Fundamentals of
Food Process Engineering

Third Edition

Fundamentals of
Food Process Engineering
Third Edition

Romeo T. Toledo

University of Georgia
Athens, Georgia

 Springer

Romeo T. Toledo
Department of Food Science
   and Technology
University of Georgia
Athens, GA 30602
USA
cmsromeo@uga.edu

Library of Congress Control Number: 2005935292

ISBN-10: 0-387-29019-2   e-ISBN-10: 0-387-29241-1
ISBN-13: 978-0-387-29019-5   e-ISBN-13: 978-0-387-29241-0

Printed on acid-free paper.

springer.com

Preface

Since the publication of the first edition in 1981 and the second edition in 2001, this textbook has been widely adopted for Food Engineering courses worldwide. The author expresses his gratitude to colleagues who have adopted this textbook and to those who have made constructive criticisms on the material. This new edition not only incorporates changes suggested by colleagues, but additional material has been added to include facilitated problem solving using a computer, and new food processing and food product technologies. New sections have been added in most of the chapters reflecting the current state of the technology. The expanded coverage may result in not enough time available in a school term to cover all areas; therefore, instructors are advised to carefully peruse the book and select the most appropriate sections to cover in a school term. The advantage of the expanded coverage is the elimination of the need for a supplementary textbook.

The success of this textbook has been attributed to the expansive coverage of subject areas specified in the Institute of Food Technologists model curriculum for food science majors in the United States of America and the use of examples utilizing conditions encountered in actual food processing operations. This theme continues in the third edition. In addition to the emphasis on problem solving, technological principles that form the basis for a process are presented so that the process can be better understood and selection of processing parameters to maximize product quality and safety can be made more effective. The third edition incorporates most of what was in the second edition with most of the material updated to include the use of computers in problem solving. Use of the spreadsheet and macros such as the determinant for solving simultaneous linear equations, the solver function, and programming in Visual BASIC are used throughout the book. The manual problem-solving approach has not been abandoned in favor of the computer approach. Thus, users can still apply the concepts to better understand a process rather than just mechanically entering inputs into a pre-programmed algorithm.

Entirely new sections include enthalpy change calculations in freezing based on the freezing point depression, evaporative cooling, interpretation of pump performance curves, determination of shape factors in heat exchange by radiation, unsteady-state heat transfer, kinetic data for thermal degradation of foods during thermal processing, pasteurization parameters for shelf-stable high-acid foods and long-life refrigerated low-acid foods, high-pressure processing of fluid and packaged foods, concentration of juices, environmentally friendly refrigerants, modified atmosphere packaging of produce, sorption equations for water activity of solid foods, the osmotic pressure and water activity relationships, vacuum dehydration, new membranes commercially available for food processing and waste treatment, and supercritical fluid extraction.

This edition contains much new hard-to-find data needed to conduct food process engineering calculations and will be very useful as a sourcebook of data and calculation techniques for practicing food engineers.

Athens, Georgia   Romeo T. Toledo

Contents

CHAPTER 1

Review of Mathematical Principles and Applications in Food Processing

1.1 GRAPHING AND FITTING EQUATIONS TO EXPERIMENTAL DATA

1.1.1 Variables and Functions

A variable is a quantity that can assume any value. In algebraic expressions, variables are represented by letters from the end of the alphabet. In physics and engineering, any letter of the alphabet and Greek letters are used as symbols for physical quantities. Any symbol may represent a variable if the value of the physical quantity it represents is not fixed in the statement of the problem. In an algebraic expression, the letters from the beginning of the alphabet often represent constants; that is, their values are fixed. Thus, in the expression $ax = 2by$, x and y represent variables and a and b are constants.

A function represents the mathematical relationship between variables. Thus, the temperature in a solid that is being heated in an oven may be expressed as a function of time and position using the mathematical expression $T = F(x, t)$. In an algebraic expression, $y = 2x + 4$, $y = F(x)$, and $F(x) = 2x + 4$.

Variables may be dependent or independent. Unless defined, the dependent variable in a mathematical expression is one that stands alone on one side of an equation. In the expression $y = F(x)$, y is the dependent and x is the independent variable. When the expression is rearranged in the form $x = F(y)$, x is the dependent and y is the independent variable. In physical or chemical systems, the interdependence of the variables is determined by the design of the experiment. The independent variables are those fixed in the design of the experiment, and the dependent variables are those that are measured. For example, when determining the loss of ascorbic acid in stored canned foods, ascorbic acid concentration is the dependent variable and time is the independent variable. On the other hand, if an experiment involves taking a sample of a food and measuring both moisture content and water activity, either of these two variables may be designated as the dependent or independent variable. In statistical design, the terms "response variable" and "treatment variable" are used for the dependent and independent variables, respectively.

1.1.2 Graphs

Each data point obtained in an experiment is a set of numbers representing the values of the independent and dependent variables. A data point for a response variable that depends on only one independent variable (univariate) will be a number pair, whereas with response variables that depend on several independent variables (multivariate), a data point will consist of a value for the response variable and one value each for the treatment variables. Experimental data are often presented as a table of numerical values of the variables or as a graph. The graph traces the path of the dependent variable as the values of the independent variables are changed. For univariate responses, the graph will be two-dimensional, and multivariate responses will be represented by multidimensional graphs.

When all variables in the function have the exponent of one, the function is called first order and will be represented by a straight line. When any of the variables has an exponent other than one, the graph will be a curve in rectangular coordinates.

The numerical values represented by a data point are called the "coordinate" of that point. When plotting experimental data, the independent variable is plotted on the horizontal axis or "abscissa" and the dependent variable is plotted on the vertical axis or "ordinate." The rectangular or Cartesian coordinate system is the most common system for graphing data. Both abscissa and ordinate are in the arithmetic scale and the distance from the origin measured along or parallel to the abscissa or ordinate to the point under consideration is directly proportional to the value of the coordinate of that point. Scaling of the abscissa and ordinate is done such that the data points, when plotted, will be symmetrical and centered within the graph. The Cartesian coordinate system is divided into four quadrants with the origin in the center. The upper right quadrant represents points with positive coordinates, the left right quadrant represents negative values of the variable on the abscissa and positive values for the variable on the ordinate, the lower left quadrant represents negative values for both variables, and the lower right quadrant represents positive values for the variable on the abscissa and negative values for the variable on the ordinate.

1.1.3 Equations

An equation is a statement of equality. Equations are useful for presenting experimental data because they can be mathematically manipulated. Furthermore, if the function is continuous, interpolation between experimentally derived values for a variable may be possible. Experimental data may be fitted to an equation using any of the following techniques:

1. Linear and polynomial regression: Statistical methods are employed to determine the coefficients of a linear or polynomial expression involving the independent and dependent variables. Statistical procedures are based on minimizing the sum of squares for the difference between the experimental values and values predicted by the equation.
2. Linearization, data transformation, and linear regression: The equation to which the data is being fitted is linearized. The data is then transformed in accordance with the linearized equation, and a linear regression will determine the appropriate coefficients for the linearized equation.
3. Graphing: The raw or transformed data is plotted to form a straight line, and from the slopes and intercept the coefficients of the variables in the equation are determined.

1.1.4 Linear Equations

Plotting of linear equations can be facilitated by writing the equation in the following forms:

1. The slope-intercept form: $y = ax + b$, where $a =$ the slope, and $b =$ the y-intercept, or the point on the ordinate at $x = 0$. The slope is determined by taking two points on the line with coordinates (x_1, y_1) and (x_2, y_2), and solving for $a = (y_2 - y_1)/(x_2 - x_1)$.
2. The point-slope form: $(y - b) = a (x - c)$, where $a =$ slope, and b, c represent coordinates of a point (c, b) through which the line must pass. When linear regression is used on experimental data, the slope and the intercept of the line are calculated. The line must pass through the point that represents the mean of x and the mean of y. A line can then be drawn easily using either the point-slope or the slope-intercept forms of the equation for the line.

The equations for slope and intercept of a line obtained by regression analysis of N pairs of experimental data are

$$a = \frac{\sum xy - (\sum x \sum y)/N}{\sum x^2 - [(\sum x)^2 /N]}; \quad b = \frac{\sum y \sum x^2 - \sum x \sum xy}{N(\sum x^2 - [(\sum x)^2 /N])}$$

The process of regression involves minimizing the square of the difference between value of y calculated by the regression equation and y_i, the experimental value of y. In linear regression, $\sum(ax + b - y)^2$ is called the explained variation, and $\sum(y_i - y)^2$ is called the random error or unexplained variation.

The ratio of the explained and unexplained variation is called the correlation coefficient. If all the points fall exactly on the regression line, the variation of y from the mean will be due to the regression equation, therefore explained variation equals the unexplained variation, and the correlation coefficient is 1.00. If there is too much data scatter, the random or unexplained variation will be very large, and the correlation coefficient will be less than 1.00. Thus, regression analysis not only determines the equation of a line that fits the data points, but it can also be used to test if a predictable relationship exists between the independent and dependent variables. The formula for the linear correlation coefficient is

$$r = \frac{N \sum xy - \sum x \sum y}{[[N \sum x^2 - (\sum x)^2][N \sum y^2 - (\sum y)^2]]^{0.5}}$$

r will have the same sign as the regression coefficient a. Values for r that is much different from 1.0 must be tested for significance of the regression. The student is referred to statistics textbooks for procedures to follow in testing significance of regression from the correlation coefficient.

Example 1.1. The protein efficiency ratio (PER) of a protein is defined as the weight gain of an animal fed a diet containing the test protein per unit weight of protein consumed. Data is collected by providing feed and water to the animal so the animal can feed at will, determining the amount of feed consumed, and weighing each animal at designated time intervals. The PER may be calculated from the slope of the regression line for weight of the animal (y) against cumulative weight of protein consumed (x). The data expressed as (x, y) where x is the amount of feed consumed and y is the weight are as follows: (0, 11.5), (0, 12.2), (0, 14.0), (0, 13.3), (0, 12.5), (2.0, 16.8), (2.2, 16.7), (1.8, 15.2), (2.5, 18.4), (1.8, 16.8), (3.4, 22.8), (4.2, 22.5), (3.7, 20.7), (4.6, 25.3), (4.0, 23.5), (6.5, 28.0), (6.3, 29.5), (6.8, 31.0), (5.8, 28.5), (6.6, 29.0).
Perform a regression analysis and determine the PER.

Solution:

The sum and sums of squares of the x and y are $\Sigma x = 62.2$; $\Sigma x^2 = 307.00$; $\Sigma y = 408.2$; $\Sigma y^2 = 9138.62$; $\Sigma xy = 1568.28$; $N = 20$. The mean of $x = \Sigma x/N = 62.2/20 = 3.11$.

$$a = \frac{408.2 - (62.2)(408.20/20)}{307.00 - (62.20)^2/20} = 2.631$$

$$b = \left(\frac{1}{20}\right)\left[\frac{(408.20)(307.00) - (62.20)(1568.28)}{307.00 - (62.20)^2/20}\right] = 12.23$$

The mean of $y = \Sigma y/N = 408.2/20 = 20.41$. Thus the best-fit line will go through the point (3.11, 20.41).

The correlation coefficient "r" is calculated as follows:

$$r = \frac{20(1568.28) - 62.2(408.20)}{[[20(307.00) - (62.20)^2][20(9138.62) - (408.20^2]]^{0.5}}$$

$$r = 0.9868$$

The correlation coefficient is very close to 1.0, indicating very good fit of the data to the regression equation. The regression and graphing can also be performed using a spreadsheet as discussed later in this chapter. The PER is the slope of the line, 2.631.

1.1.5 Nonlinear Equations

Nonlinear monovariate equations are those where the exponent of any variable in the equation is a number other than one. The polynomial: $y = a + bx + cx^2 + dx^3$ is often used to represent experimental data. The term with the exponent 1 is the linear term. that with the exponent 2 is the quadratic term, and that with the exponent 3 is the cubic term. Thus b, c, and d are often referred to as the linear, quadratic, and cubic coefficients, respectively. Linear regression analysis is used to determine the coefficients of a polynomial that fits the experimental data. Although the polynomial is nonlinear, linear regression analysis is used because the first partial derivative of the function with respect to any of the coefficients is a constant. The objective of polynomial regression is to determine the coefficients of the polynomial such that the sum of the squares of the difference between experimental and predicted value of the response variable is a minimum. Polynomial regression is more difficult to perform manually than linear regression because of the number of coefficients that must be evaluated. Stepwise regression analysis may be performed, that is, additional terms are added to the polynomial, and the contribution of each additional term in reducing the error sum of squares is evaluated. To illustrate the complexity of polynomial compared with linear regression, the equations that must be solved to determine the coefficients are as follows:

For linear regression, $y = ax + b$:

$$\Sigma y = aN + b\Sigma x$$
$$\Sigma xy = aN\,\Sigma x + b\Sigma x^2$$

For a second-order polynomial, $y = a + bx + cx^2$:

$$\Sigma y = aN + b\Sigma x + c\Sigma x^2$$
$$\Sigma xy = aN\Sigma x + b\Sigma x^2 + c\Sigma x^3$$
$$\Sigma x^2 y = aN\Sigma x^2 + b\Sigma x^3 + c\Sigma x^4$$

Thus, evaluation of coefficients for the linear regression is relatively easy, involving the solution of two simultaneous equations. On the other hand, polynomial regression involves solving $n + 1$ simultaneous equations to evaluate coefficients of an nth order polynomial. Determinants can be used to determine the constants for an nth order polynomial. Techniques for solving determinants manually and using a spreadsheet program are discussed later in this chapter. For the second-order polynomial (quadratic) equation, the constants a, b, and c are solved by substituting the values of N, Σx, Σx^2, Σx^3, Σx^4, Σxy, and $\Sigma x^2 y$, into the three equations above and solving them simultaneously.

1.2   LINEARIZATION OF NONLINEAR EQUATIONS

Nonlinear equations may be linearized by series expansion, but the technique is only an approximation and the result is good only for a limited range of values for the variables. Another technique for linearization involves mathematical manipulation of the function and transformation and/or grouping such that the transformed function assumes the form:

$$F(x, y) = aG(x, y) + b$$

where a and b are constants whose values do not depend on x and y.

Example 1.2. $xy = 5$.

Linearized form: $y = 5\left(\dfrac{1}{x}\right)$

A plot of y against $(1/x)$ will be linear.

Example 1.3.

$$y = (y^2/x) + 4.$$
$$y^2 = xy - 4x \qquad y^2 = x(y - 4) \quad \text{A plot of x}$$

against $y^2/(y - 4)$ will be linear

Example 1.4. The hyperbolic function $y = 1/(b + x)$.

$$\frac{1}{y} = b + x$$

A plot of $1/y$ against x will be linear.

Example 1.5. The exponential function $y = ab^x$.

$$\log y = \log a + x \log b$$

A plot of log y against x will be linear.

Example 1.6. The geometric function $y = ax^b$.

$$\log y = \log a + b \log x$$

A plot of log y against log x will be linear.

1.3 NONLINEAR CURVE FITTING

Linearizing an equation and fitting the linearized equation to the data has the advantage of simplicity but will require several replicates of entire data sets in order to be able to obtain reliable estimates of confidence limits for the equation parameters. Linearization also introduces complex errors particularly when two measured variables both appear in a linearized term. Nonlinear curve-fitting techniques permits determination of parameter estimates and their confidence interval from a single data set consisting of numerous data points. There are several nonlinear curve-fitting routines available. One commonly used software is Systat. To use Systat for data analysis, the data must be entered or imported into a Systat worksheet and saved as a Systat file.

To use Systat, first access the program and open the Systat main menu. Select *Window* and on the pop-up menu, select *Worksheet*. Data may then be entered in the worksheet. The first row should be the variable's name, and the values are entered in the column corresponding to the variables. Data may then be saved by selecting *File* and *Save*. Exit the worksheet by choosing the "X" (exit) button and return to the Systat main menu. To use data files saved in the Systat directory, chose *Open* in the *Worksheet* menu. Enter the *Filename* with the *.sys* extension and chose *Edit*. The system will return to the Systat Main menu and the following message is displayed: "Welcome to Systat. Systat variables available to you are." If a printout of the confidence interval of the parameter estimates is desired, select *Data* in the main menu and select *Format* in the pop-up menu. Then select *Extended (Long)* and *OK* to get back to the main menu. The Systat toolbar then becomes active. Select *Stats* in the Systat Main menu and select *Nonlin* in the pop-up menu. Follow the prompts. First select *Loss Function* and enter Loss function that is to be minimized. Usually this will be the sum of squares of the value of the dependent variable and the estimate. Although the sum of squares is the default, sometimes the program does not do the required iterations if nothing is entered for the loss function. Then select *OK* and when the display returns to the Systat Main menu, select *Stats* again, select *Nonlin* in the pop-up menu, and select *Model*. Enter the model desired for fitting into the data. Enter initial values of the coefficients separated by commas. Enter number of iterations. Select *OK* and Systat will return values of the parameter estimates and the loss function.

Example 1.7. Data on degradation of neoaxanthin, a carotenoid pigment in olives [*J. Agric. Food Chem.* (1994) 42:1551–1554] is as follows [Days, Conc. (in mg/kg)]: (0, 1.41), (4, 1.29), (8, 1.18), (14, .98), (19, 0.80), (20, 0.76), (26, 0.62), (33; 0.51), (54, 0.13). The change of concentration with time is first order, therefore the logarithm of concentration when plotted against time is linear. Fit the logarithmic equation $\ln(C) = kt + b$ by linear regression to obtain parameter estimates of k and b. Also fit the equation $C = [e]^{kt+b}$ and obtain parameter estimates of k and b and their confidence limits using nonlinear curve fitting.

Solution:

Enter the data into the worksheet, save and exit. The Systat main menu will indicate that the following variables are available: "Days and Neo." Select *Data*, then *Format*, then *Extended (long)*, and *OK*.

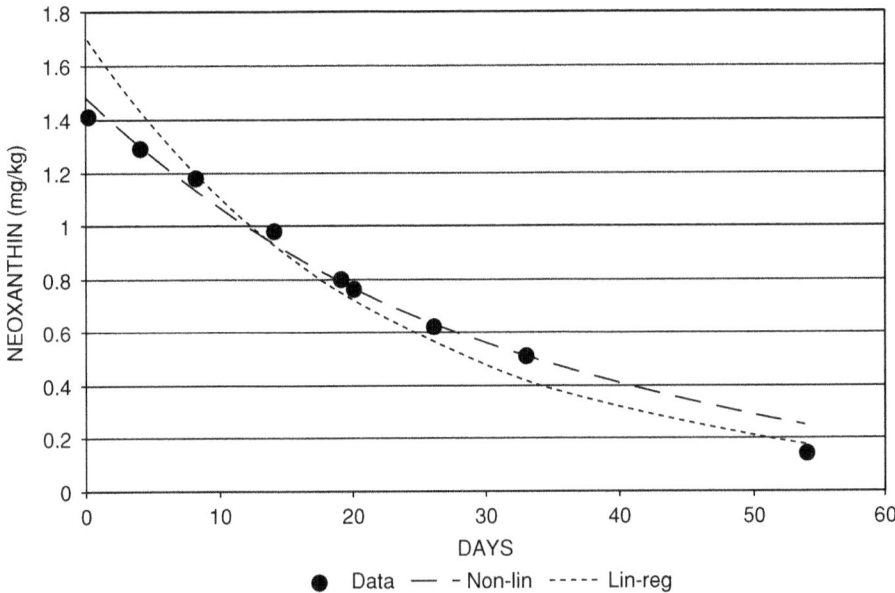

Figure 1.1 Graph showing fit to experimental data of a first-order equation with model parameters determined using linearization and linear regression (Lin-reg) and nonlinear curve fitting (Non-lin).

Back into the Systat main menu, select *Stats* then *Nonlin* and *Loss Function*. Enter "(Neo–estimate) ^2" in the loss function expression box and select *OK*. Back in the Systat main menu, select *Stats*, then *Nonlin*, then *Model*. Enter "neo = exp(k*days + b)" in the *Model* expression box, −1, 1 in the *Start* box, and 20 in the *Iterations* box. Select *OK*. Parameter estimates k = −0.033 ∀ .005 and b = 0.387 ∀ .066 and a loss function of .023 are displayed. To ensure that this is not a local minimum for the loss function, select *Stats*, then *Nonlin*, then *Resume*. Enter −.1 and 0.5 in the *Start* box and 20 in the *Iterations* box. Select *OK*. Displayed values of k and b are the same as above.

To fit a linearized form of the first-order equation, use ln(neo) = k*days + b. Transform the values for concentration of neoxanthin into their natural logarithms and perform a linear regression. This may be done using the *Regrn* function of Systat or the Statistics routine in Excel. Using Systat, enter the values of log(neo) at indicated days in the worksheet, and save. The Systat main menu then appears. Select *Regrn*. Select logneo as the dependent and days as the independent variable. Select *OK*. Systat displays −0.043 as the slope and 0.521 as the constant. The correlation coefficient is 0.958 showing reasonably good fit of the linearized equation to the data.

Figure 1.1 shows a plot of the experimental data and the fitted equations. The nonlinear curve-fitted parameters show closer values to the experimental data than the linearized transformed variable fitted parameters. Linearization forced the function to be strongly influenced by the last data point resulting in underestimation of the middle and overestimation of the first few data points. Nonlinear curve fitting is recommended over linearization, when possible.

The solver feature of Microsoft Excel may also be used to do the curve fitting. An example of how Excel may be used for curve fitting to determine kinetic parameters is shown in the section "Determining Kinetic Parameters" in Chapter 8.

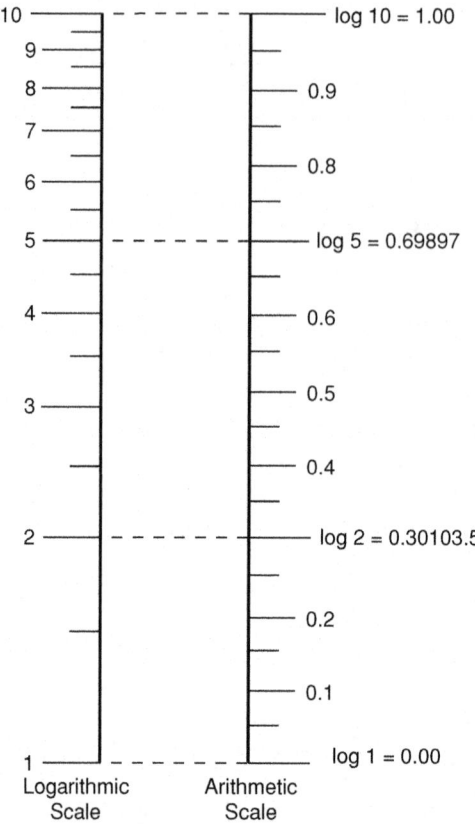

Figure 1.2 Scaling of logarithmic scale used on the logarithmic axis of semi-logarithmic or full logarithmic graphing paper.

1.4 LOGARITHMIC AND SEMI-LOGARITHMIC GRAPHS

Graphing paper is available in which the ordinate and abscissa are in the logarithmic scale. A full logarithmic or log-log graphing paper has both abscissa and ordinate in the logarithmic scale. A semi-logarithmic graphing paper has the ordinate in the logarithmic scale and the abscissa in the arithmetic scale. Full logarithmic graphs are used for geometric functions as in Example 1.6 above, and semi-logarithmic graphs are used for exponential functions as in Example 1.5. The distances used in marking coordinates of points in the logarithmic scale are shown in Fig. 1.2. Each cycle of the logarithmic scale is marked by numbers from 1 to 10. Distances are scaled on the basis of the logarithm of numbers to the base 10. Thus, there is a repeating cycle with multiples of 10. One cycle semi-logarithmic and full logarithmic graphing paper is shown in Fig. 1.3.

When plotting points on the logarithmic scale, label the extreme left and lower coordinates of the graph with the multiple of 10 immediately below the magnitude of the least coordinate to be graphed. Thus, if the least magnitude of the coordinate of the point to be plotted is 0.025, then the extreme left or lower coordinate of the graph should be labeled 0.01. The number of cycles on the logarithmic scale of the graph to be used must be selected such that the points plotted will occupy most of the graph

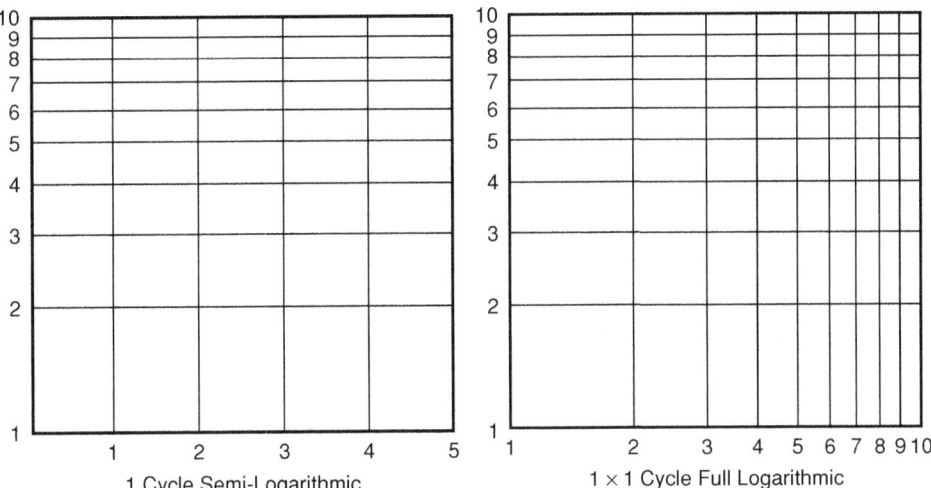

Figure 1.3 One cycle semi-logarithmic and full logarithmic graphing paper.

after plotting. Thus, if the range of numbers to be plotted is from 0.025 to 3.02, three logarithmic cycles will be needed (0.01 to 0.1; 0.1 to 1; 1 to 10). If the range of numbers is from 1.2 to 9.5, only one cycle will be needed (1 to 10).

Numerical values of data points are directly plotted on the logarithmic axis. The scaling of the graph accounts for the logarithmic relationship. Thus, points, when read from the graph, will be in the original rather than the logarithmically transformed data.

$$\text{Slope} = \frac{\log y_2 - \log y_1}{\log x_2 - \log x_1}$$

Slopes on log-log graphs are determined using the following formula:
Coordinates of points (x_1, y_1) and (x_2, y_2), which are exactly on the line drawn to best fit the data points, are located. Enough separation should be provided between the points to minimize errors. At least one log cycle separation should be allowed on either the ordinate or abscissa between the two points selected.

Slopes on semi-logarithmic graphs are calculated according to the following formula:

$$\text{Slope} = \frac{\log y_2 - \log y_1}{x_2 - x_1}$$

A separation of at least one log cycle, if possible, should be allowed between the points (x_1, y_1) and (x_2, y_2). Figure 1.4 shows the logarithmic scale relative to the arithmetic scale that would be used if the data is transformed to logarithms prior to plotting. The determination of the slope and intercept is also shown.
The following examples illustrate the use of semi-log and log-log graphs:

Example 1.8. An index of the rate of growth of microorganisms is the generation time (g). In the logarithmic phase of microbial growth, number of organisms (N) change with time of growth (t) according to:

$$N = N_0[2]^{t/g}$$

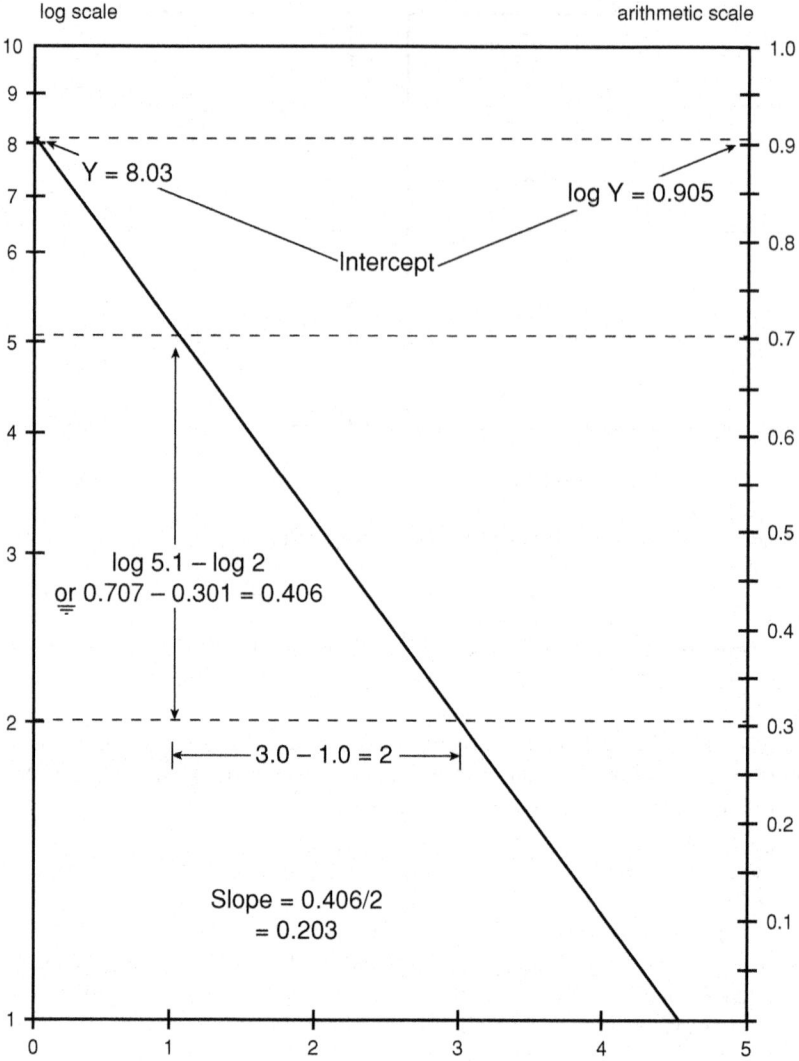

Figure 1.4 Graph showing the logarithmic relative to the arithmetic scale, and how the slope and intercept are determined on a semi-logarithmic graph.

Find the generation time of a bacterial culture that shows the following numbers with time of growth:

Numbers (N)	Time of growth, (t), in minutes
980	0
1700	10
4000	30
6200	40

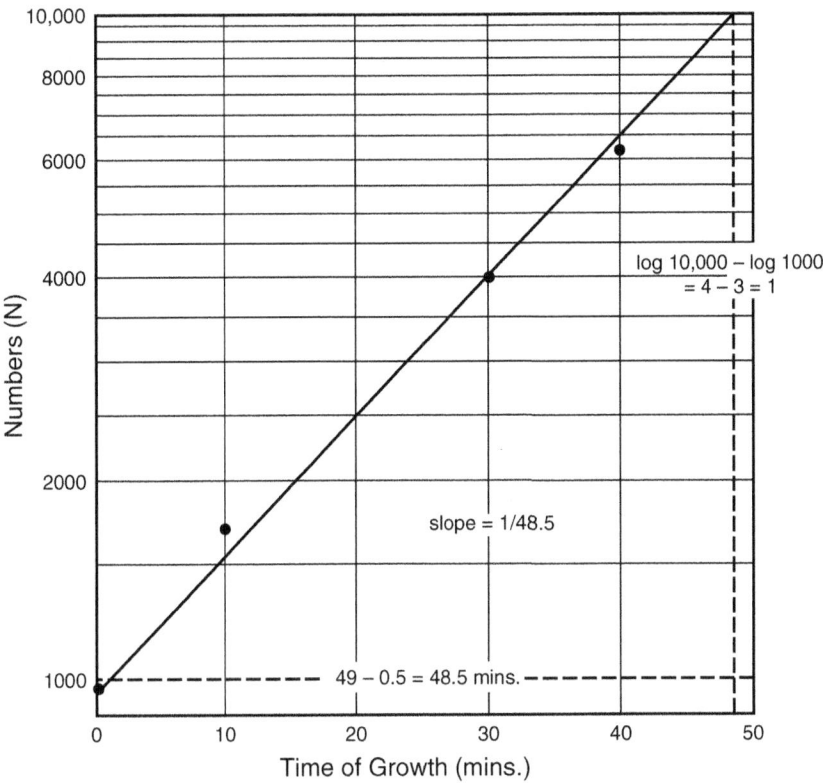

Figure 1.5   Semi-logarithmic plot of microbial growth.

Solution:

Taking the logarithm of the equation for cell numbers as a function of time:

$$\text{Slope} = \frac{\log 2}{g}$$

$$\log N = \log No + (t/g)\log 2$$

Plotting log N against t will give a straight line. A semi-logarithmic graphing paper is required for plotting. The slope of the line will be from Fig. 5, the two points selected to obtain the slope are (0,1000) and (48.5,10000). The two points are separated by one log cycle on the ordinate. The slope is $1/48.5 = 0.0206$ min^{-1}. The generation time $g = \log 2/\text{slope} = 14.6$ minutes.

Regression eliminates the guesswork in locating the position of the best-fit line among the data points. Let $\log(N) = y$ and $t = x$. The sums are $\Sigma x = 80$, $\Sigma y = 13.616$, $\Sigma x^2 = 2600$, $\Sigma y^2 = 46.740$, $\Sigma xy = 292.06$.

$$a = \frac{292.06 - 80(13.616)/4}{2600 - (80)^2/4} = 0.01974$$

$$b = \frac{13.616(2600) - 80(292.06)}{4[2600 - (80)^2/4]} = 3.0092$$

The graph is shown in Fig. 1.5. A best-fit line is drawn by positioning the straight edge such that points below the line balance those above the line. Although the equation for N suggests that any two data points may be used to determine g, it is advisable to plot the data to make sure that the two points selected lie exactly on the best-fitting line.

The correlation coefficient is:

$$r = \frac{4(292.06 - 80(13.616)}{\left([4(2600) - (80)^2][4(46.740) - (13.616)^2]\right)^{0.5}} = 0.9981$$

The correlation coefficient is very close to 1.0, indicating good fit of the data to the regression equation. The slope is 0.01974. $g = \log (2)/0.01974 = 15.2$ minutes.

The parameter estimate for g by nonlinear curve fitting using Systat and the model [$N = 980*10^{\wedge}(time/g)$] returns a parameter estimate for g of 14.834 ∀ 0.997.

Example 1.9. The term "half-life" is an index used to express stability of a compound and is defined as the time required for the concentration to drop to half the original value. In equation form:

$$C = C_o[2]^{-t/t_{0.5}}$$

where Co is concentration at $t = 0$, C is concentration at any time t, and $t_{0.5}$ is the half-life.

Ascorbic acid in canned orange juice has a half-life of 30 weeks. If the concentration just after canning is 60 mg/100 mL, calculate the concentration after 10 weeks. When labeling the product, the concentration declared on the label must be at least 90% of the actual concentration. What concentration must be declared on the label to meet this requirement at 10 weeks of storage?

Graphical Solution:

A logarithmic transformation of the equation for concentration as a function of time results in:

$$\log C = \log C_o - \left[\frac{\log 2}{t_{0.5}}\right] t$$

A plot of C against t on semi-logarithmic graphing paper will be linear with a slope of $-(\log 2)/t_{0.5}$.

Figure 1.6 is a graph constructed by plotting 60 mg/100 mL at $t = 0$ and half that concentration (30 mg/100 mL) at $t = 30$ weeks and drawing a line connecting the two points. At $t = 10$ weeks, a point on the line shows a concentration of 47.5 mg/100 mL. Thus, a concentration of 0.9(47.5) or 42.9 mg/100 mL would be the maximum that can be declared on the label.

Analytical Solution:

Given: $C_o = 60$; $t_{0.5} = 30$; at $t = 10$, C_{10} = concentration and the declared concentration on the label, $C_d = 0.9\, C_{10}$. Solving for C_d:

$$C_d = 0.9(60)[2]^{-10/30}$$
$$= 0.9(60)(0.7938) = 42.86 \text{ mg/mL}$$

Example 1.10. The pressure-volume relationship that exists during adiabatic compression of a real gas is given by $PV^n = C$, where P is absolute pressure, V is volume, n is the adiabatic expansion factor,

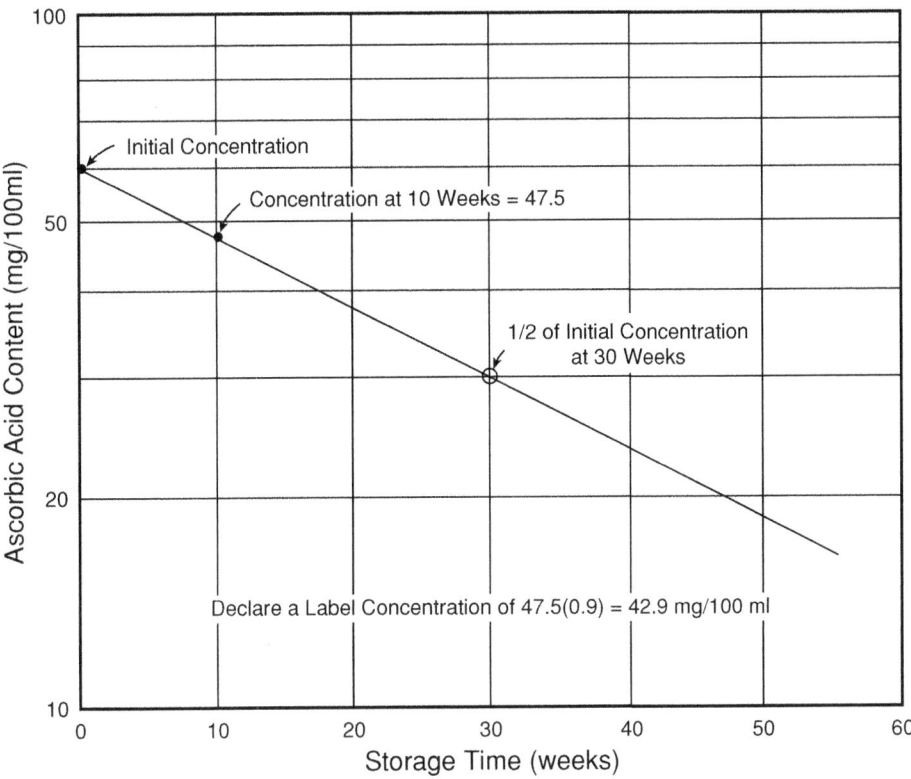

Figure 1.6 Graphical representation of the half-life illustrated by ascorbic acid degradation with time of storage.

---

and C is a constant. Calculate the value of the adiabatic expansion factor, n, for a gas that exhibits the following pressure-volume relationship:

Volume (ft^3)	Absolute Pressure (lb$_f$/in.^{2})
53.3	61.2
61.8	49.5
72.4	37.6
88.7	28.4
118.6	19.2
194.0	10.1

Solution:

The equation may be linearized and the value of n determined from the linear plot of the data. Taking the logarithm:

$$\log P = -n \log V + \log C$$

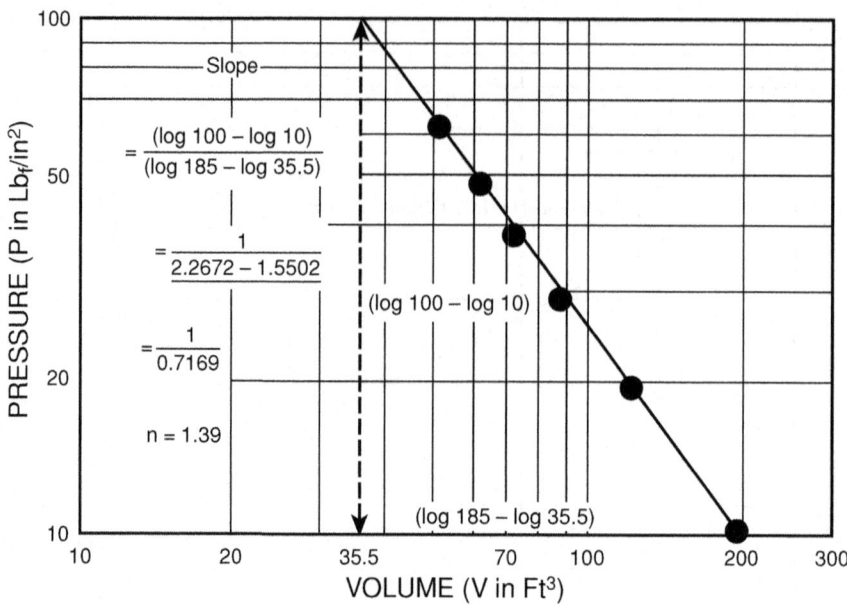

Figure 1.7 Log-log graph of pressure against volume during adiabatic expansion of a gas to determine the adiabatic expansion coefficient from the slope.

Thus, the value of n will be the negative slope of the best-fitting line line drawn through the data points in a log-log graph. The log-log graph of the data is shown in Fig. 1.7. The slope is -1.39, therefore the value of n is 1.39. This problem can be solved using linear regression analysis after transforming the data such that $x = \log(V)$ and $y = \log(P)$. The sums are $\Sigma x = 11.6953$; $\Sigma y = 8.7975$; $\Sigma x^2 = 23.006$; $\Sigma y^2 = 13.3130$; $\Sigma xy = 16.8544$

$$a = \frac{16.8544 - 8.7975(11.6953)/6}{23.006 - (11.6953)^2/6} = -1.40$$

$$b = \frac{8.7975(23.006) - (11.6953)(16.8544)}{6[23.006 - (11.6953)^2/6]} = 4.2033$$

$$r = \frac{6(16.8544) - (11.6953)(8.7975)}{[[6(23.006) - (11.6953)^2][6(8.7975) - (13.3130)^2]]^{0.5}}$$

$r = -0.9986$. r is very close to 1.0, indicating good fit of the data points to the linear regression equation. The value of n equals the slope, therefore, $n = 1.40$

1.5   INTERCEPT OF LOG-LOG GRAPHS

The y-intercept of a log-log graph is determined at a point where $\log x = 0$. On a log-log plot, $\log x = 0$ when $x = 1$. Therefore, the y-intercept is read from the graph at a point where the line crosses $x = 1$. In Example 1.10, $\log C$ is the y-intercept of the line. Figure 1.8 is drawn by

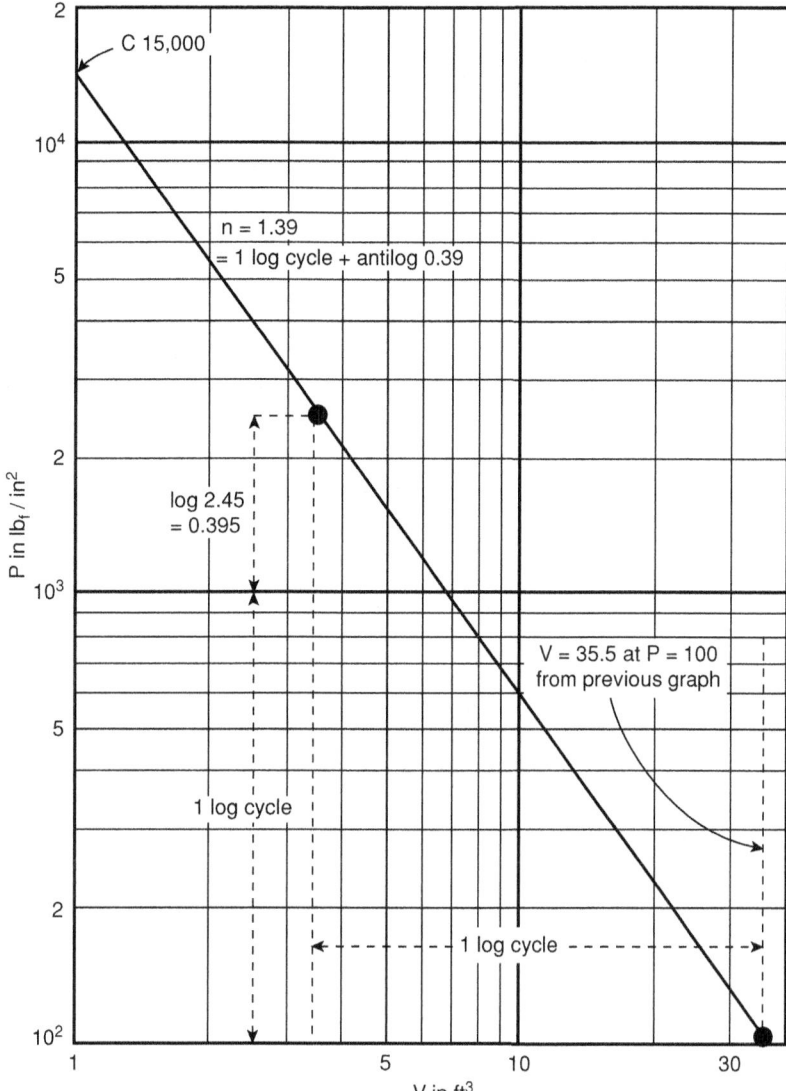

Figure 1.8 Plotting a linear log-log graph based on one point and slope, and intercept of a log-log graph.

extending the graph in Fig. 1.7 to include $V = 1$ in order to show how C may be evaluated from the intercept. The line passes through the point $V = 35.5$ and $P = 100$.

The slope, 1.39, is used to generate the other point on the line by reducing V one log cycle to a point with coordinate 3.55 and going up 1.39 log cycles; that is, 100 to $1000(10^{0.39})$ or 2450. Thus, the coordinate of the second point is 3.55, 2480. Joining the two points by a straight line and extrapolating the line to $V = 1$, the value of the intercept, which is 15,000, may be obtained. Thus, $C = 15,000$. From the regression, the intercept $= 4.2033$. $C = 10^{4.2033} = 15,500$.

1.6 ROOTS OF EQUATIONS

The roots of an equation $F(x) = 0$ are the points where the function crosses the abscissa. The roots of a system of equations are values of the variables that satisfy the equations and represent a point in the graph where the equations intersect. The following techniques are used to determine roots of equations.

1.6.1 Polynomials

A polynomial expression will have as many roots as the order of the equation. A root may be real or imaginary. Examples of imaginary roots are negative numbers raised to a fractional power. In this book, only real roots will be considered. The following techniques can be used in evaluating roots of polynomials.

1.6.1.1 Quadratic Equation

Equations with 2 as the highest power of the variable are called quadratic equations, and the root is obtained using the quadratic formula. The equation:

$$ax^2 + bx + c = 0$$

will have the following roots:

$$x = \frac{-b \pm \sqrt{(b^2 - 4ac)}}{2a}$$

Example 1.11. Determine the roots of the expression:

$$2x^2 + 3x - 2 = 0$$

Using the quadratic equation, $a = 2, b = 3, c = -2$

$$x = \frac{-3 \pm \sqrt{(3)^2 - (4)(2)(-2)}}{2(2)}$$

$x = 0.686; \; x = -2.186$

1.6.1.2 Factoring

Equations may be factored and the roots of the individual factors calculated. Thus:

$$F(x) = (ax + b)(cx + d)(ex + f) = 0$$
$$x = -b/a; \qquad x = -d/c; \qquad x = -f/e$$

All three values of x satisfy the equality $F(x) = 0$.

Example 1.12. Determine the roots of the equation: $2x^3 + 5x^2 - 11x + 4 = 0$. Dividing by $2x - 1$, the quotient is $x^2 + 3x - 4$, which when further divided by $x - 1$ will give a quotient of $x + 4$. Thus the factors $(2x - 1)(x - 1)(x + 4) = 2x^3 + 5x^2 - 11x + 4 = 0$. The roots are $x = 1/2; x = 1$; and $x = -4$.

1.6.1.3   Iteration Technique

This is a trial-and-error method involving the substitution of values of the variable into the equation and testing if equality expressed by the function is satisfied. Inspection will usually identify a range of values of the variable that gives a negative value for the function at one end of the range and a positive value at the other end. Substitution of values within that range and plotting will identify the value of the variable when the function crosses the abscissa.

Another method for iteration involves the calculus, and is called the Newton-Raphson iteration procedure. In this procedure, a value of the variable (x_1) is assumed and the next value (x_2) can be calculated as follows:

$$x_2 = x_1 - \frac{F(x)}{F'(x)}$$

The iteration is continued until $F(x) = 0$. $F'(x)$ is the value of first derivative of the function evaluated at the assumed value of x. The derivative is discussed in Section 1.3.

Example 1.13. Determine the roots of the function F(x): $2x^3 + 5x^2 - 11x + 4 = 0$.

This is the same function as the previous example, therefore the results of the iteration method can be verified. Because this is a cubic expression, three roots are to be expected. Using differential calculus, the derivative of the function is determined.

$$F'(x) = 6x^2 + 10x - 11$$

If this derivative is equated to zero, the values of x where the function exhibits a maximum and minimum can be determined. The roots of $6x^2 + 10x - 11 = 0$ can be determined using the quadratic equation as follows:

$$x = \frac{-10 \pm \sqrt{100 - 4(6)(-11)}}{2(6)}; \quad x = -2.432; \quad x = 0.755$$

Because these two points represent a peak and a valley in the curve, it would be expected that one root might exist at $x < -2.423$, one at $-2.423 < x < 0.755$, and another root at $x > 0.755$. To illustrate the Newton-Raphson iterative technique, consider the root at the region $-2.423 < x < 0.755$. First, Let $x = -1$; $F(x) = 18$; $F'(x) = -15$; $x_2 = -1 - 18/(-15) = +0.2$. x_2 is assumed to be the new value of x in the next iteration. Let $x = 0.2$; $F(x) = 2.016$; $F'(x) = -8.76$. $x_2 = 0.2 - (2.016/-8.76) = 0.43$. This is used as the new value of x in the next iteration.

Let $x = 0.43$; $F(x) = 0.353$; $F'(x) = -5.59$:

$$x_2 = 0.43 - \frac{0.0353}{-5.59} = 0.493$$

Using the new value of x in the next iteration: Let $x = 0.493$; $F(x) = 0.032$; $F'(x) = -4.612$:

$$x_2 = 0.492 - \frac{0.032}{-4.612} = 0.4999$$

The iteration is terminated when a critical value of $|x - x_2|$ is reached. For example, if the critical $|x - x_2|$ is 0.0001, another iteration is needed with $x = 0.4999$. $F(x) = 0.00045$; $F'(x) = -4.5016$ and $x_2 = 0.4999 + 0.00045/4.5016$; $x_2 = 0.5000$. Thus, the critical $|x - x_2|$ of 0.0001 is reached and the iteration is stopped. The value of x is the last one computed, which is 0.5000.

The root in this region of the function as shown in the previous example is 0.5. A similar process can be used to determine the other two roots.

1.7 PROGRAMMING USING VISUAL BASIC FOR APPLICATIONS IN MICROSOFT EXCEL

BASIC stands for "Beginner's All-purpose Symbolic Instruction Code." It is a program that interprets symbols and codes and convert them into machine language that the computer can process. The usefulness of BASIC is reinforced by bundling of the Visual BASIC interpreter in one of the more popular spreadsheet programs, Excel. Although spreadsheets may be used to solve equations, there are situations when BASIC is more efficient to use than spreadsheets. Formulas used to calculate values in spreadsheet programs are coded using BASIC syntax. The format of BASIC is very similar to processing a problem manually. The computer executes the commands in sequence. For example, when an iterative procedure is used to solve the function $F(x) = 0$, an estimated value of the variable is substituted into the equation for $F(x)$, $F(x)$ is calculated, and the process is repeated until $F(x) = 0$. The repeated calculations are done rapidly by the computer. The Newton-Raphson iteration technique in BASIC facilitates determining roots of equations. Visual BASIC is the latest version of the language. In Visual BASIC, the program statements are executed consecutively, therefore make sure that variables have been defined and values are known before any variable is used in a program line. Data may be entered using the "*inputbox*" command, by defining values of variables using an equality statement or by using a dimension statement, for example, *Dim* y(i), to enter data as an array. Answers are displayed using the *Msgbox* function. The following example illustrates the use of Visual BASIC in an iterative procedure for determining the root of a function.

Example 1.14. Determine the positive root of the function $x^2 + 109.3x^{1.35} - 20,000 = 0$. The solution is based on substitution of various values of x into the equation. To illustrate Visual BASIC programing, the program inside Microsoft Excel will be used.. Start Excel. From the main menu, access the BASIC interpreter by selecting *View*, then select *Toolbar* in the pop-up menu and select *Visual BASIC* in the secondary menu. The Visual BASIC message box appears. Select the "Visual Basic Editor" icon, which brings up the Visual Basic Main Window. Select *View* and *Code* in the pop-up menu. This opens the Visual BASIC window. The window is now available for coding the program. First type the subject **Sub** and the title of the subject, for example, **Root1**. Pressing the "Enter" key automatically displays "End Sub" in the window. The program code can then be typed in the space between "Sub" and "End Sub." Displays can be formatted by declaring a named variable to be displayed; for example, **Display** = and the variables and values to be displayed. Include the "Chr(10)" and "Chr(13)," the linefeed and return codes, to separate the lines in the display. At the point in the program where the values to be displayed are created, write the code "MsgBox Text." The program and display are shown in Fig. 1.9. The output of the program is as follows:

The first line in the program defines the type (Sub or Function), name and arguments, and is the procedure header. The last line in the program is the procedure footer. These two lines in the program define the procedure's limits in Visual BASIC.

The program is run by clicking on the "Run" icon. The screen will display each set of x and corresponding fx values in one line as defined by the Chr(10), a linefeed code, and Chr(13), a carriage return code, in statement to be printed. The value of x can be set at the beginning from a wider range (e.g., from 1 to 100 in step of 1), then narrowed down to a smaller range to obtain more significant figures in the value of the root.

The values of fx changes from negative (-6.259) at $x = 43.96$ to positive (0.1673) at $x = 43.97$. Thus, the root must be just below and close to 43.97. A more accurate value for the root may be obtained using the Newton-Raphson iteration technique.

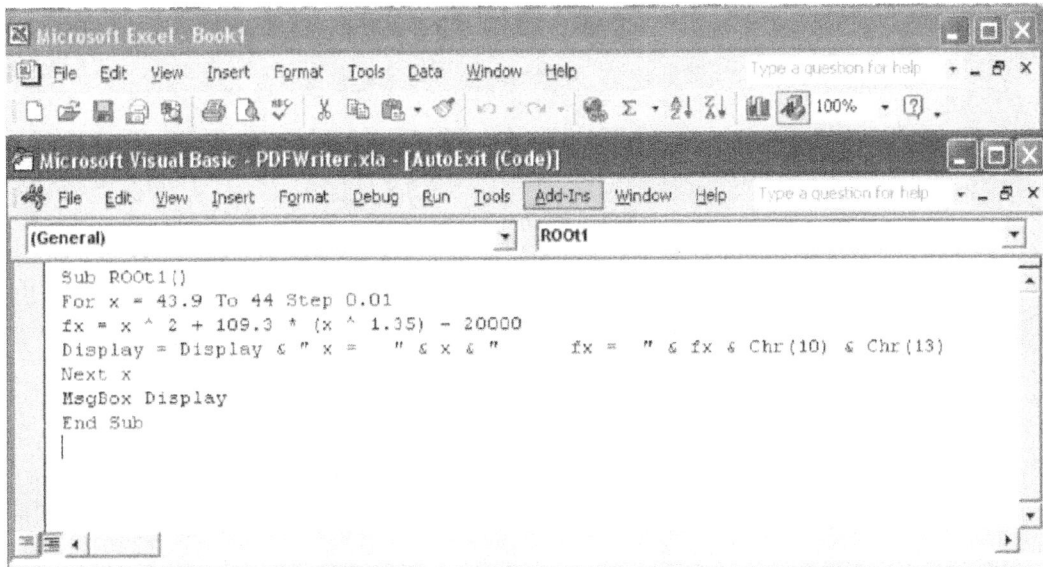

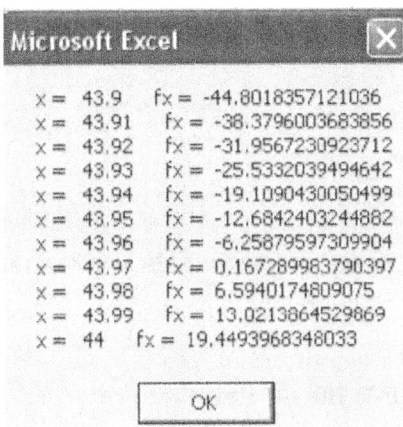

Figure 1.9 Coding of Visual BASIC program for calculating the root of a function and the display on running the program.

Solution:

Using the Newton-Raphson iteration technique, $F = (x) = 2x + 147.555x^{0.35}$.
The Visual BASIC program is shown in Fig. 1.10.
The program, when run, will output $x = 43.96973968$ as shown in Fig. 1.10.

Note that the message box is displaying only one line and is outside of the loop; there is no need to code the display as a variable as in the previous example.

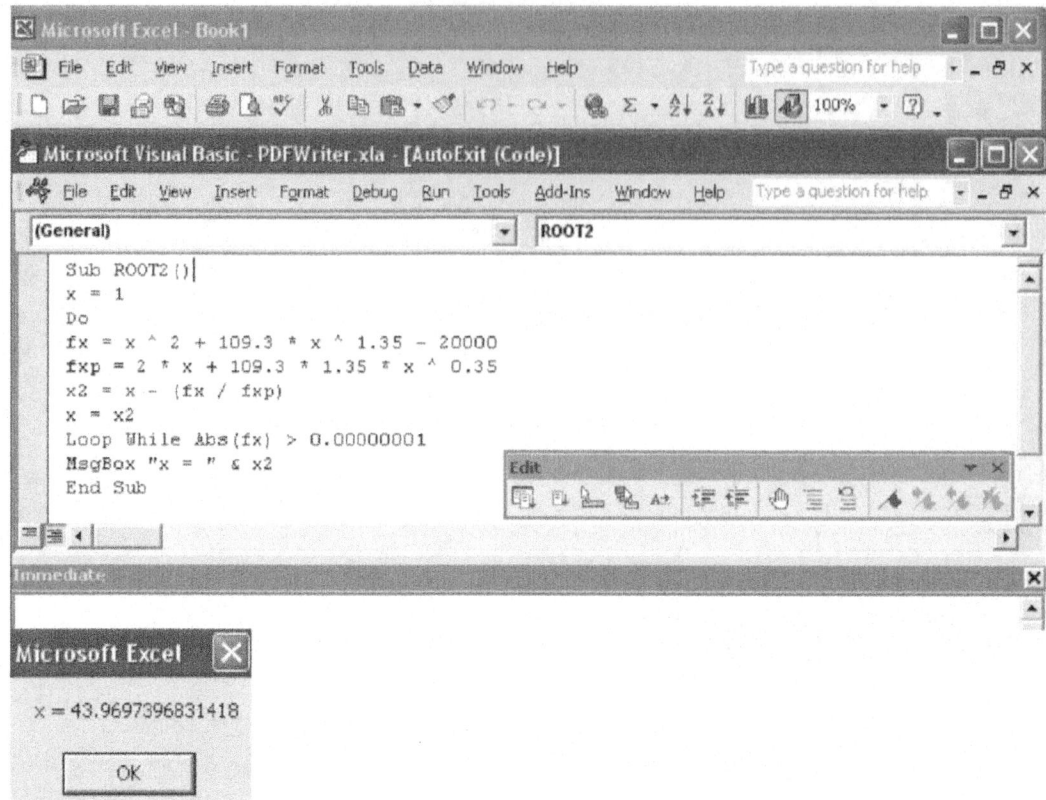

Figure 1.10 Coding of Visual BASIC program for calculating the root of a function using Newton-Raphson iteration and the display on running the program.

1.8 USE OF SPREADSHEETS TO SOLVE ENGINEERING PROBLEMS

There are three major spreadsheet programs available: Excel, Quatro Pro, and Lotus 1, 2, 3. Although there are similarities in navigating within the spreadsheet and programming formulas in cells, there are differences in syntax, therefore it is best to study one particular software rather than switch around among the different types. Excel will be used to demonstrate the concepts. Each cell in the spreadsheet may be filled by typing a title or a number. Values calculated using a formula can also be entered. *Formulas* are created with the syntax of BASIC, which are as follows:

Multiplication: $xy = x*y$ Logarithms: $\log_{10} x = \log(x); \log_e x = \ln(x)$
Division: $x) y = x/y$ Exponentiation: $x^a = x^{\wedge}a$

To *navigate between cells*, use the arrow keys. *Correcting entries in cells* may be done by typing a new entry completely inside a cell. Moving cursor to another cell will replace old content in a cell with new one. An alternative is to press F2, correct part of the contents of a cell, and press "Enter."

Copying formulas form one cell to other cells is done by locating the cursor on the cell to be copied, left clicking on the "copy" icon, and dragging the mouse to range of cells where you want the cell content to be copied. Then left click on the "paste" icon to complete the entry into the other cells. When copying contents of one cell to be used on all calculations in the block of cells to which a formula is copied, make sure to use the *absolute addressing* method. This is done by placing the $ sign before the column and row designation for a cell. Thus, if the value is in C2, an absolute cell address would be $C$2. Values in either column or row may be frozen. For example, values in column A from row 1 to 20 will be used in a formula in column D, E, and F. The cell address may be written as $C1, $C2, and so forth, in either column D, E, or F, and values from column C will always be used in the calculation. On the other hand, if relative addressing is used, and formulas in column D are copied to column E, there will be an error because the column E formulas will be using values in column D instead of column C. Thus addressing a cell as $A$2 means that the content of cell A2 will be used in all formulas regardless of location in the spreadsheet. Using the cell address $A2 means that the contents of cells in column A will be used in the calculations regardless of the column position in the spreadsheet. When *entering formulas*, use the = sign before starting the formula to tell the program to treat the entry as a formula rather than a label. When using formula calculated values in cells in another formula, an error message may be returned because the cell addresses may no longer be compatible with the current cell position in the spreadsheet. This may be corrected by absolute addressing or by *converting formula value in cells to numerical values*. To change formula values in a cell to a numerical value, position the cursor in the cell, double click, press F9 and enter.

Example 1.15. Calculate the root of the following equation by iteration using the Newton-Raphson iteration technique: $Y = X^2 + 109.3 \, X^{1.35} - 20,000$.

Solution:

Newton-Raphson is discussed in the preceding section.

The function is $F(x) = x^2 + 109.3 * x^{1.35} - 20,000$.
The first derivative is $F = (x) = 2 * X + 109.3 * 1.35 * x^{.35}$.
Start Excel. To determine an approximate value for x when $y = 0$, start with fairly large increments of x and calculate values for y. Label cell A1, x, and cell B1, y. Enter 10 in A2 and enter formula $= A2 + 10$ in A3. Copy formula in A3 to A4 to A6. Enter the formula $= x^2 + 109.3 * x^{1.35} - 20000$ in B2. Copy the formula in B2 to B3 to B6. Note that y changes from negative to positive between $x = 40$ and $x = 50$. Start Newton-Raphson with an initial value of $x = 40$.
In the spreadsheet D1, E1, F1 and G1, enter the labels X, F(x), and F = (x), X2.
In D2, enter a starting value of X , 40.
In E2, enter the formula $= D2^2 + 109.3 * D2^{1.35} - 20000$.
In F2, enter the formula $= 2 * D2 + 109.3 * 1.35 * D2^{.35}$.
In G2, enter the formula $= D2 - (E2/F2)$.
In D3, enter (calculated value of x2) formula $= \$G2$.
Copy formula in D3 to D4 to D6, and also corresponding formulas in columns E, F, and G. The value of x that gives 0 for the value of F(x) is the root. This value is 43.9674. The spreadsheet is shown in Fig. 1.11.

Example 1.16. Calculate the moisture content needed to lower the water activity of a gelatin candy to 0.7. The candy contains 7.5 g gelatin/40 g dextrose. The water activity a_w changes with mole fraction water, x , according to:

$$\log_{10}\left(\frac{a_w}{x}\right) = -0.7(1 - x)^2$$

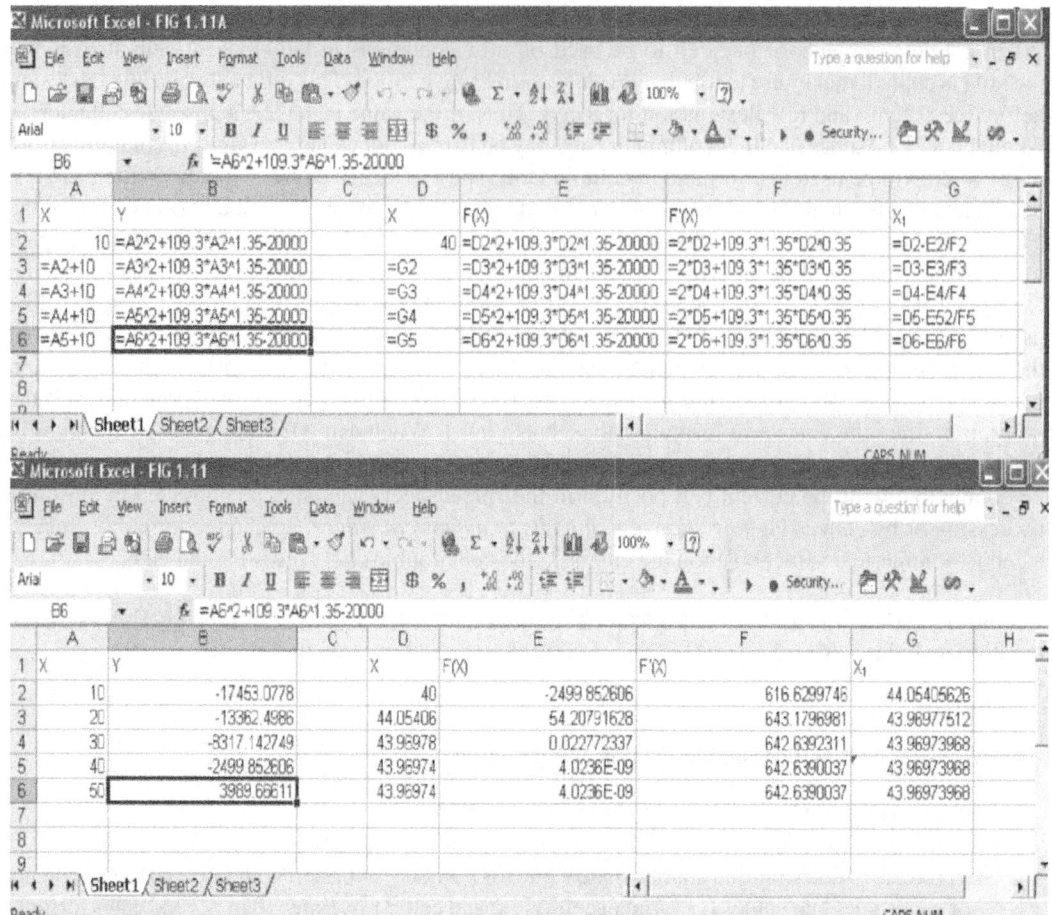

Figure 1.11 Spreadsheet solution to root of an equation using Newton-Raphson iteration.

Solution:

Use as a basis 47.5 g gelatin/dextrose mixture (7.5 g gelatin; 40 g dextrose). Let M be the mass fraction moisture in candy. The mass of water = $47.5/(1 - M) - 47.5 = 47.5M/(1 - M)$. The mole fraction of water (x) at mass fraction water M is

$$x = \frac{\dfrac{47.5M}{18(1 - M)}}{\dfrac{47.5M}{18(1 - M)} + \dfrac{40}{180}}$$

Label cell A1, M; B1, x; and C1, aw. In A2, enter M = 0.9; in A3, enter $= A2 - 0.1$. Copy to A4:A10. In B2, enter formula $= (47.5^*A2/(18^*(1 - A2)))\ /((47.5^*A2/(18^*(1 - A2))) + 40/180)$. Copy to B3:B10. In C2, enter formula $= B2^*10^{\wedge}(-.7^*(1 - B2)^{\wedge}2)$. Copy to C3:C10. Inspection of the calculated values show that M should be close to 0.3. Replace value in A2 with 0.25 and decrement by 0.01 to A6. Recalculated values show M to be between 0.21 and 0.22 for $a_w = 0.7$. Thus, the candy formula should have a moisture content greater than 21% but less than 22%. The spreadsheet is shown in Fig. 1.12.

1.9 SIMULTANEOUS EQUATIONS

Simultaneous equations are often encountered in problems involving material balances and multistage processes. The following techniques are used in evaluating simultaneous equations.

1.9.1 Substitution

If an expression for one of the variables is fairly simple, substitution is the easiest means of solving simultaneous equations. In a set of equations $x + 2y = 5$ and $2x - 2y = 3$, x may be expressed as a function of y in one equation and substituted into the other to yield an equation with a single unknown. Thus, $x = 5 - 2y$ and $2(5 - 2y) - 2y = 3$, giving a value of $y = 7/6$, and $x = 8/3$.

1.9.2 Elimination

Variables may be eliminated either by subtraction or division. Division is used usually with geometric expressions, and elimination by subtraction is used with linear expressions. When subtraction is used, equations are multiplied by a factor such that the variable to be eliminated will have the same coefficient in two equations. Subtraction will then yield an equation having one less variable than the original two equations.

Example 1.17. Calculate the value of τ and μ that would satisfy the following expressions: $\mu = 3.2 \, (\tau)^{0.75}$ and $1.5\mu = 0.35(\tau)^{0.35}$.
μ is eliminated by division.

$$\frac{1}{1.5} = \frac{3.2}{0.35}\tau^{0.75-0.35}$$

$$\tau = \frac{0.35}{(1.5)(3.2)}\frac{1}{0.4}$$

$$\tau = 0.00169; \quad \Phi = 3.2(0.00169)^{0.75} = 0.02667$$

Example 1.18. Determine x and y that satisfy the following expressions: $2x + 2y = 32; x^2 = y - 2$.

y may be eliminated by multiplying the second equation by 2 and adding the two equations.

$$2x + 2y = 32$$
$$2x^2 - 2y = -4$$

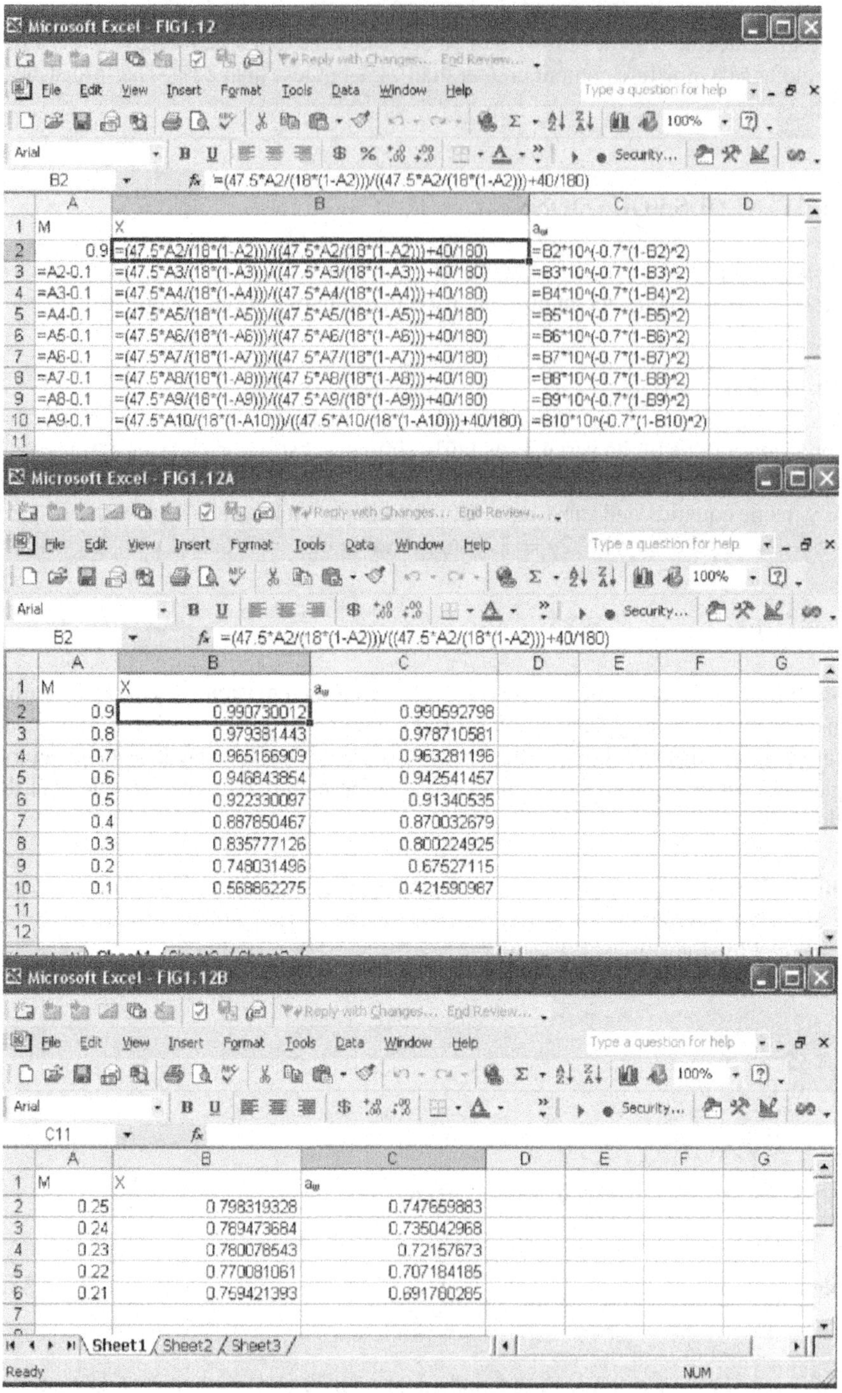

Microsoft Excel - FIG1.12

B2 f_x =(47.5*A2/(18*(1-A2)))/((47.5*A2/(18*(1-A2)))+40/180)

	A	B	C	D
1	M	X	a_w	
2	0.9	=(47.5*A2/(18*(1-A2)))/((47.5*A2/(18*(1-A2)))+40/180)	=B2*10^(-0.7*(1-B2)^2)	
3	=A2-0.1	=(47.5*A3/(18*(1-A3)))/((47.5*A3/(18*(1-A3)))+40/180)	=B3*10^(-0.7*(1-B3)^2)	
4	=A3-0.1	=(47.5*A4/(18*(1-A4)))/((47.5*A4/(18*(1-A4)))+40/180)	=B4*10^(-0.7*(1-B4)^2)	
5	=A4-0.1	=(47.5*A5/(18*(1-A5)))/((47.5*A5/(18*(1-A5)))+40/180)	=B5*10^(-0.7*(1-B5)^2)	
6	=A5-0.1	=(47.5*A6/(18*(1-A6)))/((47.5*A6/(18*(1-A6)))+40/180)	=B6*10^(-0.7*(1-B6)^2)	
7	=A6-0.1	=(47.5*A7/(18*(1-A7)))/((47.5*A7/(18*(1-A7)))+40/180)	=B7*10^(-0.7*(1-B7)^2)	
8	=A7-0.1	=(47.5*A8/(18*(1-A8)))/((47.5*A8/(18*(1-A8)))+40/180)	=B8*10^(-0.7*(1-B8)^2)	
9	=A8-0.1	=(47.5*A9/(18*(1-A9)))/((47.5*A9/(18*(1-A9)))+40/180)	=B9*10^(-0.7*(1-B9)^2)	
10	=A9-0.1	=(47.5*A10/(18*(1-A10)))/((47.5*A10/(18*(1-A10)))+40/180)	=B10*10^(-0.7*(1-B10)^2)	
11				

Microsoft Excel - FIG1.12A

B2 f_x =(47.5*A2/(18*(1-A2)))/((47.5*A2/(18*(1-A2)))+40/180)

	A	B	C	D	E	F	G
1	M	X	a_w				
2	0.9	0.990730012	0.990592798				
3	0.8	0.979381443	0.978710581				
4	0.7	0.965166909	0.963281196				
5	0.6	0.946843854	0.942541457				
6	0.5	0.922330097	0.91340535				
7	0.4	0.887850467	0.870032679				
8	0.3	0.835777126	0.800224925				
9	0.2	0.748031496	0.67527115				
10	0.1	0.568862275	0.421590967				
11							
12							

Microsoft Excel - FIG1.12B

C11 f_x

	A	B	C	D	E	F	G
1	M	X	a_w				
2	0.25	0.798319328	0.747659883				
3	0.24	0.789473684	0.735042968				
4	0.23	0.780078543	0.72157673				
5	0.22	0.770081061	0.707184185				
6	0.21	0.759421393	0.691780285				
7							

Sheet1 / Sheet2 / Sheet3 /

Ready   NUM

Figure 1.12   Spreadsheet solution to problem of candy moisture content to give a water activity of 0.70.

Adding: $2x^2 + 2x = 28$

$$x = \frac{-2 \pm \sqrt{4 - 4(2)(-28)}}{2(2)}$$

Solving by the quadratic formula:
$x = 3.275$; $y = 12.275$
$x = -4.275$; $y = 20.275$

1.9.3 Determinants

Coefficients of a system of linear equations may be set up in an array or matrix, and the matrices are resolved to determine the values of the variables. Solutions of a system of linear equations may be obtained using a spreadsheet program like Excel. For a system of three equations or more, setting up the matrix to solve the equations using Excel will be the fastest way to determine the values of the variables.

In a system of equations:

$$a_1 x + b_1 y + c_1 z = d_1$$
$$a_2 x + b_2 y + c_2 z = d_2$$
$$a_3 x + b_3 y + c_3 z = d_3$$

the array of the coefficients is as follows:

$$\begin{vmatrix} a_1 & b_1 & c_1 \\ a_2 & b_2 & c_3 \\ a_3 & b_2 & c_3 \end{vmatrix} \begin{vmatrix} d_1 \\ d_2 \\ d_3 \end{vmatrix}$$

The values of x, y, and z are determined as follows:

$$x = \frac{\begin{bmatrix} d_1 & b_1 & c_1 \\ d_2 & b_2 & c_2 \\ d_3 & b_3 & c_3 \end{bmatrix}}{\begin{bmatrix} a_1 & b_1 & c_1 \\ a_2 & b_2 & c_2 \\ a_3 & b_3 & c_3 \end{bmatrix}} \qquad y = \frac{\begin{bmatrix} a_1 & d_1 & c_1 \\ a_2 & d_2 & c_2 \\ a_3 & d_3 & c_3 \end{bmatrix}}{\begin{bmatrix} a_1 & b_1 & c_1 \\ a_2 & b_2 & c_2 \\ a_3 & b_2 & c_3 \end{bmatrix}} \qquad z = \frac{\begin{bmatrix} a_1 & b_1 & c_1 \\ a_2 & b_2 & c_2 \\ a_3 & b_3 & c_3 \end{bmatrix}}{\begin{bmatrix} a_1 & b_1 & c_1 \\ a_2 & b_2 & c_2 \\ a_3 & b_3 & c_3 \end{bmatrix}}$$

A 2×2 matrix is resolved by cross-multiplying the elements in the array and subtracting one from the other. The position of an element in the matrix is designated by a subscript ij with i representing the row and j representing the column. In order to maintain consistency in the sign of the cross-product, the element in the first column whose subscript adds up to an odd number is assigned a negative sign. Thus, the cross-product with that element will have a negative sign.

A 2×2 matrix and its value is shown below:

$$\begin{bmatrix} a_{11} & a_{12} \\ a_{21} & a_{22} \end{bmatrix} = a_{11}a_{22} - a_{21}a_{12}$$

A 3×3 matrix is evaluated by multiplying each of the elements in the first column with the 2×2 matrix left using elements in the second and third column other than those in the same row as the multiplier. As with the 2×2 matrix above, the multiplier whose subscript adds up to an odd number is assigned a negative value. The multiplier and 2×2 matrices as multiplicand are determined as follows:

$$\begin{bmatrix} a_{11} & a_{12} & a_{13} \\ a_{21} & a_{22} & a_{23} \\ a_{31} & a_{32} & a_{33} \end{bmatrix} \begin{bmatrix} a_{11} & a_{12} & a_{13} \\ a_{21} & a_{22} & a_{23} \\ a_{31} & a_{32} & a_{33} \end{bmatrix} \begin{bmatrix} a_{11} & a_{12} & a_{13} \\ a_{21} & a_{22} & a_{23} \\ a_{31} & a_{32} & a_{33} \end{bmatrix}$$

Thus, the 3×3 matrix resolves into:

$$a_{11} \begin{bmatrix} a_{22} & a_{23} \\ a_{32} & a_{33} \end{bmatrix} - a_{21} \begin{bmatrix} a_{12} & a_{13} \\ a_{32} & a_{33} \end{bmatrix} + a_{31} \begin{bmatrix} a_{12} & a_{13} \\ a_{22} & a_{23} \end{bmatrix}$$

Example 1.19. Determine the values of x, y, and z in the following equations:

$$x + y + z = 100$$
$$0.8x + 0.62y + z = 65$$
$$0.89x + 0.14y = 20$$

The array of the coefficients and constants is as follows:

$$\begin{bmatrix} 1 & 1 & 1 \\ 0.8 & 0.62 & 1 \\ 0.89 & 0.14 & 0 \end{bmatrix} \begin{bmatrix} 100 \\ 65 \\ 20 \end{bmatrix}$$

The matrix (array A) that consists of the coefficients of the variables will be the denominator of the three equations for x, y, and z.

$$\begin{bmatrix} 1 & 1 & 1 \\ 0.8 & 0.62 & 1 \\ 0.89 & 0.14 & 0 \end{bmatrix} = 1 \begin{bmatrix} 0.62 & 1 \\ 0.14 & 0 \end{bmatrix} - 0.8 \begin{bmatrix} 1 & 1 \\ 0.14 & 0 \end{bmatrix} + 0.89 \begin{bmatrix} 1 & 1 \\ 0.62 & 1 \end{bmatrix}$$
$$= 1(0 - 0.14) - 0.8(0 - 0.14) + 0.89(1 - 0.62)$$
$$= 0.3102$$

The matrix (array B) that will be the numerator in the equation for x is

$$\begin{bmatrix} 100 & 1 & 1 \\ 65 & 0.62 & 1 \\ 20 & 0.14 & 0 \end{bmatrix} = 100 \begin{bmatrix} 0.62 & 1 \\ 0.14 & 0 \end{bmatrix} - 65 \begin{bmatrix} 1 & 1 \\ 0.14 & 0 \end{bmatrix} + 20 \begin{bmatrix} 1 & 1 \\ 0.62 & 1 \end{bmatrix}$$
$$= 100(0 - .14) - 65(0 - .14) + 20(1 - .62) = 2.7$$

Thus, $x = 2.7/.3102 = 8.7$.

The matrix (array C) that is the numerator in the equation for y is

$$\begin{bmatrix} 1 & 100 & 1 \\ 0.8 & 65 & 1 \\ 0.89 & 20 & 0 \end{bmatrix} = 1 \begin{bmatrix} 65 & 1 \\ 20 & 0 \end{bmatrix} - 0.8 \begin{bmatrix} 100 & 1 \\ 20 & 0 \end{bmatrix} + 0.89 \begin{bmatrix} 100 & 1 \\ 65 & 1 \end{bmatrix}$$
$$= 1(0 - 20) - 0.8(0 - 20) + 0.89(100 - 65) = 27.15$$

Thus, $y = 27.15/0.3102 = 87.5$.

The matrix (array D) that is the numerator of the equation for z is

$$\begin{bmatrix} 1 & 1 & 100 \\ 0.8 & 0.62 & 65 \\ 0.89 & 0.14 & 20 \end{bmatrix} = 1\begin{bmatrix} 0.62 & 65 \\ 0.14 & 20 \end{bmatrix} - 0.8\begin{bmatrix} 1 & 100 \\ 0.14 & 20 \end{bmatrix} + 0.89\begin{bmatrix} 1 & 100 \\ 0.62 & 65 \end{bmatrix}$$

$$= 1(12.4 - 9.1) - 0.8(20 - 14) + 0.89(65 - 62) = 1.17$$

Thus, $z = 1.17/0.3102 = 3.8$.

Check: $x + y + z = 8.7 + 87.5 + 3.8 = 100$.

Values of determinants may be obtained using Microsoft Excel. Determinants of arrays A, B, and C above will be solved. Access Excel and enter the arrays A, B, C, and D above. For example, the array A may be entered in the block A2 to C4, array B in block E2 to G4, array C in block A8 to C10, and array D in block E8 to G10. To determine determinant of array A, block A2 to C4, move the pointer to the name box (displaying "A2") and give the array a name (e.g., ARRA). Press "Enter." Move the pointer to the cell where the answer is to be displayed (e.g., B5 for array A) and click the mouse to make B5 the active cell. Move the pointer to the formula box and click on the "=" sign. This activates the formula box. Type "Mdeterm(ARRA)" and select *OK*. The determinant of array A (.3102) will be displayed in cell B5. Repeat the process for arrays B, C, and D. Figure 1.13 shows the spreadsheet. The results are exactly the same as those solved manually above.

1.10 SOLUTIONS TO A SYSTEM OF LINEAR EQUATIONS USING THE "SOLVER" MACRO IN EXCEL

Instead of calculating determinants, solutions to a system of equations may be obtained by using the "Solver" macro in Excel. Enter the variables = coefficients into the spreadsheet as an array. Enter the row of constants in the row adjacent to the array of variables = coefficients. Then select a row of cells where the answer will be displayed. This can be the empty row above or below the array. Enter zero in each of these cells. Enter the constraints in an empty column to the right of the array. The constraints will be formulas equivalent to the equations to be solved. Position the pointer in the cell to contain the first constraint, click to make this the active cell, move the pointer to the formula box, and click on the "=" sign. This activates the formula box. Enter the formula in this cell. Use absolute cell addressing for the cells designated to contain the values of the variables. Enter all constraints. Excel will enter zero in these constraints cells when the formulas are entered. Select *Tools* then select *Solver* in the pop-up menu. The *Solver Parameters* dialogue box displays. Point the mouse to the first cell containing the constraint. This cell address will be displayed in the *Set Target Cell* box. Click *Value of* and enter the value corresponding to the constant of the first equation. Click the box under *By Changing Cells* and enter range of cell addresses of the row designated to contain values of the variables. Under *Subject to Constraints* is a box that should hold all the other constraints. Click on *Add* to enter the *Add Constraint* dialogue box. Enter the cell reference for the second constraint formula. Click on the button for the mathematical relationship of the constraint, choose "A=" and click on the box under *Constraint*. Enter the value of the second constant or enter a cell address where this constant was entered. Click *Add*. The *Add Constraint* dialogue box will reappear to accept addition of more constraints. Enter the rest of the constraints. Click *OK*. The *Solver Parameters* dialogue box reappears. Check the values and cell references entered. Click *Solve* and the calculated values of the variables will be displayed in the designated cells.

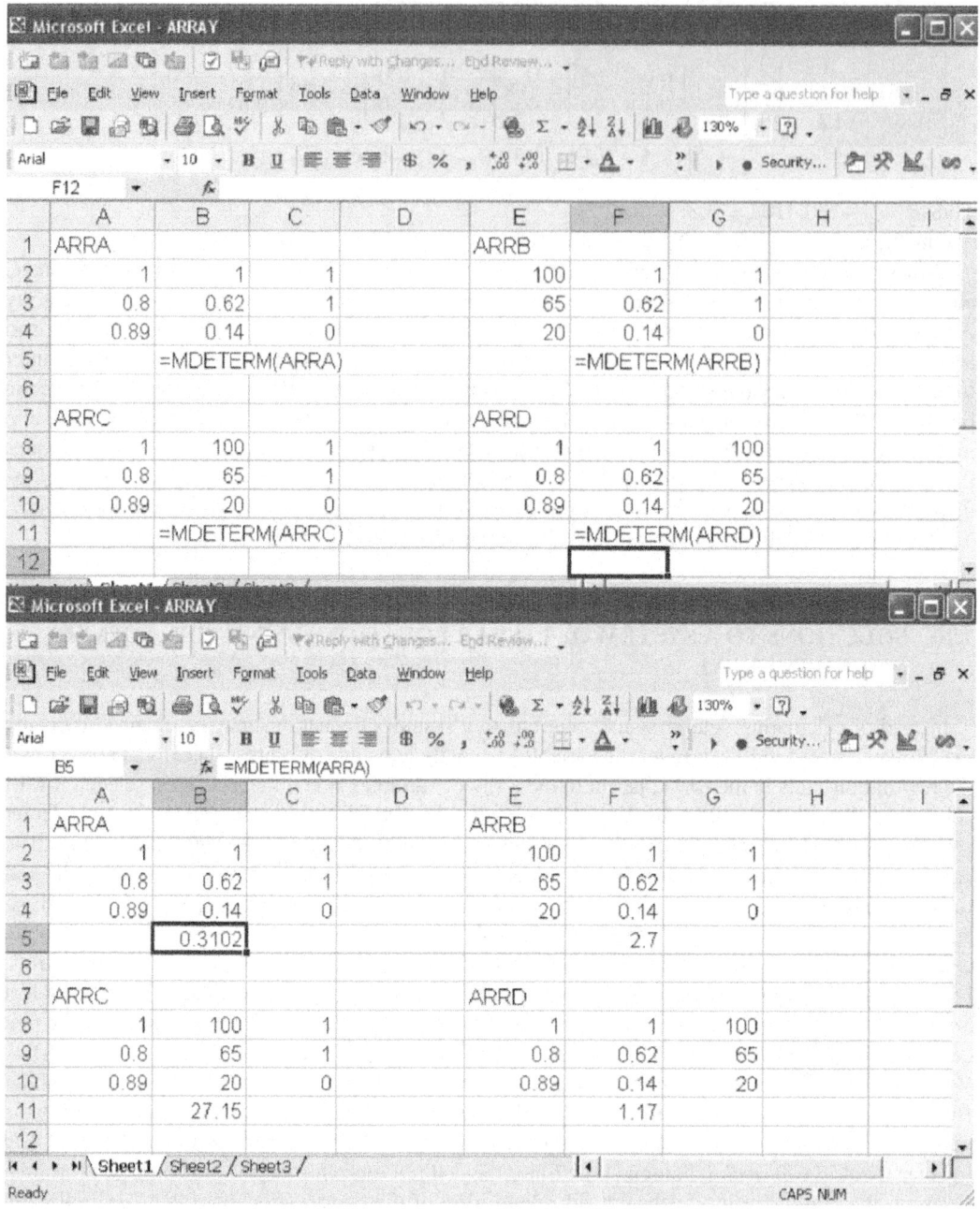

Figure 1.13 Spreadsheet solution to values of determinants using the "Mdeterm" function in Excel.

Example 1.20. Solve the system of equations in the preceding example using the "Solver" macro in Excel.

Solution:

Use the spreadsheet used for calculating determinants above. Enter the array of variables = coefficients in the block A3 to C5. Values for x, y, and z are designated to be displayed in row A8:C8. Enter the constants in column D3:D5. Enter the constraints into the column E3:E5. E3 should have the equivalent of the first equation: Cell \$A\$8 represents x; \$B\$8 represents y; and \$C\$8 represents z. Point the mouse to Cell E3 and click to make it the active cell. Move the pointer to the formula box and Click the "=" sign. Enter the first constraint \$A\$8*A3 + \$B\$8*B3 + \$C\$8*C3. Make cell E4 the active cell, activate the formula box, and enter the second constraint: \$A\$8*A4 + \$B\$8*B4 + \$C\$8*C4. Activate cell E5 and the formula box and enter the third constraint: \$A\$8*A5 + \$B\$8*B5 + \$C\$8*C5. Select *Tools* then *Solver* in the pop-up menu to open the *Solver Parameters* dialogue box. Enter \$E\$3 for the *Target Cell,* click on *Value of,* and enter the constant of the first equation 100. In the *By Changing Cell* box, enter \$A\$8:\$C\$8. In the *Subject to Constraints* box, enter the other constraints by clicking the *Add* button to display the *Add Constraints* dialogue box and enter the *Cell Reference* box \$E\$4, activate the "=" sign, and enter in the *Constraint* box the cell reference for the second equation constant, \$D\$4. Click *Add*. The *Add Constraints* dialogue box reappears and enter \$E\$5, and \$D\$5. Click *OK*. The *Solver Parameters* dialogue box reappears. Click *Solve*. Results are displayed in Fig. 1.14.

1.11 POWER FUNCTIONS AND EXPONENTIAL FUNCTIONS

Power functions and exponential functions consist of a base raised to an exponent. Although the two functions are similar, "power functions" are those that have numerical exponents, whereas the term "exponential function" is used for those that have variables as exponent. Both functions are resolved using logarithms or by taking the γth root of both sides of the equation, where r is the exponent of the variable whose value needs to be determined. The following are basic rules when working with exponents:

Multiplication: When the base is the same, add exponents

$$x^2(x^3) = x^{2+3} = x^5$$

Division: When the base is the same, subtract exponents. A negative exponent signifies division.

$$\frac{(x+2)^3}{(x+2)^{1.5}} = (x+2)^{3-1.5} = (x+2)^{1.5}$$

Exponentiation: Multiply the exponents. Extracting the γth root of a function implies exponentiation to the $1/\gamma$th power.

$$\left(\frac{P}{V}\right)^{3.5} = P^{3.5}V^{-3.5}$$

$$(x^2)^3 = x^{2(3)} = x^6$$

$$\sqrt[4]{(x^2)} = (x^2)^{1/4} = (x)^{2/4} = (x)^{1/2}$$

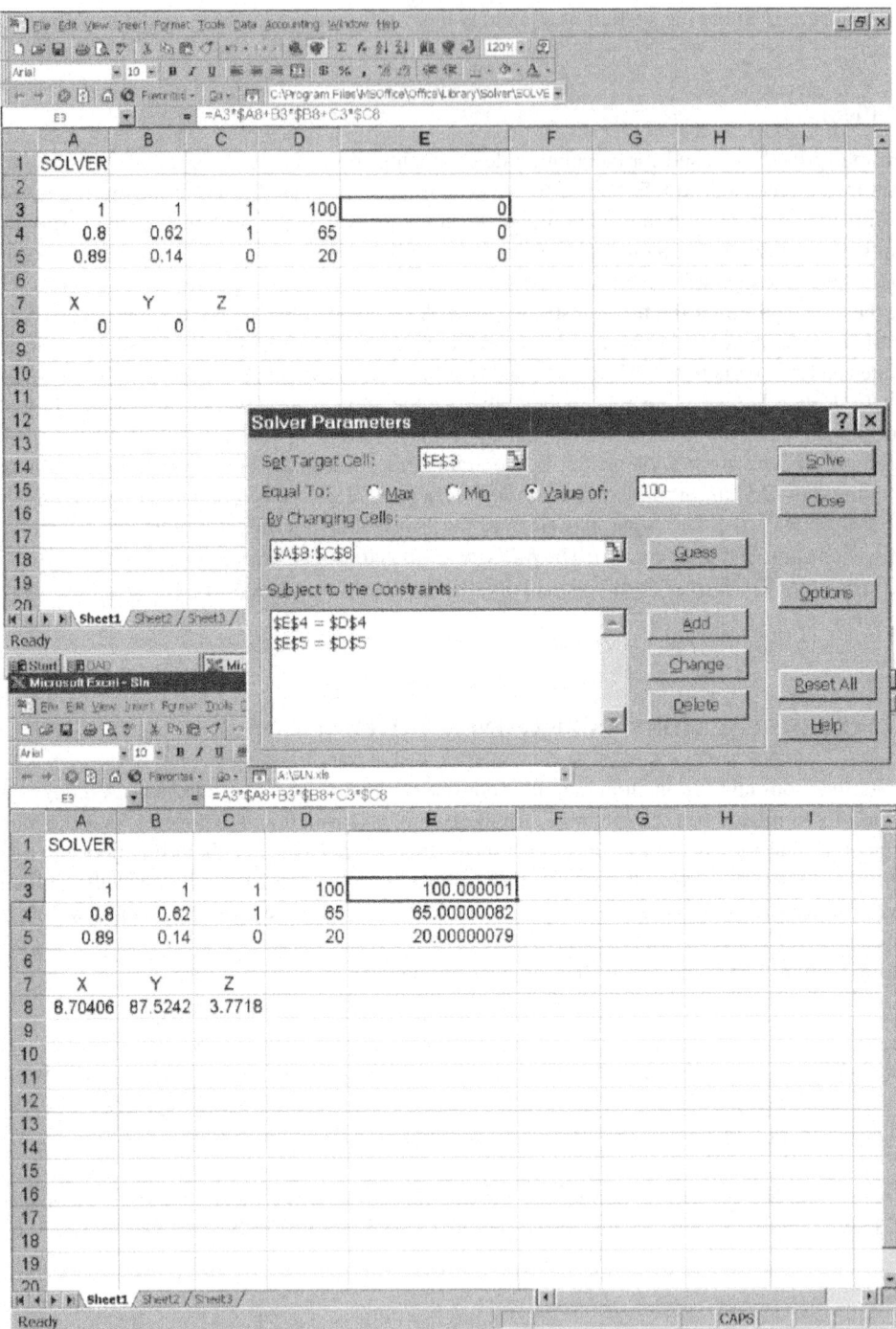

Figure 1.14   Solutions to a system of linear equations using the "Solver" macro in Excel.

1.12 LOGARITHMIC FUNCTIONS

1. Logarithm of a product = sum of the logarithms

$$\log(xy) = \log x + \log y$$

2. Logarithm of a quotient = difference of the logarithms. A negative logarithm signifies a reciprocal of the terms within the logarithm.

$$\log(x/y) = \log x - \log y$$
$$-\log x = \log(1/x)$$

3. Logarithm of a power function = exponent multiplied by the logarithm of the base.

$$\log e^{2x} = 2x \log e$$

Example 1.21. The Reynolds number of a non-Newtonian fluid is expressed as:

$$Re = \frac{8V^{2-n}R^n\rho}{k\left(\dfrac{3n+1}{n}\right)^n}$$

Calculate the value of the velocity V that would result in a Reynolds number of 2000 if $k = 1.5$ $Pa \cdot s^n$, $n = 0.775$, $\rho = 1030 \text{ kg/m}^3$, and $R = 0.0178$ m.
Substituting values:

$$2000 = \frac{8(V)^{1.225}(0.0178)^{0.775}(1030)}{1.5\left[\dfrac{2.325}{0.775}\right]^{0.775}}$$

$$V^{1.225} = \frac{2000(1.5)(2.3429)}{8(0.04406)(1030)} = 25.55$$

$$V = (25.55)^{1/1.225} = 14.08 \text{ m/s}$$

Example 1.22. The temperature (T, in EC) of a fluid flowing through a tube immersed in a constant temperature water bath at T_b changes exponentially with position along the length of the tube as follows:

$$T = T_b - (T_b - T_o)(e)^{-3.425L}$$

Calculate the length L such that when fluid enters the tube with an initial temperature $T_O = 20$ EC, the exit temperature will be 99% of the water bath temperature, T_b, which is 95 EC.

$$T = 0.99(95) = 95 - (95 - 20)(e)^{-3.425L}$$
$$(e)^{-3.425L} = [(95 - 94.05)/(75)] = .012667$$

Taking the natural logarithm of both sides and noting that ln (e) = 1:

$$-3.425L = \ln(.012667); \quad L = -4.3687/-3.425 = 1.276 \text{ m}$$

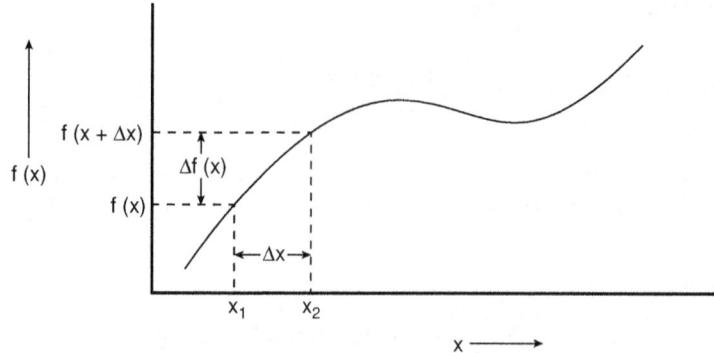

Figure 1.15   Graphical representation of the slope of a function and the derivative.

1.13   DIFFERENTIAL CALCULUS

Calculus is a branch of mathematics that deals with infinitesimal-sized segments of a whole. The concept is analogous to high-speed filming of a moving object. The action can be frozen in an infinitesimal time increment, and it will be possible to take measurements on the frozen picture frame. Analysis of a series of frames will define the nature and magnitude of changes that occur as the object moves. The calculus is particularly useful in predicting point values of variables in a system from global measurements.

Calculus is divided into differential and integral calculus. The former deals with the rate of change of variables or incremental changes in a variable. Integral calculus had its origins on earlier studies of areas of plane figures. An application of integral calculus that is particularly useful to engineers is the derivation of a function that defines a variable or processing parameter from a differential expression of the changes in the variable with respect to another.

1.13.1   Definition of a Derivative

Any function $F(x)$ may be represented by a graph shown in Fig. 1.15. A section on the x axis between x_1 and x_2 is designated Δx. The slope of the function within this section is

$$\text{Slope} = \frac{\Delta F(x)}{\Delta x} = \frac{F(x + \Delta x) - F(x)}{\Delta x}$$

If Δx is infinitesimal, Δx approaches 0, and $\Delta F(x)/\Delta x$ is the derivative of $F(x)$ with respect to x.

$$\mathop{\text{Limit}}_{\Delta x \to 0} \frac{\Delta F(x)}{\Delta x} = \frac{dF(x)}{dx} = \mathop{\text{limit}}_{\Delta x \to 0} \frac{F(x + \Delta x) - F(x)}{\Delta x}$$

If the function $F(x) = x^2$:

$$\frac{dF(x)}{dx} = \mathop{\text{limit}}_{\Delta x \to 0} \frac{(x + \Delta x)^2 - x^2}{\Delta x}$$

$$= \mathop{\text{limit}}_{\Delta x \to 0} \frac{x^2 + 2x\Delta x + (\Delta x^2) - x^2}{\Delta x}$$

The derivative of $F(x) = x^2$ is $dF(x)/dx = 2x$.
The differential form of the above derivative is

$dF(x) = 2xdx$

The symbol "d" is a differential operator, meaning that a differentiation operation has been performed on $F(x)$. Differentiation is the process of obtaining a derivative. The result of a differentiation operation is a differential equation that can then be divided by the differential term of the reference variable to obtain the derivative. The term $dF(x)/dx$ may be written as $F = (x)$.

The derivative is a rate of change, or the slope of a function. Thus, constants will have zero slope and therefore will have a derivative from zero. A linear function will have a constant slope; therefore, the derivative of a variable to the power 1 will be a constant, the coefficient of that variable.

1.13.2   Differentiation Formulas

The following are differentiation formulas that are most often used:

Constant: $d(a) = 0$
Sum: $d[F(x) + G(x)] = dF(x) + dG(x)$
Product: $d[F(x)G(x)] = F(x)dG(x) + G(x)dF(x)$
Quotient: $d(F(x)/G(x)) = (G(x)dF(x) - F(x)dG(x))/[G(x)]^2$
Power function: $d[F(x)]^n = n[F(x)]^{n-1}dF(x)$
Exponential function: $d(a)^{F(x)} = (a)^{F(x)}[dF(x)] \ln a$
Logarithmic function: $d \ln[F(x)] = dF(x)/F(x)$

$D \log[F(x)] = dF(x)/F(x)(\ln 10)$

Example 1.23.   Determine the slope of the function

$y = 2(x + 2)^3 + 2x^2 + x + 3$

at $x = 1$.

The slope is the derivative of the function. The slope is obtained by differentiating the function and solving for dy/dx. The sum and power function formulas will be used.

$dy = d[2(x + 2)^2] + d(2x)^2 + d(x) + d(3)$

The last term is the derivative of a constant and is 0.

$dy = 2(3)(x + 2)^2d(x + 2) + 2(2)(x)dx + dx = 6(x + 2)^2dx + 4xdx + dx$

$\dfrac{dy}{dx} = 6(x + 2)^2 + 4x + 1$

Substituting $x = 1$:

$\dfrac{dy}{dx} = 6(3)^2 + 4 + 1 = 54 + 5 = 59$

Example 1.24.   Differentiate the function $H = E + PV$. All terms are variables. Use the sum and the product formulas $dH = dE + PdV + VdP$.

Example 1.25. The expression for the water activity of a sugar solution is given as:
For glucose, $k = -0.7$. How much faster will the water activity (a_w) change with a change in

$$\log\left[\frac{a_w}{x}\right] = k(1 - x)^2$$

water mole fraction (x) at $x = 0.7$ compared with $x = 0.9$?

The problem requires determining da_w/dx at $x = 0.7$ and at $x = 0.9$. The ratio of the two slopes will give the relative effects of small increases in concentration around $x = 0.9$ and $x = 0.7$ on the water activity.

The expression to be differentiated after substituting $k = 0.7$ is

$$\log a_w - \log x = -0.7(1 - x)^2$$

When differentiated directly in this form, a_w will appear in the denominator of the first term involving $d(\log a_w)$. This can be eliminated by solving first for a_w before differentiation:

$$a_w = x(10)^{-0.7(1-x)^2}$$

Differentiating:

$$d(a_w) = xd[10]^{-0.7(1-x)^2} + [10]^{-0.7(1-x)^2}dx + [10]^{-0.7(1-x)^2}\ dx$$

$$= x[10]^{-0.7(1-x)^2}(-0.7)(2)(1 - x)[\ln(10)](-dx)$$

$$\frac{da_w}{dx} = [10]^{-0.7(1-x)^2}[(0.7)(2)(1 - x)(x)[\ln 10] + 1]$$

Substituting $x = 0.9$:

$$\left[\frac{da_w}{dx}\right] = [(0.7)(2)(1 - 0.9)(0.9)(2.303) + 1][10]^{-0.7(1-0.9)^2}$$

$$= 1.3226(10)^{-0.007} = 1.301$$

Substituting $x = 0.7$:

$$\left[\frac{da_w}{dx}\right] = [(0.7)(2)(1 - 0.7)(0.7)(2.303) + 1][10]^{-0.7(1-0.7)^2}$$

$$= 1.6769(10)^{-0.063} = 1.450$$

a_w will be changing faster as x is incremented at $x = 0.7$ compared to $x = 0.9$.

Example 1.26. Differentiate:

$$y = \frac{3x + 2}{x + 3}$$

Using the formula for the derivative of a quotient:

$$dy = \frac{(x+3)d(3x+2) - (3x+2)d(x+3)}{(x+3)^2}$$

$$= \frac{(x+3)(3dx) - (3x+2)dx}{(x+3)^2}$$

$$= \frac{(3x+9-3x-2)dx}{(x-3)^2}$$

$$\frac{dy}{dx} = \frac{7}{(x+3)^2}$$

Example 1.27. The growth of microorganisms expressed as cell mass is represented by the following:

$$\log\left(\frac{C}{C_0}\right) = kt$$

Determine the rate of increase of cell mass at $t = 10$ hours if it took 1.5 hours for the cell mass to double and the initial cell mass at time zero (C_0) is 0.10 g/L.

Solution:

The value of k is determined from the time required for cell mass to double. $k = (\log 2)/1.5 = 0.200$ h^{-1}. The expression to be differentiated to determine the rate is $\log(C/0.10) = 0.200t$. Differentiating using the formula for derivative of a logarithmic function:

$$d\log(C/0.10) = 0.200\, dt$$

$$\frac{dC/0.1}{(\ln 10)(C/0.1)} = 0.200\, dt; \quad \frac{dC}{dt} = 0.200C \ln(10) = 0.4605\, C$$

C from the original expression is substituted to obtain a rate expression dependent only on t. $C = C_0(10)^{0.200t}$

$$\frac{dC}{dt} = 0.4605\, C_0\, (10)^{0.200t}$$

At $t = 10$,

$$\frac{dC}{dt} = 0.4605(0.1)(10)^{2.00} = 4.605 \text{g/L} \cdot \text{h}$$

1.13.3 Maximum and Minimum Values of Functions

One of the most useful attributes of differential calculus is its use in determining the maximum and minimum values of functions. It has been shown in a previous section that determining a root of a polynomial expression is facilitated by determining where the maximum and minimum points are located. Determination of maximum and minimum values of a function can be applied in optimizing processes to identify conditions where cost is minimum, profit is maximized, or a product quality attribute is maximized.

When a function has a maximum or minimum point, the slope changes signs upon crossing the crest or valley in the curve. Thus at the maximum or minimum point, the slope is zero. After a value of the independent variable is determined at a point where the slope of the curve is zero, that point is identified as a maximum or minimum by either of two methods. (a) Determining the second derivative: As the curve approaches a maximum, the slope is positive and decreases with increasing values of the independent variable, therefore the second derivative is negative. As the curve approaches a minimum, the function has a negative slope that decreases in value, therefore the second derivative is positive. (b) For functions that have a complex second derivative, substitution of the root of the derivative equation into the original function will give the maximum and minimum value of the function. If the derivative equation has only one root, it will be possible to identify this root as the maximum or minimum point by substituting any other value of the independent variable into the original function.

Example 1.28. Plot the curve: $y = 2x^3/3 + x^2 - 6x$ and determine its maximum and minimum value.

Differentiating:

$$\frac{dy}{dx} = 2x^2 + x - 6$$

At the maximum or minimum, $dy/dx = 0$ and $2x^2 + x - 6 = 0$. The roots of the derivative are

$$x = \frac{-1 \pm \sqrt{1 - 4(2)(-6)}}{2(2)}; \quad x = -2; \quad x = 1.5$$

To determine which of these points is a maximum, take the second derivative.

$$\frac{d^2y}{dx^2} = 4x + 1$$

At $x = -2$, $d^2y/dx^2 = -7$. The point is a maximum. Substituting $x = -2$ in the expression for y:

$$y = \frac{2(-2)^3}{3} + \frac{(-2)^2}{2} - 6(-2) = 8.667$$

At $x = 1.5$, $d^2y/dx^2 = 7$. The point is a minimum. Substitute $x = 1.5$ in the expression for y:

$$y = \frac{2(1.5)^3}{3} + \frac{(1.5)^2}{2} - 6(1.5) = -5.625$$

Even without taking the second derivative, values of y show which of the roots of the derivative equation represents the maximum and minimum points.

Example 1.29. Derive an expression for the constants a and b in the equation $y = ax + b$, which is the best-fitting line to a set of experimental data points (x_i, y_i) by minimizing the sum of squares of the error $(y - y_i)$.

$$E = \Sigma(ax_i + b - y_i)^2$$

where i = 1 to n. To evaluate a and b, two independent equations must be formulated. E will be maximized with respect to b at constant a, and with respect to a at constant b. x and y are considered constants during the differentiation process with respect to either a or b. The two equations are

$$\frac{dE}{da} = 2\Sigma(ax_i + b - y_i)x_i; \quad \frac{dE}{db} = 2\Sigma(ax_i + b - y_i)$$

The second derivative of these two expressions will be $2\Sigma x_i^2$ and 2, respectively, both of which are positive quantities for all values of a or b; therefore, the root of the derivative equation will represent a point where E is minimum. The two derivative equations are then solved simultaneously.

$$2\Sigma x_i(ax_i + b - y_i) = 0; \quad 2\Sigma(ax_i + b - y_i) = 0$$

$$b = \frac{\Sigma y_i - a\Sigma x_i}{n}; \quad b = \frac{\Sigma x_i y_i - a\Sigma x_i^2}{\Sigma x_i}$$

$$a\Sigma x_i^2 + b\Sigma x_i - \Sigma y_i x_i = 0; \quad a\Sigma x_i + nb - \Sigma y_i = 0$$

Equating the two expressions for b and solving for a:

$$a = \frac{\Sigma x_i y_i - \Sigma x_i \Sigma y_i / n}{\Sigma x_i^2 - (\Sigma x_i)^2 / n}$$

Substituting the expression for a in the expression for b and solving for b:

$$\frac{\Sigma x_i y_i - \Sigma x_i y_i / n}{\Sigma x_i^2 - (\Sigma x_i)^2 / n} \Sigma x_i + nb - \Sigma y_i = 0$$

Solving for b:

$$b = \frac{1}{n}\left[\frac{\Sigma y_i \Sigma x_i^2 - \Sigma x_i \Sigma x_i y_i}{\Sigma x_i^2 - (\Sigma x_i)^2 / n}\right]$$

Example 1.30. Calculate the dimensions of a can that will hold 100 mL of material such that the amount of metal used in its manufacture is a minimum.

Let r = radius and h = height. Two independent equations are needed. One equation must involve the surface area that must be minimized. The other equation must involve the volume because the 100 mL volume requirement must be met. $V = 100 = \pi r^2 h$. $h = 100/(\pi r^2)$. $A = 2\pi rh + 2\pi r^2$. Substituting h:

$$A = \frac{2\pi(100)}{\pi r^2} + 2\pi r^2 = \frac{200}{r} + 2\pi r^2$$

$$\frac{dA}{dr} = 4\pi r - \frac{200}{r^2} = 0; \quad 4\pi r^3 = 200$$

$$r = \left[\frac{50}{\pi}\right]^{0.333} = 2.51 \text{ cm} \quad h = \frac{100}{\pi(2.51)^2} = 5.06 \text{ cm}$$

The second derivative, $d^2A/dr^2 = 4\pi + 400\pi/r^3$, will be positive for all positive values of r, therefore the root of the derivative function represents a minimum point. At r = 2.51 cm, $A = 2\pi(2.51)^2 + 200/2.51 = 119.3 \text{ cm}^2$. At r = 2.4 cm, $A = 2\pi(2.4)^2 + 200/2.4 = 119.5 \text{ cm}^2$. The value of A at r = 2.51 cm is less than at r = 2.4 cm, therefore r = 2.51 is a minimum point.

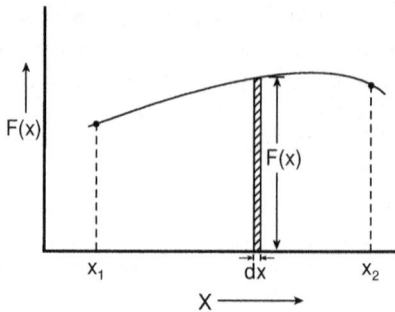

Figure 1.16   Graphical representation of an integral.

1.14   INTEGRAL CALCULUS

Integration is the inverse of differentiation. If $dF(x)$ is the differential term, the integral of $dF(x)$ is $F(x)$. Integrals are differential terms preceded by the integral sign $\int$. The terms inside the integral sign consists of a function $F(x)$ and a differential term dx. Graphically, an integral is represented in Fig. 1.16. The function $F(x)$ traces a curve and the differential term dx is represented by a small area increment with height $F(x)$ and thickness dx. When the area increments are evaluated consecutively, the sum represents the value of the integral. Limits are needed to place a definite value to the integral. The limits must correspond to the value of the variable in the differential term and is the abscissa of the curve used to plot the function within the integral sign. Graphically, the limits define which region of the curve is covered by the area summation. Figure 1.16 represents the graphical equivalent of the following integral

$$\int_{x_1}^{x_2} F(x)dx$$

When evaluating a definite integral, the limits are substituted into the integrand, and the value at the lower limit is subtracted from the value at the upper limit.

When the limits are not specified, the integral is an indefinite integral and the integrand will need a constant of integration $\int F(x) = F(x) + CI$. The value of C is determined using boundary or initial conditions that satisfy the function $F(x) + C$.

1.14.1   Integration Formulas

Only simple and most common integrals are presented here. The reader is encouraged to read a calculus textbook for techniques that are used for more complex functions.

Integral of a sum:

$$\int (du + dv + dw) = \int du + \int dv + \int dw$$

Integral of a power function:

$$\int F(x)^n dF(x) = \frac{F(x)^{n+1}}{n+1} + C$$

Integral of a quotient yielding a logarithmic function:

$$\int \frac{dF(x)}{F(x)} = \ln F(x) + C$$

Integral of an exponential function:

$$\int e^{F(x)} dF(x) = e^{F(x)} + C$$

$$\int 10^{F(x)} dF(x) = \frac{10^{F(x)}}{\ln 10} + C$$

1.14.2 Integration Techniques

1.14.2.1 Constants

Constants may be taken in or out of the integral sign. If only a constant is needed to meet the differential form needed in the above integration formulas, the integral expression may be multiplied and divided by the same constant. The multiplicand is placed within the integral to meet the required differential form, and the divisor is placed outside the integral.

$$\int F(x) dG(x) = F(x)G(x) - \int G(x) dF(x)$$

1.14.2.2 Integration by Parts

$$\int \frac{F(x)dx}{G(x)H(x)} = \int \frac{A dx}{G(x)} + \int \frac{B dx}{H(x)} + C$$

1.14.2.3 Partial Fractions

A and B are constants obtained by clearing fractions in the above equation and solving for A and B, which would satisfy the equality.

1.14.2.4 Substitution

A variable is used to substitute for a function. The substitution must result in a simpler expression that is integrable using standard integration formulas or integration by parts.

Example 1.31. $\int (x^2 + 3)^2 \, dx$.

This does not fit the power function formula because the differential term dx is not $d(x^2 + 3)$. The function can be expanded and integrated as a sum.

$$\int (x^2 + 3)^2 dx = \int (x^4 + 6x^2 + 9)\, dx = \frac{x^5}{5} + \frac{6x^3}{3} + 9x + C$$

Example 1.32.

$$\int_{0.1}^{0.3} (2x^2 + 2)^4\, x\, dx$$

This function may be integrated using the formula for a power function $d(2x^2 + 2) = 4x\, dx$. The function will be the same as in the power function formula by multiplying by 4. The whole integral is then divided by 4 to retain the same value as the original expression.

$$\int_{0.1}^{0.3} (2x^2 + 2)^4\, x\, dx = \frac{1}{4} \int_{0.1}^{0.3} (2x^2 + 2)^4 4x\, dx = \frac{1}{4} \frac{(2X^2 + 2)^5}{5} \Big|_{0.1}^{0.3} = \frac{(2.18)^5 - (2.02)^5}{20} = 0.78018$$

Example 1.33. The inactivation of microorganisms under conditions when temperature is changing is given by:

$$\frac{N_0}{N} = \int_0^t \frac{dt}{D_T}$$

N_0 is the initial number of microorganisms, and D_T is the decimal reduction time of the organism. D_T is given by:

$$D_T = D_0[10]^{(250-T)/z}$$

If $D_0 = 1.2$ minutes, $z = 18°F$, $N_0 = 10,000$, and $T = 70 + 1.1t$, where T is in °F and t is time in minutes, calculate N at $t = 250$ min. The integral to be evaluated is

$$\frac{10,000}{N} = \int_0^{250} \frac{dt}{1.2[10]^{(250-70-1.1t)/18}} = \int_0^{250} \frac{dt}{1.2[10]^{(180-1.1t)/18}}$$

$$= 0.833(10)^{-10} \int_0^{250} (10)^{0.06111t}$$

$$= \frac{1}{1.2(10)^{10}} \int_0^{250} \frac{dt}{(10)^{-0.06111t}}$$

$$= 0.833(10)^{-10} \left[\frac{(10)^{0.06111t}}{0.06111 \ln(10)} \right] \Big|_0^{250}$$

$$= \frac{0.833(10)^{-10}}{0.06111 \ln(10)} [10^{15.275} - 1]$$

The 1 in brackets is much smaller than $10^{15.275}$; therefore, it may be neglected. Therefore:

$$\frac{10,000}{N} = 5.9209(10)^{-10}(10)^{15.275}$$

$$= 5.9209(10)^{15.275-10} = 5.9209(10)^{5.275}$$

Taking logarithm of both sides:

$$\log\left(\frac{10,000}{N}\right) = 5.275 + \log 5.9209 = 6.047$$

$$N = (10)^{4-6.047} = 0.0089$$

Example 1.34.

$$\int \frac{xdx}{(2x^2 + 1)}$$

This function will yield a logarithmic function because the derivative of the denominator is $4x\,dx$. The numerator of the integral can be made the same by multiplying by 4 and dividing the whole integral by 4.

$$\int \frac{xdx}{(2x^2 + 1)} = \frac{1}{4}\int \frac{4xdx}{(2x^2 + 1)} = \frac{1}{4}\ln(2x^2 + 1) + C$$

Example 1.35. Solve for the area under a parabola $y = 4x^2$ bounded by $x = 1$ and $x = 3$ as given by the following integral:

$$A = \int_{1}^{3} 4x^2 dx = \left.\frac{4x^3}{3}\right|_{1}^{3} = \frac{1}{3}[4(27) - 4] = 34.6667$$

1.15   GRAPHICAL INTEGRATION

Graphical integration is used when functions are so complex they cannot be integrated analytically. They are also used when numerical data are available, such as experimental results, and it is not possible to express the data in the form of an equation that can be integrated analytically. Graphical integration is a numerical technique used for evaluation of differential equations by the finite difference method. Three techniques for graphical integration will be shown. Each of these will be used to evaluate the area under a parabola, solved analytically in the preceding example.

1.15.1   Rectangular Rule

The procedure is illustrated in Fig. 1.17A. The domain under consideration is divided into a sequence of bars. The thickness of the bars represent an increment of x. The height of the bars are set such that the shaded area within the curve that is outside the bar equals the area inside the bar that lies outside the curve. The increments may be unequal, but in Fig. 1.17A equal increments of 0.5 units are used.

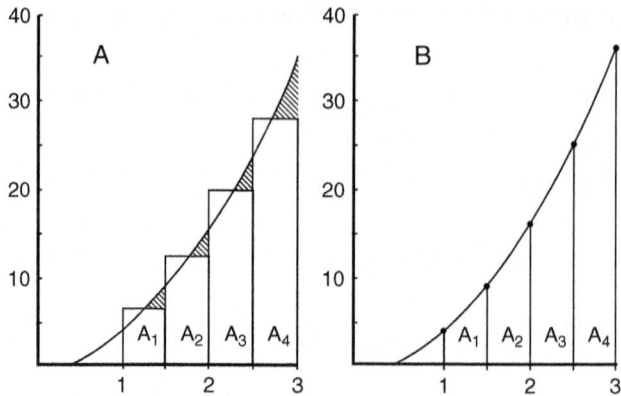

Figure 1.17   Graphical integration by the rectangular (A) and trapezoidal (B) techniques.

The height of the bars are 6.5, 12.5, 20, and 28 for A_1, A_2, A_3, and A_4, respectively. The sum of the area increments is $0.5(6.5 + 12.5 + 20 + 28) = 33.5$

1.15.2   Trapezoidal Rule

The procedure is illustrated in Fig. 1.17B. The domain under consideration is subdivided into a series of trapezoids. The area of the trapezoid is the width multiplied by the arithmetic mean of the height. For the function $y = 4x^2$, the values (height) are 4, 9, 16, 25, and 36, respectively, for x = 1, 1.5, 2, 2.5, and 3. The trapezoids A_1, A_2, A_3, and A_4 will have areas of $0.5(4 + 9)(0.5)$, $0.5(9 + 16)(0.5)$, $0.5(16 + 25)(0.5)$, and $0.5(25 + 36)(0.5)$. The sum is $3.25 + 6.25 + 10.25 + 15.25 = 35$.

1.15.3   Simpson's Rule

This procedure assumes a parabolic curve between area increments. The curve is divided into an even number of increments, therefore, including the value of the function at the lower and upper limits, there will be i values of the height of the function and $(i - 1)$ area increments. The number of increments $(i - 1)$ must be an even integer. Thickness of area increments must be uniform, therefore, for the limits x = a to x = b, $\delta x = (b - a)/(i - 1)$. Simpson's rule is as follows:

$$A = \frac{\delta x}{3}[F(a) + 4F(1) + 2F(2) + 4F(3) + 2F(4)$$
$$+ \text{---} \; 2F(i - 2) + 4F(i - 1) + F(b)]$$

Note the repeating 4, 2, 4, 2, 4 multiplier of the value of the function as succeeding increments are considered. If the index is i = 0 at x = a, the lower limit, all values of the function when the index is even are multiplied by 2 and the multiplier is 4 when the index is odd. For the example of the parabola $y = 4x^2$ from x = 1 to x = 3, setting the number of increments at 4 gives 5 values of y that must be successively evaluated for the area increment. The thickness of the area increments δx = (3 − 1)/(4) = 0.5. Values of F(x) at 1, 1.5, 2, 2.5, and 3 have been determined in the preceding examples. Substituting in the formula for Simpson's rule:

$$A = \frac{0.5}{3}[4 + 4(9) + 2(16) + 4(25) + 36] = 34.6667$$

The accuracy of graphical integration using Simpson's rule is better than the trapezoidal rule. Both the trapezoidal rule and Simpson's rule can be programmed easily in Visual BASIC. The following program in Visual BASIC illustrates how the area of a curve drawn to fit numerical data is determined using Simpson's rule. The area under the parabola $y = 4x^2$ is being analyzed. Smaller thickness of the area increment is used than in the previous example. Ten area increments are chosen, therefore 11 values of the function will be needed to include x = 1 and x = 3, the upper limit.

These values are entered into a Inputbox as variable y with dimension statements.

```
Sub test3()
Dim y (1 to 11) as double
For i = 1 To 11
y(i) = Val(InputBox("Input Data for'' & "Y('' & i & ")''"))
next i
i = 11
dx = 0.2
For n = 2 To i Step 2
sa = sa + y(n)
Next n
For n = 3 To i - 1 Step 2
sb = sb + y(n)
Next n
MsgBox "A = '' & (dx / 3)* (4* sa + 2* sb + y(1) + y(11))
End Sub
```

The program, when run, will display A = 34.66667 on the screen.

1.16 DIFFERENTIAL EQUATIONS

Differential equations are used to determine the functionality of variables from information on rate of change. The simplest differential equations are those where the variables are separable. These equations take the form:

$$\frac{dy}{dx} = F(x); \quad dy = F(x)\,dx$$

The solution to the differential equation then involves integration of both sides of the equation. Constants of integration are determined using initial or boundary conditions.

Example 1.36. If the rate of dehydration is constant at 2 kg water/h from a material that weighs 20 kg and contains 80% moisture at time = 0, derive an expression for moisture content as a function of time.

Let W = weight at any time, t = time, and x = moisture content expressed as a mass fraction of water. From the rate data, the differential equation is

$$\frac{dW}{dt} = -2$$

The negative sign arises because the rate is expressed as a weight loss, therefore W will be decreasing with time. Separating variables: $dW = -2dt$. Integrating: $W = -2t + C$. The constant of integration is determined by substituting $W = 20$ at $t = 0$ and the expression for W becomes $W = 20 - 2t$. The moisture content x is determined from W as follows:

$$W = \frac{\text{weight dry solids}}{\text{mass fraction dry solids}} = \frac{20(1 - 0.8)}{1 - x}$$

Substituting W and simplifying:

$$\frac{1}{1 - x} = 5 - 0.5t$$

Example 1.37. The venting of air from a retort by steam displacement is analogous to dilution where the air is continuously diluted with steam and the diluted mixture is continuously vented. Let C = concentration of air at any time, t = time, V = the volume of the retort, and R = the rate of addition of steam. The differential equation for C with respect to time is:

$$V\frac{-dC}{dt} = RC; \quad \frac{dC}{C} = \frac{-R}{V}\,dt$$

The integration constant will be evaluated using, as an initial condition, $C = Co$ at $t = 0$. Integrating: $\ln C = (-R/V)t + \ln C_0$.

1.17   FINITE DIFFERENCE APPROXIMATION OF DIFFERENTIAL EQUATIONS

Differential equations may be approximated by finite differences in the same manner as integrals are solved by numerical methods using graphical integration. Finite increments are substituted for the differential terms and the equation is solved within the specified boundaries. Finite difference approximations can be easily done using BASIC.

Example 1.38. A stirred tank having a volume V, in liters, contains a number of cells of microorganisms, N. The tank is continuously fed with cell-free media at the rate of R (L/h) while the volume is maintained constant by allowing media to overflow at the same rate as the feed. The number of cells in the tank will increase due to reproduction, and some of the cells are washed out in the overflow. (a) Derive an equation for the cell number at any time, as a function of the generation time, g, the feed rate, R, and the volume, V. (b) Calculate N after 5 hours, if $g = 0.568$ h, $N_0 = 10,000$, $V = 1.5$ L, and $R = 1.5$ L/h.

Solution:

The number of microorganisms in the most rapid rate of growth will increase with time of growth according to: $N = N_0(2)^{t/g}$ where g is generation time. Washout is $N(R)$. Net accumulation is dN/dt (V). Growth $= (dN/dt)_{gr}$ A balance of cell numbers will give: Cell growth = washout +

accumulation.

$$\text{Cell growth} = V\left(\frac{dN}{dt}\right)_{gr} = V\frac{d}{dt}(N_0)(2)^{t/g} = N_0(2)^{t/g}(\ln 2)V = VN\frac{\ln 2}{g}$$

The differential equation is

$$V\,N\frac{\ln 2}{g} = NR + V\frac{dN}{dt}$$

Rearranging and separating variables:

$$\frac{dN}{N} = \left(\frac{\ln 2}{g} - \frac{R}{V}\right)dt$$

Integrating and using as an initial condition $N = N_0$ at $t = 0$:

$$\ln\frac{N}{N_0} = \left(\frac{\ln 2}{g} - \frac{R}{V}\right)t$$

To solve the problem by finite difference techniques, values of V, R, and N_0 and g must be defined. If $g = 0.568$ h, $N_0 = 10,000$, $R = 1.5$ L/h, and $V = 1.5$ L, the number of organisms in the tank at $t = 5$ h is as follows: From the analytical solution:

$$\ln N = \ln 10,000 + \left[\frac{\ln 2}{0.568} - \frac{1.5}{1.5}\right](5) \quad (5)$$

$$= 9.21 + 1.101 = 10.311$$

$$N = (e)^{10.311} = 30,060$$

The differential equation can be solved using a finite difference technique utilizing Visual BASIC. The program in Visual BASIC for the differential equation is

```
Sub diff()
t = 0
no = 10000
r = 1.5
v = 1.5
g = 0.568
dt = 0.1
n = no
Do
dn = n * (Log(2) / g - (r / v)) * dt
n = n + dn
t = t + dt
Loop While t < 5
MsgBox "time='' & (t - dt) & " n = '' & n
End Sub
```

When the program is run, "Time $= 5$ N $= 30,054.757$" will be displayed on the screen.

PROBLEMS

1.1. Determine the maximum and minimum value of the following function and prove that the points are maximum or minimum.

$$C = \frac{315 + 52.5T}{(0.21T - 0.76)^{0.5} - 1.61}$$

1.2. A warehouse having a volume of 10,000 ft^3 and a floor area of 1000 ft^2 is to be built. The cost of constructing the floor is $6.00/ft^2, the cost of the roof is $10.00/ft^2, and the cost of the walls is $20.00/ft^2. If W is the width, L the length, and H the height of the building, what should the dimensions be such that the cost is minimal?

1.3. Calculate the maximum or minimum value of the following expression. Show that the calculated value is maximum or minimum.

$$q = \frac{240R}{-0.02 + 20\ln(0.5/R) + 0.02R}$$

1.4. Calculate the maximum (or minimum value) of y.

$$y = 2X^2 + 0.5R + X + 3$$
$$X = 0.5R$$

1.5. Calculate the maximum velocity of a fluid flowing inside a pipe expressed in terms of the average velocity.

The point velocity (velocity at any point r measured from the center) is

$$V = \left[\frac{\Delta p}{2\,LK}\right]^{1/n} \left[1 - \left[\frac{r}{R}\right]^{(n+1)/n}\right] \left[\frac{n}{n+1}\right] [R]^{(n+1)/n}$$

The average velocity is

$$\bar{V} = \left[\frac{\Delta P}{2\,LK}\right]^{1/n} \left[\frac{n}{3n+1}\right] [R]^{(n+1)/n}$$

1.6. The rate of evaporation of component A divided by the rate of evaporation of component B in a mixture containing A moles of A and B moles of B is directly proportional to A/B. If a mixture originally contained 5 moles of A and 3 moles of B and the rate of evaporation of A is 5 moles/h and of B is 2.6 moles/h, derive an equation for the concentration of A relative to that of B.

1.7. The rate of production of ethanol in a fermentor is directly proportional to the number of cells of yeast present. At a cell mass of 14 g yeast/liter, ethanol production rate is 18 g ethanol/ ($L \cong h$). If a batch fermentor is inoculated with 0.1 g of yeast/L, and the generation time of the yeast is 1.5 hours, what will be the ethanol production after 10 hours of fermentation. Assume no alcohol-induced inhibition of yeast growth and that yeast doubles in cell mass with each generation time.

1.8. The work required to compress a gas is given by the following integral:

$$W = \int_{V_1}^{V_2} P dV$$

Van der Waals equation of state for a gas is is follows:

$$P = \frac{nRT}{V - nb} - \frac{n^2 a}{V^2}$$

$R = 82.06$ (cm^3 $\cong$ atm)/(g mole $\cong$ K), b $= 36.6$ cm^3/g mole, and a $= 1.33 \times 10^6$ (atm $\cong$ cm^6)/(g mole)2, when P is in atm, T in Kelvin, and V in cm^3. Calculate the work done when the pressure of the gas is increased from 1 to 10 atm. There was originally 1 liter of gas at 293 K.

1.9. Write a computer program in BASIC which can be used to calculate a value for the water activity of a solution containing two sugars having mass fractions (percentage composition by weight expressed as a decimal) x_1 and x_2. x_1 is sucrose (molecular weight 342; k $= 2.7$); x_2 is glucose (molecular weight 180, k $= 0.5$). The water activity of the mixture is

$$a_w = (a_{w1})^0 (a_{w2})^0$$

where $(a_{w1})^0$ is water activity of a solution containing all the water in the mixture and component 1; and $(a_{w2})^0$ is water activity of a solution containing all the water in the mixture and component 2.

$$\log_{10}\left[\frac{(a_{wi})^0}{f_i}\right] = -k_i(1 - f_i)^2$$

where f_i is mole fraction of component i in a solution containing only component i and all the water present in the mixture.

$$f_i = 1 - \left[\frac{x_i/M_i}{(x_i/M_i) + (1 - x_1 - x_2)/18}\right]$$

1.10. Determine the slope of the following functions:

$$f(x) = 2x^4 - 3x^2 + 7 = 0 \quad \text{at } x = 2$$
$$xy = (0.5x + 3)(x + 2) \quad \text{at } x = 1$$

1.11. Determine the maximum and minimum value of the functions in Problem 10 above.

1.12. Determine the slope of the following function at the indicated point.

$$x = (0.5xy + 3)(3 + x^2) \quad \text{at } x = 1$$

1.13. Construct a spreadsheet that can be used to determine the boiling temperature of a liquid in an evaporator as it is being concentrated. The boiling point of a liquid is the temperature at which the vapor pressure equals the atmospheric pressure. A solution will exhibit a boiling point rise because the solute will lower the water activity resulting in a higher temperature to be reached before boiling occurs. The vapor pressure of water (P^0) as a function of temperature is expressed by the following equation:

$$\ln(P^0) = \left(-\frac{H}{R}\right)\left(\frac{1}{T}\right) + C$$

where P^0 is the vapor pressure in kilopascals, H/R is the ratio of the latent heat of vaporization and the gas constant, C is a constant, and T is the absolute temperature in degrees Kelvin. The value for H/R is 4950 and C is 17.86.

Atmospheric pressure is 101 kilopascals. The vapor pressure of a solution is $P = a_w P^0$. The water activity (a_w) is given by:

$$\log_{10}\left(\frac{a_w}{f_1}\right) = -k(1 - f_1)^2$$

f_1 is the mole fraction of water and is calculated by:

$$f_1 = \frac{(1 - x_1)/18}{(1 - x_1)/18 + x_1/M_1}$$

x_1 is the mass fraction of solute. Assume the solute is only sucrose with a molecular weight (M_1) of 342 and a k value of 2.7. Have the program display the value of the boiling point when the sucrose concentration is 20% and at increasing concentrations in 5% intervals to a final concentration of 60%.

1.14. Calculate a value of x that would give the minimum value for S in the following expression:

$$S = 50(9.522 \times 10^{-6}) - \frac{50}{(50X + 1)}(9.522 \times 10^{-6}) - \frac{60(0.22)X}{30(24)}$$

Show that the function is a minimum or maximum.

1.15. In the dehydration of diced potatoes, the following weights were recorded at various times in the process when dehydration rate was the slowest.

   t = 6 hours from start of drying: weight = 2350 g
   t = 8 hours from start of drying: weight is 2275 g; moisture content = 15.6%.

In this range of moisture content, the drying rate is proportional to the moisture content, expressed in mathematical form as follows:

$$-\frac{dW}{dt} = kW$$

W is the moisture content in g water/g dry matter, and k is a constant.

(a) Derive an equation for the moisture content, W, as a function of time, t, which satisfies the experimental conditions given above.

(b) How long would it take for a product to be dehydrated from 22.5% to 12.5% moisture?

$$\frac{N_0}{N} = \frac{R^2\overline{V}}{2}\left[\frac{1}{\int_0^R rV_r[10]^{-L/(V_r D)}dr}\right]$$

1.16. Write a computer program in BASIC that will solve the following integral:
where

$$V_r = \overline{V}\left[\frac{3n + 1}{n + 1}\right]\left[1 - \frac{r^{(n+1)/n}}{R}\right]$$

Given: $\overline{V} = 2$, $R = 0.02$, $n = 0.6$, $L = 10$, $D = 1$. Use the trapezoidal rule in evaluating the integral and use increments in r of 0.0001.

1.17. Linearize the following equations. In each case, indicate which function should be plotted as the independent and dependent variable to obtain a linear plot. What is the slope and intercept of each plot?

(a) $\log(a_w/X_i) = -k(1 - X_i)^2$ where a_w and X_i are variables.

(b) $(1/x) = (a + b)/2y + c/4y$ where x and y are variables and a, b, and c are constants.

(c) $N = N_0(10)^{-(t/D)}$ where N and t are variables and N_0 and D are constants.

1.18. The velocity of an enzyme-catalyzed reaction (V) expressed as a function of the substrate

$$V = \frac{V_{max}(S)}{K_m + S}$$

concentration (S) is given by the following equation:

where V_{max} is the maximum reaction rate and K_m is a constant for the reaction. Linearize the equation and determine the independent and dependent variables to be plotted to obtain V_{max} and K_m from the slope and intercept.

1.19. The temperature (T) of a refrigerant during compression in a refrigeration system increases with pressure (P) as shown in the following equation. P_1 and T_1 are reference temperature and pressure and are considered constants.

$$\frac{T}{T_1} = \left[\frac{P}{P_1}\right]^{(k-1)/k}$$

The following data are available on the temperature and pressure of a refrigerant during compression:

T(ER)	450	510	550	608	640
$P(lb_f/in^2)$	4	6	10	20	30

Linearize the above equation, perform a linear regression, and determine the constant k for this refrigerant. Calculate the value of k by nonlinear curve fitting.

1.20. The heat of respiration of leafy greens as a function of time when the temperature is changing during a cooling operation is as follows:

$$q = \int_0^t 0.009854[e]^{0.073T}\,dt$$

where q is in BTU/(lb), T is in °F, and t is time in hours. The temperature of a box of spinach containing 50 lb, originally at 110°F, will cool down exponentially when placed in a refrigerated room at 35°F according to:

$$T = 35 + (75)\exp(-t/5)$$

where t is the time in hours. Evaluate this integral using a spreadsheet and determine the total heat generated by the spinach as the box cools down from 110°F to 40°F. Use Simpson's rule to determine the value of the integral.

1.21. An immobilized enzyme reactor must have the enzyme regenerated periodically because of the decay in the activity of the enzyme. The enzyme will convert 87% of the substrate to product within the first day of operation, and this conversion changes to $0.87/(t)^{0.82}$ after the first day. If the feed rate is 150 lb of substrate per day, the amount of product formed after t days of operation will be:

$$P = 150(0.87) + \left[\frac{150(0.87)}{(t)^{0.82}} \right] (t - 1)$$

At a substrate cost of \$2.50/lb, an operating cost of \$600.00/day, a product cost of \$14.00/lb, and an enzyme replacement cost of \$600, the return from the operation will be

$$S = 14P - 600(t) - 2.5(P) - 600/t$$

Calculate the number of days t that the reactor must be operated before recharging in order that the return, S, will be maximum. Prove that S is maximum at the value of t identified.

SUGGESTED READING

Burrows, W. H. 1965. Graphical Techniques for Engineering Computation. Chemical Publishing Co., New York.

Cumming, H. G. and Anson, C. J. 1967. Mathematics and Statistics for Technologists. Chemical Publishing Co., New York.

Dull, R. W. and Dull, R. 1951. Mathematics for Engineers. McGraw-Hill Book Co., New York.

Greenberg, D. A. 1965. Mathematics for Introductory Science Courses. Calculus and Vectors with a Review of Basic Algebra. W. A. Benjamin, New York.

Feldner, R. M. and Rousseau, R. W. 1999. Elementary Principles of Chemical Processes. 2nd ed. John Wiley & Sons, New York.

Kelly, F. H. C. 1963. Practical Mathematics for Chemnists. Butterworths, London.

Mackie, R. K., Shephard, T. M., and Vincent, C. A. 1972. Mathematical Methods for Chemists. The English University Press, London.

Perry, R. H., Chilton, C. H., and Kirkpatrick, S. D. 1963. Chemical Engineers Handbook. 4th ed. McGraw-Hill Book Co., New York.

Person, R. 1997. Using Microsoft Excel 97. QUE Corporation, Indianapolis, IN.

Wilkinson, L., Hill, M., Welna, J., and Birkenbeuel, G. 1992. Statistics. Systat Inc., Evanston IL.

CHAPTER 2

Units and Dimensions

Units used to designate magnitude of a dimension have evolved based on common usage and instruments available for measurement. Two major systems for measurement have been used: the English system, which was used primarily in industry, and the metric system, which was used in the sciences. The confusion that results from the use of various terms to represent the same dimension has led to the development of a common system of units that is proposed for use in both science and industry. The *Système International d'Unites* (International System of Units) and the official international designation SI was adopted in 1960 by the General Conference on Weights and Measures. This body consists of delegates from member countries of the Meter Convention, and it meets at least once every 6 years. There are at least 44 countries represented in this convention, one of which is the United States.

The use of SI is now widespread in the scientific community. However, it is still often necessary to convert data from one system to another, as tables in handbooks may be in a different unit from what is needed in the calculations, or experimental data may be obtained using instruments calibrated in a different unit from what is desired in reporting the results.

In this chapter, the various units in SI are discussed, and techniques for conversion of units using the dimensional equation are presented. Also emphasized in this chapter is the concept of dimensional consistency of mathematical equations involving physical quantities and how units of variables in an equation are determined to ensure dimensional consistency.

2.1 DEFINITION OF TERMS

Dimension: used to designate a physical quantity under consideration (e.g., time, distance, weight).

Unit: used to designate the magnitude or size of the dimension under consideration (e.g., m for length, kg for weight).

Base unit: Base units are dimensionally independent. They are used to designate only one dimension (e.g., units of length, mass, and time).

Derived units: a combination of various dimensions. An example of a derived unit is the unit of force, which includes the dimensions of mass, length, and time.

Precision: synonymous with reproducibility, the degree of deviation of the measurements from the mean. This is often expressed as a $\pm$ term or as the smallest value of unit that can be consistently read in all determinations.

Table 2.1 Systems of Measurement.

System	Use	Length	Mass	Time	Temperature	Force	Energy
					Dimension		
English							
English absolute	Scientific	Foot	Pound mass	Second	°F	Poundal	BTU ft (poundal)
British Engineering	Industrial	Foot	Slug	Second	°F	Pound force	BTU ft (pound force)
American Engineering	U.S. industrial	Foot	Pound mass	Second	°F	Pound force	BTU ft (pound force)
Metric							
Cgs	Scientific	Centimeter	Gram	Second	°C	Dyne	Calorie, erg
Mks	Industrial	Meter	Kilogram	Second	°C	Kilogram force	Kilocalorie joule
SI	Universal	Meter	Kilogram	Second	°K	Newton	Joule

Accuracy: refers to how a measured quantity relates to a known standard. To test for accuracy of a measurement, the mean of a number of determinations is compared against a known standard. Accuracy depends on proper calibration of an instrument.

2.2 SYSTEMS OF MEASUREMENT

The various systems in use are shown in Table 2.1. These systems vary in the base units used. Under the English system, variations exist in expressing the unit of force. The chemical and food industries in the United States use the American Engineering System, although SI is the preferred system in scientific articles and textbooks.

The metric system uses prefixes on the base unit to indicate magnitude. Industry adopted the "mks" system, whereas the sciences adopted the "cgs" system. SI is designed to meet the needs of both science and industry.

2.3 THE SI SYSTEM

The following discussion of the SI system and the convention followed in rounding after conversion is based on the American National Standard, Metric Practice, adopted by the American National Standards Institute, the American Society for Testing and Materials, and the Institute of Electrical and Electronics Engineers.

2.3.1 Units in SI and Their Symbols

SI uses base units and prefixes to indicate magnitude. All dimensions can be expressed in either a base unit or combinations of base units. The latter are called derived units and some have specific names. The base units and the derived units with assigned names are shown in Table 2.2.

Table 2.2 Base Units of SI and Derived Units with Assigned Names and Symbols.

Quantity	Unit	Symbol[a]	Formula
Length	meter	m	—
Mass	kilogram	kg	—
Electric current	ampere	A	—
Temperature	kelvin	K	—
Amount of substance	mole	mol	—
Luminous intensity	candela	cd	—
Time	second	s	—
Frequency (of a periodic phenomenon)	hertz	Hz	1/s
Force	newton	N	$kg\ m/s^2$
Pressure, stress	pascal	Pa	N/m^2
Energy, work, quantity of heat	joule	J	$N \cdot m$
Power, radient flux	watt	W	J/s
Quantity of electricity, electric charge	coulomb	C	$A \cdot s$
Electric potential, potential difference, electromotive force	volt	V	W/A
Capacitance	farad	F	C/V
Electric resistance	ohm		$V \cdot A$
Conductance	siemens	S	A/V
Magnetic flux	weber	Wb	$V \cdot s$
Magnetic flux density	tesla	T	Wb/m^2
Inductance	henry	H	Wb/A
Luminous flux	lumen	1m	$cd \cdot sr^b$
Illuminance	lux	lx	$1m/m^2$
Activity (of radionuclides)	becquerel	Bq	1/s
Absorbed dose	gray	Gy	J/kg

Source: American National Standard, 1976. Metric Practice. IIEE Std. 268–1976. Institute of Electrical and Electronics Engineers, New York.
[a]Symbols are written in lowercase letters unless they are from the name of a person.
[b]sr stands for steradian, a supplementary unit used to represent solid angles.

2.3.2 Prefixes Recommended for Use in SI

Prefixes are placed before the base multiples of 10. Prefixes recommended for general use are shown in Table 2.3.

A dimension expressed as a numerical quantity and a unit must be such that the numerical quantity is between 0.1 and 1000. Prefixes should be used only on base units or named derived units. Double prefixes should not be used.

Examples:

1. 10,000 cm should be 100 m, not 10 kcm.
2. 0.0000001 m should be 1 μm.
3. 3000 m^3 should *not* be written as 3 km^3.
4. 10,000 N/m^2 can be written as 10 kPa but not 10 kN/m^2.

Table 2.3 Prefixes Recommended for Use in SI

Prefix	Multiple	Symbol[a]
tera	10^{12}	T
giga	10^{9}	G
mega	10^{6}	M
kilo	1000	k
milli	10^{-3}	m
micro	10^{-6}	μ
nano	10^{-9}	n
pico	10^{-12}	p
femto	10^{-15}	f

[a]Symbols for the prefixes are written in capital letters when the multiplying factor is 10^6 and larger. Prefixes designating multiplying factors less than 10^6 are written in lower case letters.

2.4   CONVERSION OF UNITS

2.4.1   Precision, Rounding-Off Rule, Significant Digits

Conversion from one system of units to another should be done without gain or loss of precision. Results of measurements must be reported such that a reader can determine the precision. The easiest way to convey precision is in the number of significant figures.

Significant figures include all nonzero digits and nonterminal zeroes in a number. Terminal zeroes are significant in decimals and they may be significant in whole numbers when specified. Zeroes preceding nonzero digits in decimal fractions are not significant.

Examples:

1. 123 has three significant figures.
2. 103 has three significant figures.
3. 103.03 has five significant figures.
4. 10.030 has five significant figures.
5. 0.00230 has three significant figures.
6. 1500 has two significant figures unless the two terminal zeroes are specified as significant.

When the precision of numbers used in mathematical operations is known, the answer should be rounded off following these rules recommended by the American National Standards Institute. All conversion factors are assumed to be exact.

1. In addition or subtraction, the answer shall not contain a significant digit to the right of the least precise of the numbers. Example: $1.030 + 1.3 + 1.4564 = 3.8$. The least precise of the numbers is 1.3. Any digit to the right of the tenth digit is not significant.
2. In multiplication or division, the number of significant digits in the answer should not exceed that of the least precise among the original numbers. Example: $123 \times 120 = 15,000$ if the terminal

zero in 120 is not significant. The least precise number has two significant figures and the answer should also have two significant figures.

3. When rounding-off, raise the terminal significant figure retained by 1 if the discarded digits start with 5 or larger, otherwise the terminal significant figure retained is unchanged. If the start of the digits discarded is 5, followed by zeroes, make the terminal significant digit retained even.

If the precision of numbers containing terminal zeroes is not known, assume that the number is exact. When performing a string of mathematical operations, round-off only after the last operation is performed.

Examples:

1253 rounded-off to two significant figures = 1300.
1230 rounded-off to two significant figures = 1200 (2 is even).
1350 rounded-off to two significant figures = 1400 (3 is odd).
1253 rounded-off to three significant figures = 1250.
1256 rounded-off to three significant figures = 1260.

2.5   THE DIMENSIONAL EQUATION

The magnitude of a numerical quantity is uncertain unless the unit is written along with the number. To eliminate this ambiguity, make a habit of writing both a number and its unit.

An equation that contains both numerals and their units is called a dimensional equation. The units in a dimensional equation are treated just like algebraic terms. All mathematical operations done on the numerals must also be done on their corresponding units. The numeral may be considered as a coefficient of an algebraic symbol represented by the unit. Thus, $(4m)5 = (4)^2(m)^2 = 16m5$ and

$$\left[5\frac{J}{kg\ K} \right](10\,kg)(5\ K) = 5(10)(5)\left[\frac{J\ kg\ K}{kg\ K} \right] = 250\,J$$

Addition and subtraction of numerals and their units also follow the rules of algebra; that is, only like terms can be added or subtracted. Thus, $5m - 3m = (5 - 3)m$, but $5m - 3cm$ cannot be simplified unless their units are expressed in like units.

2.6   CONVERSION OF UNITS USING THE DIMENSIONAL EQUATION

Determining the appropriate conversion factors to use in conversion of units is facilitated by a dimensional equation. The following procedure may be used to set up the dimensional equation for conversion.

1. Place the units of the final answer on the left side of the equation.
2. The number being converted and its unit is the first entry on the right-hand side of the equation.
3. Set up the conversion factors as a ratio using Appendix Table A.1.
4. Sequentially multiply the conversion factors such that the original units are systematically eliminated by cancellation and replacement with the desired units.

Example 2.1. Convert BTU/(lb $\cong$ °F) to J/(g $\cong$ K)

$$\frac{J}{g \cdot K} = \frac{BTU}{lb.°F} \times \text{appropriate conversion factors}$$

The numerator J on the left corresponds to BTU on the right-hand side of the equation. The conversion factor is 1054.8 J/BTU. Because the desired unit has J in the numerator, the conversion factor must have J in the numerator. The factor 9.48×10^4 BTU/J may be obtained from the table, but it should be entered as J/9.48×10^4 BTU in the dimensional equation. The other factors needed are 2.2046×10^{-3} lb/g or lb/453.6 g and 1.8°E/K.

The dimensional equation is

$$\frac{J}{g \cdot K} = \frac{BTU}{lb \cdot ° F} \cdot \frac{1054.8 \text{ J}}{BTU} \cdot \frac{2.2046 \times 10^{-3} lb}{g} \cdot \frac{1.8°F}{K}$$

Another form for the dimensional equation is

$$\frac{J}{g \cdot K} = \frac{BTU}{lb \cdot ° F} \cdot \frac{J}{9.48 \times 10^{-4} BTU} \cdot \frac{lb}{453.6g} \cdot \frac{1.8°F}{K}$$

Canceling out units and carrying out the arithmetic operations:

$$\frac{J}{g \cdot K} = \frac{BTU}{lb \cdot ° F} \times 4.185$$

Example 2.2. The heat loss through the walls of an electric oven is 6500 BTU/h. If the oven is operated for 2 hours, how many kilowatt hours of electricity will be used just to maintain the oven temperature (heat input = heat loss)?

To solve this problem, rephrase the question. In order to supply 6500 BTU/h for 2 hours, how many kilowatt hours are needed? Note that power is energy/time; therefore the product of power and time is the amount of energy. Energy in BTU is to be converted to J.

The dimensional equation is

$$J = \frac{6500 \text{ BTU} \cdot 2 \text{ h}}{h} \cdot \frac{1054.8 \text{ J}}{BTU}$$

W = J/s, therefore W $\cong$ s = J,

$$kW \cdot h = W \cdot s \frac{1kW}{1000W} \frac{1h}{3600s}$$

$$kW \cdot h = \frac{6500 \text{ BTU} \cdot 2 \text{ h}}{h} \cdot \frac{1054.8J}{BTU} \cdot \frac{kW}{1000W} \cdot \frac{h}{3600 \text{ s}}$$

$$kW \cong h = 3.809.$$

An alternative dimensional equation is

$$kW \cdot h = \frac{6500 \text{ BTU} \cdot 2 \text{ h}}{h} \cdot \frac{h}{60 \text{ min}} \cdot \frac{1.757 \times 10^{-2} \text{ kW}}{BTU/ \text{ min}} = 3.809$$

The conversion factors in Appendix Table A.1 expresses the factors as a ratio and is most convenient to use in a dimensional equation. Although column I is labeled "denominator" and columns II and III are labeled "numerator," the reciprocal may be used. Just make sure that the numerals in column II go with the units in column III.

2.7 THE DIMENSIONAL CONSTANT (G_c)

In the American Engineering System of measurement, the units of force and the units of mass are both expressed in pounds. It is necessary to differentiate between the two units because they have actually different physical significance. To eliminate confusion between the two units, the pound force is usually written as lb_f and the pound mass is written as lb_m.

Because the force of gravity on the surface of the earth is the weight of a given mass, in the American Engineering System, the weight of 1 lb_m is exactly 1 lb_f. Thus, the system makes it easy to conceptualize the magnitude of a $1-lb_f$.

The problem in the use of lb_f and lb_m units is that an additional factor is introduced into an equation to make the units dimensionally consistent, if both units of force and mass are in the same equation. This factor is the dimensional constant g_c.

The dimensional constant g_c is derived from the basic definition of force: Force = mass × acceleration.

If lb_f is the unit of force and lb_m is the unit of mass, substituting ft/s^2 for acceleration results in a dimensional equation for force that is not dimensionally consistent.

$$lb_f = lb_m \cdot \frac{ft}{s^2}$$

The American Engineering System of measurement is based on the principle that the weight, or the force of gravity on a pound mass on the surface of the earth, is a pound force. Introducing a dimensional constant, the equation for force becomes:

$$lb_f = lb_m \cdot \frac{ft}{s^2} \cdot \frac{1}{g_c}$$

To make lb_f numerically equal to lb_m when acceleration due to gravity is 32.174 ft/s^2, the dimensional constant g_c should be a denominator in the force equation and have a numerical value of 32.174. The units of g_c is

$$g_c = \frac{ft \cdot lb_m}{lb_f \cdot s^2}$$

The dimensional constant, g_c, should not be confused with the acceleration due to gravity, g. The two quantities have different units. In SI, the unit of force can be expressed in the base units following the relationship between force, mass, and acceleration, therefore, g_c is not needed. However, if the unit kg_f is to be converted to SI, the equivalent g_c of 9.807 kg × m/kg_f × s^2 should be used. Use of conversion factors in Appendix Table A.1 eliminates the need for g_c in conversion from the American Engineering System of units to SI.

2.8 DETERMINATION OF APPROPRIATE SI UNITS

A key feature of SI is expression of any dimension in terms of the base units of meter, kilogram, and second. Some physical quantities having assigned names as shown in Table 2.2 can also be expressed in terms of the base units when used in a dimensional equation. When properly used, the coherent nature of SI ensures dimensional consistency when all quantities used for substitution into an equation are in SI units. The following examples illustrate selection of appropriate SI units for a given quantity.

Example 2.3. A table for viscosity of water at different temperatures lists viscosity in units of $lb_m/(ft \cong h)$. Determine the appropriate SI unit and calculate a conversion factor.

The original units have units of mass (lb_m), distance (ft), and time (h). The corresponding SI base units should be $kg/(m \cong s)$. The dimensional equation for the conversion is

$$\frac{kg}{m \cdot s} = \frac{lb_m}{ft \cdot h} \cdot \text{conversion factor} = \frac{lb_m}{ft \cdot h} \cdot \frac{1kg}{2.2046\, lb_m}$$

$$\frac{3.281\,ft}{1m} \cdot \frac{1h}{3600s} \cdot \frac{kg}{m \cdot s} = \frac{lb_m}{ft \cdot h}(3.84027 \times 10^{-5})$$

Viscosity in SI is also expressed in $Pa \cong s$. Show that this has the same base units as in the preceding example.

$$Pa \cdot s = \frac{N}{m^2} \cdot s = \frac{kg \cdot m}{s^2} \frac{s}{m^2} = \frac{kg}{m \cdot s}$$

Example 2.4. Calculate the power available in a fluid that flows down the raceway of a reservoir at a rate of 525 lb_m/min from a height of 12.3 ft. The potential energy (PE) is

$$PE = m\, g\, h$$

where m = mass, g = the acceleration due to gravity = 32.2 ft/s^2, and h is the height.

From Table 2.2, the unit of power in SI is the watt and the formula is J/s. Expressing in base units:

$$P = W = \frac{N \cdot m}{s} = \frac{kg \cdot m}{s^2} \cdot \frac{m}{s} = \frac{kg \cdot m^2}{s^3}$$

$$\frac{kg \cdot m^2}{s^3} = \frac{525\,lb_m}{min} \cdot \frac{32.2\,ft}{s^2} \cdot (12.3ft) \cdot \frac{kg}{2.2046\, lb_m}$$

$$\cdot \frac{1\,min}{60s} \cdot \frac{m^2}{(3.281)^2 ft^2} = 146W$$

2.9   DIMENSIONAL CONSISTENCY OF EQUATIONS

All equations must have the same units on both sides of the equation. Equations should be tested for dimensional consistency before substitution of values of variables. A dimensional equation would easily verify dimensional consistency. Consistent use of SI for units of variables substituted into an equation ensures dimensional consistency.

Example 2.5. The heat transfer equation expresses the rate of heat transfer (q = energy/time) in terms of the heat transfer coefficient h, the area A, and the temperature difference ΔT.

Test the equation for dimensional consistency and determine the units of h.

$$q = h\, A\, \Delta T$$

The dimensional equation will be set up using SI units of the variables:

$$\frac{J}{s} = W = (-)(m^2)(K)$$

The equation will be dimensionally consistent if the units of $h = W/m^2 \cdot K$. Then m5 and K will cancel out on the right-hand side of the equation, leaving W on both sides, signifying dimensional consistency.

Example 2.6. Van der Waal's equation of state is as follows:

$$\left(P + \frac{n^2 a}{V^2}\right)(V - nb) = nRT$$

where P is pressure, n is moles, V is volume, R is gas constant, and T is temperature. Determine units of the constants a, b, and R and test for dimensional consistency.

Because the two terms in the left-hand side of the equation involve a subtraction, both terms in the first parenthesis will have units of pressure and in the second parenthesis units of volume. The product of pressure times volume will be the same units on the right side of the equation.

Using SI units for pressure of N/m^2 and volume, m^3, the dimensional equation is

$$\frac{N}{m^2}(m^3) = (kg\,mole)(- - -)(^\circ K)$$

The equation will be dimensionally consistent if the units of $R = N \cdot m/kg$ mole because that will give the same units, Nm, on both sides of the equation. The units of a and b are determined as follows. The first term, which contains a, has units of pressure:

$$\frac{N}{m^2} = \frac{(kg\,mole)^2(- - - -)}{(m^3)^2}$$

To make the right-hand side have the same units as the left, a will have units of $N(m)^4/(kg\,mole)^2$. The second term, which contains b has units of volume.

$$m^3 = (kg\,mole)(- - - -)$$

Thus, b will have units of m^3/kg mole.

2.10 CONVERSION OF DIMENSIONAL EQUATIONS

Equations may be dimensionless (i.e., dimensionless groups are used in the equation). Examples of dimensionless groups are Reynolds number (Re), Nusselt number (Nu), Prandtl number (Pr),

Fourier number (Fo), and Biot number (Bi). These numbers are defined below, and the dimensional equations for their units show that each group is dimensionless. The variables are as follows: V is velocity, D is diameter, ρ is density, μ is viscosity, k is thermal conductivity, h is heat transfer coefficient, α is thermal diffusivity, C_p is specific heat, L is thickness, and t is time.

$$Re = \frac{DV\rho}{\mu} = m\frac{m\,kg}{s\,m^3}\frac{1}{kg/(m \cdot s)}$$

$$Nu = \frac{hD}{k} = \frac{W}{m^2 \cdot K}m\frac{1}{W/(m \cdot K)}$$

$$\text{Pr} = \frac{C_\rho\,\mu}{k} = \frac{J}{kg \cdot K} \cdot \frac{kg}{m \cdot s} \cdot \frac{1}{J/(s \cdot m \cdot K)}$$

$$\text{Fo} = \frac{\propto t}{L^2} = \frac{m^2}{s} \cdot s \cdot \frac{1}{m^2}$$

$$\text{Bi} = \frac{hL}{k} = \frac{W}{m^2 \cdot K} \cdot m \cdot \frac{1}{W/(m \cdot K)}$$

An example of a dimensionless equation is the Dittus-Boelter equation for heat transfer coefficients in fluids flowing inside tubes:

$$\text{Nu} = 0.023\,(\text{Re})^{0.8}\,(\text{Pr})^{0.3}$$

Because all terms are dimensionless, there is no need to change the equation regardless of what system of units is used in determining the values of the dimensionless groups.

Dimensional equations result from empirical correlations (e.g., statistical analysis of experimental data). With dimensional equations, units must correspond to those used on the original data, and substitution of a different system of units will require a transformation of the equation.

When converting dimensional equations, the following rules will be useful:

1. When variables appear in the exponent, the whole exponent must be dimensionless, otherwise it will not be possible to achieve dimensional consistency for the equation.
2. When variables are in arguments of logarithmic functions, the whole argument must be dimensionless.
3. Constants in an equation may not be dimensionless. The units of these constants provide for dimensional consistency.

Example 2.7. The number of surviving microorganisms in a sterilization experiment is linear in a semi-logarithmic graph, therefore, the equation assumes the form:

$$\log N = -at + b$$

where N is number survivors, t is time, and a and b are constants. The equation will be dimensionally consistent if b is expressed as log N at time 0, designated $\log N_o$, and a will have units of reciprocal time. Thus, the correct form of the equation that is dimensionally consistent is

$$\log \frac{N}{N_o} = -at$$

In this equation, both sides are dimensionless.

Example 2.8. An equation for heat transfer coefficient between air flowing through a bed of solids and the solids is

$$h = 0.0128\,G^{0.8}$$

where G is mass flux of air in lb/(ft$^2 \cong$ h) and h is heat transfer coefficient in BTU/(h $\cong$ ft$^2 \cong$ F). Derive an equivalent equation in SI.

The dimensional equation is

$$\frac{BTU}{h \cdot ft^2 \cdot F} = (---)\left(\frac{lb}{ft^2 \cdot h}\right)^{0.8}$$

An equation must be dimensionally consistent; therefore, the constant 0.0128 in the above equation must have units of

$$\frac{BTU}{h \cdot ft^2 \cdot F}\left(\frac{ft^{1.6}h^{0.8}}{lb^{0.8}}\right) = \frac{BTU}{h^{0.2}ft^{0.4}lb^{0.8}F}$$

Converting the equation involves conversion of the coefficient. The equivalent SI unit for $h = W/(m^2K)$ and $G = kg/(m^2 \cong s)$. Thus, the coefficient will have units of $J/(s^{0.2}m^{0.4}kg^{0.8}K)$. The dimensional equation for the conversion is

$$\frac{J}{s^{0.2}m^{0.4}kg^{0.8}K} = \frac{0.0128\ BTU}{h^{0.2}ft^{0.4}lb^{0.8}F} \cdot \frac{1054.8\ J}{BTU} \cdot \frac{ft^{0.4}}{(0.3048)^{0.4}m^{0.4}}$$

$$\frac{h^{0.2}}{(3600)^{0.2}(s^{0.2})} \cdot \frac{(2.2048)^{0.8}lb^{0.8}}{kg^{0.8}} \cdot \frac{1.8\ F}{K} = 14.305$$

The converted equation is

$$h = 14.305\ G^{0.8}$$

where both h and G are in SI units. To check: If $G = 100\ lb/(ft^2 \cong h)$ and $h = 0.0128(100)^{0.8}$, $h = 0.5096\ BTU/(h \cong ft^2 \cong °F)$.

The equivalent SI value is

$$\frac{0.5096\ BTU}{h \cdot ft^2 \cdot F} \cdot \frac{5.678263W/(m^2 \cdot K)}{BTU/(h \cdot ft^2 \cdot °F)} = 2.893\frac{W}{(m^2 \cdot K)}$$

In SI, $G = [100lb/(ft^2 \cdot h)][ft^2/0.3048)^2 m^2](h/3600\ s)(kg/2.2046\ lb) = 0.1356$. Using converted equation,

$$h = 14.305\ (0.1356)^{0.8} = 2.893\ W/(m^2 \cong K).$$

PROBLEMS

2.1. Set up dimensional equation and determine the appropriate conversion factor to use in each of the following

$$\frac{lb}{ft^3} = \frac{lb}{gal} \times \text{conversion factor}$$

$$\frac{lb}{in^2} = \frac{lb}{ft^2} \times \text{conversion factor}$$

$$W = \frac{cal}{s} \times \text{conversion factor}$$

2.2. The amount of heat required to change the temperature of a material from T_1 to T_2 is given by:

$$q = mC_p(T_2 - T_1)$$

where q is BTU, m is mass of material in lb, C_p is specific heat of material in BTU/lb·°F, and T_1 and T_2 are initial and final temperatures in °F.
(a) How many BTUs of heat are required to cook a roast weighing 10 lb from 40°F to 130°F?
   Cp = 0.8 BTU/(lb $\cong$ °F)
(b) Convert the number of BTUs of heat in (a) into watt-hours.

(c) If this roast is heated in a microwave oven having an output of 200 watts, how long will it take to cook the roast?

2.3. How many kilowatt hours of electricity will be required to heat 100 gallons of water (8.33 lb/gal) from 60°F to 100°F? C_p of water is 1 BTU/(lb $\cong$ °F).

2.4. 4. Calculate the power requirements for an electric heater necessary to heat 10 gallons of water from 70°F to 212°F in 10 minutes. Express this in Joules/min, and in watts. Use the following conversion factors in your calculations:

Specific heat of water = 1 BTU/(lb $\cong$ °F)

3.414 BTU/(W $\cong$ h)

60 min/h

3600 s/h

8.33 lb water/gal

1.054×10^3 J/BTU

2.5. One ton of refrigeration is defined as the rate of heat withdrawal from a system necessary to freeze 1 ton (2000 lb) of water at 32°F in 24 hours. Express this in watts, and in BTU/h. Heat of fusion of water = 80 cal/g.

2.6. (a) In the equation $\tau = \mu(\gamma)$, what would be the units of τ in the equation if μ is expressed in dyne $\cong$ s/cm^2 and γ is in s^{-1}?

(b) If μ is to be expressed in lb$_m$/(ft $\cong$ s), τ is expressed in lb$_f$/ft^2, and γ is in s^{-1}, what is needed in the equation to make it dimensionally consistent?

2.7. In the equation

$$\overline{V} = \frac{1000(\rho_1 - \rho_2)}{m\rho_1\rho_2} + \frac{M}{\rho_2}$$

what units should be used for the density, ρ, such that $\overline{V}$ would have the units ml/mole?

(a) m = moles/1000 g

(b) M = g/mole

2.8. Express the following in SI units. Follow the rounding-off rule on your answer.

(a) The pressure at the base of a column of fluid 8.325 in. high when the acceleration due to gravity is 32.2 ft/s^2 and the fluid density is 1.013 g/cm^3.

P = density × height × acceleration due to gravity

(b) The compressive stress (same units as pressure) on a specimen having a diameter of 0.525 in. when the applied force is 5.62 pound force.

Stress = Force/area

(c) The force needed to restrain a piston having a diameter of 2.532 in. when a pressure of 1500 (exact) lb$_f$/in^2 is in the cylinder behind the piston.

Force = Pressure × area

2.9. An empirical equation for heat transfer coefficient in a heat exchanger is

$$h = a(V)^{0.8}(1 + 0.011T)$$

where h = BTU/(h $\cong$ ft^2 $\cong$ Δ°F), V = ft/s, and T = °F. In one experimental system, a had a value of 150. What would be the form of the equation and the value of a if h, V, and T are in SI units?

2.10. A correlation equation for the density of a liquid as a function of temperature and pressure is as follows:

$$d = (1.096 + 0.0086 \text{ T})(e)^{0.000953P}$$

where d is density in g/cm^3, T is temperature in Kelvin, P is pressure in atm, and e is the base of natural logarithms. A normal atmosphere is 101.3 kPa. Determine the form of the equation if all variables are to be expressed in SI.

2.11. The Arrhenius equation for the temperature dependence of diffusivity (D) is given by $D = D_o[e]^{-E/RT}$. R is a constant with a value of 1.987 cal/mole K. (T is temperature in degrees Kelvin). If D is in cm^2/s, determine the units of D_o and E.

2.12. The heat of respiration of fresh produce as a function of temperature is $q = a\, e^{bT}$. If q has units of BTU/(tonA24 h), and T is in °F, determine the units of a and b. The values of a and b for cabbage are 377 and 0.041, respectively. Calculate the corresponding values if q is expressed in mW/kg and T is in °C.

SUGGESTED READING

American National Standards Institute 1976. American National Standard, metric practice. Am. Natl. Stand. Inst. IIEE Std. 268–1976. IEEE, New York.

American Society for Agricultural Engineers 1978. Use of customary and SI (metric) units. American Society for Agricultural Engineers Yearbook—1978. ASAE, St. Joseph, MI.

Benson, S. W. 1971. Chemical Calculations. 3rd ed. John Wiley & Sons, New York.

Feldner, R. M. and Rousseau, R. W. 1999. Elementary Principles of Chemical Processes. 2nd ed. John Wiley & Sons, New York.

Himmelblau, D. M. 1967. Basic Principles and Calculations in Chemical Engineering. 2nd ed. Prentice-Hall, Englewood Cliffs, NJ.

Kelly, F. H. C. 1963. Practical Mathematics for Chemists. Butterworths, London.

McCabe, W. L., Smith, J. C., and Harriott, P. 1985. Unit Operations of Chemical Engineering. 4th ed. McGraw-Hill Book Co., New York.

Obert, E. and Young, R. L. 1962. Elements of Thermodynamics and Heat Transfer. McGraw-Hill Book Co., New York.

Watson, E. L. and Harper, J. C. 1988. Elements of Food Engineering. 2nd ed. Van Nostrand Reinhold Co., New York.

CHAPTER 3

Material Balances

Material balance calculations are employed in tracing the inflow and outflow of material in a process and thus establish quantities of components or the whole process stream. The procedures are useful in formulating products to specified compositions from available raw materials, evaluating final compositions after blending, evaluating processing yields, and evaluating separation efficiencies in mechanical separation systems.

3.1 BASIC PRINCIPLES

3.1.1 Law of Conservation of Mass

Material balances are based on the principle that matter is neither created nor destroyed. Thus, in any process, a mass balance can be made as follows:

Inflow = Outflow + Accumulation

Inflow may include formation of material by chemical reaction or microbial growth processes, and outflow may include material depletion by chemical or biological reactions.

If accumulation is 0, inflow = outflow and the process is at steady state. If the accumulation term is not 0, then the quantity and concentration of components in the system could change with time and the process is an unsteady state.

3.1.2 Process Flow Diagrams

Before formulating a material balance equation, visualize the process and determine the boundary of the system for which the material balance is to be made. It is essential that everything about the process that affects the distribution of components is known. The problem statement should be adequate to enable the reader to draw a flow diagram. However, in some cases, basic physical principles associated with a process may affect the distribution of components in the system but may not be stated in the problem. It is essential that a student remembers the physical principles applied in the processes used as examples. Knowing these principles not only allows the student to solve similar material balance problems but also provides information that may be used later as a basis for the design of a new process or for evaluation of parameters affecting efficiency of a process.

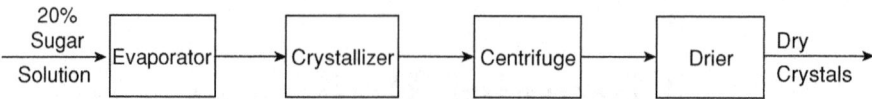

Figure 3.1 Process flow diagram for a crystallization problem.

For example, consider a problem in crystallization. The problem may be simply stated: Determine the amount of sugar (water-free basis) that can be produced from 100 kg of sugar solution that contains 20% by weight of sugar and 1% of a water-soluble uncrystallizable impurity. The solution is concentrated to 75% sugar, cooled to 20°C, centrifuged, and the crystals dried.

The problem statement is indeed adequate to draw a process flow diagram. This is shown in Fig. 3.1.

However, this flow diagram does not give a complete picture of where various streams separate and leave the system.

Figure 3.2 is a flow diagram of the same process, but after taking into consideration how components partition in various steps in the process, additional streams leaving the system are drawn in the diagram. To concentrate a 20% solution to 75% requires the removal of water. Thus, water leaves the system at the evaporator. The process of cooling does not alter the mass, therefore, the same process stream enters and leaves the crystallizer. Centrifugation separates most of the liquid phase from the solid phase, and the crystals, the solid phase containing a small amount of retained solution, enter the drier. A liquid phase leaves the system at the centrifuge. Water leaves the system at the drier.

Three physical principles involved in this problem are not stated: (1) Crystals will crystallize out of a solution when solute concentration exceeds the saturation concentration. The solute concentration in the liquid phase is forced toward the saturation concentration as crystals are formed. Given enough time to reach equilibrium, the liquid phase that leaves the system at the centrifuge is a saturated sugar solution. (2) The crystals consist of pure solute and the only impurities are those adhering to the crystals from the solution. (3) It is not possible to completely eliminate the liquid from the solid phase by centrifugation. The amount of impurities that will be retained with the sugar crystals depends on

---

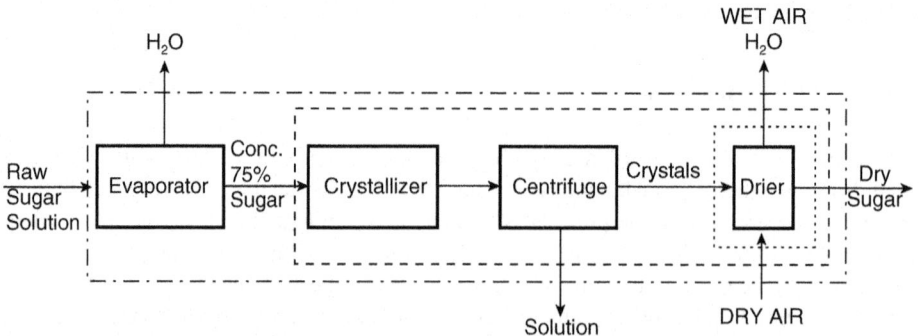

Figure 3.2 Process flow diagram for a crystallization problem showing input and exit streams and boundaries enclosing subsystems for analyzing sections of the process.

how efficiently the centrifuge separates the solid from the liquid phase. This principle of solids purity being dependent on the degree of separation of the solid from the liquid phase applies not only in crystallization but also in solvent extraction.

In order to solve this problem, the saturation concentration of sugar in water at 20°C, and the water content of the crystals fraction after centrifugation must be known.

3.1.3   System Boundaries

Figure 3.2 shows how the boundaries of the system can be moved to facilitate solving the problem. If the boundary completely encloses the whole process, there will be one stream entering and four streams leaving the system. The boundary can also be set just around the evaporator in which case there is one stream entering and two leaving. The boundary can also be set around the centrifuge or around the drier. A material balance can be carried out around any of these subsystems or around the whole system. The material balance equation may be a total mass balance or a component balance.

3.1.4   Total Mass Balance

The equation in section "Law of Conservation of Mass," when used on the total weight of each stream entering or leaving a system, represents a total mass balance. The following examples illustrate how total mass balance equations are formulated for systems and subsystems.

Example 3.1. In an evaporator, dilute material enters and concentrated material leaves the system. Water is evaporated during the process. If I is the weight of the dilute material entering the system, W is the weight of water vaporized, and C is the weight of the concentrate, write an equation that represents the total mass balance for the system. Assume that a steady state exists.

Solution:

The problem statement describes a system depicted in Fig. 3.3.
The total mass balance is

Inflow = Outflow + Accumulation
$\quad$ I = W + C (accumulation is 0 in a steady-state system)

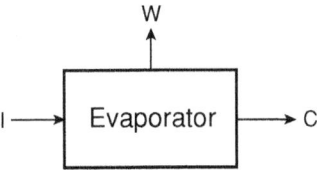

Figure 3.3   Input and exit streams in an evaporation process.

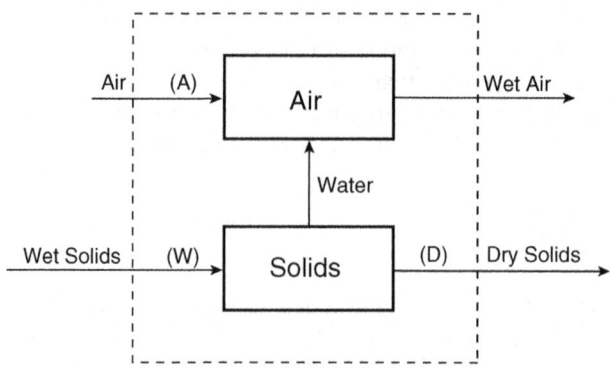

Figure 3.4 Diagram of material flow in a dehydration process.

Example 3.2. Construct a diagram and set up a total mass balance for a dehydrator. Air enters at the rate of A lb/min, and wet material enters at W lb/min. Dry material leaves the system at D lb/min. Assume steady state.

Solution:

The problem statement describes a system (dehydrator) where air and wet material enters and dry material leaves. Obviously, air must leave the system also, and water must leave the system. A characteristic of a dehydrator not written into the problem statement is that water removed from the solids is transferred to air and leaves the system with the air stream. Figure 3.4 shows the dehydrator system and its boundaries. Also shown are two separate subsystems—one for the solids and the other for air—with their corresponding boundaries. Considering the whole dehydrator system, the total mass balance is

$W + A =$ wet air $+ D$

Considering the air subsystem:

$A +$ water $=$ wet air

The mass balance for the solids subsystem is

$W =$ water $+ D$

Example 3.3. Orange juice concentrate is made by concentrating single-strength juice to 65% solids followed by dilution of the concentrate to 45% solids using single-strength juice. Draw a diagram for the system and set up mass balances for the whole system and for as many subsystems as possible.

Solution:

The problem statement describes a process depicted in Fig. 3.5. Consider a hypothetical proportionator that separates the original juice (S) to that which is fed to the evaporator (F) and that (A) which is used to dilute the 65% concentrate. Also, introduce a blender to indicate that part of the process where

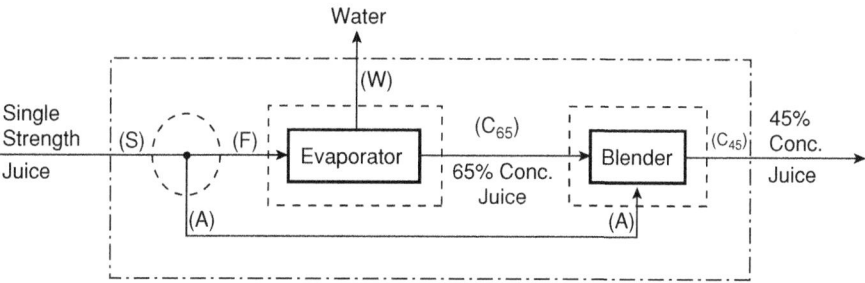

Figure 3.5 Diagram of an orange juice concentrate process involving evaporation and blending of concentrate with freshly squeezed juice.

---

the 65% concentrate (C_{65}) and the single-strength juice are mixed to produce the 45% concentrate (C_{45}). The material balance equations for the whole system and the various subsystems are:

Overall: $S = W + C_{45}$
Proportionator: $S = F + A$
Evaporator: $F = W + C_{65}$
Blender: $C_{65} + A = C_{45}$

3.1.5   Component Mass Balance

The same principles apply as in the total mass balance except that components are considered individually. If there are n components, n independent equations can be formulated; one equation for total mass balance and n − 1 component balance equations.

Because the object of a material balance problem is to identify the weights and composition of various streams entering and leaving a system, it is often necessary to establish several equations and simultaneously solve these equations to evaluate the unknowns. It is helpful to include the known quantities of process streams and concentrations of components in the process diagram in order that all streams where a component may be present can be easily accounted for. In a material balance, use mass units and concentration in mass fraction or mass percentage. If the quantities are expressed in volume units, convert to mass units using density.

A form of a component balance equation that is particularly useful in problems involving concentration or dilution is the expression for the mass fraction or weight percentage.

$$\text{Mass fraction A} = \frac{\text{mass of component A}}{\text{total mass of mixture containing A}}$$

Rearrange the equation:

$$\text{Total mass of mixture containing A} = \frac{\text{mass of component A}}{\text{mass fraction of A}}$$

Thus, if the weight of component A in a mixture is known, and its mass fraction in that mixture is known, the mass of the mixture can be easily calculated.

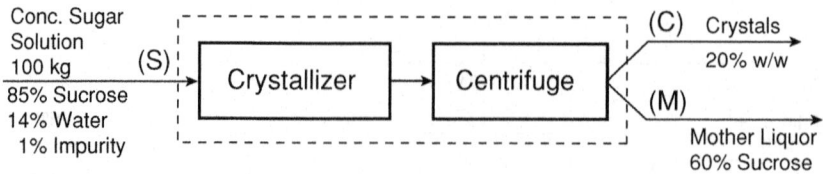

Figure 3.6   Diagram showing composition and material flow in a crystallization process.

Example 3.4. Draw a diagram and set up a total mass and component balance equation for a crystallizer where 100 kg of a concentrated sugar solution containing 85% sucrose and 1% inert, water-soluble impurities (balance, water) enters. Upon cooling, the sugar crystallizes from solution. A centrifuge then separates the crystals from a liquid fraction, called the mother liquor. The crystal slurry fraction has, for 20% of its weight, a liquid having the same composition as the mother liquor. The mother liquor contains 60% sucrose by weight.

Solution:

The diagram for the process is shown in Fig. 3.6. Based on a system boundary enclosing the whole process of crystallization and centrifugation, the material balance equations are as follows.

Total mass balance:

$$S = C + M$$

Component balance on sucrose:

$$S(0.85) = M(0.6) + C(0.2)(0.6) + C(0.8)$$

The term on the left is sucrose in the inlet stream. The first term on the right is sucrose n the mother liquor. The second term on the right is sucrose in mother liquor carried by crystals. The last term on the right is sucrose in the crystals.

Component balance on water:

Let x = mass fraction of impurity in the mother liquor

$$S(0.14) = M(0.4 - x) + C(0.2)(0.4 - x)$$

The first term on the left is water in the inlet stream. The first term on the right is water in the mother liquor. The last term on the right is water in the mother liquor adhering to the crystals.

Component balance on impurity:

$$S(0.01) = M(x) + C(0.2)(x)$$

Note that a total of four equations can be formulated but there are only three unknown quantities (C, M, and x). One of the equations is redundant.

Example 3.5. Draw a diagram and set up equations representing total mass balance and component mass balance for a system involving the mixing of pork (15% protein, 20% fat, and 63% water) and backfat (15% water, 80% fat, and 3% protein) to make 100 kg of a mixture containing 25% fat.

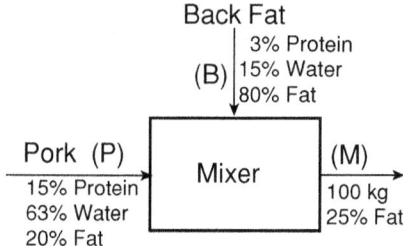

Figure 3.7 Composition and material flow in a blending process.

___

Solution:

The diagram is shown in Fig. 3.7.
Total mass balance:

$$P + B = 100$$

Fat balance:

$$0.2\,P + 0.8\,B = 0.25(100)$$

These two equations are solved simultaneously by substituting $P = 100 - B$ into the second equation.

$$0.2(100 - B) + 0.8\,B = 25$$
$$B = \frac{25 - 20}{0.8 - 0.2} = 8.33 \text{ kg}$$
$$P = 100 - 8.33 = 91.67 \text{ kg}$$

3.1.6 Basis and "Tie Material"

A "tie material" is a component used to relate the quantity of one process stream to another. It is usually the component that does not change during a process. Examples of tie material are solids in dehydration or evaporation processes and nitrogen in combustion processes. Although it is not essential that these tie materials are identified, the calculations are often simplified if it is identified and included in one of the component balance equations. This is illustrated in Example 3.6 of the section "A Steady State" where the problem is solved rather readily using a component mass balance in the solid (the tie material in this system) compared to when the mass balance was made on water. In a number of cases, the tie material need not be identified as illustrated in examples in the section "Blending of Food Ingredients."

A "basis" is useful in problems where no initial quantities are given and the answer required is a ratio or a percentage. It is also useful in continuous flow systems. Material balance in a continuous flow system is done by assuming as a basis a fixed time of operation. A material balance problem can be solved on any assumed basis. After all the quantities of process streams are identified, the specific quantity asked in the problem can be solved using ratio and proportion. It is possible to change basis when considering each subsystem within a defined boundary inside the total system.

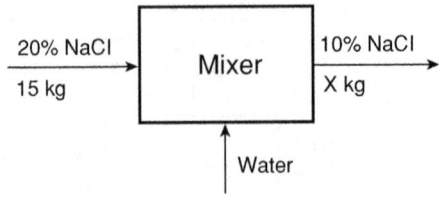

Figure 3.8   Composition and material flow for a dilution process.

3.2   MATERIAL BALANCE PROBLEMS INVOLVED IN DILUTION, CONCENTRATION, AND DEHYDRATION

3.2.1   Steady State

These problems can be solved by formulating total mass and component balance equations and solving the equations simultaneously.

Example 3.6. How many kilograms of a solution containing 10% NaCl can be obtained by diluting 15 kg of a 20% solution with water?

Solution:

The process diagram in Fig. 3.8 shows that all NaCl enters the mixer with the 20% NaCl solution and leaves in the diluted solution. Let $x = $ kg 10% NaCl solution; $y = $ kg water. The material balance equations are

Total mass: $15 = X - Y$
Component: $15(0.20) = X(0.10)$

The total mass balance equation is redundant because the component balance equation alone can be used to solve the problem.

$$x = \frac{3}{0.1} = 30 \text{ kg}$$

The mass fraction equation can also be used in this problem. Fifteen kilograms of a 20% NaCl solution contains 3 kg NaCl. Dilution would not change the quantity of NaCl so that 3 kg of NaCl is in the diluted mixture. The diluted mixture contains 10% NaCl, therefore:

$$x = \frac{3 \text{ kg NaCl}}{\text{mass fraction NaCl}} = \frac{3}{0.1} = 30 \text{ kg}$$

Example 3.7. How much weight reduction would result when a material is dried from 80% moisture to 50% moisture?

Solution:

The process diagram is shown in Fig. 3.9. Dehydration involves removal of water and the mass of solids remain constant. There are two components, solids and water, and a decrease in the concentration

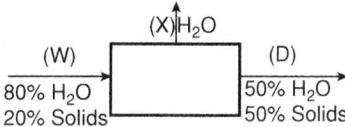

Figure 3.9   Composition and material flow in a dehydration process.

___

of water, indicating a loss, will increase the solids concentration. Let: W = mass of 80% moisture material, D = mass of 50% moisture material, and X = mass water lost, which is also the reduction in mass.

No weights are specified, therefore express reduction in weight as a ratio of final to initial weight. The material balance equations are

Total mass: $W = X + D$
Water: $0.8W = 0.5 D + X$

Solving simultaneously:

$$W = X + D; X = W - D$$
$$0.8\ W = 0.5\ D + (W - D)$$
$$0.5\ D = 0.2\ W;$$
$$\frac{D}{W} = \frac{0.2}{0.5} = 0.4$$
$$\%\text{wt. reduction} = \frac{W - D}{W}(100) = \left(1 - \frac{D}{W}\right)100$$
$$= (1 - 0.4)(100) = 60\%$$

The problem can also be solved by using as a basis 100 kg of 80% moisture material. Let W = 100.

Total mass balance: $100 = X + D$
Water balance: $0.8(100) = X + 0.5\ D$

Solving simultaneously, by subtracting one equation from the other:

$$100 - 80 = D(1 - 0.5);\ D = 20/0.5 = 40\ kg.$$
$$\%\ \text{wt. reduction} = \frac{100 - 40}{100}(100) = 60\%$$

3.2.2   Volume Changes on Mixing

When two liquids are mixed, the volumes are not always additive. This is true with most solutions and miscible liquids. Sodium chloride solution, sugar solutions, and ethanol solution all exhibit volume changes on mixing. Because of volume changes, material balances must be done on mass rather than volume of components. Concentrations on a volume basis must be converted to a mass basis before the material balance equations are formulated.

Example 3.8. Alcohol content in beverages are reported as percent by volume. A "proof" is twice the volume percent of alcohol. The density of absolute ethanol is 0.7893 g/cm^3. The density of a solution containing 60% by weight of ethanol is 0.8911 g/cm^3. Calculate the volume of absolute ethanol that must be diluted with water to produce 1 liter of 60% by weight, ethanol solution. Calculate the "proof" of a 60% ethanol solution.

Solution:

Use as a basis: 1 liter of 60% w/w ethanol.
Let X = volume of absolute ethanol in liters. The component balance equation on ethanol is:

$$X(1000)(0.7983) = 1(1000)(0.8911)(0.6)$$
$$X = 0.677 \text{ L}$$
$$\text{g water} = 1000(0.8911) - 0.677(0.7983) = 356.7 \text{ g or } 356.7 \text{ mL}$$

Total volume components = 0.677 + 0.3567 = 1.033 L

This is a dilution problem, and a component balance on ethanol was adequate to solve the problem. Note that a mass balance was made using the densities given. There is a volume loss on mixing as more than 1 liter of absolute ethanol and water produced 1 liter of 60% w/w ethanol solution.

To calculate the "proof" of 60% w/w ethanol, use as a basis 100 g of solution.

Volume of solution = 100/0.8911 = 112.22 cm^3
Volume of ethanol = 100(0.6)/0.7893 = 76.016 cm^3
Volume percent = (76.016/112.22)(100) = 67.74%
Proof = 2 (volume percent) = 135.5 proof

3.2.3 Continuous Versus Batch

Material balance calculations are the same regardless whether a batch or continuous process is being evaluated. In a batch system, the total mass considered includes what entered or left the system at one time. In a continuous system, a basis of a unit time of operation may be used and the material balance will be made on what entered or left the system during that period of time. The previous examples were batch operations. If the process is continuous, the quantities given will all be mass/time (e.g., kg/h). If the basis used is 1 hour of operation, the problem is reduced to the same form as a batch process.

Example 3.9. An evaporator has a rated evaporation capacity of 500 kg water/h. Calculate the rate of production of juice concentrate containing 45% total solids from raw juice containing 12% solids.

Solution:

The diagram of the process is shown in Fig. 3.10. Use as a basis 1 hour of operation. Five hundred kg of water leaves the system. A component balance on solids and a total mass balance will be needed to solve the problem. Let F = the feed, 12% solids juice, and C = concentrate containing 45% solids. The material balance equations are:

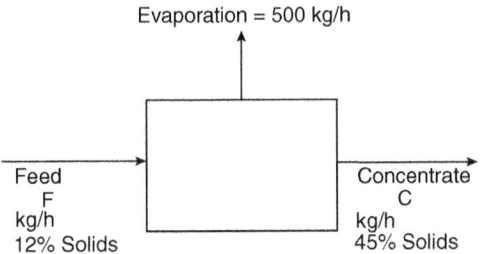

Figure 3.10   Flow diagram of an evaporation process.

Total mass:

$$F = C + 500$$

Solids:

$$0.12\,F = 0.45\,C; \quad F = \frac{0.45\,C}{0.12} = 3.75\,C$$

$$C = \frac{500}{3.75 - 1} = 181.8 \text{ kg}$$

Substituting and solving for C: $3.75\,C = C + 500$.

Because the basis is 1 hour of operation, the answer will be: Rate of production of concentrate = 181.8 kg/h

3.2.4   Recycle

Recycle is evaluated similarly as in the previous examples, but the boundaries of subsystems analyzed are moved around to isolate the process streams being evaluated. A system is defined that has a boundary surrounding the recycle stream. If this total system is analyzed, the problem may be reduced to a simple material balance problem without recycle.

Example 3.10. A pilot plant model of a falling film evaporator has an evaporation capacity of 10 kg water/h. The system consists of a heater through which the fluid flows down in a thin film and the heated fluid discharges into a collecting vessel maintained under a vacuum where flash evaporation reduces the temperature of the heated fluid to the boiling point. In continuous operation, a recirculating pump draws part of the concentrate from the reservoir, mixes this concentrate with feed, and pumps the mixture through the heater. The recirculating pump moves 20 kg of fluid/h. The fluid in the collecting vessel should be at the desired concentration for withdrawal from the evaporator at any time. If feed enters at 5.5% solids and a 25% concentrate is desired, calculate: (a) the feed rate and concentrate production rate, (b) the amount of concentrate recycled, and (c) concentration of mixture of feed and recycled concentrate.

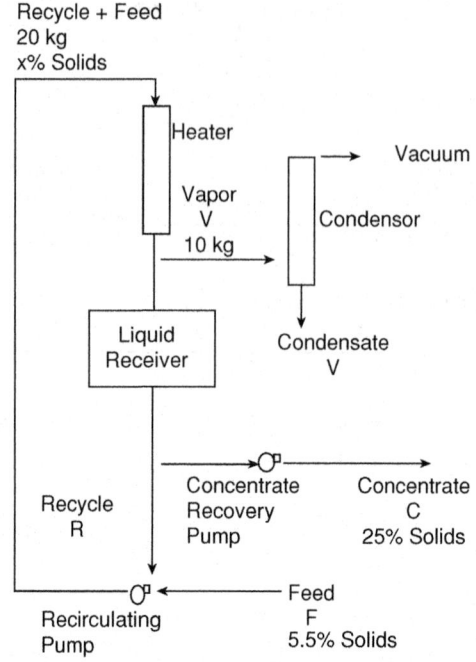

Recycle + Feed
20 kg
x% Solids

Heater

Vacuum

Vapor
V
10 kg

Condensor

Liquid
Receiver

Condensate
V

Concentrate
Recovery
Pump

Concentrate
C
25% Solids

Recycle
R

Recirculating
Pump

Feed
F
5.5% Solids

Figure 3.11   Diagram of material flow in a falling film evaporator with product recycle.

Solution:

A diagram of the system is shown in Fig. 3.11. The basis is 1 hour of operation.

A mass and solids balance over the whole system will establish the quantity of feed and concentrate produced per hour.

Total mass:

$$F = C + V; \ F = C + 10$$

Solids:

$$F(0.055) = C(0.25); \ F = C\left(\frac{0.25}{0.055}\right) = 4.545\,C$$

Substituting F:

$$4.545\,C = C + 10; \ C = \frac{10}{4.545 - 1} = 2.82 \text{ kg}$$

(a) Solving for F : $F = 4.545(2.82) = 12.82$ kg/h. The concentrate production rate is 2.82 kg/h.

(b) Material balance around the recirculating pump:

$$R + F = 20; \ R = 20 - 12.82 = 7.18 \text{ kg}$$

Recycle rate $= 7.18 \text{ kg/h}$

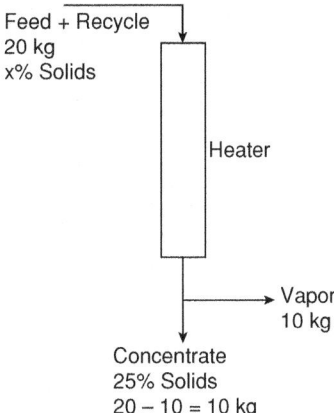

Feed + Recycle
20 kg
x% Solids

Heater

Vapor
10 kg

Concentrate
25% Solids
20 − 10 = 10 kg

Figure 3.12   Diagram of material balance around the heater of a falling film evaporator.

(c) A material balance can be made around the part of the system where the vapor separates from the heated fluid as shown in Fig. 3.12.

Solids balance: $20(x) = 10(0.25)$; $x = 2.5/20 = 0.125$.

The fluid entering the heater contains 12.5% solids.

3.2.5   Unsteady State

Unsteady-state material balance equations involve an accumulation term in the equation. Accumulation is expressed as a differential term of the rate of change of a variable with respect to time. A mass balance is made in the same manner as steady-state problems. Because a differential term is involved, the differential equation must be integrated to obtain an equation for the value of the dependent variable as a function of time.

Example 3.11. A stirred tank with a volume of 10 liters contains a salt solution at a concentration of 100 g/L. If salt-free water is continuously fed into this tank at the rate of 12 L/h, and the volume is maintained constant by continuously overflowing the excess fluid, what will be the concentration of salt after 90 min.

Solution:

This is similar to a dilution problem, except that the continuous overflow removes salt from the tank, thus reducing not only the concentration but also the quantity of salt present. A mass balance will be made on the mass of salt in the tank and in the streams entering and leaving the system.

Let x = the concentration of salt in the vessel at any time. Representing time by the symbol t, the accumulation term will be dx/dt, which will have units of g salt/(L $\cong$ min). Convert all time units to minutes in order to be consistent with the units. Multiplying the differential term by the volume will result in units of g salt/min. The feed contains no salt, therefore the input term is zero. The overflow is at the same rate as the feed, and the concentration in the overflow is the same as inside the vessel, therefore the output term is the feed rate multiplied by the concentration inside the vessel. The mass balance equation is

$$0 = Fx + V\left(\frac{dx}{dt}\right); \frac{dx}{dt} = -\frac{F}{V}x$$

The negative sign indicates that x will be decreasing with time. Separating variables:

$$\frac{dx}{x} = -\frac{F}{V}dt$$

Integrating:

$$\ln x = -\left(\frac{F}{V}\right)t + C$$

The constant of integration is obtained by substituting $x = 100$ at $t = 0$; $C = \ln(100)$; $F = 12\,L/h = 0.2\,L/min$; $V = 10\,L$

$$\ln\left(\frac{x}{100}\right) = -\frac{0.2\,t}{10}$$

At 90 min: $\ln(0.01\,x) = -1.800$.

$$x = 100(e^{-1.80}) = 16.53\,g/L$$

Example 3.12. The generation time is the time required for cell mass to double. The generation time of yeast in a culture broth has been determined from turbidimetric measurements to be 1.5 hours. (a) If this yeast is used in a continuous fermentor, which is a well-stirred vessel having a volume of 1.5 L, and the inoculum is 10,000 cells/mL, at what rate can cell-free substrate be fed into this fermentor in order that the yeast cell concentration will remain constant? The fermentor volume is maintained constant by continuous overflow. (b) If the feed rate is 80% of what is needed for a steady-state cell mass, calculate the cell mass after 10 hours of operation.

Solution:

This problem is a combination of dilution with continuous cell removal, but with the added factor of cell generation by growth inside the vessel. The diagram shown in Fig. 3.13 represents the material balance on cell mass around a fermentor. The balance on cell mass or cell numbers is as follows:

Input(from feed + cell growth) = output + accumulation

(a) The substrate entering the fermentor is cell free, therefore this term in the material balance equation is zero. Let $R =$ substrate feed rate = overflow rate, because fluid volume in fermentor is maintained constant. Let $x =$ cells/mL. The material balance on cell numbers with the appropriate time derivatives for rate of increase in cell numbers, $(dx/dt)_{gen}$, and accumulation $(dx/dt)_{acc}$ is

$$V\left[\frac{dx}{dt}\right]_{gen} = Rx + V\left[\frac{dx}{dt}\right]_{acc}$$

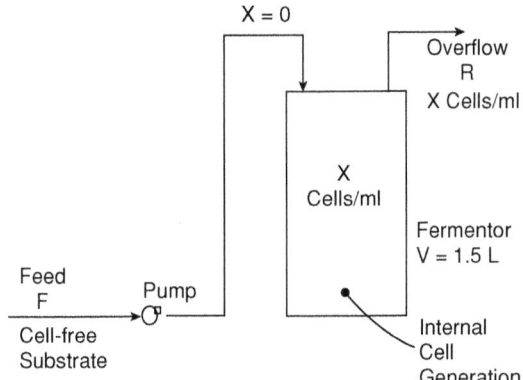

Figure 3.13   Material balance around a fermentor.

If cell mass is constant, then the last term on the right is zero. Cell growth is often expressed in terms of a generation time, g, which is the average time for doubling of cell numbers. Let $t =$ time in hours. Then:

$$x = x_0(2)^{(t/g)}$$

Differentiating to obtain the generation term in the material balance equation:

$$\left[\frac{dx}{dt}\right]_{gen} = \frac{d}{dt}\left[x_0(2)^{(t/g)}\right] = x_0(g^{-1})(2)^{(t/g)}\ln 2 = xg^{-1}\ln 2$$

Substituting in the material balance equation and dropping out zero terms:

$V \cdot x \cdot g^{-1} \ln 2 = R\,x$; x cancels out on both sides.
$$R = V(\ln 2)g^{-1}$$

Substituting known quantities:

$$R = \frac{1.5 \ln 2}{1.5} = \frac{0.693\ L}{h}$$

In continuous fermentation, the ratio R/V is the dilution rate, and when the fermentor is at a steady state in terms of cell mass, the dilution rate must equal the specific growth rate. The specific growth rate is the quotient: rate of growth/cell mass $= (1/x)(dx/dt)_{gen} = \ln 2/g$. Thus, the dilution rate to achieve steady state is strictly a function of the rate of growth of the organism and is independent of the cell mass present in the fermentor.
(b) When the feed rate is reduced, the cell number will be in an unsteady state. The material balance will include the accumulation term. Substituting the expression for cell generation into

the original material balance equation with accumulation:

$$V\frac{x\ln 2}{g} = Rx + V\frac{dx}{dt}$$

$$x\left[\frac{V\ln 2}{g} - R\right] = v\frac{dx}{dt}$$

Separating variables and integrating:

$$\frac{dx}{x} = \left[\frac{\ln 2}{g} - \frac{R}{V}\right]dt$$

$$\ln x = \left[\frac{\ln 2}{g} - \frac{R}{V}\right]t + C$$

At $t = 0$, $x = 10,000$ cells/mL; $C = \ln(10,000)$.

$$\ln\left(\frac{x}{10,000}\right) = \left[\frac{\ln 2}{g} - \frac{R}{V}\right]t$$

Substituting: $V = 1.5$ L; $R = 0.8(0.693$ L/h$) = 0.554$ L/h; $g = 1.5$ h; $t = 10$ h

$$\ln\left(\frac{x}{10,000}\right) = 0.927; \quad x = 10,000(2.5269)$$

$$x = 25,269 \text{ cells/ml}$$

3.3 BLENDING OF FOOD INGREDIENTS

3.3.1 Total Mass and Component Balances

These problems involve setting up total mass and component balances and involve simultaneously solving several equations.

Example 3.13. Determine the amount of a juice concentrate containing 65% solids and single-strength juice containing 15% solids that must be mixed to produce 100 kg of a concentrate containing 45% solids.

Solution:

The diagram for the process is shown in Fig. 3.14.

Total mass balance: $X + Y = 100$; $X = 100 - Y$
Solid balance: $0.65X + 0.15Y = 100(0.45) = 45$
Substituting $(100 - Y)$ for X:

$$0.65(100 - Y) + 0.15Y = 45$$

$$65 - 0.65Y + 0.15Y = 45$$

$$65 - 45 = 0.65Y - 0.15Y$$

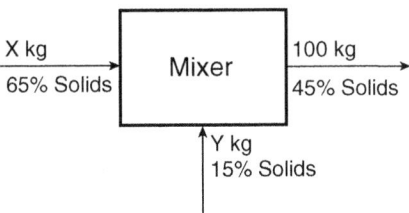

Figure 3.14 Material flow and composition in a process for blending of juice concentrates.

$20 = 0.5Y$

$Y = 40$ kg single strength juice

$X = 60$ kg 65% concentrate

Example 3.14. Determine the amounts of lean beef, pork fat, and water that must be used to make 100 kg of a frankfurter formulation. The compositions of the raw materials and the formulations are

Lean beef: 14% fat, 67% water, 19% protein.
Pork fat: 89% fat, 8% water, 3% protein.
Frankfurter: 20% fat, 15% protein, 65% water.

Solution:

The diagram representing the various mixtures being blended is shown in Fig. 3.15.

Total mass balance: $Z + X + Y = 100$
Fat balance: $0.14X + 0.89Y = 20$
Protein balance: $0.19X + 0.03Y = 15$

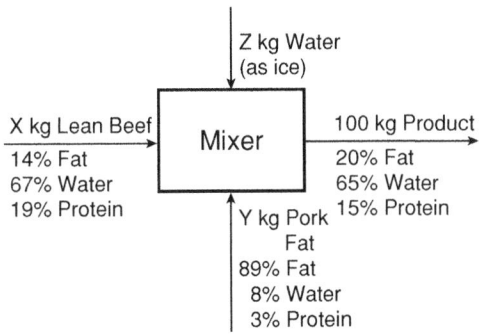

Figure 3.15 Composition and material flow in blending of meats for a frankfurter formulation.

These equations will be solved by determinants. Note that the fat and protein balance equations involve only X and Y, therefore solving these two equations first will give values for X and Y. Z will be solved using the total mass balance equation. See Section 1.9.3 for a discussion of how to set up the determinants.

The matrices for the fat and protein balance equations are

$$\begin{bmatrix} 0.14 & 0.89 \\ 0.19 & 0.03 \end{bmatrix} \begin{bmatrix} X \\ Y \end{bmatrix} = \begin{bmatrix} 20 \\ 15 \end{bmatrix}$$

$$X = \frac{\begin{vmatrix} 20 & 0.89 \\ 15 & 0.03 \end{vmatrix}}{\begin{vmatrix} 0.14 & 0.89 \\ 0.19 & 0.03 \end{vmatrix}} = \frac{(20)(0.03) - (15)(0.89)}{(0.14)(0.03) - (0.19)(0.89)}$$

$$X = \frac{-12.75}{-.1649} = 77.31$$

$$Y = \frac{\begin{vmatrix} 0.14 & 20 \\ 0.19 & 14 \end{vmatrix}}{\begin{vmatrix} 0.14 & 0.89 \\ 0.19 & 0.03 \end{vmatrix}} = \frac{(0.14)(14) - (0.19)(0.20)}{(0.14)(0.03) - (0.19)(0.89)}$$

$$Y = \frac{-1.7}{-0.1649} = 10.3 \text{ kg}$$

Total mass balance: $Z = 100 - 77.3 - 10.3 = 12.4$ kg.
The solution using the Solver Macro in Excel is shown in Fig. 3.16.

Example 3.15. A food mix is to be made that would balance the amount of methionine (MET), a limiting amino acid in terms of food protein nutritional value, by blending several types of plant proteins. Corn, which contains 15% protein, has 1.2 g MET/100 g protein; soy flour with 55% protein has 1.7 g MET/100 g protein; and nonfat dry milk with 36% protein has 3.2 g MET/100 g protein. How much of these ingredients must be used to produce 100 kg of formula that contains 30% protein and 2.2 g MET/100 g protein.

Solution:

This will be solved by setting up a component balance on protein and methionine. Let C = kg corn, S = kg soy flour, and M = kg nonfat dry milk. The material balance equations are

Total mass: $C + S + M = 100$
Protein: $0.15C + 0.55S + 0.36M = 30$

$$\text{MET: } \frac{(1.2)(0.15)}{100}C + \frac{(1.7)(0.55)}{100}S + \frac{(3.2)(0.36)}{100}M = \frac{2.2}{100}(30)$$

$$0.18C + 0.935S + 1.152M = 66$$

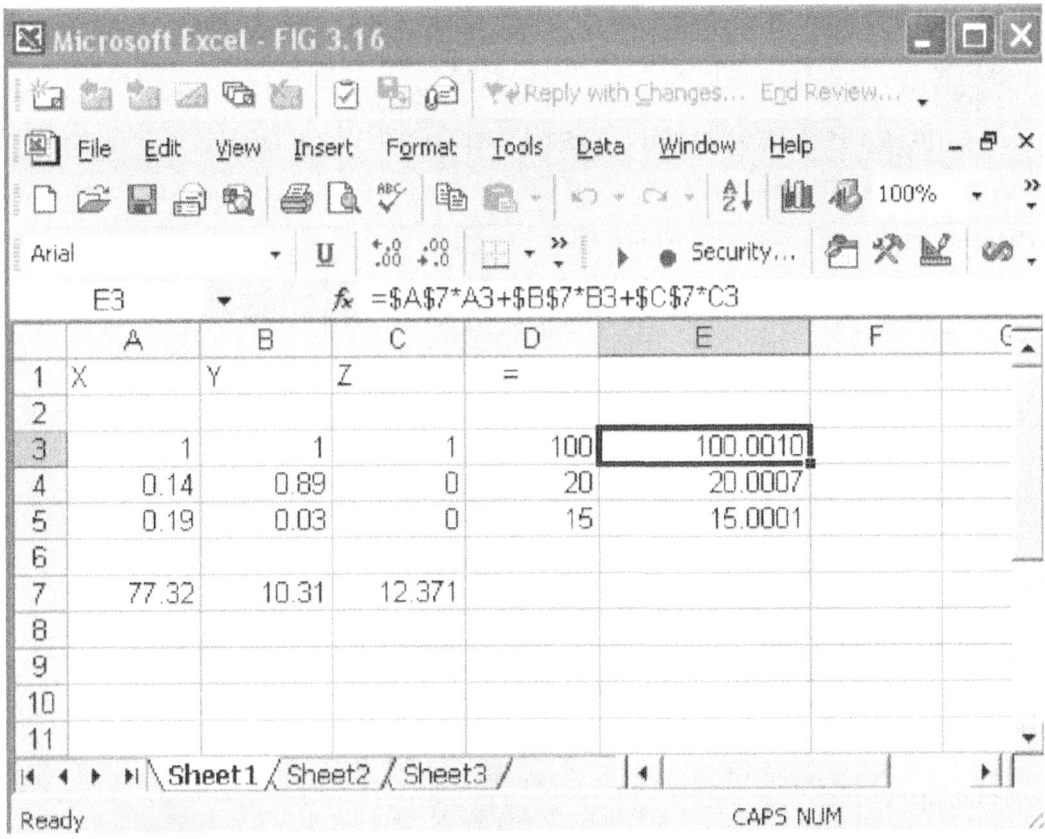

Figure 3.16   Solution to meats blending problem for frankfurter formulation using the Solver macro in Excel.

The matrices representing the three simultaneous equations are

$$\begin{bmatrix} 1 & 1 & 1 \\ 0.15 & 0.55 & 0.36 \\ 0.18 & 0.935 & 1.152 \end{bmatrix} \begin{bmatrix} C \\ S \\ M \end{bmatrix} = \begin{bmatrix} 100 \\ 30 \\ 66 \end{bmatrix}$$

Thus:

$$C = \frac{\begin{bmatrix} 100 & 1 & 1 \\ 30 & 0.55 & 0.36 \\ 66 & 0.935 & 1.152 \end{bmatrix}}{\begin{bmatrix} 1 & 1 & 1 \\ 0.15 & 0.55 & 0.36 \\ 0.18 & 0.935 & 1.152 \end{bmatrix}}$$

The denominator matrix is resolved to:

$$1\begin{bmatrix} 0.55 & 0.36 \\ 0.935 & 1.152 \end{bmatrix} - 0.15\begin{bmatrix} 1 & 1 \\ 0.935 & 1.152 \end{bmatrix} + 0.18\begin{bmatrix} 1 & 1 \\ 0.55 & 0.36 \end{bmatrix}$$

$$= 1[0.55(1.152) - 0.935(0.36)] - 0.15[1(1.152) - 1(0.935)] + 0.18[(1)(0.36) - (0.55)(1)]$$

$$= 0.297 - 0.03255 - 0.0342 = 0.23025$$

The numerator matrix resolves as follows:

$$100\begin{bmatrix} 0.55 & 0.36 \\ 0.935 & 1.152 \end{bmatrix} - 30\begin{bmatrix} 1 & 1 \\ 0.935 & 1.152 \end{bmatrix} + 66\begin{bmatrix} 1 & 1 \\ 0.55 & 0.36 \end{bmatrix}$$

$$= 100[0.55(1.152) - 0.935(0.36)] - 30[1(1.152) - 0.935(1)] + 66[1(0.36) - 0.55(1)]$$

$$= 100(0.297) - 30(0.217) + 66(-0.19) = 10.65$$

$$C = \frac{10.65}{0.23025} = 46.25 \text{ kg}$$

$$S = \frac{\begin{bmatrix} 1 & 100 & 1 \\ 0.15 & 30 & 0.36 \\ 0.18 & 66 & 1.152 \end{bmatrix}}{0.23025}$$

The numerator matrix resolves as follows:

$$1\begin{bmatrix} 30 & 0.36 \\ 66 & 1.152 \end{bmatrix} - 0.15\begin{bmatrix} 100 & 1 \\ 66 & 1.152 \end{bmatrix} + 0.18\begin{bmatrix} 100 & 1 \\ 30 & 0.36 \end{bmatrix}$$

$$= 1[30(1.152) - 66(0.36)] - 0.15[100(1.152) - 66(1) + 0.18[100(.36) - 30(1)]$$

$$= 1(10.9) - 0.15(49.2) + 0.18(6) = 4.5$$

$$S = \frac{4.5}{0.23025} = 19.54 \text{ kg}$$

The total mass balance may be used to solve for M, but as a means of checking the calculations the matrix for M will be formulated and resolved.

$$M = \frac{\begin{bmatrix} 1 & 1 & 100 \\ 0.15 & 0.55 & 30 \\ 0.18 & 0.935 & 66 \end{bmatrix}}{0.23025}$$

The numerator matrix is resolved as follows:

$$1\begin{bmatrix} 0.55 & 30 \\ 0.935 & 66 \end{bmatrix} - 0.15\begin{bmatrix} 1 & 100 \\ 0.935 & 66 \end{bmatrix} + 0.18\begin{bmatrix} 1 & 100 \\ 0.55 & 30 \end{bmatrix}$$

$$= 1[0.55(66) - 0.935(30)] - 0.15[1(66) - 0.935(100)] + 0.18[1(30) - 0.55(100)]$$

$$M = \frac{7.875}{0.23025} = 34.2 \text{ kg} = 1(8.25) - 0.15(-27.5) + 0.18(-25) = 7.875$$

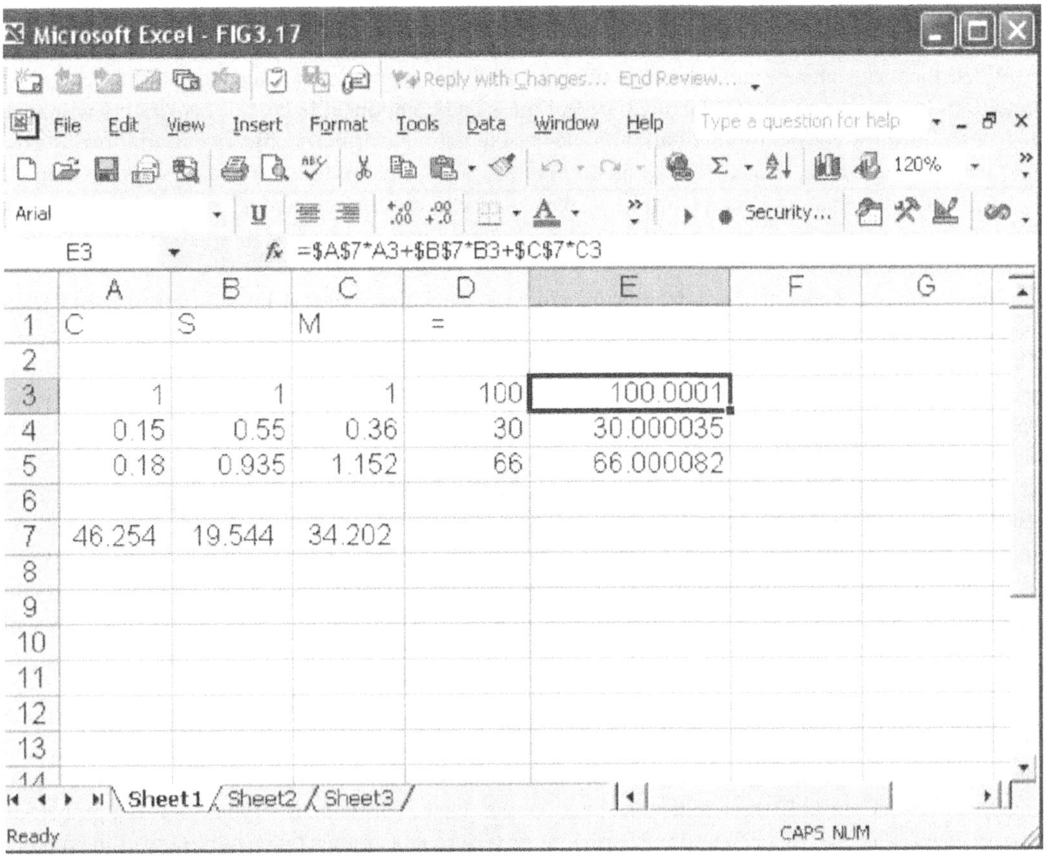

Figure 3.17 Solution to food protein blending problem using the Solver macro in Excel.

Check: C + S + M = 46.25 + 19.54 + 34.2 = 99.99.
The solution using the Solver Macro in Excel is shown in Fig. 3.17.

3.3.2 Use of Specified Constraints in Equations

When blending ingredients, constraints may be imposed either by the functional properties of the ingredients or its components, or by regulations. Examples of constraints are limits on the amount of non-meat proteins that can be used in meat products like frankfurters, moisture content in meat products, or functional properties such as fat or water binding properties. Another constraint may be cost. It is necessary that these constraints specified be included as one of the equations to be resolved.

In some cases, a number of ingredients may be available for making a food formulation. To effectively solve the quantity of each ingredient using the procedures discussed in the previous sections, the same number of independent equations must be formulated as there are unknown quantities to be solved. If there are three components that can be used to formulate component balance equations, only three independent equations can be formulated. Thus, a unique solution for components of a blend, using the solution to simultaneous equations, will be possible only if there are three ingredients to be used.

When there are more ingredients possible for use in a formulation than there are independent equations that can be formulated using the total mass and component balances, constraints will have to be used to allow consideration of all possible ingredients. The primary objective is minimizing cost and maximizing quality. Both cost and quality factors will form the basis for specifying constraints.

The constraints may not be specific (i.e., they may only define a boundary rather than a specific value). Thus, a constraint cannot be used as a basis for an equation that must be solved simultaneously with the others.

One example of a system of constraints to include as many ingredients as possible into a formulation is the "least cost formulation" concept used in the meat industry. Several software companies market "least cost formulation" strategies, and the actual algorithm within each of these computer programs may vary from one company to the other. However, the basis for these calculations is basically the same: (a) The composition must be met, usually 30% fat, and water and protein must satisfy the USDA requirement of water not to exceed 4 times the protein content plus 10% (4P + 10), or some other specified moisture range allowable for certain category of products. (b) The bind values for fat must satisfy the needs for emulsifying the fat contained within the formulation. Software companies who market least-cost sausage formulation strategies vary in how they use these fat holding capacities. (c) Water holding capacity must be maximized; that is, the water and protein content relationship should stay within the 4P + 10 allowed in the finished product. This requirement is not usually followed because of the absence of reliable water holding capacity data, and processors simply add an excess of water during formulation to compensate for the water loss that occurs during cooking.

To illustrate these concepts, a simple least-cost formulation will be set up for frankfurters utilizing a choice of five different meat types. Note that, as in the previous examples, a total mass balance and only two component balance equations can be formulated (water can be determined from the total mass balance). Thus, only three independent equations can be formulated. Product textural properties may be used as a basis for a constraint. If previous experience has shown that a minimum of the protein present in the final blend must come from a type of meat (e.g., pork) to achieve a characteristic flavor and texture, then this constraint will be one factor to be included in the analysis. Other constraints are that no negative values of the components are acceptable.

The solution will be obtained using the Solver macro in Microsoft Excel (see Chapter 1, "Simultaneous Equations").

Example 3.16. Derive a least-cost formulation involving four meat types from the five types shown. The composition and pertinent functional data, as bind values, for each of the ingredients are shown in Table 3.1. The fat content in the blend is 30% and protein is 15%.

The constraints are (1) for product color and textural considerations, at least 20% of the total protein must come from pork trim and pork cheeks; (2) the amount of beef cheek meat should not exceed 15% of the total mixture; and (3) the bind value (sum of the product of mass of each ingredient and its bind constant) must be greater than 15. These constraints are established by experience as the factors needed to impart the desired texture and flavor in the product.

Solution:

Enter data from Table 3.1 in the spreadsheet in block A5 : G10 as shown in Fig. 3.18. Designate column block H6 : H10 to hold values of the weights of the selected meat components. Designate row block B12 : H12 to hold the constraints. Enter formulas for the total mass balance in H12 and

Table 3.1 Data for the Least-Cost Formulation Problems

Meat type	Composition			Bind constant kg fat/100 kg	Cost ($/kg)
	Fat	Protein	Water		
Bull meat	11.8	19.1	67.9	30.01	1.80
Pork trm.	25.0	15.9	57.4	19.25	1.41
Beef chk.	14.2	17.3	68.0	13.96	1.17
Pork chk.	14.1	17.3	67.0	8.61	1.54
Back fat	89.8	1.9	8.3	1.13	0.22

component mass balances on fat and protein respectively in B12 and C12. Designate F12 for the resulting bind value and G12 for the resulting least cost. Enter the following formulas:

1. Total weight: Enter in H12

$$= SUM(H6 : H10)$$

and $H$12 = 100.

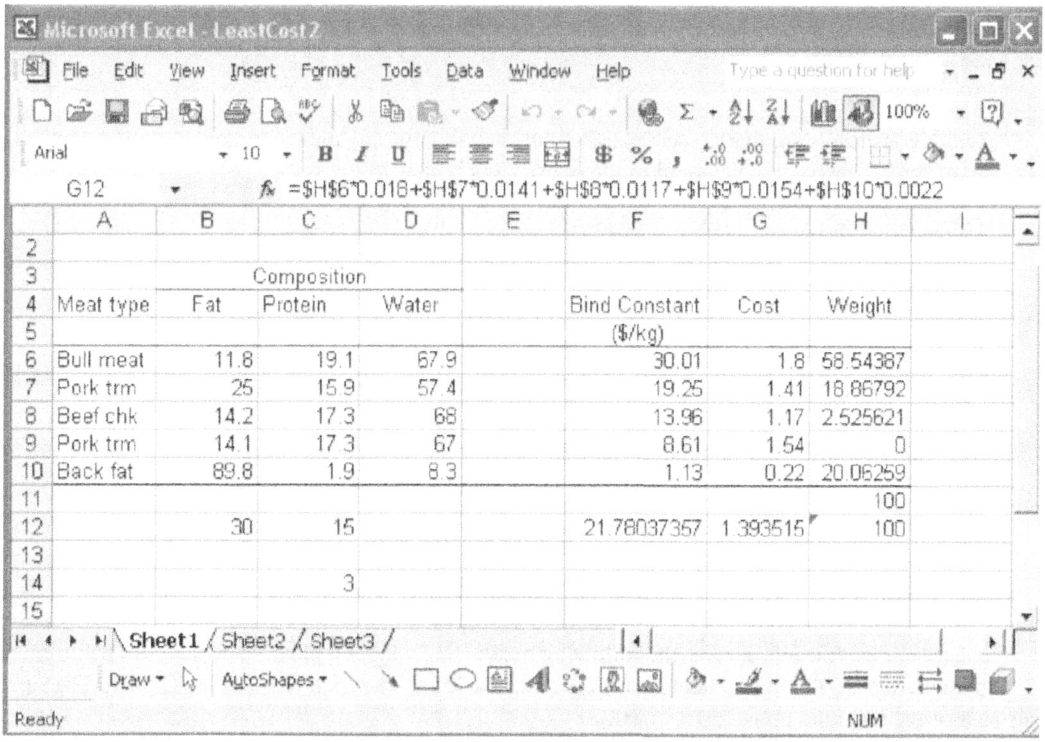

Figure 3.18 Solution to a least-cost meat formulation problem using the Solver macro in Excel.

2. Constraint 1: 15% protein in 100 kg mix will require 15 kg of protein. Enter 15 in C12. Constraint 1 specifies 20% of the 15 kg protein or 3 kg protein must come from pork trim and pork cheeks. This constraint is represented by the following equation entered in C14:

$$= \$H\$7^*0.159 + \$H\$9^*0.173 \quad \text{and} \quad \$C\$14 = 3$$

3. Because Constraint 2 refers to a meat weight (pork cheeks), the equation representing this constraint is entered in H8 as follows: $\$H\$8 < or = 15$
4. Constraint 3, the cumulative bind value of the meat components is entered in F12 as:

$$= \$H\$6^*0.3001 + \$H\$7^*0.1925 + \$H\$8^*0.1396 + \$H\$9^*0.0861$$
$$+ \$H\$10^*0.0113$$

and E12 > 15.
5. Cumulative fat from the meat components is entered in B12 as the fat balance formula:

$$= \$H\$6^*0.118 + \$H\$7^*0.25 + \$H\$8^*0.142 + \$H\$9^*0.141$$
$$+ \$H\$10^*0.898$$

and B12 = 30.
6. Cumulative protein from the meat components is entered in C12:

$$= \$H\$3^*0.191 + \$H\$4^*0.159 + \$H\$5^*0.173 + \$H\$6^*0.173$$
$$+ \$H\$7^*0.019$$

and: C12 = 15.
7. Other constraints are that no acceptable solution will have any meat component with values less than zero: These will be entered in the constraints box as follows:

H6 >= 0; H7 >= 0;   H9 >= 0; H10 >= 0

8. Formula for total cost is entered in G12:

$$= \$H\$6^*0.018 + \$H\$7^*0.0141 + \$H\$8^*0.0117 + \$H\$9^*0.0154$$
$$+ \$H\$10^*0.0022$$

The value of G12 must be minimized.

Access the Solver by clicking on Tools and selecting Solver. The Solver Parameters box will appear. In the Set Target Cell box enter $\$G\$12$ and in the row Equal To choose Min. In the By Changing Cells box enter $\$H\$6 : \$H\$10$. Add all the constraints into the "Subject to the Constraints" box. Click Solve. Results are displayed in the spreadsheet in Fig. 3.18.

The least = cost formulation has the following meats: 58.54 kg bull meat, 18.87 kg pork trimmings, 2.5 kg beef cheeks, 0 kg pork cheeks, and 20.06 kg back fat. The total mass of meats is 100 kg, and the cost of the formulation is 139.35 $/100 kg.

The least-cost formulation will change as the cost of the ingredients change, therefore, processors need to update cost information regularly and determine the current least cost.

3.4 MULTISTAGE PROCESSES

Problems of this type require drawing a process flow diagram and moving the system boundaries for the material balance formulations around parts of the process as needed to the unknown quantities. It is always good to define the basis used in each stage of the calculations. It is possible to change basis as the calculations proceed from an analysis of one subsystem to the next. A "tie material" will also be helpful in relating one part of the process to another.

Example 3.17. The standard of identity for jams and preserves specify that the ratio of fruit to added sugar in the formulation is 45 parts fruit to 55 parts sugar. A jam also must have a soluble content of at least 65% to produce a satisfactory gel. The standard of identity requires soluble solids of at least 65% for fruit preserves from apricot, peach, pear, cranberry, guava, nectarine, plum, gooseberry, figs, quince, and currants. The process of making fruit preserves involves mixing the fruit and sugar in the required ratio, adding pectin, and concentrating the mixture by boiling in a vacuum, steam-jacketed kettle until the soluble solids content is at least 65%. The amount of pectin added is determined by the amount of sugar used in the formulation and by the grade of the pectin (a 100 grade pectin is one that will form a satisfactory gel at a ratio 1 kg pectin to 100 kg sugar).

If the fruit contains 10% soluble solids and 100 grade pectin is used, calculate the weight of fruit, sugar, and pectin necessary to produce 100 kg of fruit preserve. For quality control purposes, soluble solids are those that change the refractive index and can be measured on a refractometer. Thus, only the fruit soluble solids and sugar are considered soluble solids in this context, and not pectin.

Solution:

The process flow diagram is shown in Fig. 3.19. The boundary used for the system on which the material balance is made encloses the whole process.

This problem uses data given in the problem, 100 kg fruit preserve, as a basis. The tie material is the soluble solids because pectin is unidentifiable in the finished product. The designation "fruit" is meaningless in the finished product because the water that evaporates comes from the fruit, and the soluble solids in the fruit mixes with the rest of the system.

Soluble solids balance:

$$0.1(X) + Y = 100(0.65)$$

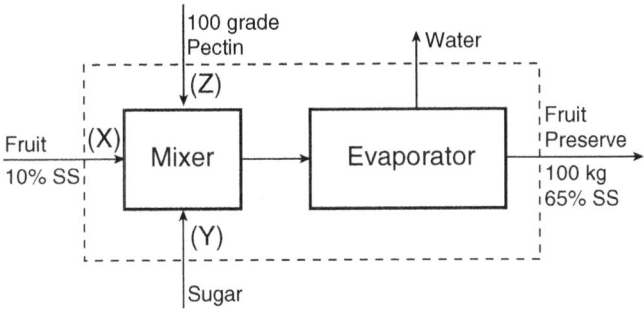

Figure 3.19 Process and material flow for manufacturing fruit preserves.

Because the ratio 45 parts fruit to 55 parts sugar is a requirement, the other equation would be

$$\frac{X}{Y} = \frac{45}{55}; \; X = \frac{45}{55} Y$$

Substituting the expression for X in the equation representing the soluble solids balance and solving for Y:

$$0.1 \left(\frac{45}{55} Y \right) + Y = 65$$

$$0.1(45)(Y) + 55Y = 55(65)$$

$$4.5Y + 55Y = 55(65)$$

$$59.5Y = 55(65)$$

$$Y = \frac{55(65)}{59.5} = 60 \text{ kg sugar}$$

Because $X = 45Y/55$, $X = 45(60)/55 = 49$ kg fruit.

$$Z = \frac{\text{kg sugar}}{\text{grade of pectin}} = \frac{60}{100} = 0.6 \text{ kg pectin}$$

The amount of pectin, Z, can be calculated from the weight of sugar used. The problem can also be solved by selecting a different variable as a basis: 100 kg fruit. Equation based on the required ratio of fruit to sugar:

$$\text{Sugar} = 100 \text{ kg fruit} \times \frac{55 \text{ kg sugar}}{45 \text{ kg fruit}} = 122 \text{ kg}$$

Let X = kg of jam produced.

Soluble solids balance: $100(0.1) + 122 = X(0.65)$

$$X = \frac{132}{0.65} = 203 \text{ kg}$$

Because 100 kg fruit will produce 203 kg jam, the quantity of fruit required to produce 100 kg jam can be determined by ratio and proportion:

$$100 : 203 = X : 100$$

$$X = \frac{100(100)}{203} = 49 \text{ kg fruit}$$

$$\text{Sugar} = 49 \times \frac{55}{45} = 60 \text{ kg}$$

$$\text{Pectin} = \frac{1}{100}(60) = 0.6 \text{ kg}$$

Example 3.18. In solvent extractions, the material to be extracted is thoroughly mixed with a solvent. An ideal system is one where the component to be extracted dissolves in the solvent, and the ratio of solute to solvent in the liquid phase equals the ratio of solute to solvent in the liquid absorbed in the solid phase. This condition occurs with thorough mixing until equilibrium is reached and if sufficient solvent is present such that the solubility of solute in the solvent is not exceeded.

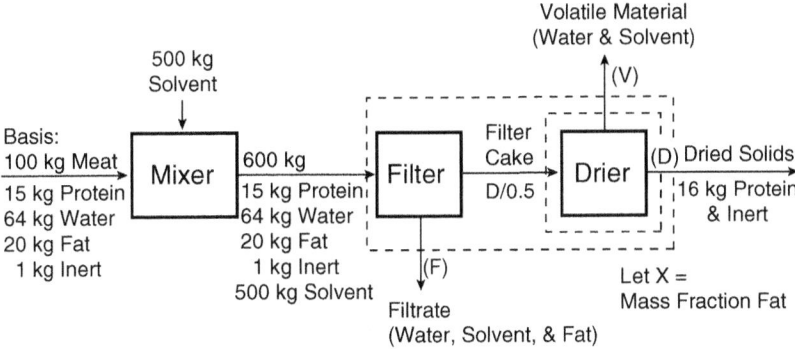

Figure 3.20 Process and material flow for a solvent extraction process involving simultaneous removal of fat and moisture.

Meat (15% protein, 20% fat, 64% water, and 1% inert insoluble solids) is extracted with five times its weight of a fat solvent that is miscible in all proportions with water. At equilibrium, the solvent mixes with water, and fat dissolves in this mixture. Assume there is sufficient solvent such that all the fat dissolves.

After thorough mixing, the solid is separated from the liquid phase by filtration, and the solid phase is dried until all the volatile material is removed. The weight of the dry cake is only 50% of the weight of the cake leaving the filter.

Assume none of the fat, protein, and inerts are removed from the filter cake by drying, and no nonfat solids are in the liquid phase leaving the filter. Calculate the fat content in the dried solids.

Solution:

The diagram of the process is shown in Fig. 3.20. Using 100 kg of meat as a basis, the ratio of solvent to meat of five would require 500 kg of solvent. Consider as a system one that has a boundary that encloses the filter and the drier. Also draw a subsystem with a boundary that encloses only the drier. Write the known weights of process stream and the weights of components in each stream in the diagram. Entering the system at the filter will be 600 kg of material containing 15 kg protein, 1 kg inerts, 64 kg water, 20 kg of fat, and 500 kg of solvent. Because all nonfat solids leave at the drier, the weight of the dried solids D includes 16 kg of the protein and inert material. From the condition given in the problem that the weight of dried solids is 50% of the weight of the filter cake, the material entering the drier should be D/0.5.

All fat dissolves in the solvent-water mixture. Fat, however, will also appear in the solids fraction because some of the fat-water-solvent solution will be retained by the solids after filtration. Although the filtrate does not constitute all of the fat-water-solvent solution, the mass fraction of fat in the filtrate will be the same as in the whole solution, and this condition can be used to calculate the mass fraction of fat in the filtrate.

$$\text{Mass fraction fat in filtrate} = \frac{\text{wt fat}}{\text{wt fat} + \text{wt solv} + \text{wt} \, H_2O}$$

$$= \frac{20}{20 + 64 + 500} = 0.034246$$

Consider the system of the filter and drier. Let x = mass fraction of fat in D. The following component balance can be made.

Fat balance:

$$F(0.034246) + D x = 20 \tag{3.1}$$

Protein and inerts balance: Sixteen kilograms of proteins and inerts enter the system and all of these leave the system with the dried solids D. Because D consists only of fat + protein + inert, and x is the mass fraction of fat, then, 1 − x would be the mass fraction of protein and inert.

$$D(1 - x) = 16; \quad D = \frac{16}{1 - x} \tag{3.2}$$

Solvent and water balance: Five hundred kilograms of solvent and 64 kg of water enter the system. These components leave the system with the volatile material V at the drier and with the filtrate F at the filter. The mass fraction of fat in the filtrate is 0.034246. The mass fraction of water and solvent = 1 − 0.034246 = 0.965754.

$$F(0.965754) + V = 564 \tag{3.3}$$

There are four unknown quantities, F, D, V, and x, and only three equations have been formulated. The fourth equation can be formulated by considering the subsystem of the drier. The condition given in the problem that the weight of solids leaving the drier is 50% of the weight entering would give the following total mass balance equation around the drier:

$$\frac{D}{0.5} = D + V; \quad D = V \tag{3.4}$$

The above four equations can be solved simultaneously. Substituting D for V in Equation (3.3):

$$F(0.965754) + D = 564 \tag{3.5}$$

Substituting Equation (3.2) in Equation (3.5):

$$F(0.965754) + \frac{16}{1 - x}(x) = 564 \tag{3.6}$$

Substituting Equation (3.2) in Equation (3.1):

$$F(0.034246) + \frac{16}{1x}(x) = 20 \tag{3.7}$$

Solving for F in Equations (3.6) and (3.7) and equating:

$$\frac{20 - 36 x}{(0.034246)(1 - x)} = \frac{548 - 564x}{(0.965754)(1 - x)}$$

Simplifying and solving for x:

$$0.965754(20 - 36x) = 0.034246(548 - 564x)$$
$$19.31508 - 34.767144x = 18.7668 - 19.314744x$$
$$x(34.76144 - 19.314744) = 19.31508 - 18.7668$$
$$x = \frac{0.548272}{15.446696} = 0.03549$$

The percentage fat in the dried solids = 3.549%.

The solution can be shortened and considerably simplified if it is realized that the ratio of fat/(solvent + water) is the same in the filtrate as it is in the liquid that adheres to the filter cake entering the drier. Because the volatile material leaving the drier is only solvent + water, it is possible to calculate the amount of fat carried with it from the fat/(solvent + water) ratio.

$$\frac{fat}{solvent + water} = \frac{20}{500 + 64} = 0.03546$$

The amount of fat entering the drier = 0.03546 V. Because D = 16 + fat; D = 16 + 0.03546 V. Total mass balance around the drier gives: D = V.

Substituting D for V gives:

$$D = 16 + 0.03546\,D$$
$$= \frac{16}{1 - 0.03546} = 16.5882$$
$$\text{wt. fat in } D = 16.5882 - 16 = 0.5882$$
$$\% \text{ fat in } D = (0.5882/16.5882)(100) = 3.546\%$$

A principle that was presented in an example at the beginning of this chapter was that in solvent extraction or crystallization, the purity of the product depends strongly on the efficiency of separation of the liquid from the solid phase. This is demonstrated here in that the amount of fat entering the drier is a direct function of the amount of volatile matter present in the wet material. Thus, the more efficient the solid-liquid separation process before drying the solids, the less fat will be carried over into the finished product.

Example 3.19. Crystallization. Determine the quantity of sucrose crystals that will crystallize out of 100 kg of a 75% sucrose solution after cooling to 15°C. The mother liquor contains 66% sucrose.

Solution:

The diagram for crystallization is shown in Fig. 3.21. A 75% sucrose solution enters the crystallizer and the material separates into a solution containing 66% sucrose and crystals of 100% sucrose.

Figure 3.21 shows sucrose appearing in the crystals and saturated solution fractions and water appears only in the saturated solution fraction. Water can be used as a tie material in this process.

The mother liquor contains 66% sucrose. The balance (34%) is water. Using the mass fraction principle, 25 kg of water enters the system and all this leaves with the mother liquor where the mass fraction of water is 0.34. Thus, the weight of the mother liquor would be 25/0.34 = 73.52 kg.

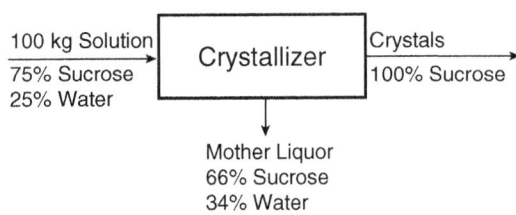

Figure 3.21 Composition and material flow in a crystallization process.

Total mass balance:

wt crystals $= 100 - 73.52 = 26.48$ kg

Example 3.20. For the problem statement and the process represented by the diagram in Fig. 3.2, calculate the yield and the purity of the sugar crystals. The mother liquor contains 67% sucrose w/w. The crystals fraction from the centrifuge loses 15% of its weight in the drier and emerges moisture free.

Solution:

The problem can be separated into two sections. First, a material balance around the evaporator will be used to determine the weight and composition of the material entering the crystallizer. The next part, which involves the crystallizer, centrifuge, and drier, can be solved utilizing the principles demonstrated in the two preceding example problems.

Material balance around the evaporator: Twenty kilograms of sucrose enters, and this represents 75% of the weight of the concentrate.

wt concentrate $= 20/0.75 = 26.66$ kg

The concentrate entering the crystallizer consists of 20 kg sucrose, 1 kg inert, and 5.66 kg water.

Consider the centrifuge as a subsystem where the cooled concentrate is separated into a pure crystals fraction and a mother liquor fraction. The weights of these two fractions and the composition of the saturated solution fraction can be determined using a similar procedure as in the preceding example. In Fig. 3.22, the known quantities indicated on the diagram.

The composition of the mother liquor fraction is determined by making a material balance around the crystallizer. The mass and composition of streams entering and leaving the crystallizer are shown in Fig. 3.23.

All the water and inert material entering the centrifuge go into the mother liquor fraction. The weight of water and inert material in the mother liquor are 5.66 and 1 kg, respectively.

The material balance around the crystallizer is

Total mass: $M + S = 26.66$
Sucrose balance: $M(0.67) + S = 20$

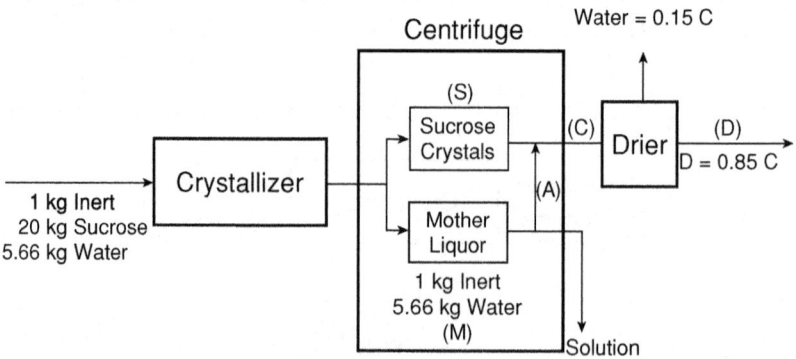

Figure 3.22   Material flow and system boundaries used for analysis of crystal purity from a crystallizer.

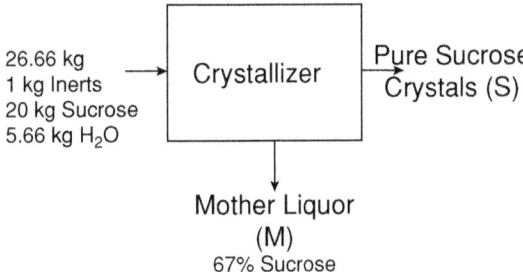

Figure 3.23   Material balance around the crystallizer to determine the quantity of mother liquor and pure crystals that separate on crystallization.

Eliminating S by subtracting the two equations:

$M(1 - 0.67) = 26.66 - 20; M = 6.66/0.33 = 20.18$ kg

From the known weights of components, the composition of mother liquor, in mass fraction, is as follows.

Sucrose $= 0.67$, as defined
Inert: 1.00 kg; mass fraction $= 1/20.18 = 0.04955$
Water: 5.66 kg; mass fraction $= 5.66/20.18 = 0.28048$

The solids fraction C leaving the centrifuge is a mixture of the mother liquor adhering to the crystals, A, and pure sucrose crystals S. Because 26.66 kg of material entered the crystallizer and 20.18 kg is in the mother liquor, the pure sucrose crystals, S=26.66 − 20.18=6.48 kg.

$C = 6.48 + A$

Water balance around the drier:
A(mass fraction water in A) $= 0.15$ C
    $A(0.28048) = 0.15$ C

Substituting C:

$A(0.28048) = 0.15(6.48 + A)$
$A(0.28048 - 0.15) = 0.15(6.48)$
$A = 7.4494$ kg
$C = 6.48 + 7.4494 = 13.9294$ kg

Wt dry sugar, $D = 0.85(6.48 + 7.4494) = 11.84$ kg
Wt inert material $=$ A(mass fraction inert in A) $= 7.4494(0.04955) = 0.3691$ kg
% inert in dry sugar $= (0.3691/11.84)(100) = 3.117\%$
The dry sugar is only 96.88% sucrose.

Example 3.21. An ultrafiltration system has a membrane area of 0.75 m² and has a water permeability of 180 kg water/h(m²) under conditions used in concentrating whey from 7% to 25% total solids at a pressure of 1.033 MPa. The system is fed by a pump that delivers 230 kg/h and the appropriate concentration of solids in the product is obtained by recycling some of the product through the membrane. The concentrate contained 11% lactose and the unprocessed when contained 5.3% lactose. There is no protein in the permeate.

Calculate (1) the production rate of 25% concentrate through the system; (2) the amount of product recycled/h; (3) the amount of lactose removed in the permeate/h; (4) the concentration of lactose in the mixture of fresh and recycled whey entering the membrane unit; and (5) the average rejection factor by the membrane for lactose based on the average lactose concentration entering and leaving the unit.

The rejection factor (F_r) of solute through a membrane is defined expressed by:

$$F_r = \frac{X_f - X_p}{X_f}$$

where X_f is solute concentration on the feed side of the membrane, which may be considered as the mean of the solute concentration in the fluid entering and leaving the membrane unit, and X_p is the solute concentration in the permeate.

Solution:

The process flow diagram is shown in Fig. 3.24.

Let L, Pr, and W represent lactose, protein, and water respectively. Let F = feed, P = permeate or fluid passing through the membrane, and R = recycled stream, which has the same composition as the concentrate, C. The fluid stream leaving the membrane unit which is the sum of the concentrate and recycle stream, is also referred to as the retentate.

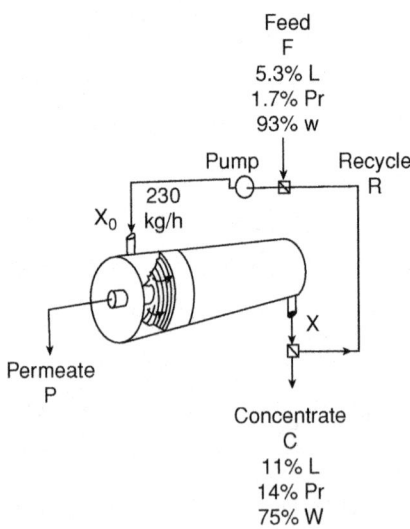

Figure 3.24 Diagram of a spiral wound membrane module in an ultrafiltration system for cheese whey.

Whole system material balance: Basis: 1 hour of operation.

$$\text{Water in permeate} = \frac{180 \text{ kg}}{\text{m}^2\text{h}}(0.75_{\text{m}^2})(1 \text{ h})$$
$$= 180(0.75) = 135 \text{ kg}$$

Water balance:

$$F(0.93) = 135 + C(0.75); \ 0.93F - 0.75C = 135$$

Protein balance:

$$F(0.017) = C(0.14); \ 0.017F - 0.14C = 0$$

Solving for F and C by determinants:

$$\begin{bmatrix} 0.93 & -0.75 \\ 0.017 & -0.14 \end{bmatrix} \begin{bmatrix} F \\ C \end{bmatrix} = \begin{bmatrix} 135 \\ 0 \end{bmatrix}$$

$$F = \frac{135(-0.14) - 0}{0.93(-0.14) - 0.017(-0.75)} = \frac{-18.9}{-0.1302 + 0.01275}$$

$$C = \frac{0 - 0.017(135)}{-0.1302 + 0.01275} = \frac{-2.295}{-0.11745} = 19.54 \text{ kg/h}$$

$$F = 160.9 \text{ kg/h}$$

1. Amount of product = 19.54 kg/h
2. Amount recycled is determined by a material balance around the pump. The diagram is shown in Fig. 3.25.

   $$R = 230 - 160.9 = 69.1 \text{ kg/h}$$

3. Lactose in permeate: Lactose balance around whole system:
   L_p = lactose in permeate = F(0.053) C(0.11)

   $$= 160.9(0.053) - 19.54(0.11) = 6.379$$

   X_p = mass fraction lactose in permeate.

   $$= \frac{6.379}{135 + 6.379} = 0.04511$$

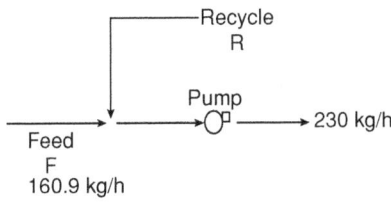

Figure 3.25 Diagram of material balance during mixing of feed and recycle streams in an ultrafiltration system.

4. Lactose balance around pump:

$$F(0.053) + R(0.11) = 230(X_{fl})$$

X_{fi} = mass fraction lactose in fluid leaving the membrane unit.

$$X_{fl} = \frac{160.9(0.053) + 69.1(0.11)}{230} = 0.0702$$

X_f = mean mass fraction lactose in the retentate side of the membrane unit.

$$X_f = 0.5(0.0702 + 0.11) = 0.09.$$

5. Rejection factor for lactose:

$$F_r = (0.09 - 0.045)/0.09 = 0.498$$

PROBLEMS

3.1. A frankfurter formulation is to be made from the following ingredients:
   Lean beef: 14% fat, 67% water, 19% protein.
   Pork fat: 89% fat, 8% water, 3% protein.
   Soy protein isolate: 90% protein, 8% water.

Water needs to be added (usually in the form of ice) to achieve the desired moisture content. The protein isolate added is 3% of the total weight of the mixture. How much lean beef, pork fat, water, and soy isolate must be used to obtain 100 kg of a formulation having the following composition: protein, 15%; moisture, 65%; fat, 20%.

3.2. If 100 kg of raw sugar containing 95% sucrose, 3% water, and 2% soluble uncrystallizable inert solids is dissolved in 30 kg of hot water and cooled to 20°C, calculate:
   (a) kilograms of sucrose that remains in solution,
   (b) crystalline sucrose,
   (c) the purity of the sucrose (in % sucrose) obtained after centrifugation and dehydration to 0% moisture. The solid phase contained 20% water after separation from the liquid phase in the centrifuge.
   A saturated solution of sucrose at 20°C contains 67% sucrose (w/w).

3.3. Tomato juice flowing through a pipe at the rate of 100 kg/min is salted by adding saturated salt solution (26% salt) into the pipeline at a constant rate. At what rate would the saturated salt solution be added to have 2% salt in the product?

3.4. If fresh apple juice contains 10% solids, what would be the solids content of a concentrate that would yield single-strength juice after diluting one part of the concentrate with three parts of water. Assume densities are constant and are equal to the density of water.

3.5. In a dehydration process, the product, which is at 80% moisture, initially has lost half its weight during the process. What is the final moisture content?

3.6. Calculate the quantity of dry air that must be introduced into an air drier that dries 100 kg/h of food from 80% moisture to 5% moisture. Air enters with a moisture content of 0.002 kg water per kg of dry air and leaves with a moisture content of 0.2 kg H_2O per kg of dry air.

3.7. How much water is required to raise the moisture content of 100 kg of a material from 30% to 75%?

3.8. In the section "Multistage Process," Example 3.18, solve the problem if the meat to solvent ratio is 1:1. The solubility of fat in the watersolvent mixture is such that the maximum fat content in the solution is 10%.

3.9. How many kilograms of peaches would be required to produce 100 kg of peach preserves? The standard formula of 45 parts fruit to 55 parts sugar is used, the soluble solids content of the finished product is 65% and the peaches have 12% initial soluble solids content. Calculate the weight of 100 grade pectin required and the amount of water removed by evaporation.

3.10. The peaches in Problem 9 come in a frozen form to which sugar has been added in the ratio of three parts fruit to one part sugar. How much peach preserves can be produced from 100 kg of this frozen raw material?

3.11. Yeast has a proximate analysis of 47% C, 6.5% H, 31% O, 7.5% N, and 8% ash on a dry weight basis. Based on a factor of 6.25 for converting protein nitrogen to protein, the protein content of yeast on a dry basis is 46.9%. In a typical yeast culture process, the growth medium is aerated to convert substrate primarily to cell mass. The dry cell mass yield is 50% of a sugar substrate. Nitrogen is supplied as ammonium phosphate.

The cowpea is a high-protein, low-fat legume that is a valuable protein source in the diet of many Third World nations. The proximate analysis of the legume is 30% protein, 50% starch, 6% oligosaccharides, 6% fat, 2% fiber, 5% water, and 1% ash. It is desired to produce a protein concentrate by fermenting the legume with yeast. Inorganic ammonium phosphate is added to provide the nitrogen source. The starch in cowpea is first hydrolyzed with amylase and yeast is grown on the hydrolyzate.

(a) Calculate the amount of added inorganic nitrogen as ammonium phosphate to provide the stoichiometric amount of nitrogen to convert all the starch present to yeast mass. Assume none of the cowpea protein is utilized by the yeast.

(b) If the starch is 80% converted to cell mass, calculate the proximate analysis of the fermented cowpea on a dry basis.

3.12. This whey is spray dried to a final moisture content of 3%, and the dry whey is used in an experimental batch of summer sausage.

In summer sausage, the chopped meat is inoculated with a bacterial culture that converts sugars to lactic acid as the meat is allowed to ferment prior to cooking in a smokehouse. The level of acid produced is controlled by the amount of sugar in the formulation. The lactic acid level in the summer sausage is 0.5 g/100 g dry solids. Four moles of lactic acid ($CH_3CHOHCOOH$) is produced from 1 mole of lactose ($C_{12}H_{22}O_{11}$). The following formula is used for the summer sausage:

3.18 kg lean beef (16% fat, 16% protein, 67.1% water, 0.9% ash)
1.36 kg pork (25% fat, 12% protein, 62.4% water, 0.6% ash)
0.91 kg ice
0.18 kg soy protein isolate (5% water, 1% ash, 94% protein)

Calculate the amount of dried whey protein that can be added into this formulation in order that, when the lactose is 80% converted to lactic acid, the desired acidity will be obtained.

3.13. Osmotic dehydration of blueberries was accomplished by contacting the berries with a corn syrup solution containing 60% soluble solids for 6 hours and draining the syrup from the solids. The solid fraction left on the screen after draining the syrup is 90% of the original weight of the berries. The berries originally contained 12% soluble solids, 86.5% water, and 1.5% insoluble solids. The sugar in the syrup penetrated the berries such that the berries themselves remaining on the screen, when washed free of the adhering solution, showed a soluble solids gain of 1.5% based on the original dry solids content. Calculate:

(a) The moisture content of the berries and adhering solution remaining on the screen after draining the syrup.

(b) The soluble solids content of the berries after drying to a final moisture content of 10%.

(c) The percentage of soluble solids in the syrup drained from the mixture. Assume none of the insoluble solids are lost in the syrup.

3.14. The process for producing dried mashed potato flakes involves mixing wet mashed potatoes with dried flakes in a 95:5 weight ratio, and the mixture is passed through a granulator before drying on a drum dryer. The cooked potatoes after mashing contained 82% water and the dried flakes contained 3% water. Calculate:

(a) The amount of water that must be removed by the dryer for every 100 kg of dried flakes produced.

(b) The moisture content of the granulated paste fed to the dryer.

(c) The amount of raw potatoes needed to produce 100 kg of dried flakes; 8.5% of the raw potato weight is lost on peeling.

(d) Potatoes should be purchased on a dry matter basis. If the base moisture content is 82% and potatoes at this moisture content cost $200/ton, what would be the purchase price for potatoes containing 85% moisture.

3.15. Diafiltration is a process used to reduce the lactose content of whey recovered using an ultrafiltration membrane. The whey is passed through the membrane first and concentrated to twice the initial solids content, rediluted, and passed through the membrane a second time. Two membrane modules in series, each with a membrane surface area of 0.5 m^2, are to be used for concentrating and removing lactose from acid whey that contains 7.01% total solids, 5.32% lactose, and 1.69% protein. The first module accomplishes the initial concentration, and the retentate is diluted with water and reconcentrated in the second module to a solids content of 14.02%. Under the conditions used in the process, the membrane has an average water permeation rate of 254 kg/(h $\cong$ m^2). The rejection factor for lactose by the membrane based on the arithmetic mean of the feed and retentate lactose concentration and the mean permeate lactose concentration is 0.2. Protein rejection factor is 1. The rejection factor is defined as: $F_r = (C_r C_p)/C_r$ where C_r is the concentration on the retentate side and C_p is the concentration on the permeate side of the membrane. Use Visual BASIC to determine:

(a) The amount of 14.02% solids delactosed whey concentrate produced from the second module per hour.

(b) The lactose content of the delactosed whey concentrate.

3.16. An orange juice blend containing 42% soluble solids is to be produced by blending stored orange juice concentrate with the current crop of freshly squeezed juice. The following are the constraints:

The soluble solids to acid ratio must equal 18; and the currently produced juice may be concentrated first before blending, if necessary. The currently produced juice contains 14.5% soluble solids, 15.3% total solids, and 0.72% acid. The stored concentrate contains 60% soluble solids, 62% total solids, and 4.3% acid. Calculate:

(a) The amount of water that must be removed or added to adjust the concentration of the soluble solids in order to meet the specified constraints.

(b) The amounts of currently processed juice and stored concentrate needed to produce 100 kg of the blend containing 42% soluble solids.

3.17. The process for extraction of sorghum juice from sweet sorghum for production of sorghum molasses, which is still practiced in some areas in the rural South of the United States, involves passing the cane through a three-roll mill to squeeze the juice out. Under the best conditions, the squeezed cane (bagasse) still contains 50% water.

    (a) If the cane originally contained 13.4% sugar, 65.6% water, and 32% fiber, calculate the amount of juice squeezed from the cane per 100 kg of raw cane, the concentration of sugar in the juice, and the percentage of sugar originally in the cane that is left unrecovered in the bagasse.

    (b) If the cane is not immediately processed after cutting, moisture and sugar loss occurs. Loss of sugar has been estimated to be as much as 1.5% within a 24-hour holding period, and total weight loss for the cane during this period is 5.5%. Assume sugar is lost by conversion to CO_2, therefore the weight loss is attributable to water and sugar loss. Calculate the juice yield based on the freshly harvested cane weight of 100 kg, the sugar content in the juice, and the amount of sugar remaining in the bagasse.

3.18. In a continuous fermentation process for ethanol from a sugar substrate, the sugar is converted to ethanol and part of it is converted to yeast cell mass. Consider a 1000 L continuous fermentor operating at a steady state. Cell-free substrate containing 12% glucose enters the fermentor. The yeast has a generation time of 1.5 hours, and the concentration of yeast cells within the fermentor is 1×10^7 cells/mL. Under these conditions, a dilution rate (F/V, where F is the rate of feeding of cell-free substrate and V is the volume of the fermentor) that causes the cell mass to stabilize at a steady state results in a residual sugar content in the overflow of 1.2%. The stoichiometric ratio for sugar to dry cell mass is 1:0.5 on a weight basis, and that for sugar to ethanol is based on 2 moles of ethanol produced per mole of glucose. A dry cell mass of 4.5 g/L is equivalent to a cell count of 1.6×10^7 cells/mL. Calculate the ethanol productivity of the fermentor in g ethanol/(L $\cong$ h).

3.19. A protein solution is to be dimineralized by dialysis. The solution is placed inside a dialysis tube immersed in continuously flowing water. For all practical purposes, the concentration of salt in the water is zero, and the dialysis rate is proportional to the concentration of salt in the solution inside the tube. The contents of the tube may be considered to be well mixed. A solution that initially contained 500 µg/mL of salt and 15 mg protein/mL contained 400 µg/mL salt and 13 mg protein/mL at the end of 2 hours. Assume no permeation of protein through the membrane, and density of the solution is constant at 0.998 g/mL. The rate of permeation of water into the membrane and the rate of permeation of salt out of the membrane are both directly proportional to the concentration of salt in the solution inside the membrane. Calculate the time of dialysis needed to drop the salt concentration inside the membrane to 10 µg/mL. What will be the protein concentration inside the membrane at this time?

3.20. A fruit juice–based natural sweetener for beverages is to be formulated. The sweetener is to have a soluble solids to acid ratio of 80. Pear concentrate having a soluble solids to acid ratio of 52 and osmotically concentrated/deacidified apple juice with a soluble solids to acid ratio of 90 are available. Both concentrates contain 70% soluble solids.

    (a) Calculate the ratio of deacidified apple and pear concentrates that must be mixed to obtain the desired sol. solids/acid ratio in the product.

    (b) If the mixture is to be diluted to 45% soluble solids, calculate the amounts of deacidified apple and pear juice concentrates needed to make 100 kg of the desired product.

3.21. A spray drier used to dry egg whites produces 1000 kg/h of dried product containing 3.5% moisture from a raw material that contains 86% moisture.

(a) If air used for drying enters with a humidity of 0.0005 kg water per kg dry air and leaves the drier with a humidity of 0.04 kg water per kg dry air, calculate the amount of drying air needed to carry out the process.

(b) It is proposed to install a reverse osmosis system to remove some moisture from the egg whites prior to dehydration to increase the drying capacity. If the reverse osmosis system changes the moisture content of the egg whites prior to drying to 80% moisture, and the same inlet and outlet air humidities are used on the drier, calculate the new production rate for the dried egg whites.

3.22. An ultrafiltration system for concentrating milk has a membrane area of 0.5 m^2, and the permeation rate for water and low molecular weight solutes through the membrane is 3000 g/(m^2 $\cong$ min). The solids content of the permeate is 0.5%. Milk flow through the membrane system must be maintained at 10 kg/min to prevent fouling. If milk containing initially 91% water enters the system, and a concentrate containing 81% solids is desired, calculate:

(a) The amount of concentrate produced by the unit/h.

(b) The fraction of total product leaving the membrane that must be recycled to achieve the desired solids concentration in the product.

(c) The solids concentration of feed entering the unit after the fresh and recycled milk are mixed.

3.23. A recent development in the drying of bluberries involves an osmotic dewatering prior to final dehydration. In a typical process, grape juice concentrate with 45% soluble solids (1.2% insoluble solids; 53.8% water) in a ratio 2 kg juice/kg berries is used to osmoblanch the berries, followed by draining the berries and final drying in a tunnel drier to a moisture content of 12%.

During osmoblanching (the juice concentrate is heated to 80°C, the berries added, temperature allowed to go back to 80°C, 5-minute hold, and the juice is drained), the solids content of blueberies increases. When analyzed after rinsing the adhering solution, the total solids content of the berries was 15%. The original berries contained 10% total solids and 9.6% soluble solids. Assume leaching of blueberry solids into the grape juice is negligible.

The drained berries carry about 12% of their weight of the solution that was drained.

Calculate:

(a) The proximate composition of the juice drained from the blueberries.

(b) The yield of dried blueberries from 100 kg raw berries.

(c) The drained juice is concentrated and recycled. Calculate the amount of 45% solids concentrate that must be added to the recycled concentrate to make enough for the next batch of 100 kg blueberies.

3.24. A dietetic jelly is to be produced. In order that a similar fruit flavor may be obtained in the dietetic jelly as in the traditional jelly, the amount of jelly that can be made from a given amount of fruit juice should be the same as in the standard pectin jelly.

If 100 kg of fruit juice is available with a soluble solids content of 14%, calculate the amount of standard pectin jelly with a soluble solids content of 65% that can be produced.

Only the sugar in jelly contribute the caloric content.

(a) What will be the caloric equivalent of 20 g (1 tsp) of standard pectin jelly?

(b) The dietetic jelly should have a caloric content 20% that of the standard pectin jelly. Fructose may be added to provide sweetness. The low methoxy pectin can be used in the same amount for the same quantity of fruit as in the standard jelly. Calculate the soluble solids content in the dietetic jelly to give the caloric reduction.

(c) In part (b), calculate the amount of additional fructose that must be used.

SUGGESTED READING

Charm, S. E. 1971. Fundamentals of Food Engineering. 2nd ed. AVI Publishing Co., Westport, CT.

Felder, R. M. and Rousseau, R. W. 1999. Elementary Principles of Chemical Processes. 2nd ed. John Wiley & Sons, New York.

Himmelblau, D. M. 1967. Basic Principles and Calculations in Chemical Engineering. 2nd ed. Prentice-Hall, Englewood Cliffs, NJ.

Hougen, O. A. and Watson, K. M. 1946. Chemical Process Principles. Part I. Material and Energy Balances. John Wiley & Sons, New York.

Sinnott, R. K. 1996. Chemical Engineering. Vol 6, 2nd ed. Butterworth-Heinemann, Oxford.

CHAPTER 4

Gases and Vapors

Gases and vapors are naturally associated with foods and food-processing systems. The equilibrium between food and water vapor determines temperatures achieved during processing. Dissolved gases in foods such as oxygen affect shelf life. Gases are used to flush packages to eliminate oxygen and prolong shelf life. Modified atmospheres in packages have been used to prolong shelf life of packaged foods. Air is used for dehydration. Gases are used as propellants in aerosol cans and as refrigerants. The distinction between gases and vapors is very loose because theoretically all vapors are gases. The term "vapor" is generally used for the gaseous phase of a substance that exists as a liquid or a solid at ambient conditions.

4.1 EQUATIONS OF STATE FOR IDEAL AND REAL GASES

Equations of state are expressions for the relationship between pressure, volume, temperature and quantity of gases within a given system. The simplest equation of state, the ideal gas equation, closely approximates the actual behavior of gases at near ambient temperature and pressure where the effect of molecular interactions is minimal. At high pressures and temperatures, however, most gases deviate from ideal behavior, and several equations of state have been proposed to fit experimental data. In this section, two equations of state will be discussed: the ideal gas equation and one of the most used equation of state for real gases, van der Waal's equation.

4.1.1 The Kinetic Theory of Gases

The fundamental theory governing the behavior of gases, the kinetic theory, was first proposed by Bernoulli in 1738 and was tested and extended by Clausius, Maxwell, Boltzman, van der Waal, and Jeans. The postulates of the kinetic theory are as follows:

1. Gases are composed of discreet particles called *molecules*, which are in constant random motion, colliding with each other and with the walls of the surrounding vessel.
2. The force resulting from the collision between the molecules and the walls of the surrounding vessel is responsible for the *pressure* of the gas.

3. The lower the pressure, the farther apart the molecules, thus, attractive forces between molecules have reduced influence on the overall properties of the gas.
4. The average kinetic energy of the molecules is directly proportional to the absolute temperature.

4.1.2   Absolute Temperature and Pressure

The pressure (P) of a gas is the force of collisions of gas molecules against a surface in contact with the gas. Because pressure is force per unit area, pressure is proportional to the number of gas molecules and their velocity. This pressure is the absolute pressure.

Pressure is often expressed as *gauge pressure* when the measured quantity is greater than atmospheric pressure, and as *vacuum* when below atmospheric. Measurement of gauge and atmospheric pressure is shown in Fig. 4.1.

Two types of pressure-measuring devices are depicted: a manometer and a Bourdon tube pressure gauge. On the left of the diagram, the pressure of the gas is counteracted by atmospheric pressure, such that if pressure is atmospheric, the gauges will read zero. The reading given by these type of gages is the *gauge pressure*. In the American Engineering System of measurement, gauge pressure is expressed as "psig" or pound force per square inch gauge. In the SI system, gauge pressure is expressed as "kPa above atmospheric."

The diagram on the right in Fig. 4.1 shows the measuring element completely isolated from the atmosphere. The pressure reading from these gauges represents the actual pressure or force of collision of gas molecules, the *absolute pressure*. In the American Engineering System, absolute pressure is expressed as "psia" or pound force per square inch absolute. In SI, it is expressed as "kPa absolute."

Conversion from gauge to absolute pressure is done using the following equation:

$$\mathbf{P}_{gauge} = \mathbf{P}_{absolute} - \mathbf{P}_{atmospheric} \qquad (4.1)$$

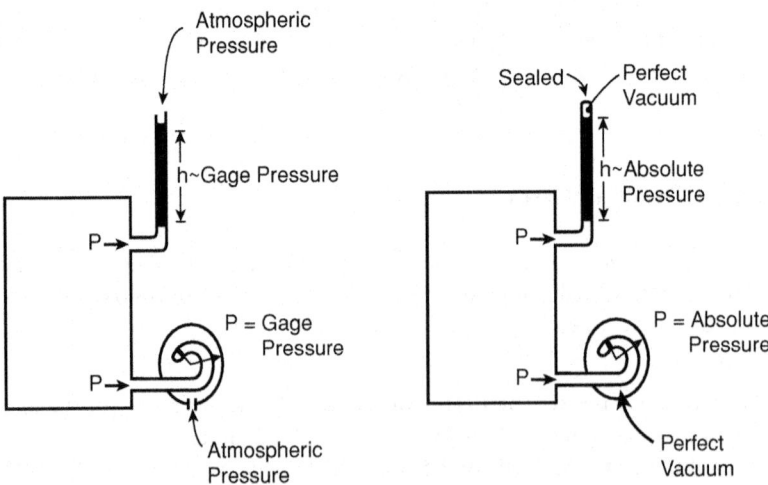

Figure 4.1   Diagram showing differences in measurement of gauge and absolute pressure.

The term "vacuum" after a unit of pressure indicate how much the pressure is below atmospheric pressure. Thus, vacuum may be interpreted as a negative gauge pressure and is related to the atmospheric pressure and absolute pressure as follows:

$$\mathbf{P}_{absolute} = \mathbf{P}_{atmospheric} - \mathbf{P}_{vacuum} \tag{4.2}$$

Pressure is sometimes expressed in *atmospheres* (atm) instead of the traditional force/area units. This has led to some confusion between a pressure specified in atmospheres and atmospheric pressure. Unless otherwise specified, the pressure term "atm" refer to a *standard atmosphere*, the mean atmospheric pressure at sea level, equivalent to 760 mm Hg, 29.921 in. Hg, 101.325 kPa, or 14.696 $lb_f/in.^2$. A *technical atmosphere* is a pressure of 1 kg_f/cm^2. Some technical articles express pressure in *Bar*, which is equivalent to 100 Pa. *Atmospheric pressure* is actual pressure exerted by the atmosphere in a particular location and varies with time and location. Atmospheric pressure must be specified if it is different from a standard atmosphere.

Example 4.1. Calculate the absolute pressure inside an evaporator operating under 20 in. Hg vacuum. Atmospheric pressure is 30 in. Hg. Express this pressure in SI and in the American Engineering System of units.

From the table of conversion factors, Appendix Table A.1, the following conversion factors are obtained:

$$\frac{0.4912\ lb_f/in.^2}{in.Hg};\ \frac{3.38638 \times 10^3\ Pa}{in.\ Hg} \mathbf{P}_{absolute} = \mathbf{P}_{atmospheric} - \mathbf{P}_{vacuum} = (30 - 20)\ in.\ Hg = 10\ in.\ Hg$$

$$= 10\ in.Hg\ \frac{0.4912\ lb_f/in.^2}{\in.Hg} = 4.912\ psia = 10\ in.Hg\ \frac{3386.38\ Pa}{in.Hg} = 33.863\ kPa\ absolute$$

Temperature (T) is a thermodynamic quantity related to the velocity of motion of molecules. The temperature scales are based on the boiling (liquid-gas equilibrium) and freezing points (solid, liquid, gas equilibrium) of pure substances such as water at 1 standard atmosphere (101.325 kPa) pressure. The Kelvin (K) is defined as the fraction, 1/273.16, of the thermodynamic temperature of the triple point of water. The absolute temperature scales are based on a value on the scale that is zero at the temperature when the internal energy of molecules is zero. These are the Kelvin, in SI, and the Rankine (R) in the American Engineering System. Conversion from the commonly used Celsius and Fahrenheit temperature scales to the absolute or thermodynamic scales are as follows:

$$K = °C + 273.16$$
$$°R = °F + 460$$

In common usage, the conversion to Kelvin is rounded off to:

$$K = °C + 273$$

The absolute temperature is used in equations of state for gases.

4.1.3 Quantity of Gases

The quantity of gases is often expressed as volume at a specific temperature and pressure. Because molecules of a gas have very weak attraction toward each other, they will disperse and occupy all space within a confining vessel. Thus, the volume (V) of a gas is the volume of the confining vessel and is not definitive of the quantity of the gas unless the pressure and temperature are also specified.

The most definitive quantification of gases is by mass, or as the number of moles, n, which is the quotient of mass and the molecular weight. The unit of mass is prefixed to mole (e.g., kgmole), to indicate the ratio of mass in kg and the molecular weight. The number of molecules in 1 gmole is the *Avogadro number*, 6.023×10^{23} molecules/gmole.

At 273 K and 760 mm Hg pressure (101.325 kPa), 1 gmole, kgmole, and lbmole occupies 22.4 L, 22.4 m^3, and 359 ft^3, respectively.

4.1.4   The Ideal Gas Equation

The ideal gas equation is the simplest equation of state and was originally derived from the experimental work of Boyle, Charles, and Guy-Lussac. The equation will be derived based on the kinetic theory of gases.

Pressure, the force of collision between gas molecules and a surface, is directly proportional to temperature and the number of molecules per unit volume. Expressing the proportionality as an equation, and using a proportionality constant, R:

$$P\alpha\frac{n}{V}T; \quad P = R\frac{n}{V}T$$

Rearranging:

$$PV = nRT \tag{4.3}$$

Equation (4.3) is the *ideal gas equation*. R is the gas constant and has values of 0.08206 L(atm)/(gmole $\cong$ K); or 8315 N(m)/(kgmole $\cong$ K) or 1545 ft(lb$_f$)/(lbmole $\cong$ °R).

4.1.4.1   P-V-T Relationships for Ideal Gases

When a fixed quantity of a gas that follows the ideal gas equation (Eq. 4.3) undergoes a process where the volume, temperature, or pressure is allowed to change, the product of the number of moles n and the gas constant R is a constant and:

$$\frac{PV}{T} = \text{constant}$$

If initial temperature, pressure, and volume are known and designated by subscript 1, these properties at another point in the process will be expressed by:

$$\frac{PV}{T} = \frac{P_1V_1}{T_1} \tag{4.4}$$

Ideality of gases predicted by Equations (4.3) and (4.4) occur only under conditions close to ambient or under processes that span a relatively narrow range of temperature and pressure. Thus, these equations are useful in problems usually encountered when using gases in food processing or packaging.

Example 4.2. Calculate the quantity of oxygen entering a package in 24 hours if the packaging material has a surface area of 3000 cm^2 and an oxygen permeability 100 $cm^3/(m^2)(24\ h)$ STP (standard temperature and pressure = $0°C$ and 1 standard atmosphere of 101.325 kPa).

Solution:

Solving for the volume of oxygen permeating through the package in 24 hours:

$$V = \frac{100\ cm^3}{m^2(24\ h)}(24\ h)\frac{1\ m^2}{(100)^2\ cm^2}(3000\ cm^2)$$

$$= 30\ cm^3$$

Using Equation (4.3): n = PV/RT
Use R = 0.08206 L(atm)/(gmole $\cong$ K)

$$n = \frac{1\ atm(30\ cm^3)}{[0.08206\ L(atm)/(gmole \cdot K)](273\ K)}\frac{1L}{1000\ cm^3} = 0.001339\ gmoles$$

Example 4.3. Calculate the volume of CO_2 in ft^3 at $70°F$ and 1 atm, which would be produced by vaporization of 1 lb of dry ice.

Solution:

Dry ice is solid CO_2 (mol. wt. = M = 44 lb/lbmole). T = $70°F + 460 = 530°R$.

$$n = \frac{W}{M} = \frac{1\ lb}{44lb\ /\ lbmole} = 0.02273\ lbmole$$

$$P = 1\ atm = \frac{14.7\ lb_f}{in.^2}\frac{144\ in^2}{ft^2} = 2116.8\ lb_f/ft^2$$

Substituting in the ideal gas equation: V = nRT/P

$$V = 0.02273\ lbmole\frac{1545\ ft\ lb_f}{lbmole(°R)}\frac{530°R}{2116\ lb_f/ft^2} = 8.791\ ft^3$$

Example 4.4. Calculate the density of air (M = 29) at $70°F$ and 1 atm in (a) American Engineering and (b) SI units.

Solution:

Density is mass/volume = W/V. Using the ideal gas equation:

$$Density = \frac{W}{V} = \frac{PM}{RT}$$

(a) In the American Engineering System of units: P = 2116.8 lb_f/ft^2; M = 29 lb/lbmole; R = 1545 ft lb_f/ lbmole(°R); and T = 70 + 460 = 530 ER.

$$Density = \frac{2116.8\ lb_f/ft^2}{1545\ ft\ lb_f/(lbmole \cdot R)}\frac{29\ lb/lbmole}{530°R} = 0.07498\ lb/ft^3$$

(b) in SI, P = 101325 N/m^2; T = (70 − 32)/1.8 = 21.1°C = 21.1 + 273 = 294.1 K; V = 1 m^3; M = 29 kg/kgmole; and R = 8315 Nm/(kgmole $\cong$ K).

$$\text{Density} = \frac{101,325 \text{ N/m}^2}{8315 \text{ Nm/(kgmole} \cdot \text{K)}} \frac{29 \text{ kg/kgmole}}{294.1 \text{ K}} = 1.202 \text{ kg/m}^3$$

Example 4.5. A process requires 10 m^3/s at 2 atm absolute pressure and 20°C. Determine the rating of a compressor in m^3/s at STP (0°C and 101325 Pa) that must be used to supply air for this process.

Solution:

Unless otherwise specified, the pressure term "atm" means a standard atmosphere, or 101.325 kPa. Use the P-V-T equation (Eq. 4.4). V$_1$ = 10 m^3; T$_1$ = 293 K; P$_1$ = 202650 N/m^2; P = 101325 N/m^2; T = 273 K. Solving for V:

$$V = \frac{P_1 T_1}{T_1} \frac{T}{P} = \frac{(202,650 \text{ N/m}^2)(2 \text{ m}^3)}{293 \text{ K}} \frac{273 \text{ K}}{101,325 \text{ N/m}^2} = 18.635 \text{m}^3/\text{s}$$

Example 4.6. An empty can was sealed in a room at 80°C and 1 atm pressure. Assuming that only air is inside the sealed can, what will be the vacuum after the can and contents cool to 20°C?

Solution:

The quantity of gas remains constant; therefore, the P-V-T equation can be used. The volume does not change, therefore it will cancel out of Equation (4.4). P$_1$ = 101325 N/m^2; T$_1$ = 353 K; T = 293 K. The pressures to be used in Equation (4.4) will be absolute.

$$P = \frac{101,325 \text{ N/m}^2}{353 \text{ K}} (293 \text{ K}) = 84,103 \text{ Pa absolute.}$$

$$\text{Vaccum} = 17,222 \text{ Pa} \left(\frac{1 \text{ cm Hg}}{1333.33 \text{ Pa}} \right) = 12.91 \text{ cm Hg vacuum}$$

The vacuum in Pa will be 101,325 − 84,103 = 17,222 Pa.

4.1.5   van der Waal's Equation of State

The ideal gas equation is based on unhindered movement of gas molecules within the confined space; therefore, at constant temperature when molecular energy is constant, the product of pressure and volume is constant. However, as pressure is increased, molecules are drawn closer, and attractive and repulsive forces between molecules affect molecular motion. When molecules are far apart, attractive forces exist. The magnitude of this attractive force is inversely proportional to the square of the distance between molecules. When molecules collide, they approach a limiting distance of separation when repulsive forces become effective, preventing molecules from directly contacting each other. Molecular contact may cause a chemical reaction, and this occurs only at very high molecular energy levels that exceed the repulsive force. The separation distance between molecules where repulsive forces are effective defines an exclusion zone, which reduces the total volume available for molecules to randomly move. The attractive forces between molecules also restrict molecular motion, and this will have the effect of reducing the quantity and magnitude of impact against the walls of the confining

Table 4.1   Values of Van der Waal's Constants for Different Gases

Gas	a Pa $(m^3/kgmole)^2$	b $m^3/kgmole$
Air	1.348×10^5	0.0366
Ammonia	4.246×10^5	0.0373
Carbon dioxide	3.648×10^5	0.0428
Hydrogen	0.248×10^5	0.0266
Methane	2.279×10^5	0.0428
Nitrogen	1.365×10^5	0.0386
Oxygen	1.378×10^5	0.0319
Water vapor	5.553×10^5	0.0306

Source: Calculated from values in the International Critical Tables.

vessel. These attractive forces are referred to as an internal pressure. A low pressures, molecular distance is large, the internal pressure is small, and the excluded volume is small compared with the total volume, and gas molecules obey the ideal gas equation. However, at high pressures, the pressure-volume-temperature relationship deviates from ideality. Gases that deviate from ideal gas behavior are considered real gases. One of the commonly used equations of state for real gases is the van der Waal's equation.

van der Waal proposed corrections to the ideal gas equation based on the excluded volume, nb, and a factor, n^2a/V^2, the internal pressure. The fit between experimental P-V-T relationship and calculated values using van der Waal's equation of state is very good except in the region of temperature and pressure near the critical point of the gas. For n moles of gas, van der Waal's equation of state is

$$\left(P + \frac{n^2a}{V^2}\right)(V - nb) = nRT \tag{4.5}$$

Values of the constants a and b in SI units are given in Table 4.1.

Example 4.7. Calculate the density of air at 150°C and 5 atm pressure using van der Waal's equation of state and the ideal gas equation.

Solution:

Using the ideal gas equation:

$$\text{Density} = W = \frac{PM}{RT}; \quad V = 1 \text{ m}^3 \left(P + \frac{n^2a}{V^2}\right)(V - nb) = nRT; \quad n = \frac{W}{M}; \quad V = 1$$

$$= \frac{5(101, 325)(29)}{8315(150 + 273)} = 4.1766 \text{ kg/m}^3$$

Using van der Waal's equation of state: $V = 1 \text{ m}^3$

$$\left(P + \frac{W^2a}{M^2V^2}\right)\left(V - \frac{W}{M}b\right) = \frac{W}{M}RT$$

$$W^3\left(\frac{ab}{M^3V^3}\right) - W^2\left(\frac{a}{M^2V}\right) + W\left(\frac{Pb + RT}{M}\right) - PV = 0$$

Expanding and collecting like terms:

Substituting $a = 1.348 \times 10^5$; $b = 0.0366$; $R = 8315$; $M = 29$; $P = 5(101325) = 506625$ Pa; $T = 150 + 273 = 423$ K; $V = 1$ m^3.

$$0.2023\, W^3 - 160.3\, W^2 + 121{,}923.7\, W - 506{,}625 = 0$$

Dividing through by 0.2023:

$$W^3 - 792.3875\, W^2 + 602{,}687.6\, W - 2{,}504{,}325.3 = 0$$

Solve using the Newton-Raphson iteration technique:

$$f = W^3 - 792.3875\, W^2 + 602{,}687.6\, W - 2{,}504{,}325.3$$
$$f' = 3\, W^2 - 1584.775\, W + 602{,}687.6$$

Thus, the mass of 1 m^3 of air at 150°C and 5 atm is 4.178 kg = the density in kg/m^3. Values calculated using the van der Waal's equation of state is more accurate than that obtained using the ideal gas equation. A worksheet in Microsoft Excel to solve this problem is shown in Fig. 4.2

4.1.6   Critical Conditions for Gases

When the pressure of a gas is increased by compression, the molecules are drawn closer together and attractive forces between the molecules become strong enough to restrict movement of the molecules. At a certain temperature and pressure, a saturation point of the gas is reached and an equilibrium condition between gas and liquid can exist. If the energy level in the gas is reduced by removal of the latent heat of vaporization, it will condense into a liquid. The higher the pressure, the higher the saturation temperature of the gas.

If the pressure of the gas is increased as the temperature is maintained following the saturation temperature curve, a point will be reached where gas and liquid become indistinguishable. This particular temperature and pressure is called the critical point. The property of the gas at the critical point is very similar to that of a liquid in terms of dissolving certain solutes, and this property is put into practical use in a process called supercritical fluid extraction, which will be discussed in more detail in Chapter 14.

4.1.7   Gas Mixtures

In this section, the concept of partial pressures and partial volumes will be used to elucidate P-V-T relationships of individual components in a gas mixture.

If components of a gas mixture at constant volume are removed one after the other, the drop in pressure accompanying complete removal of one component is the *partial pressure* of that component.

If P_t is the total pressure and P_a, P_b, P_c ... P_n are partial pressures of the components a, b, c, ... and n, then:

$$P_t = P_a + P_b + P_c + \ldots P_n \tag{4.6}$$

Equation (4.6) is *Dalton's law of partial pressures*. Because the same volume is occupied by all components, the ideal gas equation may be used on each component to determine the number of moles of that component, from the partial pressure.

$$P_a V = n_a RT \tag{4.7}$$

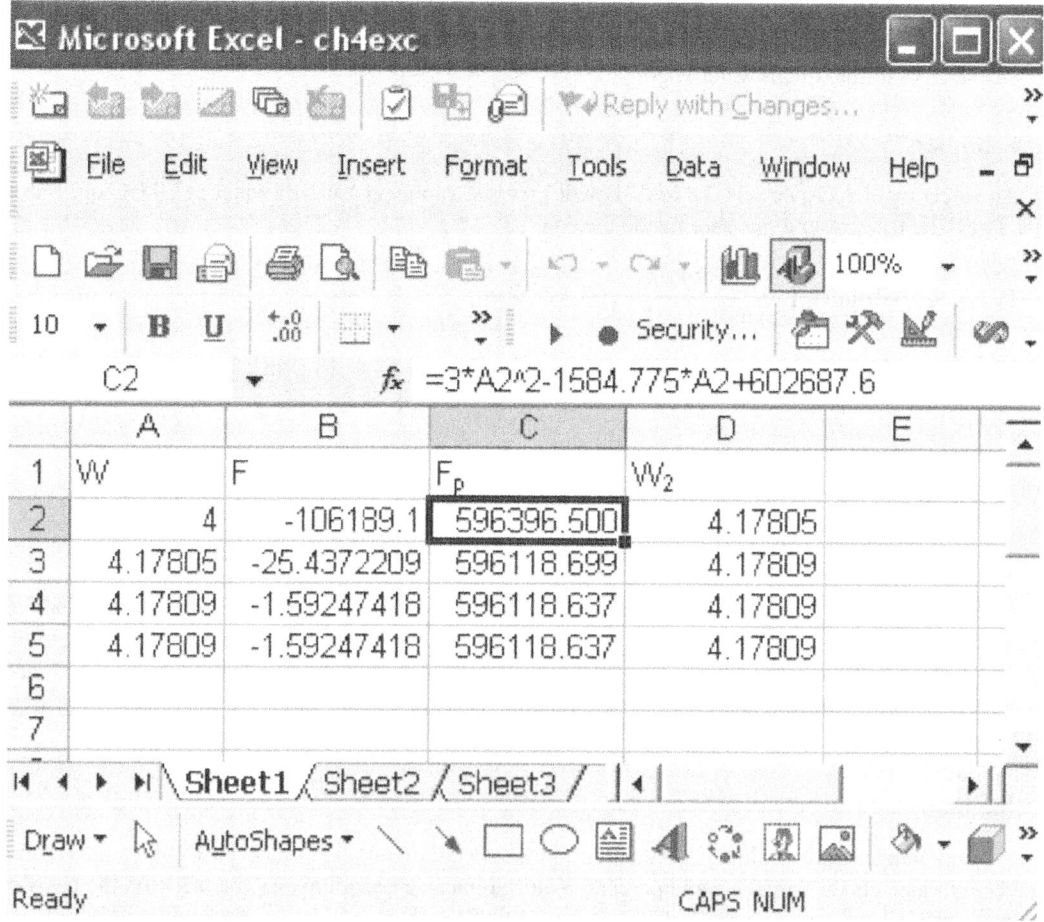

Figure 4.2   Spreadsheet for Newton-Raphson method and van der Waal's equation to calculate air density.

The *partial volume* is the change in volume of a gas mixture when each component is removed separately at constant pressure. If V_t is the total volume, and V_a, V_b, V, ... V_n are the partial volumes of the components, then:

$$V_t = V_a + V_b + V_c + \ldots V_n \tag{4.8}$$

Equation (4.8) is *Amagat's law of partial volumes*. The ideal gas equation may also be used on the partial volume of each component to determine the number of moles of that component in the mixture.

$$PV_a = n_a RT \tag{4.9}$$

Compositions of gases are expressed as a percentage of each component by volume. It may be seen from Equation (4.9) that the volume percentage of a gas is numerically the same as the mole percentage.

Example 4.8. Calculate the quantity of air in the headspace of a can at 20°C when the vacuum in the can is 10 in. Hg. Atmospheric pressure is 30 in. Hg. The headspace has a volume of 16.4 cm³. The headspace is saturated with water vapor.

Solution:

The steam tables (Appendix Table A.4) will give the vapor pressure of water at 20 EC of 2336.6 Pa. Let P_t be the absolute pressure inside the can.

$$P_t = (30 = 10)\text{in.Hg}\left(\frac{3386.38 \text{ Pa}}{\text{in. Hg}}\right) = 67,728 \text{ Pa}$$

Using the ideal gas equation: $V = 16.4 \text{ cm}^3(10^{-6})\text{m}^3/\text{cm}^3 = 1.64 \times 10^{-5} \text{ m}^3; T = 20 + 273 = 293 \text{ K}.$

$$P_{air} = P_t - P_{water}$$
$$n_{air} = \frac{P_{air}V}{RT} = \frac{(65,391.4\text{N/m}^3)(1.64 \times 10^{-5}\text{m}^3)}{8315/(293)} = 67728 - 2336.6 = 65391.4 \text{ Pa}$$
$$= 4.40 \times 10^{-7} \text{ kgmoles}$$

Example 4.9. A gas mixture used for controlled atmosphere storage of vegetables contains 5% CO_2, 5% O_2, and 90% N_2. The mixture is generated by mixing appropriate quantities of air and N_2 and CO_2 gases. 100 m³ of this mixture at 20°C and 1 atm is needed per hour. Air contains 21% O_2 and 79% N_2. Calculate the rate at which the component gases must be metered into the system in m³/h at 20°C and 1 atm.

Solution:

All percentages are by volume. No volume changes occur on mixing of ideal gases. Because volume percent in gases is the same as mole percent, material balance equations may be made on the basis of volume and volume percentages. Let X = volume O_2, Y = volume CO_2, and Z = volume N_2, fed into the system per hour.

Oxygen balance: 0.21(X) = 100(0.05); X = 23.8 m³
CO_2 balance: Y = 0.05(100); Y = 5 m³
Total volumetric balance: X + Y + Z = 100
Z = 100 − 23.8 − 5 = 71.2 m³

4.2 THERMODYNAMICS

Thermodynamics is a branch of science that deals with energy exchange between components within a system or between a system and its surroundings. A *system* is any matter enclosed within a boundary. The boundary may be real or imaginary and depends solely upon the part of the process under study. Anything outside the boundary is the *surroundings*. The *properties* of a system determines its state just as a state may imply certain properties. Properties may be *extrinsic* (i.e., it is measurable) or it may be intrinsic (i.e., no measurable external manifestations of that property exist). Changes in the intrinsic properties, however, may be measured through changes in energy associated with the change.

Equilibrium is a fundamental requirement in thermodynamic transformations. When a system reaches a condition where properties remain constant, a state of equilibrium is attained. The science of thermodynamics deals with a system in equilibrium. Thermodynamics cannot predict how long it will take for equilibrium to occur within a system, it can only predict the final properties at equilibrium.

4.2.1 Thermodynamic Variables

Energy involved in thermodynamic transformations is expressed in terms of *heat* (Q), the energy that crosses a system's boundaries due to a difference in temperature, and *work* (W), the energy associated with force displacement. The term "internal energy" (E) is used to define an intrinsic property, which is energy not associated with either work or heat. Internal energy is not measurable, but changes in internal energy can be measured. An intrinsic property *entropy* (S) is also possessed by a system and is a measure of the disorder that exists within a system. The first and second laws of thermodynamics are based on the relationships of the above main thermodynamic variables.

If δQ and δW are small increments of heat and work energy that crosses a system's boundaries, the accompanying differential change in the internal energy of the system is

$$dE = \delta Q - \delta W \tag{4.10}$$

Equation (4.10) is also expressed as:

$$\Delta E = Q - W \tag{4.11}$$

Equations (4.10) or (4.11) are expressions for the *first law of thermodynamics*, a fundamental relationship also known as the law of conservation of energy. The symbol δ, called "del," operating on W and Q is not a true differential but rather a finite difference. Although it is considered in mathematical operations as a differential operator, integration produces an increment change in a function, rather than an absolute value. This principle is used when integrating functions such as E and S, which do not have absolute values but for which increment change associated with a process can be calculated.

Entropy cannot be measured, but the change in entropy is defined as the ratio of the reversible energy (Q_{rev}) which crossed a system's boundaries, and the absolute temperature.

$$dS = \frac{\delta Q_{rev}}{T}; \quad \Delta S = \frac{Q_{rev}}{T} \tag{4.12}$$

The *second law of thermodymanics* states that any process that occurs is accompanied by a positive entropy change, and the entropy change approaches zero for reversible processes. If the process is reversible, any change in entropy in a system is compensated by a change in the entropy of the surroundings such that the net entropy change for the system and its surroundings is zero.

The concept of the entropy or state of disorder of a system has several interpretations. One interpretation is that no truly reversible process is possible, because any process produces a more disordered system than what existed before the change. This change in the state of order requires the application of or a loss of energy. Another interpretation of the concept of entropy involves the spontaneity of processes. Spontaneous processes will occur only when there is enough energy available initially to overcome the requirement for increasing the entropy. Thus, process spontaneity always requires a change from a higher to a lower energy state.

Another intrinsic thermodynamic variable is the enthalpy (H) defined as:

$$H = E + PV \tag{4.13}$$

In differential form:

$$dH = dE + PdV + VdP \qquad (4.14)$$

Because $PdV = \delta W$; $dE + \delta W = \delta Q$ and:

$$dH = \delta Q + VdP \qquad (4.15)$$

For a constant pressure process, $dP = 0$ and:

$$\delta Q = dH; \delta H = Q \qquad (4.16)$$

A specific heat at constant pressure may be defined as follows:

$$C_p = \left.\frac{dQ}{dT}\right|_p \qquad (4.17)$$

and as:

$$\Delta H = C_p dT \qquad (4.18)$$

Equation (4.18) shows why the enthalpy is referred to as the heat content. In a constant-volume process, work is zero, and Equation (4.10) becomes:

$$dE = \delta Q; \quad \Delta E = Q \qquad (4.19)$$

A specific heat at constant volume may be defined as follows:

$$C_v = \left.\frac{dQ}{dT}\right|_v \qquad (4.20)$$

and as:

$$\Delta E = C_v dT \qquad (4.21)$$

4.2.2   The Relationship Between C_p and C_v for Gases

The relationship between C_p and C_v for gases is derived as follows. Substituting $W = PdV$ for work in Equation (4.10) for a constant presssure process:

$$dE = dQ - PdV \qquad (4.22)$$

Taking the derivative with respect to temperature:

$$\frac{dE}{dT} = \left.\frac{dQ}{dT}\right|_p - P\frac{dV}{dT} \qquad (4.23)$$

Equation (4.17) gives $dQ/dT = C_p$ and Equation (4.21) gives $dE/dT = C_v$. From the ideal gas equation, for 1 mole of gas, $PV = RT$, and dV/dT at constant pressure is R/P. Substituting in Equation (4.23):

$$C_v = C_p - R \qquad (4.24)$$

A useful property in calculating P-V-T and other thermodynamic variables involved in expansion and compression of a gas is the specific heat ratio, C_p/C_v, designated by the symbol Q. The ratio C_p/R is expressed in terms of Q as follows:

$$\frac{C_p}{R} = \frac{\gamma}{\gamma - 1} \qquad (4.25)$$

4.2.3 P-V-T Relationships for Ideal Gases in Thermodynamic Processes

Adiabatic processes are those where no heat is added or removed from the system, therefore, $\delta Q = 0$. Equation (4.15) then becomes: $dH = V\,dP$. Because for 1 mole of an ideal gas, $V = RT/P$, and because $dH = C_p dT$,

$$C_p dT = RT \left(\frac{dP}{P} \right) \qquad (4.26)$$

$$\ln \left[\frac{P-2}{P_1} \right] = \frac{\gamma}{\gamma - 1} \ln \left[\frac{T_2}{T_1} \right]$$

$$\frac{C_p}{R} \int_{T_1}^{T_2} \frac{dT}{T} = \int_{P_1}^{P_2} \frac{dP}{P} \qquad (4.27)$$

Integrating and substituting $C_p/R = \gamma / (\gamma - 1)$.

The ideal gas equation may be used to substitute for P in Equation (4.27) to obtain an expression for V as a function of T in an adiabatic process. Substitute: $P_1 = nRT_1/V_1$ and $P_2 = nRT_2/V_2$ in Equation (4.27) and simplifying:

$$\frac{P_2}{P_1} = \left[\frac{T_2}{T_1} \right]^{\frac{\gamma}{\gamma - 1}} \qquad (4.28)$$

$$\frac{V_1}{V_2} = \left[\frac{T_2}{T_1} \right]^{\frac{1}{\gamma - 1}} \qquad (4.29)$$

Equations (4.28) and (4.29) can be used to derive:

$$\left[\frac{V_1}{V_2} \right]^{\gamma} = \left[\frac{P_2}{P_1} \right] \qquad (4.30)$$

Isothermal processes are those where the temperature is maintained constant. Thus, $P_1 V_1 = P_2 V_2$. *Isobaric processes* are those where the pressure is maintained constant. Thus, $V_1/T_1 = V_2/T_2$. *Isocratic processes* are those where the volume is maintained constant. Thus, $P_1/T_1 = P_2/T_2$.

4.2.4 Changes in Thermodynamic Properties, Work, and Heat Associated with Thermodynamic Processes

Adiabatic:

$$\Delta Q = 0; \ \Delta S = 0; \ \Delta E = W = \int P dV; \ \Delta H = \int V\,dP$$

Isothermal:

$$Q = W = \int P\,dV; \ \Delta S = \frac{Q}{T}; \ \Delta E = 0; \ \Delta H = O$$

Isobaric:

$$Q = \int C_p dT = \Delta H; \quad \Delta S = \int C_p \frac{dT}{T}; \quad W = P\Delta V; \quad \Delta E = Q - W$$

Isocratic:

$$Q = \Delta E = \int C_v dT; \quad \Delta S = \int C_v \frac{dT}{T}; \quad W = 0; \quad \Delta E = Q$$

4.2.5 Work and Enthalpy Change on Adiabatic Expansion or Compression of an Ideal Gas

Work and enthalpy changes are important in determining power input into compressors. Adiabatic compression is what occurs during the compression of a refrigerant in a refrigeration cycle.

In adiabatic compression or expansion:

$$W = \int P dV; \quad \Delta H = \int V \, dP \tag{4.31}$$

Equation (4.30) is rearranged, substituting P and V for P_2 and V_2.
Solving for V:

$$V = [(P_1)^{\frac{1}{\gamma}} V_1](P)^{\frac{-1}{\gamma}}$$

Differentiating with respect to P:

$$dV = \frac{-1}{\gamma}[(P_1)^{\frac{1}{\gamma}} V_1][P]^{\frac{-1}{\gamma}-1}$$

Substituting dV in the expression for work and integrating:

$$W = \int_{P_1}^{P_2} -\frac{1}{\gamma}[(P_1)^{\frac{1}{\gamma}} V_1][P]^{\frac{-1}{\gamma-1}+1}$$

The integral is

$$-\frac{1}{\frac{1}{\gamma}+1}[[P_2]^{\frac{1}{\gamma}+1} - [P_1]^{\frac{1}{\gamma}+1}]$$

$$= -\frac{(P_1)^{\frac{1}{\gamma}} V_1}{\gamma} \int_{P_1}^{P_2}(P)^{\frac{-1}{\gamma}} dP$$

Combining the integral with the multiplier and simplifying:

$$W = \frac{P_1 V_1}{1 - \gamma}\left[\left[\frac{P_2}{P_1}\right]^{\frac{(\gamma-1)}{\gamma}} - 1\right] \tag{4.32}$$

The enthalpy change is determined by evaluating the following integral:

$$\Delta H = V_1 \int_{P_1}^{P_2}\left[\frac{P_1}{P}\right]^{\frac{1}{\gamma}} dP \tag{4.33}$$

4.2.6 Work and Enthalpy Change on Isothermal Expansion or Compression of an Ideal Gas

$$\Delta H = \left[\frac{\gamma}{\gamma - 1}\right](P_1 V_1)\left[\left[\frac{P_2}{P_1}\right]^{\frac{(\gamma-1)}{\gamma}} - 1\right]$$

When an ideal gas is subjected to an isothermal process

$$Q = W = \int P\,dV \text{ and } \Delta H = \int P\,dV + \int V\,dP$$

Because T is constant, $P = nRT/V$ and $dP = -(nRTV^{-2})\,dV$

$$Q = W = \int_{V_1}^{V_2} nRT\frac{dV}{V}$$

$$Q = nRT \ln\left[\frac{V_2}{V_1}\right]$$

$$\Delta H = W + \int_{V_1}^{V_2} -\left[nRT\frac{dV}{V}\right] \tag{4.34}$$

The second term in the expression for ΔH is exactly the same as the expression for W but has a negative sign, therefore, in an isothermal process, $\Delta H = 0$.

Example 4.10. N_2 gas trapped inside a cylinder with a movable piston at 80 atm pressure is then allowed to expand adiabatically (no heat added or removed) until the final pressure is 1 atm. The gas is initially at 303 K and the initial volume is 1 liter. Calculate (a) the work performed by the gas, (b) the entropy change, (c) the enthalpy change, and (d) change in internal energy. The gas has a specific heat ratio, γ, of 1.41.

Solution:

(a) Equation (4.31) will be used to calculate the work. $V_1 = 0.001$ m^3; $P_1 = 80(101325) = 8106000$ Pa; $P_2 = 101325$ Pa. From Equation (4.31):

$$W = \frac{(8,106,000\text{Pa})(0.001\text{m}^3)}{-0.41}\left[\left(\frac{101,325}{8,106,0000}\right)^{\frac{0.41}{1.41}} - 1\right] = -14241.8\text{J}$$

(b) $\Delta S = 0$

$$\Delta H = \left[\frac{1.41}{0.41}(8,106,000 \text{ Pa})(0.001 \text{ m}^3)\right]\left[\left[\frac{101,325}{8,106,000}\right]^{\frac{0.41}{1.41}} - 1\right]$$

(c) From Equation (4.32):

$$= -20,080.9 \text{ J}$$

(d) $\Delta E = W = -14241.7 \text{ J}$

4.3 VAPOR-LIQUID EQUILIBRIUM

The molecular attraction that holds liquid molecules together is not strong enough to prevent some molecules from escaping, therefore, some molecules will escape in the gaseous form. If the volume of the system containing the liquid and vapor is held constant, eventually equilibrium will be attained when the rate of escape of molecules from the liquid phase to become vapor equals the rate at which the vapor molecules are recaptured by the liquid phase. This condition of equilibrium exists in all liquids. A liquid will always maintain an envelope of vapor around its surface, and the pressure exerted by that vapor is known as the *vapor pressure*. The pressure exerted by molecules of vapor in equilibrium with a liquid is a function only of temperature. In the absence of any other gas exerting pressure on the liquid surface (e.g., when the liquid is introduced into a container which is under a perfect vacuum), the pressure of the vapor at equilibrium is the vapor pressure. If the total pressure above a liquid is maintained constant at the vapor pressure, heating will not increase the temperature but will cause more liquid molecules to enter the vapor phase. The temperature at a pressure that corresponds to the vapor pressure of a liquid is the *boiling point*, a condition where the whole atmosphere over the liquid surface consists only of gaseous molecules of that liquid. The heat added to the system to generate a unit mass of vapor from liquid at the boiling point is the *heat of vaporization*. When the pressure above the liquid is higher than the vapor pressure, some other gas molecule (e.g., air) would be exerting that pressure in addition to the vapor molecules, therefore the vapor pressure will be the partial pressure of vapor in the gas mixture surrounding the liquid.

4.3.1   The Clausius-Clapeyron Equation

The temperature dependence of the vapor pressure is expressed by Equation (4.34), the Clausius-Clapeyron equation:

$$\ln\left[\frac{P}{P_1}\right] = \left[\frac{\Delta H_v}{R}\right]\left[\frac{T - T_1}{TT_1}\right] \tag{4.35}$$

where ΔH_v is heat of vaporization; P is vapor pressure at temperature T; R is gas constant; and P_1 is vapor pressure at temperature T_1. Equation (4.34) assumes constant heat of vaporization, but in reality, this quantity changes with temperature. Thus, the most useful function of Equation (4.34) is in interpolating between values of vapor pressure listed in abbreviated tables where temperature interval between entries in the table is quite large. The steam tables (Appendix Tables A.3 and A.4) give values of the vapor pressure for water at different temperatures. Because temperature intervals in these tables are rather close, a linear interpolation may be used, instead of a logarithmic function interpolation as suggested by Equation (4.34).

4.3.2   Liquid Condensation from Gas Mixtures

When the partial pressure of component in a gas mixture exceeds the vapor pressure of that component, condensation will occur. If the mixture is in a closed container where the volume is constant, condensation will result in a decrease in total pressure. This principle is responsible for the vacuum that results when a hot liquid is filled into a can and sealed. Replacement of air in the headspace by vapors from the product itself or by steam in a system where steam is flushed over the can headspace prior to sealing results in a reduced pressure after the can cools adequately to condense the vapors in the headspace.

Example 4.11. A canned food at the time of sealing is at a temperature of 80°C, and the atmospheric pressure is 758 mm Hg. Calculate the vacuum (in mm Hg) formed inside the can when the contents cools down to 20°C.

Solution:

Assume there are no dissolved gases in the product at the time of sealing, therefore the only gases in the headspace are air and water vapor. From Appendix Table A.3, the vapor pressure of water at 20°C and 80°C are 2.3366 and 47.3601 kPa, respectively. In the gas mixture in the headspace, air is assumed to remain at the same quantity in the gaseous phase, while water condenses on cooling.

$$P_{air} = P_t - P_{water}$$

At 80°C,

$$P_t = 758 \text{ mm Hg}\left(\frac{1 \text{ cm Hg}}{10 \text{ mm Hg}}\right)\left(\frac{1333.33 \text{ P}_a}{\text{cm Hg}}\right) = 101,066 \text{ P}_a$$

Let P = pressure in the headspace at 20°C.

$$n_{air} = \frac{PV}{RT} = \frac{(101,064 - 47,360.1)V}{8315(80 + 273)} = 0.018296 \text{ V kgmole}$$

At 20°C,

$$n_{air} = \frac{PV}{8315(293)} = 4.1046 \times 10^{-7} PV \text{ kgmole}$$

Because the number of moles of air trapped in the headspace is constant:

$$4.1046 \times 10^{-7} \text{ P V} = 0.018296 \text{ V}$$
$$P = 44575 \text{ Pa absolute} = 334 \text{ mm Hg absolute}$$
$$\text{Vacuum} = 758 - 334 = 424 \text{ mm Hg vacuum}$$

Example 4.12. Air at 5 atm pressure is saturated with water vapor at 50°C. If this air is allowed to expand to 1 atm pressure and the temperature is dropped to 20°C, calculate the amount of water that will be condensed per m³ of high pressure air at 50°C.

Solution:

The vapor pressure of water at 50°C and 20°C are 12.3354 and 2.3366 kPa, respectively. Basis: 1 m³ air at 5 atm pressure and 50°C. The number of moles of air will remain the same on cooling.

$$n_{air} = \frac{[5(101,325) - 12,335.4]1}{8315(50 + 273)} = 0.1840 \text{ kgmole}$$

At 20°C:

$$n_{air} = \frac{(101,325 - 2336.6)V}{8315(20 + 273)} = 0.04063 \text{ V}$$

Equating: 0.04063 V = 0.1840; V = 4.529 m³ at 20°C and 1 atm.
At 50°C,

$$n_{water} = \frac{12,335.4(1)}{8315(323)} = 0.004593$$

At 20°C,

$$n_{water} = \frac{2336.6(4.529)}{8315(293)} = 0.004344$$

Moles water condensed = 0.004593 - 0.004344 = 0.000249 kg moles.

Example 4.13. The partial pressure of water in air at 25°C and 1 atm is 2.520 kPa. If this air is compressed to 5 atm total pressure to a temperature of 35°C, calculate the partial pressure of water in the compressed air.

Solution:

Increasing the total pressure of a gas mixture will proportionately increase the partial pressure of each component. Using the ideal gas equation (Equation 4. 3) for the mixture and for the water vapor, let V_1 = the volume of the gas mixture at 25°C and 1 atm; P_t = total pressure; P_w = partial pressure of water vapor.

The total number of moles of air and water vapor is

$$n_t = \frac{P_1 V_1}{RT_1} = \frac{P_2 V_2}{RT_2}$$

The number of moles of water vapor is

$$n_{water} = \frac{P_{w1} V_1}{RT_1} = \frac{P_{w2} V_2}{RT_2}$$

Assuming no condensation, the ratio, n_t/n_w will be the same in the low-pressure and high-pressure air, therefore:

$$\frac{P_1}{P_{w1}} = \frac{P_2}{P_{w2}}; \quad P_{w2} = P_{w1} \frac{P_2}{P_1}$$

$$P_{w2} = \frac{5 \text{ atm}}{1 \text{ atm}} (2.520 \text{ kPa}) = 12.60 \text{ kPa}$$

The temperature was not used in calculating the final partial pressure of water. The temperature is used in verifying the assumption of no condensation, comparing the calculated final partial pressure with the vapor pressure of water at 35°C. From the steam tables, Appendix Table A.3, the vapor pressure of water at 35°C is the pressure of saturated steam at 35°C, which is 5.6238 kPa. Because the calculated partial pressure of water in the compressed air is greater than the vapor pressure, condensation of water must have occurred and the correct partial pressure of water will be the vapor pressure at 35°C. Thus: $P_{w2} = 5.6238$ kPa.

PROBLEMS

4.1. Air used for dehydration is heated by burning natural gas and mixing the combustion products directly with air. The gas has a heating value of 1050 BTU/ft^3 at 70°F and 1 atm pressure. Assume the gas is 98% methane and 2% nitrogen.

    (a) Calculate the quantity of natural gas in ft^3 at 70 EF and 1 atm. needed to supply the heating requirements for a dryer that uses 1500 lb dry air per hour at 170°F and 1 atm. Assume the products of combustion will have the same specific heat as dry air, of 0.24 BTU/(lb $\cong$ °F).

    (b) If the air used to mix with the combustion gases is completely dry, what will be the humidity of the air mixture entering the dryer.

4.2. A package having a void volume of 1500 cm^3 is to be flushed with nitrogen to displace oxygen prior to sealing. The process used involved drawing a vacuum of 700 mm Hg on the package, breaking the vacuum with nitrogen gas, and drawing another 700 mm Hg vacuum before sealing. The solids in the package prevents total collapse of the package as the vacuum is drawn, therefore the volume of gases in the package may be assumed to remain constant during the process. If the temperature is maintained constant at 25°C during the process, calculate the number of gmoles of oxygen left in the package at the completion of the process. Atmospheric pressure is 760 mm Hg.

4.3. Compression of air in a compressor is an adiabatic process. If air at 20°C and 1 atm pressure is compressed to 10 atm pressure, calculate:

    (a) The temperature of the air leaving the compressor.;

    (b) The theoretical compressor horsepower required to compress 100 kg of air. The specific heat ratio of air is 1.40; the molecular weight is 29.

4.4. Air at 25°C and 1 atm that contains water vapor at a partial pressure that is 50% of the vapor pressure at 25°C (50% relative humidity) is required for a process. This air is generated by saturating room air by passing through water sprays, compressing this saturated air to a certain pressure, P, and cooling the compressed air to 25°C. The partial pressure of water in the cooled saturated air that leaves the compressor is the vapor pressure of water at 25°C. This air is allowed to expand to 1 atm pressure isothermally. Calculate P such that after expansion, the air will have 50% relative humidity.

4.5. An experiment requires a gas mixture containing 20% CO_2, 0.5% O_2, and 79.5% N_2 at 1 atm and 20°C. This gas mixture is purchased premixed and comes in a 10 L tank at a pressure of 130 atm gauge. The gas will be used to displace air from packages using a packaging machine that operates by drawing a vacuum completely inside a chamber where the packages are placed, displacing the vacuum with the gas mixture, and sealing the packages. The chamber can hold four packages at a time, and the total void volume chamber with the packages in place is 2500 cm^3. How many packages can be treated in this manner before the pressure in the gas tank drops to 1 atm gauge.

4.6. A vacuum pump operates by compressing gases from a closed chamber to atmospheric pressure in order that these gases can be ejected to the atmosphere. A vacuum drier operating at 700 mm Hg vacuum (atmmospheric pressure is 760 mm Hg) and 50°C generates 500 g of water vapor per minute by evaporation from a wet material in the dryer. In addition, the leakage rate for ambient air infiltrating the dryer is estimated to be 1 L/h at 1 atm and 20°C.

    (a) Calculate the total volume of gases that must be removed by the vacuum pump per minute.

    (b) If the pump compresses the gas in an adiabatic process, calculate the theoretical horsepower required for the pump. The specific heat ratio for water is 1.30, and that for air is 1.40.

4.7. The mass rate of flow of air (G) used in correlation equations for heat transfer in a dryer is expressed in kg air/m^2(h). Use the ideal gas equation to solve for G as a function of the velocity of flow (V, in m/h) of air at temperature T and 1 atm pressure.

4.8. Use van der Waal's equation of state to calculate the work done on isothermal expansion of a gas from a volume of 10 to 300 m^3 at 80°C. The initial pressure was 10 atm. Calculate the entropy change associated with the process.

4.9. A supercritical CO_2 extraction system is being operated at 30.6 Mpa and 60°C in the extraction chamber. The volume of CO_2 leaving the system measured at 101.3 kPa and 20°C is 10 L/min. If the extraction chamber is a tube having a diameter of 50.6 mm and a length of 45 cm., calculate the residence time of the CO_2 in the extraction chamber. Residence time = volume of chamber/volumetric rate of flow in the chamber.

SUGGESTED READING

Feldner, R. M. and Rosseau, R. W. 1999. Elementary Principles of Chemical Processes. 2nd ed. John Wiley & Sons, New York.

Himmelblau, D. M. 1967. Basic Principles and Calculations in Chemical Engineering. 2nd ed. Prentice-Hall, Englewood Cliffs, NJ.

Martin, M. C. 1986. Elements of Thermodynamics. Prentice-Hall, New York.

Peters, M. S. 1954. Elementary Chemical Engineering. McGraw-Hill Book Co., New York.

CHAPTER 5

Energy Balances

"Energy" used to be a term that everybody took for granted. Now, to a layman, energy has been added to the list of the basic necessities of life. Increasing energy costs have forced people to recognize and appreciate the value of energy more than ever before. Energy conservation is being stressed not only in industrial operations but also in almost all aspects of an individual's daily activities.

Energy is not static; it is always in a stage of flux. Even under steady-state conditions, an object absorbs energy from its surroundings and at the same time emits energy to its surroundings at the same rate. When there is an imbalance between the energy absorbed and emitted, the steady state is altered, molecular energy of some parts of the system may increase, new compounds may be formed, or work may be performed. Energy balance calculations can be used to account for the various forms of energy involved in a system.

Energy audits are essential in identifying effectiveness of energy conservation measures and in identifying areas where energy conservation can be done. The technique is also useful in the design of processing systems involving heating or cooling to insure that fluids used for heat exchange are adequately provided and that the equipment is sized adequately to achieve the processing objective at the desired capacity. When energy exchange involves a change in mass due to evaporation or condensation, energy balances can be used during formulation such that after processing, the product will have the desired composition.

5.1 GENERAL PRINCIPLES

An energy balance around a system is based on the first law of thermodynamics: the law of conservation of energy. Mechanical (work), electrical, and thermal energy can all be reduced to the same units. Mechanical input into a system to overcome friction, electrical energy, or electromagnetic energy such as microwaves will be manifested by an increase in the heat content of the system. Defining the surroundings of the system on which to make an energy balance is done similar to material balance calculations in Chapter 3. The basic energy balance equation is

Energy in = Energy out + Accumulation   (5.1)

If the system is in a steady state, the accumulation term is zero, whereas an unsteady-state system will have the accumulation term as a differential expression. When Equation (5.1) is used, all energy terms known to change within the system must be accounted for. The heat contents are expressed as enthalpy based on increase in enthalpy from a set reference temperature. Mechanical, electrical, or electromagnetic inputs must all be accounted for if their effects on the total heat content are significant. If the system involves only an exchange of energy between two components, the energy balance will be:

Energy gain by component 1 = Energy loss by component 2   (5.2)

Either Equation (5.1) or (5.2) will give similar results if Equation (5.2) is applicable. Equation (5.1) is a general form of the energy balance equation.

5.2   ENERGY TERMS

The unit of energy in SI is the Joule. Conversion factors in Appendix Table A.1 can be used to convert mechanical and electrical energy units to the SI equivalent. Microwave and radiant energy absorbed by a material are usually expressed as a rate of energy flow, energy/time. Energy from ionizing radiation is expressed as an absorbed dose, the Gray (Gy), which has the base units J/kg. Another accepted form for reporting absorbed ionizing radiation is the rad, 100 rd = 1 Gy.

5.2.1   Heat

Sensible heat is defined as the energy transferred between two bodies at different temperatures, or the energy present in a body by virtue of its temperature. Latent heat is the energy associated with phase transitions, heat of fusion, from solid to liquid, and heat of vaporization, from liquid to vapor.

5.2.2   Heat Content, Enthalpy

Enthalpy, as defined in Chapter 4, is an intrinsic property, the absolute value of which cannot be measured directly. However, if a reference state is chosen for all components that enter and leave a system such that at this state the enthalpy is considered to be zero, then the change in enthalpy from the reference state to the current state of a component can be considered as the value of the absolute enthalpy for the system under consideration. The reference temperature (T_{ref}) for determining the enthalpy of water in the steam tables (Appendix Table A.3 and A.4) is 32.018°F or 0.01°C. The enthalpy of any component of a system that would be equivalent to the enthalpy of water obtained from the steam tables, at any temperature T is given by:

$$H = C_p(T - T_{ref})$$ (5.3)

C_p in Equation (5.3) is the specific heat at constant pressure, previously defined in Chapter 4.

5.2.3   Specific Heat of Solids and Liquids

The specific heat (C_p) is the amount of heat that accompanies a unit change in temperature for a unit mass. The specific heat, which varies with temperature, is more variable for gases compared with liquids or solids. Most solids and liquids have a constant specific heat over a fairly wide temperature range. The enthalpy change of a material with mass m is:

$$q = m \int_{T_1}^{T_2} C_p \, dT \tag{5.4}$$

Handbook tables give specific heats averaged over a range of temperature. When average specific heats are given, Equation (5.4) becomes:

$$q = m \, C_{avg} \, (T_2 - T_1) \tag{5.5}$$

For solids and liquids, Equations (5.3) and (5.5) are valid over the range of temperatures encountered in food processing systems.

Table 5.1 shows the average specific heats of various solids and liquids.

For fat-free fruits and vegetables, purees and concentrates of plant origin, Siebel (1918) observed that the specific heat varies with moisture content and that the specific heat can be determined as the weighted mean of the specific heat of water and the specific heat of the solids.

For a fat-free plant material with a mass fraction of water M, the specific heat of water above freezing is 1 BTU/(lb $\cong$ °F) or 4186.8 J/(kg $\cong$ K), and that of non-fat solids is 0.2 BTU/(lb $\cong$°F) or 837.36 J/(kg $\cong$ K). Because the mass fraction of non-fat solids is $(1 - M)$, the weighted average specific heat for unit mass of material above freezing is

$$C'_{avg} = 1(M) + 0.2(1 - M) = (1 - 0.2)M + 0.2 = 0.8 \, M + 0.2 \quad \text{in BTU/(lb °F)} \tag{5.6}$$

In SI:

$$C_{avg} = 3349 \, M + 837.36 \quad \text{in J/(kg K)} \tag{5.7}$$

Equations (5.6) and (5.7) are different forms of Seibel's equation, which has been used by Ashrae (1965) in tabulated values for specific heat of fruits and vegetables.

When fat is present, the specific heat above freezing may be estimated from the mass fraction fat (F), mass fraction solids non-fat (SNF), and mass fraction moisture (M), as follows:

$$C_{avg} = 0.4 \, F + 0.2 \, SNF + M \quad \text{in BTU/(lb °F)} \tag{5.8}$$

$$C'_{avg} = 1674.72 \, F + 837.36 \, SNF + 4186.8 \, M \quad \text{in J/(kg K)} \tag{5.9}$$

Below freezing, it is not suitable to use specific heats for the whole mixture beause the amount of frozen and unfrozen water vary at different temperatures. It will be necessary to consider the latent heat of fusion of water, and sensible heats of liquid water and ice should be evaluated separately. Refer to the section "Enthalpy Changes in Foods During Freezing" later in this chapter.

Equations (5.8) and (5.9) are general and can be used instead of Equations (5.6) and (5.7).

Example 5.1. Calculate the specific heat of beef roast containing 15% protein, 20% fat, and 65% water.

Solution:

$$C'_{avg} = 0.15(0.2) + 0.20(0.4) + 0.65(1) = 0.76 \, \text{BTU/(lb °F)}$$

$$C_{avg} = 0.15(837.36) + 0.2(1674.72) + 0.65(4186.8) = 3182 \, \text{J/(kg K)}$$

Table 5.1   Specific Heat of Food Products.

Product	% H_2O	C_{pm}
Dairy products		
Butter	14	2050
Cream, sour	65	2930
Milk, skim	91	4000
Fresh meat, fish, poultry, and eggs		
Codfish	80	3520
Chicken	74	3310
Egg white	87	3850
Egg yolk	48	2810
Pork	60	2850
Fresh fruits, vegetables, and juices		
Apples	75	3370
Apple juice	88	3850
Apple sauce	—	3730
Beans, fresh	90	3935
Cabbage, white	91	3890
Carrots	88	3890
Corn, sweet, kernels	—	3320
Cucumber	97	4103
Mango	93	3770
Orange juice, fresh	87	3890
Plums, fresh	76.5	3500
Spinach	87	3800
Strawberries	91	3805
Other products		
Bread, white	44	22720
Bread, whole wheat	48.5	2850
Flour	13	1800

Source: Adapted from Polley, S. L. Snyder, O. P., and Kotnour, P. A. Compilation of thermal properties of foods. *Food Technol.* 36(1):76, 1980.

Example 5.2. Calculate the specific heat of orange juice concentrate having a solids content of 45%.

Solution:

Using Siebel's equation: Calculating a weighted average specific heat:

$$C'_{avg} = 0.2(0.45) + 1(0.55) = 0.64 \text{ BTU/(lb °F)}$$
$$C_{avg} = 837.36(0.45) + 4186.8(0.55) = 2679 \text{ J/(kg K)}$$

Example 5.3. Calculate the heat required to raise the temperature of a 4.535 kg (10 lb) roast containing 15% protein, 20% fat, and 65% water from 4.44°C (40°F) to 65.55°C (150°F). Express this energy in (a) BTU, (b) joules, and (c) watt-hour.

Solution:

C'_{avg} from Example 5.1 = 0.76 BTU/(lb · °F)

(a) $q = mC_{avg}(T_2 - T_1)$

$$= 10\,lb\,\frac{0.76\,BTU}{lb \cdot °F}(150 - 40)°F = 836\,BTU$$

(b) Use C_{avg} from Example 5.1 = 3182 J/(kg · K)

$q = 4.535\,kg[3182\,J/(kg \cong K)]\,(65.55 - 4.44)\,K = 0.882\,MJ$

(c) $q = 0.882\,MJ \cdot \dfrac{10^6}{MJ} \cdot \dfrac{1\,Ws}{J} \cdot \dfrac{1\,h}{3600\,s} = 245\,W \cdot h$

The specific heat calculated by Siebel's equation is used by the American Society for Heating, Refrigerating, and Air Conditioning Engineers in one of the most comprehensive tabulated values for specific heat of foodstuffs. However, it is overly simplified, and the assumption that all types of non-fat solids have the same specific heat may not always be correct. Furthermore, Siebel's equation for specific heat below the freezing point assumes that all the water is frozen, and this is most inaccurate.

Specific heats of solids and liquids may also be estimated using correlations obtained from Choi and Okos (1987). The procedure is quite unwieldy to do by hand, however, the data may be entered into a spreadsheet to facilitate repetitive calculations. The specific heats, in J/(kg · K), as a function of T (°C) for various components of foods are as follows:

Protein: $C_{pp} = 2008.2 + 1208.9 \times 10^{-3}\,T - 1312.9 \times 10^{-6}\,T^2$
Fat: $C_{pf} = 1984.2 + 1473.3 \times 10^{-3}\,T - 4800.8 \times 10^{-6}\,T^2$
Carbohydrate: $C_{pc} = 1548.8 + 1962.5 \times 10^{-3}\,T - 5939.9 \times 10^{-6}\,T^2$
Fiber: $C_{pfi} = 1845.9 + 1930.6 \times 10^{-3}\,T - 4650.9 \times 10^{-6}\,T^2$
Ash: $C_{pa} = 1092.6 + 1889.6 \times 10^{-3}\,T - 3681.7 \times 10^{-6}\,T^2$
Water above freezing: $C_{waf} = 4176.2 - 9.0864 \times 10^{-5}\,T + 5473.1 \times 10^{-6}\,T^2$

The specific heat of the mixture above freezing is

$$C_{avg} = P(C_{pp}) + F(C_{pf}) + C(C_{pc}) + Fi(C_{pfi}) + A(C_{pa}) + M(C_{waf}) \tag{5.10}$$

where P, F, Fi, A, C and M, represent the mass fraction of protein, fat, fiber, ash, carbohydrate and moisture, respectively.

A spreadsheet for calculating specific heat at one temperature using the data given in Example 5.4 is shown in Fig. 5.1.

Example 5.4. Calculate the specific heat of a formulated food product that contains 15% protein, 20% starch, 1% fiber, 0.5% ash, 20% fat, and 43.5% water at 25°C.

Solution:

The calculated values, in J/(kg $\cong$ K), are $C_{pp} = 2037.6$; $C_{pf} = 2018.0$; $C_{pc} = 1594.1$; $C_{pfi} = 1891.3$; $C_{pa} = 1137.5$; and $C_{waf} = 4179.6$ Substituting in Equation (5.14)

$$C_{pavg} = 0.15(2037.6) + 0.2(1594.1) + 0.01(1891.3) + 0.005(1137.5)$$
$$+ 0.2(2018) + 0.435(4179.6) = 2870.8\,J/(kg \cong K)$$

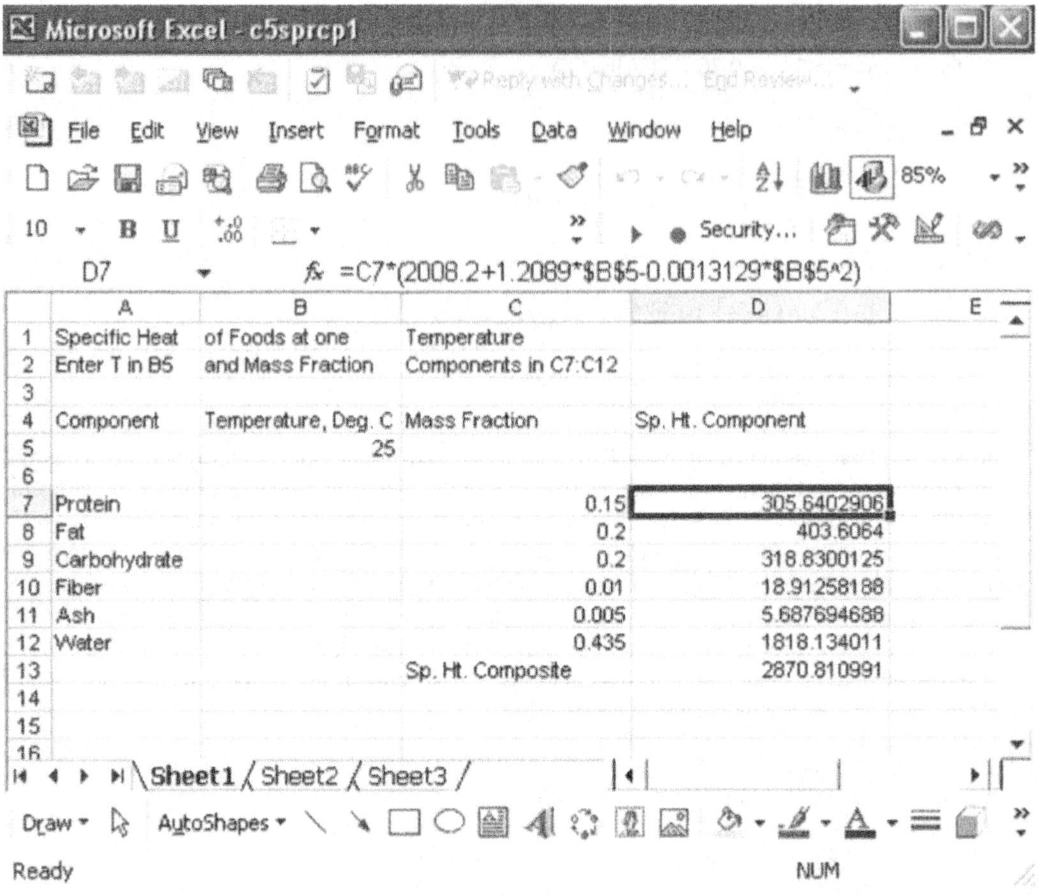

Figure 5.1 Spreadsheet for calculating specific heat using data in Example 5.4.

By comparison, using Siebel's equation:

$$C_p = 1674.72(0.2) + 837.36(0.15 + 0.01 + 0.005 + 0.2) + 4186.8(0.435) = 2462 \text{ J/(kg K)}$$

Values for C_p calculated using Choi and Okos' (1988) correlations are generally higher than those calculated using Siebel's equation at high moisture contents.

Seibel's equations have been found to agree closely with experimental values when $M > 0.7$ and when no fat is present. Choi and Okos' correlation is more accurate at low moisture contents and for a wider range of product composition because it is based on published literature values for a wide variety of foods. The simplicity of Siebel's equations however appeals to most users particularly when tolerance for error is not too stringent.

For enthalpy change calculations, Choi and Okos' equations for specific heat must be expressed as an average over the range of temperatures under consideration. The mean specific heat, C^*, over a temperature range T_1 to T_2, where $(T_2 - T_1) = \delta$, $T_2^2 - T_1^2 = \delta^2$ and $T_2^3 - T_1^3 = \delta^3$ is

$$C^* = \frac{1}{\delta} \int_{T_1}^{T_2} C_p \, dT$$

Thus, the equations for the mean specific heats of the various components over the temperature range δ, become

> Fiber: $C_{pfi}^* = [1/\delta][1845.9(\delta) + 0.9653(\delta^2) - 1550 \times 10^{-6}(\delta^3)]$
> Ash: $C_{pa}^* = [1/\delta][1092.6(\delta) + 0.9448(\delta^2) - 1227 \times 10^{-6}(\delta^3)]$
> Water: $C_{waf}^* = [1/\delta][4176.2(\delta) - 4.543 \times 10^{-5}(\delta^2) + 1824 \times 10^{-6}(\delta^3)]$
> Protein: $C_{pp}^* = [1/\delta][2008.2(\delta) + 0.6045(\delta^2) - 437.6 \times 10^{-6}(\delta^3)]$
> Fat: $C_{pf}^* = [1/\delta][1984.2(\delta) + 0.7367(\delta^2) - 1600 \times 10^{-6}(\delta^3)]$
> Carbohydrate: $C_{pc}^* = [1/\delta][1548.8(\delta) + 0.9812(\delta^2) - 1980 \times 10^{-6}(\delta^3)]$

$$C_{avg}^* = P(C_{pp}^*) + F(C_{pf}^*) + C(C_{pc}^*) + Fi(C_{pfi}^*) + A(C_{pa}^*) + M(C_{waf}^*) \tag{5.11}$$

Example 5.5. Calculate the mean specific heat of the formulated food product in example 4, in the temperature range 25°C to 100°C.

Solution:

$\delta = 75°C$, $\delta^2 = 9375°C^2$, $\delta^3 = 984375°C^3$. The mean specific heats of the components in J/(kg $\cong$ K), are $C_{pp}^* = 2078$; $C_{pf}^* = 2055.3$; $C_{pc}^* = 1645.4$; $C_{pfi}^* = 1948.7$; $C_{pa}^* = 1137.5$; $C_{waf}^* = 4200$. The mean specific heat: $C_{pavg}^* = 2904J/(kg.K)$. The spreadsheet for calculating average specific heat over a temperature range using the data in Example 5.5 is shown in Fig. 5.2.

5.3   ENTHALPY CHANGES IN FOODS DURING FREEZING

5.3.1   Correlation Equations Based on Freezing Points of Food Products Unmodified from the Natural State

When considering the heat to be removed during freezing of a food product, a change in phase is involved, and the latent heat of fusion must be considered. Water in a food does not all change into ice at the freezing point. Some unfrozen water exists below the freezing point, and therefore Siebel's equations for specific heat below the freezing point is very inaccurate. The best method for determining the amount of heat that must be removed during freezing, or the heat input for thawing, is by calculating the enthalpy change. One method for calculating enthalpy change below the freezing point (*good only for moisture contents between 73% and 94%*) is the procedure of Chang and Tao (1981). In this correlation, it is assumed that all water is frozen at 227 K (−50° F).

A reduced temperature (T_r) is defined as:

$$T_r = \frac{T - 227.6}{T_f - 227.6} \tag{5.12}$$

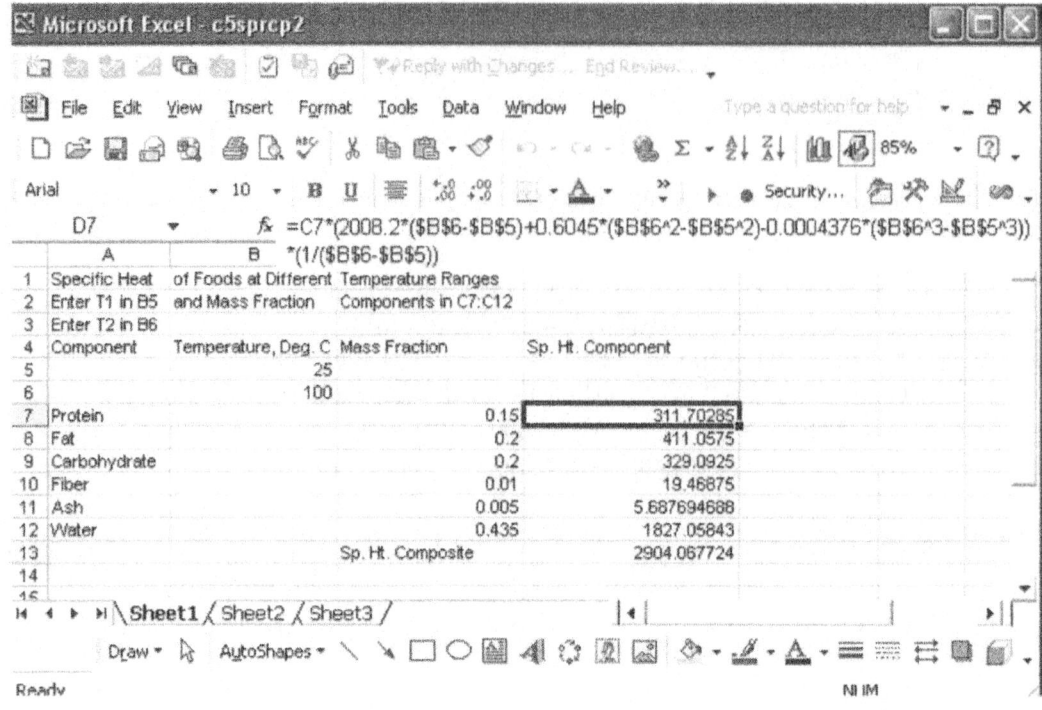

Figure 5.2 Spreadsheet for calculating specific heat as a function of temperature using data in Example 5.5.

where T_f is the freezing point temperature and T is the temperature at which the enthalpy is being determined. Two parameters, a and b, have been calculated for different products as a function of the mass fraction of moisture in the product, M. The correlation equations are as follows.
Meats:

$$a = 0.316 - 0.247(M - 0.73) - 0.688(M - 0.73)5 \qquad (5.13)$$
$$b = 22.95 + 54.68(a - 0.28) - 5589.03(a - 0.28)5 \qquad (5.14)$$

Vegetables, fruits, juices:

$$a = 0.362 + 0.0498(M - 0.73) - 3.465(M - 0.73)^2 \qquad (5.15)$$
$$b = 27.2 - 129.04(a - 0.23) - 481.46(a - 0.23)^2 \qquad (5.16)$$

The freezing point (T_f), in K is

$$\text{Meats}: T_f = 271.18 + 1.47\,M \qquad (5.17)$$

$$\text{Fruits and vegetables}: T_f = 287.56 - 49.19\,M + 37.07\,M^2 \qquad (5.18)$$
$$\text{Juices}: T_f = 120.47 + 327.35\,M - 176.49\,M^2 \qquad (5.19)$$

The enthalpy at the freezing point, H_f, in J/kg, relative to 227.6 K is

$$H_f = 9792.46 + 405096\,M \qquad (5.20)$$

The enthalpy at temperature T relative to 227.6 K is determined by:

$$H = H_f \left[a \, T_r + (1-a) T_r^{\ b} \right] \tag{5.21}$$

Example 5.6. Calculate the freezing point and the amount of heat that must be removed in order to freeze 1 kg of grape juice containing 25% solids from the freezing point to $-30°C$.

Solution:

$Y = 0.75$.
Using Equation (5.19) for juices:

$$T_f = 120.47 + 327.35(0.75) - 176.49(0.75)^2 = 266.7 \text{ K}$$

Using Equation (5.20):

$$H_f = 9792.46 + 405,096(0.75) = 313,614 \text{ J}$$

Using Equation (5.15):

$$a = 0.362 + 0.0498(0.02) - 3.465(0.02)^2 = 0.3616$$

Using Equation (5.16):

$$b = 27.2 - 129.04(0.1316) - 481.46(0.1316)^2 = 1.879$$

$$T_r = (-30 + 273 - 227.6)/(266.7 - 227.6) = 0.394$$

Using Equation (5.21):

$$H = 313{,}614 \left[(0.3616)0.394 + (1 - 0.3616)(0.394)^{1.879} \right] = 79{,}457 \text{ J/kg}$$

The enthalpy change from T_f to -30 EC is

$$\Delta H = 313{,}614 - 79{,}457 = 234{,}157 \text{ J/kg}$$

5.3.2 Enthalpy Changes During the Freezing of Foods Calculated from Molality of Liquid Water Fraction of the Food

When pure water is frozen, a phase change from liquid water to ice occurs at the freezing point. The cooling curve for water, a plot of temperature against time, will show the temperature sloping toward the freezing point followed by a constant temperature at the freezing point if the rate of cooling is slow enough to permit ice crystals adequate time to develop. The constant temperature is due to the release of energy associated with the phase change; all energy removed from the system at the freezing point serves as an energy sink for the latent heat of fusion. There will be no sensible heat loss (no temperature reduction) until all water has been converted to ice.

In foods, water exists as a solution. All water-soluble compounds in the food contribute to depress the freezing point. Because water crystals consist of pure water, transformation of water to ice is accompanied by an increase in solute concentration in the unfrozen water. Because solutes lower the freezing point, ice formation in foods occurs over a range of temperature. The freezing point is only the start of ice crystal formation. A typical freezing curve of temperature versus time exhibits a section where temperature is almost constant or drops much slower than the rest of the cooling curve. This section of the freezing curve is often referred to as the "ice crystal zone." Generally, a high initial moisture and low heat transfer rate extends the time temperature stays in the ice crystal zone. Food temperature will drop below the freezing point even when liquid water is still present.

5.3.3 Freezing Point Depression by Solutes

When the solution is ideal or when the solute concentration is low, the freezing point depression is

$$\Delta t_f = K_f M$$

where ΔT_f is the freezing point depression relative to the freezing point of water, $0°C$; K_f = cryoscopic constant = 1.86 for water. M = molality = mole solute/(g water/1000).

Let n = gmoles solute contained in w grams water. If the freezing point of a food is known, n can be calculated from the moisture content and the freezing point depression. Otherwise, the freezing point can be calculated from concentration of solutes and the moisture content. For highly ionized solutes such as salts of sodium and potassium, multiply the actual moles of solute by 2 to obtain n.

$$M = \frac{n(1000)}{w}$$

n will be constant during the freezing process, whereas w changes as ice is formed.

Let T_f = freezing point; $\Delta T_f = 0 - T_f = (-T_f)$

$$M = \frac{(-T_f)}{1.86}$$

5.3.4 Amount of Liquid Water and Ice at Temperatures Below Freezing

Basis: 1 kg food. Let w_o = original water in the mixture before freezing = weight fraction water × 1000.

$$n = \left[\frac{(-T_f)}{1.86}\right]\left[\frac{w_o}{1000}\right] = \frac{w_o(-T_f)}{1860}$$

At any temperature below the freezing point T; $\Delta Tf = 0 - T = (-T)$

$$M = \frac{-(T)}{1.86} = \frac{1000n}{w}$$

$$w = \frac{1000(n)(1.86)}{(-T)} = \frac{(1000)(w_o)(-T_f)}{1860}\left(\frac{1.86}{(-T)}\right) = \frac{w_o(-T_f)}{(-T)}$$

w = the amount of liquid water in the food at temperature T. The amount of ice = $I = w_o - w$

$$I = w_o - w_o\frac{(-T_f)}{(-T)} = w_o\left(1 - \frac{(-T_f)}{(-T)}\right)$$

5.3.5 Sensible Heat of Water and Ice at Temperatures Below the Freezing Point

For an increment of temperature dT, water will lose sensible heat according to:

$$dq = w\, C_{pl}\, dT$$

The change in sensible heat for liquid water (q_{sl}) from T_f to T is determined by:

$$q_{sl} = \int_{T_f}^{T} C_{pl}\, w\, dT = C_{pl}\, w_0 \int_{T_f}^{T} \frac{(-T_f)}{(-T)}\, dT = C_{pl} w_0 (-T_f) Ln\left[\frac{(-T)}{(-T_f)}\right] \tag{5.22}$$

The change in sensible heat of ice (q_{si}) from T_f to T is

$$q_{si} = \int_{T_f}^{T} C_{pi} w_0 \left[1 - \frac{(-T_f)}{(-T)}\right] dT = C_{pi} w_0 \left[(T_f - T) - (-T_f) \ln \frac{(-T)}{(-T_f)}\right] \tag{5.23}$$

Because the specific heat for water or ice are in J/kg.K, w_0 in the equations for q should be the mass in kilograms.

When considering the energy change associated with a temperature change from any temperature T_1 below the freezing point to T, substitute the amount of liquid water at T_1 for w_0 and T_1 for T_f.

5.3.6   Total Enthalpy Change

The total enthalpy change will consist of: sensible heat of fat; sensible heat of non-fat solids; sensible heat of ice; sensible heat of liquid water; and the latent heat of fusion of ice.

The total enthalpy change from the freezing point to any tempeature T is

$$q = \Delta H = F\, C_{pf}(T_f - T) + SNF\, C_{psnf}\, (T_f - T) + q_{sw} + q_{si} + I\,(334,860)$$

Values for the specific heats of water, ice, fat and non-fat solids were given earlier in this chapter.

Example 5.7. Boneless broiler breast meat contains 70.6% water, 24.0% protein, 1.2% ash, and 4.2% fat. The freezing point is $-1.2°C$. If this meat is marinated by adding salt solution to obtain a weight gain over the unmarinated meat of 15% and a net salt (NaCl) content of 1.0%, calculate (a) the new freezing point and (b) the enthalpy change as the marinated meat is frozen to $-18°C$ from the new freezing point per kg of marinated meat.

Basis: 1 kg original meat

$M = 1.2/1.86 = 0.645;$ g water $= 706$
$n = M\,(w_0)/1000 = .645(706)/1000 = 0.455$ gmoles

For marinated meat:

mass $= 1 + 0.15\,(1) = 1.15$ kg
Salt $= NaCl = 0.01(1.15) = 0.015$ kg
Water $= 1.15 - 0.015 - (1 - 0.706) = 0.841$ kg $= 841$ g
Consider NaCl to have two moles solute/mole NaCl; the Na^+ and Cl^- component.
The molecular weight of NaCl $= 35.5$.
Total gmoles solute in the marinated meat $= 2(0.015)(1000)/35.5 + 0.455 = 1.301.$
The molality of solutes in marinated meat:

$M = 1.301(1000)/841 = 1.547$

(a) $T_f = 0 - 1.547(1.86) = -2.9°C$

(b) At $-18°C$:

$w_o = 841 = 0.841$ kg

$w = 841(2.9/18) = 135.5$ g $= 0.1355$ kg

$I = 841 - 135.5 = 705.5$ g $= 0.1355$ kg

$q_{sl} = 4186.8 (0.841) (2.9) \ln (18/2.9) = 18,642$ J

$q_{si} = 2093.4 (.841)[\{-2.9 -(-18)\} - 2.9 \ln (18/2.9)] = 17,263$ J

Total q = $18,642 + 17,263 +$ SNF $(837.36)(18 - 2.9) + F(1674.72)(18 - 2.9) + I (334.860)$

Because the original fat content was 4.2% and there was no fat addition during marination, F = 0.042 kg. The solids non-fat will consist of protein (24%), ash (1.2%), and added salt (0.015 kg). SNF $= 0.24 + 0.012 + 0.015 = 0.267$ kg.

$$\text{Total} \quad q = 18,642 + 17,263 + .267(837.36)(18 - 2.9)$$
$$+ .042(1674.72)(18 - 29) + 0.1355(334.860)$$
$$= 84,910 \text{ J}$$

Although the original basis is 1 kg of unmarinated meat, note that the amount of water, ice, fat, and solids non-fat are those present in the marinated product, therefore the calculated q will be the enthalpy change per kg of marinated meat.

5.3.7 Specific Heats of Gases and Vapors

From Chapter 4, the specific heat of gases depends upon whether the process is carried out at constant pressure or at constant volume. If a gas is heated by blowing air across heating elements, the process is a constant pressure process. The specific heat is designated by C_p, the specific heat at constant pressure. The heat required to raise the temperature of a gas with mass, m, at constant pressure equal the change in enthalpy, ΔH. The enthalpy change associated with a change in temperature from a reference temperature T_0 to T_2 is:

$$\Delta H = m \int_{T_0}^{T_2} C_p \, dT = mC_{pm}(T_2 - T_0) \tag{5.24}$$

C_{pm} is the mean specific heat over the temperature range T_0 to T_2, and C_p is the expression for specific heat as a function of T.

If a gas is heated from any temperature T_1 to a final temperature T_2, the change in enthalpy accompanying the process must be calculated as follows:

$$\Delta H = q = mC_{pm}(T_2 - T_0) - mC'_{pm}(T_1 - T_0) \tag{5.25}$$

where C'_{pm} is mean specific heat from the reference temperature T_0 to T_1. Tabulated values for the mean specific heat of gases are based on ambient temperature of 77°F or 25°C, as the reference temperature. Table 5.2 lists mean specific heats of common gases in the American Engineering System of units, and Table 5.3 lists specific heats of the same gases in SI. Values for C_{pm} in Tables 5.2 and 5.3 change very little at temperatures ordinarily used in food processes, therefore, C_{pm} based on the temperature range from T_0 to T_2 can be used for ΔH between T_1 and T_2 with very little error in comparison with the use of Equation (5.25).

Table 5.2 Mean Heat Capacities of Various Gases from 77°F to Indicated Temperatue.

Temperature (°F)	Mean specific heat in BTU/lb(°F)				
	O_2	N_2	CO_2	H_2O vapor	Air
77	0.219	0.248	0.202	0.446	0.241
100	0.219	0.248	0.203	0.446	0.241
200	0.221	0.249	0.209	0.448	0.241
300	0.223	0.249	0.216	0.451	0.242
400	0.224	0.250	0.221	0.454	0.242
500	0.226	0.250	0.226	0.457	0.243
600	0.228	0.251	0.231	0.461	0.245
700	0.230	0.253	0.236	0.456	0.246

Source: Calculated from Mean molal heat capacity of gases at constant pressure. *In* Harper, J. C. 1976. *Elements of Food Engineering*. AVI Publishing Co., Westport, Conn.

Example 5.8. Calculate the heating requirement for an air drier that uses 2000 ft^3/min of air at 1 atm and 170°F, if ambient air at 70°F is heated to 170°F for use in the process.

Solution:

$$q = mC_{pm}(170 - 70)$$

m, the mass of air/min going through the drier, in pounds, will be calculated using the ideal gas

Table 5.3 Mean Heat Capacities of Various Gases from 25°C to Indicated Temperature.

Temperature (°C)	Mean specific heat J/kgK				
	O_2	N_2	CO_2	H_2O vapor	Air
25	916	1037	847	1863	1007
50	919	1039	858	1868	1008
100	926	1041	879	1880	1011
125	930	1043	889	1886	1012
150	934	1044	899	1891	1014
175	936	1045	910	1897	1015
200	938	1047	920	1903	1017
250	945	1049	941	1914	1020
300	952	1052	962	1925	1023
350	960	1054	983	1937	1026

Source: Calculated from Mean molal heat capacity of gases at constant pressure. *In* Harper, J.C. 1976 *Elements of Food Engineering*. AVI Publishing Co., Westport, Conn.

equation.

$$PV = \frac{m}{M}RT; \quad M = mol.wt.air = 29 \text{ lb/lb mole}$$
$$= 126.14 \text{ lb/min}$$

From Table 5.2, C_{pm} is constant at 0.241 BTU/(lb $\cong$ °F) up to 200°F, therefore it will be used for the temperature range 70°F to 170°F.

$$= 126.13 \text{ lb/min}$$

$$m = \frac{PVM}{RT} = \frac{14.7(144)\,\text{lb/ft}^2[2000\,\text{ft}^3/\min][29\,\text{lb/lbmole}]}{[1545\,\text{ft lb}_f/\text{lbmole °R}](460+170)\text{°R}}$$

$$q = 126.13\frac{\text{lb}}{\min} \times \frac{0.241\,\text{BTU}(170-70)\text{°F}}{(\text{lb °F})} = 3039.8\,\text{BTU/min}$$

Example 5.9. How much heat would be required to raise the temperature of 10 m³/s of air at 50°C to 120°C at 1 atm?

Solution:

Using Equation (5.25):

C'_{pm} at $T_1 = 50°C = 1008$ J/(kg $\cong$ K)

C'_{pm} at $T_2 = 120°C = 1012$ J/(kg $\cong$ K)

The mean specific heat from T_o to 50°C and from T_o to 120°C are different enough to require the use of Equation (5.25) to determine ΔH. The reference temperature, T_o, for C_{pm} in Table 5.3 is 25°C. The mass is determined using the ideal gas equation.

$$m = \frac{PVM}{RT} = \frac{1\,\text{atm}(10\text{m}^3)/\text{s}[29\,\text{kg/kg mole}]}{0.08206[\text{m}^3\,\text{atm/kg mole K}](50+273)(\text{K})} = 10.94\,\text{kg/s}$$

$$q = 10.94(1012)(120-25) - 10.94(1008)(50-25) = 776.08\,\text{kJ/s} = 776.08\,\text{kW}$$

5.4   PROPERTIES OF SATURATED AND SUPERHEATED STEAM

Steam and water are the two most used heat transfer mediums in food processing. Water is also a major component of food products. The steam tables that list the properties of steam are a very useful reference when determining heat exchange involving a food product and steam or water. At temperatures above the freezing point, water can exist in either of the following forms.

Saturated Liquid:. Liquid water in equilibrium with its vapor. The total pressure above the liquid must be equal to or be higher than the vapor pressure. If the total pressure above the liquid exceeds the vapor pressure, some other gas is present in the atmosphere above the liquid. If the total pressure above a liquid equals the vapor pressure, the liquid is at the boiling point.

Saturated Vapor: This is also known as saturated steam and is vapor at the boiling temperature of the liquid. Lowering the temperature of saturated steam at constant pressure by a small increment will cause vapor to condense to liquid. The phase change is accompanied by a release of heat. If heat is removed from the system, temperature and pressure will remain constant until all vapor is converted to liquid. Adding heat to the system will change either temperature or pressure or both.

Vapor-Liquid Mixtures: Steam with less than 100% quality. Temperature and pressure correspond to the boiling point; therefore, water could exist either as saturated liquid or saturated vapor. Addition of heat will not change temperature and pressure until all saturated liquid is converted to vapor. Removing heat from the system will also not change temperature and pressure until all vapor is converted to liquid.

Steam Quality: The percentage of a vapor-liquid mixture that is in the form of saturated vapor.

Superheated Steam: Water vapor at a temperature higher than the boiling point. The number of degrees the temperature exceeds the boiling temperature is the *degrees superheat*. Addition of heat to superheated steam could increase the superheat at constant pressure or change both the pressure and temperature at constant volume. Removing heat will allow the temperature to drop to the boiling temperature where the temperature will remain constant until all the vapor has condensed.

5.4.1 The Steam Tables

The steam tables are tabulated values for the properties of saturated and superheated steam.

5.4.1.1 The Saturated Steam Table

The saturated steam table consists of entries under the headings of temperature, absolute pressure, specific volume, and enthalpy. A saturated steam table is in the Appendix (Tables A.3 and A.4).

The temperature and absolute pressure correspond to the boiling point, or the temperature and pressure under which steam can be saturated. The absolute pressure at a given temperature is also the vapor pressure.

The rest of the table consists of three general headings each of which are subdivided into saturated liquid, evaporation, and saturated vapor. The entries under saturated liquid give the properties of liquid water at the indicated temperature. The entries under saturated vapor give the properties of steam at the boiling point. The entries under evaporation are changes due to the phase transformation and are the difference between the properties of saturated vapor and saturated liquid.

Specific volume is the reciprocal of the density. It is the volume in cubic feet occupied by 1 lb of water or steam under the conditions given.

Enthalpy is the heat content of a unit mass of steam or water at the indicated temperature and pressure. Enthalpy values in the steam tables are calculated from a base temperature of 0°C. The energy change associated with a change in temperature or pressure of steam is the difference in the initial and final enthalpies. The following examples illustrate the use of the steam tables. In these examples, atmospheric pressure is specified and may not be a standard atmosphere. Please refer to the section on "Absolute Pressure and Temperature" in Chapter 4 for a discussion of the difference between atmospheric pressure and a standard atmosphere.

Example 5.10. At what vacuum would water boil at 80°F? Express this in (a) inches of mercury vacuum (given: atm pressure = 30 in. Hg), (b) absolute pressure in pascals.

Solution:

From steam tables, Appendix Table A.3: The boiling pressure for water at 80°F is 0.50683 psia.

(a) Inches Hg absolute pressure

$$= 0.5068 \frac{lb}{in.^2} \left[\frac{2.035 \text{ in. Hg}}{lb/in.^2} \right] = 1.03 \text{ in. Hg absolute}$$

Vacuum $= 30$ in. Hg $- 1.03$ in. Hg $= 28.97$ in. Hg vacuum

(b) Pressure

$$= \frac{0.50683 lb_f}{in.^2} \left[\frac{6894.757 Pa}{lb/in.^2} \right]$$

$$= 3.494 \text{ kPa absolute pressure}$$

Example 5.11. If 1 lb of water at 100 psig and 252°F is allowed to expand to 14.7 psia, calculate (a) the resulting temperature after expansion and (b) the quantity of vapor produced.

Solution:

The absolute pressure $= 100 + 14.7 = 114.7$ psia. At 252°F, water will not boil until the pressure is reduced to 30.9 psia. The water therefore is at a temperature much below the boiling point at 114.7 psia and it would have the properties of liquid water at 252°F.

(a) After expansion to 14.7 psia, the boiling point at 14.7 psia is 212°F. Part of the water will flash to water vapor at 212°F and the remaining liquid will also be at 212°F.
(b) The enthalpy of water at 252°F is (h_f at 252°F) 220.62 BTU/lb. Basis: 1 lb H_2O. Heat content $= 220.62$ BTU. When pressure is reduced to 14.7 psia, some vapor will be formed, but the total heat content of both vapor and liquid at 212°F and 14.7 psia will still be 220.62 BTU.

If $x = $ wt vapor produced, $1 - x = $ wt water at 212°F and 14.7 psia:

$$x(h_g) + (1 - x)(h_f) = 220.62$$

$h_g = 1150.5$ BTU/lb; $h_f = 180.17$ BTU/lb

$$x(1150.5) + (1 - x)(180.17) = 220.62$$

$$x = \frac{220.62 - 180.17}{1150.5 - 180.17} = \frac{40.45}{970.33} = 0.0417 \text{ lb } H_2O$$

Example 5.12. If water at 70°F is introduced into an evacuated vessel, initially at 0 psia, what would be the pressure inside the vessel when equilibrium is finally attained at 70°F?

Solution:

Because the vessel is completely evacuated, the gaseous phase after introduction of water will be 100% water vapor. Upon introduction, water will vaporize until the pressure of water in the space above the liquid equals the vapor pressure.

Pressure = vapor pressure of water at 70°F

From steam tables, pressure at 70°F $= 0.36292$ psia.

Example 5.13. If the vessel in Example 5.11 had initially 14.7 psia absolute pressure and contained completely dry air, what would be the absolute pressure after introducing the water, assuming that none of the original air had escaped during the process?

Solution:

Pressure = partial pressure of air + partial pressure of water(volume is constant)

$\quad\quad$ = original pressure of dry air + vapor pressure of water

$\quad\quad$ = 14.7 + 0.36292 = 15.063 psia

Example 5.14. How much heat would be given off by cooling steam at 252°F and 30.883 psia to 248°F, at the same pressure?

Solution:

First, check the state of water at 30.883 psia and 252°F and 248°F. From steam tables, the boiling point of water at 30.883 psia is 252°F. Therefore, steam at 252°F and 30.883 psia is saturated vapor. At 30.883 psia and 248°F, water will be in the liquid state, because 248°F is below the boiling temperature at 30.883 psia.

Heat given off = q = h_g at 252°F − h_f at 248°F

From steam tables, Appendix Table A.4

$\quad\quad h_g$ at 252°F = 1164.78 BTU/lb
$\quad\quad h_f$ at 248°F = 216.56 BTU/lb
$\quad\quad q$ = 1164.78 − 216.56 = 948.22 BTU/lb

Saturated steam is a very efficient heat transfer medium. Note that for only a 2°F change in temperature, 948 BTU/lb of steam is given off. The heat content of saturated vapors come primarily from the latent heat of vaporization, and it is possible to extract this heat simply by causing a phase change at constant temperature and pressure.

5.4.1.2 The Superheated Steam Tables

A superheated steam table is in Appendix Table A.2. Both temperature and absolute pressure must be specified to accurately define the degree of superheat. From the temperature and absolute pressure, the specific volume v in ft³/lb and the enthalpy h in BTU/lb can be read from the table. Example problems on the use of the superheated steam tables are as follows.

Example 5.15. How much heat is required to convert 1 lb of water at 70°F to steam at 14.696 psia and 250°F?

Solution:

First determine the state of steam at 14.696 psia and 250°F. At 14.696 psia, the boiling point is 212°F. Steam at 250°F and 14.696 psia is superheated steam. From the superheated steam table, h at

250°F is 1168.8 BTU/lb.

Heat required = h_g at 250°F and 14.696 psia $- h_f$ at 70°F
$$= 1168.8 \text{ BTU/lb} - 38.05 \text{ BTU/lb}$$
$$= 1130.75 \text{ BTU/lb}$$

Example 5.16. How much heat would be given off by cooling superheated steam at 14.696 psia and 500°F to 250°F at the same pressure?

Solution:

Basis: 1 lb of steam.

Heat given off = q = h at 14.696 psia and 500°F $- h_g$ at 14.696 psia and 250°F
$$= 1287.4 - 1168.8 = 118.6 \text{ BTU/lb}$$

Superheated steam is not a very efficient heating medium. Note that a 250°F change in temperature is accompanied by the extraction of only 118.6 BTUs of heat.

5.4.1.3   Double Interpolation from Superheated Steam Tables

Because the entries in the table are not close enough to cover all conditions, it may be necessary to interpolate between entries to obtain the properties under a given set of conditions. In the case of superheated steam where both temperature and pressure are necessary to define the state of the system, a double interpolation is sometimes necessary. The following example shows how the double interpolation is carried out.

Example 5.17. Calculate the enthalpy of superheated steam at 320°F and 17 psia.

Solution:

Entries in Appendix Table A.2 show enthalpies at 15 and 20 psia and 300°F and 350°F. The tabular entries and the need for interpolation are as follows:

	Enthalpy	
P (psia)	300°F	350°F
15	1192.5	1216.2
17	?	?
20	1191.4	1215.4

Enthalpies at 300°F and 350°F at 17 psia are obtained by interpolating between 15 and 20 psia at each temperature.

At 300°F: P = 15, h = 1192.5; P = 20, h = 1191.4; at P = 17, h = ?

$$h_{(300°F \wedge 17 \text{ psia})} = 1191.4 + \frac{(1192.5 - 1191.4)}{(20 - 15)}(20 - 17) = 1192.06 \text{ BTU/lb}$$

At 350°F: P = 15, h = 1216.2; P = 20, h = 1215.4; at P = 17, h = ?

$$h_{(350°F \wedge 17 \text{psia})} = 1215.4 + \frac{(1216.2 - 1215.4)}{(20 - 15)}(20 - 17) = 1215.88 \text{ BTU/lb}$$

Now it is possible to interpolate between 300°F and 350°F to obtain the enthalpy at 17 psia and 320°F.

$$h = 1215.88 - \frac{(1215.88 - 1192.06)}{(350 - 300)}(350 - 320) = 1215.88 - 14.29 = 1201.59 \text{ BTU/lb}$$

5.4.2 Properties of Steam Having Less Than 100% Quality

If steam is not 100% vapor, the properties can be determined on the basis of the individual properties of the component.

If x = % quality

$$v = xv_g + (1 - x)v_f$$
$$h = xh_g + (1 - x)h_f$$

Example 5.18. Calculate the enthalpy of steam at 252°F having 80% quality.

Solution:

From the saturated steam tables, Appendix Table A.3: At 252°F, saturated steam or water has the following properties:

$h_f = 220.62 \text{ BTU/lb}$

$h_g = 1164.78 \text{ BTU/lb}$

$h = 1164.78(0.8) + 220.62(0.2) = 931.82 + 44.12 = 975.94 \text{ BTU/lb}$

Note that only the temperature of steam is given in this problem. If either the temperature or pressure is given, but not both, steam is at the boiling point.

5.5 HEAT BALANCES

Heat balance calculations are treated in the same manner as material balances. The amount of heat entering a system must equal the amount of heat leaving a system, or:

Heat in = heat out + accumulation

At a steady state, the accumulation term is zero and heat entering the system must equal what leaves the system.

Heat balance problems are facilitated by using diagrams that show process streams bringing heat and taking heat out of a system.

Example 5.19. Calculate the amount of water that must be supplied to a heat exchanger that cools 100 kg/h of tomato paste from 90°C to 20°C. The tomato paste contains 40% solids. The increase in water temperature should not exceed 10°C while passing through the heat exchanger. There is no mixing of water and tomato paste in the heat exchanger.

Solution:

The diagram of the system is shown in Fig. 5.3.

This problem may be solved by assuming a datum from which enthalpy calculations are made. This datum temperature is the lowest of the process stream temperatures. Let T_1 = inlet water

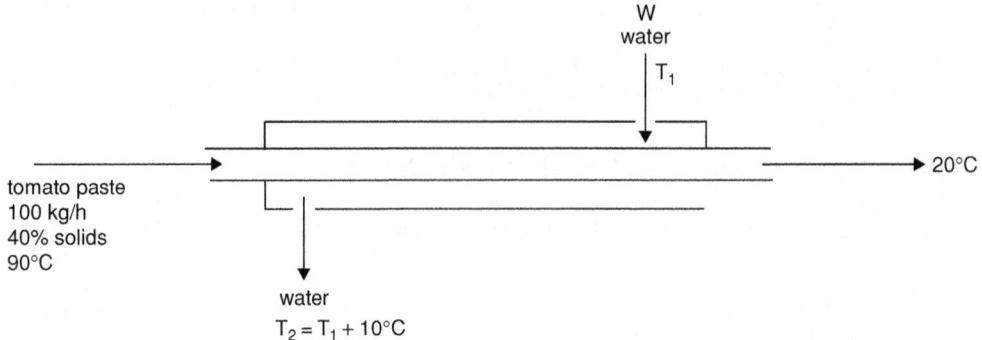

Figure 5.3   Diagram of heat exchange during cooling of tomato paste in Example 5.19.

temperature $= 20°C$, and $T_2 =$ exit water temperature $= 20 + 10 = 30°C$. Let $20°C$ be the reference temperature for enthalpy calculations. The specific heat of water $= 4187$ J/(kg $\cong$ K), and that of tomato paste is obtained, using Equation (5.7).

$C_{avg} = 3349(0.6) + 837.36 = 2846.76$ J/(kg $\cong$ K)

Heat content of entering tomato paste:

$q_1 = (100$ kg$)$ [2846.76 J/(kg $\cong$ K)]$(90 - 20)$ K$= 19.927$ MJ

Heat content of tomato paste leaving system:

$q_2 = 100$kg[2846.76 J/(kg $\cong$ K)]$(20 - 20)$ K $= 0$

Let $W =$ kg water entering the system

$q_3 =$ Wkg[4187 J/(kg $\cong$ K)]$(20 - 20)$ K $= 0$

$q_4 =$ heat content of water leaving the system

$= W$ kg [4187 J/(kg $\cong$ K)]$(30 - 20)$ K $= 41,870(W)$ J

The heat balance is

$q_1 + q_3 = q_2 + q_4$

Because q_2 and $q_3 = 0$, $q_1 = q_4$, and:

$$(19.927 \text{ MJ})\frac{10^6 \text{ J}}{\text{MJ}} = 41,871(W) \text{ J}$$

$$W = \frac{19.927 \times 10^6}{41870} = 475.9 \text{ kg}$$

The heat balance may also be expressed as follows:

Heat gain by water $=$ heat loss by the tomato paste

$$100 \text{ kg}\left(2846.76\frac{\text{J}}{\text{kg} \cdot \text{K}}\right)(90 - 20) \text{ K} = W\left(4187\frac{\text{J}}{\text{kg} \cdot \text{K}}\right) \cdot (T_1 + 10 - T_1) \text{ K}$$

$$41870W = 100(2846.76)(70); \quad W = 475.9 \text{ kg}$$

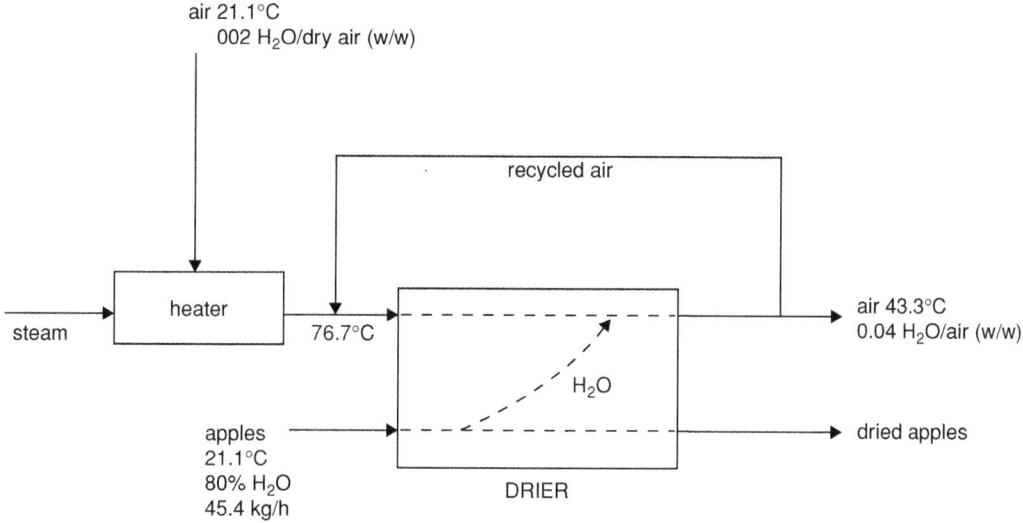

Figure 5.4 Diagram of material and energy balance in apple dehydration.

If there is no mixing and exchange of mass between process streams, calculation is simplified by equating heat gain by one process stream to heat loss by the other. However, when mixing and material transfer occurs, a heat balance based on heat content of each stream entering and leaving the system will simplify analysis of the problem.

Example 5.20. Calculate the amount of saturated steam at 121.1°C that must be supplied to a dehydrator per hour. Steam condenses in the heater, which heats the drying air from steam to water at 121.1°C. The dehydrator is operated as follows: Apples at 21.1°C enter the dehydrator with 80% moisture and leave the dehydrator at 37.7°C and 10% moisture. One hundred pounds per hour of fresh apples enter the drier. Fresh air at 21.1°C and a humidity of 0.002 kg H_2O/kg dry air enter the drier, mixes with recycled hot air until the humidity is 0.026 kg H_2O/kg dry air, and is heated to 76.7°C using steam in a finned heat exchanger. Hot air leaves the drier at 43.3°C and a humidity of 0.04 kg H_2O/kg dry air.

Solution:

The diagram for the process is shown in Fig. 5.4. Use as a basis 1 h of operation.

The system looks complicated with the hot air recycling, but if the boundary of the system is enlarged as shown in Fig. 5.4, solving the problem is considerably simplified.

The problem will be solved by separating air into dry air and water vapor, apples into dry matter and water, and heat contents of each component entering and leaving the system will be calculated. The lowest temperature, 21.1°C may be used as the base temperature for heat content calculations as in the preceeding example, but because the steam tables will be used to determine the enthalpy of water and steam, it will be more consistent to use the reference temperature for the steam tables of 0°C.

The following enthalpies are to be calculated:

Heat input $= q$

$q_1 =$ enthalpy of H_2O in entering air (vapor at $21.1°C$)

$q_2 =$ enthalpy of dry air entering at $21.1°C$

$q_3 =$ enthalpy of H_2O in apples entering (liquid at $21.1°C$)

$q_4 =$ enthalpy of dry matter in apples entering at $21.1°C$

$q_5 =$ enthalpy of H_2O in exit air (vapor at $43.3°C$)

$q_6 =$ enthalpy of dry air leaving at $43.3°C$

$q_7 =$ enthalpy of H_2O in apples leaving (liquid at $37.7°C$)

$q_8 =$ enthalpy of dry matter in apples leaving at $37.3°C$

The heat balance is

$$q + q_1 + q_2 + q_3 + q_4 = q_5 + q_6 + q_7 + q_8$$

q_2 and q_6 are calculated using the quantity of dry air that enters the system per hour. The latter is obtained by performing a material balance on water.

Amount of water lost by apples $=$ amount of water gained by air

Solids balance for apples:

$$x = \frac{45.4(0.2)}{0.9} = 10.09 \text{ kg/h}$$

$$45.4(0.2) = x(0.9); \quad x = \text{wt dried apples}$$

$$\frac{\text{Water gained by air}}{\text{kg dry air}} = (0.04 - 0.002) = 0.038$$

Water lost by apples $= 45.4 - 10.09 = 35.31$ kg/h

Let $w =$ mass of dry air:

Total water gained by air $= 0.038\ w$

Material balance on water: $0.038\ w = 35.31$

$w = 929.21$ kg dry air/h

The mean specific heat of air is obtained from Table 5.3. These values are

$25°C$: $C_{pm} = 1008$ J/(kg $\cong$ K)

$50°C$: $C_{pm} = 1007$ J/(kg $\cong$ K)

These mean specific heat will be used for determining enthalpies of air at $21.1°C$ and $43.3°C$.

From the section "Specific Heat of Solids and Liquids";
$$C_p \text{for solids non-fat in apples} = 837.36 \text{ J/(kg} \cong \text{K)}$$

Use steam tables for heat contents of water and steam. Calculating the enthalpies:

$q = m\ C_{pm}(T - 0)$ because $0°C$ is used as the reference temperature.

$$q_1 = (929.21 \text{ kg dry air})\frac{0.002 \text{ kg } H_2O}{\text{kg dry air}}(h_g \text{ at } 21.1°C)$$

From Appendix Table A.4, h_g at 21.1°C C by interpolation = 2.54017 MJ/kg.

$q_1 = 929.21(0.002)(2.54017 \times 10^6) = 4.7207$ MJ

$q_2 = (929.21)$ kg dry air$(1008)\dfrac{J}{kg} \cdot K(21.1 - 0) = 19.7632$ MJ

$q_3 = 45.4(0.8)(h_f$ at 21.1°C)

From Appendix Table A.4, h_f at 21.1°C = 0.08999 MJ/kg

$q_3 = 45.4(0.8)(0.08999 \times 10^6) = 3.2684$ MJ

$q_4 = (45.4)(0.2)$ kg dry solids $\left(\dfrac{837.36\ J}{kg \cdot K}\right)(21.1 - 0)°C = 0.16043$ MJ

$q_5 = (929.21$ kg dry air$)\left(0.04\dfrac{kg\ H_2O}{kg\ dry\ air}\right)(h_g$ at 43.3°C)

From Appendix Table A.4, h_g at 43.3°C = 2.5802 M/kg

$q_5 = (929.21)(0.04)(2.5802 \times 10^6) = 95.9019$ MJ

$q_6 = (929.21$ kg dry air$)\dfrac{1007\ J}{kg\ K}(43.3 - 0)$

$\quad = (929.21)(1007)(43.3) = 40.5164$ MJ

$q_7 = (10.09)(0.1)$ kg $H_2O(h_f$ at 37.7°C)

From Appendix Table A.4, h_f at 37.7°C = 0.15845 MJ/kg

$q_7 = (10.09)(0.1)(0.15845) = 0.15987$ MJ

$q_8 = (10.09)\,(0.9)$ kg dry matter $\left(\dfrac{837.36\ J}{kg\ F}\right)(37.7 - 0)$

$\quad = (10.09)(0.9)(837.36)(37.7) = 0.28667$ MJ

The heat balance with q in MJ is

$q + [4.7207 + 19.7632 + 3.2684 + 0.16043]$
$\quad = [95.9019 + 40.5164 + 0.15988 + 0.28667]$
$q = 136.8648 - 27.91273 = 108.952$ MJ

Because the basis is 1 h of operation, the heating requirement for the process is 108.952 MJ/h. From Appendix Table A.4, the heat of vaporization of steam at 121.1°C is 2.199144 MJ/kg.

$$\text{The amount of steam supplied/h} = \frac{108.952\ MJ/h}{2.199144\ MJ/kg} = 49.543\ kg/h$$

Alternative solution:

The problem will be solved using a balance of heat sinks and heat source in the system. Consider the system as a whole, enclosing the recycle stream within the system boundary. Entering the system are air and apples, which both gain sensible heat through the process. Also entering the system is steam, which is the only heat source. Incoming air with the water vapor may be combined, and using an average specific heat, the sensible heat gain is easily calculated. Additional moisture in the air leaving the system is accounted for by moisture evaporation from apples, which may be assumed to occur at the temperature the apples leave the system. Thus energy gain by vaporized water is the the difference between the the enthalpy of the vapor at 43.3°C and the enthalpy of liquid at 37.7°C. The sensible heat gain by apples will be asessed with the initial weight and specific heat. In equation form, the energy balance is $m_{ma} C_{pm} (43.3 - 21.1) + m_{ap} C_p (37.7 - 21.1) + m_v (h_{g@43.3 C} - h_{f@37.7 C}) = m_s(h_{fg@121.1 C})$ where m_{ma}, m_{ap}, and m_s are the mass of moist air, wet apples, and steam, respectively. C_{pm} is the average specific heat of the moist air, and C_p is the specific heat of the wet apples. The specific heat of apples $= 0.8(4186.8) + 0.2(837.36) = 3516.9$ J/kg $\cong$ K. Mass of vapor is the same as the amount removed from apples and is determined by a material balance:

$$m_v = 45.4 - \frac{45.4(0.2)}{0.9} = 35.31 \frac{kg}{h}$$

Air will gain $(0.04 - 0.002)$ or 0.038 kg water/kg dry air. Thus, the mass of dry air entering the system is $35.31/0.038 = 929.21$ kg. The mass of moist air is

$$m_{ma} = \left[\frac{929.21 \text{ kg dry air}}{h} \right] \left[\frac{1.002 \text{ kg moist air}}{\text{kg dry air}} \right] = 931.07 \frac{kg}{h}$$

The specific heat of air and water, respectively at 50°C are 1008 and 1868 J/kg $\cong$ K. Thus, the mean specific heat for moist air is $C_{pm} = [1008 (1) + 1868 (0.002)](1/1.002) = 1009.7$ J/kg $\cong$ K.

The enthalpy of vapor at 43.3°C ($h_{g@43.3 C}$) from Appendix Table A.4 by interpolation between 42.5°C and 45°C is

$$2.5788 + [(2.5832 - 2.5788)/2.5](43.3 - 42.5) = 2.58021 \text{ MJ/kg}.$$

The enthalpy of liquid at 37.7°C $= 0.15816$ MJ/kg. The latent heat of evaporation at 121.1°C ($h_{fg@121.1 C}$) from Appendix Table A.4 by interpolation is

$$2.20225 - [(2.20225 - 2.19519)/(122.5 - 120)](121.1 - 120) = 2.19914 \text{ J/kg}.$$

Solving for m_s:

$$m_s = \frac{(931.07)(1009.7)(22.2) + (45.4)(3516.9)(16.6) + (35.31)(2.42236 \times 10^6)}{2.19914 \times 10^6}$$

$$= 49.58 \frac{kg}{h}$$

Example 5.21. Calculate the amount of steam at 121.1°C (250°F) that must be added to 100 kg of a food product with a specific heat of 3559 J/(kg $\cong$ K) to heat the product from 4.44°C (40°F) to 82.2°C (180°F) by direct steam injection.

Solution:

The diagram of the system is shown in Fig. 5.5. Let $x =$ kg of steam required. From steam table, the enthalpies of water and steam are as follows:

For steam at 121.1°C or 250°F, h_g from Table A.3 $= 1164.1$ BTU/lb $= 2.70705$ MJ/kg

$$\frac{\text{Heat loss steam}}{kg} = h_g \text{ at } 121.1°C - h_f \text{ at } 82.2°C$$

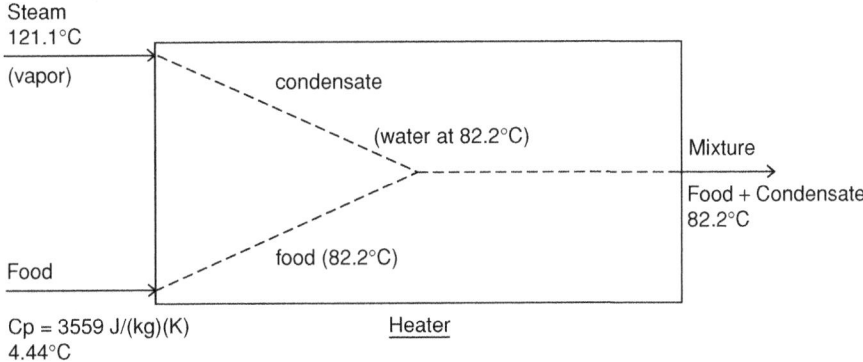

Figure 5.5   Diagram of a direct steam injection process for heating a food product.

For water at 82.2°C or 180°F, h_f = 148.00 BTU/lb = 0.34417 MJ/kg = 2.70705 − 0.34417 = 2.36288 MJ/kg

Total heat loss steam = x(2.36288) MJ

Heat gain by product = 100kg[3559 J/(kg $\cong$ K)](82.2 − 4.44) K = 27.67478 MJ

$$x(2.36288) = 27.67478$$
$$x = 11.71 \text{ kg steam required}$$

PROBLEMS

5.1. What would be the pressure generated when milk is heated to 135°C in a closed system? If the system is not pressurized, can this temperature be attained?

5.2. A method for heating food with saturated steam at temperatures below the boiling point of water is by carrying out the process under a vacuum. At what vacuum should a system be operated to heat a material with saturated steam at 150°F.

5.3. If a retort indicates a pressure of 15 psig but mercury in glass thermometer registers only 248°F, what does this indicate? Assume that both instruments are in working order and have been properly calibrated.

5.4. An evaporator is operated at 15 in. Hg vacuum. What would be the temperature of the product inside the evaporator? Assume product has the same boiling point as water.

5.5. How much heat is required to convert 1 kg water at 20°C to steam at 120°C?

5.6. How much heat must be removed to convert 1 lb steam at 220°F to (a) water at 220°F; (b) water at 120°F?

5.7. One pound of steam at 260°F contains 80% steam and 20% liquid water. How much heat will be released by this steam when it is allowed to condense to water at 200°F?

5.8. At what temperature would water be expected to boil at 10 in. Hg vacuum? Atm pressure = 14.696 psia.

5.9. How much steam at 250°F would be required to heat 10 lb of water from 70°F to 210°F in a direct steam injection heater?

5.10. How much heat will be required to convert steam at 14.696 psig to superheated steam at 600°F at the same pressure?

5.11. Ten pounds of water at a pressure of 20 psig is heated to a temperature of 250°F. If this water is allowed to empty into an open vessel at atmospheric pressure, how much of the water will remain in the liquid phase?

5.12. (a) If water at 70°F is introduced into an evacuated vessel where the original pressure is 0 psia, what would be the pressure inside the vessel at equilibrium? Assume no change in the temperature of water. (b) If the original pressure is 14.696 psia, what will be the final pressure?

5.13. Determine the heat content in BTU/lb for water (it could be liquid, saturated steam, or superheated steam) under the following conditions: (a) 180°F and 14.696 psia pressure, (b) 300°F and 14.696 psia pressure, (c) 212.01°F and 14.696 psia pressure.

5.14. In the formulation of a pudding mix, it is desired that the solids content of the product would be 20%. The product leaving the batch tank has a temperature of 26.67°C (80°F) and this is preheated to 90.56°C (195°F) by direct steam injection using culinary steam (saturated) at 104.4°C (220°F) followed by heating in a closed system to sterilizing temperatures. There is no further loss or gain of moisture in the rest of the process. What should be the solids content of the formulation in the batch tank such that after the direct steam injection heating the final solids content of the product will be 20%. Use Siebel's equation for calculating the specific heat of the product.

5.15. A fruit juice at 190°F is allowed to flash into an essence recovery system maintained at a vacuum of 29 in. of mercury. Atmospheric pressure is 29.9 in. The vapors that flash-off are rectified to produce an essence concentrate, and the juice after being stripped of the aromatic constituents is sent to an evaporator for concentration.

Assuming sufficient resident time for the juice in the system to allow equilibrium in temperature between the liquid and the vapor, calculate:

(a) The temperature of the juice leaving the essence recovery system.
(b) The solids content of the juice leaving the system if the original solids content is 10%. Assume no additional heat input and the latent heat for vaporization is derived from the loss in sensible heat of the liquid. The specific heat of the solids is 0.2 BTU/(lb ≅ °F).
(c) The quantity of water vaporized per 100 lb of juice entering the system.

5.16. An evaporator has a heat transfer surface area that would allow the transfer of heat at the rate of 100,000 BTU/h. If this evaporator is concentrating apple juice from 10% to 45% solids, under a vacuum of 25 in. Hg (atmospheric pressure is 30 in. of mercury), how much apple juice can be processed per hour?

5.17. Orange juice concentrate at 45% total solids leaves the evaporator at 50°C. This is frozen into slush in swept surface heat exchangers until half of the water is in the form of ice crystals prior to filling into cans, and the cans are frozen at −25°C. Assume that the sugars are all hexose sugars, and that the freezing point reduction can be determined using $\Delta T_f = K_f m$, where $K_f =$ the cryoscopic constant = 1.86 and m = the molality. Calculate:

(a) The total heat that must be removed from the concentrate in the swept surface heat exchangers per kg of concentrate processed.
(b) The amount of heat that must be further removed from the concentrate in frozen storage.
(c) The amount of water still in the liquid phase at −25°C.

Note: The moisture content is beyond the range where Chang and Tao's correlation is applicable. Determine the freezing point by calculating the freezing point depression: $\Delta T_b = K_f m$. The specific heat of the solids is the same below and above freezing. The specific heat of ice is 2093.4 J/(kg$K). The heat of fusion of ice is 334860 J/kg. The juice contains 42.75% soluble solids.

5.18. In a falling film evaporator, fluid is pumped to the top of a column and the fluid falls down as a film along the heated wall of the column increasing in temperature as it drops. When the fluid emerges from the column, it is discharged into a vacuum chamber where the fluid drops in temperature by flash evaporation until it reaches the boiling temperature at the particular vacuum employed. If juice containing 15% solids is being concentrated to 18% solids in one pass through the heated column, and the vacuum in the receiving vessel is maintained at 25 in. Hg, calculate the temperature of the fluid as it leaves the column such that when flashing occurs the desired solids content will be obtained.

5.19. When sterilizing foods containing particulate solids in the Jupiter system, solids are heated separately from the fluid components by tumbling the solids in a double cone processing vessel with saturated steam contacting the solids. The fluid component of the food is heated, held until sterile, and cooled using conventional fluid heating and cooling equipment. The cooled sterile liquid is pumped into the double cone processing vessel containing the hot solids, which cools the latter and drops the pressure to atmospheric. After allowing the mixture to cool by cooling the walls of the processing vessel, the sterile mixture is filled aseptically into sterile containers

    (a) Meat and gravy sauce is being prepared. Beef cubes containing 15% solids non-fat, 22% fat, and 63% water are heated from 4°C to 135°C, during which time condensate accumulates within the processing vessel with the meat. Saturated steam at 135°C is used for heating. Calculate the total amount of meat and condensate at 135°C.

    (b) The gravy mix is of equal weight as the raw meat processed and consists of 85% water and 15% solids non-fat. Calculate the temperature of the mixture after equilibration if the gravy mix is at 20°C when it is pumped into the processing chamber containing the meat at 135°C.

5.20. The chillers in a poultry processing plant cool broilers by contacting the broilers with a mixture of water and ice. Broilers enter at 38°C and leave the chillers at 4°C. USDA requires an overflow of 0.5 gallons of water per broiler processed, and this must be replaced with fresh water to maintain the liquid level in the chiller. Melted ice is part of this overflow requirement. If a plant processes 7000 broilers/h, and the broilers average 0.98 kg with a composition of 17% fat, 18% solids non-fat, and 65% water, calculate the ratio by weight of ice to fresh water that must be added into the chiller to meet the overflow requirement and the cooling load. Fresh water is at 15°C, and the overflow is at 1.5°C.

5.21. Saturated steam at 280°F is allowed to expand to a pressure of 14.696 psia without a loss of enthalpy. Calculate: (a) the temperature, (b) the weight of high pressure steam needed to produce 100 m^3/min of low pressure steam at 14.696 psia and the temperature calculated in (a).

5.22. In one of the systems for ultra-high-temperature sterilization, milk enters a chamber maintained at 60 psia and 800°F in an atmosphere of superheated steam where it discharges from a plenum into vertical tubes where it falls down in a thin film while exposed to the steam. The milk will be at the boiling temperature at 60 psia on reaching the bottom of the heating chamber. After a sterilizing hold time at constant temperature, the milk is discharged into a vacuum chamber for rapid cooling. If the vacuum chamber is at 15 in. Hg vacuum, calculate:

    (a) The temperature of the milk leaving the flash chamber.

    (b) The total solids content. Raw milk enters the heater at 2°C, and contains 89% water, 2% fat, and 9% solids non-fat. Given: The enthalpy of superheated steam at 60 psia and 800°F is 1431.3 BTU/lb. Saturated steam temperature at 60 psia is 292.7°F. The enthapy of saturated liquid (h$_f$) at 292.7°F is 260.7 BTU/lb.

5.23. (a) Calculate the freezing point of fresh strawberries that contain 8.5% soluble solids, 1% insoluble solids, and 90.5% water. Assume that the average molecular weight of the soluble solids is 261.

(b) If sucrose is added to the above strawberries in the ratio 1 part of sugar to 3 parts strawberries by weight, calculate the new freezing point.

(c) Calculate the change in enthalpy of the sugared strawberries from 20°C to −20°C.

5.24. A food mix containing 80% solids non-fat and 20% fat on a dry basis is to be extruded. Water is added continuously along with the product into the extruder and the mixture temperature increases to 135°C at a pressure of 600 kPa at the die entrance. Extrudate leaves the die and immediately expands to atmospheric pressure, releasing vapor as it exits the die. Assume that the temperature of the extrudate is 100°C immediately after leaving the die. It is desired that the moisture content of the extrudate will be 18% after the pressure reduction. Calculate the rate of moisture addtion to the extruder if the solid feed originally contains 10% water and is fed at the rate of 30 kg/h.

5.25. A food product that contains 15% solids non-fat, 2% fat, and 83% water is to be pasteurized by heating to 95°C by direct steam infusion using 90% quality steam at 15 psig. After heating to 95°C, the product is held in a holding tube and cooled in a heat exchanger; therefore, there is no opportunity for removing added water from the steam condensate later in the process. Calculate the moisture content of the raw product entering the heater such that the desired moisture content will be obtained after heating and cooling.

SUGGESTED READING

American Society for Heating, Refrigerating and Air Conditioning Engineers, 1967. Guide and Data Book, Applications for 1966 and 1967. ASHRAE, New York.

American Society of Mechanical Engineers. 1967. American Society of Mechanical Engineers 1967 steam tables. Properties of saturated and superheated steam from 0.08865 to 15,500 lb per sq in. absolute pressure. Combustion Engineering Inc., Windsor, CT.

Chang, H. D. and Tao, L. C. 1981. Correlation of enthalpy of food systems. J. Food Sci. 46:1493.

Choi, Y. and Okos, M. R. 1987. Effects of temperature and composition on thermal properties of foods. In: Food Engineering and Process Applications. M. Le Maguer and P. Jelen, Eds. Vol. I. Elsevier, New York, pp. 93–102.

Charm, S. E. 1971. Fundamentals of Food Engineering. 2nd ed. AVI Publishing Co., Westport, CT.

Felder, R. M. and Rousseau, R. W. 1999. Elementary Principles of Chemical Processes. 2nd ed. John Wiley & Sons, New York.

Himmelblau, D. M. 1967. Basic Principles and Calculations in Chemical Engineering. 2nd ed. Prentice-Hall, Englewood, Cliffs, NJ.

Hougen, O. A. and Watson, K. M. 1946. Chemical Process Principles. Part I. Material and Energy Balances. John Wiley & Sons, New York.

McCabe, W. L., Smith, J. C., and Harriott, P. 1985. Unit Operations of Chemical Engineering. 4th ed. McGraw-Hill Book Co., New York.

Seibel, J. E. 1918. Compend of Mechanical Refrigeration and Engineering. 9th ed. Nickerson and Collins, Chicago.

Sinnott, R. K. 1996. Chemical Engineering. Vol. 6. 2nd ed. Butterworth-Heinemann, Oxford, England.

Watson, E. L. and Harper, J. C. 1988. Elements of Food Engineering. 2nd. ed. Van Nostrand Reinhold, New York.

CHAPTER 6

Flow of Fluids

Fluids are substances that flow without disintegration when pressure is applied. This definition of a fluid includes gases, liquids, and certain solids. A number of foods are fluids. In addition, gases such as compressed air and steam are also used in food processing and they exhibit resistance to flow just like liquids. In this chapter, the subject of fluid flow will be discussed from two standpoints: the resistance to flow and its implications in the design of a fluid handling system, and evaluation of rheological properties of fluid foods.

6.1 THE CONCEPT OF VISCOSITY

Viscosity is a measure of resistance to flow of a fluid. Although molecules of a fluid are in constant random motion, the net velocity in a particular direction is zero unless some force is applied to cause the fluid to flow. The magnitude of the force needed to induce flow at a certain velocity is related to the viscosity of a fluid. Flow occurs when fluid molecules slip past one another in a particular direction on any given plane. Thus, there must be a difference in velocity, *a velocity gradient*, between adjacent molecules. In any particular plane parallel to the direction of flow, molecules above and below that plane exert a resistance to the force that propels one molecule to move faster than the other. This resistance of a material to flow or deformation is known as stress. *Shear stress* (τ) is the term given to the stress induced when molecules slip past one another along a defined plane. The velocity gradient ($-dV/dr$ or γ) is a measure of how rapidly one molecule is slipping past another, therefore, it is also referred to as the *rate of shear*. The position from which distance is measured in determining the shear rate is the point in the flow stream where velocity is maximum, therefore, as distance r increases from this point of reference, V decreases and the velocity gradient is a negative quantity. Because shear stress is always positive, expressing the shear rate as $-dV/dr$ satisfies the equality in the equation of shear stress as a function of shear rate. A plot of shear stress against shear rate for various fluids is shown in Fig. 6.1.

Fluids that exhibit a linear increase in the shear stress with the rate of shear (Eq. 6.1) are called *Newtonian fluids*. The proportionality constant (μ) is called the *viscosity*.

$$\tau = \mu\left(-\frac{dV}{dr}\right) \tag{6.1}$$

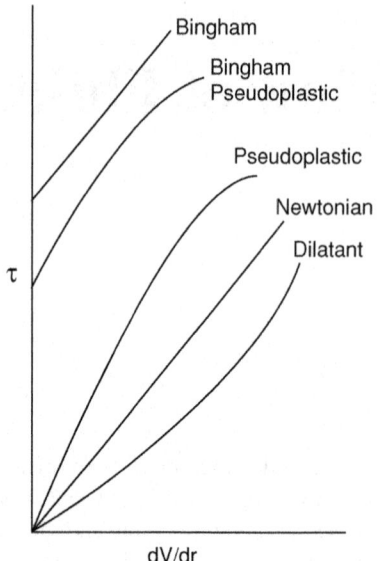

Figure 6.1   A plot showing the relationship between shear stress and shear rate for different types of fluids.

Newtonian fluids are those that exhibit a linear relationship between the shear stress and the rate of shear. The slope μ is constant, therefore the viscosity of a Newtonian fluid is independent of the rate of shear. The term viscosity is appropriate to use only for Newtonian fluids. Fluids with characteristics deviating from Equation (6.1) are called *non-Newtonian* fluids. These fluids exhibit either shear thinning or shear thickening behavior, and some exhibit a yield stress (i.e., a threshold stress that must be overcome before the fluid starts to flow). The two most commonly used equations for characterizing non-Newtonian fluids are the power law model (Eq. 6.2) and the Herschel-Bulkley model for fluids (Eq. 6.3):

$$\tau = K(\gamma)^n \tag{6.2}$$

$$\tau = \tau_0 + K(\gamma)^n \tag{6.3}$$

Equation (6.2) can fit the shear stress versus shear rate relationships of a wide variety of foods. Equation (6.2) can also be used for fluids which exhibit a yield stress as represented by Equation (6.3), if τ_0 is very small compared to τ, or if the shear rate is very high. n is the *flow behavior index* and is dimensionless. K is the *consistency index* and will have units of pressure multiplied by time raised to the *n*th power.

Equations (6.2) and (6.3) can be rearranged as follows:

$$\tau = [K(\gamma)^{n-1}]\gamma \tag{6.4}$$

$$\tau - \tau_0 = [K(\gamma)^{n-1}]\gamma \tag{6.5}$$

Equations (6.4) and (6.5) are similar to Equation (6.1), and the factor that is the multiplier of γ is the apparent viscosity defined as follows:

$$\mu_{app} = K(\gamma)^{n-1} \tag{6.6}$$

For fluids having characteristics which fit Equations. 6.4 and 6.5, the apparent viscosity can also be expressed as:

$$\mu_{app} = \tau/\gamma \tag{6.7}$$

$$\mu_{app} = (\tau - \tau_o)/\gamma \tag{6.8}$$

Substitution of Equation (6.2) for τ in Equation (6.7) also yields Equation (6.6).

The apparent viscosity has the same units as viscosity but the value varies with the rate of shear. Thus, when reporting an apparent viscosity, the shear rate under which it was determined must also be specified.

Equation (6.6) shows that the apparent viscosity will be decreasing with increasing shear rate if the flow behavior index is less than 1. This type of fluids exhibits "shear thinning" behavior and are referred to as *shear thinning fluids* or *pseudoplastic fluids*.

When the flow behavior index of a fluid is greater than 1, μ_{app} increases with increasing rates of shear and the fluid is referred to as *shear thickening* or *dilatant* fluids. Shear thinning behavior is exhibited by emulsions and suspensions where the dispersed phase has a tendency to aggregate following the path of least resistance in the flow stream, or the particles align themselves in a position that presents the least resistance to flow. Shear thickening behavior will be exhibited when the dispersed phase swell or change shape when subjected to a shearing action, or when the molecules are so long that they tend to cross-link with each other, trapping molecules of the dispersion medium. Shear thickening behavior is seldom observed in foods, although it has been reported in suspensions of clay and high molecular weight organic polymers.

Fluids that exhibit a yield stress and a flow behavior index of 1, are referred to as *Bingham plastics*. Some foods exhibit a yield stress and a flow behavior index less than 1, and there is no general term used for these fluids, although the terms *Herschel-Bulkley* and *Bingham pseudoplastic* have been used.

6.2 RHEOLOGY

Rheology is the science of flow and deformation. When a material is stressed, it deforms and the rate and nature of the deformation that occurs characterizes its rheological properties. The science of rheology can be used on solids or fluids. Some food materials exhibit both fluid and solid behavior and these materials are called *viscoelastic*. In general, *flow* indicates the existence of a velocity gradient within the material, and this characteristic is exhibited only by fluids. The subject of rheology in this section covers only flow properties of fluids. In foods, rheology is useful in defining a set of parameters that can be used to correlate with a quality attribute. These parameters can also be used to predict how the fluid will behave in a process and in determining energy requirements for transporting the fluid from one point in a processing plant to another.

6.2.1 Viscometry

Instruments used for measuring flow properties of fluids are called viscometers. Newtonian viscosity can be easily measured since only one shear rate needs to be used and therefore viscometers for this purpose are relatively simple compared to those for evaluating non-Newtonian fluids. Viscometers require a mechanism for inducing flow that should be measurable, a mechanism for measuring the applied force, and the geometry of the system in which flow is occurring must be simple in design such that the force and the flow can be translated easily into a shear stress and shear rate.

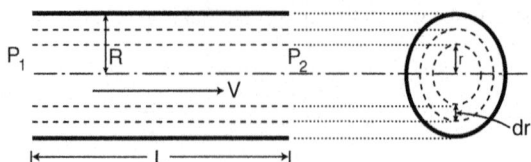

Figure 6.2   Differential control element for analysis of fluid flow through a tube.

6.2.1.1   Viscometers Based on Fluid Flow Through a Cylinder

These viscometers are called capillary or tube viscometers, depending on the inside diameter. The principle of operation is based on the Poiseuille equation, if the fluid is Newtonian. The Rabinowitsch-Mooney equation applies when the fluid is non-Newtonian. The Poiseuille equation will be derived based on the Newtonian flow equation (Eq. 6.1), and the Rabinowitsch-Mooney equation will be derived based on the power law equation (Eq. 6.2).

6.2.1.2   Derivation of the Poiseuille Equation

Figure 6.2 shows a tube of length L and radius R. A pressure P_1 exists at the entrance and P_2 exists at the end of the tube. At a distance r from the center, a ring of thickness dr is isolated, and at this point, the fluid velocity is V, the point velocity. The shear stress at this point is force resisting flow/area of the fluid undergoing shear. The force resisting flow on the fluid occupying the area of the ring is the pressure drop times area of the ring, and the area of the plane where the fluid is sheared is the circumferential area of the cylinder formed when the ring is projected through length, L.

$$\tau = \frac{(P_1 - P_2)(\pi r^2)}{2\pi r L} = \frac{(P_1 - P_2)\, r}{2L} \tag{6.9}$$

Substituting Equation (6.9) in Equation (6.1):

$$\frac{(P_1 - P_2)\, r}{2L} = \mu\left(-\frac{dV}{dr}\right)$$

Separating variables and integrating:

$$\int dV = \frac{(P_1 - P_2)}{2L\mu}\int -r\, dr$$

$$V = \frac{(P_1 - P_2)}{2L\mu}\left(\frac{-r^2}{2}\right) + C$$

The constant of integration can be determined by applying the boundary condition: at $r = R$, $V = 0$. $C = (P_1 - P_2)(R^2)/4L\mu$.

$$V = \frac{(P_1 - P_2)}{4L\mu}(R^2 - r^2) = \frac{\Delta P}{4L\mu}(R^2 - r^2) \tag{6.10}$$

Equation (6.10) gives the point velocity in a flow stream for any fluid flowing within a cylindrical tube. The point velocity is not very easily measured. However, an average velocity defined as the volumetric flow rate/area can be easily measured. Equation (6.10) will be expressed in terms of the

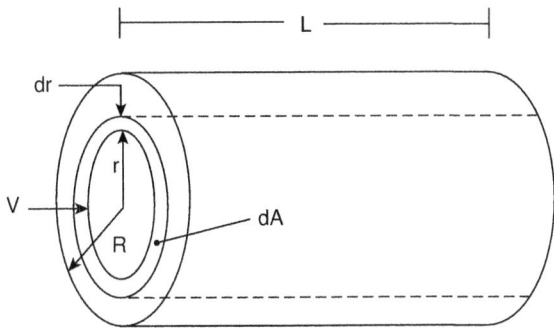

Figure 6.3   Pipe showing ring of thickness dr used as a control element for analysis of fluid flow.

---

average velocity. Consider a pipe of radius R and a control volume that is the walls of a hollow cylinder of thickness dr within the pipe, shown in Fig. 6.3.

The cross-sectional area of the ring of thickness dr is $dA = \pi[(r + dr)^2 - r^2)] = \pi(r^2 + 2rdr + (dr)^2 - r^2) = 2\pi rdr + (dr)^2$. Because dr is small $(dr)^2$ is negligible therefore, $dA = 2\pi rdr$. The volumetric rate of flow (volume/time) through the control volume is $VdA = 2\pi rVdr$. The total volume going through the pipe will be the integral of the volumetric rate of flow through the control volume from $r = 0$ to $r = R$. Substituting Equation (6.10) for V:

$$\bar{V}(\pi R^2) = \frac{(P_1 - P_2)}{4L\mu}(2\pi)\int_0^R (R^2 - r^2)r\,dr$$

Rearranging and integrating:

$$\bar{V} = \frac{(P_1 - P_2)}{2L\mu R^2}\left[\frac{R^2r^2}{2} - \frac{r^4}{4}\right]\Bigg|_0^R$$

Substituting limits, combining terms, and substituting ΔP for $(P_1 - P_2)$:

$$\bar{V} = \frac{(P_1 - P_2)\,R^2}{8L\mu} = \frac{\Delta PR^2}{8L\mu} \tag{6.11}$$

Equation (6.11) is the Poiseuille equation and can be used to determine the viscosity of a Newtonian fluid from pressure drop data when the fluid is allowed to flow through a tube or a capillary.

When Equation (6.11) is used to determine the viscosity of non-Newtonian fluids, the viscosity obtained will be an apparent viscosity. Thus a viscosity obtained from measurements using a single rate of flow is:

$$\mu_{app} = \frac{\Delta PR^2}{8L\bar{V}}$$

Equations (6.10) and (6.11) may be combined to obtain an expression for $V/\underline{V}$ which can be differentiated to obtain a rate of shear as a function of the average velocity.

$$V = \bar{V}\left[\frac{\Delta P}{4L\mu}\right](R^2 - r^2)\left[\frac{8L\mu}{\Delta PR^2}\right]$$

Simplifying:

$$V = 2\bar{V}\left[1 - \left(\frac{r}{R}\right)^2\right]$$ (6.12)

Equation (6.12) represents the velocity profile of a Newtonian fluid flowing through a tube expressed in terms of the average velocity. The equation represents a parabola where the maximum velocity is $2\,\bar{V}$ at the center of the tube ($r = 0$). Differentiating Equation (6.12):

$$\frac{dV}{dr} = 2\bar{V}\left[\frac{2r}{R^2}\right]$$

The shear rate at the wall (r = R) for a Newtonian fluid is

$$-\frac{dV}{dr}\bigg|_w = \frac{4\bar{V}}{R}$$ (6.13)

6.2.1.3   Velocity Profile and Shear Rate for a Power Law Fluid

Equations (6.2) and (6.9) can be combined to give:

$$\frac{(P_1 - P_2)\,r}{2L} = K\left(-\frac{dV}{dr}\right)^n$$

Re-arranging and using ΔP for $(P_1 - P_2)$:

$$\frac{dV}{dr} = \left[\frac{\Delta P}{2LK}\right]^{1/n} (r)^{1/n}$$

Integrating and substituting the boundary condition, V = 0 at r = R:

$$V = \left[\frac{\Delta P}{2LK}\right]^{1/n}\left[\frac{1}{(1/n) + 1}\right]\left[R^{(1/n)+1} - r^{(1/n)+1}\right]$$ (6.14)

Equation (7.14) represents the velocity profile of a power law fluid. The velocity profile equation will be more convenient to use if it is expressed in terms of the average velocity. Using a procedure similar to that used in the section "Derivation of the Poiseuille Equation," the following expression for the average velocity can be derived:

$$\bar{V}(\pi R^2) = \int_0^R 2\pi r\left[\frac{\Delta P}{2LK}\right]^{1/n}\left[\frac{n}{n + 1}\right]\left[R^{(1/n)+1} - r^{(1/n)+1}\right]dr$$

Integrating and substituting limits:

$$\bar{V} = \left[\frac{\Delta P}{2LK}\right]^{1/n}[R]^{(n+1)/n}\left[\frac{n}{3n + 1}\right]$$ (6.15)

The velocity profile in terms of the average velocity is

$$V = \bar{V}\left[\frac{3n + 1}{n + 1}\right]\left[1 - \left[\frac{r}{R}\right]^{(n+1)/n}\right]$$ (6.16)

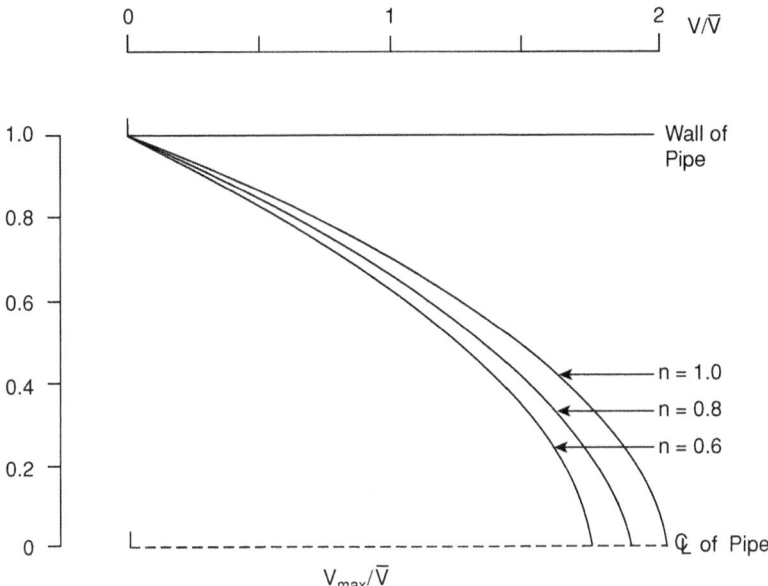

Figure 6.4 Plot of V/V̄ as a function of position in the pipe for fluids with different values of the flow behavior index, n.

Figure 6.4 shows V/V̄ as a function of r/R for various values of n. When $n = 1$, the velocity profile is parabolic and Equations (6.12) and (6.16) give the same results. The velocity profile flattens out near the center of the pipe as n decreases.

Differentiating Equation (6.16) and substituting $r = R$ for the shear rate at the wall:

$$-\frac{dV}{dr}\Big|_w = \bar{V}\left[\frac{3n+1}{n+1}\right]\left[\frac{n+1}{n}\right][R]^{-(n+1)/n}[R]^{1/n}$$

Simplifying:

$$-\frac{dV}{dr}\Big|_w = \bar{V}\left(\frac{3n+1}{n}\right)\left(\frac{1}{R}\right) \tag{6.17}$$

Equation (6.17) is a form of the Rabinowitsch-Mooney equation used to calculate shear rates for non-Newtonian fluids flowing through tubes. Converting into a form similar to Equation (6.13):

$$-\frac{dV}{dr}\Big|_w = \frac{4\bar{V}}{R}\left[\frac{3}{4} + \frac{1}{4n}\right] \tag{6.18}$$

Equation (6.18) shows that the shear rate at the wall for a non-Newtonian fluid is similar to that for a Newtonian fluid except for the multiplying factor $(0.75 + 0.25/n)$.

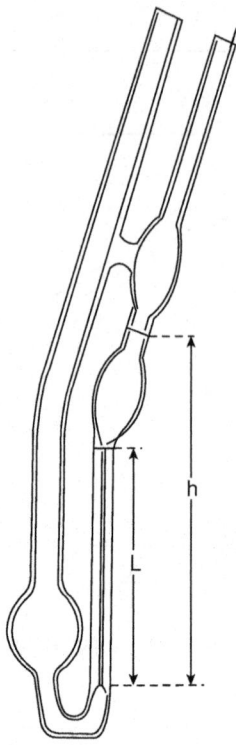

Figure 6.5 Diagram of a glass capillary viscometer showing the components and parameters used in fitting viscometer data to the Poiseuille equation.

6.2.1.4 Glass Capillary Viscometers

The simplest viscometer operate on gravity induced flow, and are commercially available in glass. Figure 6.5 shows a Cannon-Fenske type viscometer. These viscometers are available with varying size of the capillary. The size of capillary is chosen to minimize the time of efflux for viscous fluids. The Poiseuille equation (Eq. 6.11) is used to determine the viscosity in these viscometers. The pressure drop needed to induce flow is

$$\Delta P = \rho g h$$

Where ρ is the density of the fluid and h is the height available for free fall in the viscometer (Fig. 6.5). The viscosity is then calculated as using the above expression for ΔP in Equation (6.13) as follows:

$$\frac{\mu}{\rho} = \frac{g h R^2}{8 L \bar{V}} \tag{6.19}$$

The ratio μ/ρ is the *kinematic viscosity*.

The viscosity is obtained from the time of efflux of the fluid through the viscometer. To operate the viscometer, fluid is pipetted into the large leg of the viscometer until the lowermost bulb is about half full. Fluid is then drawn into the small leg to about half of the uppermost bulb. When the suction on the small leg is released, fluid will flow and timing is started when the fluid meniscus passes through the first mark. When the fluid meniscus passes the second mark, timing is stopped and the efflux time is recorded. The lower portion of the small leg of the viscometer is a capillary of radius R and length L. If t_e is the time of efflux, the average velocity $\bar{V}$ will be L/t_e. Substituting in Equation (6.18):

$$\frac{\mu}{\rho} = \frac{ghR^2}{8L}t_e \qquad (6.20)$$

For each viscometer, the length and diameter of capillary, and height available for free fall, are specific, therefore, these factors can be grouped into a constant , k_v, for a particular viscometer. The kinematic viscosity can then be expressed as:

$$\frac{\mu}{\rho} = k_v t_e \qquad (6.21)$$

The viscometer constant is determined from the efflux time of a fluid of known viscosity and density.

Example 6.1. A glass capillary viscometer when used on a fluid with a viscosity of 10 centipoises allowed the fluid to efflux in 1.5 minutes. This same viscometer used on another fluid allowed an efflux time of 2.5 minutes. If the densities of the two fluids are the same, calculate the viscosity of the second fluid.

Solution:

Using Equation (6.21) to solve for the viscometer constant:

$$k_v = \frac{\mu}{\rho t_e} = \frac{10}{1.5\rho}$$

For the second fluid:

$$\mu = \rho \left[\frac{10}{1.5\rho} \right] (2.5) = 16.67 \text{ centipoises}$$

6.2.1.5 *Forced Flow Tube or Capillary Viscometry*

For very viscous fluids, gravity induced flow is not sufficient to allow measurement of viscosity. Forced flow viscometers are used for these fluids. In order to obtain varying flow rates in viscometers, some means must be provided to force a fluid through the viscometer at a constant rate. This may be obtained by using a constant pressure and measuring the flow rate that develops, or by using a constant flow rate and measuring the pressure drop over a length of test section. Flow rates through glass capillary viscometers may also be varied by applying pressure instead of relying solely on gravity flow.

The flow properties of a fluid are the constants in Equations (6.1) to (6.3), which can be used to characterize the relationship between the shear stress and the rate of shear. These are the viscosity, if the fluid is Newtonian, the flow behavior index, the consistency index, and the yield stress. Equation (6.3) is most general. If a fluid has a yield stress, the test is carried out at very high rates of shear in order that τ_0 is very small in comparison to τ and Equation (6.3) simplifies to Equation (6.2). A fluid with properties that fits Equation (6.1) is a special case of a general class of fluids which fits Equation (6.2), when n = 1. Evaluation of τ_0 in Equation (6.3), from data at low shear rates is simple, once n is established from data at high shear rates. The shear rate at the wall may be determined using Equations (6.13) or (6.18). These equations show that the shear rate is the product of some factor which is independent of the rate of flow, and another factor which is a function of the rate of flow, i.e the average velocity, $\bar{V}$, the volumetric rate of flow Q, or even the speed of a piston V_p which delivers fluid to the tube. In equation form:

$$\gamma_w = F_1\bar{V} = F_2Q = F_3V_p$$

The shear stress at the wall is determined by substituting R for r in Equation (6.9).

$$\tau_w = \frac{\Delta PR}{2L} \tag{6.22}$$

Equation (6.22) shows that τ_w is a product of a factor which is independent of the rate of flow, and the pressure drop. In equation form: $\tau_w = F_{p1}(\Delta P) = F_{p2}$(height of manometer fluid) $= F_{p3}$(transducer output). Substitution of any of the above expressions for τ_w and Δ_w into Equation (6.2):

$$PF_{p1} = [\bar{V}F_1]^n$$

P and $\bar{V}$ are, respectively, measurements from which pressure or velocity can be calculated. Taking logarithms of both sides:

$$\log P + \log F_{p1} = n \log \overline{linbe}V + n \log F_1$$

Rearranging:

$$\log P = n \log \bar{V} + (n \log F_1 - \log F_{p1})$$

Thus, a log-log plot of any measure of pressure against any measure of velocity can give the flow behavior index n for a slope.

The shear rate can then be calculated using either Equation (6.13) or (6.18), and the shear stress can be calculated using Equation (6.22). Taking the logarithm of Equation (6.2): $\log \tau_w = \log K + n \log \gamma_w$. The intercept of a log-log plot of τ_w against γ_w will give K.

Example 6.2. A tube viscometer having an inside diameter of 1.27 cm and a length of 1.219 m is used to determine the flow properties of a fluid having a density of 1.09 g/cm^3. The following data were collected for the pressure drop at various flow rates measured as the weight of fluid discharged from the tube. Calculate the flow behavior and consistency indices for the fluid.

Data with pressure drop (in kPa)	
$(P_1 - P_2)$	Flow rate (g/s)
19.197	17.53
23.497	26.29
27.144	35.05
30.350	43.81
42.925	87.65

Solution:

Figure 6.6 shows a plot of log ΔP against log (mass rate of flow). The slope, n = 0.5, indicates that the fluid is non-Newtonian. Equation (6.18) will be used to solve for γ_w, and Equation (6.22) for τ_w.

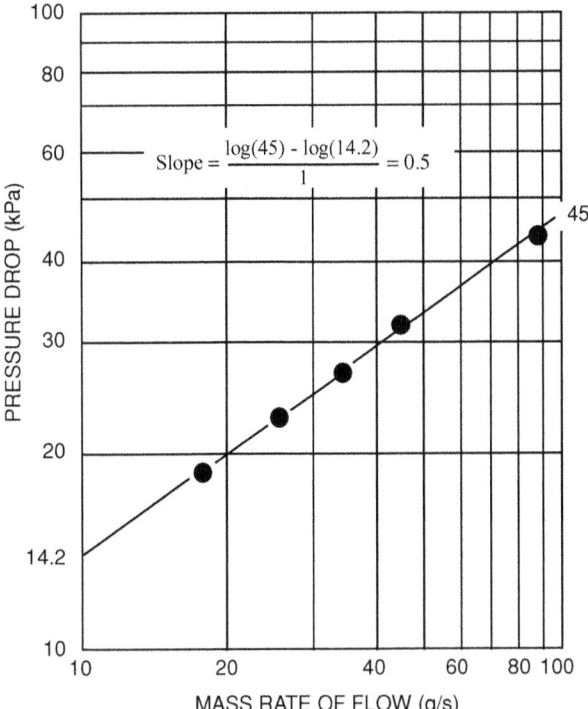

Figure 6.6 Log-log plot of pressure drop against mass rate of flow to obtain the flow behavior index from the slope.

Solving for τ_w: R = 0.5(1.27 cm)(0.01 m/cm) = 0.00635 m; L = 1.219 m.

$$\tau_w = [0.00635(0.5)/1.219]\,\Delta P = 0.002605\,\Delta P\,\text{Pa}$$

The average velocity $\bar{V}$ in m/s can be calculated by dividing the mass rate of flow by the density and the cross-sectional area of the tube. Let q = mass rate of flow in g/s.

$$\bar{V} = q\frac{\text{g}}{\text{s}}\frac{\text{cm}^3}{1.09\,\text{g}}\frac{\text{m}^3}{(100)^3\,\text{cm}^3}\frac{1}{\pi(0.00635)^2\,\text{m}^2}$$
$$= 0.007242\,q\;\text{m/s}$$

Equation (6.18) is used to calculate the shear rate at the wall. n = 0.5.

$$\gamma_w = \frac{4\bar{V}}{R}\left(0.75 + \frac{0.25}{n}\right) = \frac{4(0.007242\,q)}{0.00635}(1.25)$$
$$= 5.7047\,q$$

The shear stress and shear rates are

τ_w (Pa)	γ_w (1/s)
50.008	99.966
61.209	149.92
70.710	199.87
79.062	249.83
111.82	499.83

Equation (6.2) will be used to determine K. The yield stress τ_o is assumed to be much smaller than the smallest value for τ_w measured. Thus, Equation (6.3) reduces to the same form as Equation (6.2), the logarithm of which is as follows:

$$\log \tau_w = \log K + n \log \gamma_w$$

A log-log plot of shear stress against the shear rate is shown in Fig. 6.7. The intercept is the consistency index,

$$k = 5 \text{ Pa} \cdot \text{s}^n.$$

Example 6.3. A fluid induces a pressure drop of 700 Pa as it flows through a tube having an inside diameter of 0.75 cm and a length of 30 cm at a flow rate of 50 cm^3/s.

(a) Calculate the apparent viscosity defined as the viscosity of a Newtonian fluid that would exhibit the same pressure drop as the fluid at the same rate of flow through this tube.
(b) This same fluid flowing at the rate of 100 cm^3/s through a tube 20 cm long and 0.75 cm in inside diameter, induces a pressure drop of 800 Pa. Calculate the flow behavior and consistency indices for this fluid.
(c) What would be the shear rates at 50 and 100 cm^3/s?

Solution:

Equation (6.11) will be used to calculate the apparent viscosity from the pressure drop of a fluid flowing through a tube.

(a) $R = (0.0075)(0.5) = 0.00375 \text{ m}; L = 0.3 \text{ m}; \Delta P = 700 \text{ Pa}.$

$$\bar{V} = 50 \frac{\text{cm}^3}{\text{s}} \frac{1 \text{ m}^3}{(100)^3 \text{ cm}^3} \frac{1}{\pi (0.00375)^2} = 1.1318 \text{ m/s}$$

$$\mu_{app} = \frac{\Delta P R^2}{8 L \bar{V}}$$

$$\mu_{app} = \frac{700(0.00375)^2}{8(0.3)(1.1318)} = 0.003624 \text{ Pa} \cdot \text{s}$$

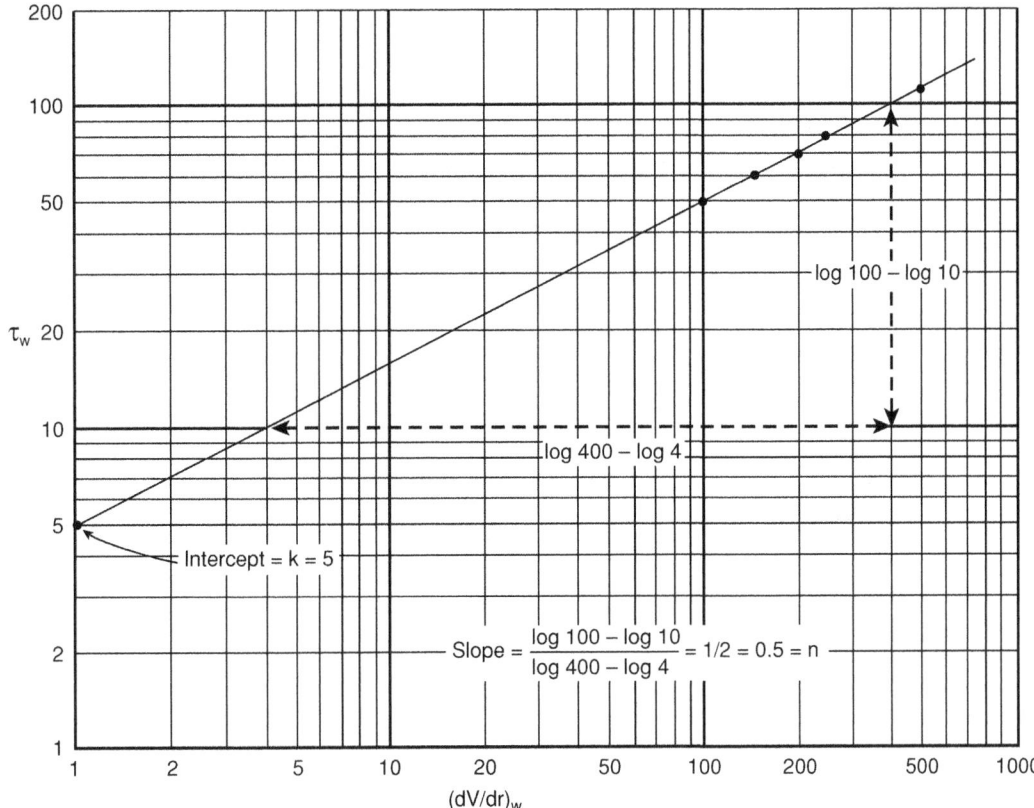

Figure 6.7 Log-log plot of shear stress against shear rate to obtain the consistency index from the intercept.

(b) $\bar{V} = 100 \frac{\text{cm}^3}{\text{s}} \dfrac{1\,\text{m}^3}{(100)^3\,\text{cm}^3} \dfrac{1}{\pi(0.00375)^2} = 2.2635\ \text{m/s}$

$\mu_{\text{app}} = \dfrac{800(0.00375)^2}{8(0.2)(2.2635)} = 0.003106\ \text{Pa}\cdot\text{s}$

Equation (6.6) will be used to solve for K and n.

Let $\varphi = 0.75 + 0.25/\text{n}$

Equation (6.18): $\gamma_w = \dfrac{4\bar{V}}{R}(\phi)$

Equation (6.6) : $\mu_{\text{app}} = K(\gamma)^{n-1}$

Substituting ϕ : $\mu_{\text{app}} = K\left[\dfrac{4\phi}{R}\right]^{n-1}\bar{V}^{(n-1)}$

Using the subscripts 1 and 2 for the velocity and apparent viscosity at 50 and 100 cm^3/s, respectively:

$$\frac{\mu_{app}1}{\mu_{app}2} = \frac{(K(4\phi/R)^{n-1}\bar{V}_1)^{n-1}}{(K(4\phi/R)^{n-1}\bar{V}_2)^{n-1}} = \left(\frac{V_1}{V_2}\right)^{n-1}$$

$$n = 1 + \frac{\log(1.16674)}{\log(0.5009)}$$

$$\log(1.16674) = (n-1)\log(0.5009)$$

$$\log\left(\frac{0.003624}{0.003106}\right) = (n-1)\log\left(\frac{1.1318}{2.2635}\right) = 0.777$$

$$\phi = 0.75 + \frac{0.25}{0.777} = 1.072$$

Using Equation (6.6) on either of the two apparent viscosities:

$$\gamma_w = \frac{4(1.1317)}{0.00375}(1.072) = 1294 \text{ s}^{-1}$$

Using Equation (6.6):

$$K = \frac{0.003624}{(1294)^{0.777-1}} = 0.0178 \text{ Pa} \cdot \text{s}^n$$

(c) At 50 cm^3/s, $\gamma_w = 1294$ s^{-1}

At 100 cm^3/s, $\gamma_w = \dfrac{4(2.2635)}{0.00375}(1.072) = 2414$ s^{-1}

6.2.1.6 Evaluation of Wall Effects in Tube Viscometry

The equations derived in the previous sections for evaluating flow properties of fluids by tube or capillary viscometry are based on the assumption that there is zero slip at the wall. This condition may not always be true for all fluids, particularly suspensions.

To determine if slip exists at the wall, it will be necessary to conduct experiments on the same fluid using viscometers of different radius. Kokini and Plotchok (1987) defined a parameter, β_c, which can be used to correct for slip at the tube wall. β_c is the slope of a plot of $Q/(\pi R^3 \tau_w)$ again $1/R^2$ obtained on the same fluid using different viscometers of varying radius at flow rates which would give a constant shear stress at the wall, τ_w. Q = the volumetric rate of flow, and R is the viscometer radius. With β_c known, a corrected volumetric flow rate Q_c is calculated as follows:

$$Q_c = Q - \pi R \tau_w \beta_c \tag{6.23}$$

The corrected flow rate Q_c is then used instead of the actual flow rate in determining the mean velocity and shear rates in the determination of the flow behavior index and the consistency index of fluids.

The experiments will involve a determination of pressure drop at different rates of flow on at least four tubes, each with different radius. The value of β_c may vary at different values of τ_w, therefore, for each tube, data are plotted as log Q against log τ_w. The four linear plots representing data obtained on the four tubes are drawn on the same graph, and at least four values of τ_w are selected. Points on each line at a constant value of τ_w, are then selected. At each value of τ_w, a plot is made of $Q/(\pi\,R^3\tau_w)$ against $1/R^2$. The plot will be linear with a slope, β_c. The value of β_c may be different at different τ_w, therefore, plots at 4 different values of τ_w are needed. After determining β_c values at each τ_w, Equation (6.23) is used to determine Q_c, which is then used to determine Q_w. An example is given in Kokini and Plotchok's (1987) article.

6.2.1.7   Glass Capillary Viscometer Used as a Forced Flow Viscometer

Fluids that exhibit non-Newtonian behavior but have low consistency indices are usually difficult to study using tube viscometers because of the low pressure drop. A glass capillary viscometer may be used as a forced flow viscometer by attaching a constant pressure source to the small leg of the viscometer. A simple set-up is shown in Fig. 6.8.

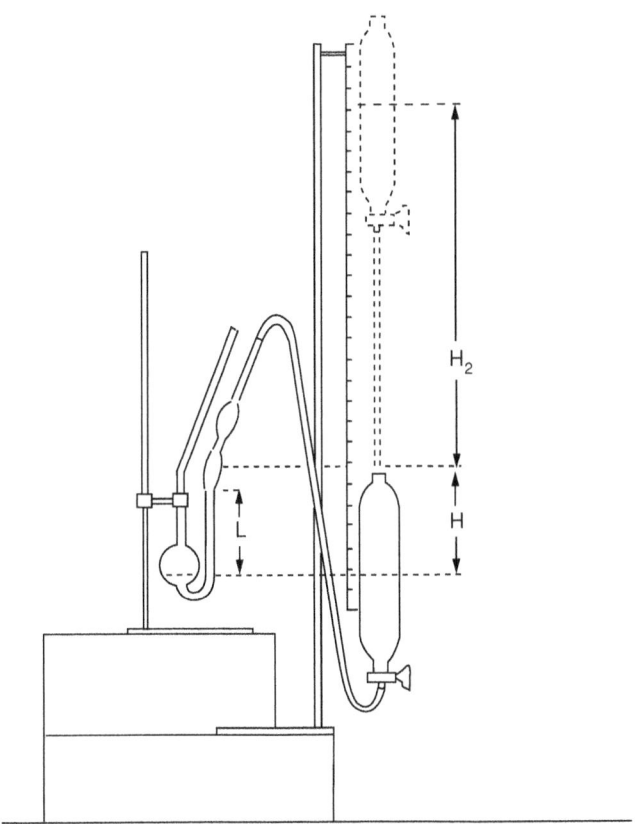

Figure 6.8   Diagram of a system used to adapt a glass capillary viscometer for forced flow viscometry.

The pressure source is a water column of movable height. To operate, the fluid is pipetted into the large leg of the viscometer. The fluid level in the u-tube is made level with the fluid in the large bulb of the viscometer by adjusting the height of the dropping funnel and the stopcock is shut-off. The viscometer is disconnected from the pressure source and fluid is drawn to the top of the second bulb on the small viscometer leg by suction. After reconnecting the viscometer to the pressure source, the height of the dropping funnel is raised and pressure is applied by opening the stopcock. The time of efflux is then measured as the time for the fluid meniscus to pass through the first and second mark on the viscometer. Pressure is discontinued by closing the stopcock and the difference in fluid level between the dropping funnel and the large bulb of the viscometer is measured. The flow behavior and consistency indices will be calculated from the efflux times at different levels of fluid in the pressure source.

In a capillary viscometer, the volume of fluid which passes through the capillary is the volume which fills the section between the two marks on the viscometer. Two other critical factors are the length of the capillary (L) and the height available for free fall (H). These measurements are made on the viscometer as indicated in Fig. 6.8. The radius of the capillary may be calculated from the viscometer constant using Equation (6.19). When used as a forced flow viscometer, the pressure drop will be the pressure equivalent to the height of the forcing fluid as shown in Fig. 6.8 and the pressure equivalent to the height available for free fall of the fluid in the viscometer. The calculations are shown in the following example.

Example 6.4. A Cannon-Fenske type glass capillary viscometer has a height available for free fall of 8.81 cm and a length of capillary of 7.62 cm. When used on distilled water at 24°C, an efflux time of 36 seconds was measured. The volume of fluid which drains between the two marks on the viscometer was 3.2 cm^3.

(a) Calculate the radius of the capillary.
(b) When used as a forced flow viscometer on a concentrated acid whey containing 18.5% total solids at 24°C, the following data were collected:

Height of (H_2) forcing fluid (cm)	Efflux time (t_e) (s)
0	92
7.1	29
20.8	15
34.3	9

The whey has a density of 1.0763 g/cm^3. Calculate the flow behavior and consistency indices.

Solution:

(a) At 24°C, the density and viscosity of water are 947 kg/m^3 and 0.00092 Pa · s, respectively. Using Equation (6.11):

$$\mu = \frac{\Delta PR^2}{8L\bar{V}}; \Delta P = \rho gH$$

$$\mu = \frac{\rho gHR^2 \pi R^2 t_e}{3.2 \times 10^{-6}(8)L} \quad R^4 = \frac{(3.2 \times 10^{-6})\, 8\, L\mu}{\rho gH\pi t_e}$$

$$\bar{V} = \frac{\text{Volume}}{\pi}R^2 t_e$$

Substituting known values and solving for R:

$$R = \left[\frac{0.00092\,(3.2 \times 10^{-6})(8)(0.0762)}{997\,(9.8)(0.0881)(\pi)(36)} \right]^{0.25}$$

$$= 3.685 \times 10^{-4} \text{ m}$$

(b) The total pressure forcing the fluid to flow is: $\Delta P = g(\rho_1 H + \rho_2 H_2)$

$$\Delta P = 9.8\,[(1076.3)(0.0881) + (997)(H_2)] = 9.8(94.822 + 997H_2) \text{ Pa}$$

Let q = volumetric rate of flow.

$$q = \frac{3.2}{t_e} \text{ cm}^3/\text{s}$$

The pressure inducing flow, and the volumetric rate of flow calculated using the above equations are

ΔP (Pa)	q (cm^3/s)
929	0.0564
1623	0.110
2961	0.213
4280	0.355

A log-log plot of ΔP against q shown in Fig. 6.9 gives $n = 0.84$.
   The consistency index will be calculated from the rates of shear and the shear stress.

$$\bar{V} + \frac{q \times 10^{-6}}{\pi(R^2)} = 2.3466\,q$$

$$\gamma_w = \frac{4\bar{V}}{R}\left[0.75 + \frac{0.25}{n}\right] = 26699\,q$$

$$= \frac{4(2.3466)\,q}{3.683 \times 10^{-4}}\left[0.75 + \frac{0.25}{0.84}\right]$$

The shear stress is calculated from ΔP. Substituting known values in Equation (6.22):

$$\tau_w = \frac{\Delta PR}{2L} = \frac{\Delta P(3.683 \times 10^{-4})}{2(0.00762)} = 24.1666 \times 10^{-4}\Delta P$$

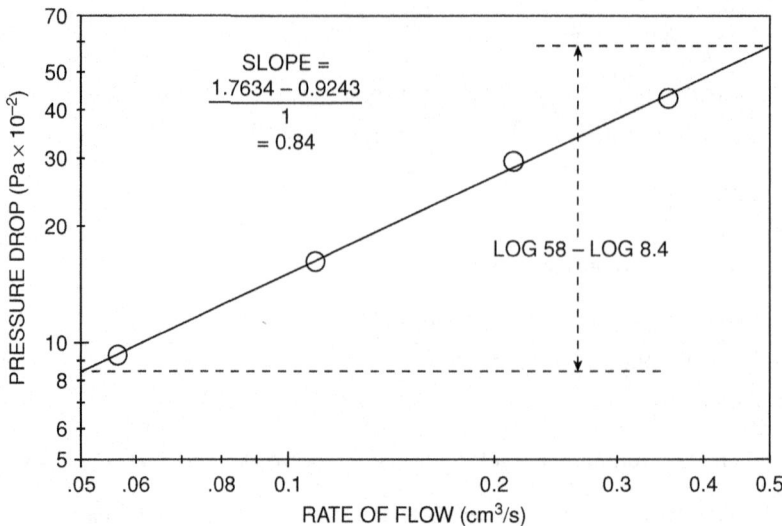

Figure 6.9 Log-log plot of pressure drop against the rate of flow to determine the flow behavior index, n, from the slope.

The shear stress and shear rates are

τ_w (Pa)	γ_w (1/s)
2.245	1505
3.922	2936
7.155	5687
10.343	9478

Figure 6.10 shows a log-log plot of shear stress against shear rate. A point on the line can be used to calculate K. The first data point (2.245,1505) falls directly on the line.

$$K = \frac{\tau_w}{(\gamma_w)^n} = \frac{2.245}{(1505)^{0.84}} = 0.0048 \; \text{Pa} \cdot \text{s}^n$$

6.2.2 Effect of Temperature on Rheological Properties

Temperature has a strong influence on the resistance to flow of a fluid. It is very important that temperatures be maintained constant when making rheological measurements. The flow behavior index, n, is relatively constant with temperature, unless components of the fluid undergo chemical changes at certain temperatures. The viscosity and the consistency index on the other hand, are highly temperature dependent.

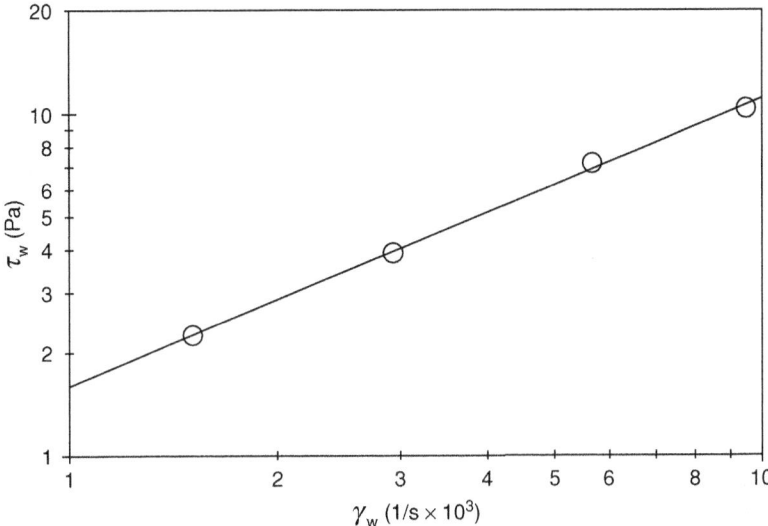

Figure 6.10   Log-log plot of shear stress against shear rate to determine a point on the line that can be used to calculate the consistency index, K.

The temperature dependence of the viscosity and the consistency index can be expressed in terms of the Arrhenius equation:

$$\ln\left(\frac{\mu}{\mu_1}\right) = \frac{E_a}{R}\left(\frac{1}{T} - \frac{1}{T_1}\right) \tag{6.24}$$

where μ = the viscosity at absolute temperature T; μ_1 = viscosity at temperature T_1; E_a = the activation energy, J/gmole; R = the gas constant, 8.314 J/(gmole K); and T is the absolute temperature.

Equation (6.24) is useful in interpolating between values of μ at two temperatures. When data at different temperatures are available, Equation (6.24) may be expressed as:

$$\ln\mu = A + \frac{B}{T} \tag{6.25}$$

where the constants B and A are slope and intercept respectively of the plot of $\ln(\mu)$ against $1/T$. The same expressions may be used for the temperature dependence of the consistency index, K.

There are other expressions for the dependence of the viscosity or consistency index on temperature such as the Williams-Landel-Ferry (WLF) equation and the Fulcher equation (Rao et al. 1987), but these also have limitations for general use and determination of the parameters from experimental data requires specialized curve-fitting computer programs.

The WLF equation however, has implications in the flow behavior of polymer solutions and the viscoelastic properties of gels, therefore, and the basis for the model will be briefly discussed. The model is based on the concept of a free volume. Any polymer, by virtue of the size of the molecule, contains a free volume. The free volume may include the space within the helix in a protein or starch molecule, or the space formed when sections of a random coil mesh. The magnitude of the free volume is temperature dependent. At a specific temperature called the glass transition temperature, the free volume is essentially zero, and this will be manifested by a drastic change in the thermophysical properties of the material. The free volume is best manifested by the specific volume. A plot of the specific volume with temperature above the glass transition temperature is continuous and the slope defines the thermal expansion coefficient. The point in the curve where a discontinuity occurs, is the glass transition temperature. The WLF model relates the ratio, μ/μ_g, where μ = the viscosity at temperature T and μ_g = the viscosity at the glass transition temperature, T_g, to a function of $T - T_g$. The Arrhenius equation is the simplest to use and is widely applied in expressing the temperature dependence of a number of factors.

Example 6.5. The flow behavior index of applesauce containing 11% solids is 11.6 and 9.0 Pa · s^n, at 30°C and 82°C respectively. Calculate the flow behavior index at 50°C.

Solution:

Using Equation (6.24): T = 303 K; T_1 = 355 K

$$\frac{E_a}{R} = \frac{\ln(11.6/9.0)}{1/303 - 1/355} = 524.96$$

$$K = 11.6[e]^{524.96(1/323 - 1/303)}$$

Using Equation (6.24) with k_1 = 11.6, T_1 = 303 K, and T = 323 K.

$$K = 11.6(.898) = 10.42 \, \text{Pa} \cdot \text{s}^n$$

6.2.3   Back Extrusion

Back extrusion is another method for evaluating the flow properties of fluids. It is particularly useful with materials that have the consistency of a paste or suspensions that have large particles suspended in them, since the suspended solids tend to intensify wall effects when the fluid is flowing in a small tube. When a rotational viscometer is used, pulsating torque readings are usually exhibited when a lumpy fluid is being evaluated.

The key to accurate determination of fluid flow properties using back extrusion is the assurance of annular flow, i.e. the plunger must remain in the center of the larger stationary cylinder all throughout the test. Figure 6.11 shows a back extrusion device designed for use on an Instron universal testing machine. A stanchion positioned at the center of the large cylinder which is machined to a close tolerance to fit an opening at the center of the plunger, guides the movement of the plunger and ensures concentric positioning of the plunger and outer cylinder at all times during the test.

The equations that govern flow during back extrusion have been derived by Osorio and Steffe (1987).

Let F_t = the force recorded at the maximum distance of penetration of the plunger L_p. The height of the column of fluid which flowed through the annulus, L is

$$L = \frac{A_p L_p}{A_a}$$

where A_p is the area of the plunger which displaces fluid, and A_a is the cross-sectional area of the annulus.

$$A_p = \pi(R_p^2 - R_s^2); \quad A_a = \pi(R_c^2 - R_p^2)$$

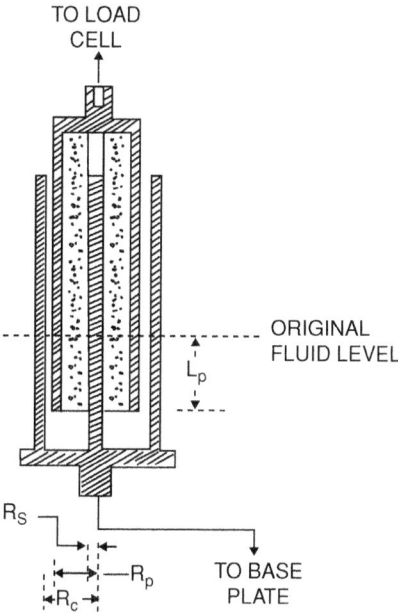

TO LOAD
CELL

ORIGINAL
FLUID LEVEL

L_p

R_S

R_p

R_c

TO BASE
PLATE

Figure 6.11   Diagram of a back extrusion cell suitable for measuring flow properties of fluids with lumpy or paste-like consistency.

where R_c is the radius of the cylinder, R_p is the radius of the plunger, and R_s is the radius of the stanchion which positions the plunger in the center of the cylinder. L_p cannot easily be measured on the plunger and cylinder assembly itself since the fluid level in the cylinder is not visible outside, but it can easily be read from the force tracing on the Instron chart, by measuring the distance from the initiation of the rise in force and the point of maximum force on the chart. Figure 6.12 shows a typical back extrusion force-displacement curve.

Part of the total force F_t counteracts the force of gravity on the mass of fluid in the annulus.

Force of gravity on annular fluid $= \rho g\, L\pi(R_c{}^2 - R_p{}^2)$

The net force, F_n, which is exerted by the fluid to resist flow through the annulus, is the total force corrected for the force of gravity.

$$F_n = F_t - \rho gL\pi(R_c^2 - R_p^2)$$

where ρ = density of the fluid.

The flow behavior index, n, is the slope of a log-log plot of F_n/L, against the velocity of the plunger, V_p.

Determination of the shear rate at the plunger wall and the shear stress at this point involves a rather complex set of equations which were derived by Osorio and Steffe (1987). The derivation is based on a dimensionless radius, λ, which is the ratio r_m/R_c, where r_m = radius of a point in the flow stream where velocity is maximum. The deferential momentum and mass balance equations were solved simultaneously to obtain values of λ for different values for the flow behavior index of fluids, and

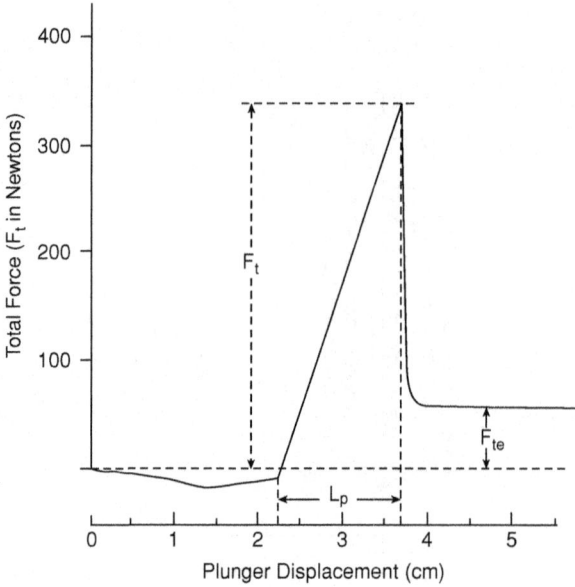

Figure 6.12 Typical force deformation curve obtained for back extrusion showing how total force, plunger displacement, and yield force are obtained.

different annular gaps. The authors presented results of the calculations as a Table of λ for different n and annular gap size expressed as σ, the ratio of plunger to cylinder radius (R_p/R_c). Table 6.1 shows these values. Figure 6.13 related the value of σ to a parameter θ for fluids with different n. The parameter θ is used to calculate the shear rate at the wall from the plunger velocity, V_p.

Table 6.1   Values of λ for Different Values of σ and n.

	n									
σ	0.1	0.2	0.3	0.4	0.5	0.6	0.7	0.8	0.9	1.0
0.1	0.4065	0.4889	0.5539	0.6009	0.6344	0.6586	0.6768	0.6907	0.7017	0.7106
0.2	0.5140	0.5680	0.6092	0.6398	0.6628	0.6803	0.6940	0.7049	0.7138	0.7211
0.3	0.5951	0.6313	0.6587	0.6794	0.6953	0.7078	0.7177	0.7259	0.7326	0.7382
0.4	0.6647	0.6887	0.7068	0.7206	0.7313	0.7399	0.7469	0.7527	0.7575	0.7616
0.5	0.7280	0.7433	0.7547	0.7636	0.7705	0.7761	0.7807	0.7846	0.7878	0.7906
0.6	0.7871	0.7962	0.8030	0.8082	0.8124	0.8158	0.8186	0.8209	0.8229	0.8246
0.7	0.8433	0.8480	0.8516	0.8544	0.8566	0.8584	0.8599	0.8611	0.8622	0.8631
0.8	0.8972	0.8992	0.9007	0.9019	0.9028	0.9035	0.9042	0.9047	0.9052	0.9055
0.9	0.9493	0.9498	0.9502	0.9504	0.9507	0.9508	0.9510	0.9511	0.9512	0.9513

From: Osorio and Steffe (1987); used with permission.

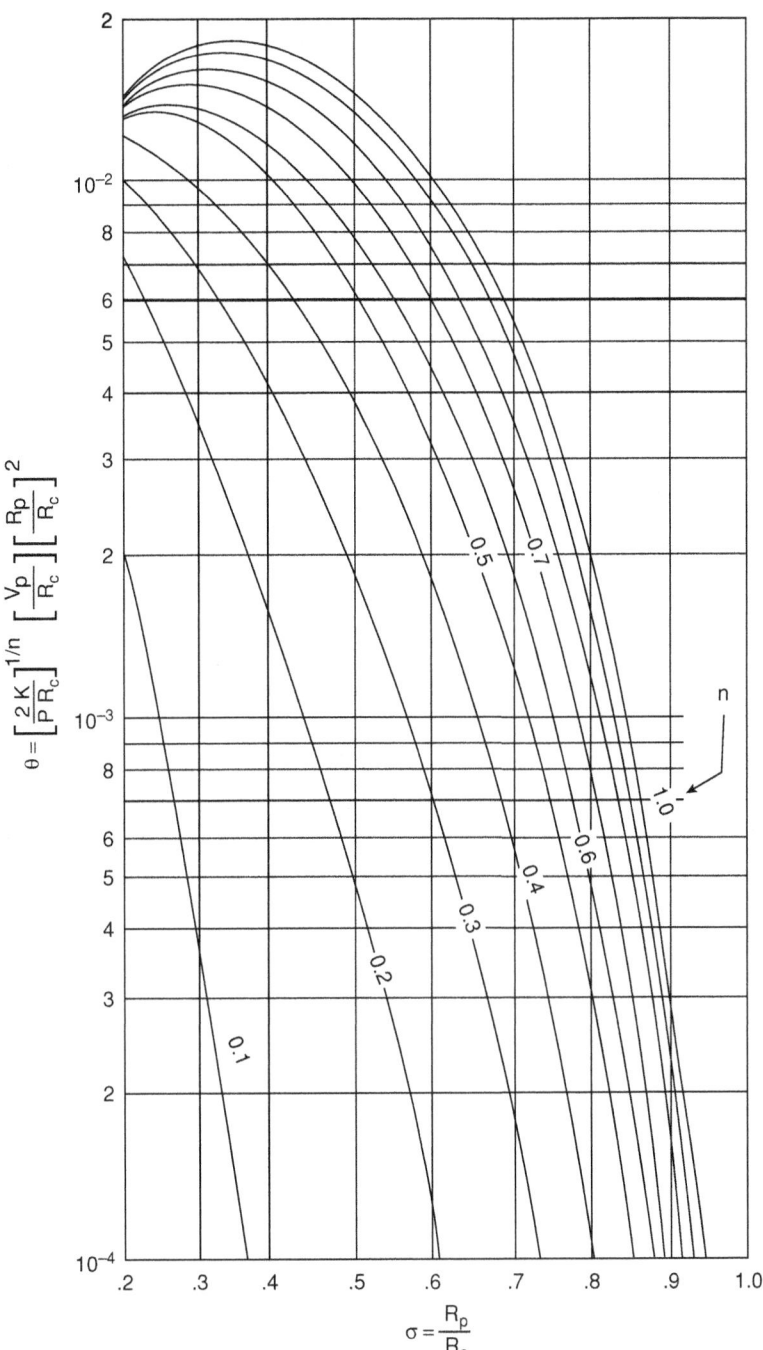

Figure 6.13 Graph of the parameter θ used to determine the shear rate from the plunger velocity. (Source: Osorio, F. A. and Steffe, J. F. 1987. Back extrusion of power law fluids. J. Tex. Studies. 18:43. Used with permission.)

Let : $\sigma = R_p/R_c$; $\sigma_s = R_s/R_c$; $\lambda = r_m/R_c$

The parameter θ in Fig. 6.12 is a dimensionless parameter which satisfies the expressions for the integral of the dimensionless point velocity of fluid through the annulus and the quantity of fluid displaced by the plunger. The terms V_p, P, and K in the expression of θ in Fig. 6.13 are V_p = plunger velocity, P = pressure drop per unit distance traveled by the fluid, and k = the fluid consistency index.

Knowing the flow behavior index of the fluid, and the dimensionless plunger radius, σ, a value for λ is obtained from Table 6.1. Figure 6.13 is then used to obtain a value for θ. The shear rate at the plunger wall is calculated using Equation (6.26).

$$-\frac{dV}{dr}\bigg|_{R_p} = \frac{V_p}{R_c\theta}(\sigma^2 - \sigma_s{}^2)\left[\frac{\lambda^2}{\sigma} - \sigma\right]^{\frac{1}{n}} \tag{6.26}$$

The shear stress at the plunger wall is calculated from the net force, F_n, exerted on the base of the column of fluid to force it up the annulus.

The shear stress at the wall, τ_w, is calculated using Equation (6.27). Equations (6.26) and (6.27) are derived to account for the fact that part of the plunger area being occupied by the stanchion which positions the plunger in the center of the concentric cylinder, does not displace fluid up through the annular gap.

$$\tau_w = \frac{F_n}{2\pi LR_p}\left(\frac{\lambda^2 - \sigma^2}{\lambda^2 - \sigma^2}\right) \tag{6.27}$$

The consistency index of the fluid can then be determined from the intercept of a log-log plot of shear stress and shear rate.

The yield stress can be calculated from the residual force after the plunger has stopped, F_{ne} corrected for the hydrostatic pressure of the height of fluid in the annulus.

$$\tau_0 = \frac{(F_{ne})}{2\pi(R_p + R_c)L} \tag{6.28}$$

where $F_{ne} = F_{te} - \rho gL\pi(R_c^2 - R_p^2)$ and F_{te} = the residual force after the plunger has stopped.

Example 6.6. The following data were collected in back extrusion of a fluid through a device similar to that shown in Fig. 6.8. The diameters were: 2.54 cm for the stanchion, 6.64 cm for the plunger, and 7.62 cm for the outer diameter. The fluid had a density of 1.016 g/cm^3. The total force, the distance penetrated by the plunger, and the plunger speed at each of the tests are as follows:

Plunger speed, V_p (mm/min)	Penetration depth, L_p (cm)	Total force, F_p (N)
50	15.2	11.44
100	13.9	12.16
150	16.0	15.36
200	13.2	13.60

Calculate the flow behavior and consistency indices for this fluid.

Solution:

The flow behavior index is calculated by determining the ratio F_n/L and plotting against V_p.

$$L = L_p \frac{[(0.0332)^2 - (0.0127)^2]}{[(0.0381)^2 - (0.0332)^2]}$$

$$= 2.6905 L_p$$

$$F_n = F_t - 1016(9.8)(2.6905 L_p)(\pi)[(0.0381)^2 - (0.0332)^2]$$

$$= F_t - 29.4028 L_p$$

Values of F_n, L, and F_n/L at different V_p are

V_p (mm/min)	L (m)	F_n (N)	F_n/L (N/m)
50	0.408	6.98	17.1
100	0.374	8.08	21.59
150	0.430	10.66	24.79
200	0.355	9.72	27.37

A log-log plot of F_n/L against V_p is shown in Fig. 6.14. The slope is 0.33, the flow behavior index, n. The value of λ for $\sigma = 0.871$, $n = 0.33$ is obtained by interpolation from Table 6.1.
The value of λ for $n = 3$, $\sigma = 0.871$ is

$$\lambda = 0.9007 + \frac{(0.9504 - 0.9019)(0.071)}{0.1} = 0.9351$$

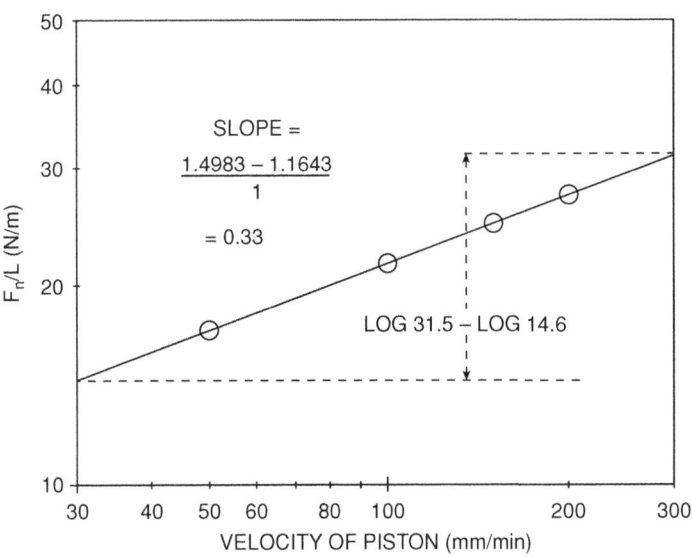

Figure 6.14 Log-log plot of corrected force to extrude fluid through an annular gap, against piston velocity, to obtain the flow behavior index, n, from the slope.

The value of λ for $n = 0.4$, $\sigma = 0.871$ is

$$\lambda = 0.9019 + \frac{(0.9504 - 0.9019)(0.071)}{0.1} = 0.9363$$

The falue of λ for $n = 0.33$ is obtained by interpolation:

$$\lambda = \frac{0.9351 + (0.9363 - 0.9351)(0.03)}{0.01} = 0.9354$$

The value of θ is obtained from Fig. 6.13. The curves for $n = 0.3$ and $n = 0.4$ do not intercept $\sigma = 0.871$ in Fig. 6.13. However, the curves are linear at low values of θ, and may be represented by the following equations:

$$n = 0.3 : \log \theta = 1.48353 - 7.5767\sigma$$
$$n = 0.4 : \log \theta = 1.9928 - 7.4980\sigma$$

Thus, at $n = 0.3$, $\theta = 7.665 \times 10^{-6}$; at $n = 0.4$, $\theta = 2.8976 \times 10^{-5}$. At $n = 0.33$, $\theta = 1.4058 \times 10^{-5}$ by interpolation. Knowing the value of n, θ, and λ, τ_w and γ_w can be calculated. Using Equation (6.26):

$$\sigma = 0.871, \lambda = 0.9354, \sigma_s = 0.3333, 1/n = 3.$$

$$\gamma_w = \frac{1}{0.381(1.4058 \times 10^{-5})}[(0.871)^2 - (0.3333)^2]\left[\frac{(0.9354)^2}{0.087}\right]^3\left(\frac{1}{60,000}\right)V_p$$

$$= 0.04796 \, V_p;$$

where V_p is in mm/min.
    Using Equation (6.27):

$$\tau_w = \frac{(0.9354)^2 - (0.871)^2}{2\pi(0.0332)[(0.9354)^2 - (0.3333)^2]}\left(\frac{F_n}{L}\right)$$

$$= 0.73(F_n/L)$$

The values of the shear stress and shear rates are

τ_w (Pa)	λ_w
12.48	2.398
15.76	4.796
18.09	7.194
19.98	9.592

The consistency index, K is determined from the intercept of the log-log plot in Fig. 6.15.

$$K = 9.2 \, Pa \cdot s^n$$

6.2.4 Determination of Rheological Properties of Fluids Using Rotational Viscometers

The most common rotational viscometer used in the food industry is the concentric cylindrical viscometer. The viscometer shown in Fig. 6.16 consists of concentric cylinders with fluid in the annular

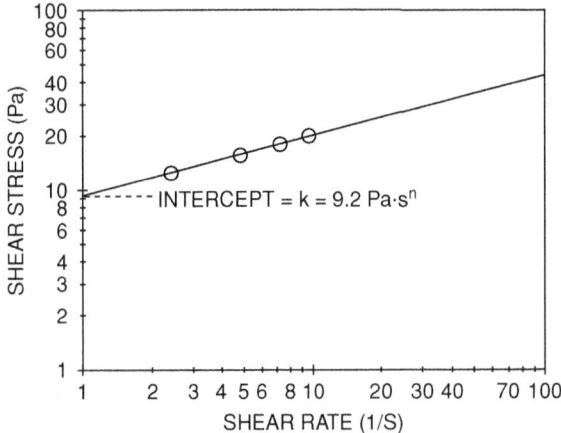

Figure 6.15 Log-log plot of shear stress against shear rate to obtain the consistency index, K, from the intercept.

space. This system is also referred to as a couette. The torque needed to rotate one of the cylinders is measured. This torque is proportional to the drag offered by the fluid to the rotation of the cylinder.

If the outer cup is stationary and if the torque measured is A, the force acting on the surface of the inner cylinder to overcome the resistance to rotation will be A/R.

The shear stress τ_w at the wall is

$$\tau_w = \frac{A}{R}\frac{1}{2\pi RL} = \frac{A}{R^2(2\pi L)} \tag{6.29}$$

For a Newtonian fluid in narrow gap viscometers where the rotating cylinder has a radius, R and the

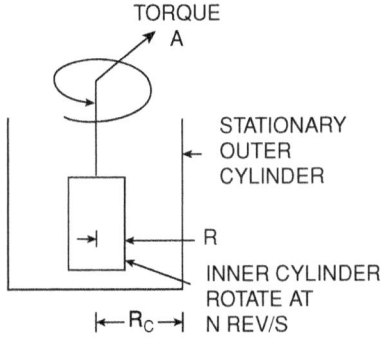

Figure 6.16 Schematic diagram of a rotational viscometer.

cup radius is R_c, the shear rate at the wall at a rotational speed of N rev/s is

$$\gamma_w = \frac{4\pi NR_c}{(R_c^2 - R^2)} \tag{6.30}$$

If the fluid is non-Newtonian with a flow behavior index, n, the shear rate at the wall has been derived by Kreiger and Maron (1954) for a couette type viscometer ($R_c/R < 1.2$) as follows:

$$\gamma_w = N\left[\frac{4\pi R_c^2}{(R_c^2 - R^2)} + 2\pi\left[1 + \frac{2}{3}\ln\left(\frac{R_c}{R}\right)\right]\left(\frac{1-n}{n}\right) + \frac{2\pi}{3}\ln\left(\frac{R_c}{R}\right)\left[\frac{1-n}{n}\right]^2\right] \tag{6.30a}$$

Equation (6.30) is a special case of Equation (6.30a) when the flow behavior index $= 1$. The factor in brackets which is a multiplier for N in Equation (6.30a), is easily calculated using a spreadsheet program.

Example 6.7. A rotational viscometer with a spring constant equivalent to 673.7 dyne $\cong$ cm full scale is used on a narrow gap viscometer with a rotating cylinder diameter of 29.5 mm and a cup diameter of 32 mm. The cylinders are 44.5 mm high. Assume end effects are negligible. This couette type viscometer is used to measure the flow properties of whole eggs and the data are shown in the spreadsheet (Figure 6.17) with N in rev/min entered in cells A4 to A7 and torque in % of full scale entered in cells B4 to B7. Calculate the flow behavior and consistency indices in SI units. The term in brackets to multiply N to obtain the shear rate in Equation (6.30a) has a value of 87.2 when n $= 0.66$ and is entered in cell A12 in Figure 6.17.

Solution:

From the spreadsheet, (Figure 6.17), the slope of the regression of log(torque) versus log(RPM) is 0.66. Thus the value of n used to calculate the multiplying factor for N to obtain the shear rate from the rotational speed. The consistency index is the antilog of the intercept of a regression of log(τ) against log(γ). The intercept is -1.246 and the value of K $= 10^{-1.246} = 0.057$ Pa $\cdot$ s^n.

6.2.4.1 Wide Gap Rotational Viscometer

A wide gap rotational viscometer consists of a cylinder or a disk shaped bob on a spindle, which rotates in a pool of liquid. A torque transducer attached to the base of the rotating spindle measures the drag offered by the fluid to the rotation of the bob. Different size spindles may be used. Based on the spindle size and the rotational speed, the indicated torque can be converted to an apparent viscosity by using an appropriate conversion factor specific for the size of spindle and rotational speed. For a shear thinning fluid, the apparent viscosity increases as the rotational speed decreases. A log-log plot of the apparent viscosity against rotational speed will have a slope equivalent to n $- 1$. If one spindle is used at several rotational speeds, the unconverted torque reading is directly proportional to the shear stress, and the rpm is directly proportional to the shear rate. A log-log plot of torque reading against rpm will have a slope equivalent to the flow behavior index n.

Example 6.8. A Brookfield viscometer model RVF was used to evaluate the apparent viscosity of tomato catsup. One spindle (No. 4) gave readings within the measuring scale of the instrument at four rotational speeds. The viscometer constant was 7187 dyne $\cong$ cm full scale. The torque reading in per cent full scale at various rotational speeds in revolutions/min (rpm), respectively are as follows: (53.5, 2); (67, 4); (80.5,10); (97, 20).

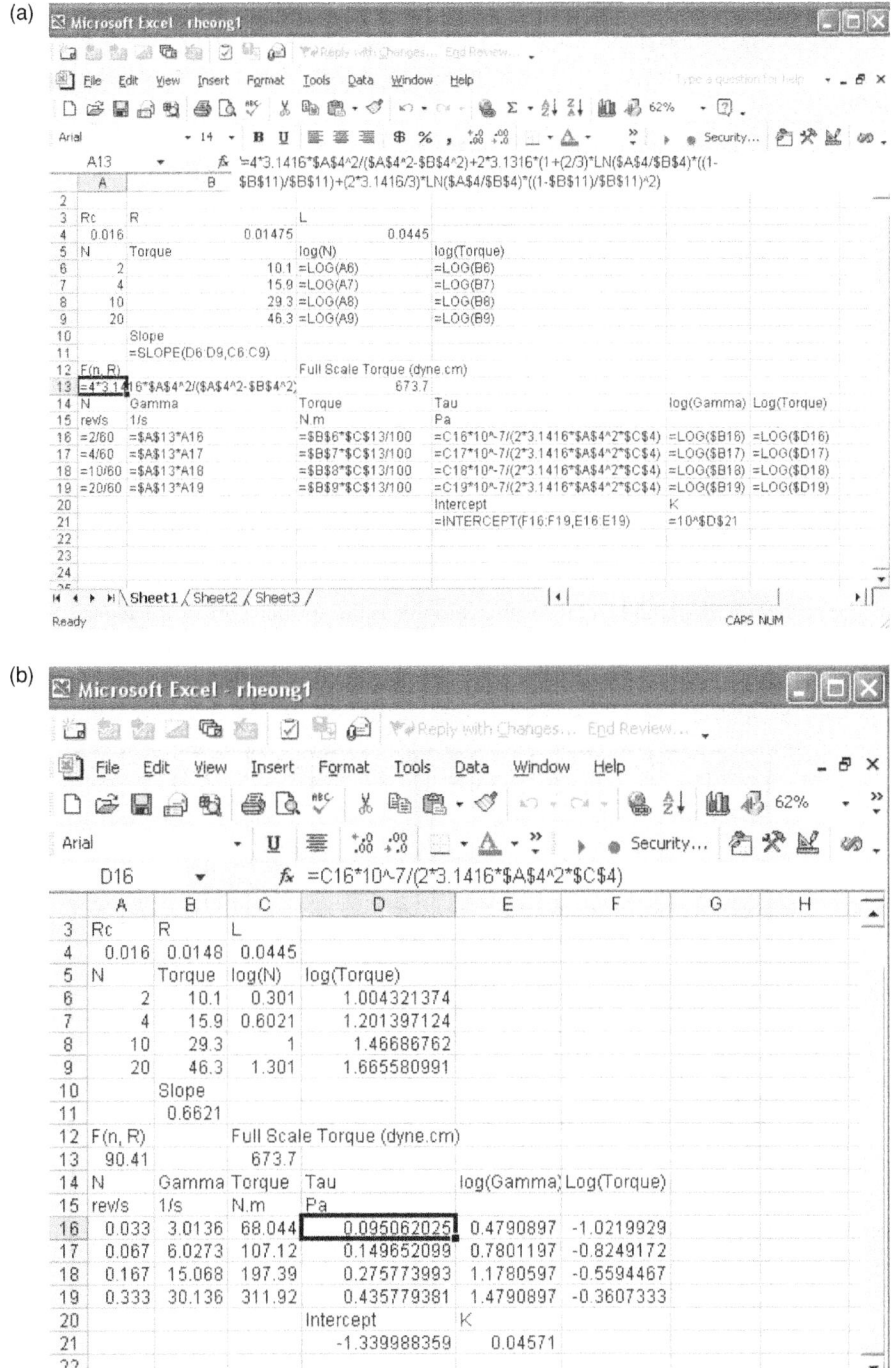

Figure 6.17 (a) Spreadsheet program in Microsoft Excel showing data entered and calculated values of shear stress, shear rate, flow behavior index, n, and consistency index, K. (b) Spreadsheet program in Microsoft Excel showing formulas in the cells used to calculate values in Fig. 6.17a.

Solution:

n is the slope of a log-log plot of torque against rpm.

Speed (rpm)	Indicator Reading (% Full Scale)	Torque (dyne-cm)
2	53.5	3845
4	67	48l5
10	80.5	5786
20	97	697l

The data plotted in Fig. 6.18 shows a slope of 0.25, which is the value of n.

6.2.4.2   Wide Gap Viscometer with Cylindrical Spindles

An important characteristic of a wide gap viscometer is that the fluid rotates in the proximate vicinity of the surface of the rotating spindle at the same velocity as the spindle (zero slip) and the fluid velocity

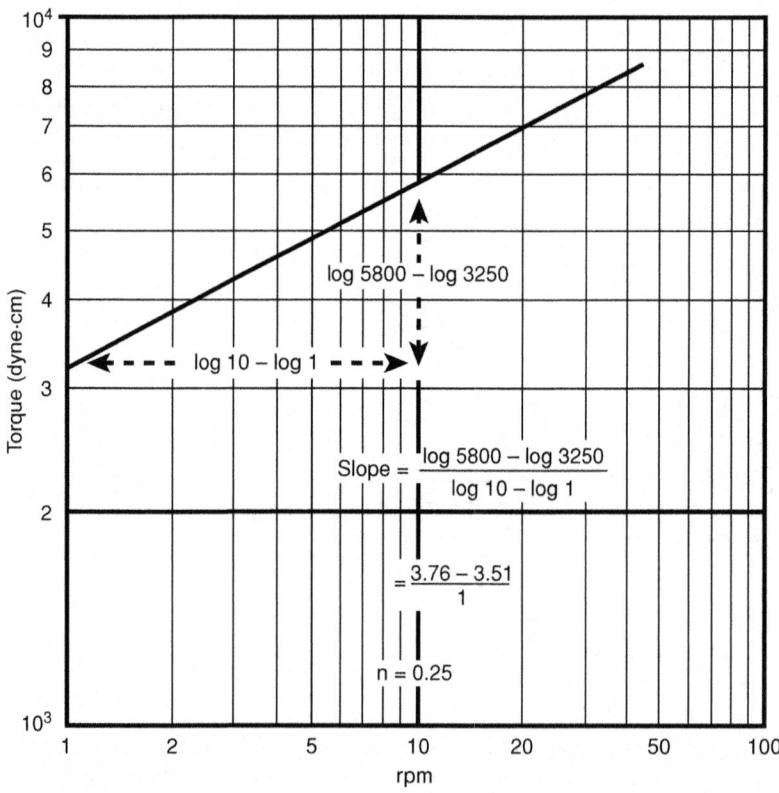

Figure 6.18   Log-log plot of torque against rotational speed to obtain the flow behavior index from the slope.

reaches zero at some point before the wall of the stationary outer cylinder. If these characteristics are met in a system, Kreiger and Maron (1952) derived an expression for the shear rate at the rotating inner cylinder wall as follows.

Let ω = angular velocity at a distance r from the center of the rotating spindle. $\gamma_w = (-dV/dr)$. Because $V = \omega r$, then: $dV = rd\omega$

$$-\frac{dV}{dr} = r d\frac{\omega}{dr}$$

The shear stress "τ" at a point in the fluid a distance "r" from the center of the rotating spindle is:

$$\tau = \frac{\text{Torque}}{2\pi r^2 h} \quad r^2 = \frac{\text{Torque}}{2\pi \tau h}$$

Differentiating the expression for shear stress

$$d\tau = \frac{\text{Torque}}{2\pi h} - 2\frac{dr}{r^3}$$

Substituting r^2

$$d\tau = \frac{\text{Torque}}{2\pi h} \left(-2\frac{dr}{r}\right)\left(\frac{2\pi h}{\text{Torque}}\right)$$

$$d\tau = \frac{-2\tau\, dr}{r}$$

Solving for dr/r : $dr/r = (-1/2)d\tau /\tau$
Because $(-dV/dr) = F(\tau)$; then $r\,(d\omega/dr) = F(\tau)$, and $d\omega = (dr/r)F(\tau)$.
Substituting $(dr/r) = d\tau/2\tau$; $d\omega = (d\tau/2\tau)F(\tau)$ and $d\omega = (1/2)F(\tau)\,d\ln(\tau)$.
At the surface of the spindle, the shear stress is τ_w and the angular velocity is $2\pi N$. Thus:

$d(2\pi N) = (1/2)\, F(\tau_w)\,d\ln(\tau_w)$. Because $(-dV/dr)_w = F(\tau_w)$ Then $2\pi dN = (1/2)(-dV/dr)\,d\ln(\tau_w)$
and: $-(dV/dr)_w = 4\pi dN/d\ln(\tau_w)$

Because $d\ln(N) = dN/N$, then $dN = N\,d\ln(N)$

$$-(dV/dr)_w = 4\pi N/[d\ln(\tau_w)/d\ln(N)]$$

The denominator is the slope of the log-log plot of torque versus N and has the same value as the flow behavior index n.

$$\gamma_w = \frac{4\pi N}{n} \tag{6.31}$$

Brookfield Engineering Laboratories gives the following equation for shear rate at a point x distant from the center of the rotating cylindrical spindle in a wide gap viscometer. R = the radius of the spindle, and Rc = the radius of the cup.

$$\gamma = \frac{4\pi N Rc^2 R^2}{x^2(Rc^2 - R^2)} \tag{6.31a}$$

If Rc is much bigger than R, then $Rc^2/(Rc^2 - R^2)$ therefore, at the surface of the spindle where x = R, the shear rate is $\gamma = 4\pi N$. Thus Equations (6.30), (6.31) and (6.31a) are equivalent when n = 1.

To use Equation (6.31), a value of n is needed. The value of n is determined by plotting log(Torque) against log(N). The slope is the value of n. Knowing n, Equation (6.31) can then be used to calculate γ_w. τ_w is calculated using Equation (6.29). A log-log plot of τ_w against γ_w will have K for an intercept.

Example 6.9. The following data were obtained in the determination of the flow behavior of a fluid using a rotational viscometer. The viscometer spring constant is 7197 dyne $\cong$ cm full scale. The spindle has a diameter of 0.960 cm and a height of 4.66 cm.

N (rev/min)	Torque (Fraction of full scale)
2	0.155, 0.160, 0.157
4	0.213, 0.195, 0.204
10	0.315, 0.275, 0.204
20	0.375, 0.355, 0.365

Calculate the consistency and flow behavior indices for this fluid.

Solution:

Equation (6.29) will be used to calculate τ_w, and Equation (6.31) for γ_w.

It will be necessary to determine n before Equation (6.31) can be used. The slope of a log-log plot of torque against the rotational speed will be the value of n. This plot shown in Fig. 6.19 shows a slope

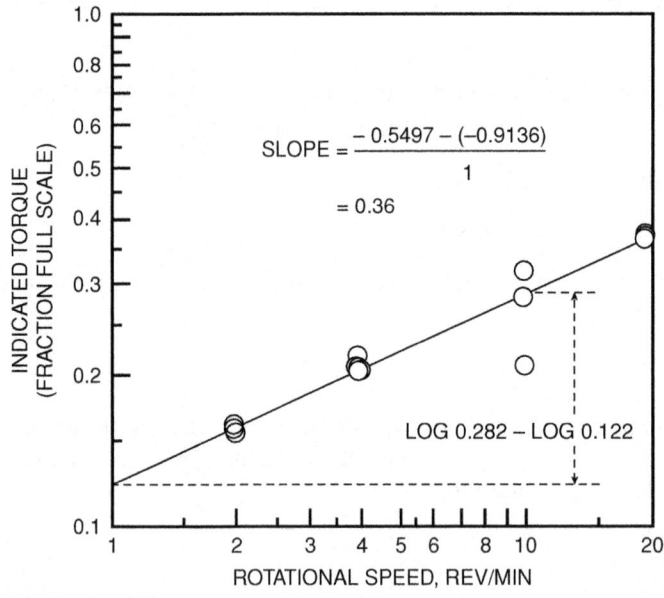

Figure 6.19 Log-log plot of indicated torque against the rotational speed to obtain the flow behavior index, n, from the slope.

of 0.36, which is the value of n. The multiplying factor to convert the rotational speed in rev/min to the shear rate at the wall of the rotating cylinder, γ_w , in s $^{-1}$ units is: 4(3.1416)/[0.36(60)] = 0.5818. The multiplying factor to convert the indicated torque as a fraction of viscometer full scale torque to the shear stress at the rotating cylinder wall, τ_w, in Pascals is

$$7187(10^{-7})/[2(3.1416)(0.0096/2)^2(0.0466)] = 106.5367.$$

The values for τ_w and Q_w are

Shear stress (τ_w) (Pa)	Shear rate (γ_w) (1/s)
16.5, 17.0, 16.7	1.16
22.7, 20.8, 21.7	2.33
33.6, 29.3, 21.7	5.82
39.9, 37.8, 38.9	11.64

The plot of τ_w against γ_w shown in Fig. 6.20 shows that $\tau_w = 15.2$ Pa at $\gamma_w = 1$; thus, K = 15.2 Pa Xs^n.

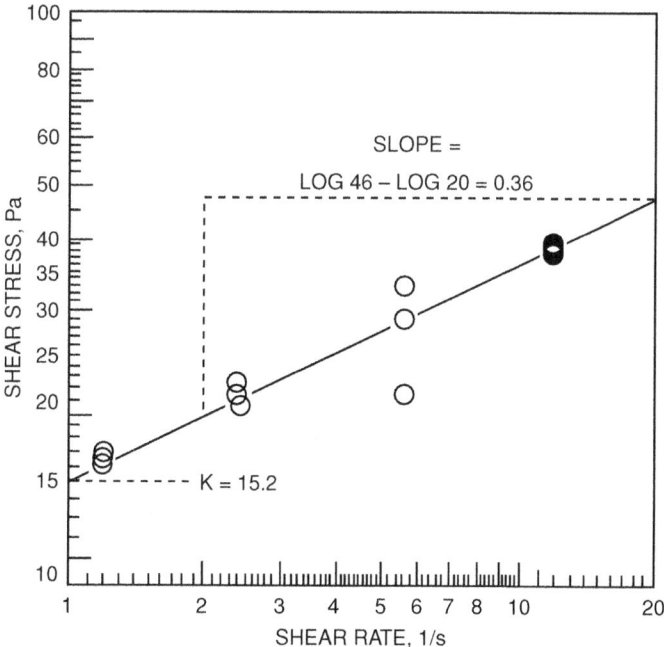

Figure 6.20 Log-log plot of shear stress against shear rate to obtain the consistency index, K, from the intercept.

6.3   CONTINUOUS VISCOSITY MONITORING AND CONTROL

In the formulation of foods where viscosity or consistency is a quality attribute, continuous monitoring of flow properties and automatic control of the feeding of ingredients is important. Two good examples of these processes are: blending of tomato paste with sugar, vinegar and flavoring in manufacturing of catsup; and blending of flour, water and flavoring ingredients to make batter for battered and breaded fried fish, poultry, or vegetables. These processes are rather simple because the level of only one ingredient in the formulation controls the consistency of the blend, and therefore will be very suitable for automatic control of flow properties. The principles for continuous viscosity monitoring are similar to those used in rheology. Some means must be used to divert a constant flow of fluid from the main pipeline to the measuring device. There are several configurations of viscosity measuring devices suitable for continuous monitoring in flowing systems. Three of the easiest to adapt on a system are described below.

6.3.1   Capillary Viscometer

A capillary viscometer may be used for continuous monitoring of consistency by tapping into the main piping, using a metering pump to deliver a constant flow through a level capillary, and returning the fluid to the main pipe downstream from the intake point. The metering pump must be a positive displacement pump to prevent flow fluctuations due to changes in the main pipeline pressure. A differential pressure transducer mounted at the entrance and exit of the capillary is used to measure the pressure drop. Poiseuille's equation (Eq. 6.11) is used to calculate the apparent viscosity from the pressure difference measured and the flow delivered by the metering pump.

6.3.2   Rotational Viscometer

Rotational viscometers could be used for continuous monitoring of viscosity in open vessels. Provision must be allowed for changes in fluid levels to ensure that the immersion depth of the spindle is appropriate. Mounting the instrument on a float device resting on the fluid surface, will ensure the same spindle immersion depth regardless of fluid level. Some models of rotational viscometers can remotely display the torque, others give a digital readout of the torque which can be easily read on the instrument.

For fluids flowing through pipes, problems may be encountered with high fluid velocities. Semicontinuous measurement may be made by periodically withdrawing fluid from the pipe through a sampling valve and depositing this sample into a cup where the viscosity is measured by a rotational viscometer. This type of measurement will have a longer response time than a truly in-line measurement.

6.3.3   Viscosity Sensitive Rotameter

This device is similar to a rotameter and operates on the principle that the drag offered by an obstruction in a field of flow is proportional to the viscosity and the rate of flow. A small float positioned within a vertical tapered cylinder will assume a particular position when fluid a fixed flow rate, and with a particular viscosity is flowing through the tube. This device needs a constant rate of flow to be effective, and usually, a metering pump draws out fluid from a tap on a pipe, delivers the fluid through the viscosity sensitive rotameter, and returns the fluid to the pipe through a downstream tap.

6.4 FLOW OF FALLING FILMS

Fluid films are formed when fluids flow over a vertical or inclined surface, or when fluid is withdrawn from a tank at a rapid rate. The differential equation for the shear stress and shear rate relationship for the fluid coupled with a differential force balance will allow a resolution of the thickness of the film as a function of fluid flow properties.

6.4.1 Films of Constant Thickness

When the rate of flow of a fluid over a vertical or inclined plane is constant, the film thickness is also constant. Figure 6.21 shows a fluid film flowing down a plane inclined at an angle α with the vertical plane. If the width of the plane is W, a force balance on a macroscopic section of the film of length L and of thickness x, measured from the film surface, is as follows:

Mass of the differential increment $= LW\rho x$
Force due to gravity on mass $= LW\rho gx$
Component of gravitational force along inclined plane $= LW\rho gx \cos(\alpha)$
Fluid resistance against the flow $= \tau WL$.

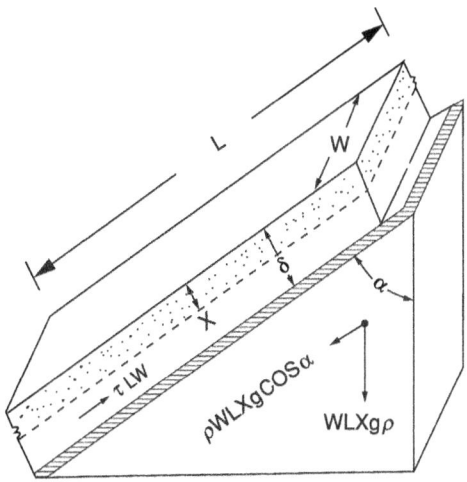

Figure 6.21   Diagram of a control element used in analysis of fluid film flowing down an inclined plane.

If the fluid follows the power law equation (Eq. 6.2), a force balance results in:

$$WLK\left[-\frac{dV}{dr}\right]^n = LW\rho gx\cos(\alpha)$$

Rearranging:

$$-\frac{dV}{dx} = \left[\frac{\rho gx\cos(\alpha)}{K}\right]^{1/n}$$

Integrating:

$$\int dV = -\left[\frac{\rho g\cos(\alpha)}{K}\right]^{1/n}\int [x]^{1/n}\,dx$$

The integrated expression is

$$V = -\left[\frac{\rho g\cos(\alpha)}{K}\right]^{1/n}\left[\frac{n}{n+1}\right][x]^{(n+1)/n} + C$$

The boundary condition, at $x = \delta$, $V = 0$, when substituted in the above expression will give the value of the integration constant, C. δ is the film thickness, and since x is measured from the surface of the film, the position $x = \delta$ is the surface of the incline along which the film is flowing. The final expression for the fluid velocity profile is

$$V = \left[\frac{\rho g\cos(\alpha)}{K}\right]^{1/n}\left[\frac{n}{n+1}\right][\delta]^{(n+1)/n}\left[1 - \left[\frac{x}{\delta}\right]^{(n+1)/n}\right]$$

The thickness of the film δ can be expressed in terms of an average velocity of flow, $\bar{V}$. The expression is derived as follows:

$$W\delta\bar{V} = \int_0^\delta VW\,dx = W\left[\frac{\rho g\cos(\alpha)}{K}\right]^{1/n}\left[\frac{n}{n+1}\right][\delta]^{(n+1)/n}\int_0^\delta\left[1 - \left[\frac{x}{\delta}\right]^{(n+1)/n}\right]dx$$

$$\bar{V} = \left[\frac{\rho g\cos(\alpha)}{K}\right]^{1/n}\left[\frac{n}{2n+1}\right][\delta]^{(n+1)/n} \tag{6.32}$$

The film thickness is

$$\delta = \left[\frac{\bar{V}(2n+1)K^{1/n}}{n[\rho g\cos(\alpha)]^{1/n}}\right]^{n/(n+1)} \tag{6.33}$$

Example 6.10. Applesauce at 80°C ($K = 9$ Pa· s^n; $n = 0.33$, $\rho = 1030$ kg/m^3 is to be de-aerated by allowing to flow as a film down the vertical walls of a tank 1.5 m in diameter. If the film thickness desired is 5 mm, calculate the mass rate of flow of applesauce to be fed into the de-aerator. At this film thickness, and 80°C, an exposure time of film to a vacuum of 25 in. Hg of 15 seconds is required to reduce dissolved oxygen content to a level needed for product shelf stability. Calculate the height of the tank needed for the film to flow over.

Solution:

Equation (6.32) will be used to calculate the average velocity of the film needed to give a film thickness of 5 mm. Since the side is vertical, $\alpha = 0$ and $\cos \alpha = 1$.

$$\bar{V} = \left[\frac{(1030)(9.8)}{9.0} \right]^{1/0.33} \left[\frac{(0.005)^{4.03}}{5.03} \right] = 0.1845\,\text{m/s}$$

The mass rate of flow, $\bar{m}$, is

$$\bar{m} = \frac{0.1845\,\text{m}}{\text{s}} (\pi)[(0.75)^2 - (0.75 - .005)^2]\text{m}^2 \left(\frac{1030\,\text{kg}}{\text{m}^3} \right)$$

$$= 4.48\,\text{kg/s}$$

The height of the vessel needed to provide the 15 seconds required exposure for de-aeration is

$$h = \frac{0.1845\,\text{m}}{\text{s}}(15\,\text{s}) = 2.77\,\text{m}$$

6.4.2 Time-Dependent Film Thickness

Problems of this type are encountered when fluids cling as a film on the sides of a storage vessel when the vessel is emptied (drainage), or when a solid is passed through a pool of fluid and emerges from the fluid coated with a fluid film (withdrawal). The latter process may be encountered when applying batter over a food product prior to breading and frying. The solution to problems of withdrawal and drainage is the same.

Figure 6.22 shows a fluid film with width W falling down a vertical surface. A control segment of the film having a thickness δ located a distance z from the top, and a length dz along the surface on which it flows, is shown. The solution will involve solving partial differential equations which involves a mass balance across the control volume.

If $\bar{V}$ is the average velocity of the fluid in the film at any point z distant from the top, the mass of fluid entering the control volume from the top is

$$M_i = \bar{V}\delta W\rho$$

The mass leaving the control volume from the lower part of the control volume is

$$M_e = \left[\bar{V}\delta + \frac{\partial}{\partial z}(\bar{V}\delta)dz \right] W\rho$$

The accumulation term due to the difference between what entered and left the control volume is

$$\frac{\partial \delta}{\partial z}\rho\,W dz$$

The mass balance across the control volume is

$$\bar{V}\delta W\rho = \left[\bar{V}\delta + \frac{\partial}{\partial z}(\bar{V}\delta)dz \right]\rho\,W + \frac{\partial \delta}{\partial z}\rho\,W\,dz$$

Simplification will produce the partial differential equation which describes the system.

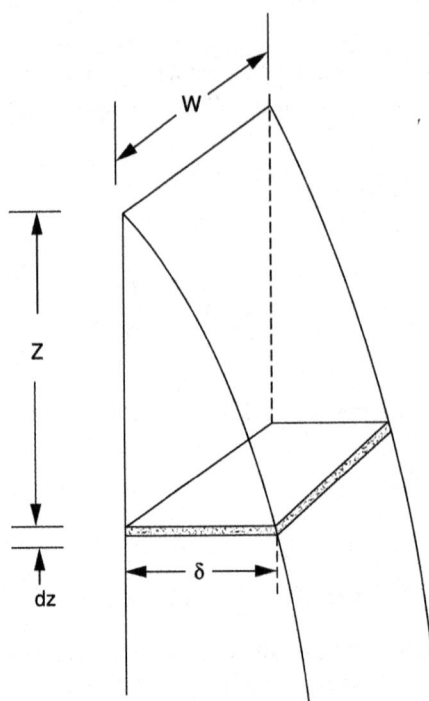

Figure 6.22 Diagram of a control element used in analysis of time dependent flow of fluid film down a surface.

$$\frac{\partial}{\partial z}(\bar{V}\delta) = -\frac{\partial \delta}{\partial t} \qquad (6.34)$$

The solution to this partial differential equation will be the product of two functions, $G(t)$ which is a function only of t, and $F(z)$ which is a function only of z.

$$\delta = F(z)G(t) \qquad (6.35)$$

Partial differentiation of Equation (6.35) gives:

$$\frac{\partial \delta}{\partial t} = F(z)\frac{\partial G(t)}{\partial t} \qquad (6.36)$$

$$\frac{\partial \delta}{\partial z} = G(t)\frac{\partial F(z)}{\partial z} \qquad (6.37)$$

Equation (6.32) represents $\bar{V}$ for a fluid film flowing down a vertical surface ($\cos \alpha = 1$). Partial differentiation of $\bar{V}\delta$ with respect to z gives:

$$\frac{\partial}{\partial z}(\bar{V}\delta) = \frac{\partial}{\partial z}\left[\left[\frac{\rho g}{K}\right]^{1/n}\left[\frac{n}{2n+1}\right]\left[\delta\right]^{(n+1)/n}\delta\right]$$

$$= \left[\frac{\rho g}{K}\right]^{1/n}[\delta]^{(n+1)/n}\frac{\partial \delta}{\partial z} \qquad (6.38)$$

Substituting Equation (6.34) for M/Mz ($\bar{V}\delta$); Equation (6.36) for Mδ/Mt; and Equation (6.37) for Mδ/Mz in Equation (6.38):

$$-F(z)\frac{\partial G(t)}{\partial t} = \left[\frac{\rho}{K}\right]^{1/n}[F(z)G(t)]^{(n+1)/n}G(t)\frac{FF(z)}{\partial z}$$

Each side of the partial differential equation contains only one variable, therefore the equation can be separated and each side integrated independently by equating it to a constant. Let the constant $= \lambda$.

The function with the time variable is obtained by integrating the left hand side of the equation, setting it equal to λ. The function with the z variable is obtained the same way, using the right side of the equation.

$$F(z) = \left[\lambda\left[\frac{K}{\rho g}\right]^{1/n}\left[\frac{n}{n+1}\right]z\right]^{n/(n+1)} + C_2$$

$$G(t) = \left[\lambda\left[\frac{n+1}{n}\right]t\right]^{-n/(n+1)} + C_1$$

The boundary conditions, $\delta = 0$ at z = 0, and t = 4 , will be satisfied when the integration constants C_1 and C_2 are zero. When expressions for F(z) and G(t) are combined according to Equation (6.35), the constant λ cancels out. The final expression for the film thickness as a function of time and position is

$$\delta = \left[\left[\frac{K}{\rho g}\right]^{1/n}\left[\frac{z}{t}\right]\right]^{n/(n+!)}\tag{6.39}$$

Example 6.11. Calculate the thickness of batter that would be attached the sides of a brick-shaped food product, at the midpoint of its height, if the product is immersed in the batter, removed, and allowed to drain for 5 seconds before breading. The batter has a flow behavior index of 0.75, a density of 1003 kg/m^3, and a consistency index of 15 Pa $\cdot$ s^n. The vertical side of the product is 2 cm high.

Solution:

z = 0.001 m; t = 5 s. Using Equation (6.38):

$$\delta = \left[\left[\frac{15}{1003(9.8)}\right]^{1/0.75}\left(\frac{0.001}{5}\right)\right]^{0.75/1.75}$$

$$= 0.000599 \text{ m}$$

Example 6.12. Applesauce is rapidly drained from a drum 75 cm in diameter and 120 cm high. At the end of 60 seconds after the fluid is drained from the drum, calculate the mass of applesauce adhering to the walls of the drum. The applesauce has a flow behavior index of 0.34, a consistency index of 11.6 Pa $\cdot$ s^n; and a density of 1008 kg/m^3.

Solution:

Equation (6.39) will be used. However, the thickness of the film will vary with the height, therefore it will be necessary to integrate the quantity of fluid adhering at the different positions over the height of the vessel. At any given time, t, the increment volume of fluid dq, over an increment of height dz will be

$$dq = \pi[r^2 - (r - \delta)^2] \, dz$$

The total volume of film covering the height of the drum will be

$$q = \int_0^z \pi(2r\delta - \delta_2) \, dz$$

Substituting the expression for δ (Eq. 6.39) and integrating over z, holding time constant:

$$q = 2\pi r \left[\frac{2}{\rho g}\right]^{1/(n+1)} \left[\frac{1}{t}\right]^{n/(n+1)} \left[\frac{n+1}{2n+1}\right][z]^{(2n+1)/(n+1)}$$

$$-\pi \left[\frac{K}{\rho}\right]^{2/(n+1)} \left[\frac{1}{t}\right]^{2n/(n+1)} \left[\frac{n+1}{3n+1}\right][z]^{(3n+1)/(n+1)}$$

Substituting values of known quantities:

$$K/\rho g = 11.6/(1008)(9.8) = 0.001174$$
$$2/(n+1) = 1.515; \, 2n/(n+1) = 0.507; \, (n+1)/(3n+1) = 0.6633$$
$$1/(n+1) = 0.7463; \, n/(n+1) = 0.254; \, (n+1)/2n+1) = 0.7976$$
$$V = 2\pi(0.375)(0.001174)^{0.7463}(1/60)^{0.254}(0.7976)$$
$$= (1.2)^{1/0.7976} - \pi(0.001174)^{1.515}(1/60)^{0.507}$$
$$= (0.6633)(1.2)^{1/0.6633}$$
$$= 0.0054289 - 0.0000125101 = 0.005416 \, \text{m}^3$$

The mass of applesauce left adhering to the walls of drum is

$$0.005416 \, \text{m}^3(1008)\text{kg/m}^3 = 5.46 \, \text{kg}$$

6.4.3   Processes Dependent on Fluid Film Thicknesses

Fluid films are involved in processes that require rapid rates of mass and/or heat transfer. Examples are as follows.:

Ultra-high temperature sterilization of fluid milk and cream: One system (DaSi) rapidly heats a fluid film as it flows down multiple tubes in a superheated steam atmosphere. The steam and fluid are introduced in parallel. The fluid is pumped to a specially constructed distributor inside a cylindrical pressure vessel where it flows down the outside periphery of the vertically oriented tube in a thin film. Heat transfer into the thin film from the superheated steam with which it is in direct contact is very rapid resulting in very short residence time to heat the product to sterilization temperatures. The short heating time minimizes adverse flavor changes on the product. The fluid flow rate and the outside surface area of the tubes must be matched to obtain the necessary film thickness to permit the rapid heating.

Falling film evaporators: Fluid flows as a film down the surface of a heated tube(s) and discharges into a plenum where a vacuum allows liquid to vaporize by flash evaporation. To maintain efficiency of heating, fluid flow must be at a rate that would permit the desired thickness of fluid film to develop without flooding the whole inside cross-sectional area of the tube.

De-aeration of food fluids: Viscous fluids at elevated temperatures are exposed to a vacuum while flowing in a thin film to facilitate removal of dissolved oxygen. One system (FMC) sprays the fluid at the top of a tall column allowing the fluid to it the walls and flow down as a falling fluid film. The column dimensions needed to successfully de-aerate the product may be calculated as shown in the example in the previous section "Films of Constant Thickness."

Stripping of volatile flavor component from foods: A system developed by Flavourtech called a "Spinning Cone Column" consists of a rotating shaft to which multiple cones arranged at even intervals along the height of the shaft are attached. The cone faces upwards with the base radiating outward from the shaft and the apex of the cone is pierced by the rotating shaft to which the cone is fastened. Another stationary cone concentric with the rotating cone but open at the apex and attached at the periphery of the base to the wall of the cylindrical column, creates a channel to guide the fluid film as it travels from one cone element to the next. Volatile flavor compounds are stripped from the fluid using an inert gas flowing countercurrent to the fluid or by vapors generated by flash evaporation when the fluid is introduced into the column at a high temperature and a vacuum is maintained in the column. The top cone is a rotating cone and fluid entering the column is deposited at the apex of this cone. The fluid film flows upwards on the wall of the rotating cone due to centrifugal force of rotation and spills over the base periphery where it falls into the top of a stationary cone. Fluid film flows by gravity over the stationary cone toward the central shaft where it drops into the apex of another rotating cone. The process repeats between the alternating rotating and stationary cones. The system has been shown to be effective in stripping volatile flavors from an aqueous slurry of tea, recovery of volatile flavors from fruit and vegetable juices and purees, de-aeration of fruit juice and purees, removing alcohol from wine or beer, and in recovery of flavors from exhaust air streams such as onion dehydrator or fruit dehydrator. The mass transfer from a large area of thin fluid film makes this system very effective compared to other types of contractors used in extraction operations.

6.5 TRANSPORTATION OF FLUIDS

The basic equations characterizing flow and forces acting on flowing fluids are derived using the laws of conservation of mass, conservation of momentum, and the conservation of energy.

6.5.1 Momentum Balance

Momentum is the product of mass and velocity and has units of $kg \cdot m/s$. When mass is expressed as a mass rate of flow, the product with velocity s the rate of flow of momentum and will have units of $kg \cdot m/s^2$, the same units as force. When a velocity gradient exists in a flowing fluid, momentum transfer occurs across streamlines, and the rate of momentum transfer per unit area, $d(mV/A)/dt$, the momentum flux, has units of $kg/m \cdot s^2$. The units of momentum flux is the same as that of stress or pressure. The pressure drop in fluids flowing through a pipe which is attributable to fluid resistance, is the result of a momentum flux between streamlines in a direction perpendicular to the direction of flow. The momentum balance may be expressed as:

Rate of momentum flow in $+ \sum F =$ Rate of momentum flow out $+$ Accumulation

$\sum F$ is the sum of external forces acting on the system (e.g., atmospheric pressure, stress on the confining vessel) or forces exerted by restraints on a nozzle discharging fluid. Because momentum is a function of velocity, which is a vector quantity (i.e., it has both magnitude and direction), all the terms in the above equation are vector quantities. The form of the equation given above implies that a positive sign on EF indicates a force applied in a direction entering the system. When making a momentum balance, the component of the force or velocity acting in a single direction (e.g., x, y, or z component)is used in the equation. Thus, in three dimensional space, a momentum balance may be made for the x, y, and z directions.

The force balance used to derive the Poiseuille equation (Eq. 6.9) is an example of a momentum balance over a control volume, a cylindrical shell of fluid flowing within a cylindrical conduit. Equation (6.9) shows that the momentum flow (momentum flux multiplied by the surface area of the control volume) equals the net force (pressure drop multiplied by the crosssectional area of the control volume) due to the pressure acting on the control element.

When the control volume is the whole pipe (i.e., the wall of the pipe is the boundary surrounding the system under consideration), it will be possible to determine the forces acting on the pipe or its restraints. These forces are significant in instances where a fluid changes direction as in a bend, when velocity changes as in a converging pipe section, when a fluid discharges out of a nozzle, or when a fluid impacts a stationary surface.

Momentum flow (rate of change of momentum, $\bar{M}$) has the same units as force, and is calculated in terms of the mass rate of flow, $\bar{m}$, kg/s, the velocity, V, or the volumetric rate of flow, q, as follows:

$$\bar{M} = \bar{m}V = AV^2\rho = qV\rho$$

A is the area of the flow stream perpendicular to the direction of flow, and is the fluid density. The momentum flux (rate of change of momentum per unit area) is the quotient, $\bar{M}/A$. A force balance over a control volume is the same as a balance of momentum flow.

Example 6.13. Calculate the force acting on the restraints of a nozzle from which fluid is discharging to the atmosphere at the rate of 5 kg/s. The fluid has a density of 998 kg/m^3. It enters the nozzle at a pressure of 238.3 kPa. above atmospheric pressure. The nozzle has a diameter of 6 cm at the inlet and 2 cm at the discharge.

Solution:

Assume that the velocity of the fluid through the nozzle is uniform (i.e., there is no velocity variation across the pipe diameter). The pipe wall may be used as a system boundary for making the momentum balance, in order to include the force acting on the restraint. The momentum balance with subscripts 1 and 4 referring to the inlet and discharge of the nozzle respectively, is

$$\bar{m}V_1 + P_1A_1 + F_x = \bar{m}V_4 + P_4A_4$$

Atmospheric pressure (P_a) acts on the inlet to the nozzle entering the control volume, and this is also the pressure at the discharge. Because P_1 is pressure above atmospheric at the inlet, the momentum balance becomes:

$$\bar{m}V_1 + P_aA_1 + P_1A_1 + F_x = \bar{m}V_4 + P_aA_4$$

Solving for F_x:

$$F_x = \bar{m}(V_4 - V_1) - P_a(A_1 - A_4) - P_1A_1$$

Substituting known quantities and solving for V_1 and V_4:

$$A_1 = 0.002827\,m^2;\ A_4 = 0.001257$$

$$V_1 = \frac{5\,kg}{s}\frac{m^3}{998\,kg}\frac{1}{0.002827\,m^2} = 1.77\,m/s$$

$$V_4 = \frac{5\,kg}{s}\frac{m^3}{998\,kg}\frac{1}{0.001257\,m^2} = 3.99\,m/s$$

$$F_X = 5(3.99 - 1.77) - 438000(0.002827) - 101{,}300(0.002827 - 0.001257)$$

$$= 11.1 - 1238.2 - 159 = -1386\,N$$

The negative sign indicates that this force is acting in a direction opposite the momentum flow out of the system, thus the nozzle is being pushed away from the direction of flow.

6.5.2   The Continuity Principle

The principle of conservation of mass in fluid dynamics is referred to as the continuity principle. This principle is applied whenever there is a change of velocity, a change in diameter of the conduit, split flow, or a change in density of compressible fluids such as gases. The continuity principle for flow in one direction, expressed in equation form is as follows:

$$\rho_1 V_1 A_1 = \rho_2 V_2 A_2 + A\frac{\partial}{\partial t}(\rho V) \tag{6.40}$$

If velocity is changing in all directions, a three-dimensional mass balance should be made. The above equation is a simplified form suitable for use in fluid transport systems where velocity is directed in one direction. At a steady state, the time derivative is zero. If the fluid is incompressible, the density is constant, and the continuity equation reduces to:

$$V_1 A_1 = V_2 A_2 \tag{6.41}$$

Example 6.14. A fluid having a density of 1005 kg/m3 is being drawn out of a storage tank 3.5 m in diameter through a tap on the side at the lowest point in the tank. The tap consists of a short length of 1.5 in. nominal sanitary pipe (ID = 0.03561 m) with a gate valve. If fluid flow out of the tap at the rate of 40 L/min, calculate the velocity of the fluid in the pipe and the velocity at which the fluid level recedes inside the tank.

Solution:

The mass rate of flow, $\bar{m}$ is constant. The mass rate of flow is the product of the volumetric rate of flow and the density. $\bar{m} = q\,\rho$

$$\bar{m} = \left[\frac{40\,L}{min} \cdot \frac{0.001\,m^3}{L} \cdot \frac{1\,min}{60\,s}\right] \cdot \left[\frac{1005\,kg}{m^3}\right] = 0.67\,kg/s$$

The velocity of fluid in the pipe is

$$V = \frac{q}{A} = \frac{\bar{m}}{\rho A}$$

$$V_{pipe} = \left[\frac{0.67\,kg}{s} \right] \cdot \frac{1\,m^3}{1005\,kg} \cdot \frac{1}{\pi(0.01781)^2\,m^2} = 0.669\,m/s$$

From the continuity principle:

$$A_{pipe}V_{pipe} = A_{Tank}V_{Tank}$$

$$V_{tank} = V_{pipe} \cdot \frac{A_{pipe}}{A_{tank}} = 0.669\frac{(0.03561)^2}{(3.5)^2} = 6.92 \times 10^{-5}\frac{m}{s}$$

6.6   FLUID FLOW REGIMES

Equation (6.10) indicates that a Newtonian fluid flowing inside a tube has a parabolic velocity profile. Fluid molecules in the center of the tube flows at the maximum velocity. The equation holds only when each molecule of the fluid remains in the same radial position as the fluid traverses the length of the tube. The path of any molecule follows a well-defined streamline, which can be readily shown by injecting a dye into the flow stream. When injected in a very fine stream, the dye will trace a straight line parallel to the direction of flow. This type of flow is called streamline or laminar flow. Figure 6.23 shows streamline flow that is observed when a dye is continuously injected into a flow stream (A), and the development of the parabolic velocity profile (D) that can be observed when a viscous dye solution is injected into a tube filled with a very viscous fluid at rest (B) and flow is allowed to develop slowly (C).

As rate of flow increases, molecular collisions occur at a more frequent rate and cross-overs of molecules across streamlines occur. Eddy currents develop in the flow stream. This condition of flow is turbulent and the velocity profile predicted by Equation (6.10) no longer holds. Figure 6.24 shows what can be observed when a dye is continuously injected into a fluid in turbulent flow. Swirling and curling of the dye can be observed. A certain amount of mixing also occurs. Thus, a particle originally introduced at the center of the tube may traverse back and forth between the wall and the center as the fluid travels the length of the tube. The velocity profile in turbulent flow is flat compared to the parabolic profile in laminar flow. The ratio of average to maximum velocity in laminar flow is 0.5 as opposed to an average ratio of approximately 0.8 in fully developed turbulent flow.

6.6.1   The Reynolds Number

The Reynolds number is a dimensionless quantity that can be used as an index of laminar or turbulent flow. The Reynolds number, represented by Re, is a function of the tube diameter, D, the average velocity $\bar{V}$, the density of the fluid ρ, and the viscosity μ.

$$Re = \frac{D\bar{V}\rho}{\mu} \tag{6.42}$$

For values of the Reynolds number below 2100, flow is laminar and the pressure drop per unit length of pipe can be determined using Equation (6.11) for a Newtonian fluid and Equation (6.5) for power law fluids. For Reynolds numbers above 2100, flow is turbulent and pressure drops are obtained using an empirically derived friction factor chart.

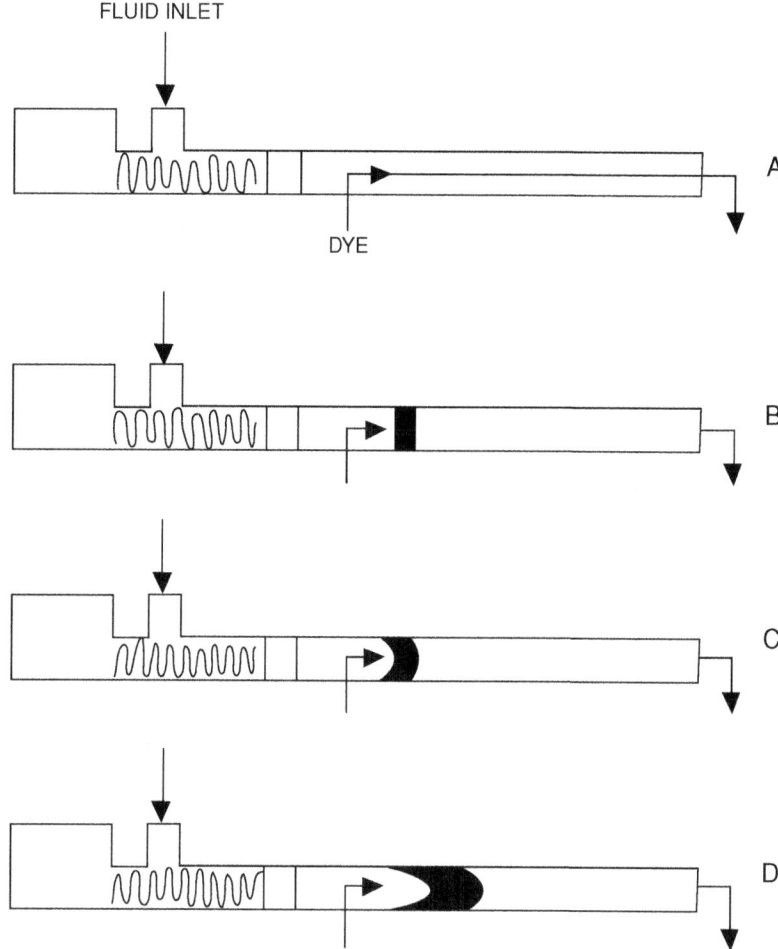

Figure 6.23   Development of laminar velocity profile during flow of a viscous fluid through a pipe.

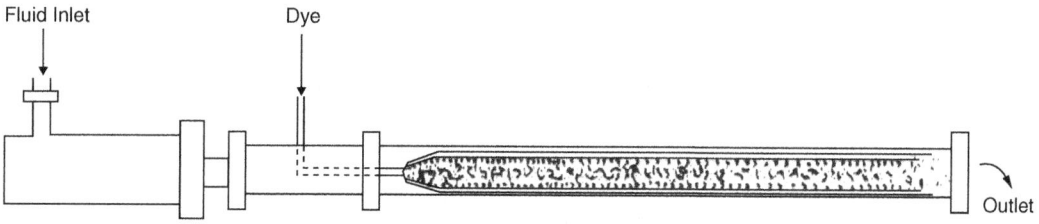

Figure 6.24   Path traced by a dye injected into a fluid flowing through a tube in turbulent flow.

Table 6.2 Pipe and Heat Exchanger Tube Dimensions (Numbers in Parentheses Represent the Dimension in Meters).

Nominal size(in.)	Steel Pipe (Sch. 40)		Sanitary Pipe		Heat Exchanger Tube (18 Ga)	
	ID in./(m)	OD in./(m)	ID in./(m)	OD in./(m)	ID in./(m)	OD in./(m)
0.5	0.622 (0.01579)	0.840 (0.02134)	–	–	0.402 (0.01021)	0.50 (0.0217)
0.75	0.824 (0.02093)	1.050 (0.02667)	–	–	0.652 (0.01656)	0.75 (0.01905)
1	1.049 (0.02664)	1.315 (0.03340)	0.902 (0.02291)	1.00 (0.0254)	0.902 (0.02291)	1.00 (0.0254)
1.5	1.610 (0.04089)	1.900 (0.04826)	1.402 (0.03561)	1.50 (0.0381)	1.402 (0.03561)	1.50 (0.0381)
2.0	2.067 (0.05250)	2.375 (0.06033)	1.870 (0.04749)	2.00 (0.0508)	–	–
2.5	2.469 (0.06271)	2.875 (0.07302)	2.370 (0.06019)	2.5 (0.0635)	–	–
3.0	3.068 (0.07793)	3.500 (0.08890)	2.870 (0.07289)	3.0 (0.0762)	–	–
4.0	4.026 (0.10226)	4.500 (0.11430)	3.834 (0.09739)	4.0 (0.1016)	–	–

ID = inside diameter.
OD = outside diameter.

6.6.2 Pipes and Tubes

Tubes are thin walled cylindrical conduits whose nominal sizes are based on the outside diameter. Pipes are thicker walled than tubes and the nominal size is based on the inside diameter. The term "sanitary pipe" used in the food industry is used for stainless steel tubing where the nominal designation is based on the outside diameter. Table 6.2 shows dimensions for steel and sanitary pipes, and heat exchanger tubes.

6.6.3 Frictional Resistance to Flow of Newtonian Fluids

Flow through a tubular conduit is accompanied by a drop in pressure. The drop in pressure is equivalent to the stress that must be applied on the fluid to induce flow. This stress can be considered as a frictional resistance to flow. Working against this stress requires the application of power in transporting fluids. An expression for the pressure drop as a function of the fluid properties and the dimension of the conduit would be useful in predicting power requirements for inducing flow over a given distance or for determining the maximum distance a fluid will flow with the given pressure differential between two points. For Newtonian fluids in laminar flow, the Poiseuille equation (Eq. 6.11) may be expressed in terms of the inside diameter (D) of a conduit, as follows:

$$\frac{\Delta P}{L} = \frac{32\bar{V}\mu}{D^2} \tag{6.43}$$

Equation (6.43) can be expressed in terms of the Reynolds number.

$$\frac{\Delta P}{L} = \frac{32\bar{V}\mu}{D^2} \cdot \frac{(D\bar{V}\rho)/\mu}{Re}$$

$$\frac{\Delta P}{L} = \frac{2\left(\dfrac{16}{Re}\right)(\bar{V})^2(\rho)}{D} \tag{6.44}$$

Equation (6.44) is similar to the Fanning equation derived by dimensional analysis where the friction factor f is 16/Re for laminar flow. The Fanning equation, which is applicable in both laminar and turbulent flow, is

$$\frac{\Delta P}{\rho} = \frac{2f(\bar{V})^2 L}{D} \tag{6.45}$$

 For turbulent flow, the Fanning friction factor f may be determined for a given Reynolds number and a given roughness factor from a friction factor chart as shown in Fig. 6.25. For smooth tubes, a plot of log f obtained from the lowest curve in Fig. 6.25 against the log of Reynolds number would give the following relationships between f and Re:

$$f = 0.048 \, (Re)^{-0.20} \quad 10^4 < Re < 10^6 \tag{6.46}$$

$$f = 0.193 \, (Re)^{-0.35} \, 3 \times 10^3 < Re < 10^4 \tag{6.47}$$

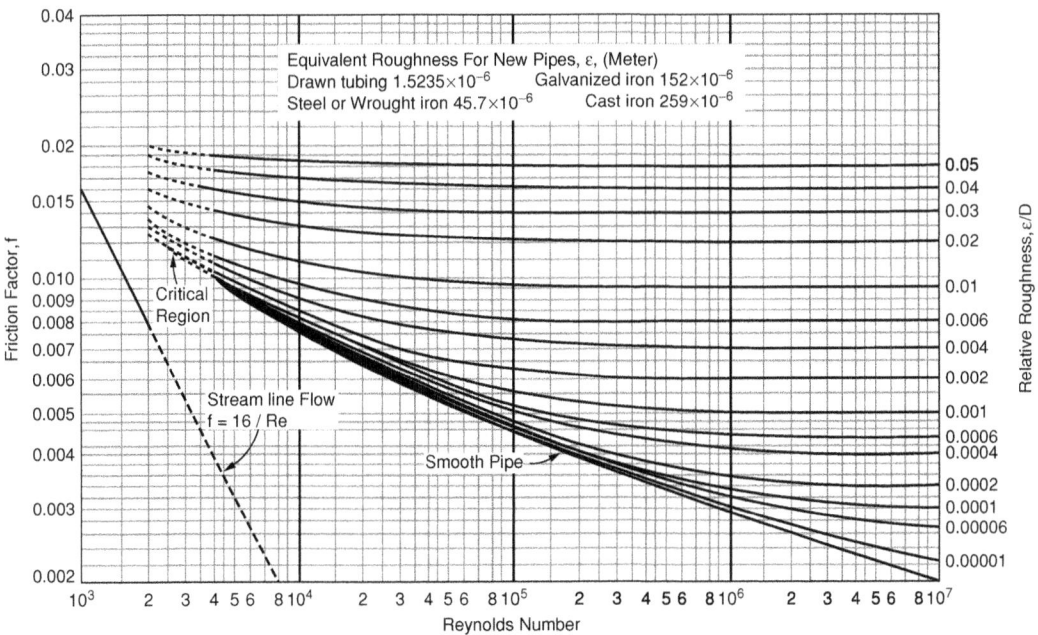

Figure 6.25 The Moody diagram for the Fanning friction factor. (Based on Moody, L. F. Friction factors for pipe flow. Trans. ASME. 66:671, 1944.)

Example 6.15. What pressure must be generated at the discharge of a pump that delivers 100 L/min of a fluid having a specific gravity of 1.02 and a viscosity of 100 centipoises? The fluid flows through a 1.5 in. (nominal) sanitary pipe, 50 m long. The pipe is straight and level, and the discharge end of the pipe is at atmospheric pressure.

Solution:

The Reynolds number is determined to determine if flow is laminar or turbulent.

$$Re = \frac{D\bar{V}\rho}{\mu}$$

Substituting units of terms in the equation for Re:

$$Re = \frac{D(m)\bar{V}(m/s)\rho(kg/m^3)}{\mu(Pa \cdot s)}$$

[Note that the Reynolds number will be dimensionless if the base units for Pa · s, kg/(m s), is substituted in the dimensional equation.]

Converting values of variables into SI units:

$$\rho = 1.02 \cdot \frac{1000\,kg}{m^3} = 1020\,kg/m^3$$

$$D = 1.402\,in. \cdot \frac{0.0254\,m}{in.} = 0.0356\,m$$

$$\bar{V} = \frac{q}{area} = \frac{0.00167\,m^3}{s} \cdot \frac{1}{\pi/4(0.0356)^2\,m^2} = 1.677\,m/s$$

$$q = 100\frac{1}{min} \cdot \frac{0.001\,m^3}{L} \cdot \frac{1\,min}{60\,s} = 0.00167\,m^3/s$$

$$\mu = (100\,cP \cdot \frac{0.001\,Pa \cdot s}{cP} = 0.1\,Pa \cdot s$$

Calculating the Reynolds number:

$$Re = \frac{D\bar{V}\rho}{\mu} = \frac{(0.0356)(1.677)(1020)}{0.1} = 609$$

Flow is laminar. Using Equation (6.44):

$$\Delta P = \frac{2(16/Re)(\bar{V})^2(\rho)(L)}{D} = \frac{2(16)(1.677)^2(1020)(50)}{609(0.0356)}$$

$$= 211,699\,Pa\,gauge$$

Because the pressure at the end of the pipe is 0 gauge pressure (atmospheric pressure), the pressure at the pump discharge will be

$$P = 211,699\,Pa\,gauge$$

Example 6.16. Milk with a viscosity of 2 centipoises and a specific gravity of 1.01 is being pumped through a 1 in. (nominal) sanitary pipe at the rate of 3 gallons per minute. Calculate the pressure drop in lb_f/in^2 per foot length of level straight pipe.

Solution:

Use SI units and Equations (6.40) or (6.41) to calculate ΔP and convert to lb_f/in^2. From Table 6.2, a 1 in (nominal) sanitary pipe will have an inside diameter of 0.02291 m.

$$q = 3\frac{gal}{min} \cdot \frac{0.00378541\,m^3}{gal} \cdot \frac{1\,min}{60\,s} = 0.00018927\,m^3/s$$

$$\bar{V} = \frac{q}{A} = 0.00018927\frac{m^3}{s} \cdot \frac{1}{\pi/4(0.02291)^2\,m^3} = 0.459\,m/s$$

$$\rho = 1.01 \cdot \frac{1000\,kg}{m^3} = 1010\,kg/m^3$$

$$Re = \frac{D\bar{V}\rho}{\mu} = \frac{0.02291(0.459)(1010)}{0.002} = 5310$$

Because Re > 2100 flow is turbulent, use Equation (6.45) for calculating ΔP and Equation (6.47) for f at Re < 10^4. Sanitary pipe is considered a smooth pipe.

$$f = 0.193(Re)^{-0.35} = 0.193(5310)^{-0.35} = 0.0095$$

f = 0.0095 can also be obtained from Fig. 6.25. The pressure drop per unit length of pipe may be calculated as $\Delta P/L$. Because the units used in the calculations are in SI, the result will be Pa/m. This can then be converted to the desired units of $(lb\ f_f/in.^2)/ft$. It may also be solved by substituting a value for L in Equation (6.45).

$$L = 1ft(0.3048m/ft) = 0.3048m.$$

$$\Delta P = \frac{2f\,\bar{V}^2 L\rho}{D} = \frac{2(0.0095)(0.459)^2(0.3048)(1010)}{0.02291}$$

$$= 53.79\,Pa$$

$$\Delta P = 53.79\,Pa \cdot \frac{1\,lb_f/in^2}{6894.757\,Pa} = 0.0078\frac{lb_f/in^2}{ft\ of\ pipe}$$

Example 6.17. Calculate the pressure drop for water flowing at the rate of 10 gal./min through 100 m of level straight wrought iron pipe having an inside diameter of 0.3579 m. Use a density for water of 62.4 lb/ft^3 and a viscosity of 0.98 centipoises.

Solution:

This problem is different from Examples 6.15 and 6.16 in that the pipe is not smooth and therefore Equations (6.46) or (6.47) cannot be used for calculating f. After calculating the Reynolds number, f can be determined from a Moody diagram (Fig. 6.25) if flow is turbulent, and ΔP can be calculated using Equation (6.45).

From Table 6.2: D = 0.03579 m

$$q = \frac{10\,\text{gal}}{\text{min}} \cdot \frac{0.00378541\,\text{m}^3}{\text{gal}} \cdot \frac{1\,\text{min}}{60\,\text{s}} = 0.0006309\,\text{m}^3/\text{s}$$

$$\text{Re} = \frac{D\bar{V}\rho}{\mu} = \frac{0.03579(0.6274)(999.7)}{0.00098} = 22{,}906$$

$$\rho = 62.4\frac{\text{lb}_m}{\text{ft}^3} \cdot \frac{(3.281)^3\text{ft}^3}{\text{m}^3} \cdot \frac{0.45359\,\text{kg}}{\text{lb}_m} = 999.7\,\text{kg/m}^3$$

$$\bar{V} = \frac{q}{A} = \frac{0.0006309\,\text{m}^3/\text{s}}{(\pi/4)(0.03579)^2(\text{m}^2)} = 0.6274\,\text{m/s}$$

$$\mu = 0.98\,\text{cP} \cdot \frac{0.001\,\text{Pa} \cdot \text{s}}{\text{cP}} = 0.00098\,\text{Pa} \cdot \text{s}$$

To use Fig. 6.25, first determine ε/D. From Fig. 6.24, $\varepsilon = 0.0000457$ m. $\varepsilon/D = 0.0000457/0.03579 = 0.00128$. For Re $= 2.29 \times 10^4$, and $\varepsilon/D = 0.00128$, f = 0.0069.

$$\Delta P = \frac{2(0.0069)(0.6274)^2(100)(999.7)}{0.03579} = 15.17\,\text{kPa}$$

6.6.4   Frictional Resistance to Flow of Non-Newtonian Fluids

For fluids whose flow behavior can be expressed by Equation (6.2) (pseudoplastic or power law fluids), the Rabinowitsch-Mooney equation (Eq. 6.18) for the shear rate at the wall was derived in the section "Velocity Profile and Shear Rate for a Power Law Fluid." Substituting the shear stress at the wall (Eq. 6.22) for τ_w and the shear rate at the wall, γ_w (Eq. 6.18), in Equation (6.2):

$$\frac{\Delta PR}{2L} = K\left[\frac{\bar{V}}{R} \cdot \frac{3n+1}{n}\right]^n$$

$$\Delta P = \frac{2LK(\bar{V})^n}{(R)^{n+1}}\left[\frac{3n+1}{n}\right]^n \tag{6.48}$$

Equation (6.44) for the pressure drop of a fluid in laminar flow through a tubular conduit may be written in terms of the radius of the tube and the Reynolds number as follows:

$$\Delta P = \frac{2(16/\text{Re})(\bar{V})^2\rho L}{2R} \tag{6.49}$$

Equation (6.49) is simply another form of Equation (6.44). These equations are very general and can be used for any fluid. In laminar flow, the friction factor, expressed as 16/Re, may be used as a means of relating the pressure drop of a non-Newtonian fluid to a dimensionless quantity, Re. Thus, Re for a non-Newtonian fluid can be determined from the flow behavior and consistency indices. Equating

Equations (6.48) and (6.49):

$$\frac{2LK(\bar{V})^n}{(R)^{n+1}}\left[\frac{3n+1}{n}\right]^n = \frac{2(16/Re)(\bar{V})^2L\rho}{2R}$$

$$Re = \frac{8(\bar{V})^{2-n}(R)^n\rho}{K\left[\dfrac{3n+1}{n}\right]^n} \tag{6.50}$$

Equation (6.50) is a general expression for the Reynolds number that applies for both Newtonian (n = 1) and non-Newtonian liquids (n ≠ 1). If n = 1, Equation (6.50) is exactly the same as Equation (6.42), when K = μ. Pressure drops for power law non-Newtonian fluids flowing through tubes can be calculated using Equation (6.49) using the Reynolds number calculated using Equation (6.50), if flow is laminar. If flow is turbulent, the Reynolds number can be calculated using Equation (6.50) and the pressure drop calculated using Equation (6.45) with the friction factor Af A determined from this Reynolds number using Equations (6.46) or (6.47). The Moody diagram (Fig. 6.25) can also be used to determine "f" from the Reynolds number.

Example 6.18. Tube viscometry of a sample of tomato catsup shows that flow behavior follows the power law equation with K = 125 dyne As/cm² and n = 0.45. Calculate the pressure drop per meter length of level pipe if this fluid is pumped through a 1-in. (nominal) sanitary pipe at the rate of 5 gallons (US)/min. The catsup has a density of 1.13 g/cm³.

Solution:

Convert given data to SI units. n = 0.45 (dimensionless).

$$K = 125\frac{\text{dyne s}^n}{\text{cm}^2} \cdot \frac{1\times10^{-5}\,\text{N}}{\text{dyne}} \cdot \frac{(100)^2\,\text{cm}^2}{\text{m}^2} = 12.5\,\text{Pa}\cdot\text{s}^n$$

From Table 6.2, D = 0.02291 m; R = 0.01146 m

$$q = \frac{5\,\text{gal}}{\text{min}} - \frac{0.00378541\text{m}^3}{\text{gal}} - \frac{1\,\text{min}}{60\,\text{s}} = 0.00031545\,\text{m}^3/\text{s}$$

$$\bar{V} = \frac{q}{A} = \frac{0.00031545\,\text{m}^3/\text{s}}{(\pi/4)(0.02291)^2\,\text{m}^2} = 0.7652\,\text{m/s}$$

$$\rho = 1.13\frac{\text{g}}{\text{cm}^3}\frac{(100)^3\,\text{cm}^3}{\text{m}^3}\frac{1\,\text{kg}}{1000\,\text{g}} = 1130\,\text{kg/m}^3$$

L = 1 m

The Reynolds number is calculated using Equation (6.50).

$$Re = \frac{8(0.7652)^{2-0.45}(0.01146)^{0.45}(1130)}{12.5\left[\dfrac{3(.45)+1}{0.45}\right]^{0.45}} = 30.39$$

Flow is laminar and pressure drop is calculated using Equation (6.45).

$$\Delta P = \frac{2\left(\dfrac{16}{30.39}\right)(0.7652^2)(1130)(1)}{0.02291} = 30.41 \text{ kPa}$$

Example 6.19. Peach puree having a solids content of 11.9% has a consistency index K of 72 dyne As^n/cm^2 and a flow behavior index, n = 0.35. If this fluid is pumped through a 1-in. (nominal) sanitary pipe at 50 gallons (US)/min, calculate the pressure drop per meter of level straight pipe. Peach puree has a density of 1.07 g/cm^3.

Solution:

D = 0.02291 (from Table 6.2); R = 0.01146 m; n = 0.35

$$K = 72 \frac{\text{dyne s}^n}{\text{cm}^2} \cdot \frac{(10)^{-5}}{\text{dyne}} \cdot \frac{(100)^2 \text{ cm}^2}{\text{m}^2} = 7.2 \text{ Pa} \cdot \text{s}^n$$

$$\rho = 1070 \text{ kg/m}^3$$

$$q = 50\frac{\text{gal}}{\text{min}} \frac{0.00378541 \text{ m}^3}{\text{gal}} \frac{1 \text{ min}}{60 \text{ s}} = 0.0031545 \text{ m}^3/\text{s}$$

$$\bar{V} = \frac{q}{A} = \frac{0.00378541 \text{ m}^3}{(\pi/4)(0.02291)^2} = 7.6523 \text{ m/s}$$

Using Equation (6.50):

$$Re = \frac{8(7.6523)^{2-0.35}(0.01146)^{0.35}(1070)}{7.2\left[\dfrac{3(0.35)+1}{0.35}\right]^{0.35}} = 3850$$

Flow is turbulent. The sanitary pipe is a smooth pipe so Equation (6.47) will be used to calculate f.

$$f = 0.193(Re)^{-0.35} = 0.193(3850)^{-0.35} = 0.010731$$

P can now be calculated using equation 42. L = 1 m.

$$\Delta P = \frac{2(0.01073)(7.6523)^2(1)(1070)}{0.02291} = 58.691 \text{ kPa/m}$$

6.6.5 Frictional Resistance Offered by Pipe Fittings to Fluid Flow

The resistance of pipe fittings to flow can be evaluated in terms of an equivalent length of straight pipe. Each type of pipe fitting has its specific flow resistance expressed as a ratio of equivalent length of straight pipe (L=) over its diameter. Table 6.3 lists the specific resistance of various pipe fittings. The equivalent length of a fitting, which is the product of L=/D obtained from Table 6.3 and the pipe diameter, is added to the length of straight pipe within the piping system to determine the total drop pressure drop across the system.

Table 6.3 Specific Resistance of Various Pipe Fittings Expressed as an Equivalent Length of Straight Pipe to Pipe Diameter Ratio.

Fitting	L'/D (dimensionless)
90° elbow std.	35
45° elbow std.	15
Tee (Used as a coupling), branch plugged	20
Tee (used as an elbow) entering the branch	70
Tee (used as an elbow) entering the tee run	60
Tee branching flow	45
Gate valve, fully open	10
Globe valve, fully open	290
Diaphragm valve, fully open	105
Couplings and unions	Negligible

Example 6.20. Calculate the pressure drop due to fluid friction across 50 m of a 1-in. (nominal) sanitary pipe that includes five 90-degree Eelbows in the piping system. Tomato catsup with properties described in Example 6.18 in the section "Frictional Resistance to Flow for non-Newtonian Fluids" flows through the pipe at the rate of 5 gallons (US)/min.

Solution:

In Example 6.18 of the section referred to above, the tomato catsup pumped under these conditions exhibited a pressure drop of 30.41 kPa/m of level pipe. The equivalent length of the five elbows is determined as follows:

L=/D from Table 6.3 for a 90-degree elbow is 35. The diameter of the pipe is 0.02291 m.
Total equivalent length of straight pipe $= 50\text{ m} + 5\,(35)(0.02291) = 54\text{ m}$.
$\Delta P = (30.41\text{ kPa/m})(54\text{m}) = 1642\text{ kPa}$.

6.7 MECHANICAL ENERGY BALANCE: THE BERNOULLI EQUATION

When fluids are transferred from one point to another, a piping system is used. Because fluids exhibit a resistance to flow, an energy loss occurs as the fluid travels downstream along the pipe. If the initial energy level is higher than the energy at any point downstream, the fluid will flow spontaneously. On the other hand, if the energy change needed to take the fluid to a desired point downstream exceeds the initial energy level, energy must be applied to propel the fluid through the system. This energy is an input provided by a pump.

An energy balance across a piping system is similar to the energy balance made in Chapter 5. Table 6.4 lists the energy terms involved in fluid flow, their units, and the formulas for calculating them. All energy entering the system including the energy input must equal that leaving the system and the energy loss due to fluid friction. The boundaries of the system must be carefully defined to identify the input and exit energies. As in previous problems in material and energy balances, the boundaries

Table 6.4 Energy Terms Involved in the Mechanical Energy Balance for fluid Flow in a Piping System, the Formulas for Calculating Them, and Their Units.

Energy Term	Formula	Dimensional Expression	Formula (basis: 1 kg)	Unit
Potential energy Pressure	$m\left(\dfrac{P}{\rho}\right)$	$\dfrac{kg(N \cdot m^{-2})}{kgm^{-3}}$	$\dfrac{P}{\rho}$	joule/kg
Elevation	mgh	$kg(m \cdot s^{-2})(m)$	gh	joule/kg
Kinetic energy	$\dfrac{1}{2}mV^2$	$kg(m \cdot s^{-1})^2$	$\dfrac{V^2}{2}$	joule/kg
Work input (from pump)	W		W	joule/kg
Frictional resistance	$\dfrac{m\Delta P_f}{\rho}$	$\dfrac{kg(N \cdot m^{-2})}{kg \cdot m^{-3}}$	$\dfrac{\Delta P_f}{\rho}$	joule/kg

may be moved around to isolate the unknown quantities being asked in the problem. The mechanical energies involved are primarily kinetic and potential energy and mechanical energy at the pump.

The potential and kinetic energy could have finite values on both the source and discharge points of a system. Work input should appear on the side of the equation opposite the frictional resistance term. A balance of the intake and exit energies in any given system including the work input and resistance terms based on a unit mass of fluid is

$$\frac{P_1}{\rho} + gh_1 + \frac{V_1^2}{2} + W_s = \frac{P_2}{\rho} + gh_2 + \frac{V_2^2}{2} + \frac{\Delta P_f}{\rho} \qquad (6.51)$$

Equation (6.51) is the Bernoulli equation.

Example 6.21. A pump is used to draw tomato catsup from the bottom of a de-aerator. The fluid level in the de-aerator is 10 m above the level of the pump. The de-aerator is being operated at a vacuum of 0.5 kg_f/cm^2 (kilogram force/cm^2). The pipe connecting the pump to the de-aerator is a 2.5-in. (nominal) stainless steel sanitary pipe, 8 m long with one 90-degree elbow. The catsup as a density of 1130 kg/m^3, a consistency index K of 10.5 Pa · s^n, and a flow behavior index n of 0.45 (dimensionless). If the rate of flow is 40 L/min, calculate the pressure at the intake side of the pump to induce the required rate of flow.

Solution:

Figure 6.26 is a diagram of the system. The problem asks for the pressure at a point in the system before work is applied by the pump to the fluid. Therefore, the term W_s in Equation (6.51) is zero. The following must be done before quantities can be substituted into Equation (6.51).
Assume atmospheric pressure is 101 kPa. P_1 must be converted to absolute pressure. The volumetric rate of flow must be converted to velocity ($\underline{V}$). The frictional resistance to flow ($\Delta P_f/\rho$ must be calculated.

$$\text{Vacuum} = 0.5 \, \frac{kg_f}{cm^2} \cdot \frac{0.8 \, m \cdot kg}{kg_f \cdot s^2} \cdot \frac{(100)^2 cm^2}{m^2} = 49 \, kPa$$

$$P_1 = \text{Atmospheric pressure} - \text{vacuum} = 101 - 49 = 52 \, kPa.$$

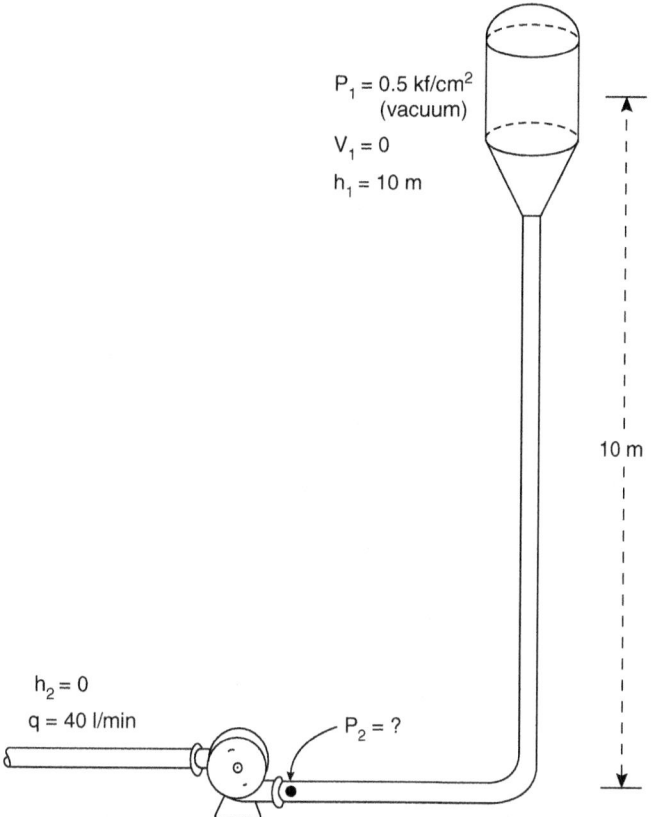

$P_1 = 0.5 \text{ kf/cm}^2$
(vacuum)

$V_1 = 0$

$h_1 = 10 \text{ m}$

10 m

$h_2 = 0$

$q = 40 \text{ l/min}$

$P_2 = ?$

Figure 6.26 Diagram of piping system for catsup from a de-aerator to a pump.

$$q = 40 \frac{\text{L}}{\text{min}} \cdot \frac{0.001 \text{ m}^3}{\text{L}} \cdot \frac{1 \text{ min}}{60 \text{ s}} = 0.0006666 \text{ m}^3/\text{s}$$

$$\text{Cross-sectional area, A} = \frac{\pi}{4}(0.06019)^2 = 0.002845 \text{ m}^2$$

From Table 6.2, $D = 0.06019$; $R = 0.03009$.

$$\bar{V} = \frac{q}{A} = \frac{0.0006666 \text{ m}^3/\text{s}}{0.002845 \text{ m}^2} = 0.2343 \text{ m/s}$$

The frictional resistance to flow can be calculated using Equation (6.45) after the Reynolds number has been calculated using Equation (6.50).

$$\text{Re} = \frac{8(0.2343)^{2-0.45}(0.03009)^{0.45}(1130)}{10.5 \left[\dfrac{3(0.45) + 1}{0.45} \right]^{0.45}} = 8.921$$

Flow is laminar and using Equation (6.45):

$$\frac{\Delta P_f}{L\rho} = \frac{2(16/8.920)(0.2343)^2}{0.060019} = 3.272 \, J/(kg \cdot m)$$

where L = length of straight pipe + equivalent length of fitting.
From Table 6.3, the equivalent length of a 90-degree elbow is 35 pipe diameters.

$$\frac{\Delta P_f}{\rho} = 3.272(10.1) = 33.0 \, J/kg$$

$$L = 8 + 1(35)(0.06019) = 10.1 \, m$$

Substituting all terms in Equation (6.51):

$$P_2 = \left[\frac{52,00}{1130} + 98 - 33.0 - 0.03\right](1130)$$

$$\frac{52,000}{1130} + 9.8(10) + 0 + 0 = \frac{P_2}{1130} + 0 + \frac{(0.2343)^2}{2} + 33$$

$$= (46.02 + 9833.00.03)(1130)$$

$$= 125.4 \, kPa \, absolute$$

This problem illustrates one of the most important factors often overlooked in the design of a pumping system; that is, to provide sufficient suction head to allow fluid to flow into the suction side of the pump. As an exercise, the reader can repeat these same calculations using a pipe size of 1.0 in. at the given volumetric flow rate. The calculated absolute pressure at the intake side of the pump will be negative, a physical impossibility.

Another factor that must be considered when evaluating the suction side of pumps is the vapor pressure of the fluid being pumped. If the suction pressure necessary to induce the required rate of flow is lower than the vapor pressure of the fluid at the temperature it is being pumped, boiling will occur, vapor lock develops, and no fluid will flow into the intake side of the pump. For example, suppose the pressure at the pump intake is 69.9 kPa. From Appendix Table A.3, the vapor pressure of water at 90°C is 70 kPa. If the height of the fluid level is reduced to 4.8 m, the pressure at the pump intake will be 69 kPa, therefore, under these conditions, if the temperature of the catsup is 90°C and above, it would not be possible to draw the liquid into the intake side of the pump.

Example 6.22. The pump in Example 6.21 delivers the catsup to a heat exchanger and then to a filler. The discharge side of the pump consists of 1.5-in. stainless steel sanitary pipe 12 m long with two elbows. Joined to this pipe is the heat exchanger, which is 20 m of steam-jacketed 1-in. stainless steel sanitary pipe with two U-bends. (Assume the resistance of a U-bend is double that of a 90-degree elbow.) Joined to the heat exchanger is another 20 m of 1.5 in. pipe with two 90-degree elbows and the catsup is discharged into the filler bowl at atmospheric pressure. The pipe elevation at the point of discharge is 3 m from the level of the pump. The piping system before the pump and the rates of flow are the same as that in Example 6.21. Calculate the power requirement for the pump. The diagram of the system is shown in Fig. 6.27.

Solution:

From Example 6.21, $h_1 = 10$ m; $V_1 = 0$; $P_1 = 52$ kPa. $\Delta P_f/\rho$ for the section prior to the pump = 33 J/kg. $\Delta P_f/\rho$ for the section at the discharge side of the pump will be calculated in two steps: for the 1.5-in. pipe and the 1-in. pipe at the heat exchanger. D =1.5-in. pipe = 0.03561 m, R = 0.0178.
Equivalent length of four 90-degree elbows = 4(35)(0.03561) = 4.9854 m.

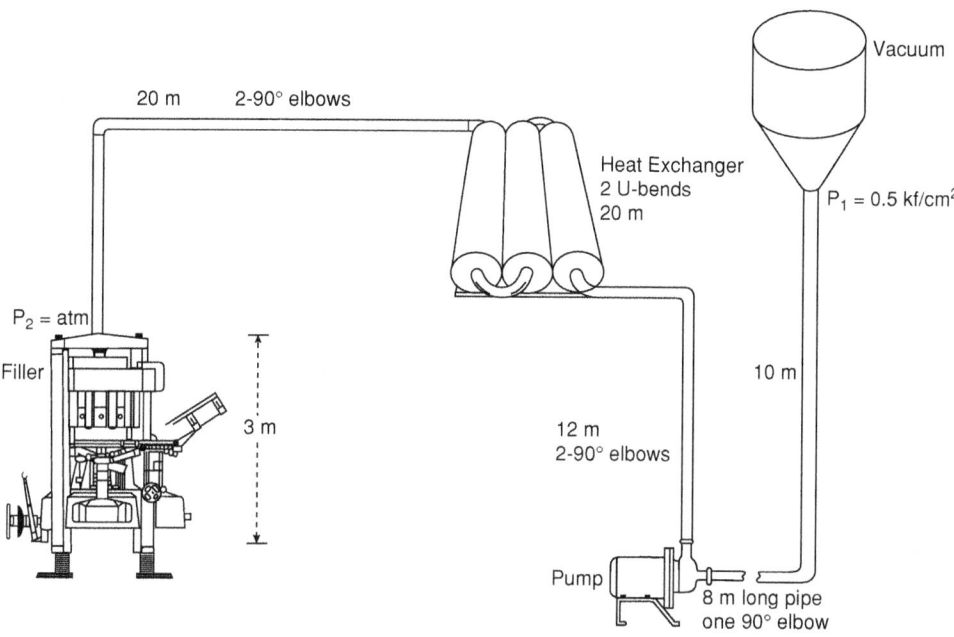

Figure 6.27 Diagram of a piping system for catsup from a de-aerator to a pump and heat exchanger and filler.

Total length $= 20 + 12 + 4.99 = 36.99$ m.

$$A = \frac{\pi}{4}(0.03561)^2 = 0.00099594 \text{ m}^2$$

$$\frac{\Delta P_f}{\rho} = \frac{2(16/35.82)(0.6693)^2(36.99)}{0.03561} = 415.7 \text{ J/kg}$$

$$\bar{V} = \frac{q}{A} = \frac{0.0006666 \text{ m}^3/\text{s}}{0.0009954 \text{ m}^2} = 0.6693 \text{ m/s}$$

$$Re = \frac{8(0.6693)^{2-0.45}(0.0178)^{0.45}(1130)}{10.5\left[\dfrac{3(0.45)+1}{0.45}\right]^{0.45}} = 35.87$$

D of 1-in. pipe $= 0.02291$; $R = 0.01146$ m.

$$A = \pi(0.01146)^2 = 0.0004126 \text{ m}^2$$

$$V = \frac{q}{A} = \frac{0.0006666 \text{ m}^3/\text{s}}{0.0004126 \text{ m}^2} = 1.616 \text{ m/s}$$

$$\frac{\Delta P_f}{\rho} = \frac{2(16/115.2)(1.616)^2(L)}{0.02291} = 31.64(L) \text{ J/kg}$$

$$Re = \frac{8(1.616)^{2-0.45}(0.01146)^{0.45}(1130)}{10.5\left[\dfrac{3(0.45)+1}{0.45}\right]^{0.45}} = 115.3$$

The equivalent length of two U-bends $= 2(2)(35)(D)$: $L = 20 + 2(2)(35)(0.02291) = 23.2$ m.

$$\frac{\Delta Pf}{\rho} = 31.64(23.27) = 734.0 \text{ J/kg}$$

Total resistance to flow $= 33.0 + 415.7 + 734.0 = 1182.7$ J/kg.
Substituting in Equation (6.51): $P_2 = 101000$ Pa; $h_2 = 3$ m; and $V_2 = 0.6697$ m/s.

$$\frac{52,000}{1130} + 9.8(10) + 0 + W_s = \frac{101,000}{1130} + 9.8(3) + \frac{(0.6697)^2}{2} + 1183.2$$

$$W_s = \frac{101,000}{1130} + 9.8(3) + \frac{(0.6697)^2}{2} + 1183.2 - \frac{52,000}{1130} - 9.8(10)$$

$$= 89.4 + 29.4 + 0.2 + 1182.7 - 46 - 98 = 1157.7 \text{ J/kg}$$

$$\text{Mass flow rate} = q\,(\rho) = 0.0006666(1130) = 0.7532 \text{ kg/s}$$

$$\text{Power} = (1157.7 \text{ J/kg})(0.7532 \text{ kg/s}) = 872 \text{ watts}$$

6.8   PUMPS

6.8.1   Types of Pumps and Their Characteristics

Pumps for transporting liquids may be classified into two general classes: *positive displacement* and *centrifugal*. Positive displacement pumps can operate effectively over a range of relatively slow drive shaft rotational speeds, and they deliver a fixed volume per revolution. These pumps can handle viscous liquids, generate high discharge pressure, are self-priming, and have some suction lift capability. Because of the relatively flat flow delivery within a moderate range of discharge pressure, positive displacement pumps are useful as metering pumps in processes where flow rate is a critical parameter for success of a process. Positive displacement pump flow must not be throttled otherwise high pressures will develop and break the pipes, destroy bearing seals in the pump drive shaft or burn up the drive motor because of the excessive load. In contrast, centrifugal pumps depend on an impeller rotating at high rotational speeds to increase the kinetic energy of fluid at the periphery of the impeller propelling the fluid at high velocity toward the pump discharge. Centrifugal pumps must be primed before they can generate any flow, or the source must be elevated above the pump to permit natural flooding of the pump inlet by gravity flow. Pressures generated are low and these pumps are not capable of pumping very viscous fluids. Fluid aeration, foaming, and fluid temperature elevation in the pump casing are other undesirable characteristics of centrifugal pumps particularly when operating at head pressures close to the maximum for the particular pump. However, when pressures against which the pump must work are low, and the fluid has relatively low viscosity, centrifugal pumps are the least expensive pump to use.

Positive displacement pumps may be classified as: reciprocating plunger or piston, gear; lobe, diaphragm, or progressing cavity pumps. Plunger or piston pumps generate the highest pressure but they deliver pulsating flow and develop pressure hammers, which might be detrimental to equipment downstream or to the piping system. Diaphragm pumps have the pumped fluid within a cavity formed by the diaphragm and the pump head cover. The diaphragm repeatedly moves forward and retracts. With each retraction of the diaphragm, the enlarged volume of the fluid cavity permits fluid to enter the cavity while forward motion of the diaphragm forces the fluid from the cavity toward the discharge. Check valves control the intake and discharge of fluid from the diaphragm cavity. A major advantage of diaphragm pumps is the complete isolation of the fluid from the surroundings while in the diaphragm cavity preventing contamination or aeration of the product. Diaphragm pumps are also available with an air drive rather than electric, and is a good feature in areas that are constantly wet. All the other positive displacement pumps have moving shafts that turn the pump rotor or gear, and these shafts must be adequately sealed to prevent fluid from leaking around the shaft. Progressing cavity pumps have the advantage of a gradual build-up of pressure as the fluid moves forward from the fluid intake at the opposite end of the pump from the discharge. Thus, the shaft seal for progressing cavity pumps only works against a fraction of the pressure generated at the pump discharge.

6.8.2 Factors to Be Considered in Pump Selection

Flow and *discharge pressure:* Pumps are sized according to the amount of fluid they deliver at designated discharge pressures. The latter is commonly called the *head*, which is defined as the height of a column of fluid that generates a pressure at its base equivalent to the pressure registered by a gauge placed at the discharge port of the pump. A graph of flow as a function of the head is called the *performance curve* of a pump and is available from pump manufacturers. Each make, model, and size of pump has a performance curve, which should be consulted to verify if a given pump can satisfactorily perform the required application.

Fluid viscosity: The viscosity of the fluid not only determines the head against which the pump must work but also the net positive suction head available at the suction inlet to the pump. For example, centrifugal pumps require a *net positive suction head* (NPSH) to be available, in the application under consideration, which is greater than that specified for a particular make and model of the pump. The NPSH is discussed im more detail later in this section. Generally, the limitation of centrifugal pumps in effectively pumping a fluid of high viscosity is the relatively high available NPSH they require compared to positive displacement pumps. However, even with positive displacement pumps, fluids with very high viscosity that would exhibit a very high pressure drop due to fluid friction in a short length of pipe connecting the suction to the fluid reservoir could prevent acceptable functioning of the pump for the desired application.

Fluid temperature affects the available NPSH. High fluid vapor pressure results in flashing of vapor to negate the suction developed by the pump and result in a condition called *vapor lock*. Fluid is prevented from entering the pump resulting in zero flow. To prevent vapor lock when pumping high temperature fluids, the fluid source must be elevated above the pump level in order that gravity flow will permit flooding of the pump inlet at the desired rate of flow.

Fragility of suspended particles: Fluids that contain fragile suspended particles require pumps that have large cavities to hold the particles without compressing them, and there should be a minimum of restrictions, sharp turns, and areas of low fluid velocities in the flow stream to prevent solids from bridging while flowing out of the pump.

Ease of cleaning: Fluids that are good substrates for microbiological growth will require a pump that can be easily cleaned. Pumps should be capable of being cleaned in place. Otherwise, the pump should be capable of being cleaned by simply removing the pump head cover without having to disconnect the pump from the connecting pipes.

Abrasiveness of suspended solids in the fluid: Suspended solids that are abrasive will require a pump where there is a minimum likelihood of having particles trapped between a stationary and moving part. Centrifugal and diaphragm pumps are ideal for this type of application.

Corrosiveness of fluid: Material of construction of any part of the pump that contacts an abrasive fluid must be of the type that will resist chemical attack of the fluid.

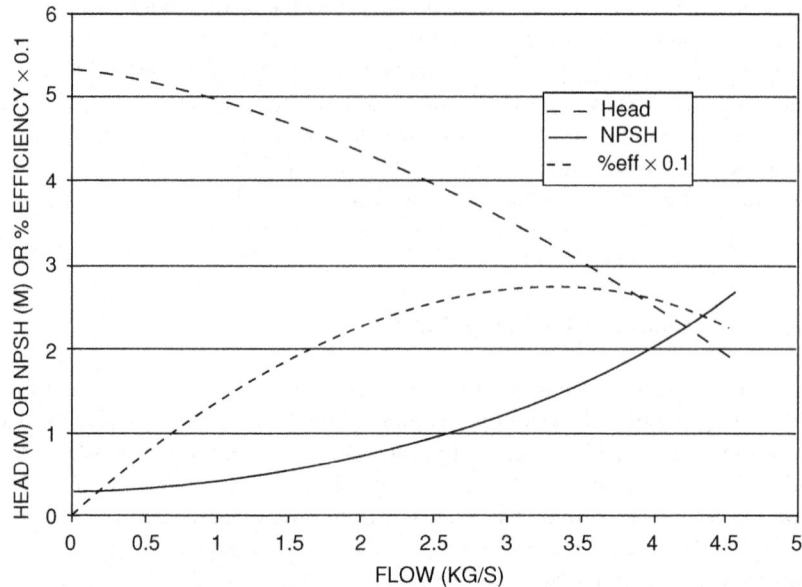

Figure 6.28   Performance curve of a volute type centrifugal pump.

6.8.3   Performance Curves of Pumps

The performance curve of a volute-type centrifugal pump is shown in Fig. 6.28. A characteristic of centrifugal pumps is the low head that they can work against. This particular pump was effectively shut off at a pressure equivalent to 5.3 m of fluid (51.9 kPa ga). Another characteristic is the required NPSH. For example, at the maximum pump efficiency, flow is 3.5 kg/s and the head is 2.8 m. This will require a NPSH of 1.5 m. The NPSH is defined as follows:

$$\text{NPSH available} = \left(\frac{1}{\rho g}\right)(P_{atm} \pm \rho\, g h - \Delta P_f - P_v)$$

where ρ, g, h, and ΔP_f are as previously defined. P_{atm} is the atmospheric pressure, and P_v is the fluid vapor pressure.

Example 6.23.   A fluid with a density of 1004 kg/m^3 and a viscosity of 0.002 Pa s is to be pumped using the pump with a performance curve shown in Fig. 6.28. If the inlet pipe is 1.5 in. sanitary pipe, the suction pipe section is 1.2 m long with one 90-degree elbow, fluid source is 1 m below pump level, and the fluid vapor pressure is the same as that of water at 35°C (5.6238 kPa), calculate the NPSH available when operating at a head of 2.8 m, and determine if the pump will be suitable for operating under the given conditions. The atmospheric pressure is 101,000 Pa.

Solution:

The pipe diameter is 0.03561 m. The equivalent length of pipe L = 1.2 + 1(35)(.03561) = 2.45 m. At a head of 2.8 m, the flow rate is 3.5 kg/s.

$$\bar{V} = 3.5 \, \frac{kg}{s} \cdot \frac{1m^3}{1004\,kg} \cdot \frac{1}{\pi \left(\dfrac{.03561}{2} \right)^2} = 3.5 \frac{m}{s}$$

Using Poiseuille's equation:

$$\Delta P_f = \frac{8\,L\mu V}{R^2} = \frac{8(2.446)(0.002)(3.5)}{\left(\dfrac{0.03561}{2} \right)^2} = 432 \text{ Pa}$$

The NPSH available is greater than the 1.5 m required by the pump, therefore it will be possible to operate the pump under the specified conditions.

$$\text{NPSH available} = \frac{1}{(1004)(9.8)} \, [101000 - 9.8(1000)(1) - 432 - 5624] = 8.65 \text{ m}$$

The performance curve of a positive displacement pump shown in Fig. 6.29 depicts a flow rate that depends on the speed of the rotor. Because a fixed volume is delivered with each rotation of the rotor, the rate of flow is proportional to the rotational speed. Note that the rate of flow at 400 rpm is almost double that at 200 rpm. High discharge pressures generated are one order of magnitude greater than those generated by centrifugal pumps. The flow reduction with an increase in head is gradual. Flow reduction with increase in head depends on the rotor design. A tight clearance between the rotor and the casing minimizes backflow of fluid at high pressures thus maintaining a fairly flat flow rate with pressure. Positive displacement pumps generally develop more constancy of flow with increasing head as the viscosity of the fluid increases because backflow is reduced as fluid viscosity increases.

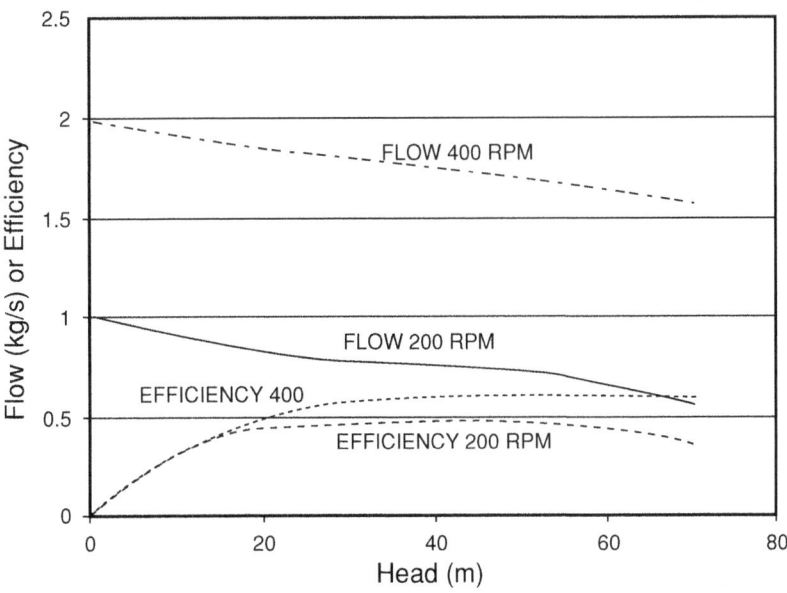

Figure 6.29   Performance curve for a two-lobe rotary positive displacement pump.

Attaining high pressures and maintaining constancy of flow with increasing pressures, however, is not desirable if these conditions are not required in an application. The reduction in backflow requires tight clearances between the rotor and casing, and to generate high pressures, there must be a tight shaft seal increasing friction thus reducing the pump efficiency.

The *pump efficiency* is defined as the ratio of the hydraulic horsepower to the brake horsepower. Let H = pressure at pump discharge expressed as height of fluid in meters, g = acceleration due to gravity in m/s^2 and $\bar{m}$ is the mass rate of flow in kg/s:

$$\text{Hydraulic Horsepower} = \frac{H \cdot g \cdot \bar{m}}{745.7}$$

The numerator is power in Watts and the denominator is 745.7 W/HP.

The brake horsepower is what the pump manufacturer specifies for the motor drive of a pump to deliver a specified rate of flow against a specified head. Efficiencies of centrifugal pumps and positive displacement pumps are rather low as shown in Figs. 6.28 and 6.29. Centrifugal pump efficiencies peak at 28% for the centrifugal pump shown in Fig 6.28, and the maximum efficiency occurs over a very narrow flow rate–head combination. In contrast, positive displacement pump efficiency (Fig. 6.29) remains relatively constant over a wide range of head. Higher rotor speeds induces higher efficiency than at lower speeds. Efficiencies for this particular pump were 50% or 60% at 200 and 400 rpm, respectively, at the optimum head and flow rate. Energy losses in pumps are attributable primarily to backflow and mechanical friction.

PROBLEMS

6.1. The following data were obtained when tomato catsup was passed through a tube having an inside diameter of 1.384 cm and a length of 1.22 m.

Flow rate (cm^3/s)	P (dynes/cm^2)
107.5	50.99 × 10^4
67.83	42.03 × 10^4
50.89	33.07 × 10^4
40.31	29.62 × 10^4
10.10	15.56 × 10^4
8.80	14.49 × 10^4
33.77	31.00 × 10^4
53.36	35.14 × 10^4
104.41	46.85 × 10^4

Determine the fluid consistency index K and the flow behavior index n of this fluid.

6.2. Figure 6.30 shows a water tower and the piping system for a small manufacturing plant. If water flows through the system at 40 L/min, what would be the pressure at point B? The pipes are wrought iron pipes. The water has a density of 998 kg/m^3 and a viscosity of 0.8 centipoise. The pipe is 1.5-in. (nominal) wrought iron pipe.

6.3. The catsup in Problem 1 (n = 0.45, K = 6.61 Pa · s^n) is to be heated in a shell and tube heat exchanger. The exchanger has a total of 20 tubes 7-m long arranged parallel inside a shell. Each tube is a 3/4-in. outside diameter, 18 gauge heat exchanger tube, (From a table

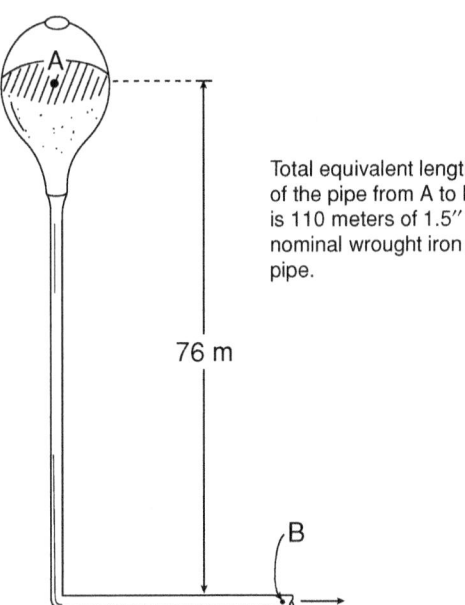

Total equivalent length of the pipe from A to B is 110 meters of 1.5″ nominal wrought iron pipe.

76 m

Figure 6.30 Water tower and piping system for Problem 2.

of thickness of sheet metal and tubes, an 18-gauge wall is 0.049 in.) It is possible to arrange the fluid flow pattern by the appropriate selection of heads for the shell, such that number of passes and the number of tubes per pass can be varied. Calculate the pressure drop across the heat exchanger (the pressure at the heat exchanger inlet necessary to push the product through the heat exchanger) if a flow rate of 40 L of product per min (density = 1013 kg/m³) is going through the system for (a) two-pass (10 tubes/pass) and (b) five-pass system (four tubes per pass). Consider only the tube resistance and neglect the resistance at the heat exchanger heads.

6.4. Figure 6.31 shows a de-aerator operated at 381 mm Hg vacuum. (Atmospheric pressure is 762 mm Hg.) It is desired to allow a positive suction flow into the pump. If the fluid has a density of 1040 kg/m³, the flow rate is 40 L/min, and the pipe is 1.5-in. sanitary pipe, calculate the height "h" the bottom of the de-aerator must be set above the pump level in order that the pressure at the pump intake is at least 5 kPa above atmospheric pressure. The fluid is Newtonian and has a viscosity of 100 centipoises.

[Note: Total length of straight pipe from first elbow below the pump to the entrance to the pump = 4 m. Distance from pump level to horizontal pipe = 0.5 m.]

6.5. Calculate the total equivalent length of 1-in. wrought iron pipe that would produce a pressure drop of 70 kPa due to fluid friction, for a fluid flowing at the rate of 50 L/min. The fluid has a density of 988 kg/m³ and a viscosity of two centipoises.

6.6. Calculate the horsepower required to pump a fluid having a density of 1040 kg/m³ at the rate of 40 L/min through the system shown in Fig. 6.32. $\Delta P_f/\rho$ calculated for the system is 120 J/kg. Atmospheric pressure is 101 kPa.

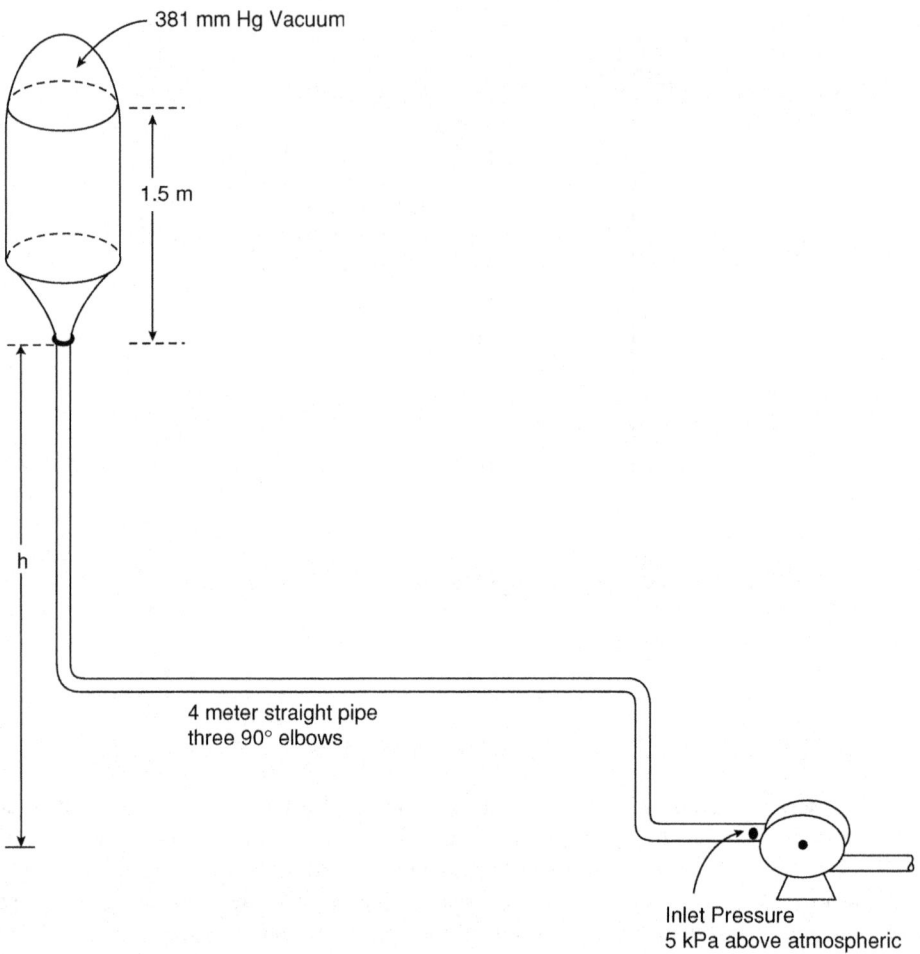

Figure 6.31   Diagram of Problem 4.

6.7. Calculate the average and maximum velocities of a fluid flowing at the rate of 20 L/min through a 1.5-in. sanitary pipe. The fluid has a density of 2030 kg/m³ and a viscosity of 50 centipoises. Is the flow laminar or turbulent?

6.8. Determine the inside diameter of a tube that could be used in a high-temperature, short-time heater-sterilizer such that orange juice with a viscosity of 3.75 centipoises and a density of 1005 kg/m³ would flow at a volumetric flow rate of 4 L/min and have a Reynolds number of 2000 while going through the tube.

6.9. Calculate the pressure generated at the discharge of a pump that delivers a pudding mix ($\rho = 995kg/m^3$; $K = 1.0\ Pa \cdot s^n$, $n = 0.6$) at the rate of 50 L/min through 50 m of a 1.5-in. straight, level 1.5-in. stainless steel sanitary pipe. What would be the equivalent viscosity of a Newtonian fluid that would give the same pressure drop?

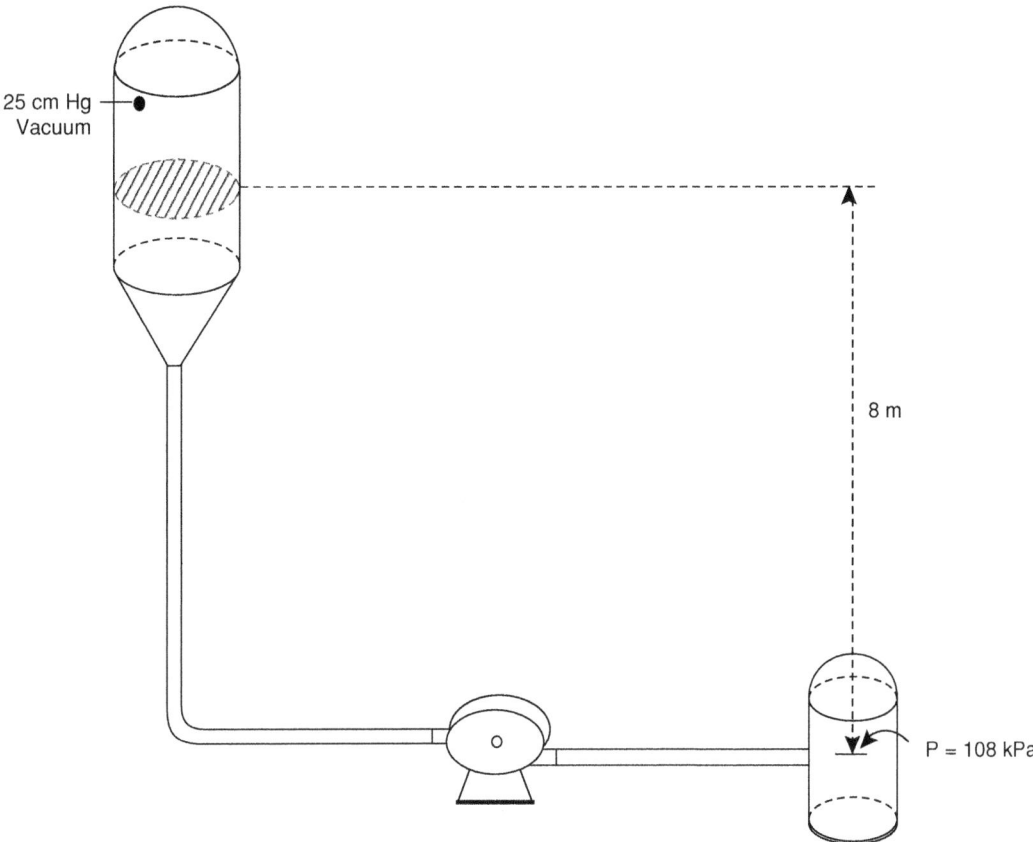

25 cm Hg
Vacuum

8 m

P = 108 kPa

Figure 6.32 Diagram for Problem 6.

6.10. What pipe diameter will give a rate of flow of 4 ft/min for a fluid delivered at the rate of 2 gal/min?

6.11. Calculate the viscosity of a fluid that would allow a pressure drop of 35 kPa over a 5-m length of $3/4$-in. stainless steel sanitary pipe if the fluid is flowing at 2 L/min and has a density of 1010 kg/m^3. Assume laminar flow.

6.12. A fluid is evaluated for its viscosity using a Brookfield viscometer. Collected data of rotational speed in rev/min and corresponding apparent viscosity in centipoises, respectively, are 20, 7230; 10, 12060; 4, 25200; 2, 39500. Is the fluid Newtonian or non-Newtonian. Calculate the flow behavior index, n, for this fluid.

6.13. A fluid having a viscosity of 0.05 lb$_{mass}$/ft (s) requires 30 seconds to drain through a capillary viscometer. If this same viscometer is used to determine the viscosity of another fluid and it takes 20 seconds to drain, calculate the viscosity of this fluid. Assume the fluids have the same densities.

6.14. Figure 6.33 shows a storage tank for sugar syrup that has a viscosity of 15.2 centiposes at 25°C, and a density of 1008 kg/m^3. Friction loss includes an entrance loss to the drain pipe

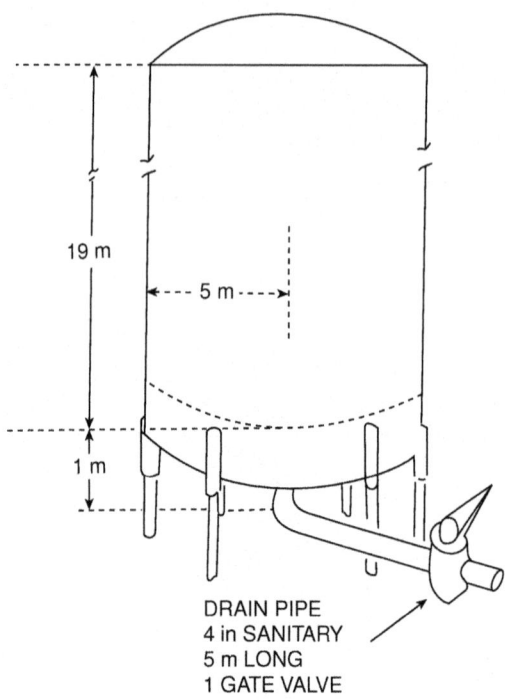

DRAIN PIPE
4 in SANITARY
5 m LONG
1 GATE VALVE

Figure 6.33   Diagram for Problem 14.

that equals the kinetic energy gain of the fluid, and resistance to flow through the short section of the drain pipe.

(a) Formulate an energy balance equation for the system.

(b) Formulate the continuity equation that represents the increment change in fluid level in the tank as a function of the fluid velocity through the drain pipe.

(c) Solve simultaneously equations formulated in (a) and (b) to calculate the time to drain the tank to a residual level 1 m from the bottom of the tank.

(d) Calculate the amount of residual fluid in the tank including the film adhering to the side of the tank.

6.15. A fluid tested on a tube viscometer 0.75 cm in diameter and 30 cm long exhibited a pressure drop of 1200 Pa when the flow rate was 50 cm^3/s.

(a) Calculate the apparent viscosity and the apparent rate of shear under this condition of flow.

(b) If the same fluid flowing at the rate of 100 cm^3/s through a viscometer tube 0.75 cm in diameter and 20 cm long exhibits a pressure drop of 1300 Pa, calculate the flow behavior and consistency indices. Assume wall effects are negligible.

6.16. Apparent viscosities in centipoises (cP) of 4000, 2500, 1250, and 850 were reported on a fluid at rotational speeds of 2, 4, 10, and 20 rev/min. This same fluid was reported to require a torque of 900 dyne cm to rotate a cylindrical spindle 1 cm in diameter and 5 cm high within the fluid at 20 rev/min.

(a) Calculate the flow behavior and consistency indices for this fluid.

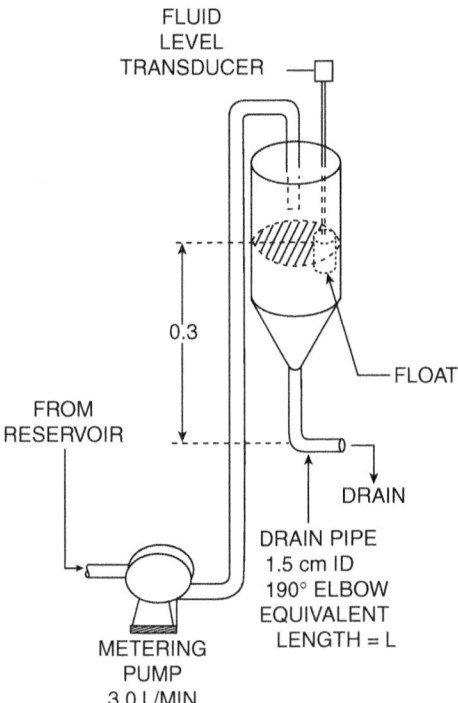

FLUID
LEVEL
TRANSDUCER

0.3

FLOAT

FROM
RESERVOIR

DRAIN

DRAIN PIPE
1.5 cm ID
190° ELBOW
EQUIVALENT
LENGTH = L

METERING
PUMP
3.0 L/MIN

Figure 6.34 Diagram for Problem 19.

(b) When exhibiting an apparent viscosity of 4000 cP, what would have been the shear rate under which the measurement was made?

6.17. The flow behavior and consistency indices of whey at a solids content of 24% has been reported to be 0.94 and 4.46×10^{-3} Pa · s^n, respectively. If a rotational viscometer having a full scale torque of 673.7 dyne Acm is to be used for testing the flow behavior of this fluid, determine the diameter and height of a cylindrical spindle to be used such that at the slowest speed of 2 rev/min. The minimum torque will be 10% of the full scale reading of the instrument. Assume a length to diameter ratio of 3 for the spindle. Would this same spindle induce a torque within the range of the instrument at 20 rev/min?

6.18. Egg whites having a consistency index of 2.2 Pa · s^n and a flow behavior index of 0.62 must be pumped through a 1.5-in. sanitary pipe at a flow rate, which would induce a shear rate at the wall of 150/s. Calculate the rate of flow in L/min, and the pressure drop due to fluid flow resistance under these conditions.

6.19. The system shown in Fig. 6.34 is used to control the viscosity of a batter formulation used on a breading machine. A float indicates the fluid level in the reservoir, and by attaching the float to an appropriate transducer, addition of dry ingredients and water into the mixing tank may be regulated to maintain the batter consistency. The appropriate fluid level in the reservoir may be maintained when a different consistency of the batter is required, by changing the length of the pipe draining the reservoir. If the fluid is Newtonian with a viscosity of 100 cP, and

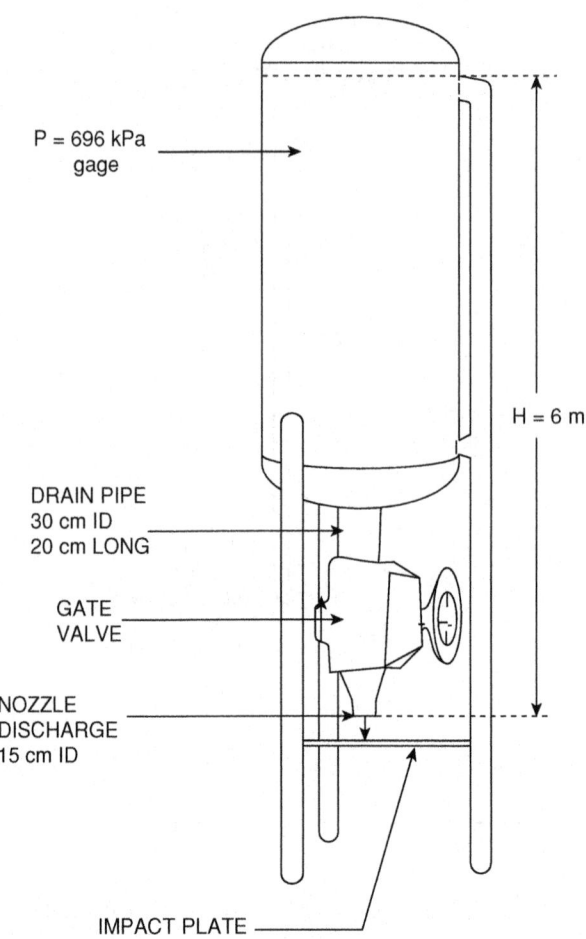

P = 696 kPa
gage

H = 6 m

DRAIN PIPE
30 cm ID
20 cm LONG

GATE
VALVE

NOZZLE
DISCHARGE
15 cm ID

IMPACT PLATE

Figure 6.35   Diagram for Problem 20.

if fluid density is 1004 kg/m^3, calculate the feed rate that must be metered into the reservoir to maintain the level shown. Assume entrance loss is negligible compared to fluid resistance through the drain pipe.

6.20. The system shown in Fig. 6.35 has been reported to be used for disintegrating wood chips in the pulp industry, after digestion. It is desired to test the feasibility of using the same system on a starchy root crop such as cassava or sweet potatoes, to disrupt starch granules for easier hydrolysis with enzymes to produce sugars for alcoholic fermentation. In simulating the system on a small scale, two parameters are of importance: the shear rate of the slurry as it passes through the discharge pipe and the impact force of the fluid against the plate. Assume that entrance loss from the tank to the discharge pipe is negligible. Under the conditions shown, calculate the shear rate through the drain pipe, and the impact force against the plate at the time the drain pipe is first opened. The slurry has flow behavior and consistency indices of 0.7 and 0.8 Pa · s^n, and a density of 1042 kg/m^3.

6.21. In a falling film direct contact steam heater for sterilization, milk is pumped into a header that distributes the liquid to several vertical pipes, and the liquid flows as a film in laminar flow down the pipe. If the fluid has a density of 998 kg/m^3, and a viscosity of 1.5 cP, calculate the flow rate down the outside surface of each of 3.7-cm outside diameter pipes in order that the fluid film will flow at a Reynolds number of 500. Calculate the fluid film thickness when flow develops at this Reynolds number.

6.22. A sauce product is being formulated to match a reference product (Product A) that has a consistency index of 12 Pa s^n and a flow behavior index of 0.55. Rheological measurements of the formulated product (Product B) on a wide gap rotational viscometer using a cylindrical spindle 1 cm in diameter and 5 cm long are as follows, with speed in rev/min and torque in % of full scale, respectively: 2, 11; 4, 18; 10, 34; 20, 56. The viscometer constant is 7187 dyne cm. Calculate the apparent viscosity of Product A and Product B at 0.5 rev/min. At this rotational speed, did the apparent viscosity of Product B match that of Product A?

6.23. An FMC de-aerator 2.97 m high and 96.5 cm in diameter is rated to de-aerate from 4.2 to 8.4 kg/s of product. If the product has a density of 1008 kg/m^3, and has a flow behavior index of 0.44 and a consistency index of 8.1 Pa s^n, calculate the film thickness and film velocity to achieve the mid-range (6.3 kg/s) of the specified capacity. If half the de-aerator height is to be covered by the fluid film, calculate the time available for any gas bubbles to leave the film into the vapor space in the chamber.

SUGGESTED READING

1. Bennet, C. O. and Myers, J. E. 1962. Momentum, Heat, and Mass Transport. McGraw-Hill Book Co., New York.
2. Charm, S. E. 1971. Fundamentals of Food Engineering. 2nd ed. AVI Publishing Co., Westport, CT.
3. Felder, R. M. and Rousseau, R. W. 1999. Elementary Principles of Chemical Processes. 2nd ed. John Wiley & Sons, New York.
4. Foust, A. S., Wenzel, L. A., Clump, C. W., Maus, L. and Andersen, L. B. 1960. Principles of Unit Operations. John Wiley & Sons, New York.
5. Geankoplis, C. J. 1993. Transport Processes and Unit Operations. 3rd ed. Prentice-Hall, Englewood Cliffs, NJ.
6. Heldman, D. R. and Singh, R. P. 1981. Food Process Engineering. AVI Publishing Co., Westport, CT.
7. Kokini, J. L. and Plutchok, G. J. 1987. Viscoelastic properties of semisolid foods and their biopolymeric components. Food Technol. 41(3):89.
8. Kreiger, I. M. and Maron, S. H. 1952. Direct determination of flow curves of non-Newtonian fluids. J. Appl. Phys. 33:147.
9. Kreiger, I. M. and Maron, S. H. 1954. Direct determination of the flow curves of non-Newtonian fluids. III. Standardized treatment of viscometric data. J. Appl. Phys. 25:72.
10. Leniger, H. A. and Beverloo, W. A. 1975. Food Process Engineering. D. Riedel Publishing Co., Boston.
11. McCabe, W. L. and Smith, J. C. 1967. Unit Operations of Chemical Engineering. 2nd ed. McGraw-Hill Book Co., New York.
12. Osorio, F. A. and Steffe, J. F. 1987. Back Extrusion of Power Flow Fluids. J. Tex. Studies 18:43.
13. Rao, M. A., Shallenberger, R. S., and Cooley, H. J. 1987. Effect of temperature on viscosity of fluid foods with high sugar content. In: Food Engineering and Process Applications. Vol. I. M. Le Maguer and P. Jelen, Eds. Elsevier, New York, pp. 23–38.

CHAPTER 7

Heat Transfer

Heat transfer is the movement of energy from one point to another by virtue of a difference in temperature. Heating and cooling are manifestations of this phenomenon, which is used in industrial operations and in domestic activities. Increasing energy costs and in some cases inadequate availability of energy will require peak efficiency in heating and cooling operations. An understanding of the mechanisms of heat transport is needed in order to recognize limitations of heating and cooling systems, which can then lead to adoption of practices that circumvent these limitations. In industrial and domestic heating and cooling, energy-use audits can be used to determine total energy use and the distribution within the process, to identify areas of high energy use, and to target these areas for energy conservation measures.

7.1 MECHANISMS OF HEAT TRANSFER

Heat will be transferred from one material to another when there is a difference in their temperature. The temperature difference is the driving force which establishes the rate of heat transfer.

7.1.1 Heat Transfer by Conduction

When heat is transferred between adjacent molecules, the process is called conduction. This is the mechanism of heat transfer in solids.

7.1.2 Fourier's First Law of Heat Transfer

According to Fourier's first law, the heat flux, in conduction heat transfer, is proportional to the temperature gradient.

$$\frac{q}{A} = -k\frac{dT}{dx} \tag{7.1}$$

In Equation (7.1), q is the rate of heat flow, and A is the area through which heat is transferred. A is the area perpendicular to the direction of heat flow. The expression q/A, the rate of heat transfer per unit area, is called the heat flux. The derivative dT/dx is the temperature gradient. The negative sign in Equation (7.1) indicates that positive heat flow will occur in the direction of decreasing temperature. The parameter k in (7.1), the thermal conductivity, is a physical property of a material. Values of the thermal conductivity of common materials of construction and insulating materials and of food products are given in Appendix Tables A.9 and A.10, respectively.

7.1.3   Estimation of Thermal Conductivity of Food Products

The thermal conductivity of materials varies with the composition and, in some cases, the physical orientation of components. Foods, being of biological origin possess highly variable composition and structure, therefore, k of foods presented in the tables is not always the same for all foods in the category listed. The effect of variations in the composition of a material on values of the thermal conductivity, has been reported by Choi and Okos (1987). Their procedure may be used to estimate k from the composition. k is calculated from the thermal conductivity of the pure component k_i and the volume fraction of each component, X_{vi}. An important assumption used in this estimation procedure is that the contribution of each component to the composite thermal conductivity is proportional to the component volume fraction as follows:

$$k = \sum (k_i X_{vi}) \tag{7.2}$$

The thermal conductivity in W/(m · K) of pure water (k_w), ice (k_{ic}), protein (k_p), fat (k_f), carbohydrate (k_c), fiber (k_{fi}), and ash (k_a) are calculated at the temperature, T in °C, using Equations (7.3) to (7.9), respectively.

$$k_w = 0.57109 + 0.0017625\,T - 6.7306 \times 10^{-6}\,T^2 \tag{7.3}$$

$$k_{ic} = 2.2196 - 0.0062489\,T + 1.0154 \times 10^{-4}\,T^2 \tag{7.4}$$

$$k_p = 0.1788 + 0.0011958\,T - 2.7178 \times 10^{-6}\,T^2 \tag{7.5}$$

$$k_f = 0.1807 - 0.0027604\,T - 1.7749 \times 10^{-7}\,T^2 \tag{7.6}$$

$$k_c = 0.2014 + 0.0013874\,T - 4.3312 \times 10^{-6}\,T^2 \tag{7.7}$$

$$k_{fi} = 0.18331 + 0.0012497\,T - 3.1683 \times 10^{-6}\,T^2 \tag{7.8}$$

$$k_a = 0.3296 + 0.001401\,T - 2.9069 \times 10^{-6}\,T^2 \tag{7.9}$$

The volume fraction X_{vi} of each component is determined from the mass fraction X_i, the individual densities ρ_i, and the composite density, ρ as follows:

$$X_{vi} = \frac{X_i \rho}{\rho_i} \tag{7.10}$$

The individual densities, in kg/m^3, are obtained using Equations (7.12) to (7.18), respectively, for water (ρ_w), ice (ρ_{ic}), protein (ρ_p), fat(ρ_f), carbohydrate (ρ_c), fiber (ρ_{fi}), and ash (ρ_a).

$$\rho = \frac{1}{[\sum (X_i / \rho_i)]} \tag{7.11}$$

$$\rho_w = 997.18 + 0.0031439\,T - 0.0037574\,T^2 \tag{7.12}$$

$$\rho_{ic} = 916.89 - 0.13071\,T \tag{7.13}$$

$$\rho_p = 1329.9 - 0.51814\,T \tag{7.14}$$

$$\rho_f = 925.59 - 0.41757\,T \tag{7.15}$$

$$\rho_c = 1599.1 - 0.31046\,T \tag{7.16}$$

$$\rho_{fi} = 1311.5 - 0.36589\,T \tag{7.17}$$

$$\rho_a = 2423.8 - 0.28063\,T \tag{7.18}$$

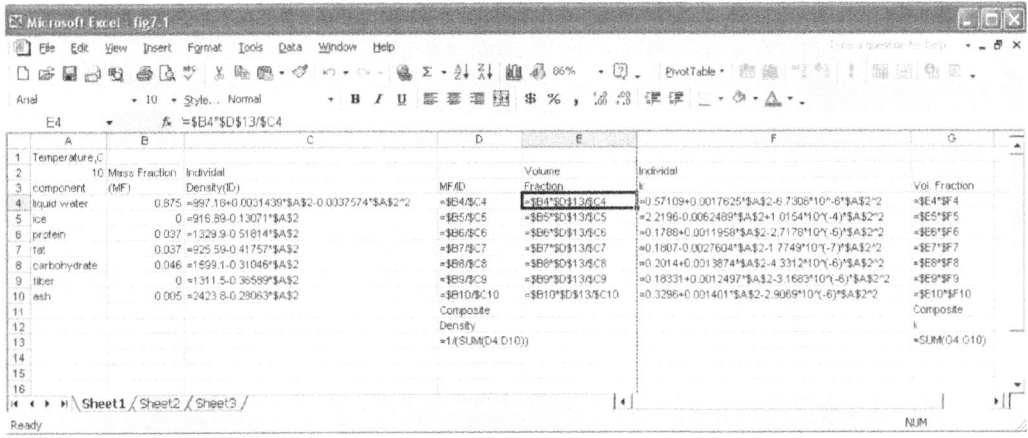

Figure 7.1   Spreadsheet program in Excel used to calculate the thermal conductivity of milk.

A Spreadsheet program in Excel shown in Fig. 7.1 may be used to calculate the thermal conductivity from the composition of a material.

To use this program, enter the temperature in cell A2 and the mass fraction of each specified component in B4 to B10. Excel will calculate required intermediate values and display the composite thermal conductivity in cell G13. Enter zero for a component whose mass fraction is not given.

Example 7.1. Calculate the thermal conductivity of lean pork containing 7.8% fat, 1.5% ash, 19% protein, and 71.7% water, at 19°C.

Solution:

The spreadsheet showing values of the component mass fraction in B4 to B10 and the composite thermal conductivity is shown in Fig. 7.2. The composite thermal conductivity is 0.4970 W/m · AK displayed in cell G13 in Fig. 7.2.

Example 7.2. Calculate the thermal conductivity of milk which contains 87.5% water, 3.7% protein, 3.7% fat, 4.6% lactose, and 0.5% ash, at 10°C.

Solution:

The spreadsheet program (Fig. 7.1) is used with the mass fraction of components and the temperature entered in the corresponding cells. Results are shown in Fig. 7.3.

The composite thermal conductivity is shown in cell G13 to be 0.5473 W/mAK.

Data on average composition of foods from USDA Handbook 8, which can be used to estimate thermophysical properties, is given in Appendix A.8.

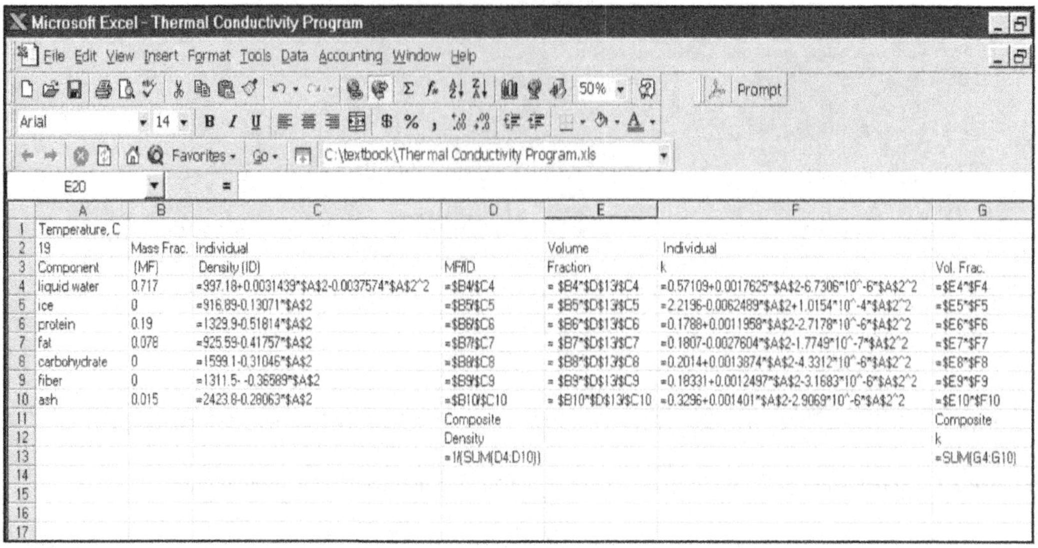

	A	B	C	D	E	F	G
1	Temperature, C						
2	19	Mass Frac.	Individual		Volume	Individual	
3	Component	(MF)	Density (ID)	MF/ID	Fraction	k	Vol. Frac.
4	liquid water	0.717	=997.18+0.0031439*$A$2-0.0037574*$A$2^2	=$B4/$C4	= $B4*$D$13/$C4	=0.57109+0.0017625*$A$2-6.7306*10^-6*$A$2^2	=$E4*$F4
5	ice	0	=916.89-0.13071*$A$2	=$B5/$C5	= $B5*$D$13/$C5	=2.2196-0.0062489*$A$2+1.0154*10^-4*$A$2^2	=$E5*$F5
6	protein	0.19	=1329.9-0.51814*$A$2	=$B6/$C6	= $B6*$D$13/$C6	=0.1788+0.0011958*$A$2-2.7178*10^-6*$A$2^2	=$E6*$F6
7	fat	0.078	=925.59-0.41757*$A$2	=$B7/$C7	= $B7*$D$13/$C7	=0.1807-0.0027604*$A$2-1.7749*10^-7*$A$2^2	=$E7*$F7
8	carbohydrate	0	=1599.1-0.31046*$A$2	=$B8/$C8	= $B8*$D$13/$C8	=0.2014+0.0013874*$A$2-4.3312*10^-6*$A$2^2	=$E8*$F8
9	fiber	0	=1311.5- -0.36589*$A$2	=$B9/$C9	= $B9*$D$13/$C9	=0.18331+0.0012497*$A$2-3.1683*10^-6*$A$2^2	=$E9*$F9
10	ash	0.015	=2423.8-0.28063*$A$2	=$B10/$C10	= $B10*$D$13/$C10	=0.3296+0.001401*$A$2-2.9069*10^-6*$A$2^2	=$E10*$F10
11				Composite			Composite
12				Density			k
13				=1/(SUM(D4:D10))			=SUM(G4:G10)

Figure 7.2 Spreadsheet program in Excel used to calculate the thermal conductivity of lean pork.

7.1.4 Fourier's Second Law of Heat Transfer

When the rate of heat transfer across a solid is not uniform (i.e., there is a difference in the rate at which energy enters and leaves a control volume), this difference will be manifested as a rate of

---

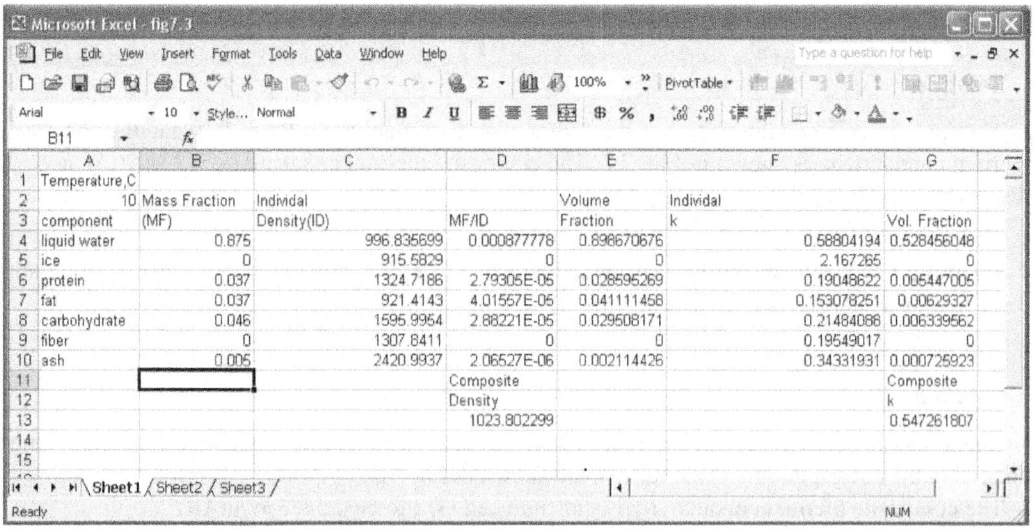

	A	B	C	D	E	F	G
1	Temperature,C						
2	10	Mass Fraction	Indivdal		Volume	Individal	
3	component	(MF)	Density(ID)	MF/ID	Fraction	k	Vol. Fraction
4	liquid water	0.875	996.835699	0.000877778	0.898670676	0.58804194	0.528456048
5	ice	0	915.5829	0	0	2.167265	0
6	protein	0.037	1324.7186	2.79305E-05	0.028595269	0.19048622	0.005447005
7	fat	0.037	921.4143	4.01557E-05	0.041111458	0.153078251	0.00629327
8	carbohydrate	0.046	1595.9954	2.88221E-05	0.029508171	0.21484088	0.006339562
9	fiber	0	1307.8411	0	0	0.19549017	0
10	ash	0.005	2420.9937	2.06527E-06	0.002114426	0.34331931	0.000725923
11				Composite			Composite
12				Density			k
13				1023.802299			0.547261807
14							
15							

Figure 7.3 Excel Spreadsheet showing the solution for Example 7.2.

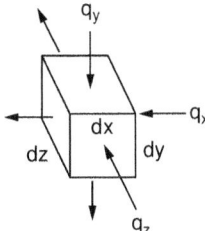

Figure 7.4 Control volume for analysis of heat transfer into a cube.

change of temperature with time, within the control volume. This type of problem is called unsteady state heat transfer. Figure 7.4 shows a control volume for analyzing heat transfer into a cube with sides dx, dy, and dz. The following are heat balance equations across the control volume in the x, y and z directions. Partial differential equations are used since in each case only one direction is being considered.

$$q = q_x + q_y + q_z = \rho \, dx \, dy \, dz \, C_p \frac{\delta T}{\delta t}$$

$$q_x = k \, dy \, dz \left[\frac{\partial T}{\partial x}\bigg|_1 - \frac{\partial T}{\partial x}\bigg|_2 \right]$$

$$q_y = k \, dx \, dz \left[\frac{\partial T}{\partial y}\bigg|_1 - \frac{\partial T}{\partial y}\bigg|_2 \right]$$

$$q_z = k \, dx \, dy \left[\frac{\partial T}{\partial z}\bigg|_1 - \frac{\partial T}{\partial z}\bigg|_2 \right]$$

Combining:

$$\rho C_p \frac{\partial T}{\partial t} = k \left[\frac{\frac{\partial T}{\partial x}\big|_1 - \frac{\partial T}{\partial x}\big|_2}{dx} + \frac{\frac{\partial T}{\partial y}\big|_1 - \frac{\partial T}{\partial y}\big|_2}{dy} \right] + k \left[\frac{\frac{\partial T}{\partial z}\big|_1 - \frac{\partial T}{\partial z}\big|_2}{dz} \right]$$

The difference in the first derivatives divided by dx, dy, or dz is a second derivative, therefore:

$$\frac{\partial T}{\partial t} = \frac{k}{\rho C_p} \left[\frac{\partial^2 T}{\partial x^2} + \frac{\partial^2 T}{\partial y^2} + \frac{\partial^2 T}{\partial z^2} \right] \tag{7.19}$$

Equation (7.19) represents Fourier's second law of heat transfer. The rate of change of temperature with time at any position within a solid conducting heat is proportional to the second derivative of the temperature with respect to distance at that particular point. The ratio $k/(\rho \cdot C_p)$ is α, the thermal diffusivity. The spreadsheet program for k (Fig. 7.1) and C_p (Chapter 3, Fig. 3.1) can be used to estimate the thermal diffusivity of foods from their composition. An Excel program for calculating α determined from food composition is given in Appendix A.11.

7.1.5   Temperature Profile for Unidirectional Heat Transfer Through a Slab

If heat is transferred under steady-state conditions, and A is constant along the distance, x, the temperature gradient, dT/dx will be constant, and integration of Equation (7.1) will result in an expression for temperature as a linear function of x. Substituting the boundary conditions, $T = T_1$ at $x = x_1$, in Equation (7.20):

$$T = -\frac{q/A}{k} \cdot x + C \tag{7.20}$$

Substituting the boundary condition, $T = T_1$ at $x = x_1$, in Equation (7.20):

$$T = \frac{q/A}{k} \cdot (x_1 - x) + T_1 \tag{7.21}$$

Equation (7.21) is the expression for the steady-state temperature profile in a slab where A is constant in the direction of x. The temperature gradient, which is constant in a slab transferring heat at a steady state, is equal to the ratio of the heat flux to the thermal conductivity.

If the temperature at two different points in the solid is known (i.e., $T = T_1$ at $x = x_1$ and $T = T_2$ at $x = x_2$), Equation (7.21) becomes:

$$T_1 - T_2 = -\frac{q/A}{k}(x_1 - x_2)$$

$$\frac{q}{A} = -k \cdot \frac{\Delta T}{\Delta x} \tag{7.22}$$

Substituting Equation (7.22) in Equation (7.21):

$$T = -\frac{\Delta T}{\Delta x} \cdot (x_1 - x) + T_1 \tag{7.23}$$

To keep the convention on the signs, and ensure that proper temperatures are calculated at any position within the solid, $\Delta T = T_2 - T_1$ with increasing subscripts in the direction of heat flow, and $\Delta x = x_2 - x_1$ is positively increasing in the direction of heat flow. Thus, if a temperature drop ΔT across any two points in a solid separated by distance Δx, is known, it will be possible to determine the temperature at any other point within that solid using Equation (7.23). Equations (7.22) and (7.23) also show that when heat is transferred in a steady state in a slab, the product, $k \cong (\Delta T/\Delta x)$, is a constant, and:

$$k\left[\frac{\Delta T}{\Delta x}\right]_1 = k\left[\frac{\Delta T}{\Delta x}\right]_2 = \cdots \tag{7.24}$$

If the solid consists of two or more layers having different thermal conductivity, the heat flux through each layer is constant, therefore the temperature profile will exhibit a change in slope at each junction between layers. A composite slab containing three materials having thermal conductivity k_1, k_2, and k_3, is shown in Fig. 7.5. The temperature profile shows a different temperature gradient within each layer. For a composite slab, Equation (7.24) becomes:

$$k_1\left(\frac{\Delta T_1}{\Delta x_1}\right) = k_2\left(\frac{\Delta T_2}{\Delta x_2}\right) = k_3\left(\frac{\Delta T_3}{\Delta x_3}\right) \tag{7.25}$$

Equation (7.25) can be used to experimentally determine the thermal conductivity of a solid by placing it between two layers of a second solid and determining the temperature at any two points within each

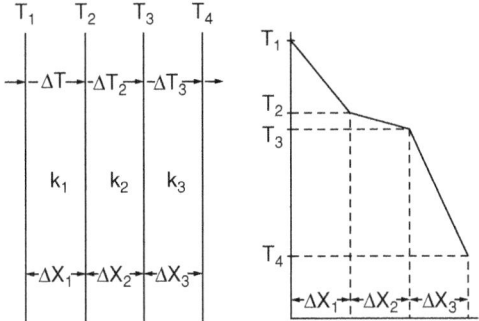

Figure 7.5   Diagram of composite slab and temperature drop at each layer.

layer of solid after a steady state has been achieved. The existence of a steady state and unidimensional heat flow can be proved by equality of the first and third terms. Two of the terms involving a known thermal conductivity k_1 and the unknown thermal conductivity k_2 can be used to determine k_2.

Example 7.3. Thermocouples embedded at two points within a steel bar, 1 and 2 mm from the surface, indicate temperatures of 100°C and 98°C, respectively. Assuming no heat transfer occurring from the sides, calculate the surface temperature.

Solution:

Because the thermal conductivity is constant, either Equations (7.23) or (7.25) can be used. Following the convention on signs for the temperature and distance differences, $T_2 = 98, T_1 = 100, x_2 = 2$ mm and $x_1 = 1$ mm. The temperature gradient $\Delta T/\Delta x = (T_2 - T_1)/(x_2 - x_1) = (98 - 100)/0.001(2 - 1) = -2000$. Equation (7.23) becomes:

$$T = -(-2000)(x_1 - x) + T_1$$

At the surface, $x = 0$, and at point $x_1 = 0.001, T_1 = 100$. Thus:

$$T = 2000(0.001) + 100 = 102°C$$

Example 7.4. A cylindrical sample of beef 5 cm thick and 3.75 cm in diameter is positioned between two 5-cm-thick acrylic cylinders of exactly the same diameter as the meat sample. The assembly is positioned inside an insulated container such that the bottom of the lower acrylic cylinder contacts a heated surface maintained at 50°C, and the top of the upper cylinder contacts a cool plate maintained at 0°C. Two thermocouples each are embedded in the acrylic cylinders, positioned 0.5 cm and 1.5 cm from the sample-acrylic interface. If the acrylic has a thermal conductivity of 1.5 W/(m · K), and the temperatures recorded at steady state are, respectively, 45° C, 43° C, 15° C, and 13°C, calculate the thermal conductivity of the meat sample.

Solution:

Equation (7.25) will be used. Proceeding along the direction of heat flow, Let $T_1 = 45, T_2 = 43, T_3 = $ temperature at the meat and bottom acrylic cylinder interface, $T_4 = $ temperature at the meat and top acrylic cylinder interface, $T_5 = 15$, and $T_6 = 13°C$. Let $x_1 = 0$ at the lowermost thermocouple location; $x_2 = 1$; $x_3 = 1.5$; $x_4 = 6.5$; $x_5 = 7$; $x_6 = 8$ cm. Because the temperatures have reached a steady state, the rate of heat transfer across any part of the system are equal therefore the temperature differences indicated by thermocouples within the acrylic should be equal if they are separated the same distances apart. Thus: $(T_2 - T_1)/(x2 - x1) = (T6 - T5)/(x6 - x5) = -2/0.01 = -200$. The heat flux through the system is

$$\frac{q}{A} = -k\left(\frac{\Delta T}{\Delta x}\right) = -1.5(-200) = 300 \text{ W}$$

Using Equation (7.23), the temperatures at the acrylic and meat sample interfaces are

$$T_3 = -(-200)(0.01 - 0.015) + 43 = 42°C$$
$$T_4 = -(-200)(0.07 - 0.065) + 15 = 16°C$$

Using Equation (7.25), with ΔT as the temperature drop across the meat sample $(T_3 - T_4)$, and Δx = the thickness of the meat sample, the thermal conductivity of the meat is

$$k = \frac{q/A}{\Delta T/\Delta x} = \frac{300}{26/0.05} = 0.5769 \text{W}/(m \cdot K)$$

7.1.6   Conduction Heat Transfer Through Walls of a Cylinder

When heat flows through the walls of a cylinder, the area perpendicular to the direction of heat flow changes with position. A in Equation (7.1) at any position within the wall is the surface area of a cylinder of radius r, and equation 1 may be expressed as:

$$q = -k(2\pi rL)\frac{dT}{dr}$$

When heat transfer occurs in a steady state, q is constant.

$$-\frac{dr}{r} = \frac{2\pi Lk}{q}dT \tag{7.26}$$

After integration and substitution of the boundary conditions: at $r = r_1$, $T = T_1$ and at $r = r_2$, $T = T_2$, the resulting equation for the rate of heat transfer expressed in Equation (7.27) is obtained.

$$q = \frac{T_1 - T_2}{[\ln(r_2/r_1)/2]} \tag{7.27}$$

Equation. (7.27) follows the convention established in Equation (7.24), that q is positive in the direction of decreasing temperature. Increasing subscripts represent positions proceeding farther away from the center of the cylinder.

7.1.7   The Temperature Profile in the Walls of a Cylinder in Steady-State Heat Transfer

At any point r, the temperature T may be obtained from Equation (7.27) by substituting r for r_2 and T for T_2.

$$T = \frac{\ln(r/r_1)q}{2\pi Lk} + T_1$$

Substituting Equation (7.27) for q:

$$T = \frac{\ln(r/r_1)}{\ln(r_2/r_1)}(T_1 - T_2) + T_1 \tag{7.28}$$

or:

$$\frac{(T - T_1)}{\ln(r/r_1)} = \frac{(T_1 - T_2)}{\ln(r_2/r_1)} \tag{7.29}$$

If the wall of the cylinder consists of layers having different thermal conductivities, the temperature profile may be obtained using the same procedure used in deriving Equations (7.28)

$$\frac{(T - T')}{\frac{1}{k_2}\ln\left(\frac{r}{r'}\right)} = \frac{(T_1 - T_2)}{\frac{1}{k_1}\ln\left(\frac{r_2}{r_1}\right)} = \frac{q}{2\pi L} \tag{7.30}$$

and (7.29), except that the thermal conductivity of the different layers are used.

In Equation (7.30), the temperatures T_1 and T_2 must transect a layer bounded by r_1 and r_2, which has a uniform thermal conductivity k_1. Similarly, the layer bounded by r and r' where the temperatures are T and T' must also have a uniform thermal conductivity, k_2. Each of the terms in Equation (7.30) is proportional to the rate of heat transfer. Equations (7.30) and (7.25) represent a fundamental relationship in steady-state heat transfer by conduction (i.e., the rate of heat flow through each layer in a multilayered solid is equal).

Example 7.5. A $^3/_4$-in. steel pipe is insulated with 2-cm-thick fiberglass insulation. If the inside wall of the pipe is at 150°C, and the temperature at the outside surface of the insulation is 35°C, (a) calculate the temperature at the interface between the pipe and the insulation and (b) calculate the rate of heat loss for one meter length of pipe.

Solution:

From Table 6.2 on pipe dimensions, the inside diameter of a $^3/_4$-in. steel pipe is 0.02093 m and the outside diameter is 0.02667 m. From Appendix Table A.9, the thermal conductivity of steel is 45 W/(m · K) and that of fiberglass is 0.035 W/(m · K). From the pipe dimensions, $r_1 = 0.01047$ m, $r_2 = 0.01334$, and $r_3 = (0.01334 + 0.020) = 0.03334$ m. $T_1 = 150$. $T_2 =$ the temperature at the pipe-insulation interface. $T_3 = 35°C$. T_1 and T_2 transect the metal wall of the pipe, with k = 45 W/(m · K). From Equation (7.30): $T = T_2$; $T' = T_s$; $r = r_3$ and $r' = r_2$; $k_1 = 0.035$ W/(m · K); and $k_2 = 45$ W/(m · K). Substituting in Equation (7.30):

$$\frac{(T_2 - T_s)k_1}{\ln(r_3/r_2)} = \frac{(T_1 - T_2)k_2}{\ln(r_2/r_1)}$$

$$0.038210\, T_2 - 1.33735 = 185.75(150) - 185.75\, T_2$$

$$\frac{(T_2 - 35)(0.035)}{\ln(0.03334/0.01334)} = \frac{(150 - T_2)(45)}{\ln(0.01334/0.01047)}$$

$$T_2(185.75 + 0.03821) = 27,864$$

$$T_2 = 149.97°C$$

7.1.8   Heat Transfer by Convection

This mechanism transfers heat when molecules move from one point to another and exchanges energy with another molecule in the other location. Bulk molecular motion is involved in convection heat transfer. Bulk molecular motion is induced by density changes associated with difference in fluid temperature at different points in the fluid, condensation, or vaporization (natural convection), or when a fluid is forced to flow past a surface by mechanical means (forced convection). Heat transfer by convection is evaluated as the rate of heat exchange at the interface between a fluid and a solid. The rate of heat transfer by convection is proportional to the temperature difference and is expressed as:

$$q = hA(T_m - T_s) = hA\,\Delta T \tag{7.31}$$

where h is the heat transfer coefficient, A is the area of the fluid-solid interface where heat is being transferred, and ΔT, the driving force for heat transfer, is the difference in fluid temperature, T_m, and the solid surface temperature, T_s. Convection heat transfer is often represented as heat transfer through a thin layer of fluid that possesses a temperature gradient, at the fluid-surface interface. The temperature, which is assumed to be uniform at T_m in the fluid bulk, gradually changes through the fluid film until it assumes the solid surface temperature past the film. Thus, the fluid film may be considered as an insulating layer that resists heat flow between the fluid and the solid. The fluid film is actually a boundary layer that has different properties and different velocity from the bulk of the fluid. The magnitude of the heat transfer coefficient varies in an inverse proportion to the thickness of the boundary layer. Conditions that result in a reduction of the thickness of this boundary layer will promote heat transfer by increasing the value of the heat transfer coefficient.

7.1.8.1   Natural Convection

Natural convection depends on gravity and density and viscosity changes associated with temperature differences in the fluid to induce convective currents. The degree of agitation produced by the convective currents depends on the temperature gradient between the fluid and the solid surface. When the ΔT is small, convective currents are not too vigorous, and the process of heat transfer is referred to as free convection. The magnitude of the heat transfer coefficient in free convection is very low, of the order 60 W/(m²· K) for air, and 60 to 3000 W/(m²· K) for water. When the surface is in contact with a liquid, and the surface temperature exceeds the boiling point of the liquid, bubbles of superheated vapor are produced at the solid-liquid interface. As these bubbles leave the surface, the boundary layer is agitated resulting in very high heat transfer coefficients. This process of heat transfer is called nucleate boiling, and the magnitude of the heat transfer coefficient for water is of the order 5000 to 50,000 W/(m²· K). When the ΔT is very high, excessive generation of vapor at the interface produces an insulating layer of vapor that hinders heat transfer. This process of heat transfer is called film boiling, and the heat transfer coefficient is much lower than that in nucleate boiling.

Another form of natural convection is the transfer of heat from condensing vapors. Condensing vapors release a large amount of energy on condensation, therefore, heat transfer coefficients are very high. When the vapors condense as droplets, which eventually coalesce and slide down the surface, the vapor is always in direct contact with a clean surface, and therefore, heat transfer coefficients are very high. This type of heat transfer is called dropwise condensation. Heat transfer coefficients of the order 10,000 W/(m²· K) are common in dropwise condensation. When vapors condense as a film of liquid on a surface, the liquid film forms a barrier to heat transfer and heat transfer coefficients are lower. This process of heat transfer is called filmwise condensation, and the magnitude of the heat transfer coefficient may be in the order of 5000 $W/(m^2 \cdot K)$.

7.1.8.2 *Forced Convection*

In forced convection heat transfer, heat transfer coefficients depend on the velocity of the fluid, its thermophysical properties, and the geometry of the surface. In general, heat transfer coefficients for noncondensing gases are about two orders of magnitude lower than that for liquids. Techniques for calculating heat transfer coefficients are discussed in section "ALocal Heat Transfer Coefficients."

7.1.9 Heat Transfer by Radiation

Heat transfer by radiation is independent of and additive to that transferred by convection. Electromagnetic waves traveling through space may be intercepted by a suitable surface and absorbed, raising the energy level of the intercepting surface. When the electromagnetic waves are of the frequency of light, the phenomenon is referred to as radiation. All bodies at temperatures higher than absolute zero emit energy in proportion to the fourth power of their temperatures. In a closed system, bodies exchange energy by radiation until their temperatures equalize. Radiation heat transfer, like convection, is a surface phenomenon, therefore the conditions at the surface determine the rate of heat transfer. Thermal radiation includes the spectrum ranging from the high ultraviolet (0.1 μm) through the visible spectrum (0.4 to 0.7 μm) to infrared (0.7 to 100 μm). Surfaces emit thermal radiation in a range of wavelengths, therefore radiation may be expressed as a *spectral intensity*, which is the intensity at each wavelength, or a *total intensity*, which is the integral of the energy emitted over a range of wavelengths. Surfaces also receive energy from the surroundings. The total energy received by a surface is called *irradiation*, which could also be considered as *spectral* or total. The irradiation received by a surface may be absorbed or reflected. The total energy leaving a surface is the sum of the emitted energy by virtue of its temperature and the reflected energy and is called the *radiosity*. The fraction of incident energy absorbed by a surface is called the *absorptivity* (α) and the fraction reflected is called the *reflectivity*(ρ). Some of the energy may be transmitted across the surface, and the fraction of incident energy transmitted is the *transmissivity* (τ). Thus, $\alpha + \rho + \tau = 1$.

7.1.9.1 *Types of Surfaces*

Surfaces may be classified according to their ability to absorb radiation, as *black* or *gray* bodies. A black body is one that absorbs all incident radiation. An example of a black body is the interior of a hollow sphere that has a small opening to admit radiation. All the energy that enters the small opening is reflected back and forth within the inside of the sphere and, eventually, it will be totally absorbed. Emissivity (ε) is a property that is the fraction of radiation emitted or absorbed by a black body at a given temperature that is actually emitted or absorbed by a surface at the same temperature. Thus black bodies have $\varepsilon = 1$. If a surface absorbs a fraction ($\varepsilon < 1$) of incident radiation equally at all wavelengths , the surface is considered gray. Table 7.1 shows emissivity of some surfaces.

7.1.9.1.1 Kirchoff's Law. A body at constant temperature is in equilibrium with its surroundings, and the amount of energy absorbed by radiation will be exactly the same as that emitted. Thus, the absorptivity of a surface (α) is exactly the same as the emissivity (ε), and these two properties may be used interchangeably.

Table 7.1 Emissivity of Various Materials

Material	Temp.('C)	Emissivity
Aluminum, bright	170	0.04
Aluminum paint	100	0.3
Chrome, polished	150	0.06
Iron, hot rolled	20	0.77
Brick, mortar, plaster	20	0.77
Glass	90	0.94
Oil paints, all colors	212	0.92–0.96
Paper	95	0.92
Porcelain	20	0.93
Stainless steel, polished	20	0.24
Wood	45	0.82–0.93
Water	32–212	0.96

Source: Bolz; R. E., and Tuve, G. L. (eds.) 1970 *Handbook of tables for Applied Engineering Science*. CRC Press, Cleveland, Ohio.

7.1.9.1.2 Stephan-Boltzman Law. The energy flux emitted by a black body is directly proportional to the fourth power of the absolute temperature.

$$\frac{q}{A} = \sigma T^4 \tag{7.32}$$

σ is the Stephan-Boltzman constant which has a value of 5.6732×10^{-8} W/(m$^2 \cdot$ K^4). For gray bodies:

$$q = A\sigma\varepsilon T^4 \tag{7.33}$$

The intensity of energy flux radiating from a black surface is a function of the wavelength, and a wavelength exists where the intensity of radiation is maximum. The energy flux from a black surface at an absolute temperature T, as a function of the wavelength λ is

$$\frac{q}{A} = \frac{C_1}{\lambda^5} \cdot \frac{1}{\left[e^{C_2/(\lambda T)} - 1\right]} \tag{7.34}$$

C_1 and C_2 are constants. The total flux over the entire spectrum will be:

$$\frac{q}{A} = \int_{-\infty}^{\infty} \left[\frac{C_1 \lambda^{-5}}{e^{C_2/(\lambda T)}} - 1\right] d\lambda \tag{7.35}$$

Equation (7.34) can be used to show that there will be a wavelength where the energy flux is maximum. The total energy flux over the whole spectrum is the energy radiated by a body at a temperature, T; thus, Equation (7.35) is equivalent to Equation (7.32).

7.1.9.1.3 Wein's Displacement Law. The wavelength for maximum energy flux from a body shifts with a change in temperature. The product of the wavelength for maximum flux intensity and absolute temperature is a constant. $\lambda_{max} \cdot T = 2.884 \times 10^{-3}$ m $\cdot$ K.

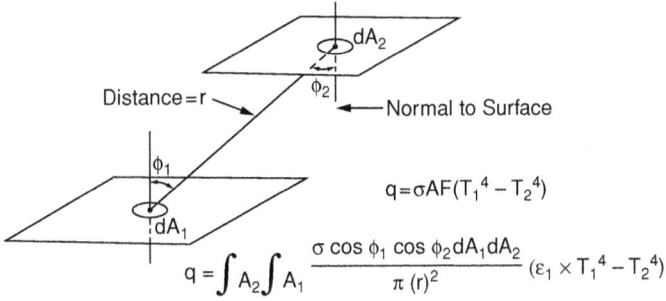

Figure 7.6 Two area elements and representation of the cosine law for heat transfer by radiation.

7.1.9.1.4 Lambert's Law. This is also called the cosine law. The energy flux over a solid angle ω in a direction φ from a normal drawn towards the surface, is a function of cos φ.

$$\frac{dq}{dA\,d\omega} = \frac{\varepsilon\sigma T^4}{\pi}\cos\phi \tag{7.36}$$

Equation (7.36) is the basis for the derivation of view factors used in calculating rate of radiation heat transfer betweem two bodies having the same emissivities. Figure 7.6 shows two area increments transferring heat by radiation and the representation of the cosine law according to Equation (7.36). The solid angle dω with which the area element dA_1 is viewed from dA_2 is

$$d\omega = \frac{\cos\phi_2 dA_2}{r^2}$$

Integrating Equation (7.36) after substituting the expression for the solid angle (dω), will give the rate of heat transfer (q_{1-2}) from area 1 to area 2.

$$q_{1-2} = \varepsilon_1\sigma(T_1{}^4 - T_2{}^4)\int\limits_{A_1}\int\limits_{A_2}\frac{\cos\phi_2 dA_2}{\pi r^2}\cdot dA_1\cos\phi_1$$

The double integral times $1/A_1$ is the view factor (F_{1-2}) used to calculate the energy transferred by area 1 and when the multiplier is $1/A_2$ it becomes the view factor (F_{2-1}) for area 2. The equations for the rate of heat transfer become:

$$q_{2-1} = (F_{2-1})A_2\varepsilon\sigma\left(T_1{}^4 - T_2{}^4\right) \tag{7.37}$$

$$q_{1-2} = (F_{1-2})A_1\varepsilon\sigma\left(T_1{}^4 - T_2{}^4\right) \tag{7.38}$$

Because in a steady state the rate of heat transfer from A_1 to A_2 is the same as that from A_2 to A_1, the product of A_1 and the view factor for A_1 is equal to the product of A_2 and the view factor for A_2. If the two surfaces have different emissivities, the effective view factor $(\bar{F})$ derived by Jacob and Hawkins (1957) is

$$\frac{1}{\bar{F}_{2-1}A_2} = \frac{1}{A_2F_{2-1}} + \frac{1}{A_1}\left(\frac{1}{\varepsilon_1}-1\right) + \frac{1}{A_2}\left(\frac{1}{\varepsilon_2}-1\right) \tag{7.39}$$

$$\frac{1}{\bar{F}_{1-2}A_1} = \frac{1}{A_1F_{1-2}} + \frac{1}{A_1}\left(\frac{1}{\varepsilon_1}-1\right) + \frac{1}{A_2}\left(\frac{1}{\varepsilon_2}-1\right) \tag{7.40}$$

The rate of heat transfer based on area A_1 or A_2 using the effective view factor is

$$q = A_1 \bar{F}_{1-2} \sigma \left(T_1^4 - T_2^4\right) = A_2 \bar{F}_{2-1} \sigma \left(T_1^4 - T_2^4\right) \tag{7.41}$$

View factors for some geometries and procedures for applying view factor algebra are available in the literature (e.g., Rohsenow and Hartnett, 1973). View factors for simple geometries are given in the succeeding section on "Radiant Energy Exchange."

7.1.9.2 *Effect of Distance Between Objects on Heat Transfer*

The total energy intercepted by an area from a point on another area is dependent on the solid angle with which the point views the area. The surface area seen from a point over a solid angle ω may be considered as the base of a cone of distance r from the point and has an area of ωr^2. Because the energy leaving the point toward this viewed area is q, the energy flux is $q/\omega r^2$, indicating that flux is inversely proportional to the square of the distance from the source. Radiant energy flux from a source weakens as a body moves away from a source.

7.1.9.3 *Radiant Energy Exchange*

View factors for simple geometries:

1. Small object (A_1) surrounded by a large object:

    $$\bar{F}_{1-2} = \varepsilon_1; \quad F_{1-2} = 1$$

2. Large parallel planes with equal areas:

    $$\frac{1}{\bar{F}_{1-2}} = \frac{1}{\varepsilon_1} + \frac{1}{\varepsilon_2} - 1 \tag{7.42}$$

3. Two parallel disks with centers directly in line (from Rohsenow and Hartnett, 1973): a = diameter of one disk and b = diameter of the other; c = distance between disks. A_1 is the area of the larger disk with diameter b (Fig. 7.7):

    $$\bar{F}_{1-2} = 0.5 \left[Z - \sqrt{(Z^2 - 4X^2Y^2)} \right] \tag{7.43}$$

    where X = a/c; Y = c/b, and $Z = 1 + (1 + X^2)Y^2$.

4. Two parallel long cylinders of equal diameters, b separated by distance 2a (From Roshenow and Hartnett, 1973) (See Figure 7.7):

    $$(\bar{F})_{1-2} = \frac{2}{\pi} \left[\sqrt{(X^2 - 1)} - X + \frac{\pi}{2} - \text{arc } \cos(1/x) \right] \tag{7.44}$$

    where X = 1 + a/b.

A spreadsheet program in Excel for solving the view factors represented by Equations (7.44) is shown in Figure 7.8.

Example 7.6. Glass bottles may be prevented from breaking on filling with hot pasteurized juice when their temperature is close to that of the juice being filled. The bottles are rapidly heated by passing through a chamber that has top, bottom, and side walls heated by natural gas. The glass bottles may be considered as an object completely surrounded by a radiating surface. The glass bottles have

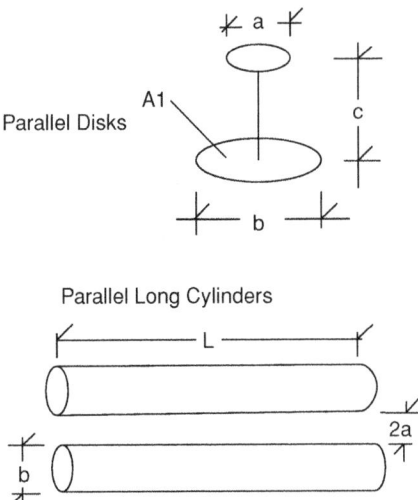

Figure 7.7 Figure of two parallel disks and two parallel long cylinders exchanging energy by radiation.

an emissivity of 0.94, a mass of 155 g each, a specific heat of 1256 J/(kg · K), and a surface area of 0.0219 m². If the glass bottles are to be heated from 15.5°C to 51.6°C in 1 minute, calculate the temperature of the walls of the chamber to achieve this average heating rate when the glass is at the midpoint of the temperature range (33.6°C).

Solution:

The required heat transfer rate is m $C_p \Delta T$:

$$q = \frac{0.155(1256)(51.6 - 15.5)}{60} = 117.1 \, \text{W}$$

Equations (7.39) and (7.41) may be used to calculate $\bar{F}_{1-2}$ and q_{1-2}. Because A_2 in Equation (7.29) is large, the third term on the right of the equation is zero, and the because $F_{1-2} = 1, \bar{F}_{1-2} = \varepsilon$. The heat transfer by radiation may also be calculated using Equation (7.38):

$$F_{1-2} = 1; \quad T_2 = 33.6 + 273 = 303.6 \, \text{K}$$

$$q = A \, \varepsilon \, \sigma (T_1^4 - T_2^4)$$

$$q = 0.0219(0.94)(5.6732e - 0.8)(T^4 - (306.6)^4)$$

$$T^4 = \frac{117.1}{0.0219(0.94)(5.6732e - 0.8)} + (306.6)^4$$

$$T = (1.091034 \times 10^{11})^{0.25} = 574.7 \, \text{K}$$

Example 7.7. Cookies traveling on a conveyor inside a continuous baking oven. occupy most of the area on the surface of the conveyor. The top wall of the oven directly above the conveyor has an emissivity of 0.92, and the cookies have an emissivity of 0.8. If the top wall of the oven has a

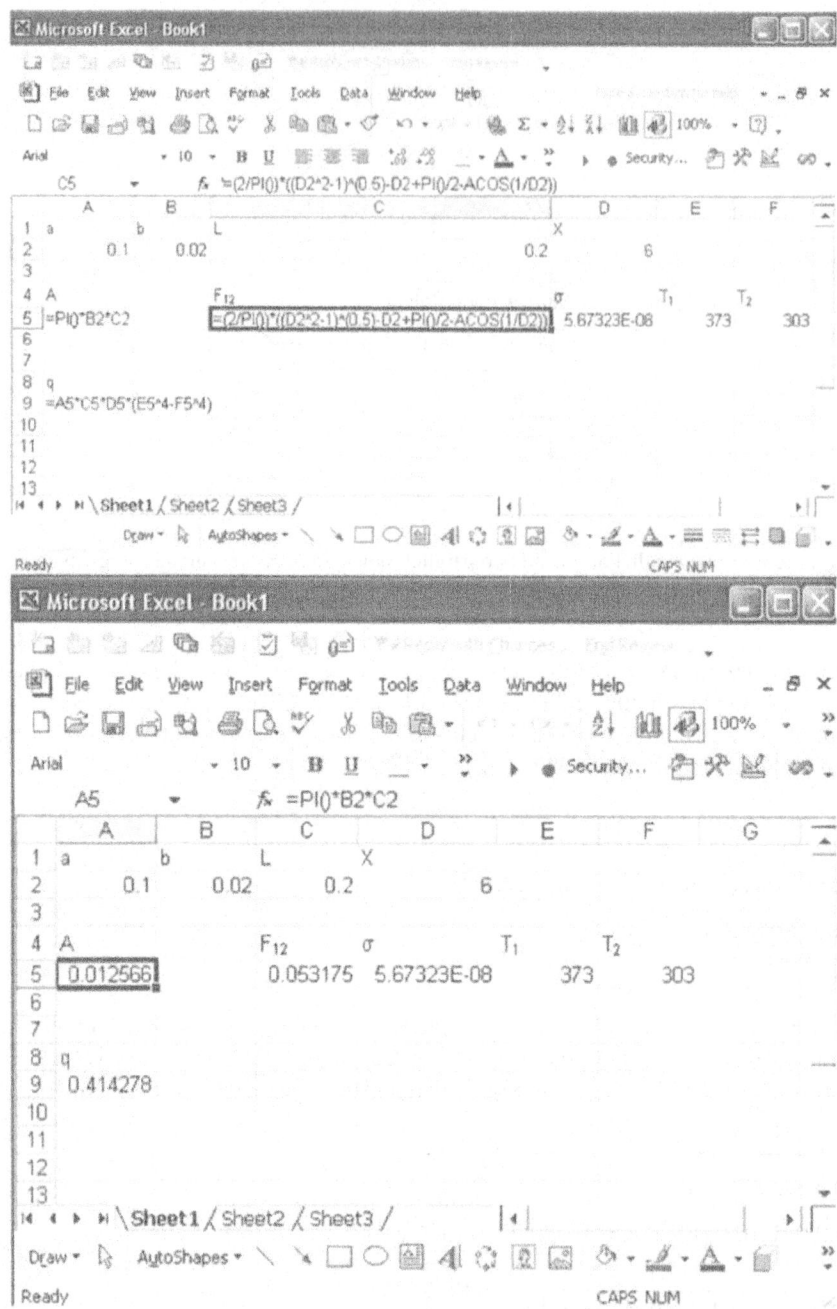

Figure 7.8 Spreadsheet program in excel to solve view factors Equations 7.43 and 7.44.

temperature of 175°C, calculate the average rate of heat transfer by radiation between the cookies per unit area on the side that faces the top wall of the oven when the cookie surface temperature is 70°C.

Solution:

Assume the layer of cookies on the conveyor and the top oven wall constitute a set of long parallel plates. The view factor $\overline{F}_{1-2}$ is calculated using Equation (7.42):

$$\frac{1}{\overline{F}_{1-2}} = \frac{1}{0.92} + \frac{1}{0.80} - 1 = 1.3369$$

$$\overline{F}_{1-2} = \frac{1}{1.3369} = 0.748$$

$$T_1 = 70 + 273 = 343K; T_2 = 175 + 273 = 448K$$

Using Equation (7.41):

$$\frac{q}{A} = 0.748(5.6732 \times 10^{-8})(448^4 - 343^4) = 1122 \text{ W/m}^2$$

Example 7.8. The temperature registered by a thermocouple in a gas stream is often the result of steady-state heat transfer by convection from the gas to the thermocouple junction. However, when radiation heat transfer from a surrounding wall is significant, the actual temperature indicated by the thermocouple will be higher than the gas temperature. Consider a superheater for steam that consists of a gas-heated pipe through which steam is flowing. The pipe wall temperature is 500°C and the heat transfer coefficient between the thermocouple junction and the superheated air is 342 W/m²K. If the thermocouple reads 200°C, what is the actual temperature of the superheated steam. Assume the thermocouple junction is a black body.

Solution:

The thermocouple junction may be considered as an area totally enclosed by another, therefore $F_{1-2} = 1$. Because the temperature of the thermocouple junction is higher than that of steam, energy will be lost by the junction due to convection. To maintain a constant temperature, the heat transfer by radiation will equal that lost by convection and the heat balance becomes:

$$hA(T - T_s) = \sigma A(T_w^4 - T^4)$$

Solving for T_s:

$$T_s = (1/h)[hT - \sigma (T_w^4 - T^4)]$$

T_w and T in the expression for radiation heat transfer is in absolute temperature of 773 and 473 K, respectively, therefore for consistency T and T_s in the expression for convection heat transfer should also be expressed in Kelvin. Solving for T_s in Kelvin:

$$T_s = (1/342)[342A\ 473 - 5.6732 \times 10^{-8}\ \{(773)^4 - (473)^4\}] = 422 \text{ K or } 149°C$$

Note that the actual steam temperature T_s will approach the indicated temperature T as h approaches infinity or if T_w approaches T.

Example 7.9. A vacuum belt dryer consists of a belt traveling over a heated plate that conveys the product and a heated plate positioned over the belt the whole length of the belt. Heat transfer from the heated top plate to the product on the conveyor is by radiation and the view factor may be considered as that for long parallel plates. The top plate is sand blasted stainless steel with $\varepsilon = 0.52$. Apple slices $\varepsilon = 0.85$ are being dried and the desired product temperature during drying equals the boiling temperature of water at an absolute pressure of 7 mm Hg, the pressure inside the vacuum dryer. If the surface area of sliced apples at 86% water is 0.731 cm^2/g, calculate the rate of evaporation of water from the apple surface due to heat transfer by radiation, expressed in g water evaporated/(h Ag dry apple solids) if the top plate temperature is maintained at 80°C.

Solution:

At 7 mm Hg absolute pressure, the pressure in Pascals is

$$P = 0.7 \text{ cm Hg} \times 1333.33 \text{ Pa/cm Hg} = 933.33 \text{ Pa}$$

From Appendix table A.4, $T = 5 + [2.5/(1.0365 - 0.8724)](0.93333 - 0.8724) = 5.9°C$
Using Equation (7.42): $\bar{F}_{1-2} = 1/[1/0.52 + 1/.85 - 1] = 0.4763$
Using Equation (7.41): $q/A_1 = 0.4763(5.6732 \times 10^{-8})[(353)^4 - (278.9)^4] = 256.07 \text{W/m}^2$
From Appendix Table A.4, the latent heat of evaporation of water at .9333 kPa is

$$h_{fg} = 2.4897 - [(2.4897 - 2.4839)/(1.0365 - 0.8724)](0.9333 - 0.8724) = 2.4918 \text{ MJ/kg}$$

Drying rate:

$$dW/dt = [256.07 \text{ J/(s A m}^2)][kg/2,491,800 \text{ J}][1000g/kg][3600 \text{ s/h}]$$
$$= 369.95 \text{ g water/(m}^2 \text{ A h)}$$

The apple surface area in m^2/g dry matter:

$$= [0.731 \text{ cm}^2/(1 - 0.86) \text{ g dry matter}][1 \text{ m}^2/10000 \text{ cm}^2] = 0.000522$$

$$dW/dt = [369.95 \text{ g water/(m}^2 \text{ A h)}][0.000522 \text{ m}^2/\text{g dry matter}]$$

$$= 0.193 \text{ g water/(h A g dry apple solids)} \text{ evaporation from the radiant heat transfer alone.}$$

7.1.10 Microwave and Dielectric Heating

Microwaves like light are also electromagnetic vibration. Heat transfer is dependent on the degree of excitability of molecules in the absorbing medium and the frequency of the field to which the medium is exposed. Dielectric heating is the term used when relatively low frequencies are used and the material is placed between two electrodes to which an electric current is passed. Frequencies from 60 Hz to 100 MHz may be used for dielectric heating. Microwave heating refers to the use of electromagnetic waves of very high frequency making it possible to transmit the energy through space. The most common frequencies used for microwave heating are 2450 MHz and 915 MHz. Domestic microwave ovens operate at 2450 MHz. The equations that govern heat transfer by microwave and dielectric systems are the same.

7.1.10.1 Energy Absorption by Foods in a Microwave Field

The energy absorbed by a body is

$$\frac{q}{V} = 0.556(10^{-12}) \text{ f E}^2\text{e tan}(\delta) \tag{7.45}$$

Table 7.2 Dielectric Properties of Food and Other Materials.

Material	Temperature $^\circ C$	e''	$tan(\delta)$
Beef (raw)	−15	5.0	0.15
Beef (raw)	25	40	0.30
Beef (roast)	23	28	0.20
Peas (boiled)	−15	2.5	0.20
Peas (boiled)	23	9.0	0.50
Pork (raw)	−15	6.8	1.20
Pork (roast)	35	23.0	2.40
Potatoes (boiled)	−15	4.5	0.20
Potatoes (boiled)	23	38.0	0.30
Spinach (boiled)	−15	13.0	0.50
Spinach (boiled)	23	34.0	0.80
Suet	25	2.50	0.07
Porridge	−15	5.0	0.30
Porridge	23	47.0	0.41
Pyrex	25	4.80	0.0054
Water	1.5	80.5	0.31
water	25	76.7	0.15
0.1 M NaCl	25	75.5	0.24

Sources: Copson, 1971. *Microwave Heating. AVI* Publishing Co. West-port, Conn.; Schmidt, W. 1960. Phillips *Tech. Rev* 3.89.

where q/V = energy absorbed, W/cm^3; f = frequency, Hz; e = dielectric constant, an index of the rate at which energy penetrates a solid; dimensionless tan(δ) = dielectric loss factor, an index of the extent to which energy entering the solid is converted to heat; dimensionless E = field strength in volts/cm^2. e and tan(δ) are properties of the material and are functions of composition and temperature. f and E are set by the type of microwave generator used. Table 7.2 shows the dielectric constant and the dielectric loss factor for foods, food components, and some packaging materials. Metal containers are opaque to microwave (i.e., microwaves are reflected from the surface therefore none passes across to food contained inside). However, an electrically conductive metal finite electrical resistance will heat up in the same manner as an electrical resistance wire will heat up when an electric current is passed through it. Similarly, electrically conductive wires will heat very rapidly in a microwave field. A continuous metal sheet with very low electrical resistance and will not heat up in a microwave field. However, a discontinuous metal sheet such as metallized plastic contains many small areas of metal that presents a large resistance to electrical current flow, therefore intense heating occurs in these materials. Such materials called absorbers or intensifiers are used in microwavable packages of frozen breaded fried foods or pizza to ensure a crispy crust when heated in a microwave. Glass and plastic are practically transparent to microwaves, i.e. they transmit microwaves and very little energy is absorbed.

The frequency of microwave power generated by a microwave generator is declared on the name plate of the unit. The power output is also supplied by the manufacturer for each unit. The coupling efficiency of a microwave unit is expressed as the ratio of power actually supplied to the unit and the actual power absorbed by the material heated. When the quantity of material being heated is large, the power generated by the system limits the power absorbed, rather than the value predicted by Equation

(7.45).The time it takes to heat a large quantity of material can be used to determine the microwave power output of a unit.

If a maximum q/v is determined by varying the quantity of food heated until further reduction results in no further increase in q/v, this value will be the limiting power absorption and will be dependent on the dielectric loss properties of the material according to Equation (7.45). If the dielectric constant and loss tangent of the material are known, it will be possible to determine the electromagnetic field strength which exists, and differences in heating rates of different components in the food mixture can be predicted using Equation (7.45).

7.1.10.2 Relative Heating Rates of Food Components

When the power output of a microwave unit limits the rate of energy absorption by the food, components having differenct dielectric properties will have different heating rates. Using subscripts 1 and 2 to represent component 1 and 2, e'' to represent the product of e and $\tan(\delta)$, and C to represent the constant, Equation (7.45) becomes:

$$q_1 = \frac{m_1}{\rho_1} Cf E^2 e_1''$$

$$q_2 = \frac{m_2}{\rho_2} Cf E^2 e_2''$$

Because $P = q1 + q2$; and $q = m C_p\, dT/dt$:

$$Cf E^2 = \frac{P}{(m_1/\rho_1)e_1'' + (m_2/\rho_2)e_2''}$$

$$\frac{dT_1}{dt} = \frac{\rho_2 e''_1 P}{C_{p1}(\rho_2 e''_1 m_1 + \rho_1 e''_2 m_2)} \tag{7.46}$$

$$\frac{dT_2}{dt} = \frac{\rho_1 e''_2 P}{C_{p2}(\rho_2 e''_1 m_1 + \rho_1 e''_2 m_2)} \tag{7.47}$$

The relative rate of heating is

$$\frac{dT_1}{dT_2} = \frac{\rho_2 e''_1 C_{p2}}{\rho_1 e''_2 C_{p1}} \tag{7.48}$$

Similar expressions may be derived for more than two components.

Example 7.10. The dielectric constant of beef at 23°C and 2450 MHz is 28 and the loss tangent is 0.2. The density is 1004 kg/m^3 and the specific heat is 3250 J/(kg · K). Potato at 23°C and 2450 MHz has a dielectric constant of 38 and a loss tangent of 0.3. The density is 1010 kg/m^3 and the specific heat is 3720 J/(kg · K).

(a) A microwave oven has a rated output of 600 W. When 0.25 kg of potatoes were placed inside the oven, the temperature rise after 1 minute of heating was 38.5°C. When 60 g of potato was heated in the oven, a temperature rise of 40°C was observed after 20 s. Calculate the average power output of the oven, and the mass of potatoes that must be present such that power output of the oven is limiting the rate of power absorption rather than the capacity of the material to absorb the microwave energy.

(b) If potatoes and beef are heated simultaneously, what would be the relative rate of heating?

Solution:

(a) Assume that 0.25 kg mass of product is sufficient to make microwave power availability the rate limiting factor for microwave absorption.

$$P = 0.25 \text{ kg} \left[\frac{3720 \text{ J}}{\text{kg} \cdot \text{K}} \right] \left[\frac{38.5 \text{ K}}{1 \text{ min}} \right] \left[\frac{1 \text{ min}}{60 \text{ s}} \right] = 596.75 \text{ W}$$

When a small amount of material is heated:

$$P = 0.06 \text{ kg} \left[\frac{3720 \text{ J}}{\text{kg} \cdot \text{K}} \right] (40 \text{ K}) \left[\frac{1}{20 \text{ s}} \right] = 446.4 \text{ W}$$

The amount absorbed with the small mass in the oven is much smaller than when a larger mass was present; therefore it may be assumed that power absorption by the material limits the rate of heating.

$$\frac{q}{V} = \left[\frac{446.4 \text{ W}}{0.06 \text{ kg} \frac{1 \text{m}^3}{1010 \text{kg}}} \right] \cdot \frac{1 \text{ m}^3}{10^{-6} \text{ cm}^3} = 7.5144 \text{ W/cm}^3$$

Because with a small load in the oven, the power absorption is only 7.5177 W/cm^3, this may be assumed to be $(q/V)_{\text{lim}}$, the maximum rate of power absorption by the material. If the power output of the oven as calculated above is 596.75 W, the mass present where the rate of power absorption equals the power output of the oven is

$$\frac{q}{V} \bigg|_{\text{lim}} = \frac{P(\rho)10^{-6}}{m}$$

$$m = \frac{P(\rho)10^{-6}}{(q/V)_{\text{lim}}} = \frac{596.75(1010 \times 10^{-6})}{7.5177} = 0.08 \text{ kg}$$

Thus, any mass greater than 0.08 kg will heat inside this microwave oven at the rate determined by the power output of the oven.

(b) Using Equation (7.48): Let subscript 1 refer to beef and subscript 2 refer to the potato.

$$\frac{dT_1}{dT_2} = \frac{1010(28)(0.2)(3720)}{1004(38)(0.3)(3250)} = 0.565$$

Thus, the beef will be heating slower than the potatoes.

7.2 TEMPERATURE MEASURING DEVICES

Temperature is defined as the degree of thermal agitation of molecules. Changes in molecular motion of a gas or liquid will change the volume or pressure of that fluid, and a solid will undergo a dimensional change such as expansion or contraction. Certain metals will lose electrons when molecules are thermally excited and when paired with another metal whose molecules could receive these displaced electrons, an electromotive force will be generated. These responses of materials to changes in temperature are utilized in the design of thermometers, electronic and mechanical temperature measuring devices. The calibration of all temperature measuring devices is based on conditions of use. Some electronic instruments have ambient temperature compensation integrated in the circuitry, and others do not. Similarly, fluid filled thermometers are calibrated for certain immersion depths, therefore partial immersion of a thermometer calibrated for total immersion will lead to errors in measurement. Thus, conditions of use that correspond to the calibration of the instrument must be known, for accurate measurements. For other conditions, the instrument will need recalibration. Thermometers filled with fluids other than mercury may have to be recalibrated after a certain time of use to ensure that breakdown of the fluid, which might change its thermal expansion characteristics, has not occurred. In the food industry, a common method of calibration involves connecting the thermometers to a manifold which contains saturated steam. The pressure of the saturated steam can then be used to determine the temperature from steam tables, and compared to the thermometer readings. Recalibration of fluid filled thermometers is normally not done. Thermometers used in critical control points in processes must be replaced when they lose accuracy. Electronic temperature measuring devices must be recalibrated frequently against a fluid-filled thermometer.

The temperature indicated by a measuring device represents the temperature of the measuring element itself, rather than the temperature of the medium in contact with the element. The accuracy of the measurement would depend on how heat is transferred to the measuring element. The temperature registered by the instrument will be that of the measuring element after heat exchange approaches equilibrium.

Thermometers will not detect oscillating temperatures if the mass of the temperature measuring element is such that the lag time for heat transfer equilibration is large in relation to the period of the oscillation. Response time of temperature measuring devices is directly proportional to the mass of the measuring element; the smaller the mass the more accurate the reading.

Measurement of the surface temperature of a solid in contact with either a liquid or a gas is not very accurate if the measuring element is simply laid on the solid surface. Depending upon the thickness of the measuring element, the temperature indication will be intermediate between that of the surface and the fluid temperature at the interface. Accurate solid surface temperatures can only be determined by embedding two thermocouples a known distance from the surface and extrapolating the temperature readings towards the surface to obtain the true surface temperature.

Measurements of temperatures of gases can be influenced by radiation. As shown in the previous section on Radiation Heat Exchange, an unshielded thermometer or thermocouple in a gas stream surrounded by surfaces of a different temperature the gas will read a temperature different from the true gas temperature because of radiation. Shielding of measuring elements is necessary for accurate measurements of gas temperatures.

7.2.1 Liquid-in-Glass Thermometers

Liquid-in-glass thermometers most commonly use mercury for general use, and mineral spirits, ethanol, or toluene for low temperature use. When properly calibrated, any temperature stable liquid may be used, and non–mercury-filled thermometers are now available for use as temperature indicators in food processing facilities.

7.2.2 Fluid-Filled Thermometers

This temperature measuring device consists of a metal bulb with a long capillary tube attached to it. The end of the capillary is attached to a device that causes a definite movement with pressure transmitted to it from the capillary. The capillary and bulb is filled with fluid that changes in pressure with changes in temperature. Capillary length affects the calibration and so does the ambient temperature surrounding the capillary. Thus, temperature measuring devices of this type must be calibrated in the field. Figure 7.9B shows a typical fluid-filled thermometer with a spiral at the end of the capillary. The movement of the spiral is transmitted mechanically through a rack and pinion arrangement or a lever to a needle which moves to indicate the temperature on a dial. Some stainless steel cased dial type thermometers are of this type. Proper performance of these instruments requires total immersion of the bulb.

7.2.3 Bimetallic Strip Thermometers

When two metals having dissimilar thermal expansion characteristics are joined, a change in temperature will cause a change in shape. For example, thermostats for domestic space cooling and heating utilize a thin strip of metal fixed at both ends to a thicker strip of another metal. An increase in temperature will make the thinner metal, which has a higher thermal coefficient of expansion, to bend and the change in position may open or close properly positioned electrical contacts, which then activates heating or cooling systems. In some configurations, the bimetallic strip may be wound in a helix or a spiral. The strip is fixed at one end, and the free end would move with changes in temperature to position an indicator needle on a temperature scale. This type of thermometer requires total immersion of the bimetallic strip for proper performance.

7.2.4 Resistance Temperature Devices (RTDs)

The principle of RTDs is based on the change in resistance of a material with changes in temperature. Figure 7.9a shows an RTD circuit. The resistance strip may be an insulated coil of resistance wire encased in a metal tube (resistance bulb) or a strip of metal such as latinum. Other materials such as semiconductors exhibit large drop in resistance with increasing temperatures. The whole instrument consists of the resistance or semiconductor element for sensing the temperature, electronic circuitry for providing the excitation voltage, and a measuring circuit for the change in resistance. Response is usually measured as a change in voltage or current in the circuit. RTDs vary in size and shape depending upon use. Total immersion of the sensing element is necessary for accurate results.

7.2.5 Thermocouples

A thermocouple is a system of two separate dissimilar metals with the ends fused together forming two junctions. When the two junctions are at different temperatures, an electromotive force is generated. This electromotive force is proportional to the difference in temperature between the two junctions. A thermocouple circuit is shown in Fig. 7.10. One of the junctions, the reference junction, is immersed in a constant temperature bath which is usually ice water. Electromotive

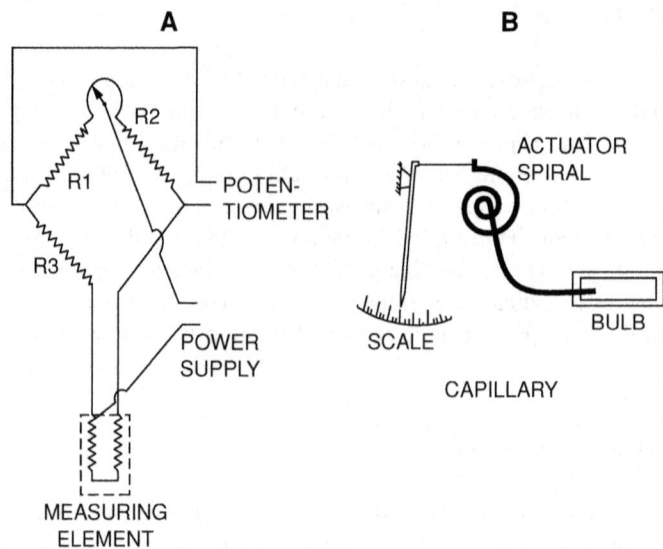

A

R2

R1

POTEN-
TIOMETER

R3

POWER
SUPPLY

MEASURING
ELEMENT

B

ACTUATOR
SPIRAL

BULB

SCALE

CAPILLARY

Figure 7.9   Schematic diagram of temperature-measuring devices. (A) RTD circuit and (B) fluid-filled thermometer.

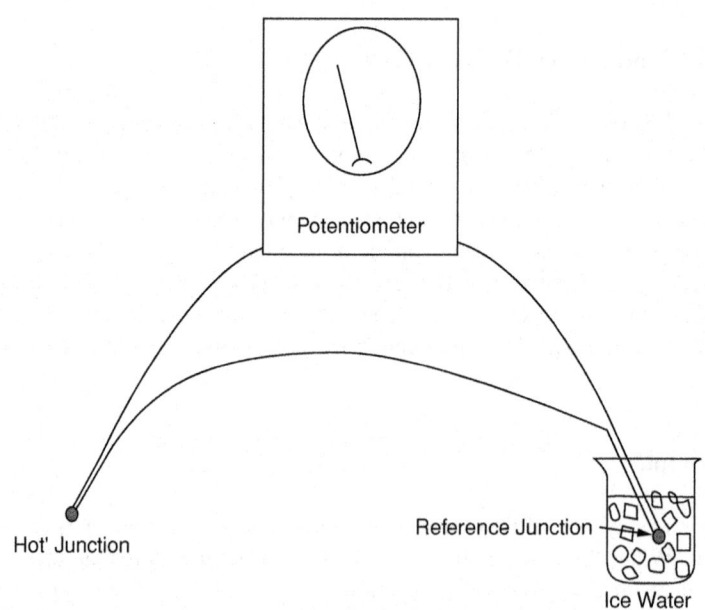

Potentiometer

Hot' Junction

Reference Junction

Ice Water

Figure 7.10   Thermocouple circuit showing ice water as the reference.

force values at various temperatures with ice water as the reference are tabulated and available in handbooks.

7.2.6 Radiation Pyrometers

This type of temperature sensing device focuses the energy received from a surface through a lens on a device such as a photovoltaic cell or a thermopile. This energy stimulates the production of an electromotive force which can then be measured. Since the amount of energy received per unit area by a receiving surface from a source, varies inversely as the distance, these types of measuring devices have distance compensation features. Distance compensation is achieved by focusing the energy received on the lens of the instrument into a smaller area having an energy sensitive surface. Radiation pyrometers are useful for sensing very high temperatures such as in furnaces, flames, and red hot or molten metals. Some, which have been calibrated for operation at near ambient temperatures, are very useful for non-contact monitoring of surface temperatures of processing equipment.

7.2.7 Accurate Temperature Measurements

Fluid in glass thermometers: The most common of these is the mercury in glass thermometer, but there are other fluids used. The temperature indicated is based on thermal expansion of the fluid in the thermometer. The temperature markings on the stem are constructed based on the volume of the bulb, the diameter of the glass capillary into which the fluid emerges from the bulb, and the coefficient of thermal expansion of the fluid. The temperature scale in the stem is marked based on a specified *minimum immersion depth*. A *total immersion* thermometer must have the bulb and the whole stem immersed in the fluid to be measured.

Partial immersion thermometers have the minimum immersion depth marked on the stem below the temperature scale. The minimum immersion depth for partial immersion thermometers is usually 76 mm from the tip of the bulb. Errors are introduced when the immersion depth is below the minimum. The magnitude of the error will depend on the length of the exposed mercury column and the ambient temperature around the exposed mercury column.

Thermocouples: Errors in temperature measurement with thermocouples come from improper contact between thermocouple wires at the measuring junction, conduction of heat along the thermocouple wires, which would make the temperature at the measuring juction different from the medium measured; interference with convection heat transfer at the measuring junction; radiation heat transfer between the measuring junction and the surroundings; resistance to current flow by the thermocouple wire, inaccurate reference junction compensation, and accuracy of the instrument that reads the electromotive force developed within the thermocouple circuit. One of the most common problems in temperature measurement is in the selection of the size of thermocouple wire. Although a very thin wire thermocouple eliminates the effect of heat conduction, a long thermocouple lead with thin wires would present a high resistance in the circuit resulting in a significant drop in the measured electromotive force. Another common problem is the joining of the thermocouple wires at the measuring junction. The two wires must have good electrical contact, and the joint must contribute minimum electrical resistance in the thermocouple circuit. With thin wires, the bare ends are twisted at least three turns before soldering or brazing, while with thicker wires, the two ends may simply be butted together and brazed or welded.

7.3 STEADY-STATE HEAT TRANSFER

7.3.1 The Concept of Resistance to Heat Transfer

Equation (7.25), derived from Fourier's first law, shows that in a steady-state system the quantity of heat passing through any part of the system must equal that passing through any other part and equal the total passing through the system. The problem of heat transfer through multiple layers can be analyzed as a problem involving a series of resistance to heat transfer. The transfer of heat can be considered as analogous to the transfer of electrical energy through a conductor. ΔT is the driving force equivalent to the voltage E in electrical circuits. The heat flux q is equivalent to the current, I.

Ohm's law for electrical circuits is

$$I = \frac{E}{R} \tag{7.49}$$

For heat transfer through a slab:

$$\frac{q}{A} = \frac{\Delta T}{[\Delta X/k]} = \frac{\Delta T}{R} \tag{7.50}$$

Thus, in comparing Equations (7.48) and (7.49), R is equivalent to $\Delta X/k$. The resistance to heat transfer is $\Delta X/k$ and is the "R" rating used in the insulation industry to rate the effectiveness of insulating materials. For multilayered materials in the geometry of a slab where A is constant in the direction of increasing x, the overall resistance to heat transfer is the sum of the individual resistance in series, and:

$$R = R1 + R2 + R3 + \ldots \ldots Rn$$

Thus:

$$R = \frac{\Delta x_1}{k_1} + \frac{\Delta x_2}{k_2} + \ldots \ldots \frac{\Delta x_n}{k_n}$$

and:

$$\frac{q}{A} = \frac{\Delta T}{[\Delta x_1/k_1 + \Delta x_2/k_2 \ldots \ldots \ldots \Delta x_n/k_n]} \tag{7.51}$$

Equation (7.50) can also be written as:

$$\frac{q}{A} = \frac{\Delta T_1}{R_1} = \frac{\Delta T_2}{R_2} = \frac{\Delta T_3}{R_3} = \ldots \ldots \frac{\Delta T_n}{R_n} = \frac{\Delta T}{R} \tag{7.52}$$

Equation (7.52) is similar to Equation (7.25) previously derived using Fourier's first law.

For heat transfer through a cylinder, the heat transfer rate in Equation (7.27) results in an expression for the heat transfer resistance as:

$$R = \frac{\ln(r_2/r_1)}{2\pi LK} \tag{7.53}$$

For resistances in series:

$$q = \frac{\Delta T}{R}$$

$$= \frac{\Delta T}{[\ln(r_2/r_1)/2\pi Lk_1 + \ln(r_3/r_2)/2\pi Lk_2 + \ldots \ln(r_{n+1}/r_n)/2\pi Lk_n]} \tag{7.54}$$

and:

$$\frac{\Delta T_1}{[\ln(r_2/r_1)2\pi Lk_1]} = \frac{\Delta T_2}{[\ln(r_3/r_2)/2\pi Lk_2]} = \cdots \frac{\Delta T_n}{[\ln(r_{n+1}/r_n)/2\pi Lk_n]} \tag{7.55}$$

For convection heat transfer:

$$q = h \, A \, \Delta T = \frac{\Delta T}{(1/hA)}$$

$$R = \frac{1}{hA} \tag{7.56}$$

7.3.2 Combined Convection and Conduction: The Overall Heat Transfer Coefficient

Most problems encountered in practice involve heat transfer by combined convection and conduction. Usually, the temperatures of fluids on both sides of a solid are known and the rate of heat transfer across the solid is to be determined. Heat transfer involves convective heat transfer between a fluid on one surface, conductive heat transfer through the solid and convective heat transfer again at the opposite surface to the other fluid. The rate of heat transfer may be expressed in terms of U, the overall heat transfer coefficient, or in terms of R, an overall resistance.

Consider a series of resistances involving n layers of solids and n fluid to surface interfaces. The thermal conductivity of the solids are $k_1 \ldots k_2 \ldots k_3 \ldots k_n$ and the heat transfer coefficients are $h_1 \ldots h_2 \ldots h_n$ with subscript n increasing along the direction of heat flow.

$$q = UA\Delta T = \frac{\Delta T}{R} \tag{7.57}$$

For a slab:
Because A is the same across the thickness of a slab:

$$R = \frac{1}{UA} \tag{7.58}$$

$$R = \sum \left[\frac{1}{h_n A}\right] + \sum \left[\frac{X_n}{k_n A}\right] = \frac{1}{UA}$$

$$\frac{1}{U} = \frac{1}{h_1} + \frac{\Delta x_1}{k_1} + \frac{\Delta x_2}{k_2} + \cdots\cdots\cdots \frac{\Delta x_n}{k_n} + \cdots\cdots\cdots \frac{1}{h_n} \tag{7.59}$$

or:

$$U = \frac{1}{\left[\sum(1/h_n) + \sum(x_n/k_n)\right]} \tag{7.60}$$

For a cylinder:

$$R = \sum \left[\frac{\ln(r_{n+1}/r_n)}{2\pi Lk_n}\right] + \sum \left[\frac{1}{h_n A_n}\right] = \frac{1}{UA}$$

If the A used as a multiplier for U is the outside area, then $U = U_o$ the overall heat transfer coefficient based on the outside area. $U_i =$ overall heat transfer coefficient based on the inside area. Using $h_i =$ inside heat transfer coefficient and $h_o =$ outside heat transfer coefficient; r_i and r_o are inside and outside radius of the cylinder, respectively.

$$(2\pi r_0 L)U_0 = \frac{1}{(1/2\pi L)\sum [\ln(r_{n+1}/r_n)/k_n] + (1/2\pi L)[1/h_o r_o + 1/h_{ir_i}]}$$

or:

$$U_o = \frac{1}{r_o \sum [\ln(r_n/r_{n-1})/k_n] + [r_o/r_i h_i] + [1/h_o]}$$

$$\frac{1}{U_o} = \frac{r_o}{r_i h_i} + \frac{r_o \ln(r_2/r_1)}{k_1} + \frac{r_o \ln(r_3/r_2)}{k_2} + \ldots\ldots \frac{r_o \ln(r_n/r_{n-1})}{k_n} + \frac{1}{h_o} \qquad (7.61)$$

$$U_i = \frac{1}{r_i \sum [\ln(r_n/r_{n-1})/k_n] + [r_i/r_o h_o] + [1/h_i]}$$

or:

$$\frac{1}{U_i} = \frac{1}{h_i} + \frac{r_i \ln(r_2/r_1)}{k_1} + \frac{r_i \ln(r_3/r_2)}{k_2} + \ldots\ldots \frac{r_i \ln(r_n/r_{n-1})}{k_n} + \frac{r_i}{r_o h_o} \qquad (7.62)$$

Example 7.11. Calculate the rate of heat transfer across a glass pane that consists of two 1.6-mm-thick glass separated by 0.8-mm layer of air. The heat transfer coefficient on one side that is at $21°C$ is 2.84 W/(m$^2 \cdot$ K) and on the opposite side that is at $-15°C$ is 11.4 W/(m$^2 \cdot$ K). The thermal conductivity of glass is 0.52 W/(m $\cdot$ K) and that of air is 0.031 W/(m $\cdot$ K).

Solution:

When stagnant air is trapped between two layers of glass, convective heat transfer is minimal and the stagnant air layer will transfer heat by conduction. There are five resistances to heat transfer. R_1 is the convective resistance at one surface exposed to air, R_2 is the conductive resistance of one 1.6-mm-thick layer of glass, R_3 is the conductive resistance of the air layer between the glass, R_4 is the conductive resistance of the second 1.6 mm thick layer of glass, and R_5 is the convective resistance of the opposite surface exposed to air. Using Equation (7.9):

$$\frac{1}{U} = \frac{1}{h_1} + \frac{x_1}{k_1} + \frac{x_2}{k_2} + \frac{x_3}{k_3} + \frac{1}{h_2}$$

$$= \frac{1}{2.84} + \frac{1.6 \times 10^{-3}}{0.52} + \frac{0.8 \times 10^{-3}}{0.031} + \frac{1.6 \times 10^{-3}}{0.52} + \frac{1}{11.4}$$

$$= 0.352 + 0.0031 + 0.0258 + 0.0031 + 0.0877 = 0.4718$$

$$U = 2.12\frac{W}{m^2 \cdot K}$$

$$\frac{q}{A} + U\Delta T = 2.12(21 - (-15)) = 76.32\frac{W}{m^2}$$

Example 7.12. (a) Calculate the overall heat transfer coefficient for a 1-in. (nominal) 16-gauge heat exchanger tube when the heat transfer coefficient is 568 W/(m^2 K) inside and 5678 W/(m$^2 \cdot$ K) on the outside. The tube wall has a thermal conductivity of 55.6 W/(m $\cdot$ K). The tube has an inside diameter of 2.21 cm and a 1.65 mm wall thickness. (b) If the temperature of the fluid inside the tube is $80°C$ and $120°C$ on the outside, what is the inside wall temperature?

Solution:

$r_i = 2.21/2 = 1.105$ cm

$r_o = 1.105 + 0.1651 = 1.2701$ cm

(a) Using Equation. (7.61):

$$\frac{1}{U_O} = \frac{1}{5678} + \frac{1.2701 \times 10^2 \ln(1.2701/1.105)}{55.6}$$

$$+ \frac{1.2701 \times 10^{-2}}{(1.105 \times 10^{-2})(568)}$$

$$= 1.76 \times 10^{-4} + 0.318 \times 10^{-4} + 20.236 \times 10^{-4}$$

$$= 22.315 \times 10^{-4}; \quad U_o = 448 W/m^2 \cdot K$$

(b) Temperature on the inside wall:

q(overall) = q(across inside convective heat transfer coefficient).

Let: T_f = temperature of fluid inside tube and T_w = temperature of the wall.

$$U_o A_o \, \Delta T = h_i \, A_i \, (T_w - T_f)$$

$$448(2\pi r_o L)(120 - 80) = 568(2\pi r_i L)(T_w - 80)$$

$$(T_w - 80) = \frac{448(r_o)(40)}{568(r_i)}$$

$$= 80 + \frac{448(1.2701)(40)}{568(1.105)}$$

$$= 80 + 36.3 = 116.3°C$$

7.4   HEAT EXCHANGE EQUIPMENT

Heat exchangers are equipment for transferring heat from one fluid to another. In its simplest form a heat exchanger could take the form of a copper tube exposed to air, where fluid flowing inside the tube is cooled by transferring heat to ambient air. When used for transferring large quantities of heat per unit time, heat exchangers take several forms to provide for efficient utilization of the heat contents of both fluids exchanging heat, and to allow for compactness of the equipment. The simplest heat exchanger to make is one of tubular design. Heat exchangers commonly used in the food industry include the following.

Swept surface heat exchangers: The food product passes through an inner cylinder and the heating or cooling medium passes through the annular space between the inner cylinder and an outer cylinder or jacket. A rotating blade running the whole length of the cylinder continuously agitates the food as it passes through the heat exchanger and at the same time, continuously scrapes the walls through which heat is being transferred (the heat transfer surface). This type of heat exchanger can be used to heat, cool or provide heat to concentrate viscous food products. Figure 7.11 shows a swept surface heat exchanger.

Double pipe heat exchanger: This heat exchanger consists of one pipe inside another. The walls of the inner pipe forms the heat transfer surface. This type of heat exchanger is usually built and

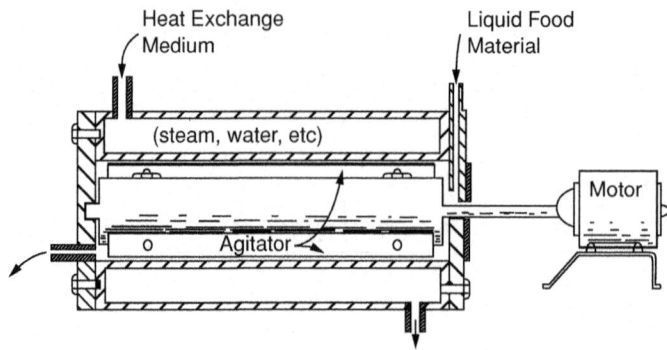

Figure 7.11   Diagram of a swept surface heat exchanger.

installed in the field. A major disadvantage is the relatively large space it occupies for the quantity of heat exchanged, compared to other types of heat exchangers. Figure 7.12 shows a double pipe heat exchanger.

Shell and tube heat exchanger: This heat exchanger consists of a bundle of tubes enclosed by a shell. The type of head arrangement allows for one tube pass or multiple passes for the product. In a one-pass arrangement, the product enters at one end and leaves at the opposite end. In a multipass arrangement, the product may travel back and forth through different tubes with each pass before finally leaving the heat exchanger. The heat exchange medium on the outside of the tubes are usually distributed using a system of baffles. Figure 7.13 shows two types of shell-and-tube heat exchanges.

Plate heat exchanger: This type of heat exchanger was developed for the dairy industry. It consists of a series of plates clamped together on a frame. Channels are formed between each plate. The product and heat transfer medium flow through alternate channels. Because of the narrow channel between the plates, the fluid flows at high velocities and in a thin layer resulting in very high heat transfer rates per unit heat transfer surface area. The plate heat exchanger is mostly used for heating fluids to temperatures below the boiling point of water at atmospheric pressure. However there are units designed for high temperature service are commercially available. Plate heat exchangers are now used in virtually any application where tubular heat exchangers were previously commonly used. Newer

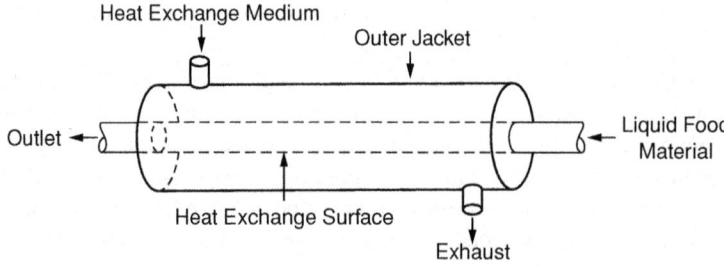

Figure 7.12   Diagram of a double-pipe heat exchanger.

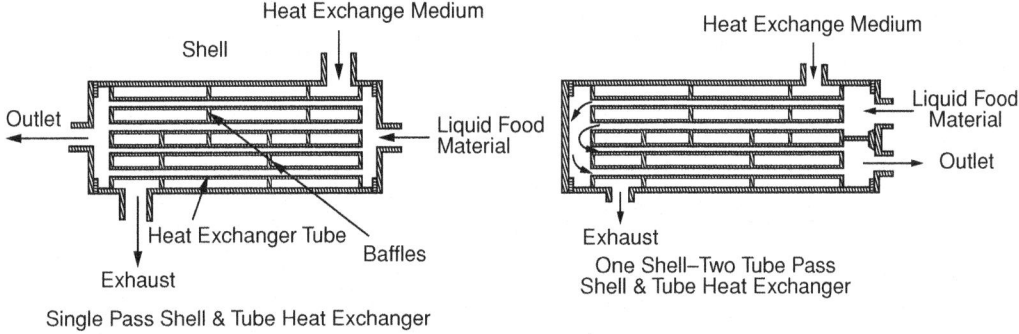

Figure 7.13 Diagram of a single-pass and a multiple-pass shell-and-tube heat exchanger.

designs have strength to withstand moderate pressure or vacuum. A major limitation is the inability to handle viscous liquids. Figure 7.14 shows a plate heat exchanger.

7.4.1 Heat Transfer in Heat Exchangers

The overall heat transfer coefficient in tubular exchangers is calculated using Equations (7.61) or (7.62). Because in heat exchangers the ΔT can change from one end of the tube to the other, a mean ΔT must be determined for use in Equation (7.57) to calculate q.

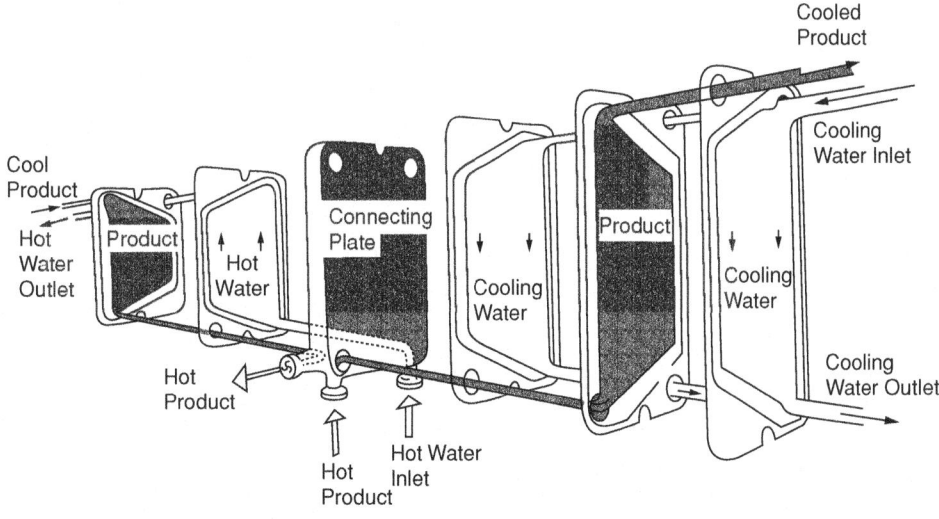

Figure 7.14 Diagram of a plat heat exchanger showing alternating paths of processed fluid and heat exchange medium.

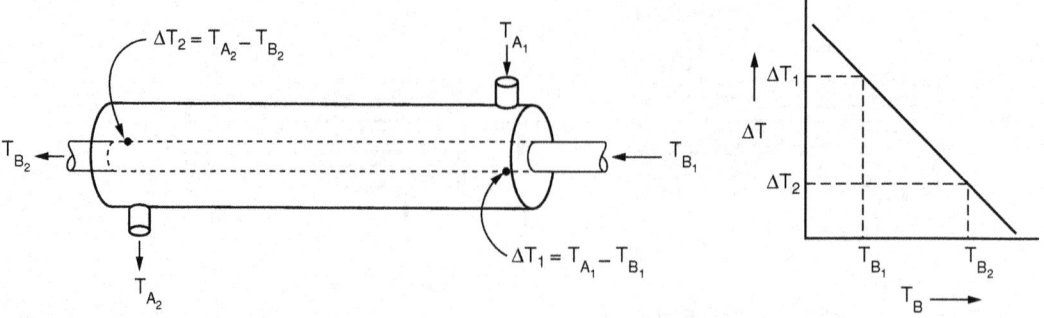

Figure 7.15   Diagram of the temperature profiles of fluid along the length of a heat exchanger tube.

7.4.2   The Logarithmic Mean Temperature Difference

Refer to Fig. 7.15 for the diagram and the representation of the symbols used in the following derivation.

If the ΔT changes linearly along the length of the heat exchanger relative to the temperature of one of the fluids, the slope of the line representing ΔT vs T_B is

$$\text{Slope} = \frac{d(\Delta T)}{dT_B} + \frac{\Delta T_2 - \Delta T_1}{T_{B2} - T_{B1}}$$

Rearranging:

$$dT_B = d\Delta T \frac{T_{B2} - T_{B1}}{\Delta T_2 - \Delta T_1}$$

The amount of heat transferred across any point in the exchanger is

$$q = U \, dA \Delta T$$

This heat transferred will cause a rise in product temperature, dT_B, and can be also expressed as:

$$q = m \, C_p \, dT_B$$

Equating:

$$U \, dA \, \Delta T = m \, C_p \, dT_B$$

Substituting the expression for dT_B:

$$U \, dA \, \Delta T = mC_p \left[\frac{d \Delta T (T_{B2} - T_{B1})}{(\Delta T_2 - \Delta T_1)} \right]$$

Rearranging:

$$mC_p(T_{B2} - T_{B1}) \frac{d\Delta T}{\Delta T} = U(\Delta T_2 - \Delta T_1)(dA)$$

Integrating:

$$mC_p(T_{B2} - T_{B1}) \int_{\Delta T_1}^{\Delta T_2} \frac{d\Delta T}{\Delta T} = U(\Delta T_2 - \Delta T_1) \int_0^A dA$$

The limits are such that at the entrance of fluid B into the heat exchanger, $\Delta T = \Delta T_1$ and $A = 0$ at this point. When fluid B leaves the heat exchanger, $\Delta T = \Delta T_2$ and $A = A$ the total heat transfer area for the heat exchanger.

$$mC_p(T_{B2} - T_{B1}) \ln\left(\frac{\Delta T_2}{\Delta T_1}\right) = UA(\Delta T_2 - \Delta T_1)$$

$mCp(T_{B2} - T_{B1}) = q$, the total amount of heat absorbed by fluid B. Thus:

$$q = UA\frac{\Delta T_2 - \Delta T_1}{\ln(\Delta T_2/\Delta T_1)} = UA\ \Delta T_L$$

Where: ΔT_L is the logarithmic mean ΔT expressed as:

$$\Delta T_L = \frac{\Delta T_2 - \Delta T_1}{\ln\left(\dfrac{\Delta T_2}{\Delta T_1}\right)} \tag{7.63}$$

When heat is exchanged between two liquids or between liquid and gas, the temperature of both fluids will be changing as they travel through the heat exchanger. If both fluids enter on the same end of the unit, flow will be in the same direction and this type of flow is cocurrent. Countercurrent flow exists when the fluids flow in opposite direction through the unit. True countercurrent or cocurrent flow would occur in double pipe, plate, and multiple tube single pass shell and tube heat exchangers. When true cocurrent or countercurrent flow exists, the rate of heat transfer can be calculated by:

$$q = U_iA_i\Delta T_L \qquad \text{or} \qquad q = U_oA_o\Delta T_L \tag{7.64}$$

The log mean temperature difference is evaluated between the inlet and exit ΔT values.

In multipass heat exchangers, the fluid inside the tubes will be traveling back ad forth across zones of alternating high and low temperatures. For this type of heat exchangers, a correction factor is used on the ΔT_L. The correction factor depends upon the manner in which shell fluid and tube fluid pass through the heat exchanger.

$$q = U_iA_iF\Delta T_L \quad \text{or} \quad q = U_oA_oF\Delta T_L \tag{7.65}$$

where F is the correction factor. The reader is referred to a heat transfer textbook or handbook for graphs showing correction factors to ΔT for multipass shell an tube heat exchangers.

Example 7.13. Applesauce is being cooled from $80°C$ to $20°C$ in a swept surface heat exchanger. The overall coefficient of heat transfer based on the inside surface area is $568\,W/m^2\cdot K$. The applesauce has a specific heat of $3187\,J/kg\cdot K$ and is being cooled at the rate of $50\,kg/h$. Cooling water enters in countercurrent flow at $10°C$ and leaves the heat exchanger at $17°C$. Calculate: (a) the quantity of cooling water required; (b) the required heat transfer surface area for the heat exchanger.

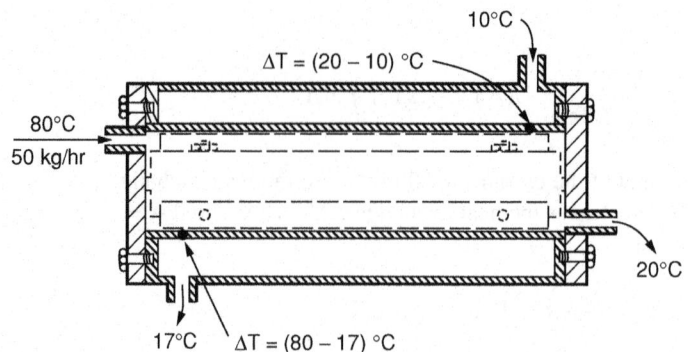

Figure 7.16   Diagram of a heat exchanger in a counterflow configuration for the fluids exchanging heat.

Solution:

The diagram of the heat exchanger in countercurrent flow is shown in Fig. 7.16.

(a) The rate of heat transfer =

$$50\frac{kg}{h} \cdot 3187\frac{J}{kg \cdot K} \cdot (80 - 20)°K = 9,561,000\frac{J}{h} = 2.656\,kW$$

Let x = quantity of water used. The specific heat of water is 4186 J/kg · K

$$x(4186)(17 - 10) = 9561000$$
$$x = \frac{9,561,000}{4186(7)} = 326\,kg/h$$

(b) $\Delta T_2 = 20 - 10 = 10°C$

$\Delta T_1 = 80 - 17 = 63°C$

Using Equation (7.63):

$$\Delta T_L = \frac{63 - 10}{\ln 63/10} = \frac{53}{\ln 6.3} = 28.8$$

The rate of heat transfer is calculated using Equation (7.64).

$$q = U_i A_i \, \Delta T_L = 568\,A_i\,(28.8)$$
$$(568)(Ai)(28.8) = 2656$$
$$A_i = \frac{2656}{568(28.8)} = 0.162\,m^2$$

7.5 LOCAL HEAT TRANSFER COEFFICIENTS

A major problem in heat transfer is the estimation of heat transfer coefficients to be used for design purposes. These values of h must be determined from the properties of the fluid and the geometry of the system.

7.5.1 Dimensionless Quantities

The use of dimensionless quantities arises from the principle of similarity. Equations that describe different systems having similar characteristics can be superimposed on each other to form a single expression suitable for all systems. Thus, if the physical characteristics of a fluid and the conditions that exist in an experiment are expressed in terms of dimensionless quantities, it will be possible to extrapolate results of an experiment to other fluids and other conditions. The principle of similarity makes it unnecessary to experimentally establish equations for heat transfer to each fluid. A general correlation equation will be suitable for all fluids. The determination of the different dimensionless quantities involved in relationships between variables describing a system is done by a method called dimensional analysis. In dimensional analysis, an equation relating the various variables is first assumed, and by performing an analysis of the base dimensions of the variables to make the equation dimensionally consistent, specific groupings of the variables are formed. The following dimensionless quantities have been identified and used in correlations involving the heat transfer coefficient.

Nusselt number (Nu): This expression involves the heat transfer coefficient (h), the characteristic dimension of the system (d), and the thermal conductivity of the fluid (k). This dimensionless expression may be considered as the ratio of the characteristic dimension of a system and the thickness of the boundary layer of fluid that would transmit heat by conduction at the same rate as that calculated using the heat transfer coefficient.

$$Nu = h\frac{d}{k} \tag{7.66}$$

Reynolds number (Re): This expression involves the characteristic dimension of the system (d), the velocity of the fluid (V), the density (ρ), and the viscosity (μ). It may be considered as the ratio of inertial forces to the frictional force. The Reynolds number has been discussed in Chapter 6.

Prandtl number (Pr): This expression involves the specific heat (C_p), the viscosity (μ), and the thermal conductivity (k). It may be considered as the ratio of rate of momentum exchange between molecules and the rate of energy exchange between molecules that lead to the transfer of heat.

$$Pr = \frac{C_P\,\mu}{k} \tag{7.67}$$

Grashof number (Gr): This quantity involves the characteristic dimension of a system (d), the acceleration due to gravity (g), the thermal expansion coefficient (β), the density of the fluid (ρ), the viscosity (μ), and the temperature difference ΔT between a surface and the fluid temperature beyond the boundary layer. This number may be considered as a ratio of the force of gravity to buoyant forces that arise as a result of a change in temperature of a fluid.

$$Gr = \frac{d^3\,g\,\beta\,\rho^2\,\Delta T}{\mu^2} \tag{7.68}$$

Peclet number (Pe): This dimensionless number is the product of the Reynolds number and the Prandtl number.

$$Pe = Re \ Pr = \frac{\rho \ V \ C_p \ d}{k} \tag{7.69}$$

Rayleigh number (Ra): This dimensionless number is the product of the Grashof number and the Prandtl number.

$$Ra = Gr \ Pr = \frac{d^3 \ g \ \beta \ C_p \rho^2 \ \Delta T}{\mu \ k} \tag{7.70}$$

Graetz number (Gz): This is similar to the Peclet number. It was derived from an analytical solution to the equations of heat transfer from a surface to a fluid flowing along that surface in laminar flow. The Graetz number is

$$Gz = \frac{\pi}{4}\left[Re \cdot Pr \cdot \left(\frac{d}{L}\right)\right] = \frac{\dot{m} \ C_p}{kL} \tag{7.71}$$

where $\overline{m}$ is the mass rate of flow, kg/s.

7.5.2   Equations for Calculating Heat Transfer Coefficients

These equations generally take the form:

$$Nu = f\left(Re, Pr, \left(\frac{L}{d}\right), Gr, \left(\frac{\mu}{\mu_w}\right)\right) \tag{7.72}$$

The Grashof number is associated with free convection, and the length to diameter ratio (L/d) appears when flow is laminar. When calculating these dimensionless quantities, the thermophysical properties of fluids at the arithmetic mean temperature at the inlet and exit are used. The viscosity of fluid at the wall (μ_w) affects heat transfer and is included in the correlation equation to account for the difference in cooling and heating processes. Most of the English literature on correlation equations for heat transfer are based on the general expression:

$$Nu = \alpha(Re)^\beta (Pr)^\gamma \left[\frac{L}{d}\right]^\delta \tag{7.73}$$

where α, β, γ, and δ are constants obtained from correlation analysis of experimental data.

A number of correlation equations suitable for use under various conditions have been published. A summary of equations suitable for use under commonly encountered conditions in food processing is given in Appendix Table A.12. In order to illustrate the use of these equations in the design of a heating or cooling system, some equations are also given in this chapter.

7.5.2.1   Simplified Equations for Natural Convection to Air or Water

For natural convection to air or water, Table 7.3 lists simplified equations based on those given by McAdams (1954).

Table 7.3 Equations for Calculating Heat Transfer
Coefficients in Free Convection from Water or Air

		Value of C for:	
Surface Conditions	*Equation*	*Air*	*Water*
Horizontal cylinder heated or cooled	$h = C\left(\frac{\Delta T}{D}\right)^{0.25}$	1.396	291.1
Fluid below heated horizontal plate	$h = C(\Delta T)^{0.25}$	2.4493	
Fluid below heated horizontal plate	$h = C(\Delta T)^{0.25}$	1.3154	
Fluid above cooled horizontal plate	$h = C(\Delta T)^{0.25}$	1.3154	
Fluid below cooled horizontal plate	$h = C(\Delta T)^{0.25}$	2.4493	
Vertical cylinder heated or cooled	$h = C\left(\frac{\Delta T}{D}\right)^{0.25}$	1.3683	127.1
Vertical plate heated or cooled	$h = c\left(\frac{\Delta T}{L}\right)^{0.25}$	1.3683	127.1

Source: Calculated from values reported in McAdams, W. H. 1954.
Heat Transmission Co., New York. 3rd ed. McGraw-Hill Book Co.,
New York.

7.5.2.2 Fluids in Laminar and Turbulent Flow Inside Tubes

Equation (7.74), used for laminar flow, was originally derived by Leveque and is discussed in a number of heat transfer textbooks. Equation (7.75) is another form of Equation (7.74). Equation (7.76), commonly referred to as the Dittus-Boelter equation, is used for turbulent flow.

$$\mathrm{Nu} = 1.615 \left[\mathrm{Re}\,\mathrm{Pr}\left(\frac{d}{L}\right) \right]^{0.33} \left(\frac{\mu}{\mu_w}\right)^{0.14} \tag{7.74}$$

Sieder and Tate (1936) introduced the use of the ratio of viscosity of fluid bulk and that at the wall as a multiplying factor to account for the difference in the heat transfer coefficient between a fluid and a heated or a cooled surface.

$$\mathrm{Nu} = 1.75(\mathrm{Gz})^{0.33} \left(\frac{\mu}{\mu_w}\right)^{0.14} \tag{7.75}$$

$$\mathrm{Nu} = 0.023(\mathrm{Re})^{0.8}(\mathrm{Pr})^{0.33} \left(\frac{\mu}{\mu_w}\right)^{0.14} \tag{7.76}$$

7.5.2.3 Heat Transfer to Non-Newtonian Fluids in Laminar Flow

Metzner et al. (1957) used Leveque's solution to heat transfer to a moving fluid in laminar flow as a basis for non-Newtonian flow heat transfer. They defined the delta function, a factor $\Delta^{0.33}$, as the ratio $\mathrm{Nu_n}/\mathrm{Nu}$ (non-Newtonian Nusselt number/Newtonian Nusselt number).

Using the delta function as a multiplying factor for the Newtonian Nusselt number, Equation (7.74) can be used for non-Newtonian fluids as follows:

$$Nu = 1.75 \Delta^{0.33} Gz^{0.33} \left(\frac{\mu}{\mu_w}\right)^{0.14} \tag{7.77}$$

When the flow behavior index is greater than 0.4 for all values of Gz, $\Delta^{0.33} = [(3n+1)/4n]^{0.33} = \delta^{0.33}$. When $n < 0.4$, $\Delta^{0.33}$ is determined as follows:

$$\Delta^{0.33} = -0.24\,n + 1.18 \quad Gz = 5, 0 < n < 0.4 \tag{7.78}$$
$$\Delta^{0.33} = -0.60\,n + 1.30 \quad Gz = 10, 0 < n < 0.4 \tag{7.79}$$
$$\Delta^{0.33} = -0.72\,n + 1.40 \quad Gz = 15, 0 < n < 0.4 \tag{7.80}$$
$$\Delta^{0.33} = -0.35\,n + 1.57 \quad Gz = 25, 0 < n < 0.4. \tag{7.81}$$

Equation (7.81) can also be used when $Gz > 25, 0.1 < n < 0.4$

For $Gz > 25$ and $n < 0.1$, the relationship between $\Delta^{0.33}$ and n is nonlinear and the reader is referred to the paper of Metzner et al. (1957) for determination of the delta function. Most food fluids will have $n > 0.1$, therefore the equations for the delta function given above should be adequate to cover most problems encountered in food processing.

7.5.2.4 Adapting Equations for Heat Transfer Coefficients to Non-Newtonian Fluids

The approach of Metzner et al. (1957) to adapting correlation equations for heat transfer coefficients derived for Newtonian fluids to non-Newtonian fluids will allow the use of practically any of the extensive correlations derived for Newtonian fluids for systems involving non-Newtonian fluids. For non-Newtonian fluids, the viscosity is not constant in the radial direction of the pipe. It will be possible, however, to use the equations derived for Newtonian fluids by using an equivalent viscosity for the non-Newtonian fluid that will cause the same pressure drop at the flow rate under consideration. A similar approach was used by Metzner et al. (1957) for heat transfer to non-Newtonian fluids in turbulent flow.

For a non-Newtonian fluid that follows the Power Law equation:

$$\tau = K\,(\gamma)^n$$

where $\tau =$ the shear stress, $\gamma =$ the shear rate, $K =$ the consistency index and $n =$ the flow behavior index.

The Reynolds number at an average rate of flow V is

$$Re = \frac{8(V)^{2-n} R^n \rho}{K[(3n+1)/n]^n}$$

The viscosity to be used in the Prandtl number and the Sieder-Tate viscosity correction term for the fluid bulk is determined as follows.

For a Newtonian fluid:

$$Re = \frac{DV\rho}{\mu}$$

Substituting the Reynolds number of the non-Newtonian fluid:

$$\mu = \frac{DV\rho}{Re}$$

$$= K\left(\frac{3n+1}{n}\right)^n\left[\frac{2RV\rho}{8V^{2-n}R^n\rho}\right]$$

$$\mu = \frac{DV\rho}{8(V)^{2-n}R^n\rho/[K(3+1/n)^n]}$$

$$\mu = \frac{K}{4}\left[\frac{3n+1}{n}\right]^n (R)^{1-n}(V)^{n-1} \tag{7.82}$$

At the wall, the viscosity μ_w used in the Sieder-Tate viscosity correction term is determined from the apparent viscosity (Eq. 6.6, Chapter 6) as follows:

$$\mu_w = K(\gamma_w)^{n-1}$$

The shear rate at the wall, γ_w, is calculated using the Rabinowitsch-Mooney equation (Eq. 6.18, Chapter 6).

$$\gamma_w = \frac{8V}{D}\left[\frac{3n+1}{4n}\right] = \frac{2V}{D}\left[\frac{3n+1}{n}\right]$$

$$\mu_w = K\left(\frac{2V}{D}\right)^{n-1}\left[\frac{3n+1}{n}\right]^{n-1} \tag{7.83}$$

Equations (7.82) and (7.83) are the viscosity terms for non-Newtonian fluids used in the correlation equations for heat transfer coefficients to Newtonian fluids.

As an example, Equation (7.74) will be used to determine an equivalent equation for non-Newtonian fluids. Because Leveque's derivation of Equation (7.74) was based on the fluid velocity profile at the boundary layer near the pipe wall, the viscosity term in the Reynolds number should be the apparent viscosity at the shear rate existing at the pipe wall. The viscosity term in the Prandtl number is based on viscous dissipation and conduction in the bulk fluid and should be the apparent viscosity based on the average velocity. Substituting Equation (7.83) for the μ term in the equation for the Reynolds number Φ and Equation (7.82) for the Φ term in the equation for the Prantl number, The Nusselt number expression of Equation (7.74) becomes:

$$Nu = 1.615\left[\left[\frac{DV\rho}{K[2V/D]^{n-1}[(3n+1)/n]^{n-1}}\right]\right.$$

$$\times\left.\left[\frac{C_p(K/4)[(3n+1)/n]^n[D/2]^{1-n}V^{n-1}}{k}\left[\frac{D}{L}\right]\right]\right]^{0.33}\left[\frac{\mu}{\mu_w}\right]^{0.14}$$

Simplifying:

$$Nu = 1.615\left[\frac{3n+1}{4n}\right]^{0.33}\left[\frac{D^2V\rho C_p}{kL}\right]^{0.33}\left[\frac{\mu}{\mu_w}\right]^{0.14}$$

Substituting the mass rate of flow calculated from the volumetric rate of flow and the density:

$$\dot{m} = \frac{\pi D^2}{4} \cdot V \cdot \rho \qquad D^2 V\rho = \frac{4}{\pi}\dot{m}$$

$$Nu = 1.615 \left[\frac{4}{\pi}\right]^{0.33} \left[\frac{\dot{m}C_p}{kL}\right]^{0.33} \left[\frac{3n+1}{4n}\right]^{0.33} \left[\frac{\mu}{\mu_w}\right]^{0.14}$$

This expression is similar to Equation (7.77) except for the value of the constant.

Example 7.14. Calculate the rate of heat loss from a 1.524 m inside diameter horizontal retort, 9.144 m long. Steam at 121°C is inside the retort. Ambient air is at 25°C. The retort is made out of steel (k = 42 W/m · K) and has a wall thickness of 0.635 cm.

Solution:

The heat transfer coefficient from the outside surface to ambient air will be controlling the rate of heat transfer; therefore, variations in the steam side heat transfer coefficient will have little effect on the answer. Assume the steam side heat transfer coefficient is 6000 W/(m² · K).

The system is a horizontal cylinder with a vertical plate at each end. The outside heat transfer coefficient can be calculated using the appropriate equations from Table 7.3.

For a horizontal cylinder to air:

$$h = 1.3196 \left[\frac{\Delta T}{D_o}\right]^{0.25}$$

For a vertical plate to air:

$$h = 1.3683 \left[\frac{\Delta T}{L}\right]^{0.25}$$

ΔT is evaluated from the outside wall temperature. The problem can be solved using a trial and error procedure by first assuming a heat transfer coefficient, calculating a wall temperature and recalculating the heat transfer coefficient until the assumed values and calculated values converge.

The magnitude of the heat transfer coefficient in free convection to air is of the order 5 W/m² · K. Equation (7.62) can be used to calculate an overall heat transfer coefficient for the cylindrical surface.

$$\frac{1}{U_i} = \frac{r_i}{r_o h_0} + \frac{r_i \ln r_o/r_i}{k} + \frac{1}{h_i} = 0.1984 + 0.00015 + 0.000167$$

$$= \frac{0.762}{0.76835(5)} + \frac{0.762 \ln(0.76835/0.762)}{42} + \frac{1}{6000}$$

$$U_i = 5.03 \, W/(m^2 AK)$$

$$U_i A_i \Delta T = h_o Ao\Delta t_w$$

$$U_i(2\pi r_i L)\Delta T = h_o(2\pi r_o L)\Delta t_w$$

$$U_i r_i \Delta T = h_o r_o \ \Delta t_w$$

$$\Delta T_w = \frac{U_i r_i \Delta T}{h_o r_o} = \frac{5.03(0.762)(121 - 25)}{5(0.76835)} = 95.78°C$$

$$h_o = 1.3196 \left(\frac{\Delta T_w}{D_o}\right)^{0.25} = 1.3196 \left(\frac{95.78}{1.5367}\right)^{0.25} = 3.71 W/(m^2 K)$$

The calculated value is less than the assumed h_o. Use this calculated value to recalculate U_i. All the terms in the expression for $1/U_i$, previously used are the same except for the first term. Assume $h_o = 3.71 \text{W}/(\text{m}^2 \cdot \text{K})$.

$$\frac{1}{U_i} = \frac{0.762}{0.76835(3.71)} + 0.00015 + 0.000167$$

$$U_i = 3.7365 \ W/(m^2\,K)$$

$$\Delta T_w = \frac{3.7365(0.762)(121 - 25)}{3.71(0.76835)} = 95.89°C$$

$$h_o = 1.3196(\frac{95.89}{1.5367})^{0.25} = 3.71 W/(m^2\,K)$$

Because the assumed and calculated values are the same, the correct h_o must be $h_o = 3.71 \text{W}/(\text{m}^2 \cdot \text{K})$ and:

$$q = U_i A_i \quad \Delta T = 3.7365(\pi)(1.524)(9.144)(121 - 25) = 15,696 \ W$$

For the ends, examination of the equation for the heat transfer coefficient reveal that h with vertical plates is approximately 4% higher than h with horizontal cylinders for the same ΔT if L is approximately the same as D. Assume $h = 3.71 (1.04) = 3.86 \text{ W}/(\text{m}^2 \cdot \text{K})$. Because the end is a vertical flat plate, U can be calculated using Equation (7.59).

$$\frac{1}{U} = \frac{1}{3.86} + \frac{0.00635}{42} + \frac{1}{6000}$$

$$U = 3.855 \ W/(m^2 \cdot K)$$

$$UA\Delta T = h_o\, A\, \Delta\, T_w$$

$$\Delta T_w = \frac{U\Delta T}{h_o} = \frac{3.855(121 - 25)}{3.86} = 95.88$$

$$h_o = 1.3683 \left[\frac{95.87}{1.5367} \right]^{0.25} = 3.846 \ W/(m^2 \cdot K)$$

The assumed and calculated values are almost the same. Use $h_o = 3.846 \text{ W}/(\text{m}^2 \cdot \text{K})$:

$$\frac{1}{U} = \frac{1}{3.846} + 0.000156 + 0.000167$$

$$U = 3.835 \ W/(m^2 \cdot K)$$

Because there are two sides, the area will be: $A = 2\pi R^2$

$$q = U\, A\, \Delta\, T = (3.835)(\pi)(1.5367)^2(2)(121 - 25) = 5463W$$

Total heat loss $= 15,704 + 5463 = 21,167 \text{ W}$.

Example 7.15. Calculate the overall heat transfer coefficient for applesauce heated from 20°C to 80°C in a stainless steel tube 5 m long with an inside diameter of 1.034 cm and a wall thickness of 2.77 mm. Steam at 120°C is outside the tube. Assume a steam side heat transfer coefficient of 6000 $\text{W}/(\text{m}^2 \cdot \text{K})$. The rate of flow is 0.1 m/s. Applesauce has a density of 995 kg/m^3 and this density is assumed to be constant with temperature. The value for n is 0.34 and the values for K are 11.6 at 30°C and 9.0 at 82° C in Pa · s units. Assume K changes with temperature according to an Arrhenius type relationship as follows:

$$\log K = A + \frac{B}{T}$$

where A and B are constants and T is the absolute temperature. The thermal conductivity may be assumed to be constant with temperature at 0.606 W/(m · K). The specific heat is 3817 J/(kg · K). The thermal conductivity of the tube wall is 17.3 W/(m · K).

First, determine the temperature dependence of K. B is the slope of a plot of log K against 1/T.

$$B = \frac{\log 11.6 - \log 9.0}{(1/303 - 1/355)} = \frac{1.064458 - 0.954243}{0.0033 - 0.00282} = 227.99$$

$$A = \log 11.6 - \frac{227.99}{303} = 1.064458 - 0.752433 = 0.3120$$

$$\log K = 0.3143 + \frac{227.99}{T}$$

The arithmetic mean temperature for the fluid is $(20 + 80)/2 = 50°C$. At this temperature, $K = \log^{-1}(0.3143 + 227.3/323) = 10.42$.

$$G = \frac{0.1 \, m}{s} \frac{1}{\pi(0.00517)^2 \, m^2} = 1190.88 \, kg/(s \cdot m^2)$$

The equivalent Newtonian viscosity is calculated using Equation (7.82).

$$\mu = \frac{10.42}{4} \left(\frac{3(0.34) + 1}{0.34} \right)^{0.34} (0.00517)^{1-0.34}(0.1)^{0.34-1}$$

$$= \frac{10.42}{4}(1.8328)(0.030967)(4.5709) = 0.676 \, Pa \cdot s$$

$$Re = \frac{0.01034(0.1)(995)}{0.676} = 1.521; \quad \text{flow is laminar}$$

Equation (7.77) must be used, because n < 0.4 and the delta function deviates from $\delta = (3n + 1)/4n$. Solving for the Graetz number:

$$\dot{m} = \pi \left(\frac{D^2}{4} \right) (V)(\rho) = \pi(0.00517)^2(0.1)(995) = 0.008355$$

Using Equation (7.79) for n < 0.4 and Gz = 10:

$$Gz = \frac{0.008355(3817)}{[(0.606)(5)]} = 10.52$$

$$\Delta^{0.33} = -0.6(0.34) + 1.30 = 1.096$$

The Nusselt number is

$$Nu = 1.75(1.096)(10.525)^{0.33} \left(\frac{\mu}{\mu_w} \right)^{0.14}$$

$$= 4.170 \left(\frac{\mu}{\mu_w} \right)^{0.14} = \frac{hD}{k}$$

Solving for h:

$$h = \frac{4.17[\mu/\mu_w]^{0.14}(0.606)}{0.01034} = 244.39\left[\frac{\mu}{\mu_w}\right]^{0.14}$$

Assume most of the temperature drop occurs across the fluid side resistance. Assume that $T_w = 116°C$.

K at $116°C = \log^{-1}(0.3126 + 227.99/389) = 7.91$ Pa $\cdot$ s

Using Equation (7.83):

$$\mu_w = 7.91\left[\frac{(2)(0.1)}{0.01034}\right]^{0.34-1}\left[\frac{3(0.34)+1}{0.34}\right]^{0.34-1} = 7.91(0.141548)(0.30849) = 0.3454$$

$$h_i = 244.39\left[\frac{0.676}{0.3454}\right]^{0.14} = 268.48 W/(m^2 \cdot \leq K)$$

Using Equation (7.62) for the overall heat transfer coefficient:

$$\frac{1}{U_i} = \frac{r_i}{r_o h_o} + \frac{r_i \ln r_o/r_i}{k} = \frac{1}{h_i}$$

$$r_i = 0.00517; \quad r_o = 0.00517 + 0.00277 = 0.00794$$

$$\frac{1}{U_i} = \frac{0.00517}{0.00794(6000)} + \frac{0.00517 \ln(0.00794/0.00517)}{17.3} + \frac{1}{268.48}$$

$$= 0.000109 + 0.000128 + 0.003725 = 0.003962; \quad U_i = 252.4$$

Check if wall temperature is close to assumed temperature.

$$U_i A_i \, \Delta T = h A_i \, \Delta T_w$$

$$\Delta T_w = \frac{U_i A_i \Delta T}{h_i A_i} = \frac{252.42}{268.48}(120-50) = 65.8$$

$$T_w = 50 + 65.8 = 115.8$$

This is close to the assumed value of $116°C$, therefore:

$$h_i = 268.48 \ W/(m^2.K) \text{ and } U_i = 252.42 \ W/(m^2 \cdot K)$$

Example 7.16. The applesauce in Example 7.14 is being cooled at the same rate from $80°C$ to $120°C$ in a double pipe heat exchanger with water flowing in the annular space outside the tube at a velocity of 0.5 m/s. The inside diameter of the outer jacket is 0.04588 m. Water enters at $8°C$ and leaves at $16°C$. Calculate the food side and the water side heat transfer coefficient and the overall heat transfer coefficient.

Solution:

First calculate the water side heat transfer coefficient. At a mean temperature of $12°C$ water has the following properties: $k = 0.58 \ W/(m \cdot K)$, $\mu = 1.256 \times 10^{-3}$ Pa $\cdot$ s; $\rho = 1000 kg/m^3$; $C_p = 4186 J/(kg \cdot K)$; and $D_o = 0.01588$ m. For an annulus, the characteristic length for the Reynolds number is 4(cross-sectional area/wetted perimeter).

$$D = \frac{4(\pi)\left(d_2^2 - d_1^2\right)/4}{(\pi)(d_1 + d_2)} = d_2 - d_1 = 0.03$$

$$Re = \frac{D_o \overline{V} \rho}{\mu} = \frac{(0.030)(0.5)(1000)}{1.256 \times 10^{-3}} = 11{,}942$$

$$Pr = \frac{C_p \mu}{k} = \frac{4186(1.256 \times 10^{-3})}{0.58} = 9.06$$

For water in turbulent flow in an annulus, Monod and Allen's equation from Appendix Table A.12 with the viscosity factor is

$$\frac{hD_o}{k} = 0.02 \, Re^{0.8} \, Pr^{0.33} \left[\frac{d_2}{d_1}\right]^{0.53} \left[\frac{\mu}{\mu_w}\right]^{0.14}$$

$$= 0.02(11942)^{0.8}(9.06)^{0.33} \left[\frac{0.04588}{0.01588}\right]^{0.53} \left[\frac{\mu}{\mu_w}\right]^{0.14} = 2565 \left[\frac{\mu}{\mu_w}\right]^{0.14}$$

Assume that $T_w = 2°C$ higher than the arithmetic mean fluid temperature. $T_w = 15°C$; $\mu_w = 1.215 \times 10^{-3}$ Pa·s.

Now, calculate the food fluid side heat transfer coefficient. Since the fluid arithmetic mean

$$h_o = \frac{0.58}{0.03}(0.02)(1827)(2.069)(1.754) \left[\frac{\mu}{\mu_w}\right]^{0.14}$$

$$h_o = 2565 \left[\frac{1.256}{1.215}\right]^{0.14} = 2577 \, W/(m^2 \cdot K)$$

temperature is the same as in Example 7.14.

$$h_i = 256.7 \left[\frac{\mu}{\mu_w}\right]^{0.14}$$

The overall ΔT between the water and the applesauce based on the mean temperature is

$$50 - 12 = 38°C. \; T_w = 15°C.$$

$$K_w = \log^{-1}\left(\frac{0.3126 + 227.99}{288}\right) = 12.69 \, Pa \cdot s$$

From Example 7.14: $\mu = 0.676$ Pa·s. Using Equation (7.83):

$$\mu_w = 12.69 \left[\frac{2(0.1)}{0.01034}\right]^{0.34-1} \left[\frac{3(0.34) + 1}{0.34}\right]^{0.34-1} = 0.554$$

$$h_i = 256.7 \left[\frac{0.676}{0.554}\right]^{0.14} = 263.95 \, W/(m^2 \cdot K)$$

Calculating for U_i using Equation (7.62):

$$\frac{1}{U_i} = \frac{0.00517}{0.00794(2577)} + \frac{0.00517 \ln(0.00794/0.00517)}{17.3} + \frac{1}{263.95}$$

$$= 0.000253 + 0.000128 + 0.003789 = 0.004170$$

$$U_i = 239.83 \, W/(m^2 \cdot K)$$

Solving for the wall temperature:

$$\Delta T_w = \frac{U_i \Delta T}{h_i} = \frac{(239.83)(38)}{263.95} = 34.5°C$$

$$T_w = 50 - 34.5 = 15.5°C$$

The small difference in the calculated and assumed wall temperatures will not alter the calculated h_i and h_o values significantly; therefore, a second iteration is not necessary. The values for the heat transfer coefficients are

> Water side: $h_o = 2577 \, W/(m^2 \cdot K)$
> Food fluid side: $h_i = 263.95 \, W/(m^2 \cdot K)$
> Overall heat transfer coefficient: $U_i = 239.83 \, W/(m^2 \cdot K)$

7.6 UNSTEADY-STATE HEAT TRANSFER

Unsteady-state heat transfer occurs when food is heated or cooled under conditions where the temperature at any point within the food or the temperature of the heat transfer medium changes with time. In this section, procedures for calculating temperature distribution within a solid during a heating or cooling process will be discussed.

7.6.1 Heating of Solids Having Infinite Thermal Conductivity

Solids with very high thermal conductivity will have a uniform temperature within the solid. A heat balance between the rate of heat transfer and the increase in sensible heat content of the solid gives:

$$mC_p \frac{dT}{dt} = hA(T_m - T)$$

Solving for T and using the initial condition, $T = T_o$ at $t = 0$:

$$\ln\left(\frac{T_m - T}{T_m - T_o}\right) = \frac{hA}{mC_p}t \tag{7.84}$$

Equation (7.84) shows that a semi-logarithmic plot of $\theta = (T_m - T)/(T_m - T_o)$ against time will be linear with a slope $-hA/mC_p$. If a solid is heated or cooled, the dimensionless temperature ratio, θ, plotted semi-logarithmically against time, will show an initial lag before becoming linear. If the extent of the lag time is defined such that the point of intersection of the linear portion of the heating curve with the ordinate is defined, it will be possible to calculate point temperature changes in a solid for long heating times, from an empirical value of the slope of the heating curve. This principle is used in thermal process calculations for foods in Chapter 8.

Fluids within a well-stirred vessel exchanging heat with another fluid that contacts the vessel walls will change in temperature as in Equation (7.84). A well-mixed steam-jacketed kettle used for cooking foods is an example of a system where temperature change will follow Equation (7.84). For a well-mixed steam-jacketed kettle, U, the overall heat transfer coefficient between the steam and the fluid inside the kettle, will be used instead of the local heat transfer coefficient h used in deriving equation 84.

Example 7.17. A steam-jacketed kettle consists of a hemispherical bottom having a diameter of 69 cm and cylindrical side 30 cm high. The steam jacket of the kettle is over the hemispherical bottom only. The kettle is filled with a food product that has a density of 1008 kg/m^3 to a point 10 cm from the rim of the kettle. If the overall heat transfer coefficient between steam and the food in the jacketed part of the kettle is 1000W/(m$^2 \cdot$ K), and steam at 120°C is used for heating in the jacket, calculate the time for the food product to heat from 20°C to 98°C. The specific heat of the food is 3100 J/(kg $\cdot$ K).

Solution:

The surface area of a hemisphere is $(1/2)\pi d^2$; the volume is $0.5(1/6)\pi d^3$. Use Equation (7.84) to solve for t. A = area of hemisphere with d = 0.69 m. A = $0.5(\pi)(d^2)$ = 0.7479 m^2.

$$\text{Volume} = 0.5(0.1666)(n)(.693) + n[(0.69)(0.5)]2(0.30 - 0.10)$$

$$= 0.08597 + 0.07478 = 0.16075 \text{ m}^3$$

$$= 0.16075 \text{m}^3(1008 \text{ kg/m}^3) = 162.04 \text{ kg}$$

$$T_m = 120; \ T_o = 20; \ U = 1000; \ C_p = 3100$$

$$\ln\left(\frac{120 - 98}{120 - 20}\right) = -\frac{1000(0.7479)}{162.04(3100)}t = -0.001489\,t$$

$$t = \frac{-\ln(0.22)}{0.001489} = 1016.9\,\text{s} = 16.95\,\text{min}$$

7.6.2   Solids with Finite Thermal Conductivity

When k is finite, a temperature distribution exists within the solid. Heat transfer follows Fourier's second law, which was derived in the section "Heat Transfer by Conduction" (Eq. 7.19) for a rectangular parallelopiped. For other geometries, the following differential equations represent the heat balance (Carslaw and Jaeger, 1959).

For cylinders with temperature symmetry in the circumferential direction:

$$\frac{\delta T}{\delta t} = \alpha \left(\frac{\delta^2 T}{\delta r^2} + \frac{\delta T}{r \delta r} + \frac{\delta^2 T}{\delta Z^2}\right) \tag{7.85}$$

For spheres with symmetrical temperature distribution:

$$\frac{\delta T}{\delta t} = \alpha \left[\frac{\delta^2(rT)}{r\delta r^2} + \frac{1}{r^2 \sin \phi} + \frac{\phi}{\delta \phi}\left(\sin \phi \frac{\delta T}{\delta \phi}\right)\right] \tag{7.86}$$

Thermal diffusivity: $\alpha = k/(\rho C_p)$. An excel program for calculating α from values of k, ρ, and C_p determined from food composition is given in Appendix A11.

These equations may be solved analytically by considering one-dimensional heat flow and obtaining a composite solution by a multiplicative superposition technique. Thus, the solution for a brick-shaped solid will be the product of the solutions for three infinite slabs, and that for a finite cylinder will be the product of the solution for an infinite cylinder and an infinite slab. Analytical solutions are not easily obtained for certain conditions in food processing, such as an interrupted process where a change in boundary conditions occurred at a midpoint in the process, or when the boundary temperature is an undetermined function of time. Equations (7.19), (7.85), and (7.86) may be solved using a finite difference technique if analytical solutions are not suitable for the conditions specified. Techniques for solving partial differential equations are discussed and analytical solutions are given in Carslaw and Yaeger (1959).

7.6.3 The Semi-Infinite Slab with Constant Surface Temperature

A semi-infinite slab is defined as one with infinite width, length, and depth. Thus, heat is transferred in only one direction, from the surface toward the interior. This system is also referred to as a "thick solid" and under certain conditions, such as at very short times of heating, the surface of a solid and a point very close to the surface would have the temperature response of a semi-infinite solid to a sudden change in the surface conditions. Equation (7.19) may be solved using the boundary conditions: at time 0, the slab is initially at T_0 and the surface is suddenly raised to temperature T_s. The temperature T at any point x, measured from the surface, expressed as a dimensionless temperature ratio, θ, is

$$\theta = \mathrm{erf}\left[\frac{x}{(4\,\alpha\,t)^{0.5}}\right] \tag{7.87}$$

$\theta = (T_s - T)/(T_s - T_O)$ and erf is the error function. The error function is a well studied mathematical function, and values are found in standard mathematical tables. Values of the error function are given in Table 7.4.

The error function approaches 1.0 when the value of the argument is 3.6. In a solid, the point where erf = 1 is undisturbed by the heat applied at the surface. Thus, a penetration depth may be calculated, beyond which the thermal conditions are undisturbed. If this penetration depth is less than the half thickness of a finite slab, the temperature distribution near the surface may be approximated by Equation (7.87). A finite body may exhibit a thick body response if the penetration depth is much less than the half thickness. The thick body response to a change in surface temperature is the temperature distribution expressed by Equation (7.87) from the surface (x = 0) to a point in the interior s distant from the surface, when $\delta = 3.8(4\alpha t)^{0.5}$ is much less than the half-thickness of a finite solid.

If the surface heat transfer coefficient is finite, the solution to Equation (7.19) with the boundary condition: $T = T_0$ at time zero, heat is transferred at the surface from a fluid at temperature T_m, with

Table 7.4 Values of the Error Function.

x	erf (x)	x	erf (x)	x	erf (x)
0	0	0.70	0.677801	1.7	0.983790
0.05	0.056372	0.75	0.711156	1.8	0.989091
0.10	0.112463	0.80	0.742101	1.9	0.992790
0.15	0.157996	0.85	0.770668	2.0	0.995322
0.20	0.222703	0.90	0.796908	2.2	0.998137
0.25	0.276326	0.95	0.820891	2.4	0.999311
0.30	0.328627	1.00	0.842701	2.6	0.999764
0.35	0.37938	1.1	0.880205	2.8	0.999925
0.40	0.428392	1.2	0.910314	3.0	0.999978
0.45	0.47548	1.3	0.934003	3.2	0.999994
0.50	0.520500	1.4	0.952790	3.4	0.999998
0.55	0.563323	1.5	0.966105	3.6	1.00000
0.60	0.603856	1.6	0.976348	3.8	1.00000
0.65	0.64202			4.0	1.00000

Source: Kreyzig, E. 1963. Advanced engineering Mathematics, John Wiley & Sons, NY.

a heat transfer coefficient h:

$$\theta = \text{erf}\left[\frac{x}{\sqrt{(4\alpha t)}} + [e]^{hx/k+(h/k)^2\alpha t}\left[\text{erfc}\left[\frac{x}{\sqrt{(4\alpha t)}} + h\text{ over }k\sqrt{(\alpha t)}\right]\right]\right] \tag{7.88}$$

where erfc[F(x)] = 1 − erf[F(x)].

Equation (7.88) becomes Equation (7.87) when h is infinite, as the complementary error function has a value of zero when the argument of the function is very large. Surface conductance plays a role along with the thermal conductivity in establishing if a body will exhibit a thick body response. Schneider (1973) defined a critical Fourier number when a body ceases to exhibit a thick body response:

$$\text{Fo}_{\text{critical}} = 0.00756\,\text{Bi}^{-0.3} + 0.02 \quad \text{for } 0.001 \leq \text{Bi} \leq 1000$$

The Fourier number is: Fo = $\alpha t/(L)^2$; L = thickness/2; Bi = Biot number = hL/k.

Example 7.18. A beef carcass at 38°C is introduced into a cold room at 5°C. The thickness of the carcass is 20 cm. Calculate the temperature at a point 2 cm from the surface after 20 min. The density is 1042 kg/m^3, the thermal conductivity is 0.44 W/(m · K), specific heat is 3558 J/(kg · K), and the surface heat transfer coefficient is 20 W/(m^2· K).

Solution:

For the short exposure time, the material might exhibit a thick body response. This condition will be affirmed by calculating the Biot number and the Fourier number.

$$L = 20(.5) = 10\text{ cm} = 0.10\text{ m}$$

$$\text{Bi} = \frac{hL}{K} = \frac{20(0.10)}{0.44} = 4.55$$

$$\text{Fo} = \frac{\alpha t}{L^2} = \frac{0.44}{1042(3558)}\frac{20(60)}{(0.10)^2} = 0.0142$$

$$\text{Fo}_{\text{critical}} = 0.00756(4.55)^{-0.3} + 0.02 = 0.024$$

The actual Fourier number is less than the critical value for the solid to cease exhibiting a thick body response, therefore, the error function solution can be used to calculate the temperature at the designated point. Using equation 88, x = 0.02 m from the surface.

$$\alpha = \frac{0.44}{(1042)(3558)} = 1.187 \times 10^{-7}$$

$$\frac{x}{(2)\sqrt{(\alpha t)}} = \frac{0.02}{(2)[(1.187 \times 10^{-7})(20)(60)]^{0.5}}$$

$$= 0.838$$

$$hx/k = 20(0.02)/0.44 = 0.909$$

$$\left(\frac{h}{k}\right)^2\alpha t = \left(\frac{20}{0.44}\right)(1.187 \times 10^{-7})(20)(60) = 0.294$$

$$= 0.542$$

$$\left(\frac{h}{k}\right)\sqrt{(\alpha t)} = \left(\frac{20}{0.44}\right)[(1.187 \times 10^{-7})(20)(60)]^{0.5}$$

$$\theta = \text{erf}(0.838) + [e]^{0.909+0.294}[\text{erfc}(0.838 + 0.542)]$$

$$= 0.7663812 + (3.3301)(.059674) = 0.9335$$

For cooling: $\theta = (T - T_m)/(T_O - T_m)$

$T = T_m + \theta(T_o - T_m) = 5 + (0.9335)(38 - 5) = 35.8°C$

7.6.4 The Infinite Slab

This is a slab with thickness 2L extending to infinity at both ends.
When h is infinite:

$$\theta = 2 \sum_{n=0}^{\infty} \left[\frac{(-1)^n}{(n+0.5)\pi}[e]^{-(n+0.5)^2(\pi^2\alpha t/L^2)} \right] \left[\cos\left[\frac{(n+0.5)\pi x}{L} \right] \right] \tag{7.89}$$

When h is finite:

$$\theta = 2 \sum_{n=1}^{\infty} [e]^{-\delta_n^2\alpha t/L^2} \left[\frac{\sin(\delta_n)\cos(\delta_n x/L)}{\delta_n + \sin(\delta_n)\cos(\delta_n)} \right] \tag{7.90}$$

δ_n are the positive roots of the transcendental equation:

$$\delta_n \tan(\delta_n) = \frac{hL}{k}$$

The center (θ_c) and surface (θ_s) temperature for an infinite solid with surface heat transfer is obtained by setting x = 0 and x = L in Equation (7.90):

$$\theta_c = 2 \sum_{n=1}^{\infty} \frac{\sin(\delta_n)[e]^{-(\delta_n/L)^2\alpha t}}{\delta_n + \sin(\delta_n)\cos(\delta_n)} \tag{7.91}$$

$$\theta_s = 2 \sum_{n=1}^{\infty} \frac{\sin(\delta_n)\cos(\delta_n)[e]^{-(\delta_n)^2\alpha t}}{\delta_n + \sin(\delta_n)\cos(\delta_n)} \tag{7.92}$$

7.6.5 Temperature Distribution for a Brick-Shaped Solid

The solution to the differential equation for a brick-shaped solid will be the product of the solution for infinite slabs of dimensions L_1, L_2, and L_3. The roots of the transcendental equation will be different for each dimension of the brick and will be designated δ_{n1}, δ_{n2}, and δ_{n3} respectively, for sides with half thickness L_1, L_2, and L_3.

Let F_o = Fourier number, $\alpha t/L^2$.

$$F(x\delta_{ni}) = \frac{\sin(\delta_{ni})\cos(\delta_{ni}x/L)}{\delta_{ni} + \sin(\delta_{ni})\cos(\delta_{ni})}$$

$$\theta = 8 \sum_{n=1}^{\infty} F(x\delta_{n1})F(x\delta_{n2})F(x\delta_{n3})[e]^{-\sum(\delta_{ni})^2 F_{o_i}} \tag{7.93}$$

Of interest to food scientists and engineers will be center temperature and surface temperature. Let:

$$F(c\delta_{ni}) = \frac{\sin(\delta_{ni})}{\delta_{ni} + \sin(\delta_{ni})\cos(\delta_{ni})}$$

$$F(s\delta_{ni}) = \frac{\sin(\delta_{ni})\cos(\delta_{ni})}{\delta_{ni} + \sin(\delta_{ni})\cos(\delta_{ni})}$$

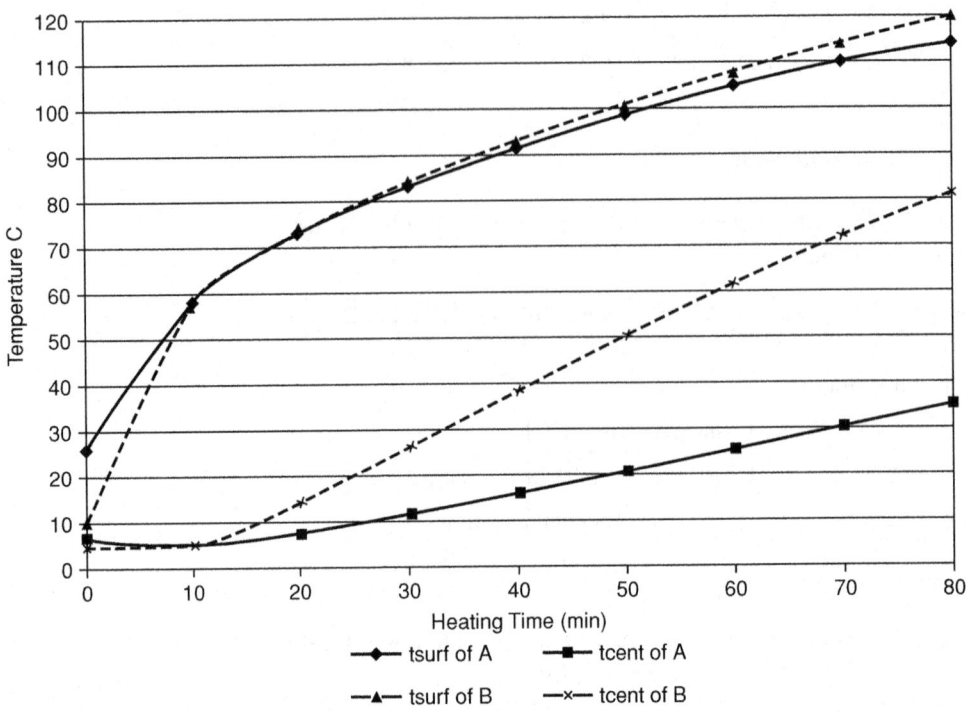

Figure 7.17   Temperature at the surface and at the geometric center of two brick-shaped solids heated in an oven at 177°C from an initial temperature of 4°C. Solid A dimensions: $20.32 \times 10.16 \times 5.08$ cm. Solid B dimensions: $30.48 \times 15.24 \times 5.08$ cm. Parameters: $\rho = 1085$ kg/m^3; $C_p = 4100$ J/(kg · K); k = 0.455 W/(m·K); h = 125 W/(m^2 K).

The center or surface temperature may be calculated using $F(c\delta_{ni})$ or $F(s\delta_{ni})$ in place of $F(x\delta_{ni})$ in Equation (7.93). Appendix Table A.13 lists a computer program in Visual BASIC for calculating the surface and center temperature of a brick shaped solid. An example of the output of the program is shown in Fig 7.17. The surface temperature will be affected by the thickness of the solid and will not assume the heating medium temperature immediately after the start of the heating process.

7.6.6   Use of Heissler and Gurney-Lurie Charts

Before the age of personal computers, calculations involving the transient temperature response of solids was a very laborious process. Solutions to the partial differential equations were plotted and arranged in a form that makes it easy to obtain solutions. Two of the transient temperature charts are the Gurney-Lurie chart shown in Fig. 7.18 and the Heissler chart shown in Figs. 7.19 and 7.20. These charts are for an infinite slab. Their use for a brick-shaped solid will involve the multiplicative superposition technique discussed earlier. Similar charts are available for cylinders and spheres.

Both charts are plots of the dimensionless temperature ratio against the Fourier number. The Gurney-Lurie chart plots the dimensionless position (x/L = n; n = 0 at the center), and 1/Biot number = m, as parameters. The Heissler chart is good for low Fourier numbers, but a chart must be made available for

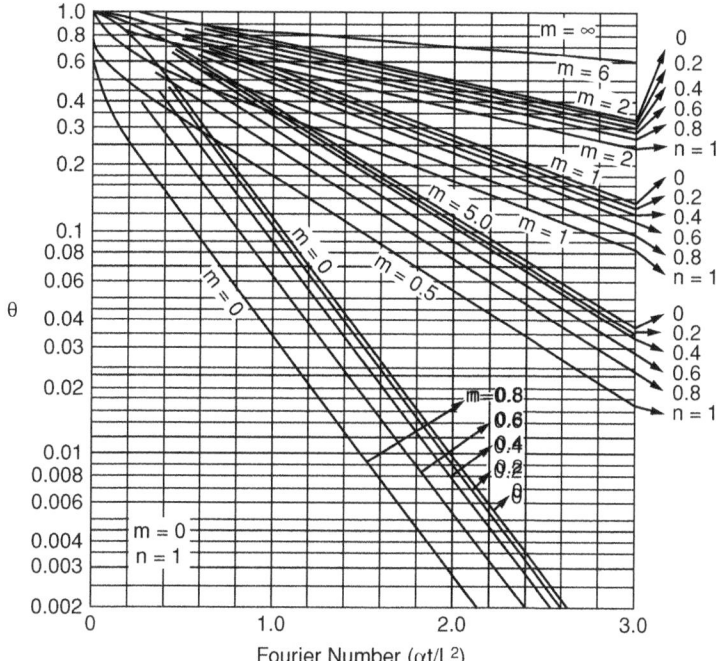

Figure 7.18 Gurney-Lurie chart for the temperature response of an infinite slab. (Source: Adapted from McAdams, W. H. 1954. Heat transmission. 3rd. ed. McGraw-Hill, New York. Used with permission of McGraw-Hill, Inc.)

---

each position under consideration. The Heissler chart shown in Fig. 7.19 is the temperature response at the surface and Fig. 7.20 is the temperature response at the center of an infinite slab.

Use of these charts involves calculating the Biot number and the Fourier number. The appropriate curve is then selected which corresponds to the Biot number and the position under consideration. The charts are used to obtain a value for the dimensionless temperature ratio, θ. If the solid is brick shaped, Fourier and Biot numbers are obtained for each direction, and the values of θ obtained for each direction are multiplied to obtain the net temperature response from all three directions.

Example 7.19. Calculate the temperature at the center of a piece of beef 3 cm thick, 12 cm wide, and 20 cm long after 30 minutes of heating in an oven. The beef has a thermal conductivity of 0.45 W/(K), a density of 1008kg/m^3, and a specific heat of 3225J/(kg · K). Assume a surface heat transfer coefficient of 20 W/(m^2 · K). The meat was originally at a uniform temperature of 5°C and it was instantaneously placed in an oven maintained at 135°C. Neglect radiant heat transfer and assume no surface evaporation.

Solution:

Consider each side separately. Consider the x direction as the thickness (L = 0.5(0.03) = 0.015 m), the y direction as the width (L = 0.5(0.12) = 0.06 m), and the z direction as the length (L = 0.5(0.2) = 0.1 m).

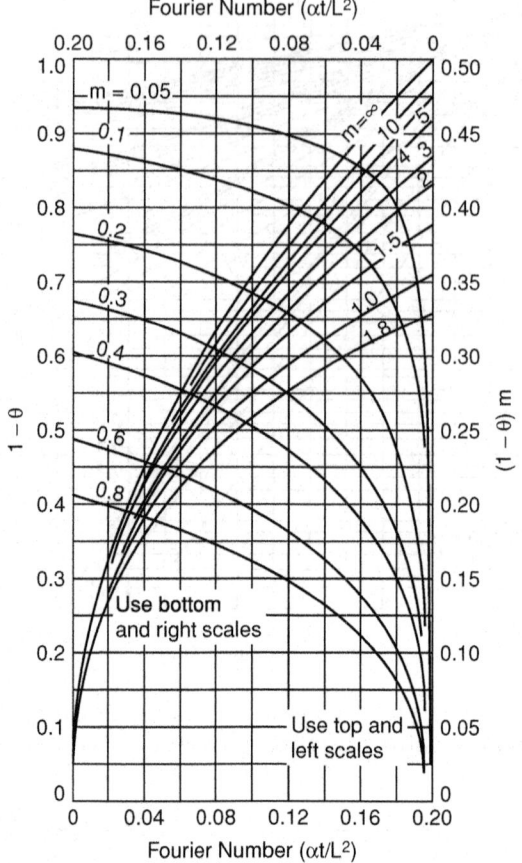

Fourier Number $(\alpha t/L^2)$

Figure 7.19   Heisler chart for the center temperature of a slab. (Adapted from Hsu, S .T. 1963. Engineering Heat Transfer. Van Nostrand Reinhold, New York.)

$$\alpha = \frac{k}{\rho C_p} = \frac{0.45}{(1008)(3225)} = 1.3843 \times 10^{-7} \, \text{m}^2/\text{s}$$

$$\text{Fo}_x = [1.3843 \times 10^{-7}(30)(60)]/(0.015)^2 = 1.107$$

$$\text{Fo}_y = [1.3843 \times 10^{-7}(30)(60)]/(0.06)^2 = 0.0692$$

$$\text{Fo} = \frac{\alpha t}{L^2}$$

$$\text{Fo}_z = [1.3843 \times 10^{-7}(30)(60)]/(0.1)^2 = 0.249$$

$$\text{Bi} = hL/k$$

$$\text{Bi}_x = 20(0.015)/0.45 = 0.667; m = 1.5$$

$$\text{Bi}_y = 20(0.06)/0.45 = 2.667; m = 0.374$$

$$\text{Bi}_z = 20(0.1)/0.45 = 4.444; m = 0.225$$

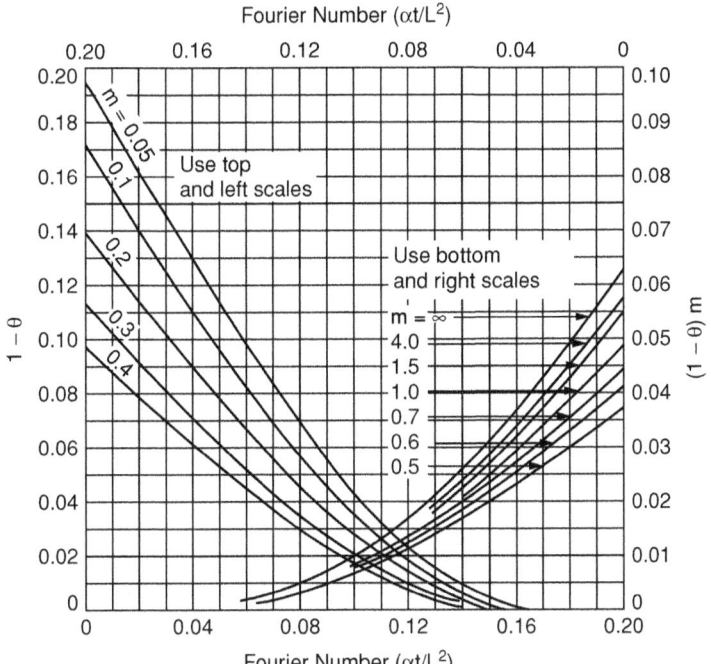

Figure 7.20 Heisler chart for the surface temperature of a slab. (Adapted from Hsu, S. T. 1963. Engineering Heat Transfer. Van Nostrand Reinhold, New York.)

At the center, n = 0

From Fig. 7.18, θ for m = 1.5 will be obtained by interpolation between values for m = 1 and m = 2. From Fig. 7.18, $Fo_x = 1.107$, n = 0, m = 1, $\theta = 0.52$, n = 0, m = 2, $\theta = 0.7$. For m = 1.5, $\theta = [0.52 + (0.70 - 0.52)(1.5 - 1)]/1 = 0.61$

The values of θ for $Fo_y = 0.0692$ and Fo_z appears to be almost 1.0 from Fig. 7.18. To verify, use Fig. 7.20 to see that at m = 0.4 and Fo = 0.07, $1 - \theta = 0$, and at m = 0.2 amd Fo = 0.01, $1 - \theta = 0$. Thus, heating for this material occurs primarily from one dimension.

$$T = T_m - \theta(T_m - T_o)$$

$$= 135 - 0.61(135 - 5) = 55.7°C.$$

7.7 CALCULATING SURFACE HEAT TRANSFER COEFFICIENTS FROM EXPERIMENTAL HEATING CURVES

If heating proceeds for a long time, the series represented by Equation (7.93) converges rapidly and the first term in the series is adequate. Equation (7.93) then becomes:

$$\theta_c = 2F(c\delta_1)F(c\delta_2)F(c\delta_3)[e]^{-Fo_1\delta_1{}^2 - Fo_2\delta_2{}^2 - Fo_3\delta_3{}^2}$$

Let $F(c\delta) = 2F(c\delta_1)F(c\delta_2)F(c\delta_3)$. Taking the logarithms of both sides of the equation:

$$(\theta_c) = \log[F(C\delta)] - [\alpha t \log(e)]\left[\frac{\delta_1^2}{L_1^2} + \frac{\delta_2^2}{L_2^2} + \frac{\delta_3^2}{L_3^2}\right] \tag{7.94}$$

Equation (7.94) shows that a plot of $\log(\theta_c)$ against t will be linear, and the surface heat transfer coefficient can be determined from the slope if α is known.

$$\text{Slope} = -[\alpha \log(e)]\left[\frac{\delta_1^2}{L_1^2} + \frac{\delta_2^2}{L_2^2} + \frac{\delta_3^2}{L_3^2}\right] \tag{7.95}$$

A computer program in BASIC shown in Appendix Table A. 14 can be used to determine the average heat transfer coefficient from the center temperature heating curve of a brick shaped solid. Similar approaches may be used for cylindrical solids.

7.8   FREEZING RATES

Temperature distribution in solids exposed to a heat exchange medium at temperatures below the solid's freezing point is complicated by the change in phase and the unique properties of the frozen and unfrozen zones in the solid. In addition, the ice front advances towards the interior from the surface, and at the interface between the two zones, a tremendous heat sink exists, in the form of the heat of fusion of water. A number of approaches have been used to mathematically model the freezing process, but the most successful in terms of simplicity and accuracy is the refinement by Cleland and Earle (1984) of the empirical equation originally developed by Plank (1913).

Plank's original equation was

$$t_f = \frac{\lambda}{T_f - T_a}\left[\frac{PD}{h} + \frac{RD^2}{k}\right] \tag{7.96}$$

where t_f is freezing time for a solid with a freezing point T_f and a thermal thickness D, which is the full thickness in the case of a slab or the diameter in the case of a sphere. P and R are shape constants (P = 1/6 for a sphere, 1/4 for a cylinder, and 1/2 for a slab; R = 1/24 for a sphere, 1/16 for a cylinder, and 1/8 for a slab). G is the latent heat of fusion per unit volume, h is heat transfer coefficient, and k is thermal conductivity of the frozen solid. Shape factors P and R have been developed for a brick shaped solid and presented as a graph based on the dimensions of the brick. This graph can be seen in Charm (1971). Plank's equation has been refined by Cleland and Earle (1984) to account for the fact that not all water freezes at the freezing point, and that the freezing process may proceed to specific final product temperature. The effect of an initial temperature different from the freezing point has also been included in the analysis. Cleland and Earle's (1984) equation is

$$t_f = \frac{\Delta H_{10}}{(T_f - T_a)(EHTD)}\left[P\frac{D}{h} + R\frac{D^2}{k_s}\right]\left[1 - \frac{1.65\,STE}{k_s}\ln\left(\frac{T_{fin} - T_a}{-10 - T_a}\right)\right] \tag{7.97}$$

ΔH_{10} = enthalpy change to go from T_f to $-10°C$ in J/m^3

T_a = freezing medium temperature

T_{fin} = final temperature

EHTD = equivalent heat transfer dimensionality, defined as the ratio of the time to freeze a slab of half thickness, D, to the time required to freeze the solid having the same D. EHTD = 1 for a slab with large width and length such that heat transfer is effectively only in one direction, and 3 for a sphere. Bricks and cylinders will have EHTD between 1 and 3.

h = surface heat transfer coefficient

k_s = thermal conductivity of the frozen solid

P, R = parameters that are functions of the Stephan number (STE) and the Plank number (PK)

$$STE = C_s \frac{(T_f - T_a)}{\Delta H_{10}}; \quad PK = C_1 \frac{(T_i - T_f)}{\Delta H_{10}}$$

C_s = volumetric heat capacity of frozen material, J/(m³ · K)

C_1 = volumetric heat capacity of unfrozen material, J/(m³ · K)

T_i = initial temperature of solid

P = 0.5 [1.026 + 0.5808 PK + STE(0.2296 PK + 0.105)]

R = 0.125 [1.202 + STE(3.41 PK + 0.7336)]

The major problem in using Equatiion (7.97) is the determination of EHTD. EHTD can be determined experimentally by freezing a slab and the solid of interest and taking the ratio of the freezing times. EHTD considers the total dimensionality of the material instead of just one dimension.

Example 7.20. Calculate the freezing time for blueberries in a belt freezer where the cooling air is at $-35°C$. The blueberries have a diameter of 0.8 cm and are to be frozen from an initial temperature of $15°C$ to a final temperature of $-20°C$. They contain 10% soluble solids, 1% insoluble solids, and 89% water. Use Choi and Okos' (1987) correlation for determining the thermal conductivity, Chang and Tao's (1981) correlation for determining the freezing point and enthalpy change below the freezing point, and Seibel's equation for determining the specific heat above and below freezing. The blueberries have a density of 1070 kg/m³ unfrozen and 1050 kg/m³ frozen. The heat transfer coefficient is 120 W/(m³ · K).

Solution:

For the specific heat above freezing, use Equation (5.11), Chapter 5:

$C_1' = 837.36(0.11) + 4186.8(0.89) = 3818.3 \, J/(kg.K)$

Converting to J/(K):

$$C_1 = \frac{3818 \, J}{kg \, K} \frac{1070 \, kg}{m^3} = 4.0856 \times 10^6 \, J/(m^3 \cdot K)$$

For the specific heat below freezing, use equation 13, Chapter 5:

$C_s' = 837.36(0.11) + 2093.4(0.89) = 1955.63 \, J/(kg \cdot K)$

Converting to J/(m³· K):

$$C_s = \frac{1955.63 \, J}{kg \cdot K} \frac{1050 \, kg}{m^3} = 2.053 \times 10^6 \, J/(m^3 \, K)$$

Use Chang and Tao's correlation for the enthalpy change fetween freezing point and $-10°C$ (see Chapter 5, the section "Enthalpy Change with a Change in Phase"). From Equation (5.20), Chapter 5:

$T_f = 287.56 - 49.19(0.89) + 37.07(0.89)^2 = 273 \, K$

Solving for the enthalpy at the freezing point: Equation (5.22), Chapter 5: $H_f = 9792.46 + 405096(0.89)$

$H_f = 370327.9 \, \text{J/kg}$

$T = -10°C = 263 \, \text{K}$

$T_r = (263 - 227.6)/(273 - 227.6) = 0.776$

$H = H_f \left[\alpha T_r + (1-a)(T_r)^b\right]$

Solving for α: Equation (5.17), Chapter 5.

$\alpha = 0.362 + 0.0498(0.89 - 0.73) - 3.465(0.89 - 0.73)^2$

$= 0.362 + 0.007968 - 0.088704 = 0.281$

Solving for b: Equation (5.18), Chapter 5:

$b = 27.2 - 129.04(0.89 - 0.23) - 481.46(0.89 - 0.23)^2$

$= 27.2 - 6.6151 - 1.265276 = 19.3196$

Solving for H at $-10°C$, $T_r = 0.776$; Equation (5.23), Chapter 5:

$H = H_f \left[(0.281)(0.776) + (1 - 0.281)(0.776)^{19.3196}\right]$

$= 0.2234 \, H_f; \quad \Delta H_{10} = H_f(1 - 0.2234) = 0.7766(370327.9)$

$\Delta H_{10} = 2.876 \times 105 \, \text{J/kg}$

The thermal conductivity will be determined using the spreadsheet program (Fig. 7.1) in the section "Estimation of Thermal Conductivity of Food Products." The amount of water unfrozen at the freezing point will be determined from the enthalpy at the freezing point using the base temperature for Chang and Tao's enthalpy correlations, of 227.3 K as the reference, and assuming that all the water is in the form of ice. The heat of fusion of ice is 334,944 J/kg.

The specific heat below freezing calculated above was 1955.23 J/(kg · K)

$H_f' = 0.89(334, 944) + 1955.23(273 - 227.6) = 386, 867$

Mass fraction unfrozen water at the freezing point

$= (386,867 - 370,327.9)/334,944 = 0.049$

Using the BASIC program in Appendix Table A.11, the following are entered with the prompts: $X_{water} = 0.049$; $X_{ice} = 0.89 - 0.049 = 0.8406$; $X_{carb} = 0.10$; $X_{fiber} = 0.01$; all other components are zero. $T = -10°C$; The output is: $k_s = 2.067 \, \text{W/(m · K)}$. Solving for freezing time using Equation (7.97):

$\Delta H_{10} = 2.876 \times 105 \, \text{J/kg} \, (1070 \, \text{kg/m}^3) = 3.077 \times 108 \, \text{J/m}^3$

$T_a = -35°C; \ T_{fin} = -20°C$

$h = 120 \, \text{W/(m}^2 \cdot \text{K)}; \ \text{EHDT} = 3 \text{ for a sphere}; \ D = 0.008 \, \text{m}$

$k_s = 2.067 \, \text{W/(m · K)}; T_i = 15 \, \text{EC}; \ C_s = 2.053 \times 10^6 \, \text{J/m}^3$

$C_l = 4.0856 \text{J/(m}^3 \, \text{AK)}$

$\text{STE} = 2.053 \times 10^6 \left[\dfrac{0 - (-35)}{6.685 \times 10^7}\right] = 1.0749$

$$PK = 4.0856 \times 10^6 \left[\frac{15 - 0}{6.685 \times 10^7} \right] = 0.9167$$

$$P = 0.5[1.026 + 0.5808(0.9167)$$
$$+ 1.0749[0.2296(0.9167) + 0.105] = 0.9488$$
$$R = 0.125[1.202 + 1.0749[(3.41)(0.91674) + 0.7336]] = 0.6688$$

Substituting values in Equation (7.97):

$$t_f = \frac{3.077 \times 10^8}{[0 - (-35)](3)} \left[0.9488 \frac{0.008}{120} + 0.6688 \frac{(0.008)^2}{2.067} \right]$$
$$= 0.029395 \times 10^8 [0.00006325 + 0.00002071](1.4383) = 353.9s$$
$$\left[1 - \frac{1.65(1.0749)}{2.067} \ln \left[\frac{-20 - (-35)}{-10 - (-35)} \right] \right]$$

PROBLEMS

7.1. How many inches of insulation would be required to insulate a ceiling such that the surface temperature of the ceiling facing the living area is within 2°C of the room air temperature. Assume a heat transfer coefficient on both sides of the ceiling of 2.84 W/(m² · K) and a thermal conductivity of 0.0346 W/(m · K) for the insulation. The ceiling is 1.27-cm-thick plasterboard with a thermal conductivity of 0.433 W/(m · K). Room temperature is 20°C and attic temperature is 49°C.

7.2. If a heat transfer coefficient of 2.84 W/(m² · K) exists on each of the two inside faces of 6.35-mm-thick glass separated by an air gap, calculate the gap that could be used such that the rate of heat transfer by conduction through the air gap would equal the rate of heat transfer by convection. What would be the rate of heat transfer if this gap is exceeded?

The thermal conductivity of air is 0.0242 W/(m · K). Solve for temperatures of 20°C and −12°C on the outside surfaces of the glass. Would the calculated gap change in value for different values of the surface temperatures? Would there be any advantage to increasing the gap beyond this calculated value?

7.3. A walk-in freezer 4 × 6 m and 3 m high is to be built. The walls and ceiling consist of 1.7 mm thick stainless steel (k = 14.2 W/(m · K)), 10 cm thick of foam insulation (k = 0.34 W/(m · K)), a thickness of corkboard (k = 0.043 W/(m · K)), and 1.27 cm thick of wood siding (k = 0.43 W/(m · K)). The inside of the freezer is maintained at −40°C. Ambient air outside the freezer is at 32°C. The heat transfer coefficient is 5 W/(m² · K) on the wood side of the wall, and 2 W/(m² · K) on the stainless steel side.

(a) If the outside air has a dew point of 29°C, calculate the thickness of the corkboard insulation that would prevent condensation of moisture on the outside wall of this freezer.

(b) Calculate the rate of heat transfer through the walls and ceiling of this freezer.

7.4. (a) Calculate the rate of heat loss to the surroundings and the quantity of steam that would condense per hour per meter of a 1.5-in. (nominal) schedule 40 steel pipe containing steam at 130°C. The heat transfer coefficient on the steam side is 11,400 W/(m² · K) and on the outside of the pipe to air is 5.7 W/m² · K. Ambient air averages 15°C in temperature for the year. The thermal conductivity of the steel pipe wall is 45 W/(m · K).

(b) How must energy would be saved in one year (365 days, 24 h/day) if the pipe is insulated with 5-cm-thick insulation having a thermal conductivity of 0.07 W/(m · K). The heat transfer coefficients on the steam and air sides are the same as in (a).

7.5. The rate of insolation (solar energy impinging on a surface) on a solar collector is 475W/m^2. Assuming that 80% of the impinging radiation is absorbed (the rest is reflected), calculate the rate at which water can be heated from 15°C to 50°C in a solar hot water heater consisting of a spiral-wound, horizontal, 1.9-cm inside-diameter polyethylene pipe 30 m long. The pipe has a wall thickness of 1.59 mm. The projected area of a horizontal cylinder receiving radiation is the diameter multiplied by the length. Assume a heat transfer coefficient of 570 W/(m^2· K) on the water side and a heat transfer coefficient of 5 W/(m^2· K) on the air side, when calculating heat loss to the surroundings after the radiant energy from the sun is absorbed. The thermal conductivity of the pipe wall is 0.3 W/m · K. Ambient temperature is 7°C.

7.6. A swept surface heat exchanger cools 3700 kg of tomato paste per hour from 93°C to 32°C. If the overall heat transfer coefficient based on the inside surface area is 855 W/m^2 · K, calculate the heating surface area required for concurrent flow and countercurrent flow. Cooling water enters at 21°C and leaves at 27°C. The specific heat of tomato paste is 3560 J/(kg · K).

7.7. Design the heating and cooling section of an aseptic canning system that processes 190 L per minute of an ice cream mix. The material has a density of 1040 kg/m^3 and a specific heat of 3684 J/(kg · K).

(a) Calculate the number of units of swept surface heat exchangers required to heat the material from 39°C to 132°C. Each unit has an inside heat transfer surface area of 0.97m^2. The heating medium is steam at 143°C. Previous experience with a similar unit on this material was that an overall heat transfer coefficient of 1700 W/(m^2 · K) based on the inside surface area may be expected.

(b) Calculate the number of units (0.9 m^2 inside surface area per unit) required for cooling the sterilized ice cream mix from 132°C to 32°C. The cooling jacket of the swept surface heat exchangers is cooled by freon refrigerant from a refrigeration system at −7°C. Under these conditions, a heat transfer coefficient of 855 W/(m^2 · K) based on the inside surface area may be expected.

7.8. A small swept surface heat exchanger having an inside heat transfer surface area of 0.11 m^2 is used to test the feasibility of cooking a slurry in a continuous system. When the slurry was passed through the heat exchanger at a rate of 168 kg/h it was heated from 25°C to 72°C. Steam at 110°C was used. The slurry has a specific heat of 3700 J/(kg · K).

(a) What is the overall heat transfer coefficient in this system?

(b) If this same heat transfer coefficient is expected in a larger system, calculate the rate at which the slurry can be passed through a similar swept surface heat exchanger having a heat transfer surface area inside of 0.75 m^2, if the inlet and exit temperatures are 25°C and 72°C, respectively, and steam at 120°C is used for heating.

7.9. A steam jacketed kettle has an inside heat transfer surface area of 0.43 m^2 that is all completely covered by the product. The product needs to be heated from 10°C to 99°C. The product contains 80% water and 20% nonfat solids. Previous experience has established that an overall heat transfer coefficient based on the inside surface area of 900 W/(m^2 · K) may be expected. The kettle holds 50 kg of product. Condensing steam at 120°C is in the heating jacket. The contents are well stirred continuously during the process.

(a) Calculate the time required for the heating process to be completed.

(b) Determine the nearest nominal size steel pipe that can be used to supply steam to this kettle if the rate of steam flow through the pipe is to average a velocity of 12 m/s.

7.10. A processing line for a food product is being designed. It is necessary to estimate the number of kettles that is required to provide a production capacity of 500 kg/h. The cooking process involving the kettles requires heating the batch from 27°C to 99°C, simmering at 99°C for 30 minutes and filling the hot product into cans. Filling the kettles and emptying requires approximately 15 minutes. The specific heat of the product is 3350 J/(kg · K). The density is 992 kg/m^3.

Available for heating are cylindrical vessels with hemispherical bottoms with the hemisphere completely jacketed. The height of the cylindrical section is 25 cm. The diameter of the vessel is 0.656 m. Assume the vessels are filled to 85% of capacity each time. The overall heat transfer coefficient based on the inside surface area averages 600 W/(m^2 · K). How many kettles are required to provide the desired production capacity? Steam condensing at 120°C is used in the jacket for heating.

7.11. A process for producing frozen egg granules is proposed where a refrigerated rotating drum that has a surface temperature maintained at −40°C contacts a pool of liquid eggs at 5°C. The eggs freeze on the drum surface and the frozen material is scraped off the drum surface at a point before the surface reenters the liquid eggs. The frozen material, if thin enough will be collected as frozen flakes. In this process, the thickness of frozen egg that forms on the drum surface is determined by the dwell time of the drum within the pool of liquid. An analogy of the process, which may be solved using the principles discussed in the section on freezing water, 7.5 h is the freezing of a slab directly in contact with a cold surface (i.e, h is infinite). It is desired that the frozen material be 2 mm thick on the drum surface.

(a) Calculate the dwell time of the drum surface within the liquid egg pool. Assume that on emerging from the liquid egg pool, the frozen material temperature will be 2°C below the freezing point.

(b) If the drum has a diameter of 50 cm and it travels 120E ($\frac{1}{3}$ of a full rotation) after emerging from the liquid egg pool before the frozen material is scraped off, calculate the rotational speed of the drum needed to satisfy the criterion stipulated in (a), and calculate the average temperature of the frozen material at the time it is scraped off the drum. The density of liquid eggs is 1012 kg/m^3, and frozen eggs 1009 kg/m^3. Calculate the thermophysical properties based on the following compositional data: 75% water, 12% protein, 12% fat, 1% carbohydrates.

7.12. In an experiment for pasteurization of orange juice, an 0.25-in. outside-diameter tube with 1/32-in.-thick wall was made into a coil and immersed in a water bath maintained at 95°C. The coil was 2 m long and when the juice was pumped at the rate of 0.2 L/min, the juice temperature changed from 25°C to 85°C. The juice contained 12% total solids. Calculate:

(a) The overall heat transfer coefficient.

(b) The inside local heat transfer coefficient if the ratio h_o/h_i is 0.8.

(c) The inside local heat transfer coefficient, h_i is directly proportional to the 0.8 power of the average velocity. If the rate at which the juice is pumped through the system is increased to 0.6 L/min, calculate the tube length needed to raise the temperature from 25°C to 90°C.

7.13. Calculate the surface area of a heat exchanger needed to pasteurize 100 kg/h of catsup by heating in a one-pass shell and tube heat exchanger from 40°C to 95°C. The catsup density is 1090 kg/m^3. The flow behavior index is 0.5, and the consistency index is 0.5 and 0.35 Pa ·s^n at 25°C and 50°C, respectively. Estimate the thermal conductivity and specific heat using correlations discussed in Chapters 5 and 7. The catsup contains 0.4% fiber, 33.8% carbohydrate, 2.8% ash, and the balance is water. The catsup is to travel within the heat exchanger at a velocity of 0.3 m/s. Steam condensing at 135°C is used for heating and, a heat transfer coefficient of

$15,000W/(m^2 \cdot K)$ may be assumed for the steam side. The heat exchanger tubes are type 304 stainless steel with an inside diameter of 0.02291 m and an outside diameter of 0.0254 m.

7.14. A box of beef that was tightly packed was originally at 0°C. It was inadvertently left on a loading dock during the summer when ambient conditions were 30°C. Assuming that the heat transfer coefficient around the box averages 20 W/(m²· K), calculate the surface temperature, and the temperature at a point 2 cm deep from the surface after 2 hours. The box consisted of 0.5-cm-thick fiberboard with a thermal conductivity of 0.2 W/(m · K), and had dimensions of $47 \times 60 \times 30$ cm. Because the box material has very little heat capacity, it may be assumed to act as a surface resistance, and an equivalent heat transfer coefficient may be calculated such that the resistance to heat transfer will be the same as the combined conductive resistance of the cardboard and the convective resistance of the surface heat transfer coefficient. The meat has a density of 1042 kg/m³, a thermal conductivity of 0.44 W/(m · K), and a specific heat of 3558 J/(kg · K).

7.15. In the operation of a microwave oven, reduced power application to the food is achieved by alternatively cutting on and off the power applied. To minimize excessive heating in some parts of the food being heated, the power application must be cycled such that the average temperature rise with each application of power does not exceed 5°C followed by a 10-second pause between power application. If 0.5 kg of food is being heated in a microwave oven having a power output of 600 W, calculate fraction of full power output that must be set on the oven controls such that the above temperature rise and pause cycles are satisfied. The food has a specific heat of 3500 J/(kg · K).

SUGGESTED READINGS

ASHRAE. 1965. ASHRAE Guide and Data Book. Fundamentals and equipment for 1965 and 1966. American Society for Heating, Refrigerating and Air Conditioning Engineers, New York.

Ball, C. O. and Olson, F. C. W. 1957. Sterilization in Food Technology. 1st ed. McGraw-Hill Book Co., New York.

Bennet, C. O. and Myers, J. E. 1962. Momentum, Heat, and Mass Transport. McGraw-Hill Book Co., New York.

Carslaw, H. S. and Jaeger, J. C. 1959. Conduction of Heat in Solids. 2nd ed. Oxford University Press, London.

Charm, S. E. 1971. Fundamentals of Food Engineering. 2nd ed. AVI Publishing Co., Westport, CT.

Chang, H. D. and Tao, L. C. 1981. Correlation of enthalpy of food system J. Food Sci. 46:1493.

Choi, Y. and Okos, M. R. 1987. Effect of temperature and composition on thermal properties of foods. In: Food Engineering and Process Applications. M. Le Maguer and P. Jelen, Eds. Vol. 1. Elsevier, New York.

Cleland, A. C. and Earle, R. L. 1984. Freezing time prediction for different final product temperature. J. Food Sci. 49:1230.

Geankoplis, C. J. 1993. Transport Processes and Unit Operations. 3rd ed. Prentice Hall, Englewood Cliffs, NJ.

Heldman, D. R. 1973. Food Process Engineering. AVI Publishing Co., Westport, CT.

Holman, J. P. 1963. Heat Transfer. McGraw-Hill Book Co., New York.

Jacob, M. and Hawkins, G. A. 1957. Heat Transfer. Vol. II. John Wiley, New York.

Leniger, H. A. and Beverloo, W. A. 1975. Food Process Engineering. D. Riedel Publishing Co., Boston.

McCabe, W. L. and Smith, J. C. 1967. Unit Operations of Chemical Engineering. 2nd ed. McGraw-Hill Book Co., New York.

McCabe, W.L., Smith, J. C., and Harriott, P. 1985. Unit Operations in Chemical Engineering. 4th ed. McGraw-Hill, New York.

McAdams, W. H. 1954. Heat Transmission. 3rd ed. McGraw-Hill, New York.

Metzner, A. B., Vaughn, R. D., and Houghton, G. L. 1957. Heat transfer to non-Newtonian fluids. AIChE J. 3:92.

Peters, M. S. 1954. Elementary Chemical Engineering. McGraw-Hill Book Co., New York.

Roshenow, W. M. and Hartnett, J. P. 1973 Handbook of Heat Transfer. McGraw Hill Book Co., New York.

Schneider, P. J. 1973. Conduction. In: Handbook of Heat Transfer, Roshenow, W. M. and Hartnett, J.P eds. McGraw-Hill, New York.

Seider, E. N. and Tate, G. E. 1936. The viscosity correction factor for heat transfer to fluids. Ind. Eng. Chem. 28:1429 (cited by McCabe and Smith, 1967, op. cit.).

Sinnott, R. K. 1996. Chemical Engineering. Vol. 6, 2nd ed. Butterworth-Heinemann, Oxford.

Watson, E. L. and Harper, J. C. 1989. Elements of Food Engineering. AVI Publishing Co., Westport, CT.

CHAPTER 8

Kinetics of Chemical Reactions in Foods

Chemical reactions occur in foods during processing and storage. Some reactions result in a quality loss and must be minimized, whereas others result in the formation of a desired flavor or color and must be optimized to obtain the best product quality. Kinetics is a science that involves the study of chemical reaction rates and mechanisms. An understanding of reaction mechanisms coupled with quantification of rate constants will facilitate the selection of the best conditions of a process or storage, in order that the desired characteristics will be present in the product.

8.1 THEORY OF REACTION RATES

Two theories have been advanced as a theoretical basis for reaction rates. The collision theory attributes chemical reactions to the collision between molecules that have high enough energy levels to overcome the natural repulsive forces among molecules. In gases, chemical reaction rates between two reactants have been successfully predicted using the equations derived for the kinetic energy of molecules and the statistical probability for collisions between certain molecules that possess an adequate energy level for the reaction to occur at a given temperature. The activation theory assumes that a molecule possesses a labile group within its structure. This labile group may be normally stabilized by oscillating within the molecule or by steric hindrance by another group within the molecule. The energy level of the labile group may be raised by an increase in temperature, to a level that makes the group metastable. Finally, a chemical reaction results that releases the excess energy and reduces the energy level of the molecule to another stable state. The energy level that a molecule must achieve to initiate a chemical reaction is called the activation energy. Both theories for reaction rates will give a reaction rate constant, which is a function of the number of reacting molecules and the temperature.

Reactions may be reversible. Reversible reactions are characterized by an equilibrium constant, which establishes steady-state concentration of product and reactants.

8.2 TYPES OF REACTIONS

8.2.1 Unimolecular Reactions

One type of chemical reaction that occurs during degradation of food components involves a single compound undergoing change. A part of the molecule may split off, or molecules may interact with each other to form a complex molecule, or internal rearrangement may occur to produce a new compound. These type of reactions are unimolecular and may be represented as:

$$A \xrightarrow{k_1} products \tag{8.1}$$

The reaction may occur in more than one step, and in some cases the intermediate product may also react with the original compound.

$$A \xrightarrow{k_1} B \xrightarrow{k_2} products \tag{8.2}$$

$$A \xrightarrow{k_1} B + A \xrightarrow{k_3} products \tag{8.3}$$

The reaction rate, r, may be considered the rate of disappearance of the reactant A or the rate of appearance of a reaction product. In reactions 8.1 and 8.2, the rate of disappearance of A, dA/dt, is proportional to a function of the concentration of A, while in reaction 8.3, dA/dt will be dependent on a function of the concentration of A and B. In reaction 8.1, the rate of formation of products will equal the rate of disappearance of the reactant, but in reactions 8.2 and 8.3, accumulation of intermediate products will result in a lag between product formation and disappearance of the original reactant. When intermediate reactions are involved, the rate of appearance of the product will depend on the rate constant k_2.

The rate constant, k, is the proportionality constant between the reaction rate and the function of the reactant concentration, $F(A)$ or $F(B)$. Thus, for reactions 8.1 and 8.2:

$$\frac{dA}{dt} = kF(A) \tag{8.4}$$

For reaction 3:

$$\frac{dA}{dt} = k_1 F(A) + k_3[F(A) + F(B)] \tag{8.5}$$

In Equations (8.4) and (8.5), A and B represent concentrations of reactants A and B, and it becomes obvious that the reaction rate will increase with increasing reactant concentration. The concentration function, which is proportional to the reaction rate, depends on the reactant and could change with the conditions under which the reaction is carried out. When studying reaction rates, it is either the rate of appearance of a product or the rate of disappearance of reactants that will be of interest. On the other hand, if intermediate reactions are involved in the process of transforming compound A into a final reaction product, and the rate constants and k_2 and k_1 are affected differently by conditions used in a process, it will be necessary to postulate rate mechanisms and evaluate an overall rate constant based on existing conditions in order to effectively optimize the process. Most of the reactions involving degradation of food nutrients are of the type shown in reaction 8.1.

8.2.2 Bimolecular Reactions

Another type of reaction involves more than one molecule. The second step of reaction 8.3 above, is one example of a bimolecular reaction. In general, a bimolecular reaction is as follows:

$$aA + bB \xrightarrow{k_3} cC + dD \tag{8.6}$$

In this type of reaction, the rate may be based on one of the compounds, either the reactant or the product, and the change in concentration of other compounds may be determined using the stoichiometric relationships in the reaction.

$$r_{3c} = \frac{dC}{dt} = -\frac{c}{b}\frac{dB}{dt} = -\frac{c}{a}\frac{dA}{dt} = \frac{c}{d}\frac{dD}{dt} = kF(A)F(B) \tag{8.7}$$

An example of the use of the relationship shown in Equation (8.7) is the expression of productivity of fermentation systems as reduction of substrate concentration, increase in product concentration, or increase in mass of the microorganism involved in the fermentation.

8.2.3 Reversible Reactions

Some reactions are reversible.

$$nA \underset{k_2}{\overset{k_1}{\rightleftharpoons}} bB + cC \tag{8.8}$$

Again, F(A), F(B), and F(C) are functions of the concentrations of A, B, and C.
The net reaction rate expressed as a net disappearance of A is

$$r = -\frac{dA}{dt} = k_1 F(A) - k_2[F(B) \cdot F(C)] \tag{8.9}$$

Expressing B and C in terms of A: $B = (n/b)(A_0 - A)$ and $C = (n/c)(A_0 - A)$, where A_0 = initial concentration of A. Let $F(A) = A$, $F(B) = B$, and $F(C) = C$ (i.e. the reaction rate is directly proportional to the concentrations of the reactants).

$$r = -\frac{dA}{dT} = k_1 A - k_2\left[\left[\frac{n}{b}(A_0 - A)\right]\left[\frac{n}{c}(A_0 - A)\right]\right] \tag{8.10}$$

At equilibrium, $r_1 = r_2$ and:

$$k_1 A = k_2\left[\left[\frac{n}{b}(A_0 - A)\right]\left[\frac{n}{c}(A_0 - A)\right]\right]$$

Clearing fractions Equation (8.11) becomes:

$$k_{eq} = \frac{k_1}{k_2} = \left[\frac{\left[\frac{n}{b}(A_0 - A)\right]\left[\frac{n}{c}(A_0 - A)\right]}{A}\right] \tag{8.11}$$

Once a constant of equilibrium is known, it will be possible to determine the concentrations of A, B, and C from the stoichiometric relationships of the reaction. An example of this type of reaction is the dissociation of organic acids and their salts.

Example 8.1. Cottage cheese whey containing initially 4.3%(w/w) lactose when treated with β-galactosidase showed maximum hydrolysis of 80% of the lactose. Calculate the equilibrium constant. In one experiment, enzyme was added to the whey and half of the lactose was hydrolyzed in 25 minutes. Calculate the rate constant for the forward and reverse reactions, and the time required to obtain 77% conversion of lactose, under the conditions given.

Solution:

The reaction involved in the hydrolysis of lactose is

$$C_{12}H_{22}O_{11} + H_2O \underset{k_2}{\overset{k_1}{\rightleftharpoons}} C_6H_{12}O_6 + C_6H_{12}O_6$$

    Lactose   Glucose   Galactose

Water is not rate limiting in this reaction, therefore its contribution to the reaction rate is ignored. Let L = the concentration of lactose, G = the concentration of glucose, and C = the concentration of galactose. The rate equations are

$$r_1 = -dL/dt = k_1L; \quad r_2 = dL/dt = k_2GC$$

At equilibrium, $r_1 = r_2$, and:

$$k_1L_{eq} = k_2G_{eq}C_{eq}$$

$$k_{eq} = \frac{k_1}{k_2} = \frac{G_{eq} \, C_{eq}}{L_{eq}}$$

Let f = fraction of lactose converted. From the stoichiometry of the reaction, $G_{eq} = C_{eq} = L_o - L_{eq} = L_o(f)$. $L_{eq} = L_o(1 - f)$.

    Basis: 1 L of whey. $L_o = 4.3(10)/342 = 0.1257$ moles/L
$L_{eq} = (1 - 0.80)(0.1257) = 0.02514$
$G_{eq} = C_{eq} = 0.80(0.1257) = 0.10056$ moles/L
$k_{eq} = (0.10056)^2/0.02514 = 0.4022$

The rate constants for the reaction will be calculated by first setting up the rate equations and integrating.
    The rate of disappearance of lactose is

$$r = -dL/dt = k_1L - k_2GC$$

$$L = L_0(1 - f); \quad dL/dt = -L_0(df/dt)$$

$$G = f \, L_0; C = f \, L_0; \quad k_1 = k_2k_{eq}$$

$$L_0df/dt = k_2k_{eq}(1 - f)L_0 - k_2f^2L_0^2$$

$$k_2 \, dt = \frac{df}{k_{eq}(1 - f) - f^2L_0}$$

The equation is evaluated by graphical integration. Using the trapezoidal rule and using a BASIC program shown in Fig. 8.1. The integral has a value of 1.814.

$$k_2 = 1.814/25 = .0726$$

$$k_1 = k_2k_{eq} = .0726(0.4022) = 0.1805$$

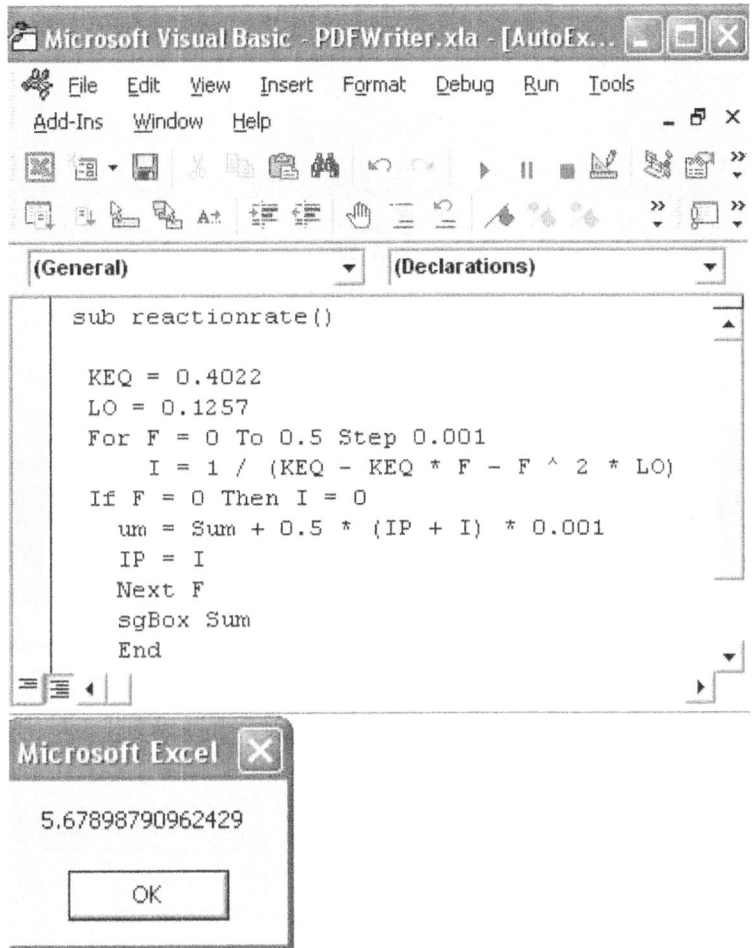

Figure 8.1 Visual BASIC program to solve kinetic parameters for example 8.1.

To obtain the time, the integral is evaluated using the limits 0 to 0.77. The same BASIC program will be used except that the second line of the program is changed to go from f = 0 to 0.77.

The value of the integral is 5.679.

t = 5.679/k2 = 5.679/(0.0726) = 78.2 minutes

8.3 ENZYME REACTIONS

Enzymatic reactions encountered in food processing occur at a rate that is limited by the concentration of enzymes present. There is usually an abundance of the substrate (reactant) so that changes in substrate concentration do not affect the reaction rate. The activity of an enzyme is defined as the rate at which a specified quantity of the enzyme will convert a substrate to product. The reaction is followed by either measuring the loss of reactant or the appearance of product. The specific activity of an enzyme is expressed as activity/mass of protein. When using an enzyme in a process, the activity of added enzyme must be known in order that the desired rate of substrate conversion will be achieved.

If there is no product inhibition:

$$-\frac{dS}{dt} = a \qquad (8.12)$$

where a is the enzyme activity. Thus, an enzymatic reaction without product inhibition suggests a linear change in substrate concentration, at the early stages of the reaction when substrate concentration is so high that enzyme concentration is rate limiting. Most enzymatic reactions however proceed in a curvilinear pattern which approaches a maximum value at infinite time. Consider an enzyme reaction where product inhibition exists. The reaction rate will be

$$-\frac{dS}{dt} = a_0 - B$$

B = enzyme bound to the product.

Assuming that B is proportional to the amount of product formed, and using ki as the constant representing the product inhibitory capacity

$$-\frac{dS}{dt} = a_0 - k_i P$$

P = product concentration.

Consider a reaction where S is converted to P. If f = the fraction of substrate converted to product: $S = S_0(1 - f)$ and $P = S_0 f$. Substituting into the rate equation, and letting S_0 = initial substrate level and a_0 = initial enzyme activity:

$$S_0 \left[\frac{df}{dt} \right] = a_0 - k_i S_0 f \qquad (8.13)$$

$$\frac{df}{dt} = \frac{a_0}{S_0} - k_i f \qquad (8.14)$$

The differential equation can be easily integrated by separation of variables if the transformation $f = a_0/S_0 - k_i f$ is used.

$$\frac{df}{dt} = -\frac{1}{k_i}\frac{df'}{dt}; \quad \frac{df'}{dt} = k_i f'$$

Integrating, substituting f' for f and using the boundary condition f = 0 at t = 0:

$$\ln\left(1 - \frac{k_i f S_0}{a_0}\right) = -k_i t; \quad f = \frac{a_0}{k_i S_0}\left(1 - e^{-k_i t}\right) \qquad (8.15)$$

A plot of the fraction of substrate converted against time, will show a curvilinear plot that will level off at a certain value of f at infinite time. The maximum conversion will be f_{max}.

$$f_{max} = \frac{a_0}{k_i S_0}; \quad \ln\left(1 - \frac{f}{f_{max}}\right) = k_i t$$

Depending on the enzyme and substrate concentration, the conversion may not immediately follow the exponential expression. For example, if the initial enzyme activity is quite high relative to substrate concentration, the fraction converted may initially be a linear function of time. However, once product accumulates and substrate concentration drops, the influence of product inhibition becomes significant and substrate conversion occurs exponentially as derived above. With product inhibition, the product

Table 8.1   Kinetic constants of reactions occurring in foods.

Factor	D_o (min)	Z (EC)	Reference
Ascorbic acid (peas)	246	50.5	Rao et al. (1981)
Browning reaction (milk)	12.5	26	Burton (1954)
Carotene (beef liver paste)	43.6	25.5	Wilkinson et al. (1981)
Chlorophyll (peas)	13.2	38.8	Rao et al. (1981)
Overall sensory	12.5	26	Lund (1977)
Thiamin	158	21	Holdsworth (1985)
Pectin methyl esterase (citrus)	0.053	14	Williams et al. (1986)

competes for active sites on the enzyme making these sites unavailable for making a complex with the substrate. Eventually, all enzyme active sites are occupied by the product and the reaction stops. Enzyme activities are determined as the slope of the substrate conversion vs. time curves at time zero (initial reaction velocity) to avoid the effect of product inhibition.

8.4   REACTION ORDER

Reaction order is the sum of the exponents of reactant concentration terms in the rate equation. Table 8.1 lists various deteriorative reactions in foods, and the order of the reaction.

8.4.1   Zero-Order Reactions

$$r = \frac{dA}{dt} = k \tag{8.16}$$
$$A = A_0 + kt$$

A characteristic of a zero-order reaction is a linear relationship between the concentration of reactant or product with time of the reaction, t.

8.4.2   First-Order Reactions

$$r = -\frac{dA}{dt} = kA \tag{8.17}$$
$$\ln\left(\frac{A}{A_0}\right) = kt$$

A_0 is the concentration of A at time $= 0$. A first-order reaction is characterized by a logarithmic change in the concentration of a reactant with time. Most of the reactions involved in the processing of foods as shown in Table 8.1 are first order reactions.

8.4.3 Second-Order Reactions

$$r = -\frac{dA}{dt} = kA^2 \tag{8.18}$$

$$\frac{1}{A} - \frac{1}{A_0} = kt$$

Second-order unimolecular reaction is characterized by a hyperbolic relationship between concentration of the reactant or product, and time. A linear plot will be obtained if $1/A$ is plotted against time. Second-order bimolecular reactions may also follow the following rate equation:

$$r = -\frac{dA}{dt} = kAB$$

where A and B are the reactants. The differential equation may be integrated by holding B constant to give:

$$\ln\left(\frac{A}{A_0}\right) = -k't \tag{8.19}$$

k' is a pseudo–first-order rate constant: $k' = kB$.

A second-order bimolecular reaction will yield a similar plot of the concentration of the reactant against time as a first-order unimolecular reaction, but the reaction rate constant will vary with different concentrations of the second reactant. An example of a second-order bimolecular reaction is the aerobic degradation of ascorbic acid. Oxygen is a reactant and a family of pseudo-first order plots will be obtained when ascorbic acid degradation is studied at different levels of oxygen availability.

8.4.4 nth-Order Reactions

$$r = -\frac{dA}{dt} = kA^n; \quad n > 1 \tag{8.20}$$

The integrated equation (8.20) is:

$$A^{1-n} - A_0^{1-n} = -(1-n)kt$$

Evaluation of reaction order is a trial and error process that involves assuming various values for n and determining which value would result in the best fit with the nth order equation above.

8.5   REACTIONS WHERE PRODUCT CONCENTRATION IS RATE LIMITING

These type of reactions are usually followed not in terms of the concentration of the reactant but by some manifestation of the completion of the reaction in terms of a physical property change. Examples are protein gelation measured as an increase of the strength of the gel, nonenzymatic browning reaction in solid foods, textural changes during cooking, sensory flavor scores during storage, and so forth. The magnitude of the attribute measured usually levels off not because of depletion of the reactants, but because the measuring technique could no longer detect any further increase in intensity. Reactions of this type could be fitted to the following:

$$\ln\left(1 - \frac{C}{C^*}\right) = \pm kt \tag{8.21}$$

where C^* is the value of the measured attribute when it remained constant at long reaction times, and C is the value at any time during the transient stage of the process.

Example 8.2. The following data shows the firmness of a protein gel as a function of time of heating. Derive an appropriate equation to fit this data and determine the rate constant for the reaction.

Time (min)	0	1	2	3	4	5	10	20
Firmness (g)	0	6.01	8.41	9.36	9.75	9.90	10.2	10.2

Solution:

The data shows firmness to reach a constant value at 10 minutes of heating. The data will be fitted to the equation:

$$\ln\left(1 - \frac{F}{F^*}\right) = -kt$$

F = firmness value and F^* = the final firmness value.

The following are the transformed data that will be analyzed by linear regression to obtain the slope, which will be the value of k.

x (time)	y [ln (1 − F/F*)]
0	0
1	−0.889
2	−1.74
3	−2.497
4	−3.121
5	−3.526

Linear regression of x and y using Microsoft Excel gives a slope of -1.3743. $k = 1.3743$ min^{-1}.

A plot of the data will show a linear fit except for the last point, which deviated slightly from linearity.

8.6 THE REACTION RATE CONSTANT

The reaction rate constant defines the reaction rate. There are several ways in which the speed of a chemical reaction can be reported for first order reactions which predominate in food systems.

Rate constant, k, for an exponential model of concentration change: This rate constant has units of reciprocal time and is the slope of a plot of $\ln(c)$ against time. This rate constant is defined for the various types of reactions in section "Reaction Order."

The D value: This method of representing the rate constant for a reaction had its origins in thermobacteriology, where the inactivation rate of microorganisms during heating is expressed as a decimal reduction time. This approach was later applied to chemical reactions, in order that the same computational scheme can be used for determining microbial inactivation and nutrient degradation during a thermal process for sterilization of foods.

The D value is defined as:

$$\log \frac{C}{C_0} = -\frac{t}{D} \tag{8.22}$$

Thus, the D value is the negative reciprocal of the slope of a plot of $\log(C)$ against t. C in the above equation is the concentration of a reactant. D is based on common logarithms, in contrast with k, which is based on natural logarithms. D and k are related as follows:

$$\ln\left(\frac{C}{C_0}\right) = \ln(10)\log\left(\frac{C}{C_0}\right) = -kt$$

Thus:

$$\frac{1}{D} = \frac{k}{\ln(10)}; \quad D = \frac{\ln(10)}{k} \tag{8.23}$$

The half-life: This method of expressing the rate of a reaction is commonly used in radioisotope decay. It is easier to visualize the rate of the reaction when expressed as a half-life rather than a rate constant based on natural logarithms. The half life is the time required for the reactant to lose half of its original concentration. The half life is related to k and D as follows:

$$\ln(10.5) = -k(t_{0.5}); \quad t_{0.5} = -\frac{\ln(0.5)}{k} \tag{8.24}$$

$$\log(0.5) = -\frac{t_{0.5}}{D}; \quad t_{0.5} = -D\log(0.5)$$

8.7   TEMPERATURE DEPENDENCE OF REACTION RATES

8.7.1   The Arrhenius Equation

The activated complex theory for chemical reaction rates is the basis for the Arrhenius equation which relates reaction rate constants to the absolute temperature. The Arrhenius equation is

$$k = A_0[e]^{-E_a/RT} \tag{8.25}$$

E_a is the activation energy, and A_0 is the rate constant as T approaches infinity. Another form of the Arrhenius equation involves the reaction rate constant at a reference temperature.

Let T_0 = the reference temperature at which $k = k_0$.

$$k_0 = A_0[e]^{-E_a/RT_0}; \quad k = A_0[e]^{-E_a/RT}$$

Taking the ratio of the two equations:

$$\frac{k}{k_0} = [e]^{(-E_a/R)(1/T - 1/T_0)} \tag{8.26}$$

The negative sign is placed on the exponent of the Arrhenius equation in order that a positive activation energy will indicate an increasing reaction rate constant with increasing temperature. Using Equation (8.26), the rate constant at any temperature can be determined from the activation energy and the rate constant ko at a reference temperature, T_0.

8.7.2   The Q_{10} Value

The Q_{10} value of a reaction is often used for reporting temperature dependence of biological reactions. It is defined as the number of times a reaction rate changes with a $10°C$ change in temperature. If a reaction rate doubles with a $10°C$ change in temperature, the $Q_{10} = 2$. For reactions such as enzymatically induced color or flavor change in foods, degradation of natural pigments, nonenzymatic browning, and microbial growth rate, the Q_{10} is usually around 2. Thus the general rule of thumb in food storage is that a $10°C$ reduction in storage temperature will increase shelf life by a factor of 2. The relationship between the Q_{10} value and the activation energy is derived as follows:

Let k_1 = rate constant at T_1 and k_2 − rate constant at T_2

From the definition of the Q_{10}:

$$k_2 = k_1[Q_{10}]^{(T_2-T_1)/10} \tag{8.27}$$

Taking the logarithm of Equation (8.27):

$$\ln\left(\frac{k_2}{k_1}\right) = \frac{T_2 - T_1}{10} \ln Q_{10} \tag{8.28}$$

Substituting k_2 for k, k_1 for k_0, T_2 for T, and T_1 for T_0 in equation 26:

$$\frac{k}{k_1} = [e]^{(-E_a/R)(1/T_2-1/T_1)} \tag{8.29}$$

Taking the logarithm of Equation (8.29):

$$\ln\left[\frac{k_2}{k_1}\right] = \frac{-E_a}{R}\left[\frac{1}{T_2} - \frac{1}{T_1}\right] \tag{8.30}$$

$$\ln\left[\frac{k_2}{k_1}\right] = \frac{-E_a}{R}\left[\frac{T_1 - T_2}{T_2T_1}\right] \tag{8.31}$$

The negative sign on E_a in Equation (8.31) drops out when the signs on T_1 and T_2 in the numerator are reversed. Equating Equations (8.28) and (8.31) and solving for E_a/R:

$$\frac{E_a}{R} = \frac{\ln(Q_{10})}{10}T_2T_1 \tag{8.32}$$

$$Q_{10} = [e]^{(E_a/R)(10/T_2T_1)} \tag{8.33}$$

The Q_{10} is temperature dependent and should not be used over a very wide range of temperature.

8.7.3   The z Value

The z value had its origins in thermobacteriology and was used to represent the temperature dependence of microbial inactivation rate. z was defined as the temperature change needed to change microbial inactivation rate by a factor of 10. The z value has also been used to express the temperature dependence of degradative reactions occurring in foods during processing and storage. The z value expressed in terms of the reaction rate constant is as follows:

$$k_2 = k_1[10]^{(T_2-T_1)/z} \tag{8.34}$$

Taking the logarithm:

$$\ln\left[\frac{k_2}{k_1}\right] = \frac{T_2 - T_1}{z}\ln(10) \tag{8.35}$$

Equating the right hand side of Equations (8.31) and (8.35):

$$\frac{\ln(10)}{z} = \left[\frac{E_a}{R}\right]\left[\frac{1}{T_2 T_1}\right] \tag{8.36}$$

$$z = \frac{\ln(10)}{(E_a/R)}T_1 T_2$$

Solving for E_a/R in Equations (8.32) and (8.36) and equating:

$$z = \frac{10\ln(10)}{\ln(Q_{10})} \tag{8.37}$$

Example 8.3. McCord and Kilara (J. Food Sci. 48:1479, 1983) reported the kinetics of inactivation of polyphenol oxidase in mushrooms to be first order and the rate constants at 50°C, 55°C, and 60°C were 0.019, 0.054, and 0.134 min^{-1}, respectively. Calculate the activation energy, the z value and Q_{10} value for the inactivation of polyphenol oxidase in mushrooms.

Solution:

The absolute temperatures corresponding to 50°C, 55°C, and 60°C are 323, 328, and 333 K respectively. A regression of $\ln(k)$ against $1/T$ gives a slope of -21009.6, thus; $-E_a/R = -21009.6$ $E_a/R = 21009\,K^{-1}$; $R = 1.987\,Cal/(gmole\$K)$; and $E_a = 41.746$ kcal/gmole.

$$\ln(Q_{10}) = 10\left(\frac{E_a}{R}\right)\left(\frac{1}{T_1 T_2}\right) = \frac{10(21,009)}{323(333)} = 1.95$$

$$Q_{10} = 7.028$$

$$z = \frac{\ln(10)}{E_a/R}T_1 T_2 = \frac{323(333)\ln(10)}{21,009} = 11.8°C$$

8.8 DETERMINATION OF REACTION KINETIC PARAMETERS

The reaction rate constant is usually determined at constant temperature by measuring changes in the concentration of reactant or concentration of reaction product with time. Because reaction rate can be affected by the presence of interfering compounds, pH, and water activity, kinetic parameters are determined with the reacting compound contained in a specific substrate. Model systems may be used to ensure that the substrate composition is constant during the determination of the kinetic parameters. However, literature data indicate variations in the value of the kinetic parameters for the same reaction in different food products.

The temperature dependence of the reaction rate constant is determined by conducting the kinetic studies at several constant temperatures and determining the z value or activation energy for the reaction.

A typical technique for determination of the kinetic parameters is to derive a linear form of the reaction rate equation and applying regression analysis on the transformed data. This approach however, has been shown to have limitations because of the smoothing out effect of the transformation used to linearize the data. One method that can be used is nonlinear curve fitting. The use of statistical software packages for determining the reaction rate constant has been shown in the section on "Nonlinear Curve Fitting" in Chapter 1.

Another approach is the use of the Solver feature in Microsoft Excel. Although Solver only allows the manipulation of one variable to minimize the least-square error, an iteration method may be employed to fit two-parameter equations to the data.

Example 8.4. Data on softening of carrots at $90°C$ from Paulus and Saguy (J. Food Sci. 45:239, 1980) given as (x, y) where y = rupture stress in kg/cm^2 and x = time in min are as follows: $(30, 0.85)$, $(40, 0.60)$, $(50, 0.40)$, $(60, 0.29)$, $(70, 0.24)$. Determine the rate constant k, defined as: $\ln(y) = kt + b$.

Solution:

Transformation of y to ln(y) and conducting a linear regression on Excel gives a slope of -0.03256 and an intercept of 0.7772. Thus, linear regression finds a k value of -0.03256.

To use the Solver feature in Microsoft Excel, enter the data in the spreadsheet. Designate two cells to contain values of k and b. For example time values are in cells A2 to A6 and force values are in B2 to B6. Designate B7 to hold the value of k and B8 to hold the value of b. Calculate y in column C as: $y = \exp(kt + b)$. Then calculate the error square in column D as the difference between the calculated and experimental value squared. For example in column D2, enter $(C2 - B2)^{\wedge}2$. Calculate the sum of squares error, e.g. in D7 enter Sum(D2.D6).

Go to Tools and choose Solver on the menu. Assume a value for k and b, e.g enter in B7 the value $-.03$, and in B8 the value 0.7772.

Solver asks to set target cell (in this case the sum of squares, error in D7) to a minimum by changing the value of k in B7. Then click solve. The sum of squares and the value of k is displayed. Repeat the process this time designating the value of b in B8 as the designated cell to change. Then click solve and Solver will find a value of b that minimizes the sum of squares of error. Results of Solver gives $k = -0.3248$ and $b = 0.7868$. Plotting the calculated y and experimental y shows better agreement with the Solver solution compared to that obtained by linear regression after data transformation.

8.9 USE OF CHEMICAL REACTION KINETIC DATA FOR THERMAL PROCESS OPTIMIZATION

Data on temperature dependence of chemical degradation reactions in foods are valuable in determining the loss or gain in product quality that might result with an elevation in processing temperature. When microbial inactivation is a major objective of the heating process, the accompanying chemical reactions become an unwanted consequence of the heating process. An acceptable process must satisfy the microbial inactivation constraint, and processing temperature and time must be selected to minimize the extent of unwanted chemical reactions. The application of kinetic data in optimizing quality factors during thermal processing is discussed further in the section "Quality Factor Degradation" in Chapter 9.

D and z values of some chemical reactions that degrade food quality and of enzyme inactivation are shown in Table 8.1. The z value of most chemical reactions associated with loss in food quality is at least two times higher than those for microbial inactivation (generally $z = 10°C$). Thus, it is a well-known

Table 8.2 Z values for quality degrading chemical reactions in foods.

Texture	Z(Celsius)	Overall Sensory	Z(Celsius)
Apples	21	Beans, green	29
Apples	28	Beets	19
Beans, black	35	Broccoli	44
Beans, Navy	37	Carrots	17
Beans, Soy	42	Corn kernels	32
Beef	4	Peas	32
Beets	40	Potatoes	26
Brussel sprouts	21	Squash	26
Carrots	18		
Potatoes	11	**Color Loss**	
Shrimp	23	Green pigment	30
		Red pigment	31
		Browning	32
Nutrients		**Enzymes**	
Carotene	25	Peroxidase	28
Thaimin	27	Catalase	8
Pyridoxine	29	Lipozgenase	9
Folic acid	20	polyphenol oxidase	8
Ascorbic acid	27	Pectin esterase	16

practice in the food industry to increase the processing temperature to shorten the processing time for microbial inactivation and minimize the extent of quality factor degradation.

When only relative quality indices are needed to optimize the temperature required for a process, the actual rate of reaction may not be needed in the calculations and only the temperature dependence of that reaction will be required. Z values for different quality factor degradation reactions are given in Table 8.2.

PROBLEMS

8.1. Nagy and Smoot (J. Agr. Food Chem 25:135, 1977) reported the degradation of ascorbic acid in canned orange juice to be first order, and the following first-order rate constants can be calculated from their data. At T = 29.4°C, 37.8°C, and 46.1°C , k in day^{-1} was 0.00112, 0.0026, 0.0087, respectively. Calculate the activation energy, Q_{10}, D value and half life at 30°C.

8.2. The following data were collected for the sensory change in beef stored while exposed directly to air at $-23°C$ (From: Gokalp et. al. J. Food Sci. 44:146, 1979). Sensory scores were 8.4, 6.2, 5.5, 5.1 at 0, 3, 6, and 9 months in storage. Plot the data and determine an appropriate form of an equation to which the data can be fitted to obtain the reaction rate constant.

8.3. Accelerated shelf life testing is often done to predict how food products would behave in the retail network. If a food product is expected to maintain acceptable quality in the retail network for 6 months at 30°C, how long should this product be stored at 40°C prior to testing in order that the results will be equivalent to 6 months at 30°C. Assume that the temperature dependence of the sensory changes in the product is similar to that for the nonenzymatic browning reaction in Table 8.1.

8.4. Ascorbic acid degradation in sweet potatoes at a water activity of 0.11 is first order with a rate constant of 0.001500 h^{-1} at 25°C. If the Q$_{10}$ for this reaction is 1.8, calculate the amount of ascorbic acid remaining in dried sweet potato stored at 30°C after 3 months in storage if the initial ascorbic acid content was 33 mg/100 g.

8.5. In the example problem on B-galactosidase action on lactose in acid whey, calculate the lactose conversion that can be expected using the same level of enzyme addition as in the example, if the whey is preconcentrated prior to treatment to have a lactose content of 12.5%, after a treatment time of 60 min.

8.6. Pectin methyl esterase in orange juice has a D value at 85°C of 8.3 min. and the z value is 14°C. Calculate the target juice temperature for pasteurization such that at least 99% of the enzyme will be inactivated after a 1-minute hold time followed by immediate cooling.

SUGGESTED READING

Burton, H. 1954. J. Dairy Res. 21:194.

Holdsworth, D.G. 1985. J. Food Eng. 4(1):89.

Holdsworth, S. D. 1990. Kinetic data, what is available and what is necessary. In: Processing and Quality of Foods. Vol I. Zeuthen, P.,Cheftel, J. C., Eriksson, C., Gormley, T. R., Linko, P., and Paulus, K. Eds. Elsevier, New York, pp 74–90.

Lund, D. B. 1977. Maximizing nutrient retention. Food Technol. 31(2):71.

Perry, R. H., Chilton, C. H., and Kirkpatrick, S. D. 1963. Chemical Engineers Handbook. 4th ed. McGraw-Hill, New York.

Rao, M. A et al. 1981. J. Food Sci. 46:636.

Skinner, G. B. 1974. Introduction to Chemical Kinetics. Academic Press, New York.

Wilkinson, S. A. et al. 1981. J. food Sci. 46:32.

Williams, D. C. et al. 1986. Food Technol. 40(6):130.

CHAPTER 9

Thermal Process Calculations

Inactivation of microorganisms by heat is a fundamental operation in food preservation. The concepts learned in this chapter are not only applicable in canning but in any process where heat is used to inactivate microorganisms and induce chemical changes that affect quality. The term "sterilization" used in this chapter refers to the achievement of commercial sterility, defined as a condition where microorganisms that cause illness, and those capable of growing in the food under normal nonrefrigerated storage and distribution, are eliminated.

9.1 PROCESSES AND SYSTEMS FOR STABILIZATION OF FOODS FOR SHELF-STABLE STORAGE: SYSTEMS REQUIREMENTS

Different systems are available for treating foods to make them shelf stable. Suitability of a system depends on the type of food processed, production rates, availability of capital, and labor costs. Product quality and economics are the major factors to be considered in system selection. Because the major production costs are overhead and labor, plants with high production capacity are inclined to use systems that have high capitalization and low labor requirements. Products with superior quality will result from systems capable of high-temperature, short-time treatments.

9.1.1 In-Can Processing

The simplest and oldest method of modern food preservation involves filling a product into a container, sealing, and heating the sealed containers under pressure. Different types of pressure vessels or retorts are used.

9.1.1.1 Stationary Retorts

These retorts are cylindrical vessels oriented vertically or horizontally. Crates are used to facilitate loading and unloading of cans. The cans are stacked vertically in the crates, and perforated metal dividers separate the layers of cans. In vertical retorts, the crates are lowered or raised using electric hoists. In horizontal retorts, the crates are of rectangular profile with dimensions to fit the cylindrical retort. The crates are mounted on a carrier with wheels, and tracks within the retort guide the wheels of the crate carrier during introduction and retrieval.

Figure 9.1   Horizontal stationary retort and crate.

Stationary retorts for processing of canned foods must be equipped with an accurate temperature controller and recording device. In addition, steam must be uniformly distributed inside the retort. A steam bleeder continuously vents small amounts of steam and promotes steam flow within the retort. A fluid-in-glass thermometer is required to provide visual monitoring of the retort temperature by the operator. At the start of the process, the retort is vented to remove air and ensure that all cans are in contact with saturated steam. Figure 9.1 shows a horizontal stationary retort and crate.

Pressure-resistant hatches for the retorts are of various design. Some are secured with hinged bolt-like locks, but more recent designs facilitate opening and closing of the retort. A wheel-type lock advances or retracts locking bars that slip into a retaining slot to secure the cover. Another cover design consists of a locking ring that can be engaged or disengaged with a turn of a lever. A locking ring type of cover assembly is shown in Fig. 9.1.

Typically, stationary retorts are operated by loading the cans, venting the retort, and processing for a specified time at a specified temperature. The time from introduction of steam to attainment of processing temperature is called the "retort come-up time." The process is "timed" when the retort reaches the specified processing temperature. A timed record of the retort temperature for each batch processed is required to be maintained in a file. Cooling may be done inside the retort. However, slow cooling cans may be removed as soon as internal pressure has dropped to just slightly above atmospheric and cooling is completed in canals, where circulating, cold chlorinated water contacts the crates, which are suspended and moved through the water by overhead conveyors.

9.1.1.2   Hydrostatic Cooker

A photograph of this type of retort is shown in Fig. 9.2. It consists of two water legs that seal steam pressure in the main processing section. When processing at 121.1°C, the absolute pressure of steam is

Figure 9.2   Hydrostatic cooker. (Courtesy of Food Machinery Corporation, Canning Machinery Division.)

205,740 Pa; therefore, if the atmospheric pressure is 101.325 Pa, a column of water 10.7 m high must be used to counteract the steam pressure. Thus, hydrostatic cookers are large structures that are often in the open. With non-agitating hydrostatic cookers, heat penetration parameters for thermal process calculations are obtained using a stationary retort, and specified processes are similar to those for a stationary retort. The specified process is set by adjusting the speed of the conveyor, which carries cans in and out of the retort such that the residence time in the steam chamber equals the specified process time.

9.1.1.3 *Continuous Agitating Retorts*

One type of continuous agitating retort consists of a cylindrical pressure vessel equipped with a rotating reel that carries cans on its periphery. When the reel rotates, cans alternately ride on the reel

Figure 9.3 Multiple-shell, continuous rotary retort. (Courtesy of Food Machinery Corporation, Canning Machinery Division.)

or roll along the cylinder wall. Figure 9.3 is a photograph of a three-shell continuous retort viewed from the end that shows the drive system for the reel. Also shown at the top of the retort is the rotary valve, which receives the cans and introduces them continuously into the retort without losing steam from the retort, and the transfer valves, which transfer cans from one retort to the other. Figure 9.4 is a cutaway view showing the reel and the automatic can transfer valve. Agitation is induced by shifting of the headspace as the cans roll. Agitation is maximum in fluid products with small suspended particles, and no agitation exists with a semi-solid product such as canned pumpkin. Agitation minimizes heat-induced changes in a product during thermal processing, when products are of low viscosity. The speed of rotation of the reel determines the rate of heating and the residence time of the cans in the retort, therefore, the heat penetration parameters must be obtained at several reel speeds to match the residence time at a given reel speed to the processing time calculated using heat penetration parameters from the same reel speed. The processing time (t) in a continuous retort is determined as follows:

$$t = \frac{N_t}{N_\rho \, \Omega} \tag{9.1}$$

where N_t is total number of cans in the retort if completely full, N_p is number of pockets around the periphery of the reel, and Ω is rotational speed of the reel.

A simulator, called the "steritort," is used to determine heat penetration parameters at different rotational speeds of the reel.

Cans enter a continuous retort and are instantaneously at the processing temperature, therefore, the process time is exactly the residence time of the cans within the retort.

9.1.1.4 Crateless Retorts

The crateless retort is one that has a labor-saving feature over conventional stationary retorts, and it appeals to processors whose level of production can not economically justify the high initial cost

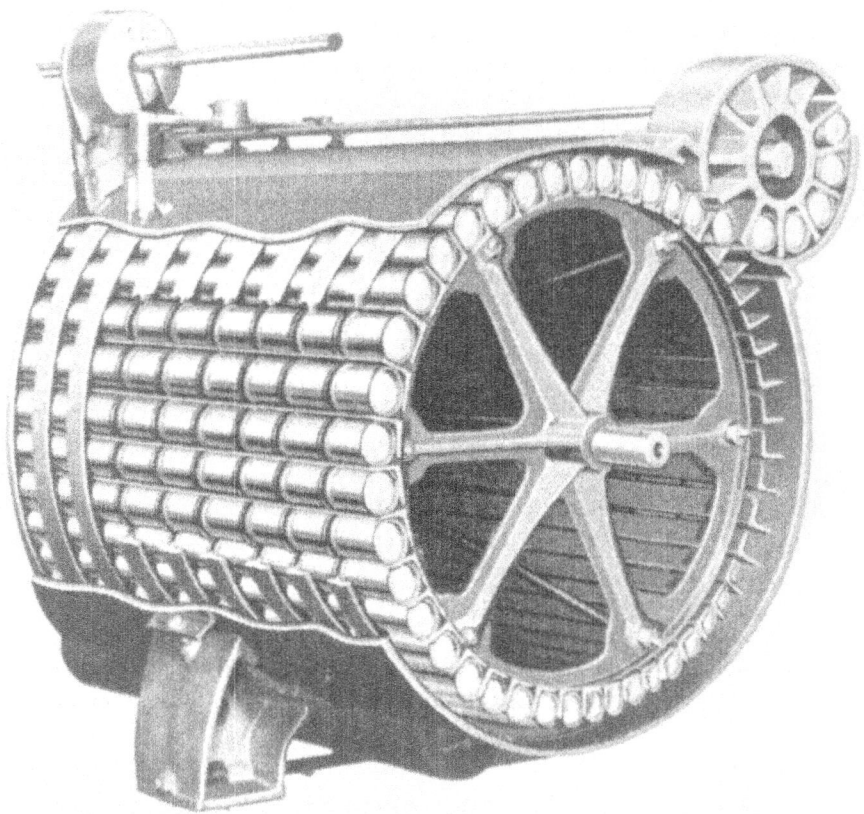

Figure 9.4 Cross section of a continuous rotary retort showing can positioning on the reel and the can transfer valve, which continuously introduces the cans into the retort without releasing the pressure. (Courtesy of Food Machinery Corporation, Canning Machinery Division.)

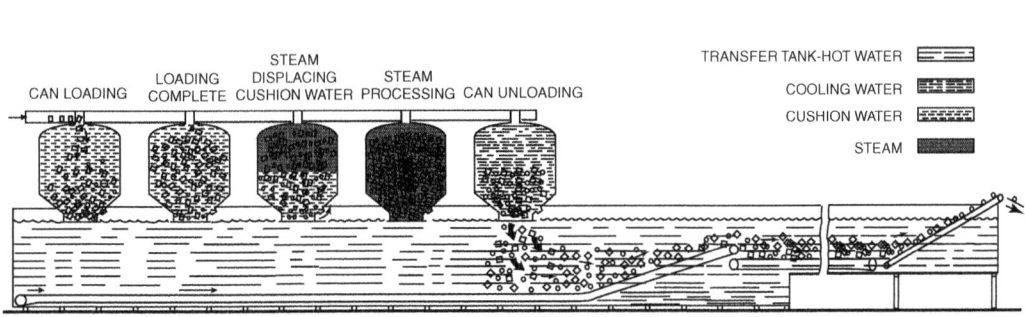

Figure 9.5 Schematic diagram of the operation of a crateless retort. (Courtesy of Food Machinery Corporation, Canning Machinery Division.)

of a hydrostatic or continuous rotational agitating retort. Figure 9.5 shows a diagram of a crateless retort. A system of pumps and hydraulic operated locks alternately opens the retort, fills it with water, receives the cans, which drop into the retort at random, seals the retort for pressure processing with steam, introduces cold water for cooling, and drops the cans and cooling water into a pool of cold chlorinated water for final cooling and retrieval by a conveyor. Energy is saved if hot water used to initially fill the retorts to receive the cans is stored and reused. Steam waste by venting is eliminated because steam displaces water at the initial phase of the process. The largest saving from this system is in labor and elimination of maintenance cost of the retort crates. Thermal process parameters are determined in the same manner as for stationary retorts.

9.1.2   Processing Products Packaged in Flexible Plastic Containers

Containers made out of plastic do not have the strength to resist sudden changes in internal pressure during thermal processing. Thus, the heating and cooling steps must be carried out slowly or air over-pressure must be applied inside the retort all the time during the process to ensure that the pressure in the retort is always greater than the pressure inside the container. Processing with air over-pressure, however, occurs under a nonsaturated steam atmosphere. Heat transfer is slower than in a saturated steam or saturated water medium. Another problem with steam/air processing medium is the possibility of large variations in the temperature within the retort and the difference in heat transfer coefficients with different concentrations of air in the medium. To solve these problems, processing may be done by complete immersion of the product in water, or water may be sprayed on the product throughout the process, or water may be cascading over the product during the process. One retort design (Fig. 9.6)

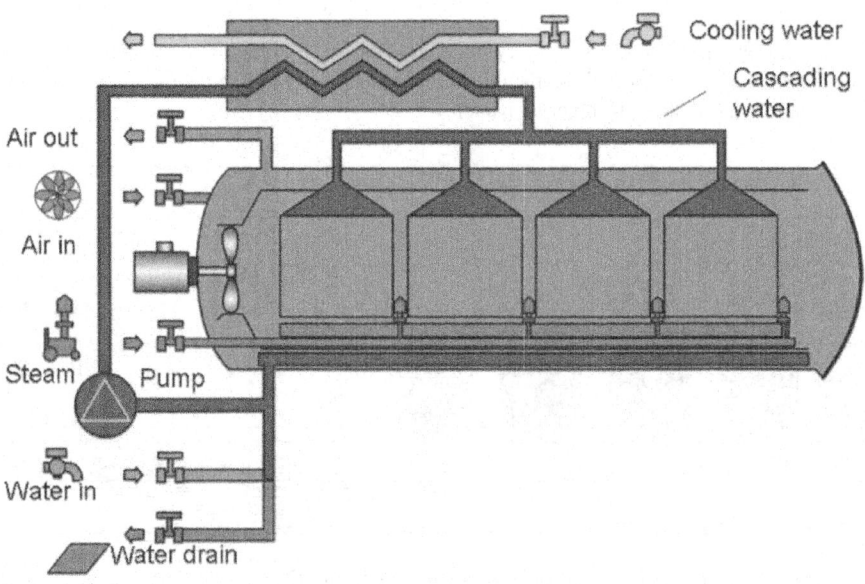

Figure 9.6   Retort system designed for steam/air mixtures as heating medium. Courtesy of Societe' Lagarde

Figure 9.7   Retort system designed for full water immersion or water spray as heating medium. (Courtesy of Stock America)

has a blower that circulates the steam/air heating medium within the retort, but water may also be sprayed on the product to improve heat transfer and minimize temperature variations at different points in the retort.

Figure 9.7 is a picture of a retort that is capable of processing by complete immersion, or spraying water on the product. In order to save energy and minimize the come-up time of the retort, a second shell is added to accommodate hot water under pressure that is inside the processing shell at the termination of the process. At the start of the process, the hot water from the upper shell is pumped into the processing vessel along with steam. At the termination of the process, steam is cut off and while air over-pressure is maintained in the processing vessel, the hot water is pumped into the upper shell. Then cold water is sprayed over the product for cooling.

9.1.3   Processing in Glass Containers

The inability of glass to withstand sudden temperature changes requires a gradual heating and cooling process. Products in glass containers are processed in stationary retorts by first filling the retort with water after the containers are loaded and heating the water by direct injection of live steam. A recirculating pump draws water from a point near the top of the water level in the retort and forces this back into the retort through the bottom. This procedure ensures uniform water temperature and uniform water velocity across all containers in the retort. Water is heated slowly thereby prolonging the come-up time and eliminating thermal shock to the glass. Cooling is accomplished by slowly introducing cold water at the termination of the scheduled process. Water temperature drops slowly eliminating thermal shock to the glass.

Evaluation of thermal precesses in hot water systems is best done using the general method for integrating process lethality. A minimum come-up time to the processing temperature, hold time at the specified temperature, and minimum cool down time must be part of the process specifications.

9.1.4   Flame Sterilization Systems

This relatively recent development in thermal processing systems is used primarily for canned mushrooms. The system consists of a conveyor that rotates the cans as they pass over an open flame. The cans themselves act as the pressure vessel, which sterilizes the contents. The fluid inside the cans must be of low viscosity, such as brine, water, or low sugar syrups because rapid heat exchange between the can walls and contents is needed to prevent scorching of product on the inner can surface. Internal vacuum at the time of filling must be at the maximum that can be achieved without paneling of the cans. This ensures that a saturated steam atmosphere will exist inside the can and internal pressure from expanding air and steam will not be too excessive during the high temperatures required for sterilization.

Thermal process determination requires a simulator that rotates the can over an open flame. Internal temperature must be monitored in the geometric center of the largest particle positioned in the geometric center of the can.

9.1.5   Continuous Flow Sterilization: Aseptic or Cold Fill

Fluids and small particle suspensions can be sterilized by heating in heat exchangers. Figure 9.8 is a schematic diagram of an aseptic processing system. The liquid phase reaches the processing temperature very rapidly, therefore the small sterilization value of the heating phase of the process is generally neglected. The specified process for sterilization in continuously flowing systems is a time

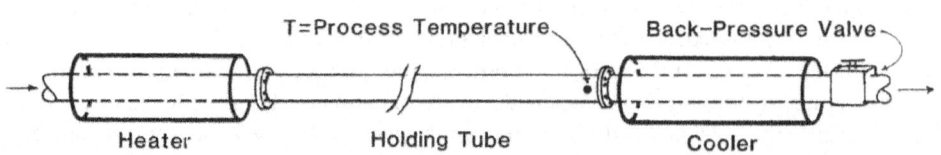

Figure 9.8   Schematic diagram of an aseptic processing system for product sterilization.

of residence in a holding tube, an unheated section of the piping system that leads the fluid from the heat exchangers for heating to the heat exchangers for cooling. A back pressure valve or a positive displacement timing pump is positioned after the cooler to maintain the pressure within the system at a level needed to keep the product boiling temperature higher than the processing temperature. After cooling, the sterile product must be handled in a sterile atmosphere, therefore the process is called *aseptic processing*. The time of residence is set by the volume of the holding tube and the rate of fluid flow delivered by a positive displacement pump.

$$t_{avg} = \frac{A_c L}{Q} \tag{9.2}$$

where t_{avg} is average fluid residence time, A_c is cross-sectional area of the holding tube, L is length of the holding tube, and Q is volumetric rate of flow.

The average velocity ($V_{avg} = Q/A_c$) may also be used to calculate the time in the holding tube.

$$t_{avg} = \frac{L}{V_{avg}} \tag{9.3}$$

In most cases, however, the time of residence of the fastest flowing portion of the fluid is used as the required hold time in the thermal process calculations. This is because the highest probability of survivors from the thermal process is contributed by the section of fluid flowing close to the geometric center of the tube. The minimum time is

$$t_{min} = \frac{L}{V_{max}} \tag{9.4}$$

The maximum velocity for Newtonian fluids in laminar flow is:

$$V_{max} = 2V_{avg} \tag{9.5}$$

For power flow fluids in laminar flow:

$$V_{max} = \frac{(3n + 1)}{(n + 1)} V_{avg} \tag{9.6}$$

For Newtonian fluids in turbulent flow, the following equation was derived by Edgerton and Jones (1970) for V_{max} as a function of the Reynolds number based on the average velocity:

$$V_{max} = \frac{V_{avg}}{0.00336 \log(Re) + 0.662} \tag{9.7}$$

An equation similar to Equation (9.7) can be derived by performing a regression analysis on data by Rothfus et al. (AIChE J. 3:208, 1957) for Reynolds number greater than 10^4.

9.1.6   Steam-Air Mixtures for Thermal Processing

A recent development in thermal processing is the use of a mixture of steam and air instead of water or saturated steam for heating. This system has been touted as ideal in processing of products in retortable pouches and glass. The advantages are elimination of a need for exhausting, and no sudden pressure changes on heating or cooling preventing breakage of the fragile containers.

Heating rates on which the scheduled process is dependent are strongly dependent on the heat transfer coefficient when steam-air is used for heating. The heat transfer coefficient is a function of velocity and mass fraction of steam. Thus, a retort designed for steam-air heating must be equipped with a blower system to generate adequate flow within the retort to maintain uniform velocity and uniform temperature. Accurate and separate controllers must be used for pressure and temperature. The mass fraction steam (X_s) in a steam-air mixture operated at a total pressure P is given by:

$$X_s = \frac{P_s}{P}\left(\frac{18}{29}\right) \tag{9.8}$$

P_s is the saturation pressure of steam at the temperature used in the process.

9.2 MICROBIOLOGICAL INACTIVATION RATES AT CONSTANT TEMPERATURE

9.2.1 Rate of Microbial Inactivation

When a suspension of microorganisms is heated at constant temperature, the decrease in number of viable organisms follows a first-order reaction.

Let N = number of viable organisms.

$$-\frac{dN}{dt} = kN \tag{9.9}$$

k is the first-order rate constant for microbial inactivation. Integrating Equation (9.9) and using the initial condition, $N = N_0$ at $t = 0$:

$$\ln\left(\frac{N}{N_0}\right) = -kt \tag{9.10}$$

Equation (9.10) suggests a linear semi-logarithmic plot of N against t. Equation (9.10) expressed in common logarithms is

$$2.303\log\left(\frac{N}{N_0}\right) = -kt; \quad \log\left(\frac{N}{N_0}\right) = \frac{-kt}{2.303}$$

or:

$$\log\left(\frac{N}{N_0}\right) = \frac{-t}{D} \tag{9.11}$$

Equation (9.11) defines D, the decimal reduction time, the time required to reduce the viable population by a factor of 10. $D = 2.303/k$. Thus, the decimal reduction time and the first-order kinetic rate constant can be easily converted for use in equations requiring the appropriate form of the kinetic parameter.

N, the number of survivors, is considered to be the probability of spoilage if the value is less than 1. Any value of N ∃ 1 means certain spoilage (probability of spoilage = 1).

9.2.2 Shape of Microbial Inactivation Curves

Microbial inactivation proceeds in a logarithmic function with time according to Equation (9.11). However, although the most common inactivation curve is the linear semi-logarithmic plot shown in Fig. 9.9A, several other shapes are encountered in practice. Figure 9.9B shows an initial rise in numbers followed by first-order inactivation. This has been observed with very heat resistant spores and may be attributed to heat activation of some spores that otherwise would not germinate and form colonies, before the heat treatment reached the severity needed to cause death to the organism. Figure 9.9C

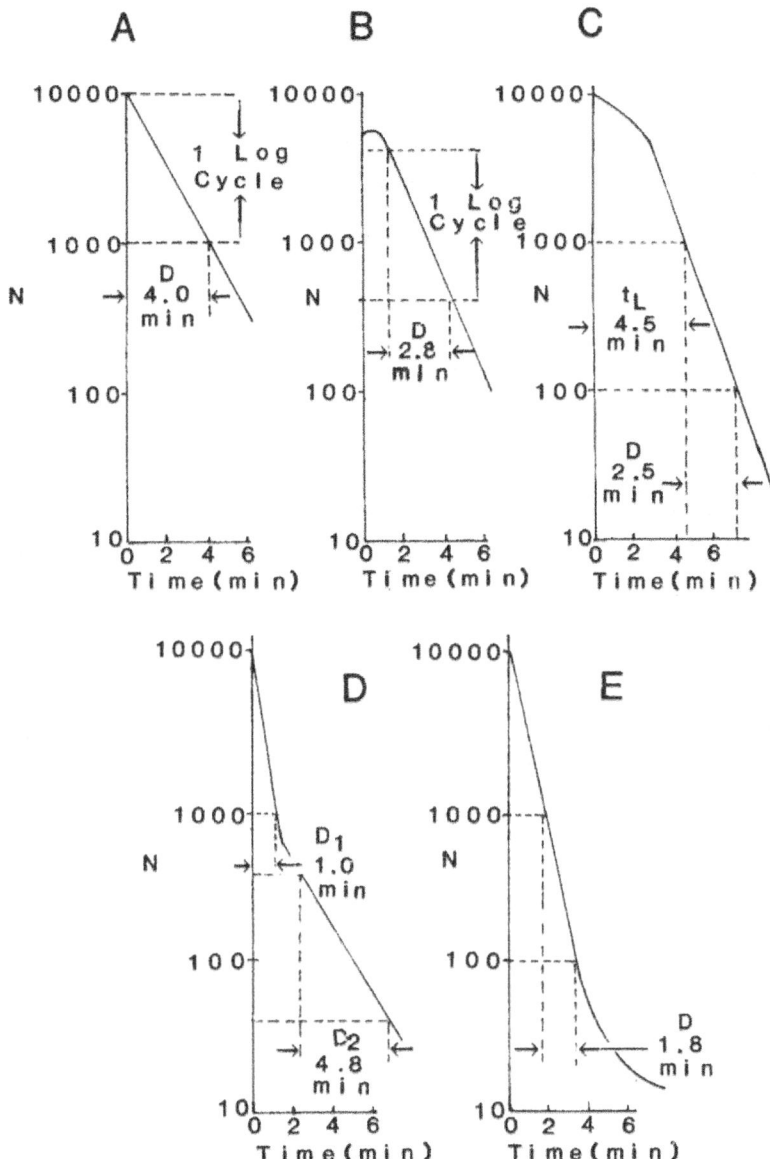

Figure 9.9 Microbial inactivation curves. (A) First-order inactivation rate. (B) Initial rise in numbers followed by first-order inactivation. (C) Initial lag in the inactivation curve. (D) Inactivation curve exhibited by a mixed culture. (E) Tailing of an inactivation curve.

shows an inactivation curve that exhibits an initial lag or induction period. Very little change in numbers occurs during the lag phase. The curve represented by Fig. 9.9C can be expressed as:

$$\log \frac{N_0}{N} = 1 + \left(\frac{t - t_L}{D}\right); \quad t > t_L \tag{9.12}$$

where t_L is the lag time, defined as time required to inactivate the first 90% of the population. In most cases, the curved section of the inactivation curve does not extend beyond the first log cycle of inactivation, therefore defining t_L as in Equation (9.12) eliminates the arbitrary selection of the lag time from the point of tendency of the curved and the straight line portion of the inactivation curve. In general, t_L approaches D as N_0 becomes smaller and as the temperature increases. When $t_L = D$, the first-order inactivation rate starts from the initiation of heating, and Equation (9.12) reduces to Equation (9.11). Equation (9.12) is not often used in thermal process calculations unless the dependence of t_L on N_0 and T are quantified. Microbial inactivation during thermal processing is often evaluated using Equation (9.11).

Figure 9.9D represents the inactivation curve for a mixed culture. The inactivation of each species is assumed to be independent of each other.

From Equation (9.11), the number of species A and B having decimal reduction times of D_A and D_B at any time are

$$N_A = N_{A0}(10)^{-(t/D_A)}; \quad N_B = N_{B0}(10)^{-(t/D_B)}$$
$$N = N_{A0}(10)^{-(t/D_A)} + N_{B0}(10)^{-(t/D_B)} \tag{9.13}$$

If $D_A < D_B$, the second term will be relatively constant at small values of t, and the first term predominates as represented by the first line segment in Fig. 9.9D. At large values of t, the first term approaches zero and microbial numbers will be represented by the second line segment in Fig. 9.9D.

The required heating time to obtain a specified probability of spoilage from a mixed species with known D values will be the longest heating time calculated using Equation (9.11) for any of the species.

Figure 9.9E shows an inactivation curve that exhibits tailing. Tailing is often associated with very high N_o values and with organisms which have a tendency to clump. As in the case of a lag in the inactivation curve, the effect of tailing is not considered in the thermal process calculation unless the curve is reproducible and the effect of initial number and temperature can be quantified.

Example 9.1. Figure 9.10 shows data on inactivation of spores of F.S. 1518 reported by Berry et al. (J. Food Sci. 50:815, 1985). When 6×10^6 spores were inoculated into a can containing 400 g of product and processed at 121.1°C, the processed product contained 20 spores/g. Calculate the equivalent heating time at 121.1°C to which the product was subjected.

Solution:

Both lines in Fig. 9.10 are parallel and the D value of 3.4 min is independent of initial number, as would be expected from either Equations (9.11) or (9.12). However, both lines show a departure from linearity at the initial stage of heating. To determine if a lag time should be considered when establishing survivors from a heating process, data from both thermal inactivation curves will be fitted to Equation (9.12) to determine if t_L is consistent with different initial numbers. A point on each plot is arbitrarily picked to obtain a value for N and t. Choosing a value of $N = 10/g$, the time required to reduce the population from N_0 to N is 20 and 16.2 min, respectively for $N_0 = 6 \times 10^6$ and 4×10^6. Using Equation (9.12): For $N_0 = 6 \times 10^6$:

$$t_L = 20 - 3.4\,[\log(6 \times 10^6/10) - 1] = 3.75\,\text{min}$$

For $N_0 = 4 \times 10^6$:

$$t_L = 16.2 - 3.4\,[\log(4 \times 10^6/10) - 1] = 0.56\,\text{min}$$

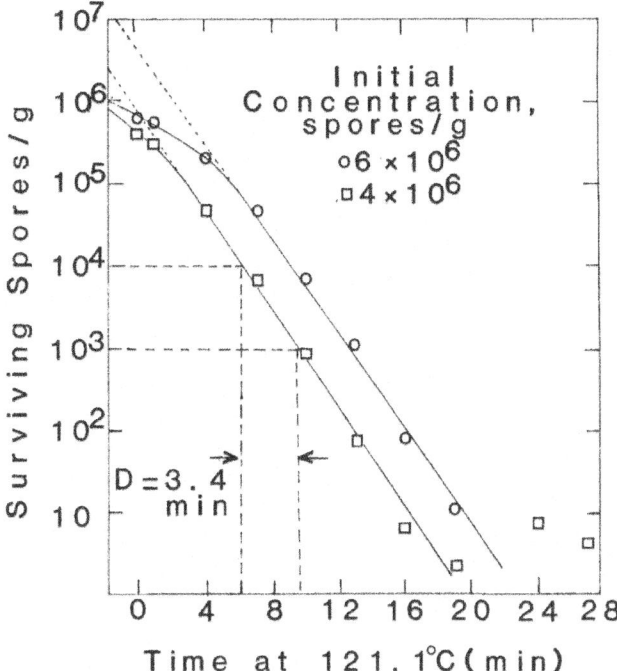

Figure 9.10   Inactivation curve for spores of FS 1518. (From Berry, M. R. et al., J. Food Sci. 50:815, 1985.)

Thus, t_L is not consistent at the two levels of N_0 tested. At $N_0 = 6 \times 10^6$, t_L is almost equal to the D value of 3.4 min, therefore inactivation was first order at the very start of heating. However, t_L was only 0.56 min when $N_0 = 4 \times 10^6$. Thus the departure from linearity at the start of heating may be considered an anomaly, and for thermal process calculations, the inactivation curve may be considered as first order from the very start of heating. Reduction of viable numbers will be calculated using Equation (9.11).

$$t = -D \log (N/N_0)$$

$$N = 20 \frac{\text{spores}}{\text{g product}} \times 400 \text{ g product} = 8000 \text{ spores}$$

$$N_0 = 6 \times 10^6; \quad t = 3.4 \log \frac{6 \times 10^6}{8000} = 9.77 \text{ min}$$

Thus, the equivalent lethality of the process is 9.77 minutes at 121.1°C.

Example 9.2. A suspension containing 3×10^5 spores of organism A having a D value of 1.5 min at 121.1°C and 8×10^6 spores of organism B having a D value of 0.8 min at 121.1°C is heated at a uniform constant temperature of 121.1°C. Calculate the heating time for this suspension at 121.1°C needed to obtain a probability of spoilage of 1/1000.

Solution:

Using Equation (9.11):

For organism A: $t = 1.5 \log(3 \times 10^5 / 0.001) = 12.72 \, \text{min}$

For organism B: $t = 0.8 \log(8 \times 10^6 / 0.001) = 7.92 \, \text{min}$

Thus, the required time is 12.72 minutes.

9.2.3 Sterilizing Value or Lethality of a Process

The basis for process lethality is Equation (9.11), the destruction of biological entities in a heated material. The following may be used as means of expressing the sterilizing value of a process:

S = number of decimal reduction = $\log N_0 / N$

F_T is process time at constant temperature T, which has the equivalent lethality of the given process. Usually F values are expressed at a reference temperature (121.1°C for sterilization processes or 82.2°C for a pasteurization process). The F and D value at 121.1°C are F_0 and D_0, respectively. In a constant temperature process, S and F values can be easily converted between each other using Equation (9.11), with F values substituted for t.

$$S = \frac{F_T}{D_T} \tag{9.14}$$

However, if the suspension is heated under changing temperature conditions such as the interior of a can during thermal processing, Equation (9.14) can be used only if the F value is calculated using the same z value as the biological entity represented by D_T. The z value is the parameter for temperature dependence of the inactivation rate and will be discussed in more detail in the section "Effect of Temperature on Thermal Inactivation of Microorganisms."

The use of S values to express process lethality is absolute (i.e., S is the expected effect of the thermal process). However, an S value represents a specific biological entity and when several must be inactivated, the S value for each entity may be calculated from the F value using Equation (9.14). The use of F values for expressing process lethality and its calculation under conditions of changing temperature during a process will be discussed in more detail in the section "Sterilizing Value of Processes Expressed as F_0."

9.2.4 Acceptable Sterilizing Value for Processes

A canned food is processed to achieve commercial sterility. Commercial sterility implies the inactivation of all microorganisms that endanger public health to a very low probability of survival. For canned foods, the critical organism is *Clostridium botulinum*. The 12D concept as a minimum process for inactivation of *C. botulinum* in canned foods is accepted in principle by regulatory agencies and the food industry. However, its interpretation has undergone a process of evolution, from a literal 12 decimal reduction, to what is now generally accepted as a probability of survival of 10^{-12}.

The latter interpretation signifies a dependence of minimum processes according to the 12D concept on initial spore loads. Thus, packaging materials that have very low spore loads will not require as

Table 9.1 N and N_o Values Used to Obtain Target ln (N_o/N) Values for Commercial Sterility of Canned Foods

Factor	N	N_o	D_{mo}
Public health	10^{-9}	General 10 Meats 10^2 Mushrooms 10^4 Packaging 10^{-5}	0.2
Mesophilic spoilage	10^{-6}	General 10 Meats 10^3	0.5
Thermophilic spoilage	10^{-2}	General 10^2	1.5

Source: From Pflug, I. V., *J. Food Protect. 50:342, 50:347, 50:528, 1987.* Reprinted from Toledo, R. T., *Food Technol. 44(2):72, 1990.*

severe a process as products such as mushrooms which may have very high spore levels. Table 9.1 shows N_0 and N values that may be used as a guide in selecting target N_0/N values for thermal processing.

Spoilage from microorganisms that pose no danger to public health is called economic spoilage. Spoilage microorganisms often have higher heat resistance than *C. botulinum*, and their inactivation is the basis for the thermal process design. A high level of spoilage will be expected from the minimum process based on the 12D concept for *C. botulinum* inactivation.

Example 9.3. The F value at 121.1°C equivalent to 99.999% inactivation of a strain of *C. botulinum* is 1.2 minutes. Calculate the D_0 value of this organism.

Solution:

A 99.999% inactivation is 5 decimal reductions (one survivor from 100,000). S = 5. Using Equation (9.14):

$$D_0 = \frac{F_0}{S} = \frac{1.2}{5} = 0.24 \, \text{min}$$

Example 9.4. Calculate F_0 based on the 12D concept using the D_0 value of *C. botulinum* in Example 9.3 and a most likely spore load in the product of 100.

Solution:

$$S = \log 100 - \log(10 - 12) = 14$$
$$F_0 = 14 \, (0.24) = 3.30 \, \text{minutes}$$

Example 9.5. The sterilizing value of a process has been calculated to be an F_0 of 2.88. If each can contained 10 spores of an organism having a D_0 of 1.5 min, calculate the probability of spoilage from this organism. Assume the F_0 value was calculated using the same z value as the organism.

Solution:

Using Equation (9.11) for a process time of 2.88 min:

$$\log \frac{N_0}{N} = \frac{2.88}{1.5}; \quad N = N_0[10]^{-(2.88/1.5)}$$
$$N = 10(10^{-1.92}) = 0.12; \quad P_{\text{spoilage}} = 12 \text{ in } 100 \text{ cans.}$$

Example 9.6. The most probable spore load in a canned food is 100 and the D_0 of the spore is 1.5 minutes. Calculate a target F_0 for a thermal process such that the probability of spoilage is 1 in 100,000. If under the same conditions *C. botulinum* type B has a D_0 of 0.2 min, would the target F_0 value satisfy the minimum 12D process for *C. botulinum*? Assume an initial spore load of 1 per can for *C. botulinum*.

Solution:

For the organism, $S = \log(100/10^{-5}) = 7$. Using Equation (9.14): $F_0 = 7(1.5) = 10.5$ minutes. For *C. botulinum*: $S = \log(1/10^{-12}) = 12$. Using Equation (9.14): $F_0 = 12(0.2) = 2.4$ minutes. The F_0 for the spoilage organism satisfies the minimum process for 12D of *C. botulinum*.

9.2.5   Selection of Inoculation Levels in Inoculated Packs

In order to be reasonably sure of the safety of a process, products may be inoculated with an organism having known heat resistance, processed, and the extent of spoilage compared to the probability of spoilage designed into a process. The organism used must have a higher heat resistance than the background micro flora in the product. The level of spoilage and the inoculation levels are set such that spoiled cans can be easily evaluated. Use of an organism that produces gas facilitates the detection of spoiled cans because spoilage will be manifested by swelled cans. If a flat sour organism is used, it will be necessary to open the cans after incubation to determine the number of spoiled cans. If the whole batch of inoculated cans is incubated, the fraction spoiled will be equivalent to the decimal equivalent of the number of surviving organisms. Inoculation levels can be calculated based on Equation (9.11).

The two examples below represent procedures used for validation of thermal processes using inoculation tests. The first example represents an incubation test, and the second example represents a spore count reduction test.

Example 9.7. A process was calculated such that the probability of spoilage from an organism with a D_0 value of 1 min is 1 in 100,000 from an initial spore load of 100. To verify this process, an inoculated pack is made. Calculate the level of inoculum of an organism having a D_0 value of 1.5 min that must be used on 100 cans such that a spoilage rate of 5 cans will be equivalent in lethality to the calculated process.

Solution:

The F_0 of the calculated process is determined using Equation (9.11) with F_0 substituted for t.

$$F_0 = 1[\log(100/1 \times 10^{-5})] = 7 \text{ minutes}$$

For the inoculum:

$$\log No - \log(5/100) = F_0/D = 7/1.5 = 4.667$$

$$N_0 = 0.05(10)^{4.667} = 2323 \text{ spores}$$

Example 9.8. In an incidence of spoilage, the isolated spoilage organism was found to have a D_0 value of 1.35 minutes. It is desired that the probability of spoilage from this organism be 1 in 100,000. Initial spore loads were generally of the order 10/can. Calculate the required F_0 for this process to achieve the desired probability of spoilage. If an inoculated pack of FS 1518 is to be made, and an initial inoculation level of 5×10^5 spores is made into cans that contained 200 g of product, what will be the spore count in the processed product such that the lethality received by the can contents will be equivalent to the desired process for eliminating spoilage from the isolated organism. The D_0 value of FS 1518 = 2.7 minutes.

Solution:

The F_0 of the process can be calculated from the heat resistance and spore reduction needed for the spoilage organism.

$$F_0 = D\,(\log N_0 - \log N) = 1.35\,[\log 10 - \log\,(1/100000)] = 8.1 \text{ minutes}$$

For FS 1518:

$$\log N - \log\,(5 \times 10^5) = -8.1/2.7 = -3.00$$

$$N = 10^{(-3.00+5.698)} = 500$$

Spore count after processing $= 500 \text{ spores}/200 \text{ g} = 2.5/\text{g}$

9.2.6   Determination of D Values Using the Partial Sterilization Technique

This technique developed by Stumbo et al. (Food Technol. 4:321, 1950) and by Schmidt (J. Bacteriol. 59:433, 1950) allows the determination of D values using survivor data at two heating times. Appropriate selection of heating times to exceed the lag time for the heated suspension to reach the specified test temperature eliminates the need to correct for temperature changes occurring during the transient period of heating. When this procedure is used, the lag time to bring the suspension to the desired temperature must be established, and the two heating times used for D value determination must exceed the lag time. If t_1 and t_2 are the heating times and N_1 and N_2 are the respective number of survivors, the D value is determined by:

$$D = \frac{t_2 - t_1}{\log\,(N_1) - \log\,(N_2)} \tag{9.15}$$

Example 9.9. Sealed tubes containing equal numbers of spores of an isolate from a spoiled canned food were heated for 10 and 15 minutes at 115.5°C. The survivors were, respectively, 4600 and 160. Calculate the D value. The lag time for heating the tubes to 115.5°C was established in prior experiments to be 0.5 minutes.

Solution:

Because the heating times are greater than the lag time for the tubes to attain the desired heating temperature, Equation (9.15) can be used.

$$D = \frac{15 - 10}{\log(4600) - \log(160)} = \frac{5}{1.458} = 3.42\,\text{min}$$

9.2.7 The Heat Resistance of Spoilage Microorganisms

The heat resistance of microorganisms is expressed in terms of a D value at a reference temperature, and the z value, the temperature dependence of the thermal inactivation rate. The use of the z value in determining D values at different temperatures from D at the reference temperature is discussed in the section "Effect of Temperature on Thermal Inactivation of Microorganisms." Reference temperatures are 121.1°C (250°F) for heat resistant spores to be inactivated in commercial sterilization processes and 82.2°C (180°F) for vegetative cells and organisms of low resistance, which are inactivated in pasteurization processes. D at 121.1°C (250°F) is D_0.

Tables 9.2 and 9.3 lists the resistance of microorganisms involved in food spoilage.

Table 9.2   Heat Resistance of Spoilage Microorganisms in Low-Acid Canned Foods

			z	
Organism	Product	D_0 (min)	(°F)	(°C)
Clostridium	Phosphate buffer (pH7)	0.16	18	10
botulinum 213-B	Green beans	0.22	22	12
	Peas	0.22	14	8
Clostridium	Phosphate buffer (pH7)	0.31	21	12
botulinum 62A	Green beans	0.22	20	11
	Corn	0.3	18	10
	Spinach	0.25	19	11
Clostridium spp.	Phosphate buffer (pH7)	1.45	21	12
PA 3679	Asparagus	1.83	24	13
	Green beans	0.70	17	9
	Corn	1.20	18	10
	Peas	2.55	19	10
	Shrimp	1.68	21	12
	Spinach	2.33	23	13
Bacillus	Phosphate buffer (pH7)	3.28	17	9
stearother-mophilus	Asparagus	4.20	20	11
FS 1518	Green beans	3.96	18	10
	Corn	4.32	21	12
	Peas	6.16	20	11
	Pumpkin	3.50	23	13
	Shrimp	3.90	16	9
	Spinach	4.94	21	12

Source: Reed, J. M., Bohrer, C. W. and Cameron, E. J., *Food Res. 16:338–408.*
Reprinted from: Toledo R. T. *1980.* Fundamentals of Food Engineering. AVI Pub. Co., Westport, CT.

Table 9.3 Heat Resistance of Spoilage Microorganisms in Acid and in Pasteurized Foods

Organism	Temperature		D (min)	Z	
	°F	°C		°F	°C
Bacillus coagulans	250	121.1	0.07	18	10
Bacillus polymyza	212	100	0.50	16	9
Clostridium pasteurianum	212	100	0.50	16	9
Mycobacterium tuberculosis	180	82.2	0.0003	10	6
Salmonella spp.	180	82.2	0.0032	12	7
Staphylococcus spp.	180	82.2	0.0063	12	7
Lactobacillus spp.	180	82.2	0.0095	12	7
Yeasts and molds	180	82.2	0.0095	12	7
Clostridium botulinum Type E	180	82.2	2.50	16	9

Source: (1) Anderson, E. E., Esselen Jr., W. B. and Fellers, C. R. *Food Res. 14:499–510, 1949. (2)* Crissley, F. D., Peeler, J. T., Angelotti, R. and Hall, H. E., *J. Food Sci. 33:133–137, 1968. (3)* Stumbo, C. R. *77termobacteriology in Food Processing,* Academic Press, New York, *1973. (4)* Townsend, C. T. *Food Res. 4:231–237, 1939. (5)* Townsend, C. T., and Collier, C. P. *Proc. Technical Session of the 48th Annual Convention of the National Canners Association (NCA). NCA information Newsl. No. 1526,* February *28, 1955. (6)* Winter, A. R., Stewart, G. F., McFarlane, V. H. and Soloway, M. *Am. J. Pub. Health 36:451–460, 1946. (7)* Zuccharo. J. B., Powers, J. J., Morse, R. E. and Mills W. C. *Food. Res. 16:3038, 1951*
Reprinted from: Toledo, *1980.* Fundamentals of Food Process Engineering, Ist. ed. AVI Pub. Co. Westport, Conn.

The type of substrate surrounding the organisms during the heating process affects heat resistance. For a more detailed discussion of techniques for determination of microbial heat resistance and the effect of various factors on thermal inactivation rates, the reader is referred to Stumbo's (1973) book *Thermobacteriology in Food Processing* and NFPA's *Laboratory Manual for Canners and Food Processors,* which are listed in the "Suggested Reading" section at the end of this chapter.

9.2.8 F_0 Values Used in Commercial Sterilization of Canned Foods

D_0, N_0, and N values that can be used as a guide for determining F_0 values for food sterilization are listed in Table 9.1. F_0 must be based on microorganisms involved in economic spoilage, as these have higher heat resistance than *C. botulinum*. Table 9.4 lists F_0 values previously used commercially for different types of foods in various size containers. Data in both Tables 9.1 and 9.4 may be used as a base for selection of F_0 values needed to calculate thermal process schedules for sterilization.

9.2.9   Surface Sterilization

Packaging materials used in aseptic packaging systems and surfaces of equipment may be sterilized using moist heat, dry heat, hydrogen peroxide, high-intensity ultraviolet, and ionizing radiation from either gamma rays or high-energy electron beams. The latter three methods have not been adopted in commercial food packaging, but various forms of heat and hydrogen peroxide combined with heat are commercially utilized.

Table 9.4   Values of F_0 for Some Commercial Canning Processes

Product	Can sizes	F_0 (min)
Asparagus	All	2–4
Green beans, brine packed	No. 2	3.5
Green beans, brine packed	No. 10	6
Chicken, boned	All	6–8
Corn, whole kernel, brine packed	No. 2	9
Corn, whole kernel, brine packed	No. 10	15
Cream style corn	No. 2	5–6
Cream style corn	No. 10	2.3
Dog Food	No. 2	12
Dog Food	No. 10	6
Mackerel in brine	301 × 411	2.9–3.6
Meat loaf	No. 2	6
Peas, brine packed	No. 2	7
Peas, brine packed	No. 10	11
Sausage, Vienna, in brine	Various	5
Chili con carne	Various	6

Source: Alstrand, D. V., and Ecklund, O. F., *Food Technol. 6(5):185, 1952.*

Dry heat includes superheated steam and hot air. Resistance of microorganisms in these heating media are shown in Table 9.5. Inactivation occurs at a slower rate in dry heat compared with moist heat at the same temperature. The z value in dry heat is also higher than in moist heat. A similar principle is utilized in evaluating microbial inactivation in dry heat as for moist heat.

The D values shown in Table 9.5, and the recommended N and N_0 values for commercial sterilization in Table 9.1, may be used to determine exposure times to the sterilant. In dry heat sterilization of

Table 9.5   Resistance of Microorganisms to Microbicidal Agents

Organism	Heating Medium	$D_{176.6°C}$ (min)	$D_{121°C}$ (min)	z (°C)
Bacillus subtilis	Superheated steam	0.57	137	23.3
Bacillus stearothermophilus (FS 1518)	Superheated steam	0.14	982	14.4
Bacillus polymyxa	Superheated steam	0.13	484	15.6
Clostridium sporogenes (P.A.3679)	Air	0.30	109	21.7
Clostridium botulinum	Air	0.21	9	33.9
Bacillus subtilis	N_z + He	0.17	285	17.2
Clostridium sporogenes (P.A.3679)	He	0.45	161	21.7
Bacillus subtilis	A + CO_2 + Oz	0.13	218	17.2

Source: Adapted from Miller, B. M., and Litskey, W. eds, 1976. *Industrial Microbiology,* McGraw-Hill. Used with permission of McGraw-Hill.

Table 9.6 Resistance of Food Spoilage Microorganisms to Inactivation in Hot Hydrogen Peroxide

Organism	$D_{80°C}$ (min)	z (°C)
Clostridium botulinum 169B	0.05	29
Bacillus subtilis ATCC 9372	0.063	41
Bacillus subtilis A	0.037	25.5
Bacillus subtilis	0.027	27
Bacillus stearothermophilus	0.07	22

Source: Toledo, R. T., *AIChE Symp. Ser. 78(218):81, 1982*. Reproduced by permission of the American Institute of Chemical Engineers, © 1982, AIChE.

surfaces, surface temperatures must be used as the basis for the process rather than the temperature of the medium. The surface heat transfer coefficient and the temperature on the opposite side of the surface sterilized determine the actual surface temperature.

Hydrogen peroxide is the only chemical sterilant allowed for use on food contact surfaces. This compound at 35% (w/w) concentration is applied to the surface by atomizing, spraying, or by dipping, in the case of packaging materials in sheet form, followed by heating to vaporize the hydrogen peroxide and eliminate residue from the surface. A maximum tolerance of residual hydrogen peroxide in the package of 0.1 parts per million is required by federal regulations in the United States. Resistance of various microorganisms in hydrogen peroxide is summarized in Table 9.6. Inactivation rate is also temperature dependent.

9.3 EFFECT OF TEMPERATURE ON THERMAL INACTIVATION OF MICROORGANISMS

Microbial inactivation is a first-order chemical reaction and the temperature dependence of the rate constant can be expressed in terms of an activation energy or a z value. Equation (8.21) of Chapter 8 expresses the rate constant for inactivation in terms of the activation energy. The activation energy is negative for reactions that increase in rate with increasing temperature.

$$\frac{k}{k_0} = [e]^{E_a/R[1/T - 1/T_0]} \tag{9.16}$$

Because $k = 2.303/D$ from Equation 8.23 Section 8.6, the temperature dependence of the D value in terms of the activation energy will be:

$$\frac{D}{D_0} = [e]^{E_a/R[1/T - 1/T_0]} \tag{9.17}$$

Because E_a is positive when reaction rates increase with temperature, Equation (9.17) represents a decrease in D value with increasing temperatures.

Thermobacteriologists prefer to use the z value to express the temperature dependence of chemical reactions. The z value may be used on the target F value for microbial inactivation, or on the D value to determine required heating times for inactivation at different temperatures. It may also be used on heating times at one temperature, to determine equivalence in lethality at a reference temperature. A semi-logarithmic plot of heating time for inactivation against temperature is called the thermal death time plot, therefore equations based on this linear semi-logarithmic plot are called the thermal death time model equations for microbial inactivation at different temperatures.

$$\log \frac{F}{F_0} = \frac{T_0 - T}{z} \tag{9.18}$$

t_0 is the equivalent heating t_T at temperature T. When $t_T = 1$, t_0 is the lethality factor, L, the equivalent heating time at 250°F for 1 minute at T.

$$\log \frac{D}{D_0} = \frac{T_0 - T}{z} \tag{9.19}$$

$$\log \frac{t_0}{t_T} = -\frac{T_0 - T}{z} \tag{9.20}$$

$$L = [10]^{T - T_0/z} \tag{9.21}$$

The inverse of L is the heating time at T equivalent to 1 minute at 250°F and is the parameter F_i used in thermal process calculations.

$$F_i = [10]^{T_0 - T/z} \tag{9.22}$$

Use of the z value is not recommended when extrapolating D values over a very large temperature range. Equation (9.17) shows that D will deviate from a linear semi-logarithmic plot against temperature when the temperature range between T and T_0 is very large.

Example 9.10. The F_0 for 99.999% inactivation of *C. botulinum* type B is 1.1 minutes. Calculate F_0 for 12D inactivation, and the F value at 275°F (135°C) when z = 18°F.

Solution:

99.999% inactivation is equivalent to S = 5. For S = 12; $F_0 = SD_0 = 12(0.22) = 2.64$ minutes. Thus the D_0 value is 1.1/5 = 0.22 minutes. F_{275} can be calculated using Equation (9.18) or using Equation (9.19) on the D value to obtain D at 275°F.

Using Equation (9.18): $(T_0 - T)/z = -25/18 = -1.389$

$$F_{275} = F_0(10^{-1.389}) = 2.64\,(0.0408) = 0.1078 \text{ min}$$

Using Equation (9.19): $D_{275} = 0.22(10^{-1.389}) = 0.00898 \text{ min}$

$$F_{275} = S\,D = 12(0.00898) = 0.1077 \text{ min}$$

Example 9.11. The D_0 for PA 3679 is 1.2 minutes and the z value is 10°C. Calculate the process time for 8D inactivation of PA 3679 at 140.5°C using the "thermal death time" model (Eq. 9.18) and the Arrhenius equation (Eq. 9.17). The z value was determined using data on D at 115.5 to 121.1°C.

Solution:

Using Equation (9.18):

$(T - T_0)/z = (140.5 - 121.1)/10 = -1.94$

$F_0 = 8(1.2) = 9.6 \text{ min}; F_{140.5} = 9.6(10^{-1.94}) = 9.6(0.01148) = 0.11 \text{ min}$

From Equation (8.36), Chapter 8:

$$z = \frac{\ln(10)R}{E_a}T_1T_2; \quad E_a = \frac{\ln(10)RT_2T_1}{z}$$

$$T_2 = 121.1 + 273 = 394.1 \text{ K}; \quad T_1 = 115.5 + 273 = 388.5 \text{ K};$$

$$R = 1.987 \text{ cal/(gmole} \cdot \text{K)}$$

$$E_a = [\ln(10)](1.987)(388.5)(394.1)/10 = 70{,}050 \text{ cal/gmole};$$

$$E_a/R = 35{,}254$$

$$T = 140.5 + 273 = 413.5 \text{ K}; \quad T_0 = 394.1 \text{ K}$$

$$(1/T - 1/T_0) = -0.000119$$

$$F_{140.5} = F_0[e]^{35254(-0.000119)} = 9.6(0.015067) = 0.1446 \text{ min}$$

These calculations show that when extrapolating D values to high sterilization temperatures using the z value based on data obtained below 250°F (121.1°C), use of the Arrhenius equation will result in a safer process compared with the TDT (Thermal Death Time) model.

9.4 INACTIVATION OF MICROORGANISMS AND ENZYMES IN CONTINUOUSLY FLOWING FLUIDS

The methods for calculating the extent of microbial inactivation that results from a heat treatment is the same regardless of the severity of the heat treatment. Mild heat treatments designed to inactivate vegetative cells of microorganisms is called *pasteurization*. Pasteurized foods will be shelf stable if they are high acid (pH ∃ 4.6) but they would require refrigeration and are perishable if the pH is > 4.6. *Sterilization* is a high-temperature heat treatment designed to inactivate heat-resistant spores to produce a shelf-stable product when stored at ambient temperature. Pasteurization is used on food products to avoid public health hazards from pathogenic microorganisms. On perishable products, pasteurization reduces the number of viable spoilage microorganisms thereby increasing product shelf life. In a system for continuous pasteurization or sterilization of flowing fluids, the product is heated to a specified temperature and held at that temperature for a specified time.

The processing system used to heat-treat fluids to destroy unwanted microorganisms or enzymes is discussed in the section "Continuous Flow Sterilization: Aseptic or Cold Fill." The process is essentially a constant temperature heating process with the residence time in the holding tube considered as the processing time and the temperature at the point of exit from the holding tube as the processing temperature.

9.4.1 Time and Temperature Used in the Pasteurization of Fluid Foods

The basis for pasteurization time and temperature for high-acid foods that will be refrigerated post-pasteurization is the resistance of pathogenic microorganisms. Liquid egg and egg product pasteurization and milk pasteurization are regulated by local and federal agencies. Data in Table 9.7 can be used to calculate minimum time/temperature combinations for safety of pasteurized low-acid products. The following data have been compiled from the literature and from discussions with processors of the different food products listed. In general, the main objective of pasteurization is the inactivation of microorganisms of public health significance. For high-acid foods, heat resistance of pathogenic and

Table 9.7 Heat resistance of spoilage and pathogenic microorganisms in pasteurized food products

Organism	Substrate	D (min) 60 °C	z °C	Reference
Low acid foods (pH ∃ 4.6)				
Aeromonas hydrophila	LWE	0.04	5.6	1
Mycobacterium tuberculosis	Milk	14.1	4.4	2
Listeria monocytogenes	Milk	2.03	5.5	3
Listeria monocytogenes	Egg yolk	1.34	9.4	4
	Egg white	2.29	9.4	4
Salmonella spp.	Egg white	0.58	4.3	4
	Egg yolk	0.91	4.3	4
E. Coli O157:H7	Milk	1.5	6.9	3
Salmonella spp.	Milk	0.52	6.9	3
Listeria monocytogenes	LWE	1.27	7.2	5
Salmonella enteritidis	LWE	0.25	6.9	5
Staphylococcus aureus	Milk	0.9	9.5	3
Aeromonas hydrophyla	Milk	0.18	7.7	3
Yersinia enterocolitica	Milk	0.51	5.8	3
High acid foods (pH < 4.6)				
E. Coli 0157:H7	Apple Juice	0.33	6.9	6
Salmonella enteritidis	Model pH4.4	0.24	6.9	3
Leuconostoc:	SSOJ	0.19	7	7
Lactobacilli:	SSOJ	1.30	7	7
Yeast:	SSOJ	1.63	6	7
Leuconostoc	42BxOJ	0.36	10	7
Lactobacilli	42BxOJ	0.091	18	7
Yeast	42BxOJ	1.02	9	7

LWE is liquid whole eggs. SSOJ is single strength orange juice and 42BxOJ is 42 brix orange juice concentrate. References: 1. Schuman JD, Sheldon BW, and Foegeding PM. (1997) J. Food Prot. 60:231; 2. Keswani J, and Frank JF (1998). J. Food Prot. 61:974; 3. ICMSF 1996 Microorganisms in Foods, Book 5. Blackie Publishers, London; 4. Schuman JD, and Sheldon BW (1997). J. Food Prot. 60:634; 5. Foegeding PM and Leasor SB (1990). J. Food Prot. 53:9; 6. Splitstoesser, DF. (1996) J. Food Prot. 59:226; 7. Murdock et al. (1953) Food Research 18:85.

aciduric spoilage microorganisms are relatively low, therefore pasteurization processes can produce a commercially sterile product.

Pasteurization processes for high-acid products (pH #4.6) or acidified products: The following time and temperature combinations are used as a guideline by the food industry for producing shelf-stable high-acid food products. Times and temperature are dependent on the product pH.

Acidified or naturally high acid products: pH < 4.0 = 1 minute at 87.8°C (190°F) ; pH 4.0 = 30 seconds at 96.1°C (205°F); pH 4.1 = 30 seconds at 100°C (212°F); pH 4.2 = 30 seconds at 102.2°C (216°F); pH > 4.2 to 4.5 = 30 seconds at 118.3°C (245°F). If sugar or starch is added to the product, the time/temperature for the next higher pH should be used. For example, if starch or sugar is a component of a product with a pH of 4.1, use a process of 30 seconds at 102.2°C (process for pH 4.2 if no sugar or starch is added).

Tomato products: The pasteurization process is based on an equivalent F value at 200°F (93.3°C). The following F_{200F} values are generally used: 1 minutes at pH 4.1; 3 minutes at pH 4.2; and 5 minutes

Table 9.7a Heat treatment conditions for long-life pasteurized products

pH	Temperature for a 40 s process	Temperature for a 20 min. Process
4.6	140°C	115°C
4.5	130	110
4.4	120	105
4.3	110	100
4.2	100	95
4.1	98	90
4.0	94	85
3.9	90	75

at pH 4.3. If temperatures other than 93.3°C is used, use a z value of 16°F (8.9°C) to obtain the equivalent process time from the F_{200F} values. For products with starch or sugar added at pH 4.3, use an F_{250F} of 0.5 minutes (z = 16°F) as a basis for the process.

Pineapple juice: The following F_{200F} values are used: pH > 4.3, F = 10 minutes; pH between 4.0 and 4.3, F = 5 minutes. Use a z value of 15°F (8.33°C) for other temperatures. Example: Juice with a pH of 4.0 processed at 210°F (98.9°C) will need a process time of 1.06 minutes.

Other juices, flash pasteurization: Flash pasteurization refers to very rapid heating to the processing temperature followed by an appropriate hold, cooling, and aseptic filling. Hold time for peach juice, pH < 4.5 is 30 seconds at 110°C; orange juice: 1 minute at 90°C or 15 seconds at 95°C; grapefruit juice: 16 seconds at 74°C or 1 second at 85°C. FDA regulations mandate a minimum pasteurization requirement of 5 log reduction of pathogenic microorganism capable of growing in the product (FDA 1998). For a typical flash pasteurization temperature of 80°C, products with pH < 4.4 will require 0.2 s to obtain 5 log reduction of *E. coli* 0157:H7 and *Salmonella* spp. Thus, the time and temperature specified above are more than adequate to meet the minimum pasteurization requirement.

Pasteurization requirements for milk: (U.S. Grade A Pasteurized milk ordinance–1978 Recommendations of the USPHS/FDA) specify the following temperature and times: 63°C (145°F) for 30 minutes; 72°C (161°F) for 15 seconds; 89°C (192.2°F) for 1 second; 90°C (194°F) for 0.1 second; 96°C (204°F) for 0.05 second.

Pasteurization requirements for liquid whole eggs: The USDA specifies a pasteurization time and temperature of 3.5 minutes at 60°C (140°F).

Comparing the process times and temperature and the heat inactivation kinetics of pathogens in Table 9.7 gives the following log reduction of pathogens. For milk, a 15-second process at 72°C = 58 log reduction of *M. tuberculosis*; = 18.7 log reduction of *L. monocytogenes*; > 100 log reduction of *E. coli* 0157:H7; > 100 log reduction of *Salmonella* spp. For liquid whole eggs, a 3.5-minute process at 60°C = 2.75 log reduction of *L. monocytogenes*; = 14 log reduction of *Salmonella enteritidis*.

Commercial pasteurization processes: Commercial processors may use processing time and temperature higher than those specified by regulations to obtain long product shelf life. For example, shelf life (defined as time for CFU to reach 10^6/mL) at 5°C of milk processed at different time and temperature (from: Kessler and Horak, Milchwissenschaft 39:451, 1984) are as follows: 21 days at 74°C/40 seconds or 78°C/15 seconds; 17 days at 74°C/15 seconds or 71°C/40 seconds; 16 days at 78°C/14 seconds or 85°C/15 seconds; 12 days at 71°C/15 seconds.

For liquid whole eggs, TetraPak recommends processing at 70°C for 90 seconds to obtain a 3-month shelf life at 5°C.

For acidified milk (flavored, yogurt-like), von Bockelman (1998) gave the heat treatment conditions to obtain a 3-month shelf life at ambient temperature storage as follows:

pH and temperature (°C) for a 40 second continuous flow process: pH 4.6, 140 °C; pH 4.5, 130 °C; pH 4.4, 120 °C; pH 4.3, 110 °C; pH 4.2, 100 °C; pH 4.1, 98 °C; pH 4.0, 94 °C; pH 3.9, 90 °C.

pH and temperature (°C) for a 20 minute batch process: pH 4.6, 115 °C; pH 4.5, 110 °C; pH 4.4, 105 °C; pH 4.3, 100 °C; pH 4.2, 95 °C; pH 4.1, 90 °C; pH 4.0, 85 °C; pH 3.9, 75 °C.

9.4.2   Microbial Inactivation in Continuously Flowing Fluids

The time and temperature for pasteurization in the previous section is based on the residence time of the fluid in a holding tube following elevation of the product temperature to the designated processing temperature. The residence time is a function of the fluid velocity. Because there is a velocity distribution in fluids flowing through a pipe, residence time of the fluid will vary at different positions in the pipe. The integrated lethality of microoganisms in the fluid leaving a hold tube must be used as the basis for the process.

The velocity distribution in a fluid flowing through a pipe has been derived for a power law fluid in laminar flow. Equation (6.16), Chapter 6 is

$$V = \overline{V}\left(\frac{3n+1}{n+1}\right)\left[1 - \left[\frac{r}{R}\right]^{(n+1)/n}\right] \tag{9.23}$$

Consider an area element of thickness dr within a tube of radius R: If N_0 is the total number of organisms entering the tube, N is the number leaving, n_0 is the organisms/unit volume entering, and n is organisms/unit volume leaving:

$$N_0 = n_0\left[2\pi \int_0^R Vr\,dr\right] \tag{9.24}$$

The expression in brackets in Equation (9.24) is the volumetric rate of flow, $nR^2\overline{V}$. Thus Equation (9.24) becomes:

$$N_0 = nR^2n_0\overline{V} \tag{9.25}$$

The residence time of fluid within the area element is L/V, and the number of survivors, $N = N_0$ $[10^{-L/(VD)}]$. Thus:

$$N = 2\pi n_0 \int_0^R Vr[10]^{-L/(VD)}dr \tag{9.26}$$

In Equation (9.23), let: A = (3n + 1)/(n + 1); B = (n + 1) /n; y = r/R. Equation (9.23) becomes:

$$V = A(1 - y^B)\overline{V} \tag{9.27}$$

The ratio $L/(\overline{V} \cdot D)$ is the lethality, S_v, based on the average velocity. Substituting Equation (9.27) into Equation (9.26), and $S_v = L/(\overline{V} \cdot D)$; r = yR; dr = Rdy. The integrated lethality, Si will be log N_0/N:

$$N = 2\pi n_0 \int_0^1 \overline{V}(A)(1 - y^B)[10]^{-S_v/[A(1-y^B)]}R^2y\,dy \tag{9.28}$$

Table 9.8   Integrated Lethality in the Holding Tube of a Continuous Sterilization System for Fluids in Laminar Flow

	Integrated Lethality, S_i Fluid flow behavior index, n						
S_v	0.4	0.5	0.6	0.7	0.8	0.9	1.0
0.1	0.093	0.093	0.092	0.091	0.091	0.091	0.091
0.5	0.426	0.419	0.414	0.409	0.406	0.401	0.401
1	0.809	0.792	0.779	0.768	0.759	0.747	0.747
2	1.53	1.49	1.46	1.44	1.42	1.40	1.38
4	2.92	2.82	2.75	2.69	2.64	2.59	2.56
6	4.27	4.11	3.99	3.89	3.81	3.74	3.68
8	5.60	5.38	5.20	5.06	4.95	4.86	4.78
10	6.92	6.63	6.41	6.23	6.08	5.96	5.86
12	8.23	7.88	7.60	7.38	7.20	7.05	6.92
14	9.54	9.12	8.79	8.52	8.31	8.13	7.98
16	10.84	10.35	9.97	9.66	9.41	9.20	9.03
18	12.14	11.58	11.14	10.79	10.51	10.27	10.07
20	13.44	12.81	12.32	11.93	11.60	11.34	11.11
22	14.73	14.03	13.49	13.05	12.69	12.40	12.15
24	16.03	15.26	14.66	14.18	13.79	13.46	13.18
26	17.32	16.48	15.83	15.30	14.87	14.52	14.22

S_v = sterilization value as number of decimal reductions based on the average velocity $(L/D \bullet \overline{V})$

Dividing Equation (9.25) by Equation (9.28):

$$\frac{N_0}{N} = \frac{\pi R^2 n_0 \overline{V}}{2 \pi n_0 \int_0^1 \overline{V}(A)(1 - y^B)[10]^{-S_v/[A(1-y^B)]} R^2 y \, dy}$$

Canceling $\pi R^2 n_0 \overline{V}$, taking the logarithm of both sides, and making the logarithm of the denominator of the right-hand side negative to bring it to the numerator:

$$\log\left(\frac{N_0}{N}\right) = -\log\left[\int_0^1 \left[2A(1 - y^B)[10]^{-S_v/[A(1-y^B)]} y \, dy\right]\right] \tag{9.29}$$

$\log(N_0/N)$ in Equation (9.29) is the integrated lethality, S_i.

Equation (9.29) can be evaluated by graphical integration. The integrated sterility in the holding tube is independent of the size of the tube and the average velocity if the length and average velocity are expressed as the sterilization value based on the average velocity, S_v.

Thus Equation (9.29) can be solved and the solution will yield a generalized table for the integrated lethality of heat in the holding tube when fluid is flowing through that tube in laminar flow. Values of S_i at different values of n and S_v obtained by graphical integration of Equation (9.29) are shown in Table 9.8. The data in Table 9.8 shows that S_i/S_v is approximately $\overline{V}/V_{max}$. Thus, use of the maximum velocity to calculate lethality produces results much closer to the integrated lethality than if the average velocity is used.

Example 9.12. A fluid food product with a viscosity of 5 cP and a density of 1009 kg/m³ is to be pasteurized in a continuous system that heats the food to 85°C followed by holding in a 1.5-in. sanitary pipe from which it leaves at 82.2°C. The process should give 12 decimal reduction of *Staphylococcus aureus*, which has a $D_{82.2EC}$ of 0.0063 minutes. Calculate the length of the holding tube if the flow rate is 19 L/min.

Solution:

From Chapter 6, Table 6.2, the inside diameter of a 1.5-in. sanitary pipe is 0.03561 m.

$$\overline{V} = \frac{19L}{min} \frac{1\,m^3}{1000\,L} \frac{1\,min}{60\,s} \frac{1}{\pi[0.5(0.03561)]^2 m^2}$$

$\overline{V} = 0.318$ m/s. Re $= (0.03561)(0.318)(1009)/5(0.001) = 2285$. Flow is in the transition zone from laminar to turbulent flow. For safety, assume flow is laminar, and $V_{max} = 2\overline{V} = (0.318) = 0.636$ m/s.

Using Equation (9.14): The F value is the required process time that must equal the residence time in the tube for the fastest-flowing particle. $F = t_{min} = S \cdot D = 12(0.0063) = 0.0756$ minutes.
Using Equation (9.4):

$$L = 0.0756\,min(60s/\,min)(0.636\,m/s) = 2.88\,m$$

Length based on the integrated lethality for a Newtonian fluid: From Table 9.8, by interpolation, to obtain $S_i = 12$, $n = 1$, $S_v = 22 - 0.17(12.17 - 11.13)/2 = 21.91$. Because $S_v = L/(\overline{V} \cdot D)$, $L = 21.91(0.318)(0.0063)(60) = 2.63$ m.

The integrated lethality gives a length much closer to that based on the maximum velocity compared to results using the average velocity.

At the Reynolds number in the range 3000 to 5000, the maximum velocity cannot be obtained using Equation (9.7). Thus, the safest approach to determination of holding tube length in this range of Reynolds number will be to assume laminar flow.

Example 9.13. An ice cream mix having a viscosity of 70 cP and a density of 1015 kg/m³ is being canned aseptically in a system which uses a 100-ft-long, 1.0-in. sanitary pipe for a holding tube. Flow rate through the system is 5 gal/min. The fluid temperature at the exit from the holding tube is 285°F (140.6°C).
Calculate (a) the sterilizing value of the process for PA 3679 ($D_0 = 1.83$ min; $z = 24°F$).

Solution:

From Table 6.2: $D = 0.02291$ m, $R = 0.01146$ m.
Solving for the average velocity:

$$\overline{V} = \frac{5\,gal}{min} \frac{3.78541 \times 10^{-3}\,m^3}{gal} \frac{1\,min}{60s} \frac{1}{\pi(0.01146)^2\,m^2}$$

$\overline{V} = 0.765$ m/s. Re $= 0.02291(0.765)(1015)/70(0.001) = 254$. Flow is laminar. $L = 100$ ft(1 m/ 3.281 ft) $= 30.48$ m.

(a) Solving for V_{max}:

$V_{max} = 2\overline{V} = 2(0.785) = 1.57$ m/s

The residence time of the fastest-flowing particle is

$t = L/V_{max} = 19.41$ s.

The process occurs at 285°F; therefore, the number of survivors from the process is calculated using the D value at 285°F for D in Equation (9.11). The D value at 285°F is determined using Equation (9.19):

$\log D = [\log D_0 + (T_0 - T)]/z = [\log[(1.83)(60)] + (250 - 285)]/24 = 2.0406 - 1.458$

$D = 10^{0.5826} = 3.8247$ s at 285°F

Using Equation (9.11): $S = 19.41/3.8247 = 5.075$. The process will result in at least 5.075 decimal reduction of PA 3679.

The number of decimal reductions based on the integrated lethality is

$S_v = L/(\overline{V} \cdot D) = 30.48/[(0.765)(3.8247)] = 10.42$

From Table 9.6, $S_v = 10.42$, $n = 1$, by interpolation:

$S_i = 5.86 + 0.42(6.93 - 5.86)/2 = 6.08$

The integrated lethality is 6.08 decimal reductions of PA 3679.

9.4.3 Nutrient Degradation

Nutrient degradation can be calculated the same way as for microbial inactivation. Equations (9.24) to (9.29) are also applicable for loss of nutrients if the appropriate kinetic parameters of D and z values are used.

Nutrient retention with increasing processing temperature can be derived by simultaneously solving the rate equations for microbial inactivation and nutrient degradation. Using the subscript c and m to signify parameters for nutrient degradation and microbial inactivation respectively, Equations (9.11) and (9.19) may be combined to determine the heating time at various temperatures for microbial inactivation.

$$t = D_{mo}\left[\log\left(\frac{N_0}{N}\right)\right][10]^{(T_0 - T)/z_m} \tag{9.30}$$

A similar expression can be formulated for nutrient degradation.

$$\log\left(\frac{C}{C_0}\right) = -\frac{t}{D_{co}[10]^{(T_0 - T)/z_c}} \tag{9.31}$$

Substituting S_v for $\log(N_0/N)$ and combining Equations (9.30) and (9.31):

$$\log\left(\frac{C}{C_0}\right) = -\left[\frac{D_{mo}}{D_{co}}S_v\right][10]^{(T_0 - T)(1/z_m - 1/z_c)} \tag{9.32}$$

Equation (9.32) is based on constant temperature processes and is applicable for estimating nutrient retention in holding tubes of aseptic processing systems. The derivation is based on plug flow of fluid at the average velocity through the tube, and adjustments will have to be made to obtain the integrated lethality and nutrient degradation. Table 9.8 can be used to determine S_v needed to obtain the microbial lethality required expressed as S_i. $-\text{Log}(C/C_o)$ will be the S_v for nutrient degradation which can be converted to S_i using Table 9.8.

Example 9.14. The data for thiamin inactivation at 95°CE to 110°C from Morgan et al. (J. Food Sci. 51:348, 1986) in milk show a D_{100EC} of 3×10^4 s and a z value of 28.4°C. Chocolate milk with a flow behavior index of 0.85 and a consistency index of 0.06 Pa s ^{n} is to be sterilized at 145°C. The density is 1006 kg/m³. If the rate of flow is 40 L/min, and the holding tube is 1.5-in. sanitary pipe, calculate the holding tube length necessary to give an integrated lethality of 7 decimal reductions of an organism having a D_o value of 0.5 minutes and a z value of 10°C. Assume the z value of the organism was determined in the temperature range that included 145°C. Calculate the retention of thiamin after this process.

Solution:

Because the z value for thiamin was determined at low temperatures, Equation (9.17) will be used to extrapolate the D value to 145°C. From Equation (8.25), Chapter 8:

$$E_a = \frac{[\log(10)]R}{z}T_1 T_2 = \frac{[\log(10)](1.987)(368)(383)}{28.4}$$

$E_a = 22{,}706$; $E_a/R = 11{,}427$; $T = 145 + 273 = 418$ K; $T_r = 100 + 273 = 373$ K.

Using Equation (9.17) and the subscript c to represent chemical degradation:

$$[1/T - 1/T_r] = (1/418 - 1/373) = -0.000289$$
$$D_c = 3 \times 10^4 (e)^{11427(-0.000289)} = 1109 \text{ s}$$

Equation (9.19) is used to determine D_m at 145°C.

$$[(T_o - T)/z] = [(121.1 - 145)/10] = -2.39$$
$$D_m = 0.5(10)^{-2.39} = 0.002037 \text{ min} = 0.1222 \text{ s}$$

Equation (9.32) will be used, but because D_m and D_c are already calculated at 145°C, $T = T_0$, the exponential term $10^{(T_o - T)(1/z_m - 1/z_c)} = 1$ and :

$$\log\left(\frac{C}{C_0}\right) = -\frac{D_m}{D_c}S_{vm} \tag{9.30a}$$

S_{vm} is the sterilizing value for the microorganisms based on the average velocity.

The Reynolds number is

$$Re = \frac{8V^{2-n}R^n \rho}{K[3 + 1/n]^n}$$

From Table 6.2, $R = 0.5(0.03561) = 0.017805$ m

$$\overline{V} = \frac{40 \text{ L}}{\text{min}} \frac{1 \text{ min}}{60 \text{ s}} \frac{\text{m}^3}{1000 \text{ L}} \frac{1}{\pi(0.017805)^2 \text{ m}^2} = 0.669 \text{ m/s}$$

$$[(3n + 1)/n] = 4.176$$

$$Re = \frac{8(0.669)^{1.15}(0.017805)^{0.85}(1006)}{0.06(4.176)^{0.85}} = 816.7$$

Flow is laminar, and the integrated lethality can be evaluated using Table 9.8. The integrated sterilizing value for the microorganism, S_{mi} is the desired outcome of the process and the value is 7.0 The sterilizing value based on the average velocity must be determined to obtain the length of the holding tube.

From Table 9.8, S_{vm} is obtained for $S_i = 7$ and $n = 0.85$, by interpolation:

At $S_i = 7$, $n = 0.8$, $S_v = 10 + [2/1.11](0.91) = 11.64$

$n = 0.9$, $S_v = 10 + [2/1.1](1.04) = 11.89$

Solving for S_v at $n = 0.85$, $S_v = 11.64 + [0.25/0.1](0.05) = 11.77$

The sterilizing value based on the average velocity $= 11.77$ and the D_m value at $145\,C = 0.1222$ seconds. Average retention time at $145\,C = 11.77(0.1222) = 1.438$ seconds. The length of the holding tube, $L = \overline{V} \cdot t = 0.669(1.438) = 0.962$ m.

The thiamine retention will be calculated using Equation (9.30a): $D_c = 1109$ s, $D_m = 0.1222$ s, and $S_{vm} = 11.77$.

$$\log\left(\frac{C}{C_0}\right) = -\frac{0.1222}{1109}11.77 = -0.001317$$

$C/C_0 = (10)^{-0.001317} = 0.997$ or 99.7% retention.

Example 9.15. Tomato paste ($n = 0.5$, $k = 7.9$ Pa. s^n, $\rho = 1085$ kg/m³) is sterilized at 95°C using a holding tube with an inside diameter of 0.03561 m. The system operates at the rate of 50 L/min. Each package to be filled with the sterilized product contains 200 L and the probability of spoilage to be expected from the process is 1 in 10,000 from spores of *Bacillus polymyxa*, which has $D_{80EC} = 0.5$ minutes and $z = 9°C$

The unprocessed paste contains 4 spores/mL. Calculate the length of the holding tube necessary to achieve an integrated sterility, which is the desired spoilage probability. Calculate the extent of nonenzymatic browning that occurs during this process expressed as percentage increase over brown color before processing. Assume the D_0 value of non-enzymatic browning is 125 minutes at 80°C and the z value is 16°C.

Solution:

Determine if flow is laminar or turbulent.

$$\overline{V} = \frac{50\,L}{min}\frac{1\,min}{60\,s}\frac{m^3}{1000\,L}\frac{1}{\pi[(0.5)(0.03561)]^2\,m^2} = 0.837\,m/s$$

$$Re = \frac{8(\overline{V})^{2-n}\,R^n\rho}{K[3 + 1/n]^n}\left[\frac{(3n + 1)}{n}\right]^n = 2.24$$

$$Re = \frac{8(0.837)^{1.5}(0.017805)^{0.5}(1085)}{7.9(2.24)} = 50$$

Flow is laminar. For each 200-L package, $N_0 = 80000$. $N = 1/10,000$. $\log(N_0/N) = 9.9$. This is to be evaluated on the integrated sterility, therefore, $S_i = 9.9$. From Table 9.8, to obtain $S_i = 9$, for $n = 0.5$, S_v is obtained by interpolation:

$S_v = 14 + (2/1.24)(9.9 - 9.12) = 15.26 = L/(V - D)$

$D_{95} = D_{80}[10]^{(80-95)/9} = 0.0215(0.5) = 0.01077$ min

Substituting and solving for L:

$L = S_v\ \overline{V}\,D = 15.26(0.837\ m/s)(0.01077\ min)(60\ s/min) = 8.25$ m.

Equation (9.32) will be used for determining the increase in brown color that results from the process. Because the reaction involves the appearance of a brown color, the negative sign in Equation (9.32) is changed to positive. The reference temperature, $T_0 = 80°C$; $T = 95°C$, $z_m = 9$ and $z_c = 16$.

$$\log\left(\frac{C}{C_0}\right) = \frac{0.5}{125}(15.26)[10]^{(80-95)(1/9-1/16)} = 0.011$$

At very small values of S, $S_v = S_i$, therefore integrated C/C_0 at $S = 0.011 = 10^{0.011} = 1.026$. An increase of 2.6% in the intensity of browning will be expected in the process. Calculations of integrated lethality for fluids in turbulent flow is not possible without an expression for velocity distribution within the tube. Currently available expressions for velocity distributions in turbulent flow is too unwieldy for a generalized treatment of the integrated sterility as was done with Equation (9.27). A possible approach to determination of an integrated lethality will be to determine experimentally the fluid residence time distribution and express this as a distribution function for velocity within the tube relative to the average velocity. In the absence of velocity distribution functions, lethality in the holding tube of continuous sterilization systems in turbulent flow must be calculated using the maximum velocity (Equation 9.7, Re > 5000). Examples shown above for nutrient degradation during high-temperature, short-time sterilization demonstrate very low values for $\log(C/C_0)$ such that the integrated value approaches the value based on the average velocity. Thus, $\log(C/C_0)$ can be based on the average velocity. On the other hand, holding tube length calculations must be based on the integrated lethality.

9.4.4   High-Pressure Pasteurization

The momentum toward adoption by the food industry of high pressure as a means of food preservation has increased because reliable equipment is now available for applying the process and adequate data have been accumulated on inactivation of pathogenic and spoilage microorganisms to increase user's confidence in the process. High-pressure processes have the advantage of inactivating microorganisms with minimal exposure of the food product to heat. Thus, the potential for preserving the fresh-like quality of the preserved food is very good. High-pressure food preservation may be applied as a pasteurization or a sterilization process. High-pressure pasteurization of high-acid products may produce a product that is stable during ambient storage. Use of the process on acid products may extend refrigerated shelf-life post-treatment. High-pressure sterilization is considered by the U.S. Food and Drug Administration as a nonconventional process, therefore careful scrutiny and proof of safety must be provided before the process can be approved for commercial use. The units of pressure used in high-pressure processing are usually expressed in megapascals (MPa). Some reports expresses the pressure in Bars. A Bar is a technical atmosphere and is equivalent to 14.5 lbf/in^2 (psi) or 0.1 MPa.

9.4.4.1   High-Pressure Systems

High-pressure systems may be classified as batch or continuous processes. The batch process is also known as "Isostatic High Pressure" or "High Hydrostatic Pressure" (HHP). The HHP system consists of a pressure vessel that holds the product to be treated; a pressure intensifier that raises the pressure of the pressurizing fluid to the target operating pressure and pumps this high pressure fluid into the processing vessel, a hydraulic pump to operate the intensifier, and appropriate controls for pressurization, temperature control, and depressurization. Because time for loading and sealing the pressure vessel then unsealing and unloading postprocess are all part of the batch cycle, reducing these elements of the batch cycle time are just as important as reducing the high-pressure exposure time

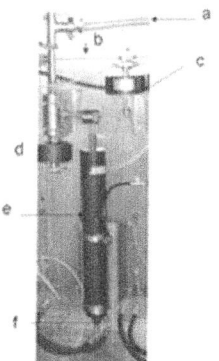

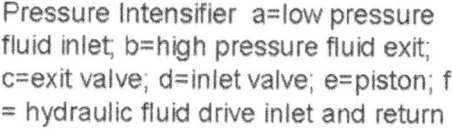

Pressure Intensifier a=low pressure fluid inlet; b=high pressure fluid exit; c=exit valve; d=inlet valve; e=piston; f = hydraulic fluid drive inlet and return

High hydrostatic pressure system.

Figure 9.11 Pressure intensifier system for generating high pressure in high-pressure processing systems and photograph of a high hydrostatic pressure processing system. High hydrostatic pressure system reprinted from Food Technology. Used with permission.

itself. HHP pressure vessels are equipped with seals that easily engage and disengage to shorten the time for loading and unloading the pressure vessel. For large systems, a product carrier is used to enable loading and unloading a full load of product rapidly. Product treated by HHP may be prepackaged and may contain large particulate material. Water is usually used as the pressurizing fluid, hence the term "Hydrostatic." Because temperature is as important as pressure in effectively inactivating microorganisms, it is important that the system incorporate temperature monitors that can measure actual temperature inside a product within the pressure vessel.

A continuous high-pressure system can be used only on homogeneous liquids. The system consists of a feed pump, a pressure intensifier, a hydraulic system to operate the intensifier pistons, and a throttling device to reduce the pressure from high pressure to atmospheric pressure. Because the throttling device also generates extremely high shear rates on the fluid and because particle size reduction or homogenization also occur during depressurization, the continuous high-pressure system may be considered a "High-Pressure Homogenization" system. High-pressure piston type homogenizers have been used for this purpose in the past, but microbial reduction has been inadequate for successful pasteurization because of the pressure cycling inherent in piston-type pumps. Recent designs of continuous flow intensifiers incorporating two separate pistons operated with programmable logic controllers to time the cycling between the two pistons have greatly reduced the amplitude of pressure oscillation. Thus, a continuous flow high-pressure system can now be used for pasteurization of fluid food products.

Figure 9.11 shows a pressure intensifier system for continuous flow pasteurization of liquids and a HHP system. An important feature of a HHP system is ease of loading and unloading. In Fig. 9.11, the pressure chamber can be lowered and moved out of the cover to expose the chamber for loading and unloading. Pressure intensifiers have the same basic features whether used to directly pasteurize fluids or used to generate the pressurizing fluid for a HHP processing vessel. The high-pressure intensifer consists of in-line dual cylinders where a drive piston is directly connected to the intensifier piston. The ratio of the area of the drive piston to that of the intensifier piston determines the pressure intensification. Standard hydraulic pumps that generate hydraulic fluid pressure of 20 MPa can be used to produce high-pressure fluid at 400 MPa if the drive and intensifer piston diameter ratio is 20:1. Each intensifier cylinder must have an inlet valve that opens to admit liquid into the intensifier and an exit valve that opens when the intensifier piston is discharging fluid at high pressure. The two valves alternately open and close, while the inlet valve is open the exit valve is closed and *vice versa*. The opening and closing of the valves are timed to the position of the drive piston. When the drive piston is at the apex of its forward travel, the exit valve closes and the inlet valve opens. The hydraulic fluid in the drive piston is released so that the drive piston is pushed back to the farthermost backward position by the pressure of the fluid entering the intensifier. At this point, the inlet valve closes and the exit valve opens and high pressure hydraulic fluid is fed into the drive piston, thus allowing the intensifier piston to advance to deliver the high pressure fluid. It is very important that the inlet and exit valves maintain positive closure when closed and that the timing of the opening and closing be properly set. A programmable logic controller may be used to control the timing of opening and closing of the valves, or a mechanical/pneumatic system may be used as a control system.

9.4.4.2 *High-Pressure Pasteurization*

Data in Table 9.9 can be used to determine the time and temperature needed for successful HHP pasteurization of food products. Microbial inactivation at high pressure has been shown to be first order and inactivation rate can be expresses as a D value as with thermal processes. Processes at 50°C will take less time than those at ambient temperature. Minimum pressure for inactivating microorganisms by HHP is 300 MPa. Increasing pressure to 700 MPa rapidly decreases the D value but a point of diminishing returns occur at pressures greater than 700 MPa. Microbial inactivation should be at least 5D for pathogenic microorganisns and 8D for spoilage microorganisms. Using this criteria, HHP processing of meat products at 500 MPa and 50°C will require 15 minutes to obtain 5D reduction of *Staphylococcus aureus*. *Staphylococcus aureus* in milk will require 12.5 minutes at 50°C and 500 MPa Orange juice will require 7.8 minutes at 500 MPa and 37°C while apple juice will require 2.3 minutes at 40°C and 450 MPa to obtain a 8 log reduction of *Saccharomyces cerevisiae*.

Microbial inactivation kinetics in continuous flow pasteurization does not follow the same trend as HHP. Hold time at high pressure prior to pressure reduction is only in the order of 0.5 seconds, yet substantial microbial inactivation results. Pressurizing to 242 MPa and releasing the pressure to 1 atm permits very rapid flow of fluid across the pressure reducing valve resulting in a breakdown of microbial cell walls killing the microorganisms. Continuous flow high-pressure pasteurization is very similar to the action of homogenizers to break down microbial cells to release cellular proteins. However, the constant pressure and controlled flow of fluid through the throttling valve results in the exposure of all suspended cells to the same high shear rates thus resulting in effective pasteurization. Cells of *Saccharomyces cerevisiae* in orange juice are reduced 8 log at 242 Mpa. Cells of *Listeria inocua* are reduced 5 log. Cells of *Lactobacillus sake* are reduced 5 log by this treatment. Orange juice processed by continuous flow pasteurization at 242 MPa retained the fresh-squeezed orange juice flavor and was stable both microbiologically and biochemically when stored for 90 days at 4°C. Temperature rise during continuous flow pasteurization must be minimized. Temperature increases instantaneously on reduction of pressure due to the conversion of potential energy at the high pressure into heat. Temperature rise is about 24°C/100 MPa pressure. To minimize exposure to high temperature, feed temperature should me kept at the lowest temperature above the freezing point and liquid must be

Table 9.9 Resistance of microorganisms important in high pressure pasteurization processes

Organism	Substrate	D	P	T	Reference
Solid medium					
Salmonella Enteritidis	meat	3	450	N/A	1
Salmonella Typhimurium	meat	1.48	414	25	2
	Meat	0.6	345	50	3
E coli	Meat	2.5	400	NA	4
S aureus	meat	3	500	50	4
l monocytogenes	meat	2.17	414	25	2
L monocytogenes	pork	1.89–4.17	414	25	5
L monocytogenes	pork	0.37–0.63	414	50	3
C. botulinum Type E (Beluga)	Crab meat	3.38	758	35	6
C. botulinum Type E (Alaska)	Crab meat	1.76	827	35	6
Liquid medium					
Salmonella Typhimurium	milk	3.0	350	N/A	1
E coli	Milk	1.0	400	50	7
E. coli 0157H7	Milk	3.0	400	50	4
S aureus	Milk	2.5	500	50	4
L. monocytogenes	Milk	3.0	375	N/A	1
L. inocua	Eggs	3.0	450	20	8
C. botulinum Type E (Alaska)	Buffer	2.64	827	35	6
Saccharomyces cerevisiae	Orange Juice	0.97	400	37	9
Saccharomyces cerevisiae	Apple juice	0.28	450	40	9

References: 1. Patterson ME, Quinn M, Simpson R, and Gilmour A. (1955) J. Food Prot. 58:524; 2. Ananth V, Dickson JS, Olson DG, and Murano EA (1998) J. Food Prot. 61:1649; 3. Kalchayanand N, Sikes A, Dunne CP, and Ray B. (1998) J. Food Prot. 61:425; 4. Patterson MF and Kilpatrick DJ. (1998) J. Food Prot. 61:432; 5. Murano EA, Murano PS, Brennan RE, Shenoy K and Moriera R. (1999). J. Food Prot. 62:480; 6. Reddy NR, Solomon HM, Fingerhut G, Balasubramanian VM, and Rodelhamel EJ. (1999) NCFST, Sumit-Argo, IL. 7. Gervilla R, Capellas M, Farragut V and Guamis B. (1997) J. Food Prot. 60:33; 8. Ponce E, Pla R, Mor-Mur M, Gervilla R and Guamis B. (1998) J. Food prot. 61:119; 9. Zook CD, Parish ME, Braddock RJ and Balaban MO. (1999) J. Food Sci. 64:533

cooled rapidly through a heat exchanger after exiting the throttling valve. This process is mild enough to permit the inactivation of 6 log of *Listeria inocua* in liquid whole eggs without coagulating the protein.

9.4.4.3 High-Pressure Sterilization

Because high pressure alone is inadequate to inactivate spores, it is necessary to raise the temperature during the high-pressure treatment to above 121.1°C. For HHP, the temperature rise that occurs with pressurization can be utilized to advantage in sterilization processes. Temperature rise on subjecting water to HHP range from 2.8°C to 4.4°C/100 Mpa pressure. The higher the initial temperature, the larger the temperature rise with increase in pressure. Foods with low density and specific heat such as oils will have higher temperature rise than water, while most fluid foods that are high in moisture content such as fruit juices have similar temperature rise as water. A food product with a temperature rise of 3°C/100 Mpa pressure will exhibit a temperature rise of 21°C going from one atmosphere to 700 Mpa. Thus, if the product initial temperature prior to pressurization is 100°C, temperature after pressurization will be 121°C, a temperature that is lethal to spore-forming microorganisms. Of interest is inactivation of heat tolerant spores of *Clostridium botulinum*. Type 62A in pH 7 buffer at 75°C and 689 MPa has a D value of 10.59 minutes. Because sterilization will require at least 12D of *C. botulinum*, a 132-minute exposure time will be required. There are currently no data on high-pressure inactivation of spores at T > 108°C. An advantage of HHP sterilization is that upon release in pressure, the temperature instantaneously drops to the temperature at the start of pressurization. The same principles may be used in continuous flow sterilization. If a fluid is heated to 80°C while under pressure at 241 Mpa, release of that pressure to atmospheric pressure will result in a temperature rise of about 55°C resulting in an exit fluid temperature of 135°C after pressure release. At this temperature, the D value for *C. botulinum* is about 0.7 second, therefore a 9-second hold will be adequate to achieve 12D inactivation of *C. botulinum*.

Sterilization by high pressure is still not an approved process for low-acid products by U.S. regulatory agencies.

9.4.5 Sterilization of Fluids Containing Discreet Particulates

Discreet particulates within a flowing fluid will be heated by heat transfer from the suspending fluid. Thus, heat transfer coefficients between the particle and the fluid play a significant role in the rate of heating. Simplified equations for heat transfer will not be applicable because the fluid temperature is not constant as the mixture passes through the heater, and temperature of fluid in an unheated holding tube may not be constant because of heat exchange between the fluid and the suspended particles. Taking a conservative approach of ignoring the heat absorbed by the particles in the heaters can result in a significant over-processing, particularly if the suspended particles have less than 0.5 cm as the thickness of the dimension with the largest area for heat transfer. Finite element or finite difference methods for solving the heat transfer equations with appropriate substitutions for changes in the boundary conditions when they occur is the only correct method to determine the lethal effect of heat in the holding tube. Residence time distribution of particles must also be considered, and as in the case of fluids in turbulent flow, the use of a probability distribution function for the residence time in the finite difference or finite element methods will allow calculation of an integrated sterility.

A discussion of the finite difference methods for evaluating heat transfer is beyond the scope of this textbook.

9.5 STERILIZING VALUE OF PROCESSES EXPRESSED AS F_0

The sterilizing value of a process expressed as the number of decimal reduction of a specific biological entity has been discussed in the section "Selection of Inoculation Levels in Inoculated Packs." When comparing various processes for their lethal effect, it is sometimes more convenient to express the lethality as an equivalent time of processing at a reference temperature. The term, the F_0, is a reference process lethality expressed as an equivalent time of processing at 121.1°C calculated using a z value of 10°C (18°F). If the z value used in the determination of F is other than 10°C, the z value is indicated as a superscript, F_0^z.

For constant temperature processes, the F_0 is obtained by calculating L in Equation (9.21), and multiplying L by the heating time at T. $F_0 = L t$.

For processes where product is subjected to a changing temperature, such as the heating or cooling stage in a canning process, a lethality is calculated over the length of the process. The process is separated into small time increments, Δt. The average temperature at each time increment is used to calculate L using Equation (9.21). The F_0 will be $3L_T\Delta t$. The z value of E used in Equation (9.21) to calculate L is 10 °C to determine F_0. If the F_0 value is to be used later to express lethality as the number of decimal reduction of a particular biological entity, the appropriate z value for that entity has to be used to calculate F_0^z.

9.6 THERMAL PROCESS CALCULATIONS FOR CANNED FOODS

When sterilizing foods contained in sealed containers, the internal temperature changes with time of heating. Lethality of the heating process may be calculated using the "general method," which is a graphical integration of the lethality-time curve, or by "formula methods," which utilize previously calculated tabular values of parameters in an equation for the required process time or process lethality. Problems in thermal process calculation can either be (I) the determination of process time and temperature to achieve a designed lethality or (II) the evaluation of a process time and temperature. These are referred to by some authors as Type I or Type II problems. Fundamental in the evaluation of thermal process schedules is heat transfer data, an equation or experimental data for temperature in the container as a function of time. Lethality may be expressed as the value achieved in a single point (i.e., the slowest heating point in the container) or it may be expressed as an integrated lethality. For microbial inactivation, where microbial numbers are nil at regions within the container nearest the wall, lethality at a single point is adequate and results in the safest process schedule, or a most conservative estimate of the probability of spoilage. However, when evaluating quality factor degradation, an integrated lethality must be determined because a finite level of the factor in question exists at all points in the container.

9.6.1 The General Method

Process lethality is calculated by graphical integration of the lethality value (Equation 9.21) using time-temperature data for the process.

$$F_0{}^z = \int_0^t L_t dt$$

Equation (9.14) may also be used to determine process lethality. However, because D is not constant, the sterilizing value expressed as the number of decimal reductions of microorganisms will be integrated over the process time, using the value of D at various temperatures in the process.

$$S = \int_0^t \frac{dt}{D_t}$$

If a process schedule is to be determined, the heating and cooling curves are used to determine the lethality curve, which is then graphically integrated to obtain either F_0 or S. The process lethality must equal the specified values for F_0 or S for the process to be adequate. If the lethality value differs from the specified, the heating time is scaled back, a cooling curve parallel to the original is drawn from the scaled back heating time, and the area is recalculated. The process is repeated until the specified and calculated values matches. Evaluation of the process lethality is done directly on the time-temperature data. Simpson's rule may be used for integration. From the section "Graphical Integration (Chapter 1 Section 1.15)" Simpson's rule is applied to thermal process determination by the general method as follows: Select time increments δt such that at the end of process time t, $t/\delta t$ will be an even number. Using i as the increment index, with $I = 0$ at $t = 0$, $I = 1$ at $t = \delta t$, $i = 2$ at $t = 2\,\delta t$; $i = 3$ at $t = 3\delta t$ ... and so forth.

$$A = \left(\frac{\delta t}{3}\right)[L_0 + 4L_1 + 2L_2 + +4L_3 + 2L_4 + \ldots\ldots 2L_{i-2} + 4L_{i-1} + L_t]$$

The area under the cooling curve may be evaluated separately from that under the heating curve.

Example 9.16. The following data represent the temperature at the slowest heating point in a canned food processed at a retort temperature of 250°F. Calculate the F_0 value for this process. What will be the required process time to have a lethality equal to an F_0 of 9 minutes.

Time (min)	Temp. (°F)	Time (min)	Temp. (°F)
0	140	55	238
5	140	60	241
10	140	65	235
15	140	70	245
20	163	75	246.3
25	185	80	247.3 (cool)
30	201	85	247.0
35	213	90	245.2
40	224	95	223.5
45	229.4	100	175
50	234.5	105	153

Solution:

From $t = 0$ to $t = 80$ minutes, $\delta t = 5$ minutes will give 16 increments. L is calculated using Equation (9.21). $T_0 = 250$°F. The values of the lethality are as follows:

$L_0 = 10^{-0.05555(110)} = 0 = L_1 = L_2 = L_3$
$L_4 = 10^{-0.05555(87)} = 1.5 \times 10^{-5}$
$L_5 = 10^{-0.05555(65)} = 2.45 \times 10^{-4}$
$L_6 = 10^{-0.05555(54)} = 0.001001$
$L_7 = 10^{-0.05555(32)} = 0.016688$
$L_8 = 10^{-0.05555(26)} = 0.035938$
$L_9 = 10^{-0.05555(20.6)} = 0.071725$
$L_{10} = 10^{-0.05555(15.5)} = 0.1377$
$L_{11} = 10^{-0.05555(12)} = 0.2155$

$L_{12} = 10^{-0.05555(9)} = 0.3163$

$L_{13} = 10^{-0.05555(6.5)} = 0.4354$

$L_{14} = 10^{-0.05555(5)} = 0.5275$

$L_{15} = 10^{-0.05555(3.7)} = 0.6229$

$L_{16} = 10^{-0.05555(2.7)} = 0.7079$

The area under the heating curve will be

$$4(L_1 + L_3 + L_5 + L_7 + L_9 + L_{11} + L_{13} + L_{15}) = 5.4498$$
$$2(L_2 + L_4 + L_6 + L_8 + L_{10} + L_{12} + L_{14}) = 2.0363$$
$$A = (5/3)(0 + 5.4498 + 2.0363 + 0.7079) = 13.57$$

The area under the cooling curve will be

$L_0 = 10^{-0.05555(2.7)} = 0.70797$

$L_1 = 10^{-0.05555(3)} = 0.61829$

$L_2 = 10^{-0.05555(4.8)} = 0.54187$

$L_3 = 10^{-0.05555(26.5)} = 0.03371$

$L_4 = 10^{-0.05555(75)} = 0$

$L_5 = 0$

$4(L_1 + L_3 + L_5) = 2.8600$

$2(L_2 + L_4 + L_6) = 1.08374$

$A = (5/3)(0.70797 + 2.8600 + 1.0837 + 0) = 7.753$

Total area $= 13.57 + 7.753 = 21.32$

The cooling curve contributed about one-third of the total lethality in this example.

The calculated total lethality is much higher than the specified F_0 of 9 minutes. Thus, it will be necessary to reduce the heating time. Reduction of heating time will result in a reduction of the can temperature prior to cooling. Let the heating time be equal to 60 minutes. There are now only 12 area increments. The can temperature at 60 minutes of heating is 241°F. The cooling curve will start at 241°F. The cooling temperature will be parallel to the cooling curve of the original process (Fig. 9.12). Using Simpson's rule on the new heating and cooling curve, Table 9.10 may be constructed.

The area under the heating curve is $(1.17479 + 0.35019 + 0.316228)(5/3) = 1.8420(5/3) = 3.07$.

The area under the cooling curve is $(1.114881 + 0.430887 + 0.316228)(5/3) = 1.96199(5/3) = 3.27$.

The total area $= F_0 = 6.3$.

Ten more minutes of heating will add approximately 3 minutes to the F_0 because one minute of heating at 241°F is equivalent to 0.31 minutes at 250°F. Thus the heating time will be 70 minutes to give an F_0 of approximately 9 minutes.

This example underscores the importance of the contribution of the cooling part of the process to the total lethal value of the process. The relative contribution of the cooling curve to total lethality increases when product characteristics or processing conditions result in a slow rate of cooling.

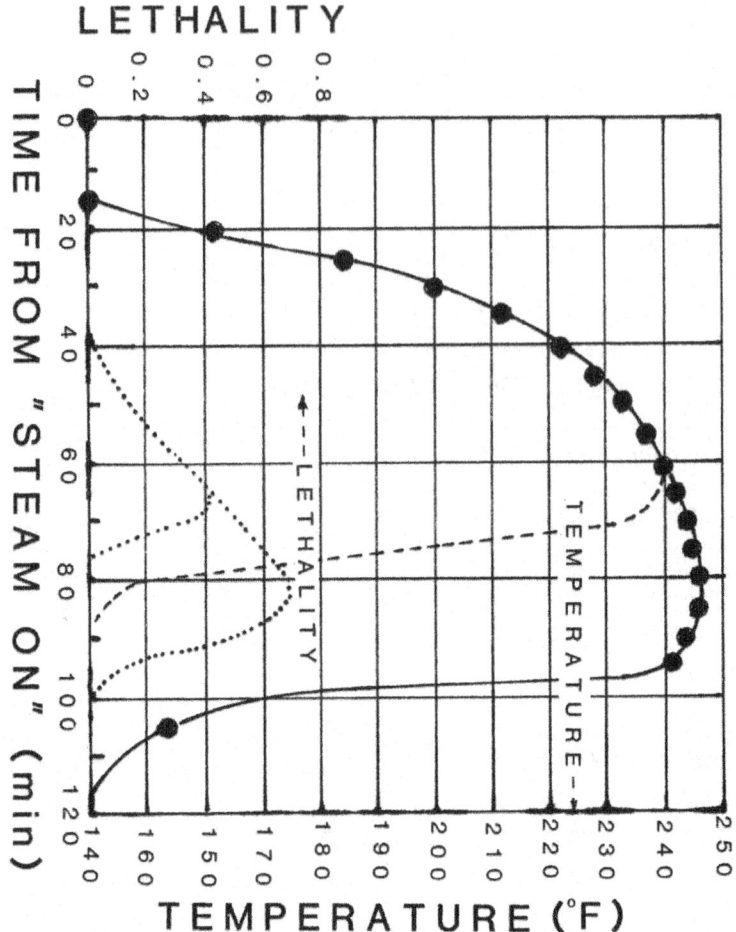

Figure 9.12 Graph showing retort temperature, lethality of the heat treatment, and the procedure for adjusting the processing time to obtain the specified process lethality.

9.6.2 Heat Transfer Equations and Time-Temperature Curves for Canned Foods

In the section "Heating of Solids Having Infinite Thermal Conductivity" Chapter 7 (Section 7.6.1), the transient temperature of a solid having an infinite thermal conductivity was derived. Equation (7.84) in Section 7.6.1 may be used to represent the temperature at a single point in a container. If the point considered is at the interior of the container, a time lag will exist from the start of heating to the time temperature at that point actually changes. The following symbols are used for thermal process heat penetration parameters.

I = initial temperature difference = $(T_r - T_0)$; T_r = heating medium temperature = retort temperature; T_0 = can temperature at the start of the heating process.

g = unaccomplished temperature difference at the end of a specified heating time = $(T_r - T)$; T = temperature at the point considered at any time, t, during the heating process.

Table 9.10 Lethality of a Process Calculated Using the General Method and Simpson's Rule for Graphical Integration

Time (min)	Temp (°F)	L	4L	2L	L
0	140	0	—	—	0
5	140	0	0	—	—
10	140	0	—	0	—
15	140	0	0	—	—
20	163	0	—	0	—
25	185	0.000245	0.000980	—	—
30	201	0.001896	—	0.003791	—
35	213	0.008799	0.035214	—	—
40	244	0.035938	—	0.071858	—
45	229.4	0.071706	0.286824	—	—
50	234.5	0.137686	—	0.275371	—
55	238	0.215443	0.861774	—	—
60	241	0.316228	—	—	0.316228
Sum			1.17479	0.351019	0.316228
60	241	0.316228	—	—	0.316228
65	240	0.278256	1.113024	—	—
70	238	0.215443	—	0.430887	—
75	190	0.000464	0.001857	—	—
80	149	0.052079	—	0	—
85	142	0	0	—	—
90	140	0	—	—	0
Sum			1.114881	0.430887	0.316228

j = lag factor, also known as the intercept index for the linear semi-logarithmic temperature versus time plot of the heating curve. jh refers to the heating curve and jc refers to the cooling curve.

f = the slope index of the linear semi-logarithmic temperature versus time plot of the heating curve. If the heating curve consists of n line segments, $f_i (i = 1$ to n) is used to represent the slope index of each line segment with 1 representing the first line segment from the start of the heating process. fc refers to the cooling curve.

Expressing Equation (7.84), Chapter 7, in terms of the above parameters:

$$\log\left(\frac{g}{jI}\right) = -\frac{t}{f_h} \tag{9.33}$$

$$\log\left(\frac{(T_r - T)}{jI}\right) = -\frac{t}{f_h} \tag{9.34}$$

$$\log(T_r - T) = \log(jI) - \frac{t}{f_h} \tag{9.35}$$

$$T = T_r - jI[10]^{-t/f_h} \tag{9.36}$$

Equation (9.35) shows that a semi-log plot of the unaccomplished temperature difference $(T_r - T)$ against time will have a slope of $-1/f_h$ and an intercept of log (jI). The latter is the reason that j is called the intercept index. j is also called the lag factor because the higher the value of j the longer it will take for the temperature at the point being monitored, to respond to a sudden change in the heating medium temperature.

The cooling curve expressed in a form similar to Equation (9.35) with T_c as the cooling water temperature is:

$$\log(T - T_c) = \log(j_c I_c) - \frac{t_c}{f_c} \tag{9.37}$$

where $I_c = (T_g - T_c)$; T_g = temperature at the end of the heating process = $(T_r - g)$.

Equation (9.37) shows that a semi-logarithmic plot of $\log(T - T_c)$ versus time t_c (with $t_c = 0$ at the start of cooling) will be linear and the slope will be $-1/f_c$. The temperature at any time during the cooling process, is

$$T = T_c + j_c I_c [10]^{-t_c/f_c} \tag{9.38}$$

Equation (9.38) represents only part of the cooling curve, and is not the critical part that contributes significantly to the total lethality. The initial segment just after the introduction of cooling water is nonlinear and accounts for most of the lethality contributed by the cooling curve. Thus, a mathematical expression which correctly fits the curved segment of the temperature change on cooling will be essential to accurate prediction of the total process lethality.

The initial curved segment of cooling curves has been represented as hyperbolic, circular, and trigonometric functions. A key parameter in any case, is the intersection of the curved and linear segments of the cooling curve. The linear segment represented by Equation (9.38) can be easily constructed from f_c and j_c and temperature at any time within this segment can be calculated easily using Equation (9.38) from the point of intersection of the curved and linear segments. Hayakawa (Food Technol. 24:1407, 1970) discussed the construction of the curved segment of the cooling curve using a trigonometric function. The equations which are valid for for $1 \le j_c \le 3$ are

$$T = T_c + [T_g - T_c]^{\cos(Bt_c)} \tag{9.39}$$

$$B = \frac{1}{t_L} \left[\arccos \left[\frac{\log(j_c I_c) - t_L/f_c}{\log(I_c)} \right] \right] \tag{9.40}$$

The cosine function in Equation (9.39) uses the value of the angle in radians as the function argument. The arccos function in Equation (9.40) returns the value of the angle in radians.

t_L is the time when the curved and linear segments of the cooling curve intersect. t_L may be derived from the intersection of a horizontal line drawn from the temperature at the initiation of cooling and the linear segment of the cooling curve represented by Equation (9.38). At the intersection, $(T - T_c) = I_c$, and $t_c = t_L$. Substituting in Equation (9.38), solving for t_L and introducing a factor k to compensate for the curvature in the cooling curve:

$$t_L = f_c \log \left(\frac{j_c}{k} \right) \tag{9.41}$$

The factor k in Equation (9.41) may be determined from the actual cooling curves, when plotting heat penetration data. $k = 0.95$ has been observed to be common in experimental cooling curves for canned foods. Equation (9.41) represents cooling data for canned foods better than the equation for t_L originally given by Hayakawa (1970).

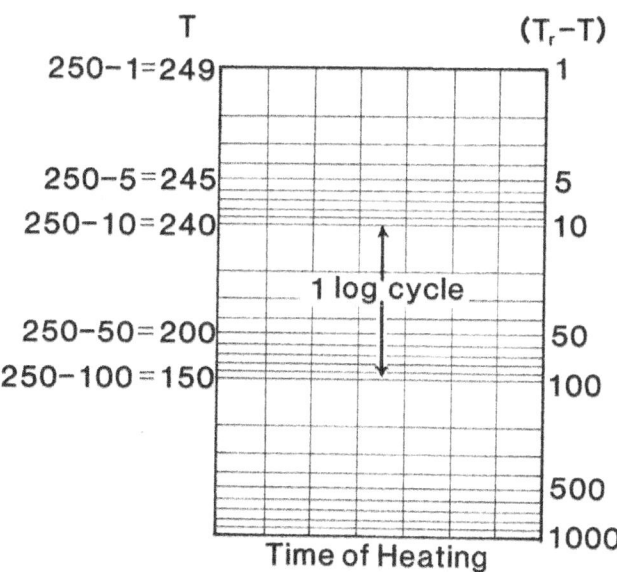

Figure 9.13 Diagram showing how the axis of semi-logarithmic graphing paper is marked for plotting heat penetration data. Retort temperature = 250°F.

9.6.3   Plotting Heat Penetration Data

Raw time-temperature data may be plotted directly on semi-log graphing paper to produce the linear plot needed to determine f_h and j, by rotating the paper 180 degrees. The numbers on the graph that mark the logarithmic scale are marked as $(T_r - T)$ and the can temperature is marked on the opposite side of the graphing paper. Figures 9.13 and 9.14 show how the can temperature is marked on 3-cycle semi-log graphing paper for retort temperature of 250°F and 240°F, respectively.

9.6.3.1   Determination of f_h and j

Can temperature is plotted on the modified graphing paper and a straight line is drawn connecting as much of the experimental data points as possible. There will be an initial curvature in the curve, but the straight line is drawn all the way to t = 0.

In any simulator used for heat penetration data collection, the retort temperature does not immediately reach the designated processing temperature. The time from introduction of steam to when processing temperature (T_r) is reached is the retort come-up time, $t_{come-up}$. Sixty percent of the retort come-up time is assumed to have no heating value, therefore heating starts from a pseudo-initial time t_{pi} which is $0.6t_{come-up}$. The pseudo-initial temperature, T_{pi}, is the intersection of the line drawn through the points and the line representing t = t_{pi}. $(T_r - T_{pi}) = jI$. The intercept index, j, is

$$j = \frac{T_r - T_{pi}}{T_r - T_0} = \frac{jI}{I} \tag{9.42}$$

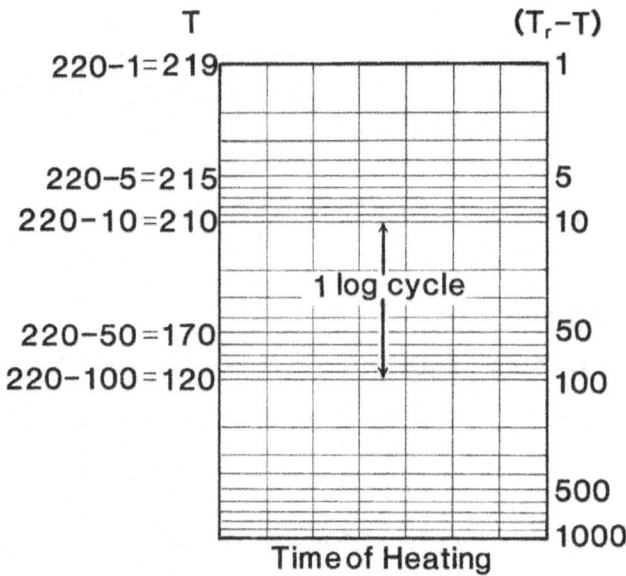

Figure 9.14   Diagram showing how the axis of semi-logarithmic graphing paper is marked for plotting heat penetration data. Retort temperature = 220°F.

The slope index, f_h, is the time for the linear section of the heating curve to traverse one log cycle on the graph.

9.6.3.2   Determination of f_c and j_c

Using 180-degree rotated semi-log graphing paper, the cooling curve is plotted on the paper with the marked side labeled $(T - T_c)$ and the opposite side labeled the can temperature, T. The abscissa is the cooling time, t_c. At $t_c = 0$, steam is shut off and cooling water is introduced. The retort is assumed to reach cooling water temperature immediately, therefore the intercept of the cooling curve is evaluated at $t_c = 0$. $j_c I_c$ is the intercept of the line drawn between the data points and $t_c = 0$. f_c is the slope index of the cooling curve and is the time for the linear section of the curve to traverse one log cycle. If the cooling data does not complete one log cycle within the graph, f_c may be evaluated as the negative reciprocal of the slope of the line. Let $(T_1 - T_c)$ and $(T_2 - T_c)$ represent the unaccomplished temperature difference at t_{c1} and t_{c2}, respectively:

$$f_c = -\frac{t_{c1} - t_{c2}}{\log(T_1 - T_c) - \log(T_2 - T_c)} \tag{9.43}$$

Example 9.17.   Determine the heat penetration parameters, j, f_h, j_c, f_c for a canned food that exhibited the following heating data when processed in a retort at 250°F. It took 3 minutes from introduction of steam to the time retort reached 250°F. Cooling water temperature is 60°F.

Time (min)	Temp. (°F)	Time (min)	Temp. (°F)
0	180	30	245
5	190	30 (cool)	245
10	210	35	235
15	225	40	175
20	235	45	130
25	241	50	101

Calculate the temperature at various times during heating and cooling of this product processed at $T_r = 251°F$ if $T_0 = 160°F$, and $t = 35$ minutes from steam introduction. $T_c = 70°F$. Retort come-up time $= 3$ minutes. Calculate the F_0 of this process using the general method and Simpson's rule for graphical integration of the lethality.

Solution:

The heating and cooling curves are plotted in Figures 9.15 and 9.16. The value for $f_h = 22$ minutes and how it is determined is shown in Fig. 9.14. $T_{pi} = 152$ minutes is read from the intersection of the line through the data points and $t_{pi} = 0.6 t_{come-up} = 1.8$ minutes. jI can be read by projecting the intersection to the axis labeled $(T_r - T)$. $jI = 98°F$, or by subtracting T_{pi} from T_r. $I = (T_r - T_0) = 250 - 180 = 70°F$. $j = 98/70 = 1.40$.

The cooling curve is shown in Fig. 9.16. The intercept of the linear portion of the curve with $t_c = 0$ is projected to the side marked $(T - T_c)$ and $j_c I_c$ is read to be $333°F$. The initial temperature difference for cooling, $I_c = 245 - 60 = 185°F$. Thus $j_c = j_c I_c / I_c = 333/185 = 1.8$. The slope index for the cooling curve is calculated from the points $(t_c = 0; (T - T_c) = 333)$ and $(t_c = 20; (T - T_c) = 41)$.

$$f_c = -\frac{0 - 20}{\log(333) - \log(41)} = 22 \, min$$

The curved section of the cooling curve intersects the linear section at $t_c = 6$ minutes. Thus, $t_L = 6$ minutes, and for this cooling curve, k in Equation (9.41) is

$$k = \frac{j_c}{[10]^{t_L/f_c}} = \frac{1.8}{[10]^{6/22}} = 0.96$$

The temperature during heating and cooling at the same point within a similar-sized container can be determined for any retort temperature, initial can temperature, or cooling water temperature once the heating and cooling curve parameters are known. Equation (9.36) is used to determine the temperature during heating. The initial heating period is assumed to be constant at T_0 until calculated values for T exceeds T_0. This assumption dose not introduce any errors in the calculation of the lethality of heat received, since at this low temperature lethality is negligible. Exceptions are rare and apply to cases where a very high initial temperature exists.

As previously discussed, when heating is carried out under conditions where a come-up time exists, the first 60% of the come-up time is assumed to have no heating value; therefore, the time variable in Equation (9.36) should be zero when $t = t_{come-up}$. If t used in Equation (9.36) is based on the time after "steam on," then 60% of $t_{come-up}$ must be subtracted from it when used in Equation (9.36).

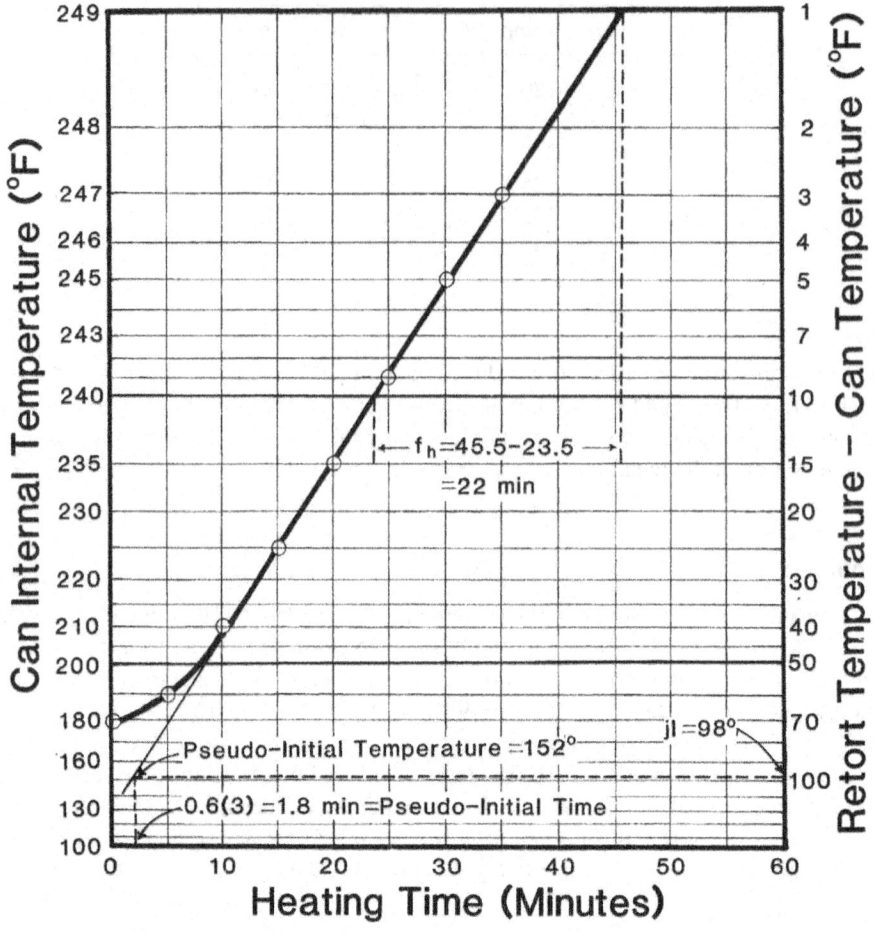

Figure 9.15   A plot of the heating curve showing how the heating curve parameters f_h and j are determined.

Let the exponential term in Equation (9.36) = A.

$$A = 10^{-\{[t-0.6(3)]/fh\}} = 10^{-[(t-1.8)/22]}$$
$$T = T_r - jIA = 251 - 1.4(251 - 160)A$$

Calculated temperatures are shown in Table 9.11.

During cooling, the curved portion of the cooling curve is constructed with the temperatures calculated using Equations (9.39), (9.40), and (9.41). The time when the linear and curved portions of the cooling curve intersect is calculated using Equation (9.39) and the previously calculated value of k of 0.96.

$$t_L = 22 \log\left(\frac{1.8}{0.96}\right) = 6.1\,\text{min}$$

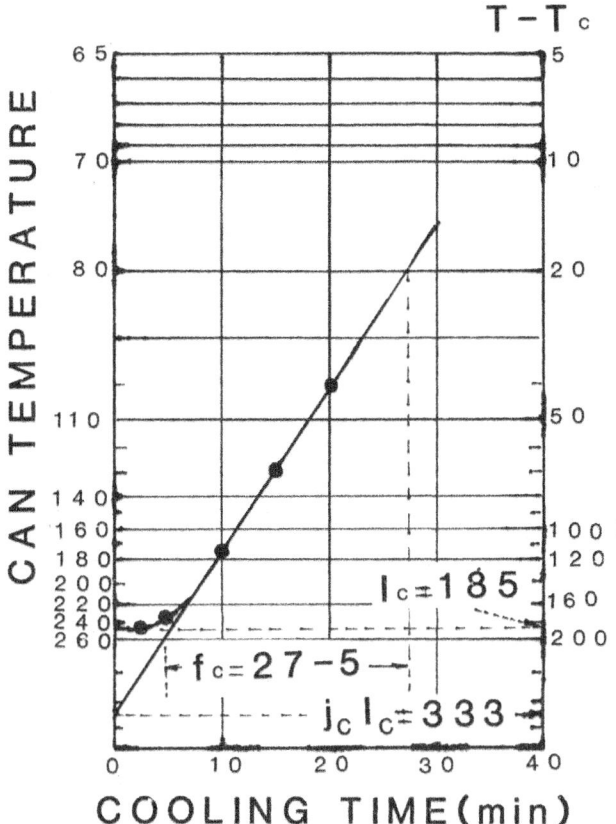

Figure 9.16   A plot of the cooling curve showing how the cooling curve parameters f_c and j_c are obtained.

Solving for parameter B in Equation (9.38):

$$B = \frac{1}{6.1}\left[\arccos\left[\frac{\ln(1.8)(174.3) - (6.1/22)}{\ln(174.3)}\right]\right]$$

$$= \frac{1}{6.1}\arccos(0.99018) = \frac{1}{6.1}\left[(8.034°)\left[\frac{2\pi \text{ rad}}{360°}\right]\right] = 0.02337$$

The temperature is then calculated using Equation (9.39):

$$T = 70 + (174.3)^{\cos[(0.02337)(tc)]}$$

Let: $E = \cos[(0.0296)(t_c)]$;

$$T = 70 + (174.3)^{E}$$

For the linear portion of the cooling curve, Equation (9.38) is used. The temperature calculated using Equation (9.30) will represent the actual can temperature only when $t_c > t_L$ because t_L is the

Table 9.11 Time and temperature during heating for Example 9.17

Time (min)	A	$T_r - jla$	Temp (°F)
0	1.207	(97)	160
2	0.979	(126)	160
4	0.794	(150)	160
6	0.644	168	168
8	0.522	184	184
10	0.423	197	197
12	0.344	207	207
14	0.279	215.5	215.5
16	0.226	222	222
18	0.184	227.6	227.6
20	0.149	232	232
22	0.121	235.6	235.6
24	0.098	238.5	238.5
26	0.079	241.6	241.6
28	0.064	242.8	242.8
30	0.052	244.3	244.3

intersection of the curved and linear portions of the cooling curve.

Equation (9.38): $T = 70 + 1.8(174.3)(10)^{(-tc/22)}$

Let $A = 10^{(-tc/22)}$; $T = 70 + (313.74)^A$

The calculated temperatures for cooling, and the lethality and area elements for area calculation using Simpson's rule, are shown in Tables 9.12 and 9.13, respectively.

Area $= F_0 = (2/3)(1.7904 + 5.5823 + 0.9698) = 5.6$ minutes

Table 9.12 Time and temperature during cooling for Example 9.17

Time (t_c)	E	A	(Temp °F) $70 + 174.3^E$	(Temp °F) $70 + (313.74)^A$
0	1	1	244.3	(383.7)
2	0.998	0.811	243.3	(324.5)
4	0.993	0.658	240.4	(276.4)
6	0.984	0.534	235.7	(237.3)
8		0.433		199.4
10		0.351		180.2
12		0.285		159.4
14		0.231		142.5
16		0.187		128.8

Table 9.13   Time, temperature, and Simpson's Rule factors for area calculations for Example 9.17

Time (min)	Temp (°F)	L	2L	4L	L
Heating					
0	160	0			
2	160	0			0
4	160	0		0	
6	169	0	0		
8	184	0.0002		0.0008	
10	197	0.0011	0.0022		
12	207	0.0042		0.0168	
14	215.5	0.0121	0.0242		
16	222.2	0.0284		0.1136	
18	227.6	0.0571	0.114		
20	232	0.1005		0.402	
22	235.6	0.1588	0.3177		
24	238.5	0.2304		0.9215	
26	241.8	0.3414	0.6829		
28	242.8	0.3977		1.5908	
30	244.3	0.4899			0.4849
Cooling					
0	244.3	0.4849			0.4849
2	243.3	0.4244		1.6976	
4	240.4	0.2929	0.5858		
6	235.7	0.1605		0.6421	
8	199	0.0015	0.0030		
10	180	0.0001		0.0004	
12	159	0	0		
14	142	0		0	
16	129	0			0
Sums			1.7297	5.3866	0.9698

The lethality of the process expressed as an equivalent heating time at 250°F is 5.3 minutes. The complete procedure used in the evaluation of the lethality of a heating process in this example is the general method. The general method may be used on data obtained experimentally or on time-temperature data reconstructed from values of heating and cooling curve parameters calculated from experimental data.

Use of a spreadsheet will greatly facilitate thermal process calculations using the general method. Time increments used in the lethality calculations can be made very small to increase the accuracy of calculated process lethality.

9.6.4   Formula Methods for Thermal Process Evaluation

Formula methods are based on tabulated values for lethality expressed as the parameter f_h/U, which have been previously calculated for various conditions of heating and cooling when unaccomplished temperature difference is expressed as the parameter "g." Two methods will be presented in this section:

Stumbo's (1973) and Hayakawa's (1970). The purpose of presenting both methods is not to compare their accuracy but to provide a means for selecting the most convenient method to use for certain conditions.

Stumbo's f_h/U versus g tables combine lethalities of both heating and cooling. A major assumption used in the calculation of lethality is that $f_h = f_c$. When actual conditions do not meet this assumption, Stumbo recommends using the general method. Hayakawa (1970) presented lethality of the heating and cooling stages in the process in separate tables thus allowing the use of his method even under different rates for heating and cooling. Hayakawa's tabular values allow substitutions for different values of z, simplifying calculation of specific F_0^z values for different z.

In this section, process calculations for products that exhibit simple heating curves will be discussed. Calculations for broken heating curves will be discussed in the section "Broken Heating Curves."

Both formula methods are based on the equation for the heating curve (Eq. 9.33). Let g = unaccomplished temperature difference, $(T_r - T)$ at the termination of the heating period; and B_b = the heating time at that point. B_b is the scheduled sterilization process. For products with simple heating curve, Equation (9.33) becomes:

$$B_b = f_h[\log(jI) - \log(g)] \tag{9.44}$$

$B_b = t - 0.6t_{come-up}$. t is time evaluated from steam introduction into the retort. In practice, B_b is timed from the point where the retort temperature reaches the processing temperature to avoid the probability of errors arising from the operator having to correct for the come-up time.

g is obtained from the tables, using a specified F_0 and z value. Stumbo's tables are simpler to use for thermal process determinations. Hayakawa's tables will require an iterative procedure involving an assumption of the value of g, calculating the F_0, and calculations are repeated until the calculated matches the specified F_0.

The following parameters are used in the formula methods:

$$F_i = [10]^{250-T/z} \tag{9.45}$$

$U = F_0F_i$ = time at T_r equivalent to F_0.

9.6.4.1 Stumbo's Procedure

Stumbo's tabulated f_h/U versus g with j_c as a parameter. j_c strongly influences the contribution of the cooling part of the process to the total lethality as discussed in the section "Sterilizing Value of Processes Expressed as F_0." In general, j_c values are higher than j. In the absence of j_c, j may be used and the error will be toward a longer process time or the safe side relative to spoilage. Condensed f_h/U versus g tables for z values from 14 to 22 and for z from 30 to 45 are shown in Tables 9.14 and 9.15. Table 9.14 is used for microbial inactivation and Table 9.15 is used for nutrient degradation. It is possible to interpolate between values in the table for other z values. Thermal process determinations can easily be made from specified F_0 values and product heat penetration parameters by solving for f_h/U determining the corresponding value for g and solving for B_b using Equation (9.44).

9.6.4.2 Hayakawa's Procedure

Hayakawa's tables are shown in Table 9.16 for lethality of the heating part of the process and in Tables 9.17, 9.18, 9.19, 9.20, and 9.21 for lethality of the cooling part of the process. The tables are based on a z value of 20°F. The parameter g/K_s with K_s defined as $K_s = z/20$ is tabulated against U/f_h. The latter is the reciprocal of Stumbo's f_h/U.

Table 9.14 f_h/U vs. g Table Used for Thermal Process Calculations by Stumbo's Procedure

f_h/U	$z=14$	$\Delta g/\Delta j$	$z=18$	$\Delta g/\Delta j$	$z=22$	$\Delta g/\Delta j$
0.2	0.000091	0.0000118	0.0000509	0.0000168	0.0000616	0.0000226
0.3	0.00175	0.00059	0.0024	0.00066	0.00282	0.00106
0.4	0.0122	0.0038	0.0162	0.0047	0.020	0.0067
0.5	0.0396	0.0111	0.0506	0.0159	0.065	0.0197
0.6	0.0876	0.0224	0.109	0.036	0.143	0.040
0.7	0.155	0.036	0.189	0.066	0.25	0.069
0.8	0.238	0.053	0.287	0.103	0.38	0.105
0.9	0.334	0.07	0.400	0.145	0.527	0.147
1.0	0.438	0.009	0.523	0.192	0.685	0.196
2.0	1.56	0.37	1.93	0.68	2.41	0.83
3.0	2.53	0.70	3.26	1.05	3.98	1.44
4.0	3.33	1.03	4.41	1.34	5.33	1.97
5.0	4.02	1.32	5.40	1.59	6.51	2.39
6.0	4.63	1.56	6.25	1.82	7.53	2.75
7.0	5.17	1.77	7.00	2.05	8.44	3.06
8.0	5.67	1.95	7.66	2.27	9.26	3.32
9.0	6.13	2.09	8.25	2.48	10.00	3.55
10	6.55	2.22	8.78	2.69	10.67	3.77
15	8.29	2.68	10.88	3.57	13.40	4.60
20	9.63	2.96	12.40	4.28	15.30	5.50
25	10.7	3.18	13.60	4.80	16.9	6.10
30	11.6	3.37	14.60	5.30	18.2	6.70
35	12.4	3.50	15.50	5.70	19.3	7.20
40	13.1	3.70	16.30	6.00	20.3	7.60
45	13.7	3.80	17.00	6.20	21.1	8.0
50	14.2	4.00	17.7	6.40	21.9	8.3
60	15.1	4.3	18.9	6.80	23.2	9.0
70	15.9	4.5	19.9	7.10	24.3	9.5
80	16.5	4.8	20.8	7.30	25.3	9.8
90	17.1	5.0	21.6	7.60	26.2	10.1
100	17.6	5.2	22.3	7.80	27.0	10.4
150	19.5	6.1	25.2	8.40	30.3	11.4
200	20.8	6.7	27.1	9.10	32.7	12.1

Source: Based on f_h/U vs. g tables in Stumbo, C. R. 1973. *Thermobacteriology in Food Processing,* 2nd ed. Academic Press, New York.
To use for values of j other than I, solve for g_j as follows:

$$g_j = g_{j-1} + (j-1)\left[\frac{\Delta g}{\Delta j}\right]$$

Example: g for (f_h/U) = 20 and j = 1.4 and z = 18: $g_{j=1.4}$ = 12.4 + (0.4)(4.28) = 14.11.
Reprinted from: Toledo, R. T. 1980. Fundamentals of Food Process Engineering, ist ed. AVI Pub. Co. Westport, CT.

In the tables for lethality of the cooling part of the process, j_c is used as a parameter. The tabular entry in the table for lethality of the cooling curve is $(T_g - T_c)/K_s = (T_r - g - T_c)/K_s = I_c/K_s$. U in the lethality table for the cooling curve is based on T_g, which must be converted to U at T_r before adding to the U obtained from tabular values for heating. The conversion from U' which is the value of U at T_g to the U at T_r is done using Equation (9.46):

$$U = U = (10)^{-g/z} \tag{9.46}$$

Table 9.15 f_h/U vs. g Table Used for Thermal Process Calculation by Stumbo's Procedure

f_h/U	$z = 60$		$z = 70$		$z = 80$		$z = 90$	
	$g_{j=1}$	$\dfrac{\Delta g}{\Delta j}$	$g_{j=1}$	$\dfrac{\Delta g}{\Delta j}$	$g_{j=1}$	$\dfrac{\Delta g}{\Delta j}$	$g_{j=1}$	$\dfrac{\Delta g}{\Delta j}$
0.2	0.00018	0.00015	0.000218	0.000134	0.000253	0.00017	0.000289	0.000208
0.3	0.0085	0.000475	0.0101	0.0062	0.000253	0.00017	0.0134	0.0097
0.4	0.0583	0.032	0.0689	0.0421	0.0118	0.00775	0.0919	0.0661
0.5	0.185	0.1025	0.0219	0.0134	0.0802	0.0545	0.292	0.208
0.6	0.401	0.2225	0.474	0.292	0.255	0.17	0.632	0.452
0.7	0.699	0.3875	0.828	0.510	0.552	0.3675	0.101	0.791
0.8	0.064	0.595	0.263	0.777	0.963	0.6425	0.678	1.205
0.9	1.482	0.8325	1.76	1.08	1.469	0.9775	2.34	1.68
1.0	1.94	1.075	2.30	1.42	2.05	1.45	3.06	2.19
2.0	7.04	4.025	8.35	5.19	2.68	1.775	11.03	7.88
3.0	11.63	6.65	13.73	8.58	9.68	6.475	18.0	12.8
4.0	15.40	9.00	18.2	11.4	12.92	8.65	23.6	16.7
5.0	18.70	10.75	21.9	13.7	15.85	10.65	28.2	19.7
6.0	21.40	12.50	25.1	15.6	18.5	12.5		
7.0	23.80	13.75	27.9	17.2	20.9	14.0		
8.0	26.00	15.00	30.3	18.6	23.1	15.5		
9.0	27.90	16.00	32.5	19.8	25.1	16.75		

Source: Based on f_h/U vs. g tables in Stumbo, C.R. 1973. *Thermobacteriology in Food Processing,* 2nd ed. Academic Press, New York.

where U is the process $U = F_0 F_i$ or the equivalent heating time at 250°F for the process at T_r. $U =$ is U obtained from Tables 9.17 to 9.21, the equivalent heating time at T_g for the lethality of the cooling part of the process.

Example 9.18. For the example in the previous section, which was evaluated using the general method, calculate F_0 using Stumbo's and Hayakawa's procedures and calculate a process time needed to obtain an F_0 of 8 minutes. The following heating and cooling curve parameters were previously determined: $f_h = f_c = 22$ minutes; $j = 1.4$; $j_c = 1.8$.

Solution:

The retort temperature, cooling water temperature, and process time are $T_r = 251°F$; $T_c = 70°F$; $t = 30 - 0.6(3) = 28.2$ minutes.

To use the formula methods to determine F_0, it is necessary to determine g from the process time and the heating curve parameters. The f_h/U versus g table is then used to determine f_h/U, which corresponds to g, from which a value of U and F_0 can be calculated. Let $T_g =$ can temperature at the termination of heating. Solving for T_g using Equation (9.36):

$$T_g = 251 - (251 - 160)(10)^{-28.2/22} = 244.3°F$$
$$g = T_r - T_g = 251 - 244.3 = 6.66°F.$$

Table 9.16 g/K_s vs. U/f_h tables used for calculating the lethality of the heating part of a thermal process by Hayakawa's procedure

g/K_s (°F)	U/f_h	g/K_s (°F)	U/f_h	g/K_s (°F)	U/f_h
100.0000	0.4165(−06)	33.0000	0.2095(−02)	0.35000	0.1161(01)
98.0000	0.5152(−06)	32.0000	0.2413(−02)	0.30000	0.1226(01)
96.0000	0.6420(−06)	31.0000	0.2780(−02)	0.25000	0.1303(01)
94.0000	0.8051(−06)	30.0000	0.3205(−02)	0.20000	0.1397(01)
92.0000	0.1015(−05)	29.0000	0.3699(−02)	0.15000	0.1519(01)
90.0000	0.1284(−05)	28.0000	0.4272(−02)	0.10000	0.1693(01)
88.0000	0.1632(−05)	27.0000	0.4939(−02)	0.09000	0.1738(01)
86.0000	0.2079(−05)	26.0000	0.5715(−02)	0.08000	0.1789(01)
84.0000	0.2655(−05)	25.0000	0.6620(−02)	0.07000	0.1846(01)
82.0000	0.3398(−05)	24.0000	0.7677(−02)	0.06000	0.1913(01)
80.0000	0.4356(−05)	23.0000	0.8914(−02)	0.05000	0.1992(01)
78.0000	0.5593(−05)	22.0000	0.1036(−01)	0.04000	0.2088(01)
76.0000	0.7191(−05)	21.0000	0.1206(−01)	0.03500	0.2146(01)
74.0000	0.9256(−05)	20.0000	0.1407(−01)	0.03000	0.2212(01)
72.0000	0.1193(−04)	19.0000	0.1643(−01)	0.02500	0.2291(01)
70.0000	0.1539(−04)	18.0000	0.1922(−01)	0.02000	0.2388(01)
68.0000	0.1986(−04)	17.0000	0.2254(−01)	0.01500	0.2513(01)
66.0000	0.2567(−04)	16.0000	0.2648(−01)	0.01000	0.2688(01)
64.0000	0.3321(−04)	15.0000	0.3119(−01)	0.00900	0.2734(01)
62.0000	0.4300(−04)	14.0000	0.3684(−01)	0.00800	0.2785(01)
60.0000	0.5573(−04)	13.0000	0.4365(−01)	0.00700	0.2844(01)
58.0000	0.7229(−04)	12.0000	0.5191(−01)	0.00600	0.2909(01)
56.0000	0.9388(−04)	11.0000	0.6198(−01)	0.00500	0.2989(01)
54.0000	0.1220(−03)	10.0000	0.7435(−01)	0.00400	0.3085(01)
52.0000	0.1589(−03)	9.0000	0.8970(−01)	0.00350	0.3143(01)
50.0000	0.2070(−03)	8.0000	0.1090(00)	0.00300	0.3210(01)
49.0000	0.2364(−03)	7.0000	0.1335(00)	0.00250	0.3290(01)
48.0000	0.2701(−03)	6.0000	0.1652(00)	0.00200	0.3384(01)
47.0000	0.3087(−03)	5.0000	0.2073(00)	0.00150	0.3509(01)
46.0000	0.3529(−03)	4.0000	0.2652(00)	0.00100	0.3685(01)
45.0000	0.4036(−03)	3.5000	0.3029(00)	0.00090	0.3734(01)
44.0000	0.4618(−03)	3.0000	0.3490(00)	0.00080	0.3780(01)
43.0000	0.5286(−03)	2.5000	0.4067(00)	0.00070	0.3842(01)
42.0000	0.6053(−03)	2.0000	0.4816(00)	0.00060	0.3903(01)
41.0000	0.6934(−03)	1.5000	0.5839(00)	0.00050	0.3986(01)
40.0000	0.7947(−03)	1.0000	0.7367(00)	0.00040	0.4073(01)
39.0000	0.9113(−03)	0.9000	0.7777(00)	0.00035	0.4143(01)
38.0000	0.1045(−02)	0.8000	0.8241(00)	0.00030	0.4204(01)
37.0000	0.1200(−02)	0.7000	0.8773(00)	0.00025	0.4274(01)
36.0000	0.1378(−02)	0.6000	0.9395(00)	0.00020	0.4358(01)
35.0000	0.1584(−02)	0.5000	0.1014(01)	0.00015	0.4505(01)
34.0000	0.1821(−02)	0.4000	0.1106(01)	0.00010	0.4659(01)

Values in parentheses indicate powers of 10 by which tabulated values are to be multiplied; e.g., $U_{h/f}$ for $g/K_s = 40°F$ is 0.0006646.

Source: Hayakawa, k., Food Technol. 24: 1407, 1970. Corrected table courtesy of K. Hayakawa.

Table 9.17 g/k_s vs. U/f_c Tables used for Calculating the Lethality of the Cooling Part of a Thermal Process by Hayakawa's Procedure ($g/K_s \leq 200$)

I_c/K_s (°F)	U'/f_c for $j_c = 0.40$ to 1.90								
	0.40	0.60	0.80	1.00	1.20	1.40	1.60	1.80	1.90
200.00	0.9339(−2)	0.1086(−1)	0.1220(−1)	0.1976(−1)	0.7021(−1)	0.9440(−1)	0.1112	0.1243	0.1300
195.00	0.9585(−2)	0.1114(−1)	0.1253(−1)	0.2030(−1)	0.7114(−1)	0.9565(−1)	0.1126	0.1260	0.1318
190.00	0.9844(−2)	0.1145(−1)	0.1288(−1)	0.2086(−1)	0.7211(−1)	0.9695(−1)	0.1142	0.1277	0.1335
185.00	0.1012(−1)	0.1177(−1)	0.1325(−1)	0.2145(−1)	0.7312(−1)	0.9830(−1)	0.1158	0.1295	0.1354
180.00	0.1041(−1)	0.1212(−1)	0.1364(−1)	0.2208(−1)	0.7418(−1)	0.9972(−1)	0.1174	0.1313	0.1373
175.00	0.1072(−1)	0.1248(−1)	0.1405(−1)	0.2275(−1)	0.7529(−1)	0.1012	0.1192	0.1332	0.1394
170.00	0.1104(−1)	0.1287(−1)	0.1449(−1)	0.2346(−1)	0.7645(−1)	0.1027	0.1210	0.1353	0.1415
165.00	0.1139(−1)	0.1328(−1)	0.1496(−1)	0.2422(−1)	0.7767(−1)	0.1044	0.1229	0.1374	0.1437
160.00	0.1176(−1)	0.1372(−1)	0.1546(−1)	0.2503(−1)	0.7895(−1)	0.1061	0.1249	0.1396	0.1460
155.00	0.1216(−1)	0.1418(−1)	0.1599(−1)	0.2589(−1)	0.8029(−1)	0.1078	0.1270	0.1420	0.1485
150.00	0.1258(−1)	0.1469(−1)	0.1657(−1)	0.2682(−1)	0.8172(−1)	0.1097	0.1292	0.1444	0.1510
145.00	0.1304(−1)	0.1523(−1)	0.1719(−1)	0.2781(−1)	0.8322(−1)	0.1117	0.1315	0.1470	0.1538
140.00	0.1353(−1)	0.1582(−1)	0.1785(−1)	0.2889(−1)	0.8481(−1)	0.1138	0.1340	0.1498	0.1566
135.00	0.1407(−1)	0.1645(−1)	0.1858(−1)	0.3005(−1)	0.8651(−1)	0.1160	0.1366	0.1527	0.1597
130.00	0.1465(−1)	0.1714(−1)	0.1936(−1)	0.3131(−1)	0.8831(−1)	0.1184	0.1393	0.1558	0.1629
125.00	0.1528(−1)	0.1789(−1)	0.2022(−1)	0.3268(−1)	0.9025(−1)	0.1209	0.1423	0.1591	0.1663

120.00	0.1598(−1)	0.1872(−1)	0.2116(−1)	0.3418(−1)	0.9232(−1)	0.1236	0.1454	0.1626	0.1700
115.00	0.1675(−1)	0.1963(−1)	0.2219(−1)	0.3583(−1)	0.9456(−1)	0.1265	0.1488	0.1663	0.1739
110.00	0.1760(−1)	0.2065(−1)	0.2334(−1)	0.3766(−1)	0.9698(−1)	0.1296	0.1524	0.1703	0.1781
105.00	0.1857(−1)	0.2175(−1)	0.2462(−1)	0.3970(−1)	0.9961(−1)	0.1329	0.1563	0.1747	0.1826
100.00	0.1969(−1)	0.2296(−1)	0.2606(−1)	0.4199(−1)	0.1025	0.1366	0.1605	0.1794	0.1875
95.00	0.2104(−1)	0.2433(−1)	0.2769(−1)	0.4463(−1)	0.1057	0.1405	0.1651	0.1845	0.1929
90.00	0.2276(−1)	0.2589(−1)	0.2955(−1)	0.4779(−1)	0.1093	0.1449	0.1702	0.1901	0.1987
85.00	0.2414(−1)	0.2768(−1)	0.3170(−1)	0.5103(−1)	0.1132	0.1498	0.1757	0.1962	0.2051
80.00	0.2576(−1)	0.2976(−1)	0.3420(−1)	0.5460(−1)	0.1176	0.1552	0.1819	0.2031	0.2122
75.00	0.2768(−1)	0.3221(−1)	0.3715(−1)	0.5878(−1)	0.1227	0.1612	0.1888	0.2107	0.2202
70.00	0.2999(−1)	0.3512(−1)	0.4069(−1)	0.6373(−1)	0.1285	0.1682	0.1967	0.2193	0.2291
65.00	0.3280(−1)	0.3865(−1)	0.4499(−1)	0.6967(−1)	0.1353	0.1762	0.2057	0.2291	0.2394
60.00	0.3626(−1)	0.4300(−1)	0.5032(−1)	0.7687(−1)	0.1434	0.1855	0.2161	0.2405	0.2511
55.00	0.4061(−1)	0.4846(−1)	0.5703(−1)	0.8575(−1)	0.1532	0.1966	0.2284	0.2539	0.2650
50.00	0.4616(−1)	0.5546(−1)	0.6564(−1)	0.9687(−1)	0.1652	0.2101	0.2432	0.2698	0.2814
45.00	0.5340(−1)	0.6462(−1)	0.7692(−1)	0.1111	0.1803	0.2268	0.2613	0.2892	0.3014
40.00	0.6302(−1)	0.7688(−1)	0.9197(−1)	0.1295	0.1997	0.2478	0.2840	0.3132	0.3261
35.00	0.7607(−1)	0.9362(−1)	0.1124	0.1539	0.2251	0.2750	0.3129	0.3437	0.3573
30.00	0.9415(−1)	0.1170	0.1408	0.1868	0.2591	0.3108	0.3507	0.3834	0.3978
25.00	0.1197	0.1501	0.1808	0.2322	0.3056	0.3593	0.4014	0.4361	0.4515

Values in parentheses are powers of 10 by which tabulated value should be multiplied.
Source: Hayakawa. K., *Food Technol.* 24:1407, 1970.

Table 9.18 g/K_s vs. U/f_c Tables used for Calculating the Lethality of the Cooling Part of a Thermal Process by Hayakawa's Procedure ($g/K_s \leq 200$)

I_c/K_s (°F)	U'/f_c for $j_c = 2.0$ to 2.8								
	2.00	2.10	2.20	2.30	2.40	2.50	2.60	2.70	2.80
200.00	0.1353	0.1403	0.1449	0.1492	0.1533	0.1572	0.1609	0.1645	0.1679
195.00	0.1371	0.1421	0.1468	0.1512	0.1553	0.1593	0.1630	0.1666	0.1700
190.00	0.1390	0.1440	0.1487	0.1532	0.1574	0.1614	0.1652	0.1689	0.1723
185.00	0.1409	0.1460	0.1508	0.1553	0.1596	0.1636	0.1675	0.1712	0.1747
180.00	0.1429	0.1481	0.1530	0.1575	0.1619	0.1660	0.1699	0.1736	0.1772
175.00	0.1450	0.1503	0.1552	0.1599	0.1643	0.1684	0.1724	0.1762	0.1798
170.00	0.1472	0.1526	0.1576	0.1623	0.1667	0.1710	0.1750	0.1788	0.1825
165.00	0.1495	0.1550	0.1600	0.1648	0.1693	0.1736	0.1177	0.1816	0.1853
160.00	0.1520	0.1575	0.1626	0.1675	0.1721	0.1764	0.1806	0.1845	0.1883
155.00	0.1545	0.1601	0.1653	0.1703	0.1749	0.1794	0.1836	0.1876	0.1914
150.00	0.1572	0.1629	0.1682	0.1732	0.1780	0.1825	0.1867	0.1908	0.1947
145.00	0.1600	0.1658	0.1712	0.1763	0.1811	0.1857	0.1901	0.1942	0.1982
140.00	0.1630	0.1689	0.1744	0.1796	0.1845	0.1891	0.1936	0.1978	0.2018
135.00	0.1661	0.1721	0.1777	0.1830	0.1880	0.1928	0.1973	0.2016	0.2057
130.00	0.1695	0.1756	0.1813	0.1867	0.1918	0.1966	0.2012	0.2056	0.2098
125.00	0.1730	0.1793	0.1851	0.1906	0.1958	0.2007	0.2054	0.2099	0.2142
120.00	0.1768	0.1832	0.1892	0.1948	0.2001	0.2051	0.2099	0.2145	0.2188
115.00	0.1809	0.1874	0.1935	0.1992	0.2046	0.2098	0.2147	0.2193	0.2238
110.00	0.1853	0.1919	0.1982	0.2040	0.2095	0.2148	0.2198	0.2246	0.2291
105.00	0.1900	0.1968	0.2032	0.2092	0.2148	0.2202	0.2253	0.2302	0.2349
100.00	0.1951	0.2021	0.2086	0.2148	0.2206	0.2261	0.2313	0.2363	0.2411
95.00	0.2006	0.2078	0.2146	0.2209	0.2268	0.2325	0.2378	0.2429	0.2478
90.00	0.2067	0.2141	0.2210	0.2275	0.2337	0.2395	0.2450	0.2502	0.2552
85.00	0.2134	0.2210	0.2281	0.2348	0.2412	0.2471	0.2528	0.2582	0.2634
80.00	0.2207	0.2286	0.2360	0.2429	0.2494	0.2556	0.2615	0.2670	0.2724
75.00	0.2290	0.2371	0.2447	0.2519	0.2587	0.2651	0.2711	0.2769	0.2824
70.00	0.2382	0.2467	0.2546	0.2620	0.2690	0.2757	0.2820	0.2879	0.2936
65.00	0.2488	0.2576	0.2658	0.2735	0.2808	0.2877	0.2943	0.3005	0.3064
60.00	0.2610	0.2701	0.2787	0.2868	0.2944	0.3015	0.3084	0.3149	0.3210
55.00	0.2752	0.2848	0.2937	0.3022	0.3101	0.3176	0.3248	0.3316	0.3380
50.00	0.2922	0.3022	0.3116	0.3204	0.3288	0.3367	0.3442	0.3513	0.3581
45.00	0.3127	0.3232	0.3331	0.3424	0.3512	0.3595	0.3674	0.3750	0.3821
40.00	0.3380	0.3491	0.3596	0.3694	0.3787	0.3876	0.3959	0.4039	0.4115
35.00	0.3700	0.3818	0.3929	0.4033	0.4132	0.4226	0.4315	0.4400	0.4480
30.00	0.4113	0.4239	0.4357	0.4468	0.4574	0.4674	0.4769	0.4860	0.4946
25.00	0.4659	0.4793	0.4920	0.5040	0.5153	0.5261	0.5363	0.5460	0.5553

Source: Hayakawa. K., *Food Technol.* 24:1407, 1970.

Determination of F_0 using Stumbo's procedure. Table 9.14 is used to determine a value of f_h/U which corresponds to $g = 6.66$. Tabular parameters are for $z = 18$ and $j_c = 1.8$. It is necessary to interpolate. A value of $g = 6.66$ is not obtainable directly from Table 9.14, because a tabular entry for g is available only for a value of $j_c = 1$. Under the column "$z = 18$" in Table 9.14, a g value is an interpolating

Table 9.19 g/K_s vs. U'/f_c Tables used for Calculating the Lethality of the Cooling Part of a Thermal Process by Hayakawa's Procedure ($200 < g/K_s \leq 400$)

I_c/K_s (°F)	U'/f_c for j_c = 0.40 to 1.90								
	0.40	0.60	0.80	1.00	1.20	1.40	1.60	1.80	1.90
400.00	0.4642(−2)	0.5348(−2)	0.5964(−2)	0.9644(−2)	0.4919(−1)	0.6616(−1)	0.7794(−1)	0.8721(−1)	0.9124(−1)
395.00	0.4700(−2)	0.5416(−2)	0.6041(−2)	0.9769(−2)	0.4951(−1)	0.6658(−1)	0.7844(−1)	0.8777(−1)	0.9182(−1)
390.00	0.4760(−2)	0.5486(−2)	0.6120(−2)	0.9897(−2)	0.4983(−1)	0.6702(−1)	0.7895(−1)	0.8833(−1)	0.9241(−1)
385.00	0.4822(−2)	0.5558(−2)	0.6201(−2)	0.1003(−1)	0.5016(−1)	0.6746(−1)	0.7947(−1)	0.8892(−1)	0.9302(−1)
380.00	0.4886(−2)	0.5632(−2)	0.6284(−2)	0.1016(−1)	0.5049(−1)	0.6791(−1)	0.8000(−1)	0.8951(−1)	0.9364(−1)
375.00	0.4951(−2)	0.5708(−2)	0.6370(−2)	0.1030(−1)	0.5083(−1)	0.6837(−1)	0.8054(−1)	0.9011(−1)	0.9428(−1)
370.00	0.5017(−2)	0.5786(−2)	0.6458(−2)	0.1045(−1)	0.5118(−1)	0.6884(−1)	0.8109(−1)	0.9073(−1)	0.9492(−1)
365.00	0.5086(−2)	0.5866(−2)	0.6548(−2)	0.1059(−1)	0.5154(−1)	0.6931(−1)	0.8165(−1)	0.9136(−1)	0.9558(−1)
360.00	0.5157(−2)	0.5949(−2)	0.6641(−2)	0.1074(−1)	0.5190(−1)	0.6980(−1)	0.8223(−1)	0.9200(−1)	0.9625(−1)
355.00	0.5229(−2)	0.6033(−2)	0.6737(−2)	0.1090(−1)	0.5227(−1)	0.7030(−1)	0.8281(−1)	0.9266(−1)	0.9694(−1)
350.00	0.5304(−2)	0.6121(−2)	0.6835(−2)	0.1106(−1)	0.5265(−1)	0.7081(−1)	0.8341(−1)	0.9333(−1)	0.9764(−1)
345.00	0.5381(−2)	0.6211(−2)	0.6937(−2)	0.1122(−1)	0.5304(−1)	0.7133(−1)	0.8403(−1)	0.9401(−1)	0.9836(−1)
340.00	0.5460(−2)	0.6303(−2)	0.7041(−2)	0.1139(−1)	0.5344(−1)	0.7187(−1)	0.8466(−1)	0.9472(−1)	0.9909(−1)
335.00	0.5542(−2)	0.6398(−2)	0.7149(−2)	0.1157(−1)	0.5384(−1)	0.7241(−1)	0.8530(−1)	0.9543(−1)	0.9984(−1)
330.00	0.5626(−2)	0.6497(−2)	0.7260(−2)	0.1175(−1)	0.5426(−1)	0.7297(−1)	0.8595(−1)	0.9617(−1)	0.1006
325.00	0.5713(−2)	0.6598(−2)	0.7374(−2)	0.1193(−1)	0.5468(−1)	0.7354(−1)	0.8663(−1)	0.9692(−1)	0.1014
320.00	0.5802(−2)	0.6703(−2)	0.7493(−2)	0.1213(−1)	0.5512(−1)	0.7413(−1)	0.8731(−1)	0.9769(−1)	0.1022
315.00	0.5895(−2)	0.6811(−2)	0.7615(−2)	0.1232(−1)	0.5556(−1)	0.7472(−1)	0.8802(−1)	0.9847(−1)	0.1030
310.00	0.5990(−2)	0.6923(−2)	0.7741(−2)	0.1253(−1)	0.5602(−1)	0.7534(−1)	0.8874(−1)	0.9928(−1)	0.1039
305.00	0.6089(−2)	0.7038(−2)	0.7871(−2)	0.1274(−1)	0.5648(−1)	0.7597(−1)	0.8948(−1)	0.1001	0.1047
300.00	0.6191(−2)	0.7157(−2)	0.8006(−2)	0.1296(−1)	0.5696(−1)	0.7661(−1)	0.9024(−1)	0.1010	0.1056
295.00	0.6297(−2)	0.7281(−2)	0.8146(−2)	0.1319(−1)	0.5746(−1)	0.7727(−1)	0.9102(−1)	0.1018	0.1065

(*Cont.*)

Table 9.19 *(Continued)*

I_c/K_s (°F)	U'/f_c for $j_c = 0.40$ to 1.90								
	0.40	0.60	0.80	1.00	1.20	1.40	1.60	1.80	1.90
290.00	0.6406(−2)	0.7409(−2)	0.8291(−2)	0.1342(−1)	0.5796(−1)	0.7795(−1)	0.9182(−1)	0.1027	0.1075
285.00	0.6519(−2)	0.7541(−2)	0.8441(−2)	0.1367(−1)	0.5848(−1)	0.7865(−1)	0.9264(−1)	0.1036	0.1084
280.00	0.6636(−2)	0.7679(−2)	0.8596(−2)	0.1392(−1)	0.5901(−1)	0.7937(−1)	0.9348(−1)	0.1046	0.1094
275.00	0.6758(−2)	0.7821(−2)	0.8758(−2)	0.1418(−1)	0.5956(−1)	0.8010(−1)	0.9435(−1)	0.1055	0.1104
270.00	0.6884(−2)	0.7969(−2)	0.8925(−2)	0.1445(−1)	0.6012(−1)	0.8086(−1)	0.9524(−1)	0.1065	0.1115
265.00	0.7015(−2)	0.8123(−2)	0.9099(−2)	0.1473(−1)	0.6070(−1)	0.8164(−1)	0.9616(−1)	0.1076	0.1125
260.00	0.7152(−2)	0.8283(−2)	0.9281(−2)	0.1503(−1)	0.6130(−1)	0.8244(−1)	0.9710(−1)	0.1086	0.1136
255.00	0.7293(−2)	0.8449(−2)	0.9469(−2)	0.1533(−1)	0.6192(−1)	0.8327(−1)	0.9807(−1)	0.1097	0.1148
250.00	0.7441(−2)	0.8623(−2)	0.9666(−2)	0.1565(−1)	0.6255(−1)	0.8412(−1)	0.9908(−1)	0.1108	0.1159
245.00	0.7595(−2)	0.8803(−2)	0.9870(−2)	0.1599(−1)	0.6320(−1)	0.8500(−1)	0.1001	0.1120	0.1171
240.00	0.7755(−2)	0.8992(−2)	0.1008(−1)	0.1633(−1)	0.6388(−1)	0.8591(−1)	0.1012	0.1132	0.1184
235.00	0.7923(−2)	0.9188(−2)	0.1031(−1)	0.1669(−1)	0.6458(−1)	0.8685(−1)	0.1023	0.1144	0.1197
230.00	0.8098(−2)	0.9394(−2)	0.1054(−1)	0.1707(−1)	0.6530(−1)	0.8782(−1)	0.1034	0.1157	0.1210
225.00	0.8281(−2)	0.9609(−2)	0.1079(−1)	0.1747(−1)	0.6604(−1)	0.8882(−1)	0.1046	0.1170	0.1224
220.00	0.8472(−2)	0.9835(−2)	0.1104(−1)	0.1788(−1)	0.6682(−1)	0.8985(−1)	0.1058	0.1184	0.1238
215.00	0.8673(−2)	0.1007(−1)	0.1131(−1)	0.1832(−1)	0.6762(−1)	0.9093(−1)	0.1071	0.1198	0.1253
210.00	0.8884(−2)	0.1032(−1)	0.1159(−1)	0.1878(−1)	0.6845(−1)	0.9204(−1)	0.1084	0.1212	0.1268
205.00	0.9106(−2)	0.1058(−1)	0.1189(−1)	0.1926(−1)	0.6931(−1)	0.9320(−1)	0.1097	0.1228	0.1284

Values in parentheses are powers of 10 hy which tabulated value should be multiplied.
Source: Hayakawa, K., *Food Technol.* 24:1407, 1970.

Table 9.20 g/K_s vs. U/f_c Tables used for Calculating the Lethality of the Cooling Part of a Thermal Process by Hayakawa's Procedure ($200 < g/K_s \leq 400$)

I_c/K_s (°F)	U'/f_c for $j_c = 2.00$ to 2.80								
	2.00	2.10	2.20	2.30	2.40	2.50	2.60	2.70	2.80
400.00	0.9497(−1)	0.9844(−1)	0.1017	0.1048	0.1077	0.1105	0.1131	0.1156	0.1180
395.00	0.9557(−1)	0.9907(−1)	0.1024	0.1054	0.1084	0.1112	0.1138	0.1164	0.1188
390.00	0.9619(−1)	0.9971(−1)	0.1030	0.1061	0.1091	0.1119	0.1145	0.1171	0.1195
385.00	0.9682(−1)	0.1004	0.1037	0.1068	0.1098	0.1126	0.1153	0.1179	0.1203
380.00	0.9747(−1)	0.1010	0.1044	0.1075	0.1105	0.1134	0.1161	0.1186	0.1211
375.00	0.9813(−1)	0.1017	0.1051	0.1083	0.1113	0.1141	0.1168	0.1194	0.1219
370.00	0.9880(−1)	0.1024	0.1058	0.1090	0.1120	0.1149	0.1176	0.1203	0.1228
365.00	0.9948(−1)	0.1031	0.1065	0.1097	0.1128	0.1157	0.1184	0.1211	0.1236
360.00	0.1002	0.1038	0.1073	0.1105	0.1136	0.1165	0.1193	0.1219	0.1245
355.00	0.1009	0.1046	0.1080	0.1113	0.1144	0.1173	0.1201	0.1228	0.1254
350.00	0.1016	0.1053	0.1088	0.1121	0.1152	0.1182	0.1210	0.1237	0.1263
345.00	0.1024	0.1061	0.1096	0.1129	0.1161	0.1190	0.1219	0.1246	0.1272
340.00	0.1031	0.1069	0.1104	0.1138	0.1169	0.1199	0.1228	0.1255	0.1281
335.00	0.1039	0.1077	0.1113	0.1146	0.1178	0.1208	0.1237	0.1265	0.1291
330.00	0.1047	0.1085	0.1121	0.1155	0.1187	0.1217	0.1246	0.1274	0.1301
325.00	0.1055	0.1094	0.1130	0.1164	0.1196	0.1227	0.1256	0.1284	0.1311
320.00	0.1064	0.1102	0.1139	0.1173	0.1206	0.1237	0.1266	0.1294	0.1321
315.00	0.1072	0.1111	0.1148	0.1183	0.1215	0.1247	0.1276	0.1305	0.1332
310.00	0.1081	0.1120	0.1157	0.1192	0.1225	0.1257	0.1287	0.1315	0.1343
305.00	0.1090	0.1130	0.1167	0.1202	0.1236	0.1267	0.1297	0.1326	0.1354
300.00	0.1099	0.1139	0.1177	0.1212	0.1246	0.1278	0.1308	0.1337	0.1365
295.00	0.1109	0.1149	0.1187	0.1223	0.1257	0.1289	0.1319	0.1349	0.1377
290.00	0.1118	0.1159	0.1197	0.1234	0.1268	0.1300	0.1331	0.1360	0.1389
285.00	0.1128	0.1170	0.1208	0.1244	0.1279	0.1312	0.1343	0.1373	0.1401
280.00	0.1139	0.1180	0.1219	0.1256	0.1290	0.1323	0.1355	0.1385	0.1414
275.00	0.1149	0.1191	0.1230	0.1267	0.1302	0.1336	0.1367	0.1398	0.1427
270.00	0.1160	0.1202	0.1242	0.1279	0.1315	0.1348	0.1380	0.1411	0.1440
265.00	0.1171	0.1214	0.1254	0.1292	0.1327	0.1361	0.1393	0.1424	0.1454
260.00	0.1183	0.1226	0.1266	0.1304	0.1340	0.1374	0.1407	0.1438	0.1468
255.00	0.1194	0.1238	0.1279	0.1317	0.1354	0.1388	0.1421	0.1452	0.1483
250.00	0.1207	0.1251	0.1292	0.1331	0.1367	0.1402	0.1435	0.1467	0.1498
245.00	0.1219	0.1264	0.1305	0.1344	0.1381	0.1417	0.1450	0.1482	0.1513
240.00	0.1232	0.1277	0.1319	0.1359	0.1396	0.1432	0.1466	0.1498	0.1529
235.00	0.1245	0.1291	0.1333	0.1373	0.1411	0.1447	0.1482	0.1514	0.1546
230.00	0.1259	0.1305	0.1348	0.1389	0.1427	0.1463	0.1498	0.1531	0.1563
225.00	0.1274	0.1320	0.1363	0.1404	0.1443	0.1480	0.1515	0.1548	0.1580
220.00	0.1288	0.1335	0.1379	0.1421	0.1460	0.1497	0.1532	0.1566	0.1599
215.00	0.1301	0.1351	0.1396	0.1438	0.1477	0.1515	0.1551	0.1585	0.1617
210.00	0.1320	0.1368	0.1413	0.1455	0.1495	0.1533	0.1569	0.1604	0.1637
205.00	0.1336	0.1385	0.1430	0.1473	0.1514	0.1552	0.1589	0.1624	0.1657

Values in parentheses an powers of 10 by which tabulated value should be multiplied.
Source: Hayakawa, K., *Food Technol.* 24:1407, 1970.

Table 9.21 g/K_s vs. U/f_c Tables used for Calculating the Lethality of the Cooling Part of a Thermal Process by Hayakawa's Procedure ($g/K_s \leq 400$)

I_c/K_s (°F)	U'/f_c for		I_c/K_s (°F)	U'/f_c for	
	$j_c = 2.90$	$j_c = 3.00$		$j_c = 2.90$	$j_c = 3.00$
400.00	0.1204	0.1226	200.00	0.1711	0.1742
395.00	0.1211	0.1234	195.00	0.1733	0.1765
390.00	0.1219	0.1242	190.00	0.1757	0.1789
385.00	0.1227	0.1250	185.00	0.1781	0.1813
380.00	0.1235	0.1258	180.00	0.1806	0.1839
375.00	0.1243	0.1266	175.00	0.1833	0.1866
370.00	0.1252	0.1275	170.00	0.1860	0.1894
365.00	0.1260	0.1284	165.00	0.1889	0.1923
360.00	0.1269	0.1293	160.00	0.1919	0.1954
355.00	0.1278	0.1302	155.00	0.1951	0.1986
350.00	0.1287	0.1311	150.00	0.1985	0.2020
345.00	0.1297	0.1321	145.00	0.2020	0.2056
340.00	0.1306	0.1331	140.00	0.2057	0.2094
335.00	0.1316	0.1341	135.00	0.2096	0.2134
330.00	0.1326	0.1351	130.00	0.2138	0.2177
325.00	0.1337	0.1361	125.00	0.2183	0.2222
320.00	0.1347	0.1372	120.00	0.2230	0.2270
315.00	0.1358	0.1383	115.00	0.2281	0.2321
310.00	0.1369	0.1394	110.00	0.2335	0.2377
305.00	0.1380	0.1406	105.00	0.2393	0.2436
300.00	0.1392	0.1417	100.00	0.2456	0.2500
295.00	0.1404	0.1430	95.00	0.2525	0.2570
290.00	0.1416	0.1442	90.00	0.2600	0.2646
285.00	0.1428	0.1455	85.00	0.2683	0.2730
280.00	0.1441	0.1468	80.00	0.2774	0.2823
275.00	0.1455	0.1481	75.00	0.2876	0.2926
270.00	0.1468	0.1495	70.00	0.2991	0.3043
265.00	0.1482	0.1509	65.00	0.3120	0.3174
260.00	0.1497	0.1524	60.00	0.3269	0.3325
255.00	0.1511	0.1539	55.00	0.3442	0.3501
250.00	0.1527	0.1555	50.00	0.3645	0.3707
245.00	0.1543	0.1571	45.00	0.3889	0.3954
240.00	0.1559	0.1587	40.00	0.4187	0.4256
235.00	0.1576	0.1604	35.00	0.4557	0.4631
230.00	0.1593	0.1622	30.00	0.5028	0.5107
225.00	0.1611	0.1640	25.00	0.5641	0.5725
220.00	0.1630	0.1659			
215.00	0.1649	0.1679			
210.00	0.1669	0.1699			
205.00	0.1689	0.1720			

Source: Hayakawa, K., *Food Technol.* 24:1407, 1970.

factor, $\Delta g /\Delta j = 1.59$. The value of g for $f_h/U = 5$ and for $j = 1.8$ is

$$g_{j=1.8} = 5.4 + 0.8(1.59) = 6.672$$

The value $g = 6.672$ exceeds 6.66, the specified g; therefore, a lower value of $f_h/U = 4$ is chosen, a corresponding g value for $j_c = 1.8$ is calculated, and by interpolation, a value of f_h/U which corresponds to $g = 6.66$ is calculated. For $f_h/U = 4$, $g_{j=1} = 4.41$; $\Delta g/\Delta j = 1.34$; and $g_{j=1.8} = 4.41 + 0.8(1.34) = 5.482$. Interpolating:

$$\left(\frac{f_h}{U}\right)_{g=6.66} = 4 + \left(\frac{1}{6.672 - 5.482}\right)(6.66 - 5.482) = 4.99$$

$$U = \frac{f_h}{(f_h/U)_{g=6.66}} = \frac{22}{4.99} = 4.41$$

At $251°F$, $F_i = (10)^{-1/18} = 0.8799$

$$F_0 = U/F_i = 4.41/0.8799 = 5.01$$

This value compares with 5.6 minutes calculated using the general method in the previous section. Stumbo's procedure for determining the process time B_b can be done directly without the need for iteration. To obtain an F_0 of 8.0 minutes a value of g is now required, and this value is obtained from Table 9.14 to correspond to a value of f_h/U. U is calculated as :

$$U = F_0 Fi = 8(0.8799) = 7.0392$$
$$f_h/U = 22/7.0392 = 3.1253$$

From Table 9.14, for $z = 18$, $j_c = 1.8$, and $f_h/U = 3$:

$$g_{j=1} = 3.26; \Delta g/\Delta j = 1.05; g_{j=1.8} = 3.26 + 0.8(1.05) = 4.10$$

For $f_h/U = 4$:

$$g_{j=1} = 4.41; \quad \Delta g/\Delta j = 1.34; \quad g_{j=1.8} = 4.41 + 0.8(1.34) = 5.482$$

Interpolating to obtain g for $f_h/U = 3.1253$:

$$g = 4.10 + \left(\frac{5.482 - 4.10}{1}\right)(3.1253 - 3.0) = 4.273$$

B_b is calculated using Equation (9.44):

$$B_b = 22[\log(1.4)(251 - 160) - \log(4.273)] = 32.4 \text{ minutes}$$

Determination of F_0 using Hayakawa's procedure: For $g = 6.658$; $K_s = 18/20 = 0.900$; $g/K_s = 7.398$.
From Table 9.16:

$$g/K_s = 7; \qquad U/f_h = 0.1252$$
$$g/K_s = 8; \qquad U/f_h = 0.1020$$

Interpolating to obtain U/f_h for $g/K_s = 7.398$:

$$U/f_h = 0.1252 - [(0.1252 - 0.1020)/1][7.398 - 7] = 0.1159$$
$$U = 22(0.1159) = 2.551$$

For the cooling curve, use Table 9.17 to 9.21. $T_g = 251 - 6.658 = 244.3$; $I_c = 244.3 - 70 = 174.3$: $I_c/K_s = 174.3/0.900 = 193.71$. The appropriate table is Table 9.17, because $I_c/K_s < 200$ and $j_c < 1.9$. Values of $I_c/K_s = 190$ and 195 can be read in Table 9.17. A value for U/f_c corresponding to $I_c/K_s = 193.71$ is obtained by interpolation. For $j_c = 1.8$, $I_c/K_s = 190$, $U = /f_c = 0.1277$; $I_c/K_s = 195$ and $U = /f_c = 0.1260$. Interpolating:

$$\frac{U'}{f_c} = 0.1277 - \left(\frac{0.1277 - 0.1260}{5}\right)(193.71 - 190) = 0.1264$$

Solving for $U =$ for the cooling part of the process:

$$U = 22(0.1264) = 2.7816.$$

$U =$ is converted to U using Equation (9.46): $U = 2.7816(10)^{-6.66/18} = 1.187$ for the cooling part of the process. Total U is the sum of U for heating and U for cooling.

$$U = 2.551 + 1.187 = 3.738$$

The process F_0 is then determined using $U = F_0 F_i$.

F_i was previously calculated at 251°F to be 0.8799. Therefore, $F_0 = U/F_i = 3.738/0.8799 = 4.248$ minutes.

Determination of B_b using Hayakawa's procedure. A value of g is first assumed. Let $g = 3.7$°F. Because $K_s = 18/20 = 0.9$, $g/K_s = 3.7/0.9 = 4.111$°F. For the heating part of the process, Table 9.16 is used. The value of U/f_h corresponding to $g/K_s = 4.111$ will be obtained by interpolation.

From Table 9.16, for $g/K_s = 4$ and $U_h/f_h = 0.2514$; for $g/K_S = 5$ and $U_h/f_h = 0.1958$:

$$\frac{U}{f_h} = 0.2514 - \left(\frac{0.2514 - 0.1958}{1}\right)(4.1111 - 4) = 0.2452$$

Solving for U, $U = 0.2452(22) = 5.39$ for the heating portion of the process.

For the cooling curve: $I_c = 251 - 3.7 - 70 = 177.3$. $I_c/K_s = 177.3/0.9 = 197$. The appropriate table to be used is Table 9.17, because $I_c/K_s < 200$ and $j_c < 1.9$. From Table 9.17, for $j_c = 1.8$, $U=/f_c$; for $I_c/K_s = 197$ will be obtained by interpolation.

$$\frac{I_c}{K_s} = 195; \frac{U_g}{f_c} = 0.1260 = 200; = 0.1243$$
$$\frac{U'}{f_c} = 0.1260 - \left(\frac{0.1260 - 0.1243}{5}\right)(197 - 195) = 0.1253$$

Because $f_c = 22$ minutes, $U = 0.1243(22) = 2.756$. $U =$ is converted to U using Equation (9.46): $U = 2.756(10)^{-3.7/18} = 1.716$ for the cooling portion of the process. The total U is the sum of U for the heating portion and U for the cooling portion.

$$U = 5.394 + 1.716 = 7.11 \text{ min} = F_0 F_i$$

F_i was previously determined to be 0.8799. $F_0 = U/F_i = 7.11/0.8799 = 8.08$ minutes. This is close to the specified F_0 value of 8.0 min; therefore required value of g is 3.7°F. If the calculated F_0 for the assumed g is not close enough to the specified F_0, it will be necessary to assume another value of g and to repeat the calculations. When selecting another value of g, keep in mind that a smaller g will result in a larger calculated F_0 value.

The selected g of 37°F, which resulted in an F_0 value close to the specified F_0 of 8.0 minutes, is used to solve for the process time. Solving for B_b using Equation (9.44):

$$B_b = 22[\log(1.4)(251 - 160) - \log(3.7)] = 33.8 \text{ minutes}$$

9.6.5   Evaluation of Probability of Spoilage from a Given Process

This procedure is used to determine if a process that deviated from specifications will give a safe product. The procedures discussed in this section will also be useful in cases of spoilage outbreaks where a spoilage organism is isolated, its heat resistance determined, and it is desired to determine if process schedule adjustment is necessary to prevent future occurrences of spoilage. Another useful application of these procedures is the conversion of standard F_0 values to F_0^z values for specific microorganisms.

9.6.5.1   Constant Temperature Processes

A process time at a constant retort temperature and an initial temperature are given. The procedure is similar to example. A g value is calculated using Equation (9.33). Tables 9.14 or 9.16 to 9.20 are then used to determine U at a specified z value, from which F_0^z is calculated. The probability of spoilage is then calculated by substituting F_0^z for t in Equation (9.11).

Example 9.19. The following data represents the heating characteristics of a canned product. $f_h = f_c = 22.5$ minutes; $j = j_c = 1.4$. If this product is processed for 25 minutes at 252°F from an initial temperature of 100°F, calculate (a) the F_0 and (b) the probability of spoilage if an organism with a D_0 value of 0.5 minutes and a z value of 14 is present at an initial spore load of 10/can.

Solution:

g is determined from the process time, using Equation (9.33):

$$g = [10]^{\log(jI) - t/f_h}$$
$$= [10]^{[\log[(1.4)(252 - 100)] - 25/22.5]}$$
$$= [10]^{1.2168} = 16.5°F$$

(a) Because $f_h = f_c$, Stumbo's procedure is used. The F_0 is determined using Table 9.14 for z = 18°F and $j_c = 1.4$. Inspection of Table 9.14 reveals that to obtain g = 16.5 when j = 1, f_h/U has to be between 40 and 45. However, the interpolation factor $\Delta g/\Delta j$ is about 6; therefore, because $j_c = 1.4$, g in the table will increase by 0.4(6) or 2.4. Thus, the entry for $f_h/U = 30$ will be considered, and after calculating g at $j_c = 1.4$, the other entry to be used in the interpolation will be selected.

$$\frac{f_h}{U} = 30; \quad g_{j=1.4} = 14.60 + 0.4(5.3) = 16.72$$

The value of g is greater than 16.5; therefore, the next lower value of f_h/U in the tables will be used to obtain the other value of g to use in the interpolation.

$$\frac{f_h}{U} = 25; \quad g_{j=1.4} = 13.6 + 0.4(4.8) = 15.52$$

Interpolating between g = 16.72 and g = 15.52 to obtain f_h/U corresponding to g = 16.5:

$$\frac{f_h}{U} = \frac{25 + 5(16.5 - 15.52)}{(16.72 - 15.52)} = 29.1$$

$$U = \frac{f_h}{(\frac{f_h}{U})_{g=16.5}} = \frac{22.5}{29.1} = 0.7736 = F_0 F_i$$

$$F_i = (10)^{-2/18} = 0.774$$

$$F_0 = \frac{0.773}{0.774} = 0.999 \, \text{min}$$

(b) To evaluate lethality to the organism with z = 14°F, F_0^{l4} must be determined. Using Table 9.14 for z = 14°F:

$$\frac{f_h}{U} = 50; \quad g_{j=1.4} = 14.2 + 0.4(4.00) = 15.8$$

For g = 16.5:

$$\frac{f_h}{U} = 60; \quad g_{j=1.4} = 15.1 + 0.4(4.3) = 16.82$$

$$\frac{f_h}{U} = \frac{50 + 10(16.5 - 15.8)}{(16.82 - 15.8)} = 56.9$$

$$U = \frac{f_h}{(f_h/U)_{g=16.5}} = \frac{22.5}{56.9} = 0.395 = F_0 F_i$$

$$f_i = (10)^{-2/14} = 0.7197$$

$$F_0^{l4} = \frac{U}{F_i} = \frac{0.395}{0.7197} = 0.549 \, \text{min}$$

The number of survivors will be

$$N = 10[10]^{-F_0/D_0} = 10(10)^{0.549/0.5} = 0.798$$

The probability of spoilage is 79.80%.

9.6.5.2 *Process Temperature Change*

When the process temperature changes, errors in the formula method are magnified because evaluation of f_h and j is based on an original uniform initial temperature, while the starting temperature distribution with in-process temperature deviations is no longer uniform. The most accurate method for evaluating the effect of process temperature changes is by using finite difference methods for evaluation of temperature at the critical point and using the general method for determining process lethality. If process deviation occurs before the temperature at the critical point exceeds 200°F, and the deviation simply involves a step change in processing temperature at $t = t_1$ from T_{r1} to T_{r2} and remains constant for the rest of the process, lethality may be approximated by the formula methods. The part of the process before the step temperature change is considered to have negligible lethality (if the temperature at the critical point did not exceed 200°F), and the temperature at t_1 is considered the initial temperature for a process at T_{r2}. Procedures for evaluation will be the same as in example 9.19.

9.7 BROKEN HEATING CURVES

Broken heating curves are those that exhibit a break in continuity of the heating rate at some point in the heating process. Thus, two or more line segments will be formed when the heat penetration data are plotted on semi-logarithmic graphing paper. This type of heating behavior will occur when the product inside the can undergoes a physical change that changes the heat transfer characteristics. A typical broken heating curve is shown in Figure 9.17. The slope indices of the curve are designated as

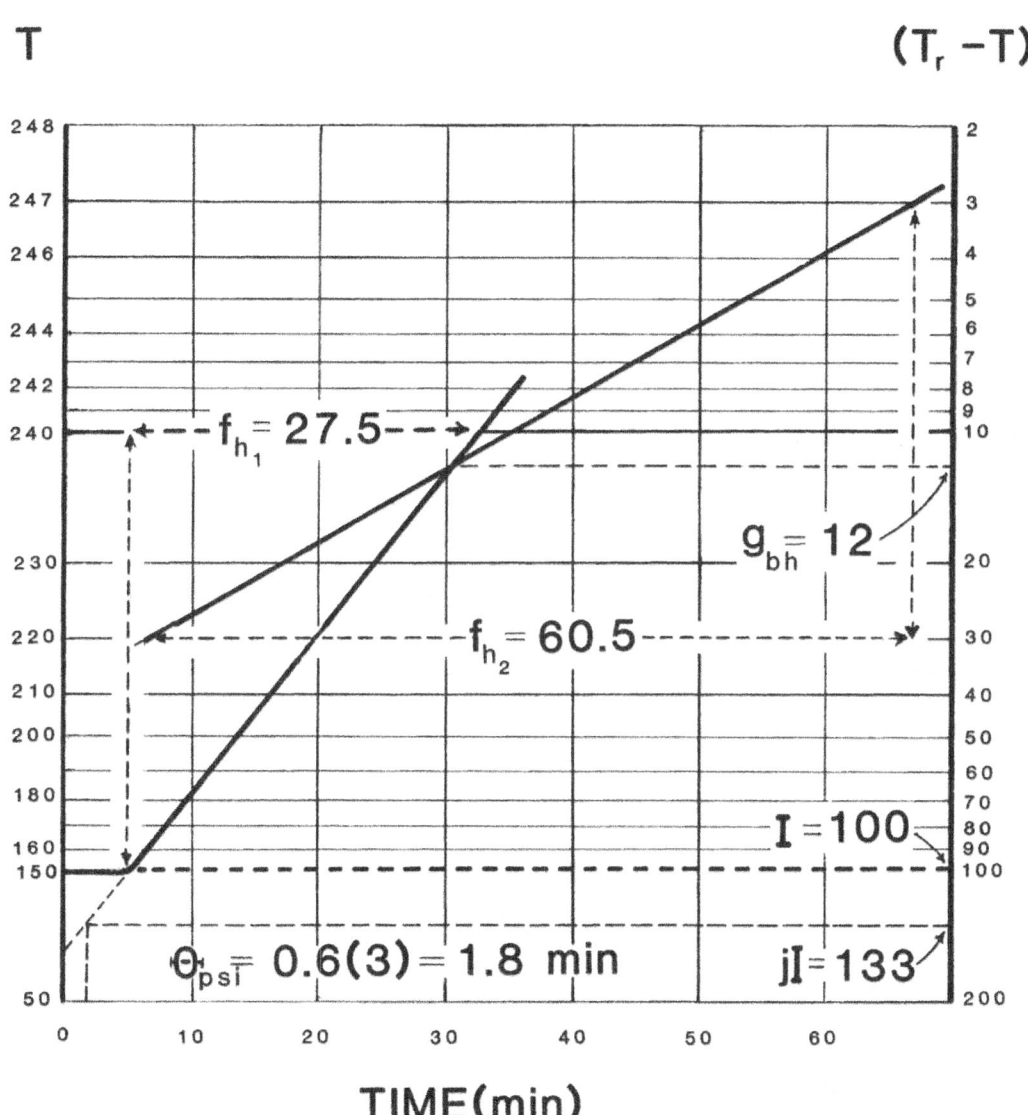

Figure 9.17   Diagram of a broken heating curve showing the heating curve parameters.

f_{h1} for the first line segment and f_{h2} for the second line segment. The retort-can temperature difference at the point of intersection of the first and second line segments is designated gbh. The rest of the parameters of the heating curve is the same as for a simple heating curve.

The equation of the first line segment is

$$\log\left(\frac{jI}{g_{bh}}\right) = \frac{t_{bh}}{f_{h1}} \tag{9.47}$$

The equation of the second line segment is

$$\log\left(\frac{g_{bh}}{g}\right) = \frac{t - t_{bh}}{f_{h2}} \tag{9.48}$$

Combining Equations (9.46) and (9.47):

$$t = f_{h1} \log\left(\frac{jI}{g_{bh}}\right) + f_{h2} \log\left(\frac{g_{bh}}{g}\right) \tag{9.49}$$

Equation (9.49) is used to calculate a process time to obtain g. An expression for g can be obtained by rearranging Equation (9.49).

$$g = [10]^{1/f_{h2}[f_{h1}\log(jI) - (f_{h1} - f_{h2})\log(g_{bh}) - t]} \tag{9.50}$$

Hayakawa's procedure involves using the lethality tables to determine U for each segment of the heating curve. However, because integration of the lethality in the U-tables is carried from the start of the process, the lethality under the second and succeeding line segments must be corrected by subtracting lethality up to times preceding the shift to the current line segment under consideration.

If there is one break in the heating curve, the following parameters define the line segments: j, I, f_{h1}, g_{bh}, f_{h2}, and g. Tabular values are obtained for U/f_h from Table 9.16 for g_{bh}/K_S and for g/K_s.

$$U = f_{h1}\left[\frac{U}{f_h}\right]_{g_{bh}} + f_{h2}\left[\left[\frac{U}{f_h}\right]_g - \left[\frac{U}{f_h}\right]_{g_{bh}}\right] \tag{9.51}$$

If there are two breaks in the heating curve, the following parameters define the line segments: j, I, f_{h1}, g_{bh1}, f_{h2}, g_{bh2}, f_{h3}, and g. Tabular values from Table 9.15 can be obtained for U/f_h at g_{bh1}/K_S, g_{bh2}/K_S, and g/K_S.

$$U = f_{h1}\left[\frac{U}{f_h}\right]_{g_{bh1}} + f_{h2}\left[\left[\frac{U}{f_h}\right]_{g_{bh2}} - \left[\frac{U}{f_h}\right]_{g_{bh1}}\right]$$
$$+ f_{h3}\left[\left[\frac{U}{f_h}\right]_g - \left[\frac{U}{f_h}\right]_{g_{bh2}}\right] \tag{9.52}$$

Evaluation of lethality under the cooling curve is the same as in the section "Formula Methods for Thermal Process Evaluation."

Stumbo's procedure involves evaluation of lethality of individual segments of the heating curve. Because the f_h/U versus g tables were developed to include the lethality of the cooling part of the process, a correction needs to be made for the lethality of cooling attributable to the first line segment of the heating curve, which does not exist. The "r" parameter was used to express the fraction of the total process lethality attributed to the heating part of the process.

For the first line segment which ends when $(T_r - T) = g_{bh}$:

$$U_1 = r\frac{f_{h1}}{[f_h/U]_{g_{bh}}} \tag{9.53}$$

The second line segment begins when $(T_r - T) = g_{bh}$ and ends when $(T_r - T) = g$. The lethality from the f_h/U tables considers the heating process with the same f_h value starting from time zero, therefore the effective lethality up to $(T_r - T) = g_{bh}$ must be subtracted from the total.

$$U_2 = \frac{f_{h2}}{(f_h/U)_g} - r\frac{f_{h2}}{(f_h/U)_{g_{bh}}} \tag{9.54}$$

Thus, the total U for the process is

$$U = \frac{f_{h2}}{(f_h/U)_g} + \frac{r(f_{h1} - f_{h2})}{(f_h/U)_{g_{bh}}} \tag{9.55}$$

The denominator, $(f_h/U)_g$ or $(f_h/U)_{g\,bh}$ in Equations (9.52) to (9.55) represent tabular values for f_h/U corresponding to g or g $_{bh}$. The parameter "r" is a function of g. Figure 9.18 can be used to obtain r corresponding to g.

Example 9.20. For the product that exhibited the heating curve shown in Fig. 9.17, assume $f_c = f_{h2}$ and $j_c = j$. This product is processed for 50 minutes at 248°F, from an initial temperature of 140°F. Calculate F_0 and the probability of spoilage from an organism having a D_0 value of 1.5 minutes and a z value of 16°F in cans given this process. The initial spore load is 100/can. The cooling water temperature is 60°F.

Solution:

Note that the conditions under which the product is processed are different from those under which the heat penetration parameters were derived. Figure 9.17 is simply used to determine the heat penetration parameters, and these parameters are utilized in the specific process to evaluate the process lethality. From Fig. 9.17: $f_{h1} = 27.5$ minutes; $f_{h2} = 60.5$ minutes; $g_{bh} = 12°F$; $I = 248 - 140 = 108°F$; $j = j_c = 1.33$. Solving for g using Equation (9.50):
The exponent is:

$$g = [10]^{\left(\frac{1}{f_{h1}} \log(jI) - (f_{h1} - f_{h2}) \log(g_{bh}) - t\right)}$$

$$\text{Exponent} = \frac{27.5 \log(1.33)(108) - (27.5 - 60.5) \log(12) - 50}{60.5}$$

$$\text{Exponent} = \frac{[27.5(2.157) + 33(1.0792) - 50]}{60.5} = 0.743$$

$$g = (10)^{0.743} = 5.53°F$$

Using Stumbo's procedure: The F_0 value for the process is determined using Table 9.14 for z = 18°F and $j_c = 1.33$. The value of f_h/U corresponding to g = 5.53°F and that corresponding to $g_{bh} = 12°F$ will have to be determined by interpolation. Inspection of Table 9.12 shows that for $j_c = 1$, g = 5.40 corresponds to f_h/U and the interpolating factor $\Delta g/\Delta j = 1.59$. Thus:

$$\frac{f_h}{U} = 5; g_{j=1.33} = 5.40 + 1.59(0.33) = 5.925$$

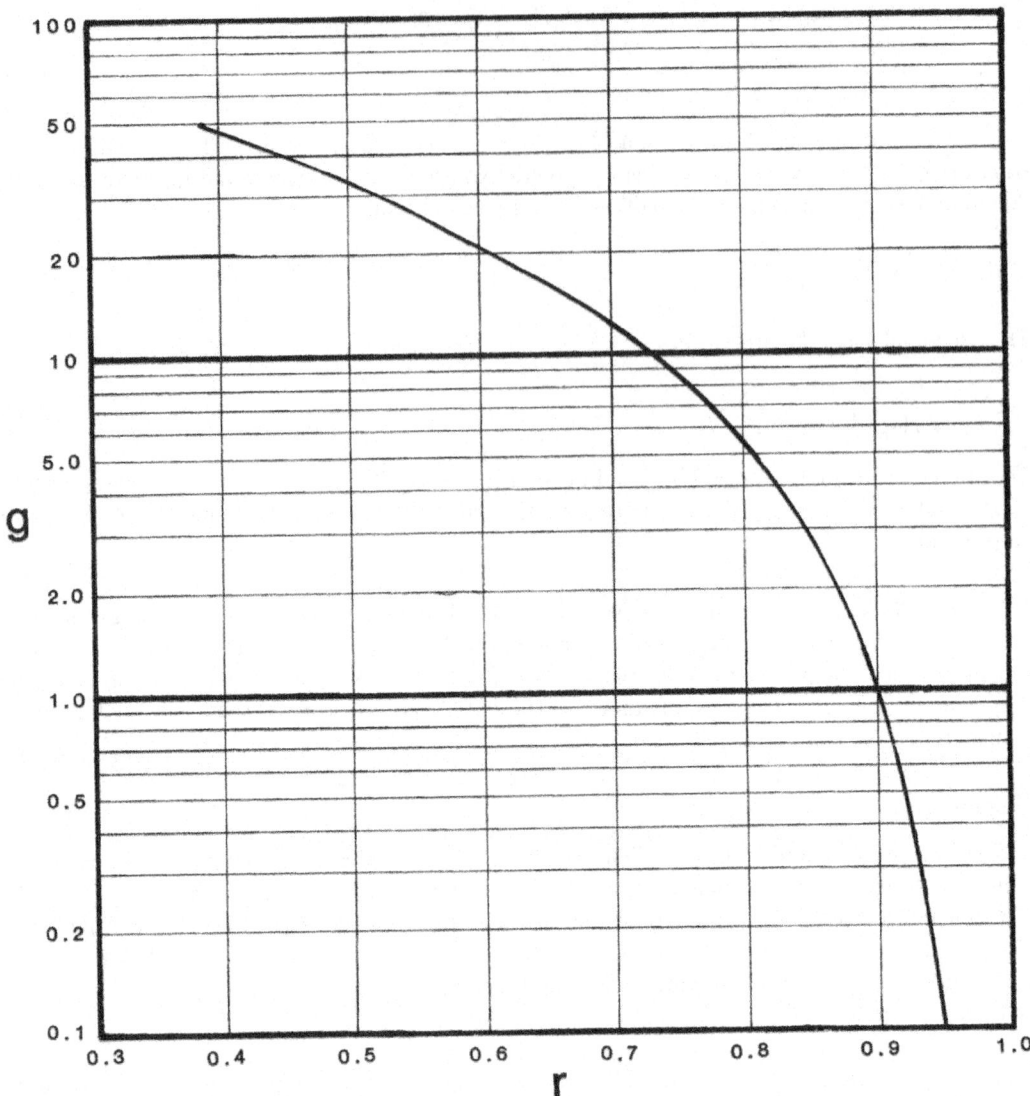

Figure 9.18 Values of the parameter r corresponding to the value of g. (Source: Anonymous. 1952. Calculation of process for canned food. American Can Company, Technical Services Division, Maywood, IL.)

This value is higher than 5.53; therefore, a lower value of f_h/U is picked for the second set of values to use in the interpolation. Thus:

$$\frac{f_h}{U} = 4; g_{j=1.33} = 4.41 + 1.34(0.33) = 4.852$$

The value of f_h/U corresponding to $g = 5.53$ is calculated by interpolating between the above two tabular values of g:

$$\left(\frac{f_h}{U}\right)_{g=5.53} = 4 + \frac{(5-4)(5.53-4.852)}{(5.925-4.852)} = 4.632$$

Next, the value of f_h/U corresponding to g_{bh} needs to be evaluated. g_{bh} is entered into Table 9.14 as a value of g, and the corresponding f_h/U is determined.

Inspection of Table 9.14 shows that for $j_c = 1$, $z = 18$, and $f_h/U = 15$ corresponds to $g = 10.88$ with the interpolating factor $\Delta g/\Delta j = 3.57$. Thus:

$$\frac{f_h}{U} = 15; \quad g_{j=1.33} = 10.88 + 0.33(3.57) = 12.058$$

The value of g is greater than 12; therefore, the next lower tabular entry is selected as the other set of values used in the interpolation.

$$\frac{f_h}{U} = 10; \quad g_{j=1.33} = 8.78 + 0.33(2.69) = 9.668$$

Interpolating between these two values to obtain f_h/U corresponding to $g = 12$:

$$\left(\frac{f_h}{U}\right)_{g_{bh}} = \left(\frac{f_h}{U}\right)_{g=12} = 10 + 5(12 - 9.668)/(12.058 - 9.668) = 14.88$$

Equation (9.55) is used to determine U from values of $(f_h/U)_g$ and $(f_h/U)_{g\,bh}$. r needs to be evaluated from Fig. 9.18 to correspond to $g_{bh} = 12$. From Fig. 9.18, $r = 0.71$. Using Equation (9.55):

$$U = \frac{60.5}{4.622} + \frac{0.71(27.5 - 60.5)}{14.88} = 13.08 - 1.574 = 11.51$$

Using Equation (9.45):

$$F_i = (10)^{2/18} = 1.291$$
$$F_0 = U/F_i = 11.51/1.291 = 8.91 \text{ minutes}$$

The probability of spoilage: Because the organism has a z value of 16°F, F_0 cannot be used in Equation (9.11) to determine the probability of spoilage. It is necessary to calculate F_0^{16} for the conditions used in the process. Because the process is the same, g calculated in the first part of this problem is the same; $g = 5.53$. Evaluating f_h/U corresponding to this value of g, however, should be done using $z = 16°F$ in Table 9.14. A double interpolation needs to be done. Values for f_h/U corresponding to $g = 5.53$ are determined for $z = 14$ and $z = 18$; and the two values are interpolated to obtain f_h/U at $z = 14$.

$f_h/U = 6$; $g_{j=1.33,z=14} = 4.63 + 0.33(1.56) = 5.145$
$g_{j=1.33,z=18} = 6.5 + 0.33(1.82) = 6.850$
$f_h/U = 6$; $g_{j=1.33,z=16} = 5.145 + [(18-16)(6.850-5.145)]/(18-14) = 5.997$
$f_h/U = 5$; $g_{j=1.33,z=14} = 4.02 + 0.33(1.32) = 4.455$
$f_h/U = 5$; $g_{j=1.33,z=18} = 5.40 + 0.33(1.59) = 0.925$
$f_h/U = 5$; $g_{j=1.33,z=16} = 4.455 + (18-16)(5.925-4.455)/(18-14) = 5.190$

Interpolating between the two values of f_h/U at $z = 16$, which straddles $g = 5.53$:

$$(f_h/U)_{g=5.53,z=16} = 5 + (6-5)(5.53-5.190)/(5.997-5.190) = 5.42$$

The same procedure is used to determine f_h/U corresponding to $g_{bh} = 12$

$f_h/U = 20$; $g_{j=1.33,z=14} = 9.63 + 0.33(2.96) = 10.61$
$f_h/U = 20$; $g_{j=1.33,z=18} = 12.4 + 0.33(4.28) = 13.81$
$f_h/U = 20$; $g_{j=1.33,z=16} = 10.61 + (16 - 14)(13.81 - 10.61)/(18 - 14) = 12.21$
$f_h/U = 15$; $g_{j=1.33,z=18} = 10.88 + 0.33(3.57) = 12.06$
$f_h/U = 15$; $g_{j=1.33,z=14} = 8.29 + 0.33(2.68) = 9.17$
$f_h/U = 15$; $g_{j=1.33,z=16} = 9.7 + (16 - 14)(12.06 - 9.17)/(18 - 14) = 11.14$

Interpolating between the two values of f_h/U at $z = 16$, which correspond to values of g, which straddle $g_{bh} = 12$:

$(fh/U)_{g\,b\,h} = 15 + (12 - 11.14)(20 - 15)/(12.21 - 11.14) = 19.02$

Because r is dependent only on g_{bh}, the same value as before is obtained from Fig. 9.18. For $g_{bh} = 12$, $r = 0.71$. Using Equation (9.55):

$$U = \frac{60.5}{5.41} + \frac{0.71(27.5 - 60.5)}{19.02} = 11.182 - 1.232 = 9.95$$
$$F_i = (10)^{2/16} = 1.333$$
$$F_0^{16} = U/F_i = 9.95/1.333 = 7.465$$

The number of survivors can now be calculated by substituting F_0^{16} for t in Equation (9.11). Using Equation (9.11):

$$\log(N/100) = -F_0/D_0 = -7.465/1.5 = -4.977$$
$$N = 10^{-4.977} = 0.0011$$

The probability of spoilage is 11 in 10,000.

Hayakawa's procedure: The solution for F_0 will not be presented here. The procedure will the same as for the determination of F_0^z, which is shown below. As an exercise, the reader can determine F_0. The calculated value is 8.16 minutes. The probability of spoilage is determined by calculating F_0^{16}.

For the heating part of the process, U is calculated using Table 9.16 to obtain $(f_h/U)_{gbh}$, which is substituted in Equation (9.51). The parameter for the table entry in Table 9.16 is $g/K_s = 5.53/0.8 = 6.91$. g was previously calculated for this problem to be $5.53°F$ and $K_s = 16/20 = 0.8$. $g_{bh}/K_s = 12/0.8 = 15.00$.

From Table 9.16: $(f_h/U)_{g/Ks=6} = 0.1555$ and $(f_h/U)_{g/Ks=7} = 0.1252$. f_h/U for $g/K_s = 6.91$ is obtained by interpolating between the above values.

$(f_h/U)_g = 0.1555 - (0.1555 - 0.1252)(6.91 - 6)/(7 - 6) = 0.1279$

From Table 9.16, for $g_{bh}/K_s = 15.00$:

$(f_h/U)_{gbh} = 0.02706$

There is only one break in the heating curve; therefore, Equation (9.51) is used to determine U for the heating portion of the process.

$U = 27.5(0.02706) + 60.5(0.1279 - 0.02706) = 6.84$ minutes

The lethality of the cooling curve is determined using Tables 9.17 to 9.21. Tabular entry is done using I_c/K_s. Because $g = 5.53°F$, $T_g = 248 - 5.53 = 242.47$. $I_c = 242.47 - 60 = 182.47$;

$I_c/K_s = 182.47/0.8 = 228$. Table 9.19 is the appropriate table to use, because $I_c/K_s > 200$ and $j_c < 1.9$. From Table 9.19:

$(U = /f_c)_{Ic/Ks=225,jc=1.2} = 0.06604$

$(U = /f_c)_{Ic/Ks=225,jc=1.4} = 0.08862$

Interpolating for $j_c = 1.33$:

$(U = /f_c) = 0.06604 + [(0.08862 - 0.06604)(1.22 - 1.2)]/0.2 = 0.8072$

From Table 9.19:

$(U = /f_c)_{Ic/Ks = 230,jc = 1.2} = 0.06530$

$(U = /f_c)_{Ic/Ks = 230,jc = 1.4} = 0.08782$

Interpolating for $j_c = 1.33$:

$(U = /fc) = 0.06530 + [(0.08782 - 0.06530)(1.33 - 1.2)]/0.2 = 0.07994$

Solving   for $(U = /f_h)_{Ic/Ks=228,jc=1.33} = 0.07994 + [(0.08782 - 0.07994) \times (230 - 228)]/5 = 0.0831$

$U = f_c(U/f_c)_{Ic/Ks = 228, jc = 1.33} = 60.5(0.0831) = 5.027$

$U =$ is then converted to U of the cooling part of the process using Equation (9.46).

$U = 5.027|10|^{-5.53/16} = 5.027(0.452) = 2.27$ minutes

The total U is the sum of U for heating and U for cooling.

$U = 2.27 + 6.86 = 9.13$ minutes

$F_0^{16} = 9.13|10|^{248-250/16} = 6.85$ minutes

The number of survivors is calculated using Equation (9.11) by substituting F_0^{16} for t:

$\text{Log} (N/100) = -6.85/1.5 = -4.567$

$N = 100(10)^{-4.567} = 0.0027$

The probability of spoilage $= 27/10,000$.

Hayakawa's procedure results in a higher probability of spoilage than Stumbo's procedure. F_0 calculated using Hayakawa's procedure is lower than that calculated using Stumbo's procedure, which in turn is lower than that calculated using the general method, as shown in the examples in the section "Determination of f_c and j_c" and "Formula Method for Thermal Process Evaluation." The safety factor built unto the formula methods is primarily responsible for the success with which these thermal process calculation techniques have served the food industry over the years in eliminating the botulism hazard from commercially processed canned foods.

9.8   QUALITY FACTOR DEGRADATION

Quality factor degradation has to be evaluated on the basis of integrated lethality throughout the container. Unlike microbial inactivation, which leaves practically zero survivors at regions in the can near the surface, there is substantial nutrient retention in the same regions. Quality factor degradation can be determined by separating the container into incremental cylindrical shells, calculating the temperature at each shell at designated time increments, determining the extent of degradation, and summing the extent of degradation at each incremental shell throughout the process. At the termination of the process, the residual concentration is calculated by integrating the residual concentration at each incremental cylindrical shell throughout the container. The procedure is relatively easy to perform using a computer, but the calculations can be onerous if done by hand.

For cylindrical containers, Stumbo (1973) derived an equation for the integrated residual nutrient based on the following observations on the temperature profiles for conduction heat transfer in cylinders.

(a) In a container, an isotherm exists where the j value at that point, designated j_v, is 0.5 j. The g at that point at any time, designated g_v, is 0.5 g, and the volume enclosed by that isotherm is 19% of the total volume.

(b) If v is the volume enclosed by the isotherm, and if the F value at the isotherm and at the critical point are F_v and F, respectively, then the difference, $(F_v - F)$, is proportional to $\ln(1 - v)$.

The following expression was then derived:

$$\overline{F} = F + D \log \left(\frac{D + 10.92(F_v - F)}{D} \right) \tag{9.56}$$

Equation (9.56) is an expression for the integrated lethal effect of the heating process $(\overline{F})$ on nutrients based on the lethality at a the critical point, F, and the lethality F_v evaluated at a point where $g_v = 0.5$ g and $j_v = 0.5$ j. Equation (9.54) has been found to be adequate for estimating nutrient retention in cylindrical containers containing foods that heat by conduction.

Example 9.21. A food product has a j value of 1.2, a j_c value of 1.4, and $f_h = f_c = 35$ minutes. This product is processed at 255°F from an initial temperature of 130°F to an F_0 of 5.5. Calculate the residual ascorbic acid remaining in this product after the process if the initial concentration was 22 Φg/g. The D_0 value for ascorbic acid in the product is 248 minutes, and the z value is 91°F.

Solution:

Stumbo's procedure will be used. The F_0 value is used to determine a value of g using the f_h/U tables for z = 18 and $j_c = 1.4$. From this value of g, a new f_h/U is determined for a z value of 91°F. The F_0^{91} obtained is the value of F in Equation (9.56). F_v is determined from f_h/U, which corresponds to $g_v = 0.5g$ and $j_v = 0.5j_c$.

$$U = F_0 F_i = 5.5(10)^{-5/18} = 2.901 \text{ minutes}$$

$$f_h/U = 35/2.901 = 12.06$$

From Table 9.14, for z = 18:

$$\frac{f_h}{U} = 10; \ g_{jc=1.4} = 8.78 + 2.69(0.4) = 9.956$$

$$\frac{f_h}{U} = 15; \ g_{jc=1.4} = 10.88 + 3.57(0.4) = 12.308$$

For $f_h/U = 12.06$:

$$g = 9.936 + (12.308 - 9.956)(12.06 - 10)/(15 - 10) = 10.925$$

For $g = 10.925$, using Table 9.15 , $z = 90°F$:

$$\frac{f_h}{U} = 1; g_{jc=1.4} = 3.06 + 0.4(2.19) = 3.936$$

f_h/U for $g = 10.925$:

$$\frac{f_h}{U} = 2; g_{jc=1.4} = 11.03 + 0.4(7.88) = 14.182 = 1 + \frac{(2-1)(10.925 - 3.936)}{(14.182 - 3.936)} = 1.682$$

U at the geometric center $= 35 / 1.682 = 20.81$ minutes.
$F_i = 10^{-5/90} = 0.8799$.
F at the geometric center $= U/F_i = 23.65$ minutes.
At a point where $g_v = 0.5$ g and $j_{vc} = 0.5j_c$, $g = 0.5(10.925) = 5.463$; $j_{vc} = 0.5(1.4) = 0.7$.
Using Table 9.15, $z = 90°F$:

$$\frac{f_h}{U} = 1; g_{jc=0.7} = 3.06 - 0.3(2.19) = 2.403$$

f_h/U for $g_v = 5.463$:

$$\frac{f_h}{U} = 2; g_{jc=0.7} = 11.03 - 0.3(7.88) = 8.666 = 1 + \frac{(2-1)(5.463 - 2.403)}{(8.666 - 2.403)} = 1.488$$

U at point where $g_v = 0.5$ g $= 35 /1.488 = 23.52$ minutes.
$F_v = U/F_I = 23.52 / 0.8799 = 26.73$ minutes.
Substituting in Equation (9.56):

$$\overline{F} = 23.65 + 248 \log \left[\frac{248 + 10.92(26.73 - 23.65)}{248} \right]$$

$$= 23.65 + 248(0.055233) = 37.348$$

$$\log \left(\frac{C}{C_0} \right) = -\frac{F}{D} = \frac{-37.348}{248} = -0.1506$$

$$\frac{C}{C_0} = (10)^{-0.1506} = 70.6$$

The percent retention of ascorbic acid is 70.6%.
The residual ascorbic acid content is $0.706(22) = 15.55$ Φg/g.

Hayakawa's tables may also be used to determine F and F_v to use in Equation (9.56). U for the heating and cooling portions of the process has to be evaluated separately, as was done for microbial inactivation. When values of z are not the same as the tabulated values in Table 9.16, use of Hayakawa's tables is recommended because interpolation across large values of z in Table 9.16 may introduce too much of an error.

PROBLEMS

9.1. Calculate the D value of an organism that shows 30 survivors from an initial inoculum of 5 $\times 10^{6}$ spores after 10 minutes at 250°F.

9.2. What level of inoculation of PA 3679 ($D_0 = 1.2$ minutes) is required such that a probability of spoilage of 1 in 100 attributed to PA 3679 would be equivalent to 12D inactivation of *C. botulinum*? Assume the same temperature process and the same z values for both organisms. The D_0 value of *C. botulinum* is 0.22 minutes.

9.3. Calculate the length of a holding tube in high-temperature processing in an aseptic packaging system that would be necessary to provide a 5D reduction of spores of PA 3679 ($D_{250} = 1.2$ minutes) at 280°F. Use a z value of 20°F. The rate of flow is 30 gal/min, density is 65 lb/ft³, and viscosity is 10 cp. The tube has 1.5-in. outside diameter and has a wall thickness of 0.064 in.

9.4. If the same system were used on another fluid having a density of 65 lb/ft³ and a viscosity of 100 cp, calculate the probability of spoilage when the process is carried out at 280°F (z = 20). The initial inoculum is 100 spores/can ($D_{250} = 1.2$ minutes). The rate of flow is 30 gal/min on a 1.5-in. outside diameter tube (wall thickness 0.064 in.).

9.5. If an initial inoculum of 10 spores/g of produce ($D_{250} = 1.2$ minutes) and a spoilage rate of 1 can in 100,000 is desired, calculate an F value for the process that would give the desired level of inactivation. Calculate the F_{280} for a z value of 18°F.

9.6. If an organism has a D value of 1.5 at 250°F and a z value of 15°F, calculate the F_{240} for a probability of spoilage of 1 in 10,000 from an initial inoculum of 100 spores/can.

9.7. Figure 9.19 shows an air sterilization system that supplies sterile air to a process. Calculate the length of the holding tube necessary to sterilize the air. The most heat resistant organism that must be avoided requires 60 minutes of heating at 150°C sterilization and has a z value of 70°C. The inside diameter of the holding tube is 0.695 in. Assume plug flow ($V_{max} = V_{avg}$).

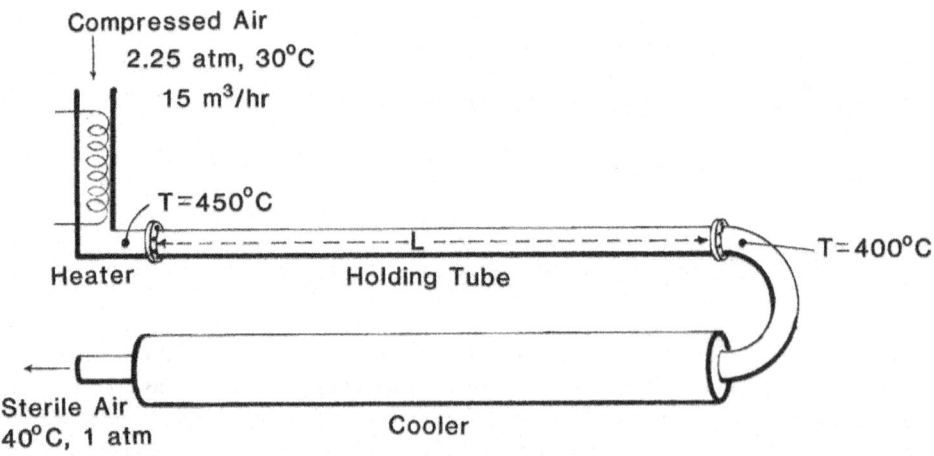

Figure 9.19 Diagram of an air sterilization system by heat (for Problem 7).

9.8. (a) A food product in a 303 × 407 can has an $f_h = 5$ and a $j = j_c = 0.8$. For an initial temperature of 80°F and a retort temperature of 250°F, calculate the process time B_b. Use an $F_0 = 4$ minutes and $z = 18°F$.

(b) The product in part (a) is processed in a stationary retort and it takes 4 minutes for the retort to reach 250°F from the time the steam was turned on. How many minutes after turning the steam on should the steam be turned off?

(c) In one of the retorts where the cans were processed, there was a mis-process and the record on the retort temperature chart showed the following:

Time (min)	Retort temperature °F
0	70
3	210
10	210
Sudden jump from 210°FE to 250°F at 10 minutes	
15	250
16	Steam off, cooling water on

What is the F_0 of this process. The can temperature at time 0 was 80°F.

9.9. In a given product, PA 3679 has a D value of 3 minutes at 250°F and a z value of 20°F. If the process was calculated at 280°F for a z value of 18°F, how many minutes of heating is required for a 5D process? What would be the actual probability of spoilage of PA 3679 if N_0 is 100 spores/can?

9.10. The following heat penetration data were obtained on Chili Con Carne processed at 250°F in a retort having a come-up time of 3 minutes. Assume $j = j_c$.

(a) Calculate the f_h and j values and processes at:

250°F, $z = 18$, $F_0 = 8$ (initial temperature = 120°F)
240°F, $z = 18$, $F_0 = 8$ (initial temperature = 120°F)
260°F, $z = 18$, $F_0 = 8$ (initial temperature = 120°F)

(b) Calculate the probability if spoilage from FS 1518 that might occur from the process calculated at 240°F, $z = 18$, $F_0 = 8$ if FS 1518 has a D value at 250 of 4 minutes and a z value of 22°F for an initial spore load of 50/can.

Heat penetration data

Time (min)	Temp. (°F)	Time (min)	Temp. (°F)
0	170	35	223
5	170	40	228
10	180	45	235
15	187	50	236
20	200		
25	209		
30	216		

9.11. A canned food having an f_h of 30 and a $j = j_c$ of 1.07 contains a spore load of 56 organisms per can and this organism had a D_{250} value of 1.2 minutes. A process with an $_0$ of 6 minutes was calculated for this product at a retort temperature of 250°F and an initial temperature of

150°F. Subsequent analysis revealed that the spores actually have a z value of 14°F instead of 18°F. If the same time as the above process was used at a retort temperature of 248°F, calculate the probability of spoilage.

9.12. A process for a pack of sliced mushrooms in 303 × 404 cans on file with the FDA specifies a processing time at 252°F for 26 minutes from an initial temperature of 110°F. A spoiled can from one pack was analyzed microbiologically and was found to contain spore-forming organisms. Data on file for similar products show j values ranging from 0.98 to 1.15 and f_h values ranging from 14 to 18 minutes.

(a) Would the filed process be adequate to provide at least a 12D reduction in spores of C. *botulinum* (D = 0.25 min; z = 14°F)?

(b) If the spores of the spoilage organism have D_0 of 1.1 minutes and z of 16°F, what would have been the initial number of organisms in the can to result in a probability of spoilage of 1 in 10,000 after the process.

(c) If you were evaluating the process, would you recommend a recall of the pack for in-adequate processing? Would you be calling for additional technical data on the process before you made a recommendation? Explain your action and provide as much detail to convince a non- technical person (i.e., lawyers and judges) that your action is the correct way to proceed.

9.13. The following data were collected in a heat penetration test on a canned food for thermal process determination.

Time (min)	Temp. (°F)	Time (min)	Temp. (°F)
0	128	35 (cool)	245
3	128	40	243
5	139	45	240
10	188	50	235
15	209	55	185
20	229	60	145
25	238	65	120
30	242	70	104

The processing temperature was 250°F and the retort come-up time was 2 minutes. Cooling water temperature was 60°F.

Calculate:

(a) The values of f_h, f_c, j_h, and j_c.

(b) If this product is processed from an initial temperature of 150°F at 248°F, how long after retort temperature reaches 248°F must the process be carried out before the cooling water is turned on if the final can temperature at the time of cooling must reach to within 2 degrees of the retort temperature.

(c) If the process is to be carried out at 252°F from an initial temperature of 120°F, calculate a process time such that an organism with a D_0 value of 1.2 minutes and a z of 18°F will have a probability of spoilage of 1 in 10,000 from an initial spore load of 100/can.

9.14. Beef stew is being formulated for canning. The marketing department of the company wants large chunks of meat in the can and they stipulate that the meat should be 5-cm cubes.

The current product utilizes 2-cm cubes of all vegetables (carrots and potatoes) and meat, and the process time used is 50 minutes at 250°F from an initial temperature of 150°F.

Marketing thinks that the change can be made without major alteration of the current process.

Heat penetration data for the current product obtained from the files did not specify if the thermocouple was embedded in a particle during the heating process. The f_h value was reported to be 35 minutes and j of heating was 1.55. There was no data available on j of cooling. There was, however, an inoculated pack done where an inoculum of 1000 spores of an organism having a D_0 value of 1.2 minutes and a z value of $18°F$ injected into a single meat particle in each can resulted in a spoilage rate of 3 cans in 1000.

(a) Calculate the F_0 of the process used on the current product based on the heat penetration data available.

(b) Calculate the F_0 of the process based on the survivors from the inoculated microorganisms. Does the inoculated pack data justify the assumption of a safe process on the current product?

(c) Is it likely that the heat penetration data was obtained with the thermocouple inside a particle or was it simply located in the fluid inside the can? Explain your answer.

(d) Calculate the most likely value for the f_h if the thermocouple was located in the center of a particle, based on the inoculated pack data, assuming that the j value would be the theoretical j for a cube of 2.02.

(e) If the f_h varies in direct proportion to the square of the cube size, and j remains the same at 2.02, estimate the process time for the 5-cm size cube such that the F_0 value for the process will be similar to that based on the inoculated pack data on the present product.

9.15. A biological indicator unit (BIU), which consists of a vial containing a spore suspension and installed at the geometric center of a can, was installed to check the validity of a thermal process given a canned food. The canned food has an f_h value of 30 minutes and a j value of 1.8. Of the 1000 spores originally in the BIU, an analysis after the process showed a survivor of 12 spores. The spores in the BIU has a D_0 of 2.3 min and a z value of $16°F$. The process was carried out at $248°F$ from an initial temperature of $140°F$. Calculate:

(a) The F_0 value received by the geometric center of the can.

(b) The process time.

(c) The sterilizing value of the process expressed as a number of decimal reductions [(log N_0/N) of *C. botulinum* having a D_0 value of 0.21 minutes and a z value of $18°F$].

9.16. A canned food with an f_h of 30 minutes and j of 1.2 is to be given a process at $250°F$ with an F_0 of 8 minutes and a z of $18°F$. In order to verify the adequacy of the process, an inoculated pack is to be performed using an organism with a D_0 of 1.5 minutes and a z of $22°F$. If the process to be used on the inoculated pack is at $250°F$ from an initial temperature of $130°F$, calculate the number of spores that must be inoculated per can such that a spoilage rate of 10 in 100 will be equivalent in lethality to the process F_0 desired. Assume j of heating and cooling are the same.

SUGGESTED READING

Aiba, S., Humphrey, A. E., and Millis, N. F. 1965. Biochemical Engineering. Academic Press, New York.

Anon. 1952. Calculation of Process for Canned Food. American Can Co., Technical Services Division, Maywood, IL.

Ball, C. O. and Olson, F. C. W. 1957. Sterilization in Food Technology. lst ed. McGraw-Hill Book Co., New York.

Charm, S. E. 1971. Fundamentals of Food Engineering. 2nd ed. AVI Publishing Co., Westport, CT.

Cleland, A. C. and Robertson, G. L. 1986. Determination of thermal process to ensure commercial sterility of food in cans. In: Developments in Food Preservation—3. S. Thorn, Ed. Elsevier, New York.

Edgerton, E. R. and Jones, V. A. 1970. Effect of process variables on the holding time in an ultra high temperature steam injection system. J. Dairy Sci. 53:1353–1357.

Hayakawa, K. 1970. Experimental formulas for accurate estimation of transient temperature of food and their application to thermal process evaluation. Food Technol. 24:1407.

Institute of Food Technologists. 2001. Kinetics of microbial inactivation for alternative food processing technologies. J. Food Sci. 66(Suppl).

Leniger, H. A. and Beverloo, W. A. 1975. Food Process Engineering. D. Riedel Publishing Co., Boston.

Lewis, M. and Heppell W. 2000. Continuous Thermal Processing of Food. Aspen Publishers Inc., Gaithersburg, MD.

Richardson, P. 2004. Improving theThermal Processing of Foods. CRC Press, Boca Raton, FL.

Ruthfus, R. R., Archer, D. H., Klimas, I. C., and Sikchi, K. G. 1957. Simplified flow calculations for tubes and parallel plates. AIChE J. 3:208.

Saravacos, G. D. and Kostaropoulos, A. E. 2002. Handbook of Food Processing Equipment. Kluwer Acdemic/Plenum Publishers, New York.

Stumbo, C. R. 1973. Thermobacteriology in Food Processing. 2nd ed. Academic Press, New York.

Zeuthen, P, Chefter, J. C., Eriksson, C., Gormley, T. T., Linko, P., and Paulus, K., Eds. 1990. Processing and Quality of Foods. Vol. I. Elsevier, New York.

CHAPTER 10

Refrigeration

Cooling is a fundamental operation in food processing and preservation. Removal of heat could involve either the transfer of heat from one fluid to another or from a solid to a fluid, or it could be accomplished by vaporization of water from a material under adiabatic conditions. Knowledge of the principles of heat transfer is an essential prerequisite to the understanding of the design and operation of refrigeration systems.

Maintaining temperatures lower than ambient inside a system requires both the removal of heat and prevention of incursion of heat through the system's boundaries. The rate of heat removal from a system necessary to maintain the temperature is the refrigeration load. Refrigeration systems must be sized to adequately handle the refrigeration load. When heat has to be removed from a system continuously, at temperatures below ambient and for prolonged periods, a mechanical refrigeration system acts as a pump that extracts heat at low temperatures and transfers this heat to another part of the system where it is eventually dissipated to the surroundings at a higher temperature. The operation requires energy, and a well-designed system will allow the maximum removal of heat at minimum energy cost.

10.1 MECHANICAL REFRIGERATION SYSTEM

10.1.1 Principle of Operation: The Heat Pump

The second law of thermodynamics mandates that heat will flow only in the direction of decreasing temperature. In a system that must be maintained at a temperature below ambient, heat must be made to flow in the opposite direction. A refrigeration system may be considered as a pump that conveys heat from a region of low temperature to another region that is at a high temperature.

The low temperature side of a refrigeration system is maintained at a lower temperature than the system it is cooling to allow spontaneous heat flow into the refrigeration system. The high temperature side must have a temperature higher than ambient to allow dissipation of the absorbed heat to the surroundings. In some instances, this absorbed heat is utilized as a heat source for use in heating processes.

Maintaining a high and a low temperature in a refrigeration system is made possible by the use of a refrigerant fluid that is continuously recirculated through the system. A liquid's boiling or condensation temperature is a function of the absolute pressure. By reducing the pressure, a low boiling temperature is made possible, allowing for absorption of heat in the form of the heat of the refrigerant's vaporization

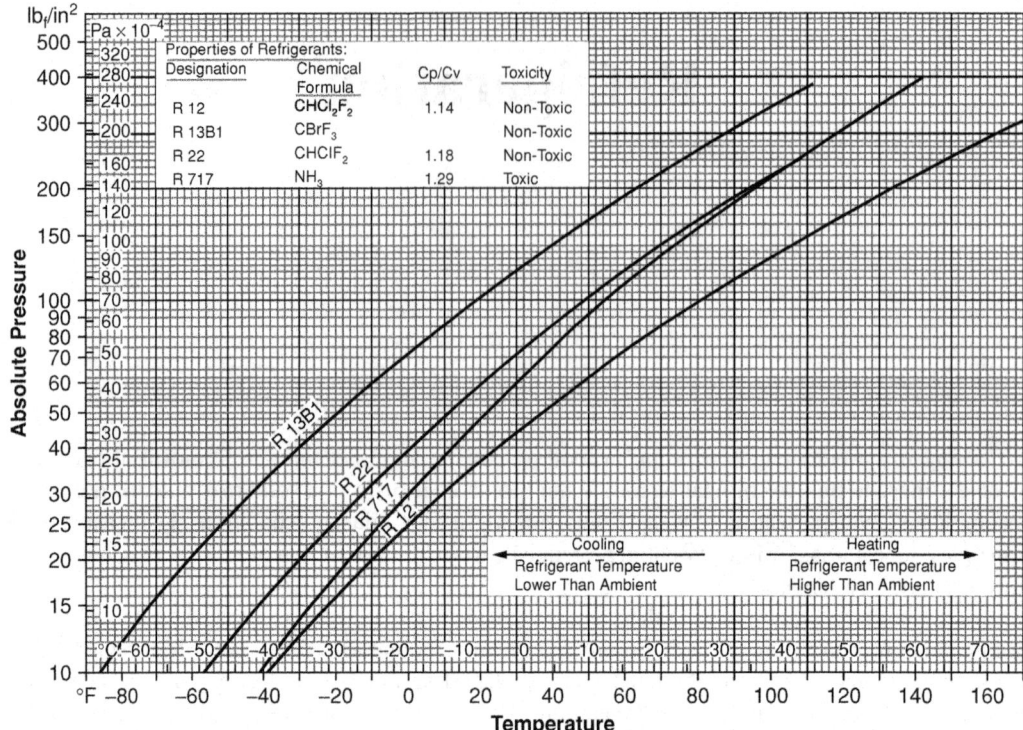

Figure 10.1   Vapor pressure of commonly used refrigerants as a function of temperature.

as it is vaporized at the low pressure and temperature. The vapors, when compressed to a high pressure, will condense at the high temperature and the absorbed heat will be released from the refrigerant as it condenses back into liquid at the high temperature and pressure. Figure 10.1 shows the vapor pressure versus temperature of commonly used refrigerants. The diagram also illustrates how this pressure and temperature relationship is utilized for cooling and heating.

10.1.2   Refrigerants

Atmospheric contamination of refrigerants that contain chlorine and fluorine in the molecule (chlorofluorocarbon; CFC) has been shown to contribute to global warming and cause ozone depletion in the upper atmosphere. Thus by international agreement, manufacturing of CFCs was stopped after 1996. CFCs have the highest ozone depletion potential (ODP) among the refrigerants and also has high global warming potential (GWP). Manufacturing of hydrochlorofluorocarbon (HCFC) refrigerants is to be phased out in 2030. HCFCs have lower ODP but also contribute to global warming. Because no new CFCs are being manufactured, replacement in existing refrigeration systems must come from recovered CFCs or the refrigeration unit must be recharged with a completely new refrigerant. Non-CFC refrigerants must be used in new refrigeration systems. Existing refrigeration systems may be

Table 10.1 Refrigerants Used in the Food Industry

Regrigerant	Composition	Boiling Point at 1°C (°F)
R12	CCl_2F_2	−30 (−22)
R22	$CHClF_z$	−41 (−41)
R32	$C11_z1^7_z$	−52 (−62)
R115	$CClF_zCF_3$	−39 (−38)
R124	$CHClFCF_3$	−12 (10)
R125	CHF_zCF_3	−49 (−56)
RI34A	$CH_z FCH_3$	−26 (−15)
R142B	CH_3CCl_zF	−10 (14)
R143A	CH_3CF_3	−25 (−13)
R152A	CH_3CHF_2	−25 (−13)
R290	$CH_3CH_zCH_3$	−42 (−44)
C318	CA (Cyclic)	−6 (23)
R401A	R22, R152A, R124(53:13:34)	
R401B	R22, R152A, R124(61:11:28)	
R402A	R125, R290, R22 (60:2:38)	
R402B	R125, R290, R22 (38:2:60)	
R404A	R125, R143A, R134A (44:52:4)	
R405A	R22, R152A, R142B, C318 (45:7:5.5:42.5)	
R407A	R32, R125, R134A (20:40:40)	
R407B	R32, R125, R134A (10:70:20)	
R407C	R32, R125.R134A (23:25:52)	
R410A	R32, R125 (50:50)	
R502	R22, R115 (48.8:51.2)	−45 (−49)
R507A	R125, R143A (50:50)	−46.7 (−52.1)

R 502 and R507A are azeotropes, while the R400 series are zeotropes

Source: ASHRAE Standards 2000. Standards for Designation and Safety of Refrigerants.

upgraded by utilizing a drop-in refrigerant (i.e., the new refrigerant is added after the original CFC refrigerant has been removed).

Refrigerants now used in the food industry include R502 for transport refrigeration, R502 and R22 for retail display cases and retail central storage, R502 for cold storage, R22 for refrigerated storage and refrigerated vending machines, and R717 for large freezers, frozen storage warehouses, and large refrigerated warehouses. R502 contains R12, which is no longer manufactured. R22 can still be used but will be phased out later. Thus, alternatives to these refrigerants must be selected for retrofits to existing installations or for new installations. For long-term retrofit, alternatives are R407A, R407B, and R507; R407C and R410A and R134A for R502, R22, and R12, respectively.

Table 10.1 lists the refrigerants used in the food industry and their boiling points. Also listed are alternative non-CFC refrigerants. Most alternatives are mixtures that are either azeotropes or zeotropes. Azeotropes are mixtures with a constant boiling point and where the vapor and liquid composition are the same. Zeotropes are mixtures where vapor and liquid compositions are different at different temperatures and there is a large gap between the dew point (the temperature where liquid first starts to form) and the bubble point (the temperature where the mixture is completely in the liquid phase).

The use of alternative refrigerants may require different compressor lubricant, so this has to be taken into consideration in retrofits.

10.1.3   The Refrigeration Cycle

Figure 10.2 shows a schematic diagram of a mechanical refrigeration system. The heart of the system is the *compressor*.

When the compressor is operating, refrigerant gas is drawn into the compressor continuously. Low pressure is maintained at the suction side and because of the low pressure, the refrigerant can vaporize at a low temperature. In the compressor, the refrigerant gas is compressed increasing in both pressure and temperature during the process. The hot refrigerant gas then flows into a heat exchange coil called the *condenser* where heat is released in the process of condensation at constant pressure and temperature. From the condenser, the liquid refrigerant flows into a liquid refrigerant holding tank. In small systems, there may be no holding tank, and the refrigerant just continuously cycles through the system.

When there is a demand for cooling, the liquid refrigerant flows from the holding tank to the low pressure side of the refrigeration system through an *expansion valve*. The drop in pressure that occurs as the refrigerant passes through the expansion valve does not change the heat content of the refrigerant. However, the temperature drops to the boiling temperature of the liquid at the low pressure. The cold liquid refrigerant then flows to another heat exchange coil called the *evaporator* where the system performs its cooling function, and heat is absorbed by the refrigerant in the process of vaporization at constant temperature and pressure. From the evaporator, the cold refrigerant gas is drawn into the suction side of the compressor thus completing the cycle.

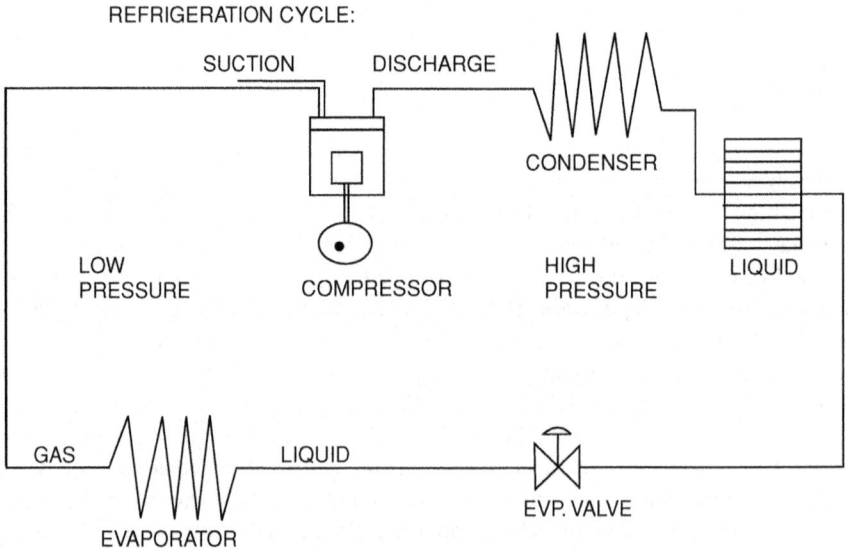

Figure 10.2   Schematic diagram of a refrigeration system.

A refrigeration system is usually equipped with low-pressure and high-pressure cut-off switches that interrupt power to the compressor when either the high-pressure set-point is exceeded (this occurs if cooling capacity of the condenser is inadequate) or when the pressure drops below the low-pressure set-point (this occurs when the compressor is running but the cooling demand is much less than the capacity of the refrigeration system). The low-pressure set-point can be used to control the evaporator temperature. In some systems, liquid refrigerant flow through the expansion valve is thermostatically controlled, interrupting flow when the evaporator temperature is lower than the set-point. In addition, refrigeration systems are equipped with thermostats that interrupt power to the compressor when the temperature of the refrigerated room reaches a set temperature.

A refrigeration system may also be used for heating. A system that alternates heating and cooling duty is called a *heat pump*. These units are used extensively for domestic heating or cooling in areas where winter temperatures are not too severe. In these units, either a low-pressure or a high-pressure refrigerant could go through the heat exchange coils that constitute the evaporator and condenser; thus, either could act as a heating or cooling coil depending upon the duty expected of the system. The ability of heat pumps to deliver heat with low power consumption is due to the fact that power is used only to pump energy from a low temperature to a higher temperature. The heat is not derived completely from the power supplied but rather it is extracted from cooler air from the surroundings. The efficiency of heat pumps in transforming applied power to heat varies inversely with the temperature differential between the temperature of the medium from which heat is extracted and the temperature of the system that is being heated. Heat pump systems can be used as a means of recovering energy from low temperature heat sources for use in low temperature heat applications such as dehydration.

10.1.4   The Refrigeration Cycle as a Series of Thermodynamic Processes

Starting from the compressor in Fig. 10.2, low-pressure gas is compressed adiabatically to high pressure, which should allow condensation at ambient temperature. Work is required to carry out this process, and this energy is supplied in the form of electrical energy to drive the compressor motor. The gas also gains in enthalpy during this compression process. At the condenser, the gas condenses and transfers the latent heat of condensation to the surroundings. There is a loss of enthalpy in a constant pressure process. At the expansion valve, the liquid expands to low pressure at constant enthalpy, while at the evaporator, the liquid evaporates at constant pressure and gains in enthalpy. The two processes crucial to the efficiency of the refrigeration system are the adiabatic compression process where energy is applied and the isobaric expansion process where energy is extracted by the refrigerant from the system. Work and enthalpy change in an adiabatic compression process has been derived in Chapter 4 in the section "Work and Enthalpy Change on Isothermal Expansion or Compression of an Ideal Gas." The ratio of enthalpy change to work in adiabatic compression of an ideal gas is equal to the specific heat ratio.

$$\frac{\Delta H}{W} = -\gamma; \quad W = \frac{-\Delta H}{\gamma} \tag{10.1}$$

The negative sign indicates that energy is being used on the system.

10.1.5   The Refrigeration Cycle on the Pressure/Enthalpy Diagram for a Given Refrigerant

The thermodynamic properties of refrigerants when plotted on a pressure-enthalpy diagram result in a plot that is useful in determining capacities and power requirement for a refrigeration system.

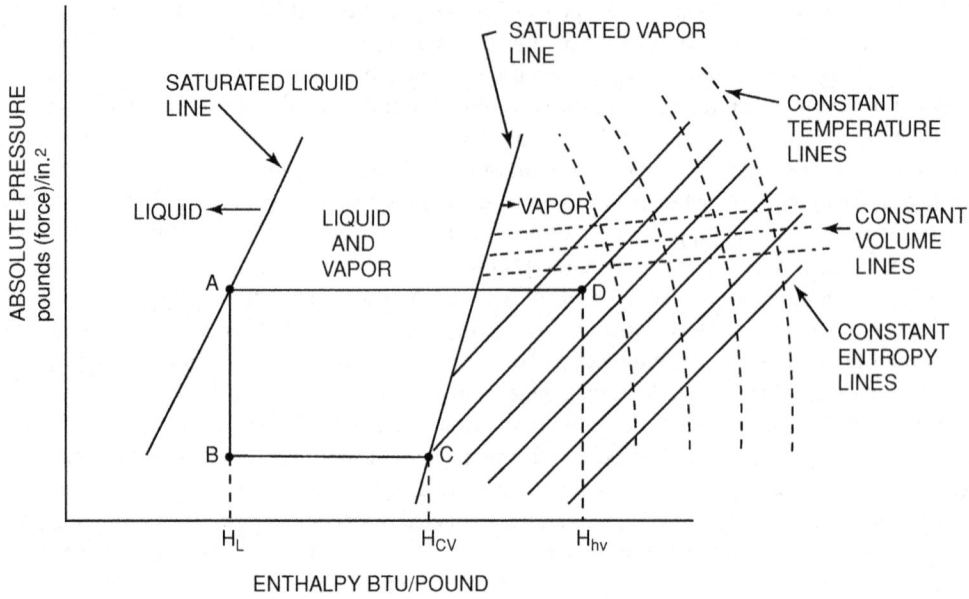

Figure 10.3   The refrigeration cycle on a pressure-enthalpy diagram.

Figure 10.3 is a schematic diagram of a refrigeration cycle on a pressure-enthalpy diagram. The diagram consists of lines representing the vapor and liquid pressure-enthalpy relationship, lines representing change in enthalpy with pressure during adiabatic compression (constant entropy), and in some charts, lines representing specific volumes at various pressures and enthalpies.

Point A in Fig. 10.3 represents the liquid refrigerant at high pressure entering the expansion valve. The refrigerant expands at constant enthalpy (H_L) as it goes through the expansion valve and leaves the valve as a mixture of liquid and vapor at a lower pressure, represented by point B. As the refrigerant absorbs heat in the evaporator, it gains in enthalpy represented by the line BC. The refrigerant leaves the evaporator as saturated vapor (represented by point C) having the enthalpy H_{cv}. The compressor raises the pressure and the change is represented by the line CD that parallels the lines of constant entropy. As the compressed refrigerant gas leaves the compressor at point D, it will have an enthalpy represented by H_{hv}. At the condenser, heat is dissipated resulting in a drop in enthalpy represented by line AD. The liquid refrigerant leaves the condenser with the pressure and enthalpy represented by point A.

The cooling capacity of the refrigeration system is represented by the length of line BC.

$$\text{Cooling capacity} = H_{cv} - H_L \qquad (10.2)$$

The condenser heat exchange requirement or condenser duty is represented by the length of line AD.

$$\text{Condenser duty} = H_{hv} - H_L \qquad (10.3)$$

The change in enthalpy due to compression ΔH_c is

$$\Delta H_c = H_{hv} - H_{cv} \tag{10.4}$$

From Equation (10.1), the work required for compression is

$$W = \frac{-\Delta H_c}{\gamma} = \frac{-(H_{hv} - H_{cv})}{\gamma} \tag{10.5}$$

If M is the mass of refrigerant recirculated through the system per unit time, the power requirement can be calculated as follows:

$$P = \frac{W}{time} = M\frac{-(H_{hv} - H_{cv})}{\gamma} \tag{10.6}$$

The negative sign on the work and power indicates that work is being added to the system.

The efficiency of a refrigeration system is also expressed in terms of a *coefficient of performance* (COP), which is a ratio of the cooling capacity over the gain in enthalpy due to compression.

$$COP = \frac{H_{cv} - H_L}{H_{hv} - H_{cv}} \tag{10.7}$$

The power requirement (P) can be expressed in terms of the coefficient of performance using Equations (10.6) and (10.7).

$$P = \frac{H_{cv} - H_L}{\gamma(COP)} \bullet M \tag{10.8}$$

The refrigeration capacity is expressed in tons of refrigeration, the rate of heat removal sufficient to freeze 1 ton (2000 lb) of water in 24 hours. Because the heat of fusion of water is 144 BTU/lb, this rate of heat removal is equivalent to 12,000 BTU/h. The refrigeration capacity in tons is

$$(tons)_r = \frac{(H_{cv} - H_L)(M)}{12,000} \tag{10.9}$$

Substituting Equation (10.9) in Equation (10.8):

$$P = \frac{(tons)_r(12,000)}{\gamma(COP)} \left[\frac{BTU}{h(ton)_r} \right]$$

Expressing the power requirement in horsepower $(HP)/(ton)_r$:

$$\frac{HP}{(tons)_r} = \frac{12,000}{\gamma(COP)} \frac{BTU}{h(ton)_r} \frac{1HP}{2545\ BTU/h} = \frac{4.715}{\gamma(COP)} \tag{10.10}$$

Because there is a certain degree of slippage of refrigerant past the clearance between the cylinder and the piston, particularly at the high pressures and also because some frictional resistance occurs between the piston and the cylinder, the actual work expended would be higher than that determined using Equation (10.10). The ratio between the theoretical horsepower as calculated from Equation (10.10) and the actual horsepower expended is the efficiency of compression. The efficiency of compression depends upon the ratio of high side to low side pressure across the compressor and will be discussed later in the text.

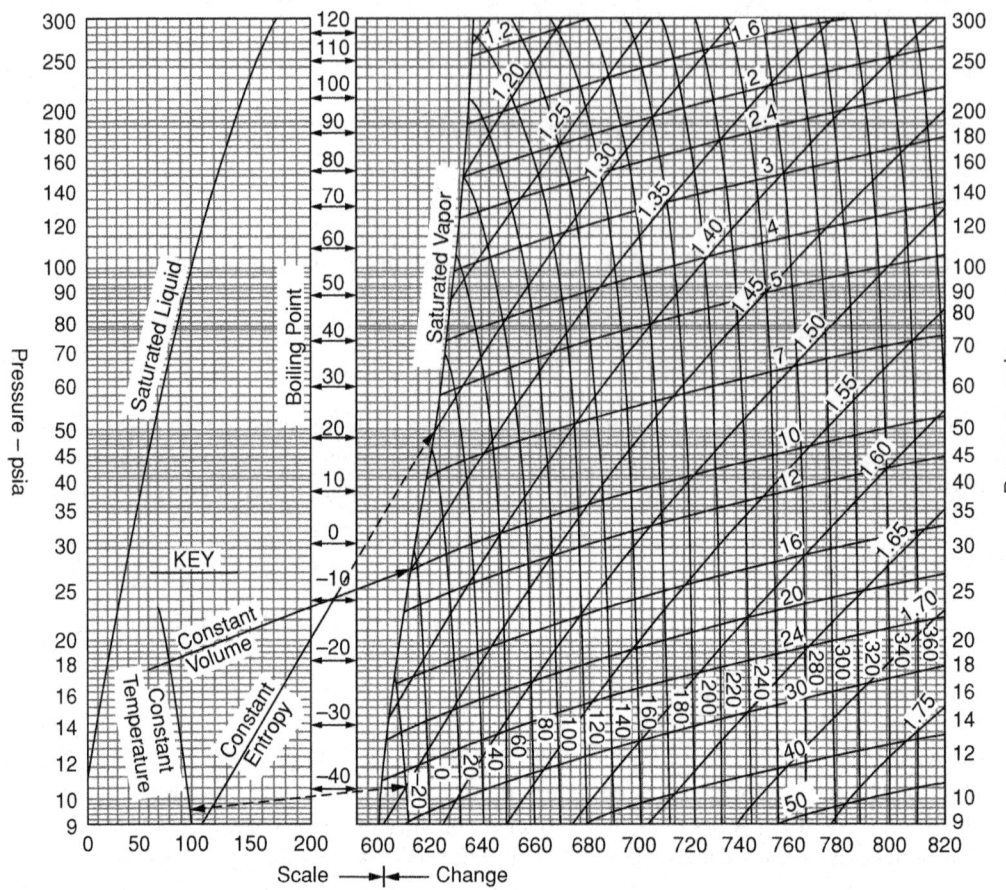

Figure 10.4   Pressure-enthalpy diagram for refrigerant 717 (NH$_3$). (Adapted from ASHRAE Guide and Data Book, Fundamentals and Equipment, 1965.)

Pressure-enthalpy diagrams for ammonia, Freon 12 and Freon 22 are shown in Figs. 10.4, 10.5, and 10.6.

10.1.5.1   *Example Problems on the Use of Refrigerant Charts*

Example 10.1.   A refrigerated room is to be maintained at 40°F (4.44°C) and 80% relative humidity. The contents of the room vaporize moisture requiring removal of moisture from the air to maintain the desired relative humidity. Assuming that air flow over the evaporator coil is such that the bulk mean air temperature reaches to within 2°F (1.11°C) of the coil temperature, determine the low side pressure for an ammonia refrigeration system such that the desired humidity will be maintained.

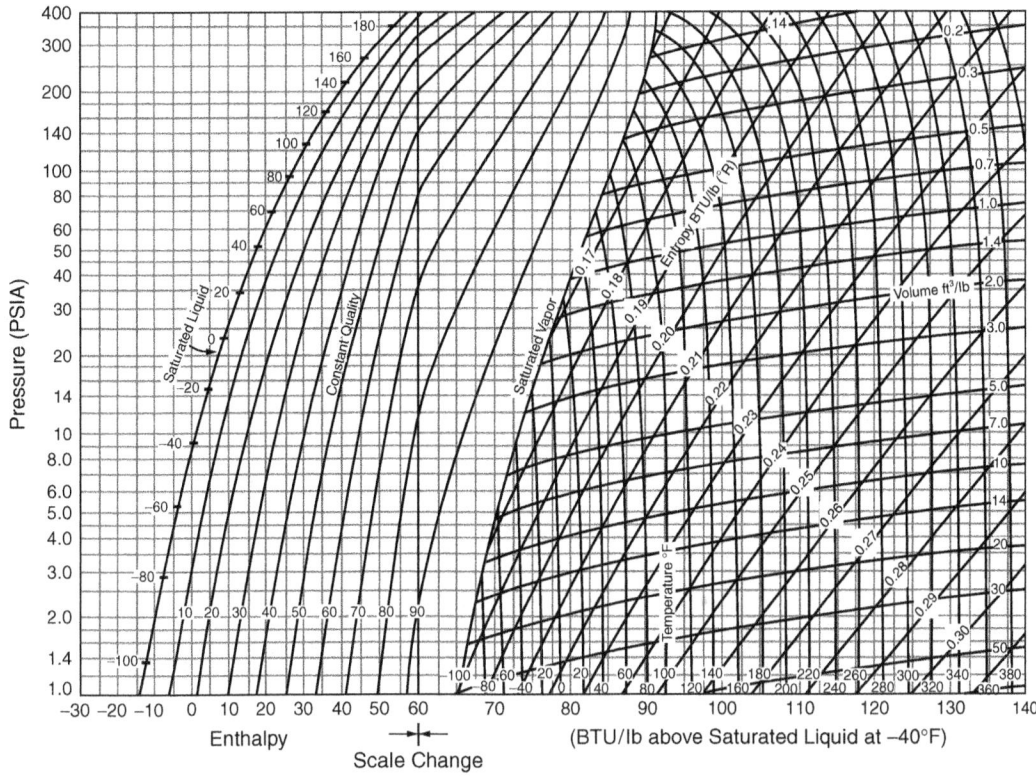

Figure 10.5   Pressure-enthalpy diagram for refrigerant 12 (CCl_2f_2). (Courtesy of E. I. Du Pont de Nemoirs Inc.)

Solution:

From a psychrometric chart, the dew point of air at 40°F and 80% RH is 33°F (0.56°C). When room air is cooled to 33°F passing over the evaporator coils, any moisture present in excess of the saturation humidity at this temperature will condense. If air flow over the coil is sufficient to allow enough moisture removal through condensation to equal the rate of moisture vaporization into the air, then the desired relative humidity will be maintained. The coil temperature should be 31°F (-0.56°C). From Fig. 10.1 for ammonia, the pressure corresponding to a temperature of 31°F is 61 psia (420 kPa absolute).

Example 10.2.   A heat pump is proposed as a means of heating a cabinet drier. Air is continuously recycled through the system passing through the condenser coil at the air inlet to the drying chamber where the air is heated. Prior to recycling, the hot moist air leaving the drying chamber is passed through the evaporator coil where moisture is removed by condensation. Inlet air to the drying chamber is at 140°F (60°C) and 4% relative humidity. Assuming that the evaporator and condenser coil design and air flow are such that air temperature approaches 5°F (2.78°C) of the evaporator coil temperature and log mean ΔT between condenser coil and air is 20°F (5.56°C), determine the high and low side pressure for the refrigeration system using refrigerant 22 as the refrigerant.

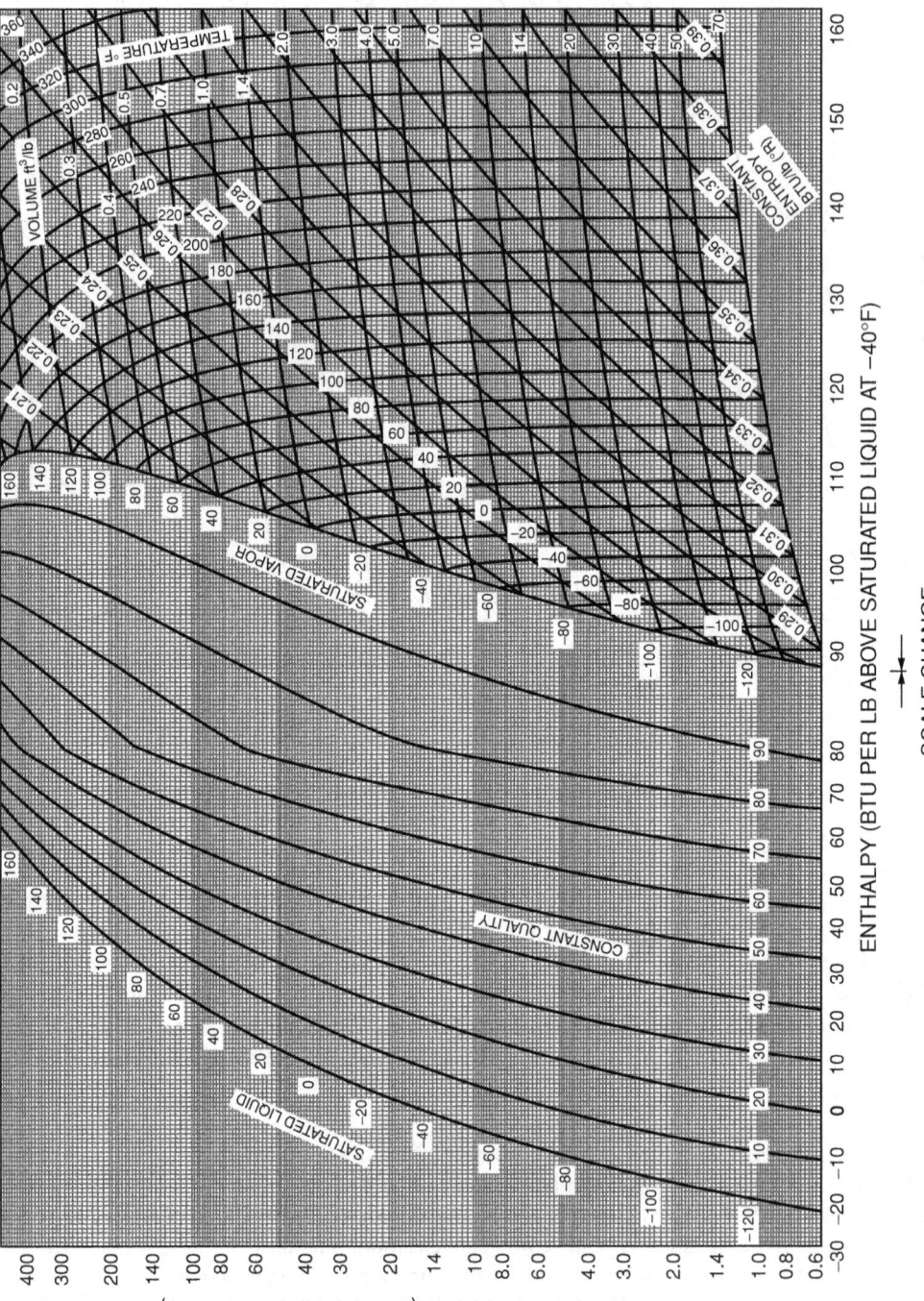

Figure 10.6 Pressure–enthalpy diagram for refrigerant 22 (CHClF$_2$). (Courtesy of E. I. Du Pont de Nemoirs Inc.)

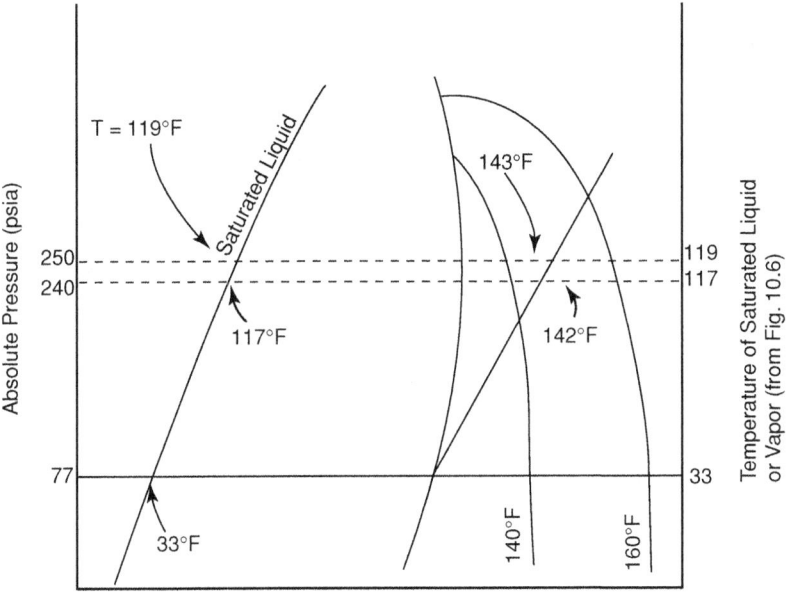

Figure 10.7 Pressure-enthalpy diagram of the heat pump system in Example 10.2.

Solution:

From a psychrometeric chart, the dew point of air at 140°F (60°C) and 4% RH is 38°F (3.33°C). Thus, the evaporator coil temperature should be at 33°F (0.56°C). The condenser coil temperature and pressure will be calculated from the specified log mean ΔT.

From Fig. 10.1, for refrigerant 22, the low side pressure corresponding to 33°F (0.56°C) is 77 psia (530 kPa absolute).

At the condenser, air enters at 38°F and leaves at 140°F. If the system is set-up in countercurrent flow, the log mean ΔT can be calculated as follows:

$$\overline{\Delta T} = \frac{(T_1 - 38) - (T_g - 140)}{\ln(T_1 - 38)/(T_g - 140)}$$

where T_1 and T_g refer to the temperature of refrigerant liquid and gas, respectively.

Figure 10.7 shows the pressure-enthalpy diagram for the system under the conditions given.

The high pressure will be determined by trial and error from Fig. 10.6. Following the constant entropy line from a pressure of 77 psia, a high pressure of 250 psia will give a hot refrigerant gas temperature of 143°F and a liquid temperature of 119°F. The log mean ΔT is

$$\overline{\Delta T_1} = \frac{(119 - 38) - (143 - 140)}{\ln 92/3} = 23.6$$

This is higher than the 20°F ΔT specified. At a pressure of 240 psia, the hot refrigerant gas temperature is 142°F and the liquid temperature is 117°F.

$$\overline{\Delta T_1} = \frac{(117 - 38) - (142 - 140)}{\ln 79/2} = 20.9$$

This calculated value for the log mean ΔT is almost equal to the specified $20°F$; therefore, the high side pressure for the system should be at 240 psia (1650 kPa).

Example 10.3. A refrigeration system is to be operated at an evaporator coil temperature of $-30°F$ ($-34.4°C$), and a condenser temperature of $100°F$ ($37.8°C$) for the liquid refrigerant. For (a) refrigerant 12 and (b) refrigerant 717, determine:

1. the high side pressure,
2. the low side pressure,
3. refrigeration capacity per unit weight of refrigerant,
4. the coefficient of performance,
5. the theoretical horsepower of compressor per ton of refrigeration, and
6. the quantity of refrigerant circulated through the system per ton of refrigeration.

Solution:

(a) For refrigerant 12:
   The high and low side pressures (1 and 2) are determined from Fig. 10.1. At $-30°F$ ($-34.4°C$), pressure is 12.3 psia (85 kPa). The high side pressure corresponding to a refrigerant liquid temperature of $100°F$ ($37.8°C$) is 133 psia (910 kPa).
   Knowing the high and low side pressures, Fig. 10.5 is used to construct the pressure-enthalpy diagram for the refrigeration cycle. This diagram is shown in Fig. 10.8.

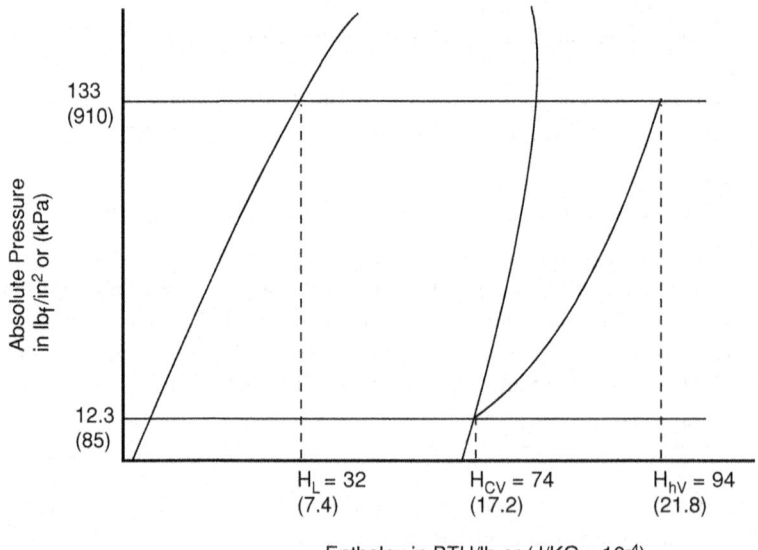

Figure 10.8   Pressure-enthalpy diagram of refrigerant 12 for the problem in Example 10.3.

The refrigeration capacity (Question 3) is calculated using Equation (10.2):

Refrigeration capacity $= (74 - 32) = 42$ BTU/lb $= (17.2 - 7.4) \times 10^4 = 98,000$ J/kg

The coefficient of performance (4) is calculated using Equation (10.7):

$$COP = \frac{74 - 32}{94 - 74} = 2.1$$

The theoretical horsepower per ton of refrigeration (5) is calculated using Equation (10.10):

$$\frac{C_p}{C_v} \text{ for Freon } 12 = \gamma = 1.14$$

$$\frac{HP}{(ton)_r} = \frac{4.715}{(\gamma)(COP)} = \frac{4.715}{(1.14)(2.1)} = 1.97$$

One ton of refrigeration (6) is equivalent to 12,000 BTU/h or 3517 W. Weight of refrigerant circulated/h is

$$\begin{aligned} \text{Weight} &= \frac{\text{coolong capacity/ton of refrigeration}}{\text{coolong capacity/unit weight iof refrigerant}} \\ &= \frac{12,000\,\text{BTU/h}}{42\,\text{BTU/lb}} = 286 \text{ lb refrigerant per hour} \\ &= \frac{3517\,\text{J/s}}{98,000\,\text{J/kg}} = 0.0359\,\text{kg/s} = 129\,\text{kg/h} \end{aligned}$$

(b) For refrigerant 717

The high and low side pressures (1 and 2) corresponding to $-30°$F $(-34.4°$C$)$ and $100°$F $(37.8°$C$)$ are determined from Fig. 10.1. The low side pressure is 13.7 psia (94 kPa). The high side pressure is 207 psia (1440 kPa).

The pressure-enthalpy diagram (Question 3) is then constructed knowing the high and low side pressures using Fig. 10.4. Figure 10.9 shows this diagram.

Refrigeration capacity $= 600 - 155 = 445$ BTU/lb $= (139 - 36) \times 10^4 = 1,030,000$ J/kg

The coefficient of performance (4):

$$COP = \frac{600 - 155}{784 - 600} = 2.42$$

The theoretical horsepower per ton of refrigeration (5) is calculated using a C_p/C_v ratio of 1.29.

$$\frac{HP}{(ton)_r} = \frac{4.715}{1.29(2.42)} = 1.51$$

Weight of refrigerant circulated per hour $(ton)_r$ (6) is

$$\begin{aligned} \text{Weight} &= \frac{12,000\,\text{BTU/(h)}(ton)_r}{445\,\text{BTU/lb}} = 26.97 \text{ lb/h}(ton)_r = \frac{1317\,\text{J/s}(ton)_r}{1,030,000\,\text{J/kg}} \\ &= 0.00341\,\text{kg/s}(ton)_r = 12.3\,\text{kg/h}(ton)_r \end{aligned}$$

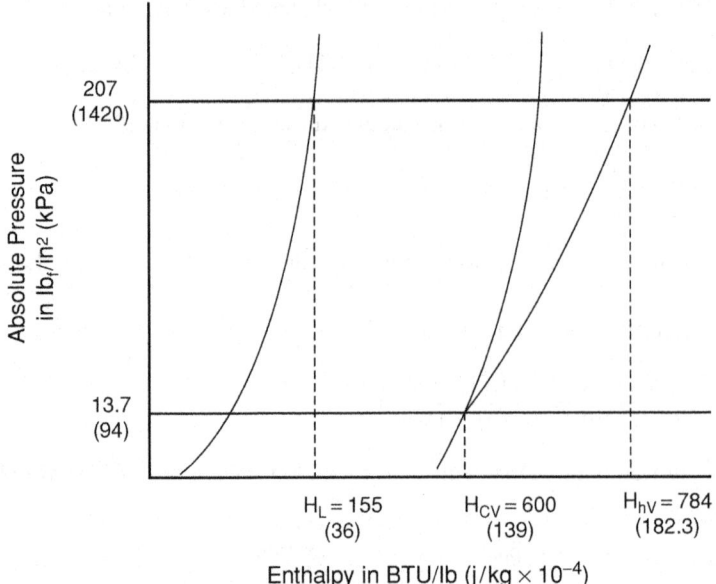

Figure 10.9 Presure-enthalpy diagram of ammonia for the problem in Example 10.3.

10.1.6   The Condenser and Evaporator

Most refrigeration systems transfer heat between the refrigerant and air. Because heat transfer coefficients to air are usually very low, the air film resistance controls the rate of heat transfer. Very large heat transfer areas would be required to achieve the necessary heat transfer rate. To accomplish the necessary heat transfer rate and still have equipment that is reasonably sized, the heat transfer surface area of the tubes that comprise the evaporator or condenser coil is increased by the use of fins.

Finned heat exchange units are sized in terms of an effective heat transfer surface area which is the sum of the area of the bare tube A_t and the effective area of the fin, which is the product of the fin surface area A_s and the fin efficiency η.

$$A_{eff} = \eta A_f + A_t$$

The fin efficiency, η, decreases as the base area of the fin decreases and as the height of the fin increases. Furthermore, as the ratio of the heat transfer coefficient to air over the thermal conductivity of the metal that constitutes the fin (h_o/k) approaches zero, 6 approaches unity. Figure 10.10 shows fin efficiency for the most common systems. The fins in these units often consist of a stack of very thin sheets of metal pierced by a bundle of tubes. The equivalent fin surface area, A_f, is calculated as follows:

$$A_f = 2\pi(r_f^2 - r^2)(n_t)(n_f)$$

where n_t is number of tubes, and n_f is number of fins along the length of each tube.

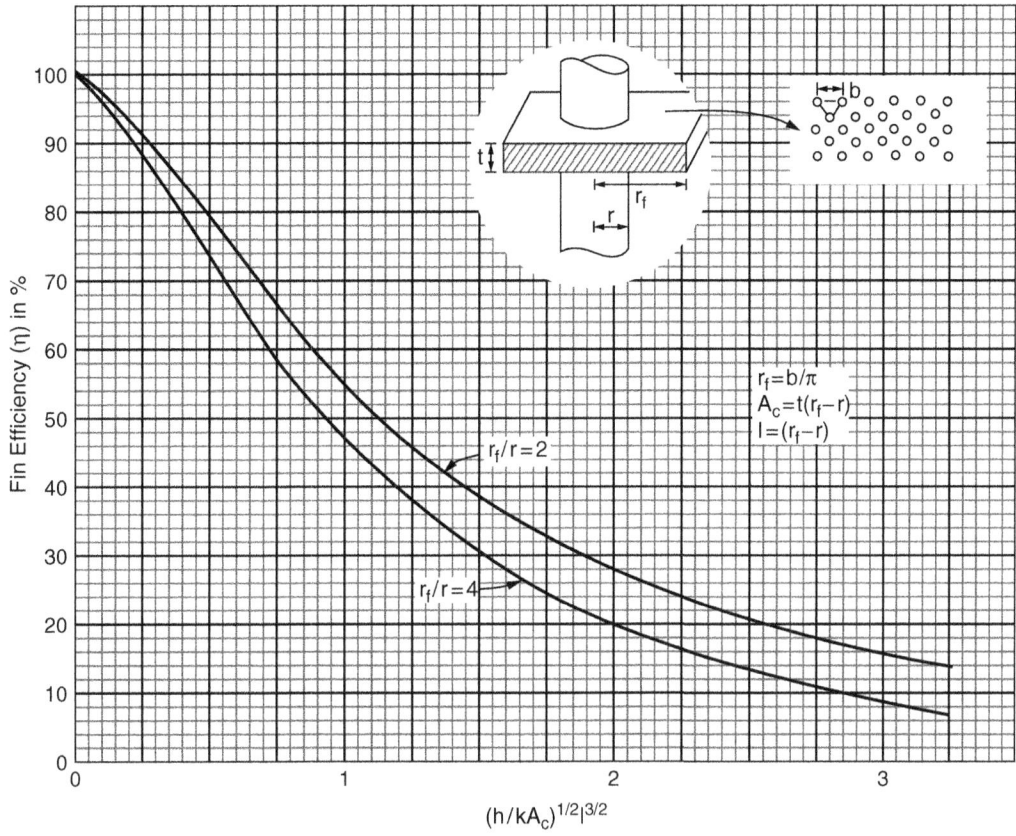

Figure 10.10 Fin efficiency of fins consisting of a stack of sheet metal each sheet with thickness l, pierced by tubes of radius r. The tubes are arranged with centers on an equilateral triangle having distance b between centers.

The capacity of a condensing or evaporating unit depends upon the surface area available for heat transfer and the mean temperature difference between the surfaces of the unit and the air going through the unit.

10.1.6.1 Problems with Heat Exchange in Systems Using Zeotropic Refrigerants

The following examples are based on systems using a single compound or an azeotrope as the refrigerant. When an azeotrope is used as refrigerant, provision must be given to the fact that local heat transfer coefficient will be reduced on the refrigerant side of the heat exchanger because the composition of liquid refrigerant on the heat exchange surface will not be the same at different points in the heat exchanger. The thermal properties of the condensate will vary at different points in the heat exchanger and the vapor phase of the other refrigerant components may also interfere with the heat transfer. Thus, the usual method of calculating the log mean ΔT in the heat transfer calculations is no longer correct. Depending upon the refrigerant, errors could go as high as 11% in the determined heat transfer surface area compared to using the standard heat transfer coefficient from condensing liquids and conventional log mean ΔT calculations. This error could result in undersized evaporator or condenser units.

Example 10.4. A condensing unit of a refrigeration system consists of 13 rows of 38.1-cm-long, 1.27-cm outside diameter copper tubes with 1.651-mm wall stacked four deep on 2.54-cm centers. The fin consists of 0.0254-cm thick aluminum sheets spaced 10 per 2.54 cm. If the rate of air flow through this condenser is such that the outside heat transfer coefficient, h, is 9 BTU/(h $\cong$ ft^2 $\cong$ °F) or 51.1 W/(m^2 $\cong$ K), calculate the effective heat transfer surface area for this unit.

Solution:

Refer to Fig. 10.10. Because the tubes are spaced on 2.54-cm centers, b = 2.54 cm. The thermal conductivity of aluminum (from Appendix Table A.9) is 206 W/(m $\cong$ K). Figure 10.10 can now be used to determine the fin efficiency η.

The tube radius, r = 0.00635 m

Equivalent fin radius, $r_f = \dfrac{b}{\sqrt{\pi}} = \dfrac{0.0254}{1.772} = 0.014334\,m$

Fin cross-sectional area $A_c = t\,(r_f - r)$

$= 0.000254\,(0.014334 - 0.00635) = 2.0279 \times 10^6$

The height of the fin, $P = (r_f - r)$

$= (0.014334 - 0.00635) = 0.007984 - m$

Values for the abscissa in Fig. 10.10 can now be calculated:

$$\left[\frac{h}{kA_c}\right]^{0.5} (\ell)^{1.5} = \left[\frac{51.1}{206(2.0279 \times 10^{-6})}\right]^{0.5} (0.007984)^{1.5} = (349.7)\,(0.000713) = 0.2495$$

$$r_f/r = 0.014334/0.00635 = 2.26$$

From Fig. 10.10, $\eta = 0.9$. The equivalent fin area is calculated as follows:

$n_t = 13(4) = 52\,tubes$

$n_f = 38.1\,cm\left[\dfrac{10\,fins}{2.54\,cm}\right] = 150$

$A_f = 2\pi(r_f^2 - r^2)\,(n_t/n_f) = 2\pi[(0.014334)^2 - (0.00635)^2](52)(150) = 8.094\,m^2$

The area of the bare tube A_t is the total tube area minus the area covered by the fins. Let $P_t = $ length of the tube.

$A_t = (2\pi r l_t)(n_t) - (2\pi r)(t)(n_t)(n_f)$

$t = $ fin thickness $= 0.000254\,m$

$A_t = 2\pi(0.00635)(0.381)(52) - 2\pi(0.00635)(0.000254)(52)(150)$

$= 0.790 - 0.0790 = 0.711\,m^2$

Effective area $A_{eff} = 0.91(8.094) + 0.711 = 8.0765\,m^2$

Example 10.5. If the conditions are such that a mean temperature difference ΔT of 45°F (25°C) exists in the condensing system and the refrigeration unit has a coefficient of performance of 2, how many tons of refrigeration can be supplied by the refrigeration system fitted with the condensing unit in Example 10.4? Assume the outside coefficient of heat transfer controls the heat transfer rate, $U = h_0 = 51.1$ W/(m$^2 \cong$ K).

Solution:

The rate of heat transferred through the unit is

$$q = U A_{eff} \Delta T = 51.1(8.0765)(25) = 10{,}318 \text{ W}$$

From Equation (10.7), let $R_r =$ refrigeration capacity.

$$COP = \frac{R_r}{H_{hv} - H_{cv}}$$

Where:

$$R_r = H_{cv} - H_L$$
$$H_L = H_{cv} - R_r$$

Condenser load:

$$C = H_{hv} - H_L = H_{hv} - H_{cv} + R_r$$
$$(H_{hv} - H_{cv}) = C - R_r$$
$$(\text{tons})_r = \frac{6878 \text{ W}}{3515.88 \text{ W/ton}} = 1.956 \text{ tons}$$

The coefficient of performance can be expressed in terms of the condenser load C and the refrigeration capacity R_r.

$$COP = \frac{R_r}{C - R_r}$$

Solving for R_r:

$$R_r = \frac{C(COP)}{1 + COP}$$

Substituting values for COP and C:

$$R_r = \frac{10{,}318(2)}{3} = 6878 \text{ W}$$

A similar procedure may be used to evaluate evaporators. The examples illustrate the dependence of the capacity of a refrigeration system on the heat exchange capacity of the condensing or evaporator unit in the system.

10.1.7 The Compressor

The compressor must be adequate to compress the required amount of refrigerant per unit time between the high side and low side pressures, to provide the necessary refrigeration capacity. Compressor capacities are determined by the displacement and the volumetric efficiency of the unit. The

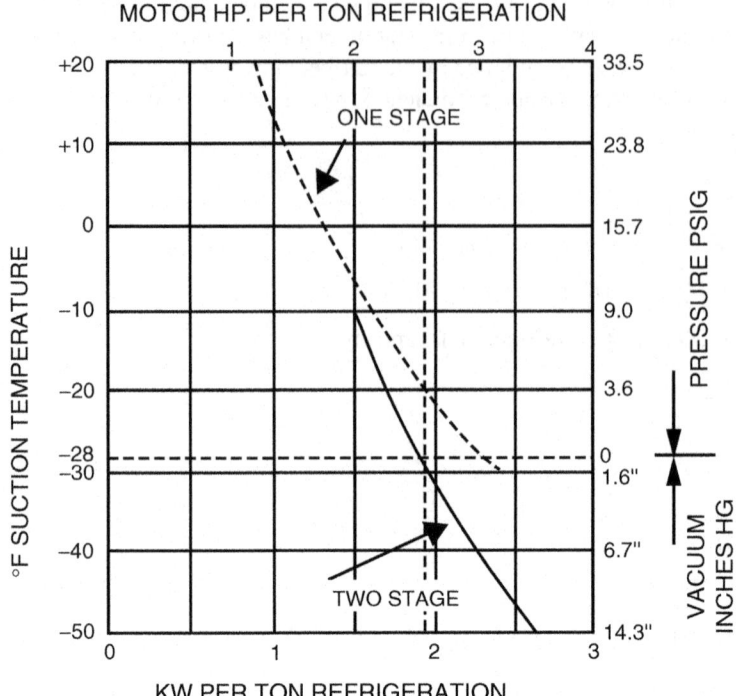

Figure 10.11 Motor horsepower and kW/Ton of refrigeration for one and two stage compressors. (From Food Eng. 41(11):91, 1969.)

---

displacement is the volume displaced by the piston or rotary unit per unit time. For a reciprocating compressor, the displacement is calculated as follows:

Displacement = (area of bore) (stroke) (cycles/min)

Theoretically, the quantity of gas passing through the unit per unit time should be independent of the pressure and should be a function only of the displacement of the unit. However, at high pressures, rotary units allow a certain amount of slip and reciprocating units leave a certain amount of gas at the space between the cylinder head and the piston at the end of each stroke. Thus, the actual amount of gas passed through the unit per cycle is less than the displacement. The volumetric efficiency of a unit is the ratio of the actual volume delivered to the displacement. The volumetric efficiency decreases with increasing ratios of high side to low side pressures.

Large units that operate at very low pressures at the evaporator (e.g., units used for freezing) often are of a multistage type, where the refrigerant is compressed to an intermediate pressure, then partially cooled before it enters a second stage where it is compressed to the next higher pressure, and so forth, until the desired high side pressure is reached. The staging is designed such that equal work is accomplished in each stage and the volumetric efficiency remains high because of the low ratio of high side to low side pressure in each of the stages.

Figure 10.11 shows a comparison between the power requirement per ton of refrigeration between one stage and two stage units operating at various suction side pressures. For an ammonia compressor operating at a suction side pressure equivalent to a temperature of $-30°F$, the motor horsepower per ton of refrigeration is 3.2 for a one stage unit and is only 2.5 for a two stage unit. In Example 10.3, the theoretical horsepower per ton of refrigeration under these conditions is only 1.51. Thus, the efficiency of the compressors represented by the graph in Fig. 10.11 is only 47% for a one-stage unit and 60% for a two-stage unit.

The presence of air in the refrigerant line reduces the efficiency of the compressor if expressed in terms of horsepower per ton of refrigeration as the same amount of power will be required to compress the gases to the required pressure. Yet, less refrigerant is passed through because part of the gas compressed is air.

10.2 REFRIGERATION LOAD

The amount of heat that must be removed by a refrigeration system per unit of time is the refrigeration load. The load may be subdivided into two categories: the *unsteady-state* load, which is the rate of heat removal necessary to reduce the temperature of the material being refrigerated to the storage temperature within a specific period of time, and the *steady-state* load, which is the amount of heat removal necessary to maintain the storage temperature. For food products, the temperature has to be reduced to the storage temperature in the shortest time possible to prevent microbiological spoilage and quality deterioration. Therefore, if large quantities need to be introduced into a refrigerated storage room intermittently, these materials are usually precooled to the storage temperature in a smaller precool or chill room or by other means prior to introduction into the large storage warehouse. This practice prevents the necessity of having to install oversized refrigeration units on the large storage warehouse that would operate at full capacity only during the short periods that materials are being introduced into the warehouse.

The unsteady-state load includes the sensible heat of the product, the heat of respiration if the product is fresh produce and the heat of fusion of the water in the product if the product is to be frozen. The steady-state load includes heat incursion through the walls of the enclosure, through cracks and crevices and through doors, latent heat of condensation of moisture infiltrating into the room, and heat generated inside the room.

10.2.1 Heat Incursion Through Enclosures

Heat transfer rates through composite solids and techniques for determination of heat transfer coefficients were discussed thoroughly in Chapter 7. The ASHRAE guide and data book for 1965 recommends the design heat transfer coefficients to be used for various surfaces as shown in Table 10.2.

10.2.2 Heat Incursion Through Cracks and Crevices

A majority of heat transferred is due to fluctuations in pressure caused by cycling of temperature in the room. At the high temperature point in the cycle, cold air will be expelled from the room. When the pressure drops at the low temperature point in the cycle, warm air will be drawn into the room. The amount of air admitted into the room because of temperature cycling can be calculated using the principles discussed in Chapter 4.

Table 10.2   Heat Transfer Coefficient to Air from Various
Surfaces under Various Conditions

	Heat Transfer Coefficients	
	BTU	W
Surface	$h(ft^2)(°F)$	$m^2(K)$
Inside wall (still air)	1.5	8.5
Outside wall or roof	5.9	33.5
15 miles/hr or 24 km/h wind		
Outside wall or roof	4.0	22.7
7.5 miles/hr or 12 km/h wind		
Horizontal surface, still air,	1.7	9.7
upward heat **flow**		
Horizontal surface, still air,	1.1	6.25
downward heat **flow**		

Source: Adapted from ASHRAE, 1965. Design heat transmission coefficient.
*ASHRAE Guide and Data Book. Fundamentals and Equipment for 1985
and 1986.* American Society of Heating, Refrigeration, and Air Conditioning
Engineers, Atlanta, GA.

10.2.3   Heat Incursion Through Open Doors

Opening doors allow entry of warm outside air and expulsion of cold air. The rate of heat incursion is dependent upon the size of the door and the temperature differential between the inside and outside. Data on rate of heat transfer through doors of refrigerated rooms have been determined empirically. The equation for rate of heat loss calculated from data published in *Food Engineering Magazine* [Food Eng. 41(11):91, 1969] for values of ΔT between 40°F and 120°F (22.2°C and 66.7°C) is

$$q = 2126\,W[e]^{0.0484(\Delta T)}(h)^{1.71} \tag{10.11}$$

where q is rate of heat incursion into the room in watts, W is the width of the door in m, ΔT is the temperature difference in °C inside and outside the room, and h is the height of the door in m.

Example 10.6.   The door to a refrigerated room is 3.048 m high and 1.83 m wide. It is opened and closed at least five times each hour with the door remaining open for at least 1 min at each opening. Calculate the refrigeration load due to the door opening if the room is maintained at 0°C and ambient temperature is 29.4°C.

Solution:

h = 3.048 m, W = 1.83 m, ΔT = 29.4°C. Substituting into Equation (10.11):

$$q = 2126(1.83)(e)^{0.0484(29.4)}(3.048)^{1.71} = 108.6\,KW$$

The total time the door was opened in 1 hour is 300 seconds. Refrigeration load = 108.6 KW (300 s) = 35.58 MJ.

10.2.4   Heat Generation

Motors inside a refrigerated room generate heat at the rate of 1025.5 W/hp. This rate drops to 732.48 W/hp if only the motor is inside and the load it drives is outside the refrigerated room. Workers inside a room generate approximately 293 W per person. The rate of heat generation by personnel increases with decreasing room temperatures. Heat dissipated by light bulbs is the same as the wattage of the lamp.

Fruits and vegetables respire and the heat of respiration adds to the refrigeration load. The heat of respiration of fruits and vegetables is a function of temperature and can be calculated using the expression:

$$q = a(e)^{bT} \tag{10.12}$$

q is the rate of heat generation (mW/kg in SI). Values for a and b for various fruits and vegetables are shown in Table 10.3.

During cooling, heat of respiration will decrease as temperature decreases. An average heat of respiration may be calculated between two temperatures T_1 and T_2 knowing the time Δt for the temperature to change between T_1 and T_2 and assuming that the temperature change during this period is linear.

$$q = \frac{1}{\Delta t} \int_0^{\Delta t} a(e)^{bT}\, dt \tag{10.13}$$

Assuming a linear temperature change:

$$T_1 - T = \frac{T_1 - T_2}{\Delta t} t \tag{10.14}$$

Substituting Equation (10.14) for T in Equation (10.13):

$$q = \frac{a}{\Delta t} \int_0^{\Delta t} [e]^{bT_1 - \frac{T_1 - T_2}{\Delta t} bt}\, dt$$

Integrating and substituting limits:

$$q = \frac{ae^{bT_1}}{b(T_1 - T_2)} \left[1 - (e)^{-b(T_1 - T_2)} \right] \tag{10.15}$$

Equation (10.15) is the average rate of heat generation over the time Δt. The units will be the same as the units of a, which in SI, is mW/kg as tabulated in Table 10.3. Multiplying q in Equation (10.15) by Δt will give the total heat generated for the time period under consideration.

Although the temperature at any given point in a material during the process of cooling usually changes exponentially with time, materials that are cooled in bulk (e.g., head lettuce or boxed products) would have a temperature gradient with the interior parts at a higher temperature than the external sections exposed to the cooling medium. Thus, if the bulk mean temperature is considered, the deviation from the linear temperature change assumption would be minimal.

Example 10.7. It is desired to cool cabbage from 32.2°C to 4.44°C in 4 hours. Calculate the heat of respiration during this cooling period. Assume a linear temperature change.

Solution:

$T_1 = 32.2°C$; $T_2 = 4.44°C$; $\Delta t = 4$ h $= 14,400$ seconds. From Table 10.3 for cabbage, a $= 16.8$ and b $= 0.074$ in SI units. Examination of Equation (10.12) will reveal that a will have the same units

Table 10.3 Heat of Respiration of Fruits and Vegetables in Air Values of the Constants a and b in the Equation $q = a(e)^{bT}$

| Product | T in °F
q in $\frac{BTU}{ton\ (24\ hr)}$ | | T in °C
q in $\frac{mW}{kg}$ | |
	a	b	a	b
Apples	213	0.06	19.4	0.108
Asparagus	2779	0.048	173.0	0.086
Beans (green or snap)	829	0.064	86.1	0.115
Beans (lima)	376	0.071	48.9	0.128
Beets (topped)	1054	0.031	38.1	0.056
Broccoli	854	0.067	97.7	0.121
Brussels sprouts	1845	0.045	104.0	0.081
Cabbage	337	0.041	16.8	0.074
Cantaloupes	128	0.07	16.1	0.126
Carrots (topped)	498	0.046	29.1	0.083
Celery	237	0.58	20.3	0.104
Corn (sweet)	2465	0.043	131.0	0.077
Grapefruit	171	0.051	11.7	0.092
Lettuce (head)	416	0.049	26.7	0.088
Lettuce (leaf)	1188	0.041	59.1	0.074
Onions	89	0.055	6.92	0.099
Oranges	151	0.059	13.4	0.106
Peaches	104	0.074	14.8	0.133
Pears	42	0.096	12.1	0.173
Peas (green)	1249	0.059	111.0	0.106
Peppers (sweet)	796	0.040	33.4	0.072
Spinach	473	0.073	65.6	0.131
Strawberries	568	0.059	50.1	0.106
Sweet potatoes	796	0.034	31.7	0.061
Tomatoes	159	0.057	13.2	0.103
Turnips	590	0.037	25.8	0.067

Source: ASHRAE, 1974. Approximate rates of evolution of heat by certain fruits and vegetables when stored at temperatures indicated. In: *ASHRAE Handbook and Product Directory—1974 Applications*. American Society of Heating, Refrgeration, and Air Conditioning Engineers, Atlanta, GA.
[1] q calculated using the constants *a* and *b* are maximum values in the range reported. Minimum values average 67% of the maximum.

as q, and b will have units of 1/EC.

$$q = \frac{16.8\,e^{0.074(32.2)}}{(0.074)(32.2 - 4.44)}\left[1 - (e)^{-0.074(32.2-4.44)}\right] = 77.253\ \text{mW/kg}$$

Heat generated over the time period of 14400 s is Q:

$$Q = (77.253 \times 10^{-3}\ \text{W/kg})(14{,}400\,\text{s}) = 1112\ \text{J/kg}$$

Example 10.8. Calculate the refrigeration load due to the heat of respiration of spinach at a constant temperature of 3.33°C.

Solution:

T = 3.33°C. From Table 10.3, a = 65.6 mW / kg and b = $0.131°C^{-1}$.

$q = 65.6 \, e^{0.131(3.35)} = 101.74 \, mW/kg$

10.2.5 The Unsteady-State Refrigeration Load

Procedures for calculating the heat capacity and sensible heat gain or loss are discussed Chapter 5. When a change in phase is involved, the latent heat of fusion of the water must be considered. The heat of fusion of ice is 144 BTU/lb, 80 cal/g, or 0.334860 MJ/kg.

Example 10.9. Calculate the refrigeration load when 100 kg/h of peas needs to be frozen from 30°C to −40°C. The peas have a moisture content of 74%. The freezing point is −0.6°C.

Solution:

The refrigeration load above freezing will be calculated using Siebel's equation. From Equation (5.9), Chapter 5, the specific heat above freezing for a material that contains 74% water and 26% solids non fat is

$C_p = 837.36(0.26) + 4186.8(0.74) = 3315.9 \, J/(kg \, K)$.

The refrigeration load required for cooling 1 kg peas from 30°C to −0.6°C = Q_1

$Q_1 = 1(3315.9)(30.6) = 101,467 \, J$

The refrigeration load below freezing ted using the concepts and equations discussed in the section "Enthalpy Changes in Food Freezing" in Chapter 5. Molality of solutes in unfrozen water at different temperatures below freezing will be used as the basis for the calculations.

$T_f = -0.6°C$; $T_2 = -40°C$; $K_f = 1.86$; $\Delta H_f = 334,860 \, J$
The specific heat of ice = 2093.4 J/(kg K)
The specific heat of water is 4186.8 J/(kg K)
The specific heat of solids non-fat = 837.36 J/(kg K)
The initial moisture before freezing = $W_o = 0.740$ kg
The liquid water at −40°C = W = $W_o(-T_f/-T_2) = 740(0.6/40) = 0.0111$ kg
Ice at −40°C = 0.740 − 0.0111 = 0.7289 kg
Change in sensible heat for ice from −0.6°C to −40°C
(From Eq. 5.23, Chapter 5)
$Q_{ice} = I \, C_{pi} \, W_o \, [(T_f - T_2) - (-T_f ln(-T_2/-T_f)] = 0.0111 \, (2093.4)[39.4 - 0.6 \, ln(40/0.6)] = 856.9 \, J$
Change in sensible heat of water from −0.6°C to −40°C
(From Eq. 5.22, Chapter 5)
$Q_{water} = C_{pw} \, W_o(-T_f) \, ln \, (-T_2/-T_f) = 4186.8(0.7289)[0.6 \, ln(40/0.6)] = 7689.9 \, J$

$Q_{snf} = 0.26\ (837.36)[-0.6-(-40)] = 8577.9$ J
The amount of ice that is formed $= 0.7289$ kg
$\Delta H_f = 0.7289\ (334,860) = 244,079$ J
Total refrigeration load/kg $= 101,467 + 856.9 + 8577.9 + 7689.9 + 244,079 = 362,670$ J/kg
For 100 kg, the refrigeration load will be 36.267 MJ.

10.3   COMMODITY STORAGE REQUIREMENTS

Most food products would benefit from a reduction in the storage temperature provided that no freezing occurs. Reduced temperature reduces the rate of chemical reactions that deteriorate the product and also reduce microbiological activity. Freezing damages the cellular structure of fruits and vegetables and severely affects acceptability. Meat pigments darken irreversibly upon freezing. Acceleration of the development of a strong fishy flavor has been observed when fish were frozen at or near the freezing point and thawed. Thus, for perishable foods not specifically prepared for frozen storage, the freezing temperature should be the lowest acceptable limit for storage. Some fruits and vegetables are susceptible to chill injury at temperatures above the freezing point. The lowest safe temperatures have been defined for these commodities. Table 10.4 shows the temperature and humidity recommended for storage of various fruits and vegetables.

10.4   CONTROLLED ATMOSPHERE STORAGE

10.4.1   Respiration

Fruits and vegetables continue their metabolic activity after harvest. Maintenance of this metabolic activity is essential in the preservation of quality. Metabolic activity is manifested by respiration, a process where the material consumes oxygen and evolves carbon dioxide. A major nutrient metabolized during respiration is carbohydrate. Because respiration depletes nutrients in a product after harvest, the key to prolonging shelf life of fruits and vegetables is in reducing the rate of respiration. Early studies on the relationship between respiration and quality have established that storage life is inversely related to the rate of respiration and that for two products respiring at different rates, when the total quantity of CO_2 evolved by each product is equal, the products would have reached a comparable stage in their storage life.

Reduction of temperature is an effective means of reducing the rate of respiration. However, for some products subject to chill injury, respiration rate may still be quite high even at the lowest safe storage temperature. Controlled atmosphere storage has been developed as a supplement to refrigeration in prolonging the storage life of actively respiring fruits and vegetables.

The reaction involved in respiration is primarily the oxidation of carbohydrates:

$$C_6H_{12}O_6 + 6O_2 = 6CO_2 + 6H_2O$$

The respiratory quotient (RQ) defined as the ratio of CO_2 produced to O_2 consumed is 1 for the reaction. Most products show respiratory quotients close to 1. The heat evolved during respiration has also been shown to be related to the quantity of CO_2 produced in the same manner as that released during the combustion of glucose (10.7 J/mg CO_2). The rate of oxygen consumption and CO_2 evolution can be calculated from the heat of respiration and vice versa.

Table 10.4 Recommended Storage Conditions for Fruits and Vegetables

(1) Storage temperatures 30°–32°F (−1.11°–0°C). Highest Freezing point 28°F (−2.22°C). 85–90% RH.

apricots	pears
cherries	peaches
grapes	plums

(2) Storage temperatures 32°F (0°C)

75%RH	90% RH	95% RH	95% RH	95% RH
garlic				
onions	mushrooms	artichokes	sweet corn	carrots
	oranges	asparagus	endives[1]	lettuce[1]
	tangerines	lima beans	escarole[1]	parsnips
		beets	leafy greens	rutabagas
		broccoli	parsley	turnips
		brussels sprouts	green peas	
		cabbage	radishes	
		cauliflower	rhubarb	
		celery[2]	spinach[1]	

(3) Storage temperature 36°F (2.22°C). 95% RH.
   apples

(4) Storage temperature 45°F (7.22°C). 90% RH. Subject to chill injury at temperatures below 45°F
   green or snap beans
   ripe tomatoes

(5) Storage temperature 50°F (10°C). Subject to chill injury at temperatures below 50°F

85% RH	90–95% RH
melons	
potatoes	cucumbers
pumpkin	eggplants
squash	sweet peppers
green tomatoes	okra

(6) Storage temperatures 58°–60°F (14.4–15.6°C). 85–90% RH
   bananas
   grapefruit
   lemons

Source: ASHRAE. 1974 ASHRAE Handbook and Product Directory—1974. Applications. American Society of Heating, Refrigeration, and Air Conditioning Engineers, New York.
[1]Highest freezing point 31.9°F (−40.06°C).
[2]Highest freezing point 31.1°F(−0.5°C).
All others have freezing points below 31°F(−0.56°C).

Controlled atmosphere (CA) storage is based on the premise that increasing the CO_2 level and decreasing the O_2 level in the storage atmosphere will result in a reduction of the rate of respiration. Indeed, calorimetric determination of heat of respiration of products in continuously flowing CA of the optimum composition for the given product has shown that CA could reduce the respiration rate to approximately 1/3 of the respiration rate in air at the same temperature.

Example 10.10. One pound (0.454 kg) of head lettuce is packaged in an air-tight container with a volume of 4 L. The product occupies 80% of the volume, the rest being air. If the product is at a constant temperature of 4°C, calculate how long it will take for the oxygen content in the package to drop to 2.5%. Assume a RQ of 1.

Solution:

From Table 10.3, the constants a and b for the heat of respiration of head lettuce (in mW/kg) is 26.7 and 0.088, respectively.

$$q = 26.7(e)^{0.088(4)} = 38 \, \text{mW/kg}$$

$$\frac{\text{mg CO}_2}{\text{h}} = (38 \times 10^{-3}) \cdot \frac{\text{J}}{\text{s} \cdot \text{kg}} \cdot \frac{1 \, \text{mg CO}_2}{10.7 \, \text{J}} \cdot 3600 \frac{\text{s}}{\text{h}}(0.454 \, \text{kg}) = 5.8$$

$$\frac{\text{g moles CO}_2}{\text{h}} = \frac{\text{g moles O}_2 \, \text{depleted}}{\text{h}} = \frac{5.8 \times 10^{-3}}{44} = 1.318 \times 10^{-4}$$

Air is approximately 21% O_2 and 79% N_2. The total number of moles of air originally in the container is:

$$n_{\text{air}} = \frac{PV}{RT} = \frac{(1 \, \text{atm})(4)(0.2)1}{0.08206(1 \, \text{atm} / \text{g mole K})(273 + 4)\text{K}}$$

$$n_{\text{air}} = 0.0352 \, \text{moles}$$

Because the RQ is 1, there will be no net change in the total number of moles of gases inside the container. The number of moles of oxygen when the concentration is 2.5% is

$$n_{O_2\text{final}} = 0.025(0.0352) = 0.00088$$

The original number of moles of O_2 is

$$n_{O_2\text{initial}} = 0.21(0.0352) = 0.007392 \, \text{moles}$$

The number of moles of O_2 that must be depleted by respiration is

$$n_{O_2\text{depleted}} = 0.007392 - 0.00088 = 0.006512$$

The time required to deplete O_2 to the desired level is

$$\text{Time} = \frac{0.006512 \, \text{mole}}{0.0001318 \, \text{mole} / \text{h}} = 49.4 \, \text{h}$$

10.4.2   CA Gas Composition

CA storage has been used successfully with apples and pears. Experimental results have been very encouraging on CA storage of cabbage, head lettuce, broccoli, and brussels sprouts. Atmospheric modification of product containers during transit of vegetables is also gaining wider acceptance. CA is recommended when transit time is 5 days or more.

Reduction of oxygen concentration in the storage atmosphere to 3% or below has been shown to be most effective in reducing respiration rate, with or without CO_2. Too low an oxygen concentration, however, leads to anaerobic respiration resulting in the development of off-flavors in the product. The O_2 concentration for onset of anaerobic respiration ranges from 0.8% for spinach to 2.3% for

Table 10.5 CA Storage Conditions Suitable for Use with Various Products

Product	% CO_2	% O_2
Apples	2–5	3
Asparagus	5–10	2.9
Brussels sprouts	2.5–5	2.5–5
Beans (green or snap)	5	2
Broccoli	10	2.5
Cabbage	2.5–5	2.5–5
Lettuce (head or leaf)	5–10	2
Pears	5	1
Spinach	11	1
Tomatoes (green)	0	3

Percentages are volume or mole percent. Balance is nitrogen.
Reprinted from Toledo, R. T. 1980. *Fundamentals of Food Process Engineering.* AVI Pub. Co., Westport, Conn.

asparagus. The optimum CA composition varies from one product to another and between varieties of a given product. CA compositions found to be effective for some products are shown in Table 10.5.

To be effective, CA storage rooms must be reasonably air-tight. To test for air-tightness, the room should be pressurized to a positive pressure of 1 in. water gauge (wg) (249 Pa gauge). A room is considered air-tight if at the end of 1 hour, the pressure does not drop below 0.2 in. wg (49.8 Pa). This requirement is equivalent to an air incursion rate of 0.2% of the gas volume in the room per hour at a constant pressure differential between inside and outside of 0.5 in. wg (124 Pa). Current standards for refrigerated tractor trailers call for a maximum air incursion rate of 2 ft³/min (3.397 m³/h) at a pressure differential of 0.5 in. water (124 Pa) between the inside and outside of the trailer. This rate is approximately 15 times what would meet the specification for air-tightness.

CA in storage warehouses are maintained at the proper composition by introduction of CO_2 and/or N_2 if product respiration is insufficient to raise CO_2 levels or lower O_2 levels. In a full, tightly sealed warehouse, ventilation with fresh air and scrubbing of CO_2 from the storage atmosphere may be necessary to maintain the correct CA composition and prevent anaerobic respiration. Successful operation of a CA storage facility necessitates regular monitoring and analysis of gas composition and making appropriate corrections when necessary. Figures 10.12 and 10.13 are schematic diagrams of CA storage systems for warehouses and for in-transit systems.

Example 10.11. The rate of flow of fluids through narrow openings is directly proportional to the square root of the pressure differential across the opening. If the pressure inside a chamber of volume V changes exponentially with time from 1 in. wg (249 Pa) at time 0 to 0.2 in. wg (49.8 Pa) at 1 hour, calculate the mean volumetric rate of flow of gases out of the chamber and the mean pressure at which this rate of flow would be expected if the pressure is held constant.

Solution:

The expression for the volumetric rate of flow Q is

$$Q = k(\Delta P)^{1/2}$$

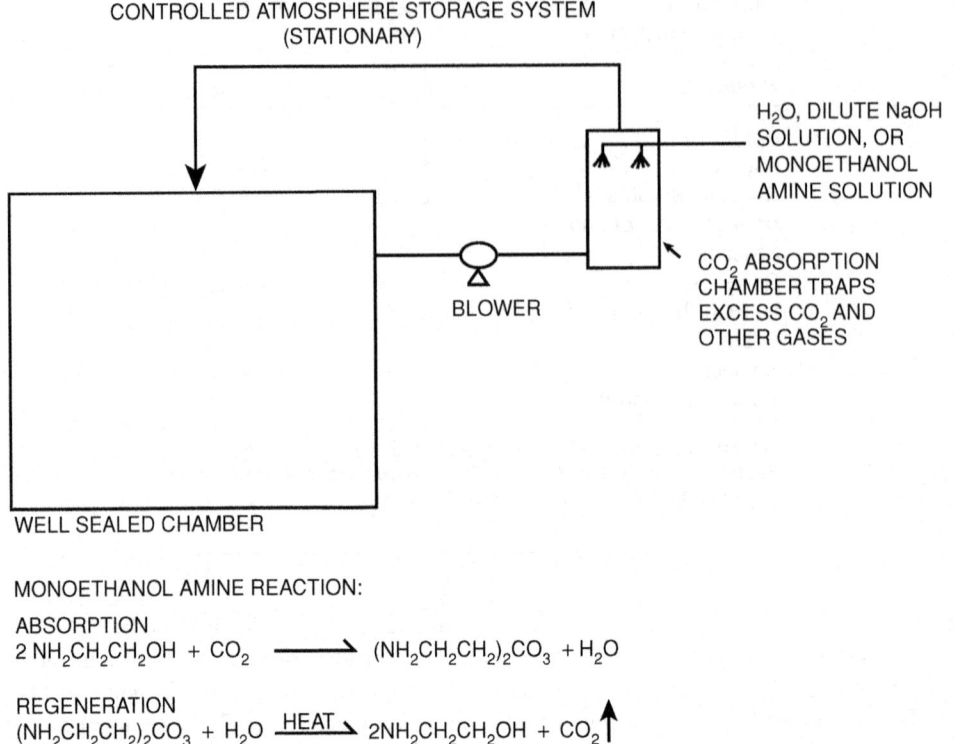

Figure 10.12   Schematic diagram of a controlled atmosphere storage system for warehouses.

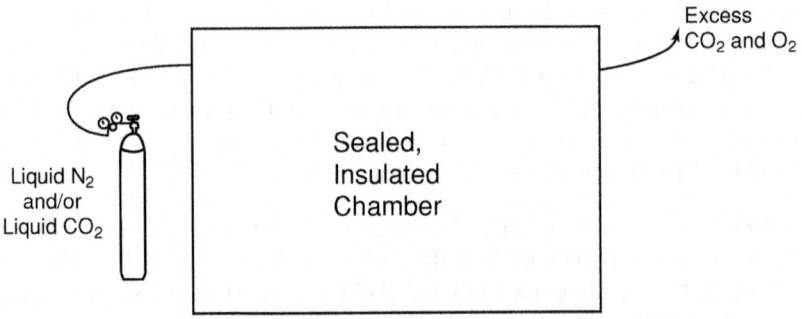

Figure 10.13   Schematic diagram of a transportable system for generating and maintaining controlled atmospheres.

The volume dQ expelled over a period of time dt is

$$dQ = k(\Delta P)^{1/2} \, dt$$

Because P is exponential with time, substituting the boundary conditions, $\Delta P = 1$ at $t = 0$, and $\Delta P = 0.2$ at $t = 1$, will give the equation for ΔP as a function of t as follows:

$$\Delta P = (e)^{-1.6094t}$$

Substituting P in the differential equation and integrating:

$$Q = \int_0^1 [k(e)^{0.5(-1.6094)t}] \, dt$$

The mean rate of flow in one hour, $\bar{q} = Q/t$

$$\bar{q} = \frac{1}{1}\left[\left(\frac{k}{-0.8047}\right)(e)^{-0.8047t}\right]_0^1 = 0.687 \, k$$

Because $\bar{q} = k(\overline{\Delta P})^{1/2}$, the mean pressure $\overline{\Delta P}$ that would give the equivalent rate of flow as $\bar{q}$ is $(0.687)^2$ or 0.472.

One atmosphere is 406.668 in. of water. Using the ideal gas equation, the number of moles gas in the chamber at pressure P_1 and P_2 are

$$n_1 = \frac{P_1 V}{RT} \text{ and } n_2 = \frac{P_2 V}{RT}$$

The number of moles lost with a drop in pressure from P_1 to P_2 would be

$$n_1 = \frac{V}{RT}(P_1 - P_2)$$

Expressing the quantity expelled as volume at standard atmospheric pressure:

$$V_1 = n_1 \frac{RT}{P} = \frac{(P_1 - P_2)V}{P}$$

Substituting $P_1 = 1$ in.; $P_2 = 0.2$ in.; and $P = 406.668$:

$$V_1 = \frac{0.8V}{406.668} = 0.00196 \, V$$

The specified conditions of a drop in pressure from 1 in. wg to 0.2 in. wg in one hour is equivalent to a rate of loss of approximately 0.2% of the chamber volume per hour at an average pressure drop of approximately 0.5 in. wg.

Example 10.12. A refrigerated tractor-trailer is loaded with head lettuce such that 20% of its total volume is air space. The trailer has a total volume of 80 m³. Assuming that the trailer meets the specifications for the maximum air incursion rate of 2 ft³/min (3.397 m³/h) at a pressure differential of 0.5 in. wg (124 Pa), what would be the rate of addition of N_2 and CO_2 necessary to maintain the atmosphere inside the truck at 2% O_2 and 5% CO_2 when the truck is traveling at the rate of 50 miles/h (80 km/h). The average temperature of the product is 1.5°C and the trailer contains 18,000 kg of lettuce.

Solution:

If the atmosphere inside the truck has been pre-equilibrated at the point of origin to the desired CA composition, the amount of nitrogen and CO_2 that must be added during transit would be that necessary to displace the air infiltrating into the trailer. Let x = rate of N_2 addition and y = rate of CO_2 addition. If n_{ai} = the rate of air infiltration, n_{rc} = the rate of CO_2 generation by the product, and n_{ro} = the rate of oxygen consumption by the product during respiration, the following material balance may be set up for CO_2 and O_2 assuming that the expelled gases are of the same composition as those in the interior of the trailer.

O_2 balance (air is 21% oxygen):

$$n_{ai}(0.21) = (n_{ai} + x + y)(0.02) + n_{ro}$$

CO_2 balance:

$$y + n_{rc} = (n_{ai} + x + y)(0.05)$$

The heat of respiration of head lettuce in mW/Kg from values of a = 26.7 and b = 0.088 from Table 10.3 and for T = 1.5°C:

$$q = 26.7(e)^{1.5(0.088)} = 30.47 \, mW/kg$$

$$\frac{mg \, CO_2}{kg \cdot h} = \frac{30.47 \times 10^{-3} \, J}{kg \cdot s} \frac{mg \, CO_2}{10.7 \, J} \frac{3600 \, s}{h} = 10.25$$

$$\frac{g \, moles \, CO_2}{kg \cdot h} = \frac{g \, moles \, O_2}{kg \cdot h} = \frac{10.25 \times 10^{-3}}{44} = 0.233 \times 10^{-3}$$

$$n_{rc} = (18,000 \, kg) \frac{0.233 \times 10^{-3} \, g \, moles}{kg \cdot h} = 4.194 \frac{g \, moles}{h}$$

$$n_{ro} = 4.194 \, g \, moles/h$$

The rate of air infiltration must now be calculated at a velocity of 80 km/h. The velocity head of a flowing fluid, $\Delta P/\rho$, is $V^2/2$. ρ is the fluid density.

$$\Delta P = \frac{(80,000)^2 \, m^2}{h^2} (0.5) \frac{\rho kg}{m^3} \frac{1 \, h^2}{(3600)^2 \, s^2} = 247\rho \, Pa$$

The specified infiltration rate of 3.397 m^3/h at a pressure of 124 Pa can be converted to the rate at the higher pressure by using the relationship that rate of flow is directly proportional to the square root of the pressure.

$$Q = 3.397 \left[\frac{247}{124} \right]^{\frac{1}{2}} = 4.79 \, m^3/h$$

Using the ideal gas equation, $n_{ai} = PV/RT$:

$$n_{ai} = \frac{(4.79 \, m^3/h)(1000 \, L/m^3)(1 \, atm)}{0.08206(L)(atm)/(gmole \, K)(274.5 \, K)} = 212.7 \, g \, moles/h$$

Substituting values for n_{ai}, n_{ro} and n_{rc} in the O_2 and CO_2 balance equations:

$$212(0.21) = (212 + x + y)(0.02) + 4.176$$

$$x + y = \frac{44.52 - 4.176 - 4.24}{0.02} = 1805$$

$$x = 1805 - y$$

$$y + 4.176 = (212 + x + y)(0.05)$$

$$0.95y = 0.05x + 6.424$$

$$0.95y - 0.05(1805 - y) = 6.424$$

$$1.0y = 6.424 + 90.25$$

$$y = 96.67 \text{ g moles } CO_2/h$$

$$x = 1708.3 \text{ g moles } N_2/h$$

Note that because of air infiltration, it would not be possible for product respiration to maintain the necessary CA composition inside the trailer. Furthermore, the respiration rate for this particular product is so low that flushing of the trailer with N_2 and CO_2 would have to be done initially to bring the CA to the desired composition within a short period of time.

10.5 MODIFIED ATMOSPHERE PACKAGING

Modified atmosphere (MA) and CA are similar, and are sometimes used interchangeably. A feature of CA is some type of control that maintains constant conditions within the storage atmosphere. MA packaging involves changing the gaseous atmosphere surrounding the product at the time of packaging and using barrier packaging materials to prevent gaseous exchange with the environment surrounding the package. The use of gas permeable packaging materials that is tailored to promote gas exchange to maintain relatively constant conditions inside the package may be considered a form of CA packaging. Reduction of respiration rates of live organisms by lowering oxygen and elevating CO_2 has been discussed in the previous section. MA packaging, as is now practiced, is primarily designed to control the growth of microorganisms and in some cases, enzyme activity, in packaged uncooked or cooked food. The atmosphere used, if the product itself is non-respiring, is usually 75% CO_2, 15% N_2, and 10% O_2. Bakery products, which are generally of lower moisture content than most foods, could be preserved longer in 100% CO_2. Dissolution of CO_2 and O_2 in the aqueous phase of the product is responsible for microbial inhibition. High CO_2 is inhibitory to most microorganisms. Although CO_2 lowers the pH, the inhibitory effect is more than from the pH reduction alone. The presence of O_2 has slight inhibitory activity against anaerobes resulting in slower growth compared with a 100% CO_2 atmosphere.

Other gas combinations and the products on which they have been used are 80% CO_2 and 20% N_2 for luncheon meats; 100% N_2 for cheese; 30% CO_2 and 70% N_2 for fresh poultry and fish, and 30% O_2, 30% CO_2 and 40% N_2 for red meats.

Typically, MA packaging involves drawing a vacuum from the product in a high barrier packaging material, displacing the vacuum with the modified atmosphere, and sealing the package.

Another approach that is suitable for engineering modeling involves the use of gas-permeable films as the primary packaging material and enclosing the packages in a larger high barrier bag filled with the desired gaseous mixture. The gas mixture composition and the package film permeability are balanced to achieve the desired constant MA inside the package. This is a rapidly growing area in packaging research. So far there are no satisfactory models that can adequately predict shelf life of MA packaged products.

PROBLEMS

10.1. An ammonia refrigeration unit is used to cool milk from 30°C to 1°C (86°F to 33.8°F) by direct expansion of refrigerant in the jacket of a shell and tube heat exchanger. The heat exchanger has a total outside heat transfer surface area of 14.58 m² (157 ft²). To prevent freezing, the temperature of the refrigerant in the heat exchanger jacket is maintained at −1°C (31.44°F).

    (a) If the average overall heat transfer coefficient in the heat exchanger is 1136 W/m² $\cong$ K (200 BTU/ h $\cong$ ft² $\cong$ °F) based on the outside area, calculate the rate at which milk with a specific heat of 3893 J/kg K(0.93 BTU/lb $\cong$ °F) can be processed in this unit.

    (b) Determine the tons of refrigeration required for the refrigeration system.

    (c) The high-pressure side of the refrigeration system is at 1.72 MPa (250 psia). Calculate the horsepower of the compressor required for the refrigeration system assuming a volumetric efficiency of 60%.

10.2. A single-stage compressor in a freon 12 refrigeration system has a volumetric efficiency of 90% at a high side pressure of 150 psia (1.03 MPa) and a low side pressure of 50 psia (0.34 MPa). Calculate the volumetric efficiency of this unit if it is operated at the same high side pressure but the low side pressure is dropped to 10 psia (68.9 kPa). Assume R12 is an ideal gas.

10.3. Calculate the tons of refrigeration for a unit that will be installed in a cooler maintained at 0°C (32°F) given the following information on its construction and operation:

    The cooler is inside a building.

    Dimensions: 4 × 4 × 3.5 m (13.1 × 13.3 × 11.5 ft)

    Wall and ceiling construction:

    3.175 mm (1/8 in.) thick polyvinyl chloride sheet inside (k of PVC = 0.173 W/m $\cong$ K or 0.1 BTU/h $\cong$ ft $\cong$ °F)

    15.24 cm (6 in.) fiberglass insulation

    5.08 cm (2 in.) corkboard

    3.17 mm (1/2 in.) PVC outside

    Floor construction:

    3.175 mm (1/8 in.) thick floor tile (k = 0.36 W/(m $\cong$ k) or 0.208 BTU/(h $\cong$ ft $\cong$ °F)

    10.16 cm (4 in.) concrete slab

    20.32 cm (8 in.) air space

    Concrete surface facing the ground at a constant temperature of 15°C (59°F)

    Door:

    1 m wide × 2.43 m high (3.28 × 8 ft)

    Design for door openings that average four per hour at 1 minute per opening

    Air infiltration rate:

    1 m³/h (35.3 ft³/h) at atmospheric pressure and ambient temperature

    Ambient conditions:

    32°C (89.6°F)

    Product cooling load:

    Design for a capability to cool 900 kg of product (C_P = 0.76 BTU/lb $\cong$ °F or 3181 J/kg $\cong$ K) from 32°C to 0°C (89.6°F to 32°F) in 5 hours. The freezing point of the product is −1.5°C (29.3°F).

10.4. The "stack effect" due to a difference in temperature between the inside and outside of a cooling room is often cited as the major reason for air infiltration. In this context, ΔP is positive at the lowest section of a cooler and is negative at the highest section, with a zone, called the neutral zone, at approximately the center of the room where the ΔP is zero. If the area of the openings at the lowest sections where ΔP is positive equals the area of the openings in the highest sections where ΔP is negative, air will enter at the top and escape at the openings in the bottom at the same volumetric rate of flow (assuming no pressure change inside the room). If the room allows air leakage at the rate of 2% of the room volume per minute at a ΔP of 0.5 in. wg (124 Pa), determine the rate of air infiltration that can be expected in a room that is 2 m (6.56 ft) high to the neutral zone if the interior of the room is at $-20°C$ ($-4°F$) and ambient temperature is 30°C (86°F). The rate of gas flow through the cracks is proportional to the square root of ΔP. Assume air is an ideal gas. ΔP due to a column of air of height h at different temperatures $= g(\rho_1 - \rho_2)h$, where ρ_1 and ρ_2 are the densities of the columns of air.

10.5. For a 1-ton refrigeration unit (80% volumetric efficiency) using refrigerant 12 at a high side pressure of 150 psia (1.03 MPa) and a low side pressure of 45 psia (0.31 Mpa), operating at an ambient temperature of 30°C (86°F), determine the effect of the following on refrigeration capacity and on HP/(ton)$_r$. Assume the same compressor displacement in each case.
   (a) Reducing the evaporator temperature to $-30°C$ ($-22°F$). High side pressure remains at 150 psia (150 MPa).
   (b) Increasing ambient temperature to 35°C (95°F). (Low side pressure remains at 45 psia [0.31 Mpa]). The high side pressure is to change such that ΔT between the hot refrigerant gas and ambient air remains the same as in the original set of conditions.
   (c) Air in the line such that the vapor phase of refrigerant always contain 10% air and 90% refrigerant by volume. Assume condensation temperature of hot refrigerant gas and temperature of cold refrigerant gas are the same as in the original set of conditions (partial pressure of refrigerant gas at the low and high side pressures are the same as in the original set of conditions, 45 and 150 psia or 0.31 and 1.03 MPa). Use R = 1.987 BTU/(lbmole $\cong$ °R) or 8318 J/(kg $\cong$ K). The specific heat ratio C_P/C_v for air is 1.4.
   (d) Oil trapped in the vapor return line such that ΔP across the constriction is 10 psi (68.9 kPa). Assume evaporator temp. $= 0°C$.

10.6. Chopped onions are frozen in a continuous belt freezer at using $-50°C$ air at high velocity. When onions with a moisture content of 86% are loaded on the belt with a thickness of 2 cm., it took 20 minutes for the temperature to drop from 10°C to $-20°C$. Onion juice is added and mixed with the chopped onions in a ratio 0.10 parts juice to 0.90 parts of the chopped onions. The freezing point of the chopped onions is $-0.5°C$. The juice contains 1.5% solids (all soluble). The juice has a freezing point of $-0.16°C$. Assuming that the rate of heat transfer is the same, calculate the time required to freeze the onions with the added juice.

SUGGESTED READING

ASHRAE. 1965. ASHRAE Guide and Data Book. Fundamentals and Equipment for 1965 and 1966. American Society for Heating, Refrigerating and Air Conditioning Engineers, Atlanta GA.

ASHRAE. 1966. ASHRAE Guide and Data Book. Applications for 1966 and 1967. American Society for Heating, Refrigerating and Air Conditioning Engineers, Atlanta, GA.

ASHRAE. 2000. ASHRAE Standard. Designation and Safety Classification of Refrigerants. American Society for Heating, Refrigerating and Air Conditioning Engineers, Atlanta, GA.

Charm, S. E. 1971. Fundamentals of Food Engineering. 2nd ed. AVI Publishing Co., Westport, CT.

Ciobanu, A. and Dincer I. 1997. Heat Transfer in Food Cooling Operations. Taylor and Francis, Washington, DC.

Green, D. W. and Mahoney, J. D. 1997. Perry's Chemical Engineers Handbook. 7th ed. McGraw-Hill Book Co., New York.

Johnson, A. T. 1999. Biological Process Engineering. John Wiley & Sons, New York.

Lascu G., Bercescu V., and Niculescu, L.1976. Cooling Technology in the Food Industry. Abacus Press, Tunbridge Wells, England.

Singh, R. P. and Heldman, D. 2001. Introduction to Food Engineering. Academic Press, San Diego.

CHAPTER 11

Evaporation

The process of evaporation is employed in the food industry primarily as a means of bulk and weight reduction for fluids. The process is used extensively in the dairy industry to concentrate milk, in the fruit juice industry to produce fruit juice concentrates, in the manufacture of jams, jellies, and preserves to raise the solids content necessary for gelling, and in the sugar industry to concentrate sugar solutions for crystallization. Evaporation can also be used to raise the solids content of dilute solutions prior to spray or freeze drying.

Evaporation is used to remove water from solutions with or without insoluble suspended solids. If the liquid contains only suspended solids, dewatering can be achieved by either centrifugation or filtration. The process of evaporation involves the application of heat to vaporize water at the boiling point. Its simplest form is atmospheric evaporation where the liquid in an open container is heated and the vapors driven off are dispersed into the atmosphere. Atmospheric evaporation is simple but it is slow and is not very efficient in the utilization of energy. Furthermore, because most food products are heat sensitive, prolonged exposure to high temperature in atmospheric evaporation causes off-flavors, color changes, or degradation of overall quality In addition, because food also contain volatile compounds, vapors produced by evaporation could generate nuisance odors, therefore they must be contained by condensation. Evaporators used on food products remove water at low temperatures by heating the product in a vacuum. Efficient energy utilization can be designed into the system by using heat exchangers to extract heat from the vapors to preheat the feed or by using multiple effects where the vapors produced from one effect are used to provide heat in the succeeding effects.

Problems in evaporation involve primarily heat transfer and material and energy balances, the principles of which have been discussed earlier.

11.1 SINGLE-EFFECT EVAPORATORS

Figure 11.1 is a schematic diagram of a single-effect evaporator. The system consists of a vapor chamber where water vapor separates from the liquid, a heat exchanger to supply heat for vaporization, a condenser to draw out the vapors from the vapor chamber as rapidly as they are formed, and a steam jet ejector for removing noncondensible gases from the system. Each vapor chamber is considered an effect.

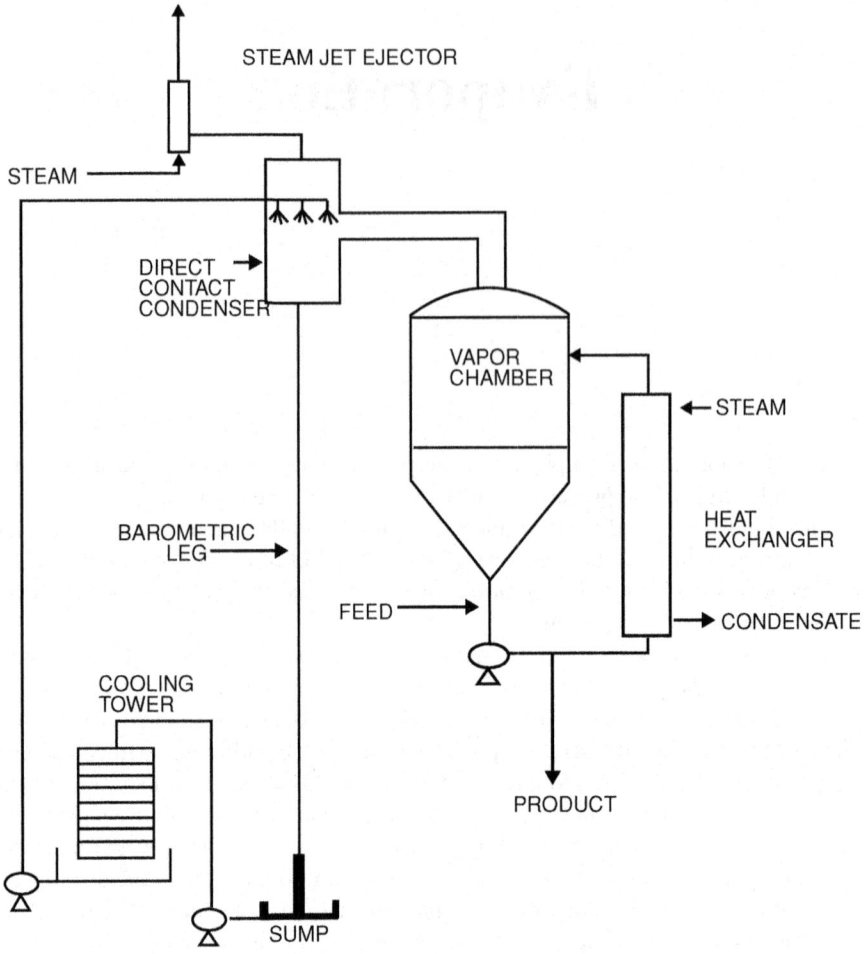

Figure 11.1   Schematic diagram of a single-effect evaporator.

11.1.1 The Vapor Chamber

This is usually the largest and most visible part of the evaporator. Its main function is to allow separation of vapor from liquid and prevent carry-over of solids by the vapor. It is also a reservoir for the product. The temperature inside an evaporator is determined by the absolute pressure in the vapor chamber. The vapor temperature is the temperature of saturated steam at the absolute pressure inside the chamber. When the liquid is a dilute solution, vapor and liquid temperatures will be the same. However, concentrated solutions exhibit a boiling point rise resulting in a higher boiling temperature than that of pure water. Thus vapors leaving the liquid will be superheated steam at the same temperature as the boiling liquid. Depending on the extent of heat loss to the surroundings around the vapor chamber, the vapor may be saturated at the absolute pressure within the vapor chamber, or superheated steam at the boiling temperature of the liquid.

In most food products the soluble solids are primarily organic compounds and the boiling point rise can be expressed as follows:

$$\Delta T_b = 0.51 \, m \tag{11.1}$$

where ΔT_b in °C is the increase in the boiling point of a solution with molality m, above the boiling point of pure water at the given absolute pressure. In addition to the boiling point rise due to the presence of solute, the pressure at the bottom of a liquid pool is higher than the absolute pressure of the vapor, and this pressure difference can add to the temperature of the liquid pool.. The pressure exerted by a column of liquid of height, h, and density, ρ, is

$$P = \rho(h)\frac{g}{g_c} \tag{11.2}$$

where $g_c = 1$ when using SI units.

Example 11.1. Calculate the boiling temperature of liquid containing 30% soluble solids at a point 5 ft (1.524 m) below the surface inside an evaporator maintained at 20 in. Hg vacuum (33.8 kPa absolute). Assume the soluble solids are hexose sugars and the density of the liquid is 62 lb/ft^3 (933 kg/m^3). Atmospheric pressure is 30 in. Hg (101.5 kPa).

Solution:

The absolute pressure in lb$_f$/in.^{2} corresponding to 20 in. Hg vacuum is $(30 - 20)(0.491) = 4.91$ psia. From the steam tables, the temperature corresponding to 4.91 psia is (by interpolation) 161.4°F (71.9°C). The molecular weight of a hexose sugar is 180. The molality of 30% soluble solids will be

$$m = \frac{\text{moles solute}}{1000 \text{ g solvent}} = \frac{0.3/180}{0.7/1000} = 2.38$$

Using Equation (11.1):

$$\Delta T_b = 0.51(2.38) = 1.21°C \text{ or } 2.2°F$$

The absolute pressure at the level considered is the sum of the absolute pressure of the vapor and the pressure exerted by the column of liquid. Expressed in lb$_f$/in.^{2}, this pressure is

$$P = 4.91 + \rho h\frac{g}{g_c} = 4.91 + 62\frac{lb_m}{ft^3}(5 \text{ ft}) \cdot \frac{lb_f}{lb_m} \cdot \frac{1 \text{ ft}^2}{144 \text{ in}^2} = 7.06 \, psia$$

From the steam tables, the boiling temperature corresponding to 7.06 psia is (by interpolation) 175.4°F (79.7°C).

The boiling temperature of the liquid will be $175.4 + 2.2 = 177.6°F$ (80.9°C).

The significance of the boiling point rise is that the liquid leaving the evaporator would be at the boiling point of the liquid rather than at the temperature of the vapor. The boiling temperature at a point submerged below a pool of liquid would have the effect of reducing the ΔT available for heat transfer in the heat exchanger if the heat exchange unit is submerged far below the fluid surface.

11.1.2 The Condenser

Two general types of condensers are used. A surface condenser is used when the vapors need to be recovered. This type of condenser is actually a heat exchanger cooled by refrigerant or by cooling water. The condensate is pumped out of the condenser. It has a high first cost and is expensive to operate. For this reason it is seldom used if an alternative is available. Condensers used on essence recovery systems fall in this category.

The other type of condenser is one where cooling water mixes directly with the condensate. This condenser may be a barometric condenser where vapors enter a water spray chamber on top of a tall column. The column full of water is called a barometric leg and the pressure of water in the column balances the atmospheric pressure to seal the system and maintain a vacuum. The temperature of the condensate-water mixture should be in the order of 5°F (2.78°C) below the temperature of the vapor in the vapor chamber to allow continuous vapor flow into the condenser. The height of the barometric leg must be sufficient to provide sufficient positive head at the base to allow the condensate and cooling water mixture to flow continuously out of the condenser at the same rate they enter. A jet condenser is one where part of the cooling water is sprayed in the upper part of the unit to condense the vapors and the rest is introduced down the throat of a venturi at the base of the unit to draw the condensed vapor and cooling water out of the condenser. The jet condenser uses considerably more water than the barometric condenser and the rate of water consumption cannot be easily controlled.

The condenser duty q_c is the amount of heat that must be removed to condense the vapor.

$$q_c = V(h_g - h_{fc}) \tag{11.3}$$

where V is the quantity of vapor to be condensed, h_g is the enthalpy of the vapor in the vapor chamber of the evaporator, and h_{fc} is the enthalpy of the liquid condensate.

For direct contact condensers, the amount of cooling water required per unit amount of vapor condensed can be determined by a heat balance:

$$W(h_{fc} - h_{fw}) = V(h_g - h_{fc}) \tag{11.4}$$

$$\frac{W}{V} = \frac{h_g - h_{fc}}{h_{fc} - h_{fw}} \tag{11.5}$$

where W is quantity of cooling water required, and h_{fw} is enthalpy of cooling water entering the condenser.

The enthalpy of the condensate-water mixture, h_{fc}, should be evaluated, in the case of barometric condensers, at a temperature 5°F (2.7°C) lower than the vapor temperature.

Example 11.2. Calculate the ratio of cooling water to vapor for a direct contact barometric condenser for an evaporator operating at a vapor temperature of 150°F (65.55°C). What would be the minimum height of the water column in the barometric leg for the evaporator to operate at this temperature? Cooling water is at 70°F (21.1°C). Atmospheric pressure is 760 mm Hg.

Solution:

At a temperature of 150°F, the absolute pressure of saturated steam is 3.7184 psia (25.6 kPa). h_g = 1126.1 BTU/lb or 2.619 MJ/kg. The condensate-cooling water mixture must be at 150 − 5 or 145°F (62.78°C). h_{fc} = 112.95 BTU/lb or 0.262 MJ/kg. The enthalpy of the cooling water h_{fw} = 38.052 BTU/lb or 0.088 MJ/kg. Basis: V = one unit weight of vapor. Using Equation (11.5):

$$\frac{W}{V} = \frac{1126.1 - 112.95}{112.95 - 38.052} = 13.52$$

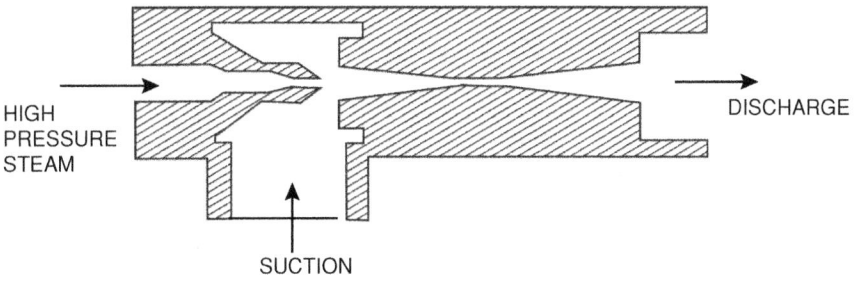

Figure 11.2 Schematic diagram of a single-stage ejector.

The atmospheric pressure is 760 mm Hg or 101.3 kPa. From the steam tables, the density of water at 145°F is l/V = 61.28 lb/ft³ or 981.7 kg/m³. The pressure that must be counteracted by the column of water in the barometric leg is the difference between barometric pressure and the absolute pressure in the system.

$$\Delta P = 101.3 - 25.6 = 75.7 \text{ kPa} = \rho g h$$

$$h = \frac{75,700 \text{ kg} \cdot m}{s^2 \cdot m^2} \frac{1}{981.7 \text{ kg} \cdot m^{-3}} \frac{1}{9.80 \text{ m} \cdot s^{-2}} = 7.868 \text{ m or } 25.8 \text{ ft}$$

11.1.3 Removal of Noncondensible Gases

A steam jet ejector is often used. Figure 11.2 is a schematic diagram of a single-stage ejector. High-pressure steam is allowed to expand through a jet, which increases its velocity. The movement of steam through the converging-diverging section at high velocity generates a zone of low pressure in the suction chamber, and noncondensible gases can be drawn into the ejector. The noncondensible gases mix with the high velocity steam and are discharged into the atmosphere. Steam jet ejectors are more effective than vacuum pumps in that water vapor present in the noncondensible gases does not interfere with its operation. If suction absolute pressures are 4 in. Hg (13.54 kPa) or lower, multistage ejectors are used. The capacity of jet ejectors is dependent on the design of the ejector, the pressure of the high-pressure steam, and the pressure differential between the suction and discharge.

Capacity charts for steam jet ejectors are usually provided by their manufacturers and the capacity is expressed as weight of air evacuated per hour as a function of suction pressure and steam pressure.

The amount of noncondensible gases to be removed from a system depends upon the extent of leakage of air into the system and the amount of dissolved air in the feed and in the cooling water. In addition to the noncondensible gases, jet ejectors also have to remove the water vapor that is present along with the noncondensible gases in the condenser. Air leakage has been estimated at 4 g air/h for every meter length of joints. The solubility of air in water at atmospheric pressure at various temperatures can be determined from Fig. 11.3. The amount of water vapor with the noncondensible gases leaving the condenser an be calculated as follows:

$$W_v = \frac{P_c(18)}{(P_v - P_c)29} \tag{11.6}$$

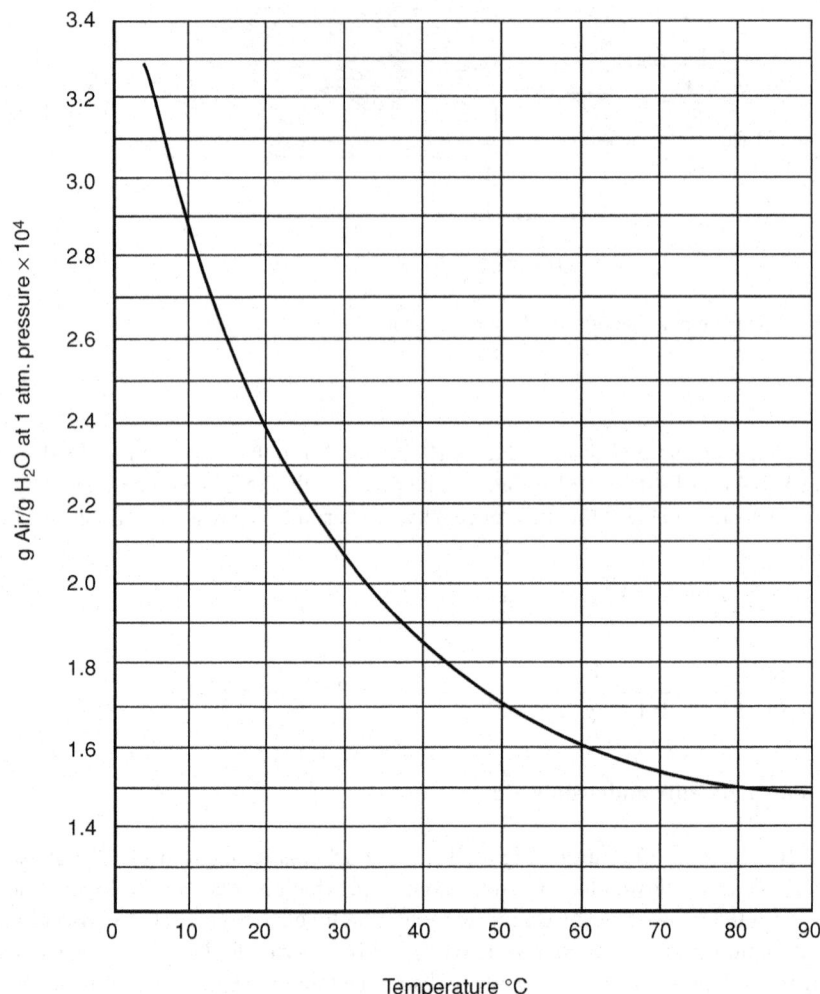

Figure 11.3 Solubility of air in water at atmospheric pressure at various temperatures. (Source: Data from Perry, et al. 1963. Chemical Engineers Handbook. McGraw-Hill, New York.)

where W_v is kg water vapor/kg air, P_c is vapor pressure of water at temperature of condensate-cooling water mixture, and P_v is absolute pressure inside the evaporator.

Example 11.3. Calculate the ejector capacity required for an evaporator that processes 100 kg/h of juice from 12% to 35% solids. The evaporator is operated at 65.6°C (150°F). Cooling water is at 21.1°C (70°F). Product enters at 35°C (89.6°F). The condenser temperature is maintained at 2.78°C (5°F) below the vapor temperature.

Solution:

From Example 11.2, the ratio of cooling water to vapors for an evaporator operated under the identical conditions specified in this problem, was determined to be 13.52 kg of water at 21.1EC. A material balance gives:

$$\text{Wt concentrate} = \frac{100(0.12)}{0.35} = 34.28 \text{ kg/h}$$

$$\text{Wt vapors} = 100 - 34.28 = 65.71 \text{ kg/h}$$

The weight of cooling water W_c is $13.52(65.71) = 888$ kg/h. From Fig. 11.3, the solubilities of air in water are 2.37×10^{-4} and 1.96×10^{-4} kg air / kg water at 21.2°C and 35°C, respectively. The amount of air, M_a, introduced with the condenser water and with the feed is:

$$M_a = 100(1.96 \times 10^{-4}) + 888(2.37 \times 10^{-4}) = 0.23 \text{ kg/h}$$

Using Equation (11.6), a vapor pressure at 62.82°C of 22.63 kPa (3.2825 psia), and a pressure of 25.63 kPa (3.7184 psia) at the vapor temperature of 65.6°C:

$$W_v = \frac{(22.63)18}{(25.63 - 22.63)(29)} = 4.68 \frac{\text{kg water}}{\text{kg air}}$$

The total ejector load excluding air leakage is equivalent to $4.68 + 0.23$ or 4.91 kg/h.
The majority of the ejector load is the water vapor that goes with the noncondensible gases.

11.1.4 The Heat Exchanger

The rate of evaporation in an evaporator is determined by the amount of heat transferred in the heat exchanger. Variations in evaporator design can be seen primarily in the manner in which heat is transferred to the product. Considerations like the stability of the product to heat, fouling of heat exchange surfaces, ease of cleaning, and whether the product could allow a rapid enough rate of heat transfer by natural convection, dictate the design of the heat exchanger for use on a given product. The schematic diagram of an evaporator shown in Fig. 11.1 shows a long tube, vertical, forced circulation evaporator. This type of design for the heat exchanger is usually used when a single effect concentrates a material that becomes very viscous at the high solids content. Because of the forced circulation, heat transfer coefficients are fairly high even at the high viscosity of the concentrate. Some evaporators would have the heat exchanger completely immersed in the fluid being heated inside the vessel that constitutes the fluid reservoir and vapor chamber. Heat is transferred by natural convection. This type of heat exchange is suitable when the product is not very viscous and is usually utilized in the first few effects of a multiple-effect-evaporator.

Heat exchangers on evaporators for food products have the food flowing inside the tubes for ease of cleaning. In very low temperature evaporation such as in fruit juice concentration, the operation has to be stopped regularly to prevent microbiological build-up and also to clean deposits of food product on the heat exchange surfaces.

The capacity of an evaporator is determined by the amount of heat transferred to the fluid by the heat exchanger. If q = the amount of heat transferred, P = mass of concentrated product, C_c = specific heat of the concentrate, V = mass of the vapor, h_g = enthalpy of the vapor, h_f = enthalpy of the water component of the feed that is converted to vapor, T_1 = feed inlet temperature, and T_2 = liquid temperature in the evaporator, a heat balance would give:

$$q = PC_c(T_2 - T_1) + V(h_g - h_f) \tag{11.7}$$

The rate of heat transfer can be expressed as:

$$Q = U A \Delta T \tag{11.8}$$

A material balance would give:

$$P = \frac{Fx_f}{x_p} \tag{11.9}$$

and

$$V = F\left(1 - \frac{x_f}{x_p}\right) \tag{11.10}$$

Equations (11.7) and (11.10) can be used to calculate the capacity of an evaporator in terms of a rate of feed F, knowing the initial solids content, x_f, the final solids content x_p, and the amount of heat transferred in the heat exchanger expressed in terms of the heat transfer coefficient U, the area available for heat transfer, A, and the temperature difference between the boiling liquid in the evaporator and the heating medium, ΔT.

In evaporators where heat transfer to the product is by natural convection, products that have tendencies to form deposits on the heat exchange surface foul the heat exchange surface and reduce the overall heat transfer coefficient, U. When evaporation rate slows down considerably such that it seriously affects production, the operation is stopped and the evaporator is cleaned. In evaporators used on tomato juice, the temperatures are high enough that microbiological build-up is not a factor, and shut-down for clean-up is usually done after about 14 days of operation. In some models of evaporators used on orange juice, on the other hand, operating time between clean-up is much shorter (2–3 days) because of the problem of microbiological build-up at the lower temperatures used.

Fouling of heat exchange surfaces is minimized with reduced ΔT across the heat exchange surface and by allowing the product to flow rapidly over the heat exchange surfaces. Although forced recirculation through the heat exchanger results in rapid heat transfer, a disadvantage is the long residence time of the product inside the evaporator.

For products that are heat sensitive and where low temperature differentials are allowable in the heat exchanger, the falling film heat exchanger is used. Figure 11.4 is a schematic diagram of a falling film type heat exchanger used extensively in the concentration of fruit juices. The product flows in a thin film down heated tubes where heat is transferred and vapor is removed. The product passes through the heat exchange tube of one effect only once, and this short time of contact with a hot surface minimizes heat induced flavor or color changes and nutrient degradation.

The coefficient of heat transfer, U, in evaporator heat exchangers is of the order 200 BTU/ $(h \cdot ft^2 \cdot {}^\circ F)$ or 1136 W/(m$^2 \cdot$ K) for natural convection and 400 BTU/ h(ft^2) (${}^\circ F$) or 2272 W/(m$^2 \cdot$ K) for forced convection. The effect of increased viscosity on heat transfer can be estimated by using the relationship: heat transfer coefficient is proportional to the viscosity raised to the power −0.44. Thus: $U_1/U_2 = (\Phi_1/\Phi_2)^{-0.44}$ where U_1 is the heat transfer coefficient corresponding to viscosity Φ_1 and U_2

Feed

Steam →

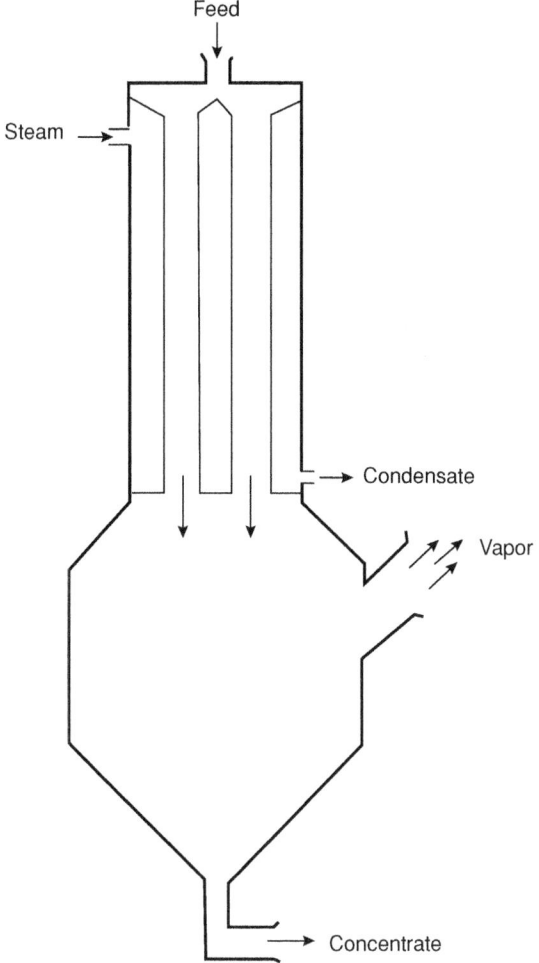

Condensate

Vapor

Concentrate

Figure 11.4 Schematic diagram of a falling film evaporator.

is the heat transfer coefficient corresponding to the viscosity Φ_2. This relationship would be useful in estimating how much of a change in evaporator performance would be expected with a change in operating conditions that would result in variations in product viscosity.

Example 11.4. A fruit juice is to be concentrated in a single-effect forced recirculation evaporator from 10% to 45% soluble solids. The feed rate is 5500 lb/h or 2497 kg/h. Steam condensing at 250°F (121.1°C) is used for heating. The vapor temperature in the evaporator should be at 130°F (54.4°C). Assume the soluble solids are hexose sugars in calculating the boiling point rise. Use Siebel's formula for calculating the specific heat of the juice. The feed is at 125°F (51.7°C). The heat transfer coefficient, U, is 500 BTU/ (h · ft ^{2} · °F) or 2839 W/(m ^{2} · K). Calculate the steam economy to be expected and the heating surface area required.

Solution:

Equations (11.7) through (11.10) will be used. The steam economy is defined as the ratio of vapor produced to steam consumed. C_c will be calculated using Siebel's formula.

$C_c = 0.8(.55) + 0.2 = 0.64$ BTU/lb($°$F) or 2679 J/(kg.K).

h_g = enthalpy of vapor at 130$°$F = 1117.8 BTU/(lb) = 2.679 MJ/(kg)

h_f = enthalpy of the water component of the feed at 125$°$F.

$\quad$ = 92.96 BTU/lb or 0.216 MJ/kg.

The temperature of the concentrate leaving the evaporator, T_2, will be the sum of the vapor temperature and the boiling point rise, ΔT_b. Using Equation (11.1), m for 45% soluble solids is

$$m = \frac{45/180}{55/1000} = 4.545 \text{ moles sugar/1000 g water}$$
$$\Delta T_b = 0.51(4.545) = 2.32°C \text{ or } 4.2°F$$
$$T_2 = 130 + 4.2 = 134.2°F \text{ or } 56.72°C$$

Substituting in Equation (11.10):

$$V = 5500 \left[1 - \frac{0.10}{0.45} \right] = 4278 \text{ lb/h or 1942 kg/h}$$

Using Equation (11.9):

$$P = 5500 \left[\frac{0.10}{0.45} \right] = 1222 \text{ lb/h or 555 kg/h}$$

Using Equation (11.7):

$$q = 1222(0.64)(134.2 - 125) + 4278(1117.8 - 92.96) = 4,391,500 \text{ BTU/h or 1.2827 MW}$$

Equation (11.8) can be used to calculate the heat transfer surface area.

$$A = \frac{q}{U\Delta T} = \frac{4,391,500}{500(250 - 134.2)} = 75.8 \text{ ft}^2 \text{ or } 7.04 \text{ m}^2$$

Note that the liquid boiling temperature was used in determining the heat transfer ΔT rather than the vapor temperature. The enthalpy of vaporization of steam at 250$°$F is 945.5 BTU/lb or 2.199 MJ/kg. Steam required = q/h_{fg} = 4,391,500/945.5 = 4645 lb/h or 2109 kg/h. Steam economy = 4278/4645 = 0.92.

11.2   IMPROVING THE ECONOMY OF EVAPORATORS

Poor evaporator economy results from wasting heat present in the vapors. Some of the techniques used to reclaim heat from the vapors include use of multiple effects such that vapors from the first effect are used to heat the succeeding effects, use of vapors to preheat the feed, and vapor recompression.

11.2.1 Vapor Recompression

Adiabatic recompression of vapor results in an increase in temperature and pressure. Figure 11.5 is a Mollier diagram for steam in the region involved in vapor recompression for evaporators. Recompression involves increasing the pressure of the vapor to increase its condensing temperature above the boiling point of the liquid in the evaporator. Compression of saturated steam would result in superheated steam at high pressure. It would be necessary to convert this vapor to saturated steam by mixing with liquid water before introducing it into the heating element of the evaporator. Superheated steam in the heat exchanger could lower the overall heat transfer coefficient. In Chapter 10, the work involved in adiabatic compression was found to be the difference in the enthalpy of the low-pressure saturated vapor and the high-pressure superheated vapor. The ratio between the latent heat of the saturated steam produced from the hot vapors and the work of compression is the coefficient of performance of the recompression system.

Example 11.5. In Example 11.4, determine the coefficient of performance for a vapor recompression system if used on the unit.

Solution:

From Fig. 11.5, the initial point for the compression is saturated vapor with an enthalpy of 1118 BTU/lb or (2.600 MJ/kg). Isentropic compression to the pressure of saturated steam at 250°F (121.1°C) would give a pressure of 29.84 psia (206 kPa) an enthalpy of 1338 BTU/lb (3.112 MJ/kg) and a temperature of 612°F (322°C). Figure 11.6 is a schematic diagram of how these numbers were obtained from Fig. 11.5.

If condensate from the heat exchanger at 250°F is used to mix with the superheated vapor after compression to produce saturated steam at 250°F, the amount of saturated steam produced will be

$$\text{Wt saturated steam} = 1 + \frac{h_{g1} - h_{g2}}{h_{fg}} = 1 + \frac{1338 - 1164}{945.5} = 1.184 \text{ lb} (0.537 \text{ kg})$$

$$\text{COP} = \frac{1.184(945.5)}{1(1338 - 1118)} = 5.09$$

The coefficient of performance in vapor recompression systems is high. COP will be higher if the ΔT is kept to a minimum. ΔT in vapor recompression systems are usually of the order 10°F (5.6°C). Although an increase in COP is achieved with the low ΔT, increased area for heat transfer in the evaporator heating unit is also required.

11.2.2 Multiple-Effect Evaporators

Steam economy can also be improved by using multiple evaporation stages and using the vapors from one effect to heat the succeeding effects. Steam is introduced only in the first effect. Figure 11.7 shows a triple effect evaporator with forward feed. This type of feeding is used when the feed is at a temperature close to the vapor temperature of the effect where it is introduced. If substantial amounts of heat are necessary to bring the feed temperature to the boiling temperature, other types of arrangements such as backward feed may be used. In a backward feed arrangement, the flow of the feed is countercurrent to the flow of vapor.

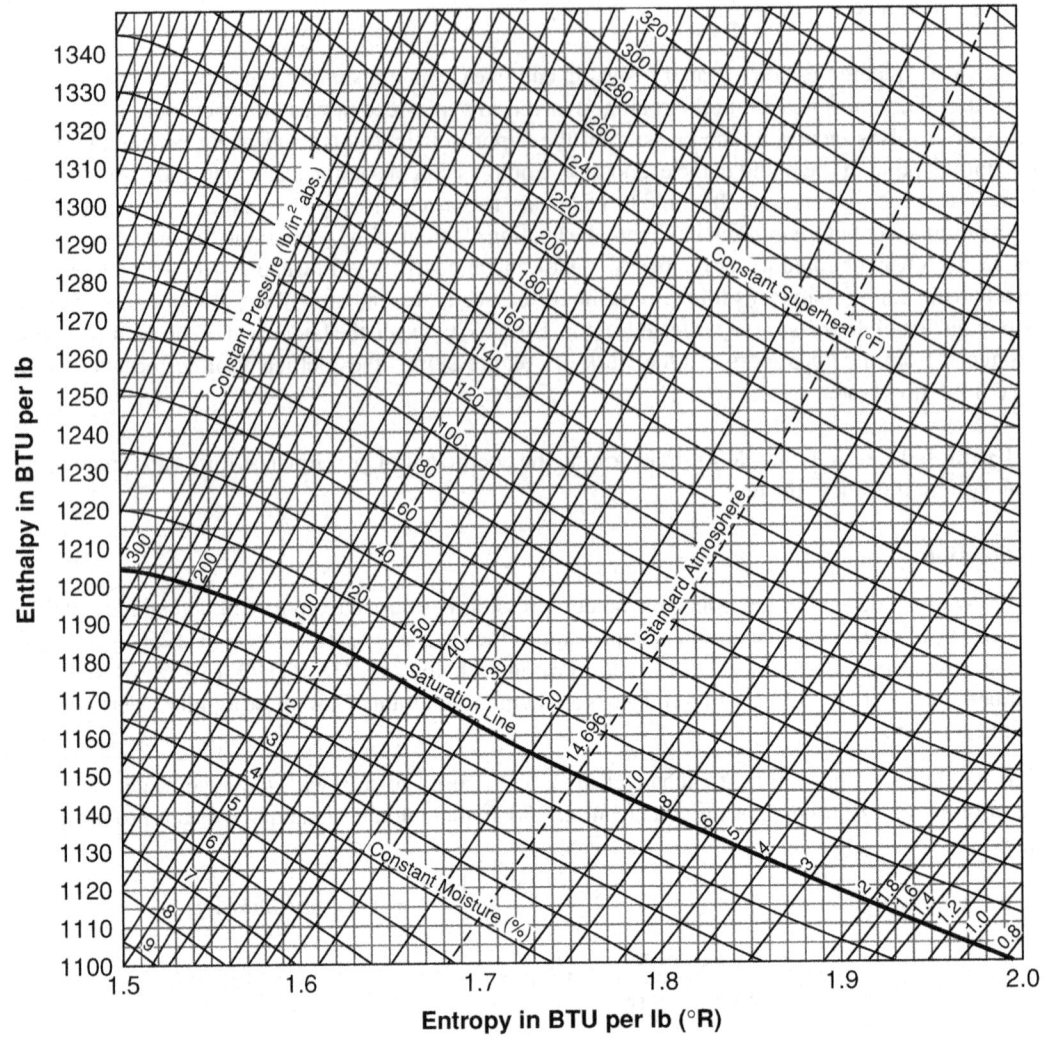

Figure 11.5 Mollier diagram for steam in the region involved in vapor recompression for evaporators. (Courtesy of Combustion Engineering Inc.)

Multiple effect evaporators are often constructed with the same heat transfer surface areas in each effect. The governing equation for evaporation rate in multiple effect evaporators is the heat transfer equation (Eq. 11.8) as in single-effect evaporators. However, the ΔT in each effect of a multiple-effect evaporator is only a fraction of the total ΔT; therefore, for the same rate of evaporation and the same total ΔT, a multiple effect evaporator with "n" effects would require approximately "n" times the heat exchange area for a single-effect evaporator. The savings in energy costs with the improvement in steam economy is achieved only with an increase in the required heat transfer surface area.

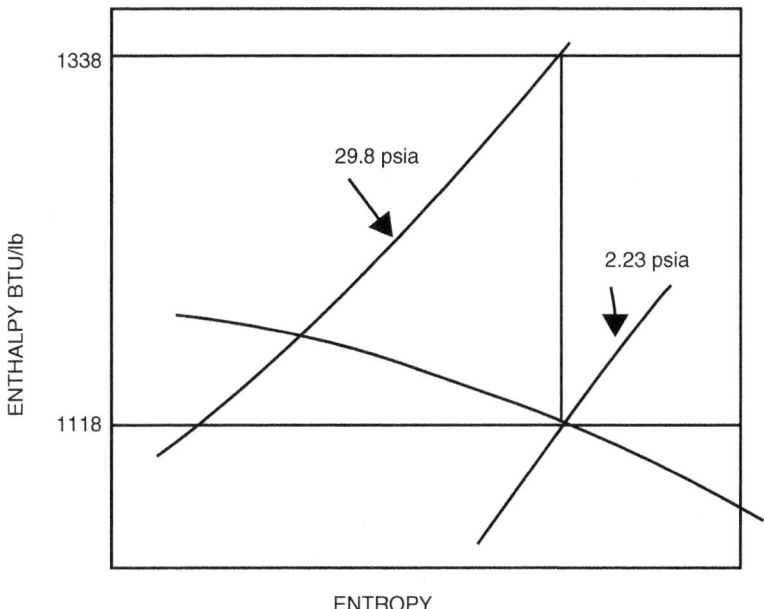

Figure 11.6 The vapor recompression process on a Mollier diagram.

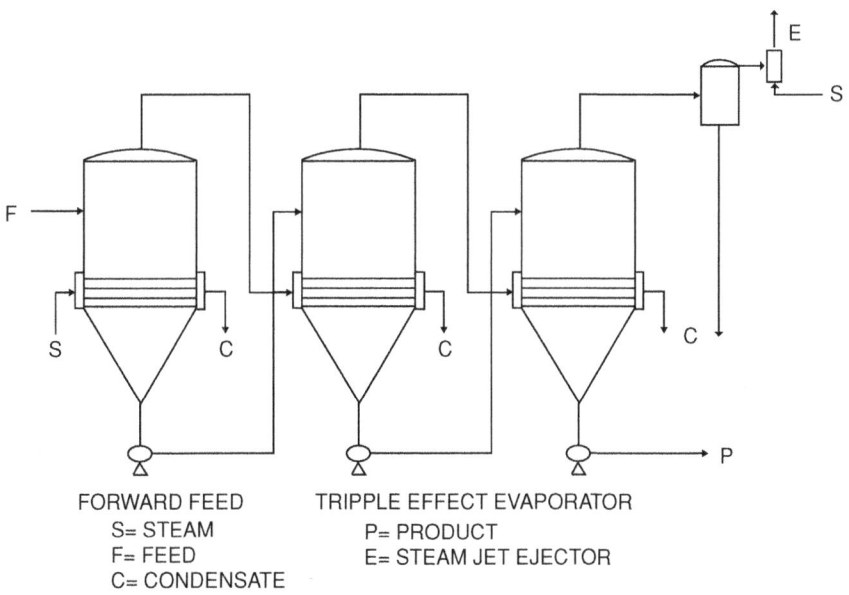

Figure 11.7 Schematic diagram of a triple effect evaporator with forward feed.

Only two parameters affecting heat transfer ΔT can be controlled in the operation of a multiple-effect evaporator. These are the temperature of the vapor in the last effect and the temperature of steam in the first effect. The vapor temperature and pressure in the first and in the intermediate effects will develop spontaneously according to the heat balance occurring within these various effects. The total ΔT in a multiple effect evaporator would be the difference between the steam temperature and the vapor temperature in the last effect. A boiling point rise would decrease the available ΔT.

$$\Delta T = T_s - T_{vn} - \Delta T_{b1} - \Delta T_{b2} - \cdots \Delta T_{bn} \tag{11.11}$$

where ΔT = the total available temperature drop for heat transfer, T_s = steam temperature, T_{vn} = vapor temperature in last effect; $\Delta T_{b1}, \Delta T_{b2} \ldots \Delta T_{bn}$ = boiling point rise in effects, 1, 2, ... n.

Multiple-effect evaporator calculations are done using a trial-and-error method. A ΔT is assumed for each effect, and by making heat and material balances in each effect, the rate of heat transfer in each effect is compared with the heat input necessary to achieve the desired evaporation rate in each effect. Adjustments are then made on the assumed ΔT's until the heat input and the heat requirement for each effect are in balance. Multiple effect evaporator calculations are tedious and time consuming. They are best done on a computer.

Reasonable approximations can be made on capacity for a given multiple effect evaporator knowing the heat transfer surface areas and the heat transfer coefficients by assuming equal evaporation in each effect and making a heat balance as if the evaporator is a single effect with a heat transfer area equal to the sum of all the areas. The overall heat transfer coefficient can be calculated as follows:

$$\frac{1}{U} = \frac{1}{U_1} + \frac{1}{U_2} + \cdots \cdots \frac{1}{U_n} \tag{11.12}$$

Example 11.6. In Example 11.4, the evaporator is a triple effect evaporator with forward feed. Assume U is the same in all effects. Calculate the approximate heat transfer surface areas in each effect and the steam economy.

Solution:

Given:

Feed rate (F) = 5500 lb/h or 2497 kg/h
Steam temperature (T_s) = 250°F (121.1°C)
Vapor temperature in last effect (T_{v3}) = 130°F (54.4°C)
Feed temperature (T_f) = 125°F (51.7°C)

$$U_1 = U_2 = U_3 = 500 \frac{BTU}{h \cdot ft^2 \cdot °F} \quad \text{or} \quad 2839 \frac{W}{m^2 \cdot K}$$

Solids content of feed (x_f) = 10%
Solids content of product (x_p) = 45%
Molecular weight of solids (M) = 180

First, determine the overall heat transfer, ΔT, using Equation (11.11). It will be necessary to determine the boiling point rise in each effect. Using Equation (11.10):

$$\text{Total evaporation (V)} = 5500 \left[1 - \frac{0.10}{0.45} \right] = 4278 \text{ lb/h or } 1942 \text{ kg/h}$$

Assuming equal evaporation, $V_1 = V_2 = V_3 = 1426$ lb/h or 647.3 kg/h. The solids content in each effect can be calculated by rearranging Equation (11.10). The subscripts n on V and F refer to the feed to, and vapor from effect n.

$$x_n = \frac{x_f F_n}{F_n - V_n} \qquad x_1 = \frac{0.10(5500)}{5500 - 1426} = 0.135$$

$$x_2 = \frac{0.135(5500 - 1426)}{5500 - 1426 - 1426} = \frac{0.135(4074)}{2648} = 0.208$$

$$x_3 = x_p = 0.45$$

The molalities are

$$m_1 = \frac{0.135(1000)}{(1 - 0.135)(180)} = 0.067 \text{ molal}$$

$$m_2 = \frac{0.208(1000)}{(1 - 0.208)(180)} = 1.459 \text{ molal}$$

$$m_3 = \frac{0.45(1000)}{(1 - 0.45)(180)} = 4.545 \text{ molal}$$

Using Equation (11.1), the boiling point rises are

$$\Delta T_{h1} = 0.51(0.067) = 0.0342°C \text{ or } 0.06°F$$

$$\Delta T_{h2} = 0.51(1.459) = 0.744°C \text{ or } 1.34°F$$

$$\Delta T_{h3} = 0.51(4.545) = 2.32°C \text{ or } 4.17°F$$

The total ΔT for heat transfer is calculated using Equation (11.11):

$$\Delta T = 250 - 130 - 4.17 - 1.34 - 0.06 = 114.4°F \text{ or } 63.57°C$$

Overall U is calculated using Equation (11.12):

$$\frac{1}{U} = \frac{1}{500} + \frac{1}{500} + \frac{1}{500} = 0.006$$
$$U = 166.67 \text{ BTU/(h} \cdot \text{ft}^2 \cdot \text{EF) or } 946.3 \text{ W/(m}^2 \cdot \text{K)}$$

From the section "The Heat Exchanger," the required heat transfer rate for this evaporator was determined to be 4,391,500 BTU/h or 1.2827 MW. Using Equation (11.8):

$$A = \frac{4,391,500}{166(114.4)} = 230.3 \text{ ft}^2 \text{ or } 21.4 \text{ m}^2$$

The heat transfer surface area for each effect will be 230.3/3 $\text{ft}^2 = 76.77 \text{ ft}^2$ or 7.13 m^2.

Calculations of steam economy can only be done using the trial-and-error procedure necessary to establish the ΔT and the vapor temperature in each effect. The steam economy of multiple effect evaporators is a number slightly less than the number of effects.

11.3   ENTRAINMENT

When the liquid to be evaporated contains suspended solids, the liquid has a tendency to foam. The level of the foam may rise much higher than the normal liquid level resulting in carry-over of solids with the vapors. Entrainment not only results in loss of valuable food solids, but the solids in the condensate may result in problems with condensate water treatment particularly if a direct contact condenser is used and the condensate is cooled in a cooling tower and recirculated.

11.4 ESSENCE RECOVERY

A major problem in concentration of fruit juices is the loss of essence during the evaporation process. With condensers where the cooling water directly contacts the vapor, it is not possible to recover the flavor components that are vaporized from the liquid. In the past, the problem of essence loss was solved in the orange juice industry by concentrating the juice to a higher concentration than is desired and diluting the concentrate with fresh juice to the desired solids concentration. The essence in the fresh juice gives the necessary flavor to the concentrate.

One method for essence recovery is by flashing the juice into a packed or perforated plate column maintained at a very low absolute pressure. Flash evaporation is a process where hot liquid is introduced into a chamber that is at an absolute pressure where the boiling point of the liquid is below the liquid temperature. The liquid will boil immediately upon exposure to the low pressure, vapor is released, and the liquid temperature will drop to the boiling point of the liquid at the given absolute pressure.

The feed is preheated to 120°F to 150°F (48.9°C to 65.6°C) and is introduced into a column maintained at an absolute pressure of approximately 0.5 psia (3.45 kPa). There is no heat input in the column; therefore, evaporative cooling drops the temperature of the liquid. The vapors rise up the packed column continually getting richer in the volatile components as they proceed up the column. A surface condenser cooled by a refrigeration system traps the volatile components. The essence concentrate recovered is blended with the concentrated product.

In a multiple-effect evaporator, a backward feed arrangement is used, and the vapors from the last effect are condensed using a surface condenser. The condensate containing the essence is flashed into the essence recovery unit.

PROBLEMS

11.1. A single-effect falling film type evaporator is used to concentrate orange juice from 14% to 45% solids. The evaporator utilizes a mechanical refrigeration cycle using ammonia as refrigerant, for heating and for condensing the vapors. The refrigeration cycle is operated at a high pressure of 200 psia (1.379 MPa) and a low side pressure of 50 psia (344.7 kPa). The evaporator is operated at a vapor temperature of 90°F (32.2°C). Feed enters at 70°F (21.1°C). The ratio of insoluble to soluble solids in the juice is 0.09 and the soluble solids may be considered as glucose and sucrose in 70:30 ratio. Consider the ΔT as the log mean ΔT between the liquid refrigerant temperature and the feed temperature at one point and the hot refrigerant gas temperature and the concentrated liquid boiling temperature at the other point. The evaporator has a heat transfer surface area of 100 ft^2 (9.29 m^2), and an overall heat transfer coefficient of 300 BTU/(h $ ft ^{2} $ °F) or 1703 W/(m^2 $K) may be expected.
Calculate:
(a) The evaporator capacity in weight of feed per hour.
(b) Tons of refrigeration capacity required for the refrigeration unit based on the heating requirement for the evaporator.
(c) Additional cooling required for condensation of vapors if the refrigeration unit is designed to provide all of the heating requirements for evaporation.

11.2. Condensate from the heating unit of one effect in a multiple effect evaporator is flashed to the pressure of the heating unit in one of the succeeding effects. If the condensate is saturated liquid at 7.511 psia (51.79 kPa) and the heating unit contains condensing steam at 2.889 psia (19.92 kPa), calculate the total available latent heat that will be in the steam produced from a unit weight of the condensate.

11.3. A single-effect evaporator was operating at a feed rate of 10,000 kg/h concentrating tomato juice at 160°F (71.1°C) from 15% to 28% solids. The ratio of insoluble to soluble solids is 0.168 and the soluble solids may be assumed to be hexose sugars. Condensing steam at 29.840 psia (205.7 kPa) was used for heating and the evaporator was at an absolute pressure of 5.993 psia (41.32 kPa). It is desired to change the operating conditions to enable the efficient use of a vapor recompression system. The steam pressure is to be lowered to 17.186 psia (118.37 kPa). Assume there is no change in the heat transfer coefficient because of the lowering of the heating medium temperature.

Calculate:

(a) The steam economy for the original operating conditions.

(b) The capacity in weight of feed per hour under the new operating conditions.

(c) The steam economy of the vapor recompression system.

Express the steam economy as the ratio of the energy required for concentration of the juice to the energy required to compress the vapor assuming a mechanical efficiency of 50% for the compressor. Assume condensate from the heating element is added to the superheated steam to reduce temperature to saturation.

SUGGESTED READING

Bennet, C. O. and Myers, J. E. 1962. Momentum, Heat, and Mass Transport. McGraw-Hill Book Co., New York.

Charm, S. E. 1971. Fundamentals of Food Engineering. 2nd ed. AVI Publishing Co., Westport, CT.

Foust, A. S., Wenzel, E. A., Clump, C. W., Maus, L., and Andersen, L. B. 1960. Principles of Unit Operations. John Wiley & Sons, New York.

Green, D. W. 1997. Perry's Chemical Engineers Handbook. 7th ed. Mcgraw- Hill Book Co., New York.

Heldman, D. R. 1973. Food Process Engineering. AVI Publishing Co., Westport, CT.

McCabe, W. L. and Smith, J. C. 1967. Unit Operations of Chemical Engineering. 2nd ed. McGraw-Hill Book Co., New York.

Perry, R. H., Chilton, C. H., and Kirkpatrick, S. D. 1963. Chemical Engineers Handbook. 4th ed. McGraw-Hill Book Co., New York.

Saravacos, G. D. and Kostaropoulos, A. E. 2002. Handbook of Food Processing Equiment. Kluwer Academic/Plenum Publishers, New York.

Singh, R. P. and Heldman, D. R. 2001. Introduction to Food Enginering. 3rd ed. Academic Press, San Diego.

CHAPTER 12

Dehydration

Dehydration is a major process for food preservation. The reduced weight and bulk of dehydrated products and their dry shelf stability reduces product storage and distribution costs. As dehydration techniques that produce good-quality convenience foods are developed, more dehydrated products will be commercially produced. At present, instant beverage powers, dry soup mixes, spices, and ingredients used in further processing are the major food products dehydrated.

12.1 WATER ACTIVITY

Dehydrated foods are preserved because water activity is at a level where no microbiological activity can occur and where deteriorative chemical and biochemical reaction rates are reduced to a minimum. Water activity (a_w) is measured as the equilibrium relative humidity (ERH), the per cent relative humidity (RH) of an atmosphere in contact with a product at the equilibrium water content. a_w is also the ratio of the partial pressure of water in the headspace of a product (P) to the vapor pressure of pure water (P^0) at the same temperature.

$$a_w = ERH = \frac{P}{P^0} \tag{12.1}$$

The relationship between a_w and the rate of deteriorative reactions in food is shown in Fig. 12.1. Reducing a_w below 0.7 would prevent microbiological spoilage. However, although microbial spoilage would not occur at $a_w = 0.7$, prevention of other deteriorative reactions needed to successfully preserve a food product by dehydration requires a_w to be reduced to $= 0.3$.

A food material may also be dehydrated only for weight or bulk reduction and finally preserved using other techniques.

12.1.1 Thermodynamic Basis for Water Activity

The first and second laws of thermodynamics are discussed in Section 4.2 –Thermodynamics. In addition to the thermodynamic variables of enthalpy (H), internal energy (E), work (W), and heat (Q) discussed under Thermodynamics in Section 4.2. Other thermodynamic variables related chemical changes will be discussed in this section.

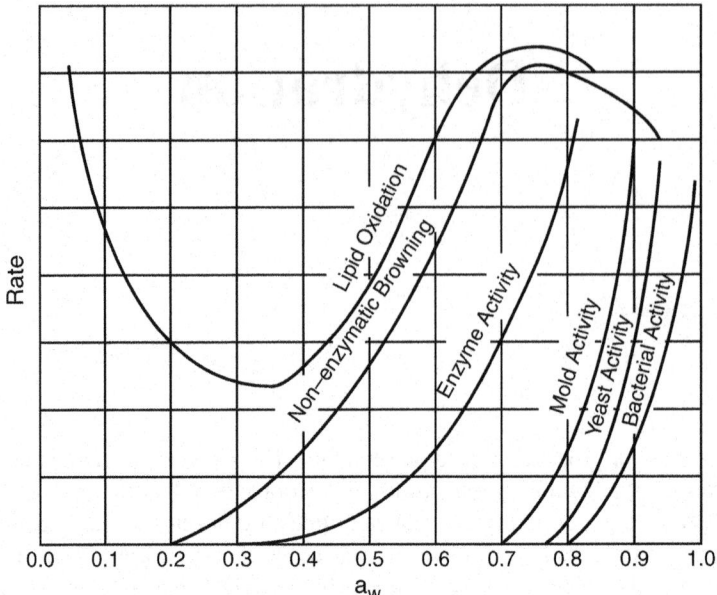

Figure 12.1 Relationship between water activity and deteriorative reaction on foods. (From Labuza T. P. et al. J. Food Sci. 37:154, 1972. Used with permission.)

F = free energy; another form of energy in a system that is different from PV work. This form of energy is responsible for chemical or electrical work and is responsible for driving chemical reactions.

$$F = H - TS \tag{12.2}$$

Φ = chemical potential. It is directly related to the free energy. In any system undergoing change, $d\Phi = dF$:

Because: $dF = dH - T\,dS - S\,dT$
and: $dH = dE + P\,dV + V\,dP$
and: $dE = T\,dS - P\,dV$
Then: $dF = V\,dP - S\,dT$;
  $d\Phi = V\,dP - S\,dT$

At constant T, $d\Phi/dP = V$.

In a system in equilibrium, the chemical potential is equal in all phases. In the gas phase, the ideal gas equation for one mole of gas is V = RT/P. Thus:

$$d\Phi = RTd(\ln P) \tag{12.3}$$

If the vapor is nonideal, the fugacity, f, may be substituted for P. At low pressures of near ambient or in a vacuum, f is almost the same as P and the two may be interchanged in use with very little error.

Thus, the chemical potential can be measured directly by the vapor pressure of a component. The activity of a component is defined as:

$$a = \frac{f}{f^0}$$

where f^0 is the fugacity of the pure component at the same temperature. If the component is water:

$$a_w = \frac{f}{f^0} \equiv \frac{P}{P^0} \tag{12.4}$$

The thermodynamic basis for the water activity is the chemical potential of water; its ability to participate in chemical reactions.

12.1.2 Osmotic Pressure

One consequence of reduced water activity is an increase in osmotic pressure, which interferes with food, water, and waste transport between a cell and its surroundings. The osmotic pressure is related to the activity coefficient as follows:

$$\pi = \frac{-RT}{V} \ln x\gamma \tag{12.5}$$

where π = osmotic pressure (atm); R = gas constant (82.06 mL $\cdot$ atm/gmole $\cdot$ K); γ = activity coefficient; V = molar volume = 18 mL/gmole for water; x = mole fraction of water; and T = absolute temperature. The argument of the logarithmic term in Equation (12.5) is the water activity. As the water activity approaches 1, the logarithmic term approaches zero and the osmotic pressure approaches zero.

12.1.3 Water Activity at High Moisture Contents

At high moisture contents, depression of the water vapor pressure by soluble solids is similar to vapor pressure depression by solutes in solution. The presence of insoluble solids is ignored and the solution phase determines the water activity.

For ideal solutions, the water activity is equal to the mole fraction of water, x_w.

$$a_w = x_w \tag{12.6}$$

For nonideal solutions, the activity coefficient, γ, corrects for deviation from nonideality.

$$a_w = \gamma x_w \tag{12.7}$$

The mole fraction of water, x_w, expressed in terms of weight fraction water (x'_w) and solute (x'_s) having molecular weights of 18 and M_s, respectively, is

$$x_w = \frac{\text{moles water}}{\text{moles water} + \text{moles solute}} \tag{12.8}$$

$$x_w = \frac{\dfrac{x'_w}{18}}{\dfrac{x'_w}{18} + \dfrac{x'_s}{M_s}}$$

Equation (12.6) is Raoult's law. In the form of Equation (12.8), Raoult's law is often cited to govern the dependence of water activity on water and solute concentration in foods. Qualitatively, Raoult's law explains adequately the dependence of water activity on water and solute content in foods. Solutes with low molecular weights also provide larger reduction of vapor pressure per unit weight of solute than those with high molecular weights. However, a_w calculated using Equation (12.8) will differ from actual a_w because the solution phase in most foods is nonideal.

The deviation of a_w from Raoult's law is related to the heat of mixing involved in the dissolution of solute. Let ΔH_{m1} = partial molal heat of mixing, defined as the change in enthalpy of mixing with the addition of or removal of 1 mole of component 1 at constant temperature and pressure. Equation (12.9) can be derived (Perry et al., 1963) by assuming that the excess entropy of mixing is zero.

$$\overline{\Delta H_{m1}} = \frac{d(\Delta H_m)}{dN_1} = RT \ln \gamma_1 \tag{12.9}$$

Solutes in food systems exhibit negative heat of mixing with water (heat is released in mixing), and γ in Equation (12.9) is less than 1. Solutions with negative heat of mixing will have a_w less than that calculated using Raoult's law.

The activity coefficient of a regular solution can be derived using Van Laar's approximations based on van der Waal's equation of state for one mole of substance:

$$P = \frac{RT}{V - b} - \frac{a}{V^2} \tag{12.10}$$

The constant a in van der Waal's equation of state represents the magnitude of attractive forces between molecules, and the constant b represents the reduction in the volume as a result of these attractive forces. In a mixture containing N_1 and N_2 moles of components 1 and 2, respectively, a for the mixture will be determined by the constant a_1 for component 1, and a_2 for component 2. Because a is associated with attraction between molecules, the rules of permutation will give N_1 ways in which N_1 molecules can interact. If each of these interactions yield a_1, the total yield of N_1^2 interactions will be $a_1 N_1^2$. The yield of all interactions between N_2 molecules of component 2 will be $a_2 N_2^2$. Molecules of component 1 and 2 also interact with each other and the yield of each of these interactions will be the geometric mean of a_1 and a_2. The number of ways in which N_1 molecules of component 1 and N_2 molecules of component 2 can interact with each other (excluding interactions between like molecules) will be $2N_1 N_2$ and the yield of these interactions will be $(a_1 a_2)^{1/2} (2N_1 N_2)$. Thus, for a mixture, the van der Waal's constant a can be expressed in terms of the constants of the components according to:

$$a = a_1 N_1^2 + a_2 N_2^2 + 2N_1 N_2 (a_1 a_2)^{1/2} \tag{12.11}$$

Constant b is associated with the volume of each molecule; therefore, b for a mixture can be expressed in terms of b_1 and b_2 is

$$b = b_1 N_1 + b_2 N_2 \tag{12.12}$$

Van Laar's first approximation assumes the ratio a/b to be the internal energy of the mixture that approximates the heat of vaporization. Furthermore, in solutions, the volume changes on mixing are small such that internal energy, E, and enthalpy, H, are equivalent. The heat of mixing will be the difference between the sum of the internal energies of pure component 1 and pure component 2 and the internal energy of the mixture.

$$-\Delta H_m = E_1 + E_2 - E_{12} \tag{12.13}$$

E_1 is a_1/b_1, E_2 is a_2/b_2, and E_{12} is a/b for the mixture determined using Equations (12.11) and (12.12). Equation (12.13)13 becomes:

$$-\Delta H_m = \frac{a_1}{b_1} + \frac{a_2}{b_2} - \frac{a_1 N_1^2 + a_2 N_2^2 + 2 N_1 N_2 (a_1 a_2)^{1/2}}{N_1 b_1 + N_2 b_2}$$

Simplifying:

$$-\Delta H_m = \frac{N_1 N_2 b_1 b_2}{N_1 b_1 + N_2 b_2} \left[\frac{a_1^{1/2}}{b_1} - \frac{a_2^{1/2}}{b_2} \right]^2$$

If the value of b for the two components are equal:

$$-\Delta H_m = \frac{N_1 N_2}{(N_1 + N_2)b} \left(a_1^{1/2} - a_2^{1/2} \right)^2 \tag{12.14}$$

The partial molal heat of mixing can be determined by differentiating Equation (12.14) with respect to N_1, keeping N_2 constant.

$$\overline{\Delta H_{m1}} = \frac{-d\Delta H_m}{dN_1} = \frac{N_2 \left(a_1^{1/2} - a_2^{1/2} \right)^2}{b} \frac{d}{dN_1} \left[\frac{N_1}{N_1 + N_2} \right] = \frac{N_2 \left(a_1^{1/2} - a_2^{1/2} \right)}{b} \left[\frac{N_2}{(N_1 + N_2)^2} \right]$$

Because a_1, a_2, and b are all constants, $k' = \frac{a_1^{1/2} - a_2^{1/2}}{b}$ and:

$$-\overline{\Delta H_m 1} = k' \left[\frac{N_2}{N_1 + N_2} \right]^2 = k = x_2^2 \tag{12.15}$$

Combining Equations (12.9) and (12.15):

$$RT \ln \gamma_1 = -k' x_2^2 \tag{12.16}$$

Because from Equation (12.7), $\gamma_w = a_w/x_w$, and in a two component system, $x_2 = 1 - x_w$. At constant temperature, Equation (12.16) can be written as:

$$\log \frac{a_w}{x_w} = -k(1 - x_w)^2 \tag{12.17}$$

Equation (12.17) shows that a plot of $\log a_w/x_w$ against $(1 - x_w)^2$ will be linear with a negative slope. Equation (12.17) has been used successfully by Norrish to predict the water activity of sugar solutions. Values of the constant k in Equation (12.17) for various solutes are shown in Table 12.1.

Example 12.1. Calculate the water activity of a 50% sucrose solution.

Solution:

From Table 12.1, the k value for sucrose is 2.7. Sucrose has a molecular weight of 342. The mole fraction of water is

$$x_w = \frac{50/18}{50/18 + 50/342} = 0.95$$

Using Equation (12.17):

$$\log a_w = \log x_w - 2.7(1 - x_w)^2 = \log 0.95 - 2.7(0.05)^2$$
$$a_w = 0.935$$

Table 12.1   Values of the Constant k for Various Solutes in Norrish's Equation for Water Activity of Solutions.

Sucrose	2,7
Glucose	0,7
Fructose	0.7
Invert sugars	0,7
Sorbitol	0.85
Glycerol	0.38
Propylene glycol	−0.12
Nacl	$15.8(x_2 < 0.02)$
	$7.9(x_2 > 0.02)$
Citric acid	6.17
d-Tartaric acid	4.68
Malic acid	1.82
Lactic acid	−1.59

Sources: Norrish, R. S., *J. Food Technol. 1:25, 1996*; Toledo, R. T., Proc. *Meat. Int Res. Conf, 1973*; Chuang, L., M.S. thesis, University of Georgia, *1974*; Chirife, J., and Ferro-Fontain, C., *J. Food Sci. 45:802, 1980*.

12.1.3.1   Gibbs-Duhem Equation

The chemical potential is a partial molal quantity that can be represented as the change in the free energy with a change in the number of moles of that component, all other conditions being maintained constant. $\Phi_1 = (dF/dN_1)$

$$dF = \left(\frac{dF}{dN_1}\right) dN_1 + \left(\frac{dF}{dN_2}\right) dN_2 = \mu_1\, dN_1 + \mu_2\, dN_2$$

But:

$$F = \Phi_1 N_1 + \Phi_2 N_2$$
$$dF = \Phi_1\, dN_1 + N_1\, d\Phi_1 + \Phi_2\, dN_2 + N_2\, d\Phi_2$$

and:

$$N_1\, d\Phi_1 + N_2\, d\Phi_2 = 0 \tag{12.18}$$

Equation (12.18) and similar expressions in Equations (12.19) and (12.20) are different forms of the Gibbs-Duhem equation.

$$x_1\left(\frac{d\ln_{a_1}}{dx_1}\right) + x_2\left(\frac{d\ln_{a_2}}{dx_1}\right) = 0 \tag{12.19}$$

or:

$$x_1\, d\ln_{a_1} + x_2\, d\ln_{a_2} = 0 \tag{12.20}$$

Let $(a_w)^0$ represent the activity of water in a system with only one solute and water in the mixture. Integration of Equation (12.20) gives

$$\ln(a_w)^0 = -\int \left(\frac{x_2}{x_1}\right) d \ln(a_2)$$

In a multicomponent system involving two solutes:

$$d \ln a_w = -\left(\frac{x_2}{x_1}\right) d \ln a_2 - \left(\frac{x_3}{x_1}\right) d \ln a_3$$

$$\ln a_w = -\int \left(\frac{x_2}{x_1}\right) d \ln a_2 - \int \left(\frac{x_3}{x_1}\right) d \ln a_3$$

$$a_w = (a_w)_2^0 (a)_3^0 \qquad (12.21)$$

The water activity of a mixture involving several components can be determined from the product of the water activity of each component separately if all the water present in the mixture is mixed with individual components. Equation (12.21) was derived by Ross (Food Technol. 3:26, 1975).

Example 12.2. Calculate the water activity of a fruit preserve containing 65% soluble solids, 2% insoluble solids, and the rest water. The soluble solids may be assumed to be 50% hexose sugars and 50% sucrose.

Solution:

Basis: 100 g of fruit preserve.

g hexose sugars $= 65(0.5) = 32.5$ g
g sucrose $= 65(0.5) = 32.5$ g
g water $= 33$ g

Consider sucrose dissolves in all of the water present.

$$x_s = \frac{32.5/342}{33/18 + 32.5/342} = 0.0492$$

$$x_w = 1 - 0.0492 = 0.9507$$

Using Equation (12.17) and a k value of 2.7 for sucrose:

$$\log(a_{w1})^0 = \log 0.9507 - 2.7(0.0492)^2$$

$$(a_{w1})^0 = 0.9365$$

Consider that hexose dissolves in all of the water present.

$$x_i = \frac{32.5/180}{32.5/180 + 33/18} = 0.0896$$

$$x_w = 1 - 0.0896 = 0.9103$$

Using Equation (12.17) and a k value for hexose of 0.7:

$$\log(a_{w2})^0 = \log 0.9103 - 0.7(0.0896)^2$$

$$(a_{w2})^0 = 0.8986$$

Using Equation (12.21), the a_w of the mixture is

$$a_w = (a_{w1})^0 (a_{w2})^0 = 0.9365\,(0.8986) = 0.841$$

12.1.3.2 Other Equations for Calculating Water Activity

12.1.3.2.1 Bromley's Equation.
For salts and other electrolytes, Bromley's equation (Bromley AIChE J. 19:313, 1073) accounts for ionic dissociation and nonideality.

$$a_w = \exp\left(-0.0183\, m_i \phi\right) \tag{12.22}$$

where ϕ = osmotic coefficient, m_i = moles of ionic species i per kilogram of solvent, $m_i = I \cdot m$, where I = number of ionized species per mole and m = molality. The osmotic coefficient, ϕ, is calculated using Equation (12.23).

$$\phi = 1 + 2.303[F_1 + (0.06 + 0.6B)\,F_2 + 0.5BI] \tag{12.23}$$

The parameter F_1 in Equation (12.23) is calculated using Equation (12.24).

$$F_1 = F_{id}[-0.017z\,I^{0.5}] \tag{12.24}$$

The parameter F_{id} in Equation (12.24) is calculated using Equation (12.25).

$$F_{id} = 3\,I^{-1.5}\left[1 + I^{0.5} - \frac{1}{1+I^{0.5}} - 2\ln\left(1+I^{0.5}\right)\right] \tag{12.25}$$

The parameter F_2 in Equation (12.23) is calculated using Equation (12.26).

$$F_2 = \frac{z}{aI}\left[\frac{1+2aI}{(1+aI)^2} - \frac{\ln(1+aI)}{aI}\right] \tag{12.26}$$

The parameter I is the ionic strength, evaluated as half the sum of the product of the molality of dissociated ions and the square of their charge; for example, for $MgCl_2$, $I = 0.5\,[m(2)^2 + 2m(-1)^2] = 3\,m$; B and a are constants for each salt obtained from a regression of activity coefficient data against ionic strength. A comprehensive listing of a and B values for different salts is given by Bromley (1973). Chirife and Ferro-Fontan (J. Food Sci. 45:802, 1980) reported a value of B for sodium lactate of 0.050 kg/gmol. z is the charge number, the ratio of the sum of the product of the number of ions and its charge to the stoichiometric number of ions; for example, for $MgCl_2$, $z = [1(2) + 2(1)]/3 = 4/3 = 1.333$. For NaCl, KCl, sodium acetate and salts with monovalent cation and anion, $z = [1(1) + 1(1)]/2 = 1.0$

The values of the parameters z, I, B, and a for various salts are as follows:

For NaCl: $z = 1$; $I = m$; $B = 0.0574$; $a = 1.5$
For KCl: $z = 1$; $I = m$; $B = 0.0240$; $a = 1.5$
For KNO_3: $z = 1$; $I = m$; $B = -0.0862$; $a = 1.5$
For $MgCl_2$: $z = 1.33$; $I = 3m$; $B = 0.01129$; $a = 1.153$

12.1.3.2.2 Lang-Steinberg Equation.
This equation (J. Food Sci. 46:670, 1981) is useful for mixtures of solids, where there are no solute–insoluble solids interaction. The moisture content of each component X_i was considered to be linear when plotted against $\log(1 - a_{wi})$, with a slope b_i and intercept a_i. a_{wi} is the water activity of component i at moisture content X_i. If X is the equilibrium

moisture content of the mixture, g water/g dry matter, and S is the total mass of dry matter in the mixture, the a_w of the mixture can be calculated from the mass of dry matter in each component, S_i as follows:

$$\log(1 - a_w) = \frac{XS - \sum(a_i S_i)}{\sum(b_i X_i)} \tag{12.27}$$

12.1.4   Water Activity at Low Moisture Contents

If the water content of a food is plotted against the water activity at constant temperature, a sigmoid curve ally results. The curve is known as the sorption isotherm for that product. Figure 12.2 shows a sorption isotherm for raw beef at 25°C.

The sorption isotherm can be subdivided into three zones, each zone representing a different mechanism for water sorption. In zone C, the influence of insoluble solids on the a_w is negligible. a_w is dependent on the solute and water content of the solution phase and can be calculated using Equations (12.17) and (12.21).

In zone B, the influence of insoluble solids on aw becomes significant. The sorption isotherm flattens out, and very small changes in moisture content are reflected by very large changes in the water activity. In this zone, water is held in the solid matrix by capillary condensation and multilayer adsorption. Some of the solutes may also be in the form of hydrates. Some of the water may still be in the liquid phase, but its mobility is considerably restricted because of attractive forces with the solid phase. The quantity of water present in the material that would not freeze at the normal freezing point usually is within this zone.

Zone A represents adsorption of water on the surface of solid particles. None of the water is in the liquid phase anymore. The heat of vaporization of water in this zone is higher than the heat of

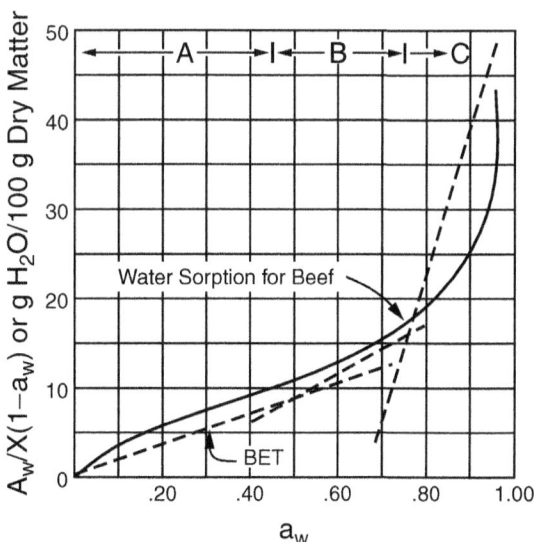

Figure 12.2   Sorption isotherm of dried raw beef at 25°C, showing the three segments of the BET plot.

vaporization of pure water because both heat of vaporization and heat of adsorption must be supplied to remove the water molecules from the solid surface.

The relationship between water activity and moisture content in zone A is best described by the Brunauer-Emmett-Teller (BET) equation. If x is the moisture content in g H_2O/g dry matter,

$$\frac{a_w}{x(1 - a_w)} = \frac{1}{x_m C} + \frac{C - 1}{x_m C}(a_w) \tag{12.28}$$

x_m in Equation (12.28) is the mass fraction of water in the material equivalent to a unimolecular layer of water covering the surface of each particle. C is a constant at constant temperature and is related to the heat of adsorption of water on the particles. C is temperature dependent.

A plot of $a_w/x(1 - a_w)$ against a_w is called the BET plot. The plot would be linear and the slope and intercept of the line can be used to determine the constant x_m, the moisture content at which the water molecules coat the surface of the solid particles in a monomolecular layer.

The region of maximum stability for a food product is usually in zone B in Fig. 12.2. When the moisture content in the product drops to a level insufficient to cover the solid molecules in a monomolecular layer, the rate of lipid oxidation increases. Determining the moisture content for maximum shelf stability of a dehydrated material would involve determining the sorption isotherm and calculating the value of x_m in Equation (12.28) from a BET plot.

The dashed lines in Fig. 12.2 are BET plots of the sorption data for raw beef. The line in zone A represents monomolecular adsorption and will be used for calculating x_m. The slope and intercept of the line are 13.7 and 0.98, respectively. From Equation (12.28), the intercept is $1/x_m C$. $x_m C = 1/0.98 = 1.02$. The slope is $(C - 1)/x_m C$.

$$\frac{C - 1}{1.02} = 13.7$$
$$C = 14.974$$
$$x_m = \frac{1.02}{14.974} = 0.068 \, g H_2O/g \text{ dry matter}$$

12.1.4.1 The GAB (Guggenheim-Anderson-de Boer) Equation

This equation gives a better fit than the BET over a wider range of moisture contents. Let X = moisture content on a dry solids basis, and X_m = the moisture content on a dry basis, equivalent to a monomolecular layer of water.
Then:

$$X = \frac{X_m C k a_w}{(1 - k a_w)(1 - k a_w + C k a_w)} \tag{12.29}$$

The GAB equation is a three-parameter equation with k, C, and X_m as constants. C and X_m have similar significance as in the BET equation. k is a third parameter that corrects for the difference in properties of adsorbed water relative to liquid water and permits the GAB equation to hold over a wider range of moisture content than the BET. Both k and C are temperature dependent.

Evaluation of X_m using Equation (12.29) is more precise than that using the BET equation, because the fit of Equation (12.29) to the data extends over a wide range of moisture contents. The narrow range of moisture contents to which Equation (12.28) fits often presents some problems in the determination of X_m using the BET equation, particularly when data points are not close enough at the lower moisture contents.

Rearranging Equation (12.29) by taking the reciprocal:

$$\frac{1}{X} = \frac{1}{X_m}\left[\frac{1}{C\,k\,a_w} - \frac{1}{C}\right]\left[1 - k\,a_w + C\,k\,a_w\right]$$

$$\frac{a_w}{X} = \left[\frac{k}{X_m\,C}\right](1 - C)\,a_w^2 + \left[\frac{C - 2}{X_m\,C}\right]a_w + \frac{1}{X_m\,C\,k}$$

Thus, a polynomial nonlinear regression of (a_w/X) against (a_w) will give values for α, the coefficient of the quadratic term, β the coefficient of the linear term, γ the intercept.

$$\beta = \frac{C - 2}{X_m\,C}; \quad \alpha = \left[\frac{k}{X_m\,C}\right]\left[1 - C\right]; \quad \gamma = \frac{1}{X_m\,C\,k}$$

These three equations can be used to evaluate C, k, and X_m.

Multiple nonlinear regression techniques must be used to fit and evaluate the parameters for this equation from moisture-water activity data.

Example 12.3. Fitting GAB Equation to Sorption Isotherm Data: The following data (a_w, X) was obtained for the sorption isotherm of potatoes at 25°C. X is the moisture content in g water/g dry matter. Calculate the GAB parameters, X_m, C, and k. Data: (0.112, 0.035), (0.201, 0.057), (0.327, 0.08), (0.438, 0.105), (0.529, 0.13), (0.577, 0.145), (0.708, 0.19), (0.753, 0.204), (0.843, 0.27), (0.903, 0.37).

Solution:

Use Excel to calculate a_w/X and tabulate against a_w. The following are the results:

Independent variable (a_w)	Dependent variable (a_w/X)
0.112	3.2
0.201	3.52
.327	4.09
.438	4.17
.529	4.07
.577	3.98
.708	3.73
.753	3.69
.843	3.12
.903	2.44

Perform a nonlinear regression (using a statistical software package, e.g., Systat) to fit the following model to the data:

$$a_w/X = \alpha(a_w)^2 + \beta(a_w) + \gamma$$

Regression results: $\alpha = -8.24$; $\beta = 7.69$; $\gamma = 2.39$. The r^2 value of the regression = 0.969 indicating a good fit throughout the full data set. Values of k, C, and X_m can then be calculated. Solving for X_m in the expressions for α, β, and γ in the GAB equation:

$$X_m = [k(1 - C)/C\alpha] = [(C - 2)/C\beta] = 1/Ck\gamma$$

Solving the equalities will result in two equations:

$$k^2(1 - C) = \alpha/\gamma; C = 2 + \beta/\gamma k$$

Substituting C: $k^2 (1 - 2 - \beta/\gamma k) = \alpha/\gamma$; $k^2 + (\beta/\gamma)k + \alpha/\gamma = 0$ Using the quadratic equation and taking the positive root: $k = 0.848$; $C = 2 + \beta/\gamma k = 5.79$; $X_m = 1/Ck\gamma = 0.0852$.

12.1.4.2 Other Equations for Sorption Isotherms of Foods

Other equations that have been used to fit the sorption isotherms of foods are Equation (12.30) to (12.39). Iglesias and Chirife (1982) fitted sorption isotherms to these equations and reported the best-fitting equations with the corresponding value of the equation parameters. Halsey's (Eq. 12.34), Henderson's (Eq. 12.35), Oswin's (Eq. 12.38), and Smith's (Eq. 12.39) was reported to fit more water sorption/desorption data among these equations.

Caurie equation:

$$\ln(1 - X) = A - r\ln(a_w) \tag{12.30}$$

X = moisture content dry basis
A and r are constants obtained from the regression equation for $\ln(1 - X)$ versus $\ln(a_w)$.
Chen equation:

$$a_w = \exp[-K\exp(-bX)] \tag{12.31}$$

$$\ln[-\ln(a_w)] = \ln(K) - bX$$

K and b are constants obtained from a linear regression of $\ln[-\ln(a_w)]$ versus X.
Chong-Pfost equation:

$$\ln(a_w) = \frac{-A}{T}\exp(-BX) \tag{12.32}$$

A and B are constants.
Day and Nelson equation:

$$a_w = 1 - \exp[P_1 T^{P_2} X^{P_3}] \tag{12.33}$$

P_1, P_2, P_3 are constants; T = absolute temperature.
Halsey equation:

$$a_w = \exp\left[\frac{-A}{T}\theta^b\right] \tag{12.34}$$

$\theta = X/X_m$; and a and b are constants.
Henderson equation:

$$1 - a_w = \exp[-aX^b] \tag{12.35}$$

a and b are constants.
Iglesias and Chirife equation:

$$X = B_1\left[\frac{a_w}{1 - a_w}\right] + B_2 \tag{12.36}$$

B_1 and B_2 are constants.
Kuhn equation:

$$X = \frac{A}{\ln(a_w)} + B \tag{12.37}$$

Oswin equation:

$$X = A\left[\frac{a_w}{1 - a_w}\right] + B \tag{12.38}$$

Smith equation:

$$X = A - B\ln(1 - a_w) \tag{12.39}$$

These equations are empirical and are usually used when the fit to the GAB equation is not very good. The advantage of the GAB and BET equations is that they are based on physical adsorption phenomena, therefore the parameters have physical meaning. Expressing sorption isotherms in equation form avoids the need for graphs in representing water activity of foods as a function of moisture content. These equations will also facilitate the calculation of equilibrium moisture content and water activity in individual particulate component when mixing particulate solids with different water activities.

12.2 MASS TRANSFER

During dehydration, water is vaporized only from the surface. The transfer of water vapor from the wet surface to a stream of moving air is analogous to convection heat transfer, therefore a mass transfer coefficient is used. Moisture flux is proportional to the driving force, which is the difference in vapor pressure on the surface and the vapor pressure of water in air surrounding the surface. At the same time that water is removed from the surface, water diffuses from the interior of a solid toward the surface. The latter is a general form of diffusion that is analogous to conduction heat transfer. The differential equations for conduction also applies for diffusion, but a mass diffusivity is used in place of the thermal diffusivity.

12.2.1 Mass Diffusion

For an infinite slab, Equation (7.89) in the section "The Infinite Slab," Chapter 7, can be used for the dimensionless moisture change with time. Expressing this equations in terms of moisture content and mass diffusivity:

$$\theta = \frac{X - X_m}{X_0 - X_m}$$

where X is moisture content at any time, dry matter basis (kg water/kg dry matter), X_0 is initial moisture content, and X_m is equilibrium moisture content. The equation for moisture content at any point y in the solid measured from the center, at any time t, when the moisture content X was originally uniform at X_0 is

$$\theta = 2\sum_{n=0}^{\infty} \frac{(-1)^n}{(n+0.5)\pi}\left[\left(e^{-(n+.5)^2\pi^2 D_m \frac{t}{L^2}}\right)\left(\cos\left(\frac{(n+0.5)\pi y}{L}\right)\right)\right] \tag{12.40}$$

Assumptions used in the derivation of this equation are constant diffusivity, and constant surface moisture at the moisture content in equilibrium with the drying air, during the process. The value of X_m is determined from the sorption isotherm of the solid at $a_w =$ decimal equivalent of per cent relative humidity of the drying air.

During the early stages of dehydration, moisture is transferred and moves from the center toward the surface by capillary action. This mechanism is more rapid than diffusion, and rate of surface evaporation controls the rate of drying. However, in the later stages of drying, diffusion controls the rate of moisture migration within the solid. Diffusivity may be constant if cells do not collapse and pack together. Firm solids such as grain may exhibit constant diffusivity, but high-moisture products such as fruits and vegetables may exhibit variable diffusivity with moisture content depending on the physical changes that occur as water is removed.

The mass diffusivity D_m has the same units as thermal diffusivity and can be used directly to substitute for a in the heat transfer equations. Of interest in dehydration processes is the average moisture content in the slab at any time during the drying process. Let W and Z represent the width and length of the slab. The total moisture in the slab is

$$\text{Total moisture} = \rho W Z \int_0^L X \, dy$$

The mean moisture $\overline{X}$ = total moisture/total mass:

$$\overline{X} = \frac{1}{L} \int_0^L X \, dy$$

The dimensionless moisture ratio $\overline{\theta}$ based on the average moisture content is:

$$\overline{\theta} = \frac{\overline{X} - X_m}{X_0 - X_m}$$

Substituting the expression for X obtained from Equation (12.40) into the above integral and integrating:

$$\overline{\theta} = 2 \sum_{n=0}^{\infty} \left[\frac{(-1)^n}{(n+0.5)^2 \pi^2} [e]^{-(n+0.5)^2 \pi^2 D_m \frac{t}{L^2}} \right] [\sin(n+0.5)\pi] \tag{12.41}$$

The squared term in the denominator of Equation (12.41) results from the integration of the cosine function of Equation (12.40).

Table 12.2 shows the values of the dimensionless mean moisture ratio $\overline{\theta}$ as a function of $(D_m t)/L^2$. L is the half thickness of the slab. For three dimensional diffusion, $\overline{\theta_{L1}}$ is obtained for $(D_m t/L_1^2)$, $\overline{\theta_{L2}}$ is obtained for $(D_m t/L_2^2)$, and $\overline{\theta_{L3}}$ is obtained for $(D_m t/L_3^2)$. The composite $\overline{\theta}$ is the product of $\overline{\theta_{L1}}$, $\overline{\theta_{L2}}$ and $\overline{\theta_{L3}}$.

When $(D_m t)/L^2 > 0.1$, the first 3 terms in the series in Equation (12.41) is adequate for series convergence. Taking logarithms of both sides of Equation (12.41):

$$\log(\overline{\theta}) = \frac{\pi^2 D_m}{L^2} t \log(e) + B$$

$$B = \log(2) + \log A_1 (e)^{0.25} + \log A_2 (e)^{2.25} + \log A_3 (e)^{6.25}$$

$$A_1, A_2, A_3 = \frac{(-1)^n \sin[(n+0.5)\pi]}{(n+0.5)^2 \pi^2} \quad \text{for} \quad n = 0, 1, 2 \tag{12.42}$$

Equation (12.42) shows that a semi-log plot of the dimensionless mean moisture ratio against time of drying will be linear if D_m is constant. The mass diffusivity, D_m, can be calculated from the slope.

Table 12.2 Dimensionless mean moisture ratio as a function of $D_m t / L^2$

$\frac{D_m t}{L^2}$	$\bar{\theta}$	$\frac{D_m t}{L^2}$	$\bar{\theta}$	$\frac{D_m t}{L^2}$	$\bar{\theta}$
0	1.0000	0.10	0.6432	1.10	0.0537
0.01	0.8871	0.20	0.4959	1.20	0.0411
0.02	0.8404	0.30	0.3868	1.30	0.0328
0.03	0.8045	0.40	0.3021	1.40	0.0256
0.04	0.7743	0.50	0.2360	1.50	0.0200
0.05	0.7477	0.60	0.1844	1.60	0.0156
0.06	0.7236	0.70	0.1441	1.70	0.0122
0.07	0.7014	0.80	0.1126	1.80	0.0095
0.08	0.6808	0.90	0.0879	1.90	0.0074
0.09	0.6615	1.00	0.0687	2.00	0.0058

Example 12.4. The desorption isotherm of apples was reported by Iglesias and Chirife (1982) to best fit Henderson's equation with the constants $a = 4.471$ and $b = 0.7131$. Experimental drying data for apple slices showed that when the average moisture content was 1.5 kg water/kg dry matter, the drying rate, which may be assumed to be diffusion controlled, was 8.33×10^{-4} kg water/[kg dry matter (s)]. The apple slices were 1.5 cm thick and 2.5 cm wide and were long enough to consider diffusion to occur from two dimensions. The drying air has a relative humidity of 5%.

(a) Calculate the mass diffusivity of water at this stage of drying.
(b) Calculate the drying rate and moisture content after 1 hour of drying from when the moisture content was 1.5 kg water/kg dry matter.

Solution:

(a) Henderson's equation will be used to calculate the equilibrium moisture content, X_m. Solving for X_m by substituting X_m for X in Henderson's equation:

$$\ln(1 - a_w) = -aX^b; \quad X_m = \left[\frac{-\ln(1 - 0.05)}{4.471} \right]^{1/0.7131}$$

$$X_m = 0.0019$$

Equation (12.42) for diffusion from two dimensions is

$$\log(\bar{\theta}) = \left[\frac{1}{(L_1)^2} + \frac{1}{(L_2)^2} \right] \pi^2 D_m \log(e)t + B \tag{12.43}$$

Differentiating Equation (12.43) with respect to t:

$$\frac{d}{dt}[\log(\bar{\theta})] = \left[\frac{1}{(L_1)^2} + \frac{1}{(L_2)^2} \right] \pi^2 D_m \log(e) \tag{12.44}$$

Differentiating the expression for $\bar{\theta}$ with respect to t:

$$\log(\bar{\theta}) = \log \left(\frac{\bar{X} - X_m}{\bar{X}_o - X_m} \right)$$

$$\frac{d}{dt}[\log(\bar{\theta})] = \frac{1}{(\bar{X} - X_m)[\ln(10)]} \frac{d\bar{X}}{dt} \tag{12.45}$$

Combining Equations (12.44) and (12.45) and solving for D_m:

$$D_m = \frac{d\overline{X}/dt}{(\overline{X} - X_m)\left[\dfrac{1}{(L_1)^2} + \dfrac{1}{(L_2)^2}\right]\pi^2} \tag{12.46}$$

Substituting known quantities:

$$D_m = \frac{8.33 \times 10^{-4}}{(1.5 - 0.0019)[(1/0.0075)^2 + (1/(0.0125)^2]\pi^2} = 2.3302 \times 10^{-9}\ m^2/s$$

(b) Solving for $D_m t/L^2$ for $t = 3600$ s:

$$L = 0.0075;\ \left(D_m t/L^2\right) = 0.149$$
$$L = 0.0125;\ \left(D_m t/L^2\right) = 0.054$$

From Table 12.2:
When $\left(D_m t/L^2\right) = 0.149$, by interpolation:

$$\overline{\theta_{L1}} = [0.6432 - (0.6432 - 0.4959)(0.049)]/0.1 = 0.571$$

When $\left(D_m t/L^2\right) = 0.054$, by interpolation:

$$\overline{\theta_{L2}} = [0.7477 - (0.7477 - 0.7236)(0.004)]/0.01 = 0.738$$

For diffusion from two directions:

$$\overline{\theta} = 0.571(0.738) = 0.4214$$
$$\overline{X} = 0.4214(1.5 - 0.019) + 0.019 = 0.633\ kg\ water/kg\ dry\ matter$$

The drying rate is $d\overline{X}/dt$. Using Equation (12.46):

$$\frac{d\overline{X}}{dt} = D_m\left(\overline{X} - X_m\right)\left[\frac{1}{(L_1)^2} + \frac{1}{(L_2)^2}\right]\pi^2$$

$$= \left[2.330 \times 10^{-9}(0.633 - 0.0019)(\pi^2)\right]\left[\frac{1}{(0.0075)^2} + \frac{1}{(0.0125)^2}\right]$$

$$= 0.000351\ kg\ water/(s \cdot kg\ dry\ matter)$$

12.2.2   Mass Transfer from Surfaces to Flowing Air

When air flows over a wet surface, water is transferred from the surface to air. The equations governing rate of mass transfer is similar to that for heat transfer. By analogy, the driving force for mass transfer is a concentration difference, and the proportionality constant between the mass flux and the driving force is the mass transfer coefficient.

$$\frac{dW_w}{A\,dt} = k_g\,M_w(a_{ws} - a_{wa}) \tag{12.47}$$

where W_w is the mass of water vapor transferred from the surface to the moving air, A is the surface area exposed to air, M_w is the molecular weight of water, a_{ws} is water activity on the surface, and a_{wa} is the water activity of the drying air. Using the dimensionless a_w difference as the driving force results in the mass transfer coefficient, k_g, having units in a general form widely used in the literature, kg moles/(m$^2 \cdot$ s).

Determination of the mass transfer coefficient is analogous to that in heat transfer, which involves the use of dimensionless groups. The equivalent of the Nusselt number in mass transfer is the Sherwood number (Sh).

$$Sh = \frac{k_g D}{D_{wm}} \tag{12.48}$$

where D is the diameter or characteristic length, and D_{wm} is the diffusivity expressed in kg mole/(m $\cdot$ s).

The equivalent of the Prandtl number in mass transfer is the Schmidt number expressed in either the mass diffusivity, D_{mX} in m^2/s, or D_{wm} in kg mole/(m $\cdot$ s). Physical property terms in the Schmidt number for dehydration processes are those for air. M_a is the molecular weight of air, 29 kg/kg mole. D_m is diffusivity of water in air $= 2.2 \times 10^{-5}$ m^2/s.

$$Sc = \frac{\mu}{\rho\, D_m} = \frac{\mu}{M_a\, D_{wm}} \tag{12.49}$$

Correlation equations for the mass transfer coefficient are similar to those for heat transfer. Typical expressions are as follows.

Gilliland and Sherwood equation:

$$Sh = 0.023\, Re^{0.81}\, Sc^{0.44} \tag{12.50}$$

Equation (12.50) was derived for vaporization from a water film flowing down a vertical tube to air flowing upwards through the tube. The Reynolds number of flowing air ranged from 2000 to 35,000, and pressure ranged from 0.1 to 3 atm.

Colburn j factor:

$$j = \frac{k_g}{G}\, Sc^{0.666} = 0.023\, Re^{-0.2} \tag{12.51}$$

G is the molar flux of air, kg moles/(m$^2 \cdot$ s) $=$ PV/RT; P $=$ pressure in Pa; V $=$ velocity, m/s, R $=$ 8315 N $\cdot$ m / (kg mole $\cdot$ K); and T $=$ absolute temperature. Equation (12.51) may be used for mass transfer into air flowing through particles in a packed bed. Equation (12.50) would be suitable for air flowing parallel to the surface of a bed of particles.

Ranz and Marshall's equation:

$$Sh = 2 + 0.6\, Re^{0.5}\, Sc^{0.33} \tag{12.52}$$

Equation (12.52) is suitable for mass transfer from surfaces of individual particles such as in fluidized beds.

In dehydration, mass transfer is not the rate limiting mechanism, particularly at high air velocities needed to maintain low humidity in the drying air.

Example 12.5. Calculate the rate of dehydration expected from mass transfer when 1 cm. carrot cubes having a density of 1020 kg/m^3 are dried in a fluidized bed with air at 2% relative humidity flowing at the rate of 12 m/s. Atmospheric pressure is 101 kPa and air temperature is 80°C. Express dehydration rate as kg/(s $\cdot$ kg dry matter) when moisture content is 5 kg water/kg dry matter.

Solution:

At 80°C, the viscosity of air from the *Handbook of Chemistry and Physics* is 0.0195 cP. The characteristic length of a cube may be calculated as the diameter of a sphere having the same surface area.

$$D = L(6/\pi)^{0.5} = 1.382\,L$$

The mass flux is calculated using the ideal gas equation by substituting air velocity, $\overline{V}$, for volume:

$$G = \frac{101{,}000(12)(29)}{8315(353)} = 11.975\frac{kg}{m^2 \cdot s}$$

$G = P\overline{V}M_a/RT \cdot M_a$ is the molecular weight of air, 29 kg/kg mole.
The Reynolds number is $DG/\Phi = 1.382\,LG/\Phi$

$$Re = \frac{1.382(0.01)(11.975)}{0.0195(0.001)} = 8487$$

The density, ρ, is obtained using the ideal gas equation.

$$\rho = \frac{PM_a}{RT} = \frac{101{,}000(29)}{8315(353)} = 0.998\frac{kg}{m^3}$$

$$Sc = \frac{\mu}{\rho\,D_m} = \frac{0.0195(0.001)}{0.998(2.2 \times 10^{-5})} = 0.888$$

$$Sh = 2 + 0.6(8487)^{0.5}(0.888)^{0.33} = 55.15$$

Equation (12.49):

$$= 7.57 \times 10^{-7}\,kgmole/(m \cdot s)$$

Equation (12.48):

$$k_g = Sh\left(\frac{D_{wm}}{D}\right)$$

$$= \frac{55.150(7.57 \times 10^{-7})}{0.01(1.382)} = 0.00302\frac{kgmole}{m^2 \cdot s}$$

$$D_{wm} = \left(\frac{\rho\,D_m}{M_a}\right) = \frac{0.998(2.2 \times 10^{-5})}{29}$$

Express area as m^2/kg dry matter (DM).

$$A = \frac{6(0.01)^2}{(0.01)^3(1020)(1/6)} = 3.529\frac{m^2}{kg\,DM}$$

Equation (12.47):

$$\frac{dW}{dt} = [3.529][0.003021(18)(1 - 0.02)] = 0.188\frac{kg}{s \cdot kg\,DM}$$

The drying rate, based on surface mass transfer, is 0.188 kg water/(s · kg DM).

12.3 PSYCHROMETRY

12.3.1 Carrying Capacity of Gases for Vapors

The mass of any component of a gas mixture can be calculated from the partial pressure using the ideal gas equation. The ideal gas equation can be written for the whole mixture or for a single component as follows:

$$P_t V = n_t RT \quad \text{for the whole mixture}$$

or:

$$P_a V = n_a RT \quad \text{for component A,} \tag{12.53}$$

where P_a and n_a are partial pressure and number of moles of component A in the mixture, respectively. If the mixture consists only of components A and B,

$$P_t = P_a + P_b \tag{12.54}$$

For component B:

$$(P_t - P_a)(V) = n_b RT \tag{12.55}$$

Dividing Equation (12.53) by Equation (12.55):

$$\frac{P_a}{P_t - P_a} = \frac{n_a}{n_b}$$

If M_a is the molecular weight of component A and M_b is the molecular weight of component B, the mass ratio of the two components can be determined from the partial pressure as follows:

$$\frac{W_b}{W_a} = \frac{n_b\, M_b}{n_a\, M_a} = \frac{P_b}{(P_t - P_b)}\frac{M_b}{M_a} \tag{12.56}$$

If component A is water and component B is air, the mass ratio is known as the "humidity" or "absolute humidity." If P_a is equal to the vapor pressure of water at the given temperature of the air, the mass ratio of components A and B is the saturation humidity. If P_a is less than the vapor pressure, the ratio P_a/P_s, where P_s is the saturation partial pressure or the vapor pressure, is the saturation ratio. This ratio expressed as a percentage is also known as the "relative humidity."

Example 12.6. Dry air is passed through a bed of solids at the rate of 1 m^3/s at 30°C and 1 atm. If the solids have an equilibrium relative humidity of 80%, and assuming that the bed is deep enough such that equilibrium is attained between the solids and the air before the air leaves the bed, determine the amount of water removed from the bed per hour. Use 29 for the average molecular weight of air. Atmospheric pressure is 101.325 kPa.

Solution:

At equilibrium, the partial pressure of water in the air should be 80% of the vapor pressure of water at 30°C. From Appendix Table A.4, the vapor pressure of water at 30°C is 4.2415 kPa. The partial pressure of water in the air is 80% of 4.215 or 3.393 kPa. The molecular weight of water is 18.

The absolute humidity can be calculated using Equation (12.56):

$$\frac{W_w}{W_a} = \frac{3.394}{(101.325 - 3.394)} \frac{18}{29} = 0.0215 \frac{\text{kg water}}{\text{kg dry air}}$$

In order to calculate the total amount of water removed in one hour, the total amount of dry air that passed through the bed per hour must be calculated.

For the dry air component:

$$PV = nRT = \frac{W_a}{M_a} RT$$

$$W_a = \frac{P_a \, M_a}{RT} V$$

Substituting values for P_b, M_b, R, and T and using $R = 8315 \, N \cdot m/(\text{kg mole}) \, (K)$:

$$W_a = \frac{(101,325 \, N \cdot m^{-2})(29)(\text{kg})(\text{kg mole})^{-1}}{8315 N \cdot m(\text{kg mole})^{-1}(K)^{-1}(30 + 273)K} \left[1\frac{m^3}{s} \right] = 1.166 \, \text{kg/s of dry air.}$$

The weight of water removed per hour is

$$W_w = 1.166 \frac{\text{kg dry air}}{s} 0.0215 \frac{\text{kg water}}{\text{kg dry air}} \frac{3600 \, s}{h} = 90.24 \, \text{kg/h of water}$$

12.3.2 The Psychrometric Chart

Psychrometry is a study of the behavior of mixtures of air and water. In the preceding section, the determination of saturation humidities from the vapor pressure and the determination of the humidity from the partial pressure of water in air was discussed. A graph of humidity as a function of temperature at varying degrees of saturation forms the main body of a psychrometric chart. For processes that involve loss or gain of moisture by air at room temperature, the psychrometric chart is very useful for determining changes in temperature and humidity.

Another main feature of a psychrometric chart is the wet bulb temperature. When a thermometer is fitted with a wet sock at the bulb and placed in a stream of air, evaporation of water from the sock cools the bulb to a temperature lower than what would register if the bulb is dry. The difference is known as the wet bulb depression and is a function of the relative humidity of the air. The more humid air allows less vaporization, thus resulting in a lower wet bulb depression.

The various quantities that can be determined from a psychrometric chart are as follows:

Humidity (absolute humidity) (H): the mass ratio of water to dry air in the mixture.
Relative humidity (% RH): the ratio of partial pressure of water in the air to the vapor pressure of water, expressed in a percentage.
Dry bulb temperature (T_{db}): actual air temperature measured using a dry temperature sensing element.
Wet bulb temperature (T_{wb}): air temperature measured using a wet sensing element that allows cooling by evaporation of water.
Dew point (T_{dp}): temperature to which a given air-water mixture needs to be cooled to start condensation of water. At the dew point, the air is saturated with water vapor. The dew point is also that temperature where the vapor pressure of water equals the partial pressure of water in the air.

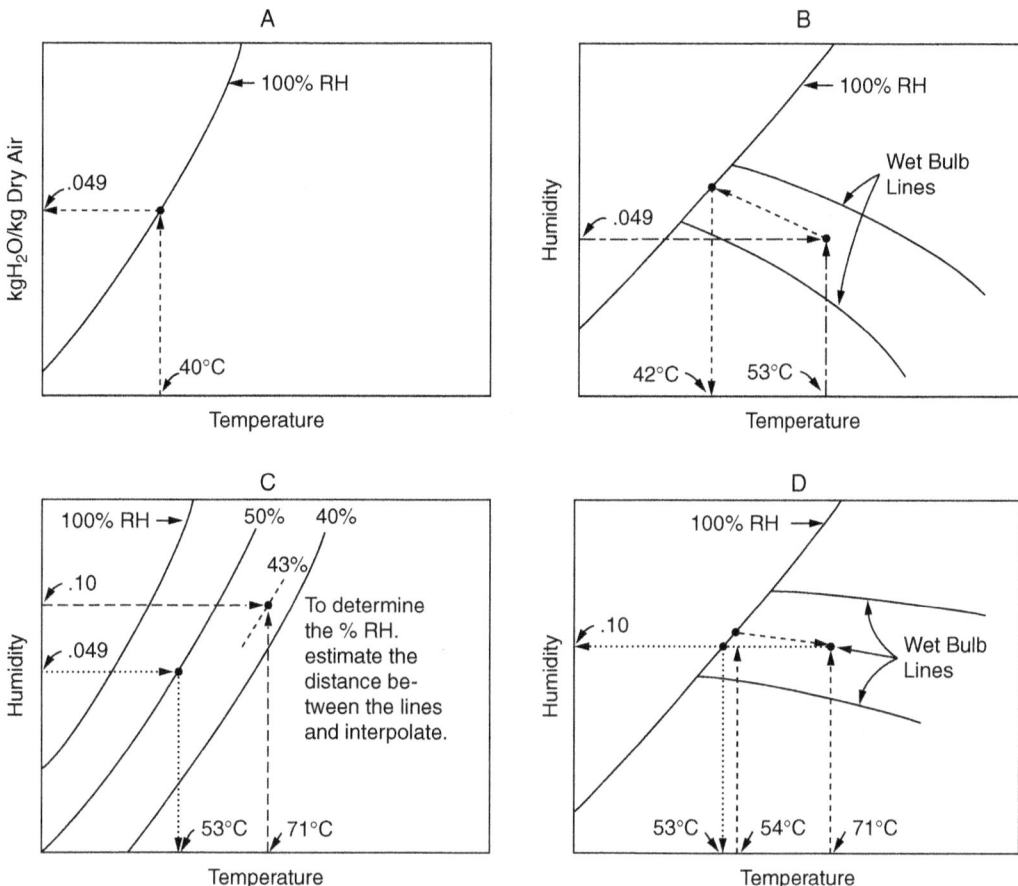

Figure 12.3   Diagram showing the use of a psychometric chart for determining properties of air-water mixtures. (Courtesy of Proctor and Schwartz, Inc. Copyrighted by Procotor and Schwartz. Inc.)

A psychrometric chart has its axes temperature on the abscissa and humidity on the ordinate. Two parameters are necessary to establish a point on the chart that represents the condition of air. These parameters could be any two from relative humidity, dry bulb temperature, wet bulb temperature, and dew point or absolute humidity. Dew point and absolute humidity are not independent, knowing one establishes the other.

Psychrometric charts in metric and English units are in the Appendix (Fig. A5 and A7). Figure 12.3 is a diagrammatic representation of how quantities are read from a psychrometric chart. Figure 12.3A shows how dew point is determined from the humidity. Figure 12.3B shows how the wet bulb temperature is determined from the humidity and the dry bulb temperature. Figure 12.3C shows how the percentage of relative humidity is determined knowing the dry bulb and the humidity, and how dry bulb temperature is determined from humidity and % relative humidity. Figure 12.3D shows how humidity and dew point are determined from the wet and dry bulb temperatures.

Example 12.7. Air has a dew point of 40°C and has a relative humidity of 50%. Determine (a) the absolute humidity, (b) the wet bulb temperature, and (c) the dry bulb temperature.

Solution:

Using a psychrometric chart, draw a vertical line through $T = 40°$. At the intersection with the line representing 100% relative humidity (% RH), draw a horizontal line and read the absolute humidity represented by this horizontal line at the abscissa (see Fig. 12.3A). (a) H = 0.049 kg water/kg dry air.

Extend the horizontal line representing the humidity until it intersects the diagonal line representing 50% RH. Draw a vertical line through this intersection and read the dry bulb temperature representing this vertical line at the abscissa (see Fig. 12.3C). (b) Dry bulb temperature = 53°C.

From the point represented by H = 0.049 and T_{tb} = 53°C, draw a line that parallels the wet bulb lines. Project this line to its intersection with the 100% RH line. Draw a vertical line at this intersection, project to the abscissa and read the wet bulb temperature at the abscissa (see Fig. 12.3B). (c) T_{tb} = 42°C.

Example 12.8. Air in a smokehouse has a wet bulb temperature of 54°C and a dry bulb temperature of 71°C. Determine (a) the humidity, (b) the dew point, and (c) the percentage of relative humidity.

Solution:

Draw vertical lines through $T = 54°C$ and the line representing 100% RH, draw a line that parallels the wet bulb lines. From the intersection of this drawn wet bulb line with the vertical line representing $T = 71°C$, draw a horizontal line and project to the abscissa. Read the humidity represented by this horizontal line (see Fig. 12.3D). (a) H = 0.1 kg water/kg dry air.

From the intersection of the horizontal line representing H = 0.1 with the 100% RH line, draw a vertical line and connect to the abscissa. (b) T_{dp} = 53°C.

From the intersection of the line representing $T = 71°C$ and that representing H = 0.1, interpolate between the diagonal lines representing 40% and 50% RH and estimate the % RH (see Fig. 12.3B). (c) % RH = 43%.

12.3.3 Use of Psychrometric Chart to Follow Changes in the Properties of Air-Water Mixtures Through a Process

Figure 12.4 shows the path of a process on a psychrometric chart for heating, cooling and adiabatic humidification.

When the temperature of air is increased at constant pressure and there is no water added or removed from the air, the process is a constant humidity process. T_{db} increases, T_{wb} also increases, but % RH decreases. The process is represented by A in Fig. 12.4. When the temperature of air is decreased above the dew point, the process is a constant humidity process represented by B in Fig. 12.4. Both T_{db} and T_{wb} decrease and % RH increases. Cooling below the dew point results in condensation of water and humidity drops. The % RH remains at 100% and the temperature and humidity drops following the line representing 100% RH. This is shown by C in Fig. 12.4.

Adiabatic humidification is a process where water is picked up by air and the heat required to vaporize the added water comes from the sensible heat loss that results from a reduction of the temperature of the air. The path traced by the temperature and humidity of the air is represented by D in Fig. 12.4, and this path parallels the wet bulb lines. The wet bulb temperature of the air remains constant, humidity and % RH increases and T_{db} decreases.

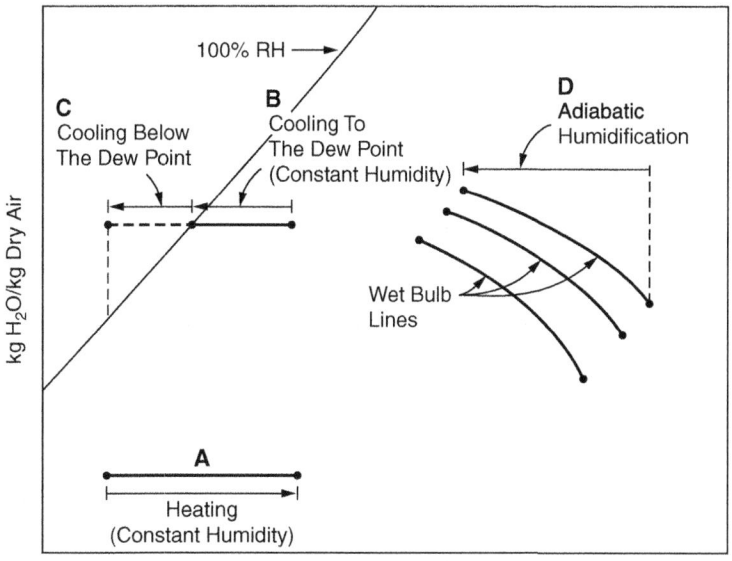

Figure 12.4 Diagram showing paths of cooling, heating, condensation, and adiabatic humidification on a psychometric chart.

This is the process that occurs when air is passed over or through a bed of wet solids in drying, or when air is passed through water sprays in a water cooling tower.

Example 12.9. Ambient air at 25°C and 50% RH is heated to 175°C. Determine (a) the % RH, and (b) the wet bulb temperature of the heated air.

Solution:

Use the psychrometric chart and draw a vertical line representing $T_{db} = 25°C$. Project this line until it intersects the diagonal line representing 50% RH. Draw a horizontal line and project to the ordinate to determine the absolute humidity. H = 0.0098 kg/water/kg dry air. Project the horizontal line representing H = 0.0098 until it intersects a vertical line representing $T_{db} = 175°C$. At the intersection, the % RH can be interpolated between the diagonal lines representing 0.15 % RH and 0.2 % RH. (a) % RH = 0.18 %.

From the intersection, draw a line parallel to the wet bulb line and project to its intersection with the line representing 100% RH. Project the intersection to the abscissa and read (b) $T_{wb} = 45°C$.

12.4 SIMULTANEOUS HEAT AND MASS TRANSFER IN DEHYDRATION

In dehydration, moisture is removed by evaporation. Heat must be transferred to equal the heat of vaporization. If the rate of mass transfer exceeds heat transfer needed to supply the heat of vaporization, the material will drop in temperature. This process is called evaporative cooling. The mass transfer equation (Eq. 12.47) can be written for a material undergoing dehydration in the form:

$$\frac{dW_w}{Adt} = k_{gw}(H_i - H) \tag{12.57}$$

dW_w/dt is mass of water transferred, H_i is the humidity at the interface where water is vaporized, and H is the humidity of the drying air. H_i is the equilibrium humidity at the interface at the temperature at the interface. When the temperature at the interface drops, H_i will also drop and the rate of mass transfer will slow down according to Equation (12.57).

Rate of heating needed to vaporize water at the rate dW_w/dt, will be

$$q = \frac{dW_w}{dt} h_{fg}$$

where h_{fg} is the enthalpy of vaporization at the temperature of the interface. Heat transfer needed for vaporization will be

$$q = k_{fg}(A)(H_i - H)(h_{fg}) \tag{12.58}$$

At equilibrium, the heat for vaporizing the mass transferred (Eq. 12.58) will equal the rate of heat transfer (Eq. 12.59). The temperature and humidity at the interface will be T_s and H_s, respectively.

$$q = hA(T - T_s) \tag{12.59}$$

Equating Equations (12.58) and (12.59):

$$(h_{fg})k_{gw}(A) (H_s - H) = hA(T - T_s)$$
$$H = H_s - \frac{hT_s}{k_{gw}(h_{fg})} + \frac{h}{k_{gw}(h_{fg})}T \tag{12.60}$$

Equation (12.60) is the equation of a wet bulb line, the relationship between equilibrium temperature and humidity when a thermometer bulb is wrapped with a wet sock and exposed to a flowing stream of air. The line will go through the point T_s and H_s the wet bulb temperature and the saturation humidity at the wet bulb temperature, respectively.

Food products during drying will follow the relationship between heat and mass transfer as expressed in Equations (12.58), (12.59), and (12.60).

The difference between the dry bulb, T, and the wet bulb temperature, T_s, is known as the wet bulb depression, an index of the drying capacity of a stream of air.

If the temperature of a product during dehydration is higher than the wet bulb temperature, then the rate of heat transfer (Eq. 12.59) exceeds evaporative cooling by mass transfer (Eq. 12.58) and mass transfer controls the drying rate. This phenomenon is often observed in the later stages of drying when the interface for vaporization is removed from the surface requiring vapor to flow through the pores of the dried material close to the surface, before mixing with the drying air at the surface.

The air temperature drops as it undergoes adiabatic humidification. Vaporization of water to increase the humidity requires energy which comes from a drop in the sensible heat of air. The heat balance is as follows.

Heat of vaporization, q_v to increase humidity from H to H_s is

$$q_v = (H_s - H)(h_{fg}) \tag{12.61}$$

The loss in sensible heat q_s for air with a specific heat C_p is

$$q_s = C_p(T - T_s) \tag{12.62}$$

Equating Equations (12.61) and (12.62):

$$C_p(T - T_s) = (H_s - H)(h_{fg}) \tag{12.63}$$

$$H = H_s + \frac{C_p}{h_{fg}} T_s - \frac{C_p}{h_{fg}} T$$

Equation (12.63) is the adiabatic humidification line; the change in humidity of air as it drops in temperature during adiabatic humidification. Equation (12.63) will be exactly equal to Equation (12.60) if the ratio of the heat transfer coefficient to the mass transfer coefficient, h/k_{gw}, equals the specific heat of air, C_p. Indeed, this has been shown to be true for water vaporizing into air at one atmosphere and at moderate temperatures. Thus, in air drying, wet bulb lines on the psychrometric chart coincide exactly with the adiabatic humidification lines.

Example 12.10. Room air at 80°F (26.7°C) and 50% RH is heated to 392°F (200°C) and introduced into a spray drier where it leaves at a temperature of 203°F (95°C).

Determine the humidity and relative humidity of the air leaving the drier. Assume adiabatic humidification in the drier.

Solution:

Using a psychrometric chart, locate the point that represents T = 80°F and 50% RH. The humidity at this point is 0.011. When air is heated, the temperature increases at constant humidity. Air leaving the heater will have a humidity of 0.011 and a temperature of 392°F (200°C). Starting from T = 392°F (200°C) and H = 0.011, draw a curve that approximates the closest wet bulb curve until a temperature of 203°F (95°C) is reached. The humidity at this point is 0.055 and the relative humidity is 10%.

12.5   THE STAGES OF DRYING

Drying usually occurs in a number of stages, characterized by different dehydration rates in each of the stages. Figure 12.5 shows the desorption isotherm and the rate of drying of apple slices in a cabinet dryer with air flowing across the trays containing the slices at 0.3 m/s, 150°F (65.6°C) and 20% RH. The drying rate curve shows a constant drying rate (line AB) from a starting moisture content of 8.7 g H_2O/g dry matter to a moisture content of 6.00 kg H_2O/kg dry matter. At this stage in the drying cycle, vaporization is occurring at the product surface and free water with a_w of 1.00 is always available at the surface to vaporize. This stage of drying is the constant rate stage. The rate of drying is limited by the rate at which heat is transferred into the material from air. The product temperature is usually at the wet bulb temperature of the drying air.

From a moisture content of 6.00 g H_2O/g dry matter, drying rate decreased linearly with the moisture content. Point B is the critical moisture content and line BC represents the first falling rate period. The first falling rate period is characterized by a slight increase in the temperature of the product, although this temperature may not be very much higher than the wet bulb temperature. Free moisture is no longer available at the surface and the rate of drying is controlled by moisture diffusion towards the surface. Most of the water in the material is still free water with a_w of 1. However, diffusion toward the surface is necessary for vaporization to occur.

From a moisture content of 1.20 g H_2O/g dry matter, the slope of the drying rate against moisture content changes and the rate of drying goes through a second falling rate period. The second falling rate period starts at point C, where the equilibrium relative humidity for the material begins to drop below

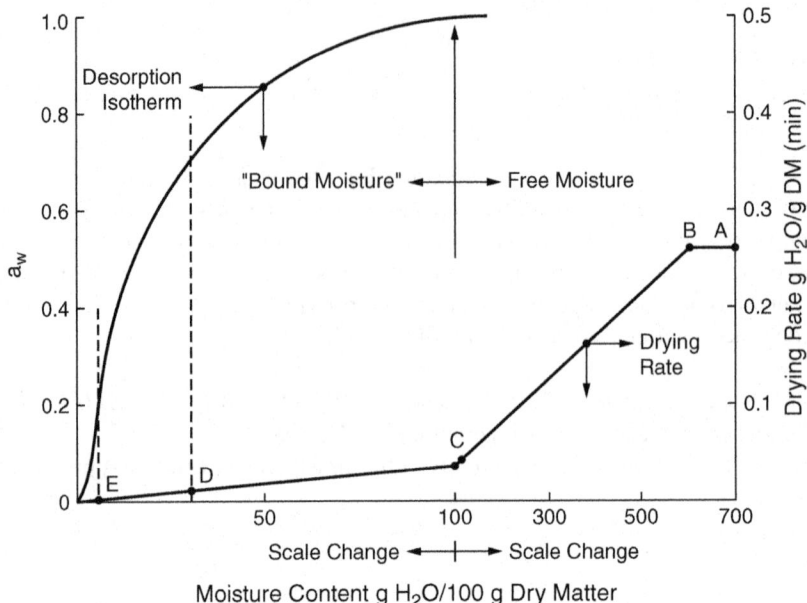

Figure 12.5 Desorption isotherm of raw apple slices and drying rates at different moisture contents.

100%. The moisture content at point C is the bound water capacity of the material. During the second falling rate period, dehydration proceeds through the portion of the sorption isotherm where water in the material is held by multimolecular adsorption and capillary condensation. Heat of vaporization of water in this stage of the dehydration process is higher than the heat of vaporization of pure water because the heats of adsorption and vaporization must be provided. Vaporization at this stage occurs at the interior rather than the surface and water vapor has to diffuse to the surface before it mixes with the flowing stream of air.

At a moisture content of 0.30 g H_2O/g dry matter, the drying rate goes through another falling rate zone. This stage of drying corresponds to the region of the sorption isotherm where water is held in mono- or multi-molecular layers. Dehydration should be terminated at any point along the line DE. Point E represents the equilibrium moisture content where dehydration stops because a_w of the surface and air are equal.

12.6 PREDICTION OF DRYING TIMES FROM DRYING RATE DATA

12.6.1 Materials with One Falling Rate Stage Where the Rate of Drying Curve Goes Through the Origin

The drying rate curve for these materials will show a constant rate R_c from the initial moisture content X_o to the critical moisture content X_c. The drying rate then falls in a linear relationship with decreasing moisture content until it becomes 0 at X = 0. The drying times required to reach a moisture content X in either of the stages of drying are

$$\text{Constant rate}: -\frac{dX}{dt} = R_c \qquad (12.64)$$

The total time, t_c, for the constant rate stage is

$$t_c = \frac{X_0 - X_c}{R_c} \qquad (12.65)$$

$$\text{At the falling rate period}: -\frac{dX}{dt} = \frac{R_c}{X_c}(X) \qquad (12.66)$$

$$\int_{t_c}^{t} dt = \frac{X_c}{R_c} \int_{X_c}^{X} \frac{dX}{X}$$

$$t - t_c = \frac{X_c}{R_c} \ln \frac{X_c}{X} \qquad (12.67)$$

The total time from X_0 to X in the falling rate stage can be calculated by substituting t_c in Equation (12.65) into Equation (12.67).

$$t = \frac{X_0 - X_c}{R_c} + \frac{X_c}{R_c} \ln \frac{X_c}{X} \qquad (12.68)$$

Equation (12.68) shows that if a material exhibits only one falling rate stage of drying and the drying rate is 0 only at $X = 0$, the drying time required to reach a desired moisture content can be determined from the constant rate R_c, and the critical moisture content X_c. For these materials, X_c is usually the moisture content when a_w starts to drop below 1.0 in a desorption isotherm.

Example 12.11. A material shows a constant drying rate of 0.15 kg H_2O/min (kg dry matter), and has an a_w of 1.0 at moisture contents >1.10 kg H_2O/kg dry matter. How long will it take to dry this material from an initial moisture content of 75% (wet basis) to a final moisture content of 8% (wet basis)?

Solution:

Converting the final and initial moisture contents from wet to a dry basis:

$$X_0 = \frac{0.75 \text{ kg water}}{0.25 \text{ kg dry matter}} = 3.0 \text{ kg water/kg dry matter}$$

$$X = \frac{0.08 \text{ kg water}}{0.92 \text{ kg dry matter}} = 0.0869 \text{ kg water/kg dry matter}$$

$X_c = 1.10$ kg water/kg dry matter

$R_c = 0.15$ kg water/min (kg dry matter)

Using Equation (12.68):

$$t = \frac{3.0 - 1.10}{0.15} + \frac{1.10}{0.15} \ln \frac{1.10}{0.0869} = 12.7 + 18.6 = 31.3 \text{ min}$$

12.6.2   Materials with More Than One Falling Rate Stage

Most food solids would exhibit this drying behavior. The drying rate curve shown in Fig. 12.5 for apple slices is a typical example. The drying time in the constant rate follows Equation (12.65). However, because the rate of drying against moisture content plot no longer goes to the origin from the point (X_c, R_c), Equation (12.66) cannot be used for the falling rate stage. If the rate versus moisture content line is extended to the abscissa, the moisture content where rate is 0 may be designated as the residual moisture content X_r and for the first falling rate period:

$$\frac{d(X - X_{r1})}{dt} = \frac{R_c}{X_{c1} - X_{r1}}(X - X_{r1})$$ (12.69)

Integrating Equation (12.69) and using Equation (12.65) for drying time in the constant rate zone, drying to a moisture content X in the first falling rate stage would take:

$$t = \frac{X_0 - X_{c1}}{R_c} + \frac{X_{c1} - X_{r1}}{R_c} \ln \frac{X_{c1} - X_{r1}}{X - X_{r1}}$$ (12.70)

where X_{c1} and X_{r1} represent the critical moisture content and the moisture content a the end of the first falling rate stage of drying. The time required to dry to moisture content X in the second falling rate stage would be

$$t = \frac{X_0 - X_{c1}}{R_c} + \frac{X_{c1} - X_{r1}}{R_c} \ln \left[\frac{X_{c1} - X_{r1}}{X_{c2} - X_{r1}}\right] + \left[\frac{X_{c1} - X_{r1}}{R_c}\right]\left[\frac{X_{c2} - X_{r2}}{R_c}\right] \ln \left[\frac{X_{c2} - X_{r2}}{X - X_{r2}}\right]$$ (12.71)

Example 12.12. Figure 12.6 shows the drying curve for apple slices blanched in 10% sucrose solution and dried in a cabinet drier using air in parallel flow at a velocity of 3.65 m/s at $_{db}$ = 170°F (76.7°C) and T_{wb} = 100°F (37.8°C) for the first 40 minutes, and T_{db} = 160°F (71.1°C) and T_{wb} = 110°F (43.3°C) for the rest of the drying period. Calculate the drying time to reach a moisture content of 0.15 kg water/kg dry matter.

Solution:

The drying rate curve was obtained by drawing tangents to the drying curve at the designated moisture contents and determining the slopes of the tangents. The drying time required to obtain a moisture content of 0.15 kg water/kg dry matter will be

$$t = \frac{5.3 - 2.5}{0.163} + \frac{2.5 - 0.35}{0.163} \ln \left[\frac{2.5 + 0.35}{1.0 - 0.35}\right] + \left[\frac{2.5 - 0.35}{0.163}\right]\left[\frac{1 - 0.1}{0.163}\right] \ln \left[\frac{1 - 0.1}{0.15 - 0.35}\right]$$

$$= 20.2 + 15.8 + 52.7 = 88.7\,\text{min}$$

The drying curve in Fig. 12.6 shows a drying time of 90 min at X = 0.15.

12.6.3   The Constant Drying Rate

The constant drying rate R_c is heat transfer controlled and can be calculated using a heat balance. Let ρ_s = the dry solids density, kg dry solids/m^3 of wet material. ρ_s = wet material density x mass fraction dry solids in the wet material.

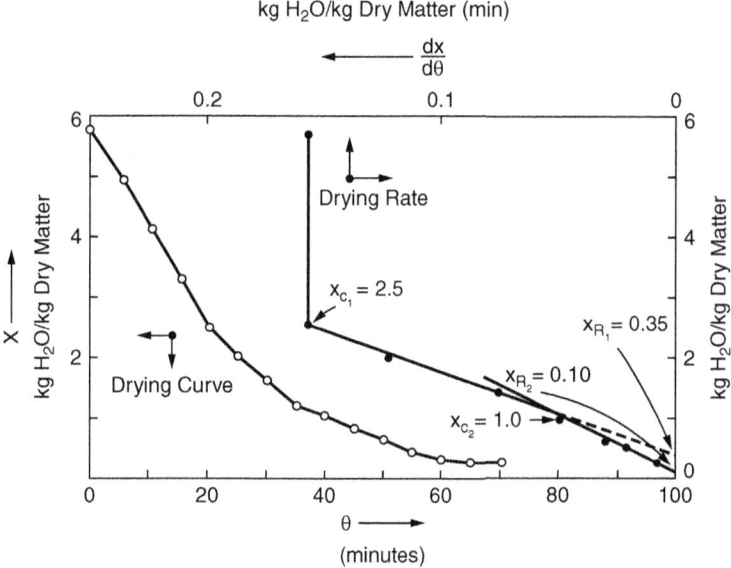

Figure 12.6 Drying curve and drying rate as a function of the moisture content of blanched apple slices, showing several breaks in the drying rate.

If L is the depth of the innermost section of the material from the drying surface (if drying occurs from both sides, L will be half the total thickness of the solid), M_s is the mass of dry solids and A is the area of the top surface of the solid, the volume V of material will be

$$V = \text{surface area (depth)} = A(L) = \frac{M_s}{\rho_s}$$

$$\frac{A}{M_s} = \frac{1}{L(\rho_s)} \tag{12.72}$$

Heat balance: latent heat of evaporation = heat transferred

$$\frac{dX}{dt} M_s\, h_{fg} = h\, A(T_a - T_s) \tag{12.73}$$

where h_{fg} = latent heat of vaporization at the surface temperature of the material, T_s, T_a = the dry bulb temperature of the air, and h is the heat transfer coefficient. The surface temperature during the constant rate period is also the wet bulb temperature (T_{wb}) of the air. $T_s = T_{wb}$. $dX/dt = R_c$.

$$R_c = \frac{h(T_a - T_s)}{h_{fg}} \cdot \frac{A}{M_s} \tag{12.74}$$

Substituting Equation (12.72) in Equation (12.74):

$$R_c = \frac{h(T_a - T_s)}{h_{fg}\, L\rho_s} \tag{12.75}$$

Equation (12.75) can be used to calculate the constant rate of drying from the heat transfer coefficient and the wet and dry bulb temperatures of the drying air for a bed of particles with drying air flowing parallel to the surface.

Expressions similar to Equation (12.75) may be derived using the same procedure as above for cubes with sides L evaporating water at all sides:

$$R_c = \frac{6h(T_a - T_s)}{h_{fg} L \rho_s} \tag{12.75a}$$

For a brick-shaped solid with sides "a" and "2a" and thickness L:

$$R_c = \frac{h(T_a - T_s)}{h_{fg} \rho_s} \left[\frac{3}{a} + \frac{2}{L} \right] \tag{12.75b}$$

For a spherical solid: $R_c = 3h(T_a - T_s)/(R h_{fg})$.

The heat transfer coefficient can be calculated using the following correlation equations (Sherwood, Ind. Eng. Chem. 21:976, 1029):

If air flow is parallel to the surface:

$$h = 0.0128 G^{0.8} \tag{12.76}$$

where h = the heat transfer coefficient in BTU/(h · ft^2·°F) and G is the mass rate of flow of air, lb$_m$/(hft^2). In SI units, Equation (12.76) is

$$h = 14.305 G^{0.8} \tag{12.77}$$

where h = W/(m^2· K) and G is in kg/(m^2· s). If flow is perpendicular to the surface:

$$h = 0.37 G^{0.37} \tag{12.78}$$

where h is in BTU/(h · ft^2·°F) and G is in lb$_m$/(ft^2· h). In SI units, Equation (12.78) is

$$h = 413.5 G^{0.37} \tag{12.79}$$

where h is in W/(m^2· K) and G is in kg/(m^2· s).

When air flows through the bed of solids, Ranz and Marshall's equation (Chem. Eng. Prog. 48(3):141, 1956; Appendix Table A.12 for particles in a gas stream) may be used for determining the heat transfer coefficient.

Example 12.13. Calculate the constant drying rate for blanched apple slices dried with air flowing parallel to the surface at 3.65 m/s. The initial moisture content was 85.4% (wet basis) and the slices were in a layer 0.5 in. (0.0127 m) thick. The wet blanched apples had a bulk density of approximately 35 lb/ft^3 (560 kg/m^3) at a moisture content of 87% (wet basis).

Dehydration proceeds from the top and bottom surfaces of the tray. Air is at 76.7°C (170°F) db and 37.8°C (100°F) wb.

$$\rho_s = \frac{560 \, \text{kg}}{\text{m}^3} \frac{0.13 \, \text{kg DM}}{\text{kg}} = 72.8 \frac{\text{kg DM}}{\text{m}^3} \text{ or } 4.55 \frac{\text{lb DM}}{\text{ft}^3}$$

$$V = 3.65 \, \text{m/s or } 12.0 \, \text{ft/s}$$

Solution:

Using the ideal gas equation: $R = 8315$ Nm/kgmole $\cdot$ K; $T = 76.7°C$, $P = 1$ atm, $= 101.3$ k Pa, $M = 29$ kg/kg mole.

$$\frac{\text{kg air}}{m^3} = \frac{P(M)}{R(T)} = \frac{(101,300)(29)}{8315(349.7)} = 1.01 \frac{\text{kg}}{m^3} \text{ or } 0.063 \frac{\text{lb}}{\text{ft}^3}$$

$$G = \frac{\text{kg air}}{m^3} \times \text{velocity} = 1.01 \frac{\text{kg}}{m^3} 3.65 \frac{m}{s}$$

$$= 3.687 \text{ kg}/(m^2 \cdot s) \text{ or } 2713 \text{ lb}/(\text{ft}^2 \cdot h)$$

Using Equation (12.77):

$$h = 14.305(3.687)^{0.8} = 40.6 \text{ W}/(m^2 \cdot K) \text{ or } 7.15 \text{ BTU}/(h \cdot \text{ft}^2 \cdot °F)$$

$$T_a - T_s = 76.7 - 37.8 = 38.9°C \text{ or } 70°F$$

$$h_{fg} = \text{heat of vaporization at } 37.8°C \ (100°F) = 1037.1 \text{ BTU/lb or } 2.4123 \text{ MJ/kg}$$

Because drying occurs on top and bottom surfaces, $L = 0.0127/2 = 0.00635$.
Using Equation (12.75):

$$R_c = \frac{40.6(38.9)}{2.4123 \times 10^6 (0.00635)(72.8)} = 0.00146 \frac{\text{kg water}}{s \cdot \text{kg DM}} = 5.098 \text{ kg water}/(h \cdot \text{kg DM})$$

12.7 SPRAY DRYING

Spray drying is a process where a liquid droplet is rapidly dried as it comes in contact with a stream of hot air. Figure 12.7 is a schematic diagram of a spray drier where the atomized feed travels concurrent with the drying air. The small size of the liquid droplets allows very rapid drying and the residence time of the material inside the spray drier is in the order of seconds. The dried material is separated from air in a cyclone separator. The dried material is continuously withdrawn and cooled. Heat could damage the product if contact with the high temperature drying air is prolonged.

While the droplet is drying, the temperature remains at the wet bulb temperature of the drying air. For this reason, very high temperatures of the drying air can be tolerated in a drier with a minimum of damage to the heat sensitive components. Furthermore, rate of degradative reactions in foods slows down at low moisture contents. Thus, the portion of the drying process where product temperatures goes higher than the wet bulb temperature does not result in severe heat damage to the product.

A major requirement of successful spray drying is the reduction of the moisture content of a liquid droplet to a dryness level that would prevent the particle from sticking to a solid surface, as the particle impinges on that surface. The rate of drying of the particles must be such that from the time the particle leaves the atomizer to the time it impinges upon the walls of the spray drier, the particle is dry. The trajectory and velocity of the particles determines the available drying time. The rate of drying and the time required to dry are dependent upon the temperature of the drying air, the heat transfer coefficient and the diameter of the droplet being dried.

A constant rate and a falling rate drying stage are also manifested in a spray drying process. As the wet droplets leave the atomizer, their surfaces rapidly lose water. Solidified solute and suspended solids rapidly form a solid crust on the surface of each particle. The diameter of the particle usually

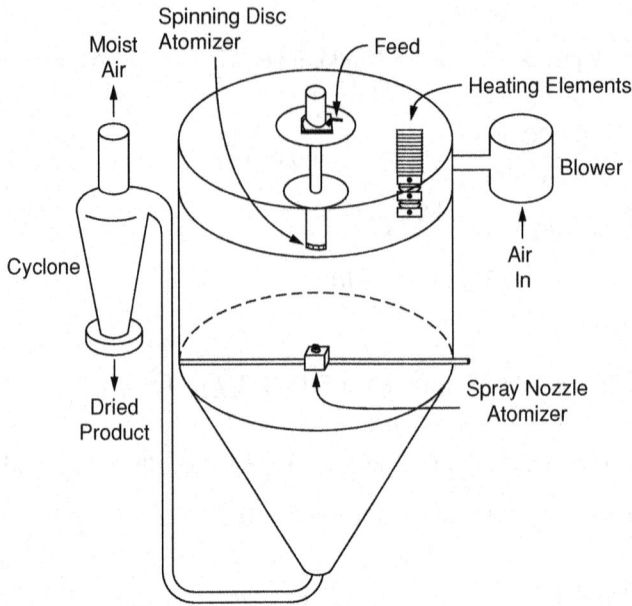

Figure 12.7 Schematic diagram of a spray drier.

decreases as drying proceeds. The formation of the solid crust constitutes the constant rate stage of drying. When the crust becomes sufficiently thick to offer considerable resistance to movement of water toward the surface, the drying rate drops, and the rate of drying is controlled by the rate of mass transfer. The temperature of the particle increases and the liquid trapped in the interior of the particle vaporizes and generates pressure. Eventually, a portion of the crust breaks and the vapor is released. Spray dried particles consist of hollow spheres or fragments of spheres. This shape of the particles is responsible for the excellent rehydration properties of spray dried powders.

12.7.1   Drying Times in Spray Drying

Drying rates in the constant rate period are generally heat transfer controlled and the heat balance is given in Equation (12.73). If ρ_L is the density of the liquid being dried, r is the radius of the droplet, and X_0 is the initial moisture content (dry basis, kg water/kg dry matter);

$$M_s = \frac{4\pi r^3 (\rho_L)}{3(1 + X_0)}$$

and Equation (12.73) becomes:

$$\frac{dX}{dt} \cdot \frac{4\pi r^3 (\rho_L)(h_{fg})}{3(1 + X_0)} = h(\pi r^2)(T_a - T_s)$$

The constant drying rate in Equation (12.75) can be determined if the heat transfer coefficient h and the radius of the liquid droplets are known. T_s during the constant drying stage would be the wet bulb temperature of the drying air. Integrating Equation (12.80), the constant rate drying time t_c is

$$\frac{dX}{dt} = \frac{3(1 + X_0)(h)(T_a - T_s)}{4r\rho_L h_{fg}} \tag{12.80}$$

$$t_c = \frac{4(X_0 - X_c)(r)(\rho_L)h_{fg}}{3(1 + X_0)(h)(T_a - T_s)} \tag{12.81}$$

For water vaporizing from a very small spherical particle in slow moving air, the following relationship has been derived for the limiting case of very small Reynolds number in Froessling's boundary later equations for a blunt-nosed solid of revolution:

$$\frac{hr}{k_f} = 1.0; \quad h = \frac{k_f}{r} \tag{12.82}$$

k_f in Equation (12.82) is the thermal conductivity of the film envelope around the particle. In spray drying, k_f may be assumed to be the thermal conductivity of saturated air at the wet bulb temperature. Substituting Equation (12.82) in Equation (12.81):

$$t_c = \frac{4(X_0 - X_c)(r^2)(\rho_L)h_{fg}}{3k_f(1 + X_0)(T_a - T_s)} \tag{12.83}$$

t_c represents the critical time in spray drying that must be allowed in a particles' trajectory before it impinges on a solid surface in the drier. The drying time in the falling rate period derived by Ranz and Marshal is

$$t_f = \frac{h_{fg}(\rho_L)(r_c^2)(X_c - X)}{3k_f(\overline{\Delta T})} \tag{12.84}$$

where r_c is the radius of the dried particle and $\overline{\Delta T}$ is the mean temperature between the drying air and the surface of the particle during the falling rate period. $\overline{\Delta T}$ may be considered as a log mean between the wet bulb depression and the difference between the exit air and product temperatures. ρ or liquids that contain a high concentration of suspended solids or crystallizable solutes, there is very little change in the droplet diameter during spray drying. Both r and r_c in Equations (12.83) and (12.84), therefore, may be approximated to be the diameter of the liquid droplet leaving the atomizer.

For centrifugal atomizers, the diameter of the droplets as a function of peripheral speed of the atomizer is shown in Fig. 12.8. For pneumatic atomizers, a graph of drop diameter as a function of atomizing air pressure at different liquid flow rates is shown in Fig. 12.9.

Example 12.14. Calculate the drying time for a liquid atomized in a centrifugal atomizer at a feed rate of 15 lb/min (6.8 kg/min) at a peripheral speed of 200 ft/s. Base the drying time for a particle size representing a diameter larger than that of 90% of the total droplets produced. Assume there is no change in droplet diameter with drying. The liquid originally has a density of 61 lb/ft^3 (993 kg/m^3) and a moisture content of 8% (wet basis) using air at 347°F (175°C) and a humidity of 0.001 H_2O/dry air. The critical moisture content is 2.00 g H_2O/g dry matter. The dried solids have a density of 0.3 g/cm^3. Exit air temperature is 220°F (104.4°C). Product exit temperature is 130°F (54.4°C).

Solution:

From Fig. 12.8, the correction factor for a 15 lb/min feed rate is 0.9. At a peripheral speed of 200 ft/s, the diameter corresponding to 90% cumulative distribution is 310 microns. Using the correction

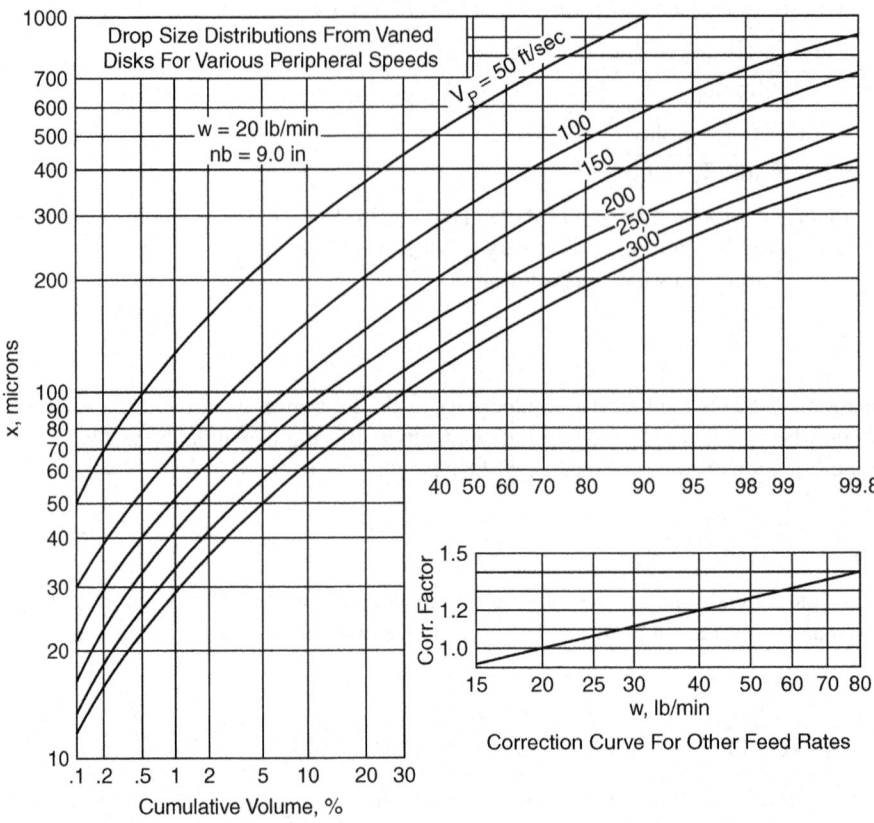

Figure 12.8 Droplet size as a function of peripheral speed of a centrifugal atomizer. (From Marshall, W. R. Jr., 1954. Chem. Eng. Prog. Monogr. Ser. 50(2):71. AIChE, New York. Used with permission.)

factor, the diameter is 279 microns. Using Equations (12.48) and (12.49):

$$X_0 = \frac{89}{11} = 8.09 \text{ kg } H_2O/\text{kg dry matter}$$

$$X_c = 2.00$$

$$X = \frac{0.08}{0.92} = 0.087 \text{ kg } H_2O/\text{kg dry matter}$$

From a psychrometric chart, $T_s = T_{wb} = 109°\,F\,(43°C)$. h_{fg} at 109°F is 1031.4 BTU/lb or 2.3999 MJ/kg. The thermal conductivity of the gas film envelope around the particle can be calculated. It is the thermal conductivity of saturated air at 109°F (43°C). The humidity is 0.056 kg H_2O/kg dry matter. The thermal conductivity of dry air is 0.0318 W/m · K and that of water vapor is 0.0235 W/m · K.

$$k_f = 0.318\frac{1}{1.056} + 0.0235\frac{0.056}{1.056} = 0.0314 \text{ W/m · K}$$

$$r = \frac{279}{2} \times 10^{-6} \text{ m} = 193.5 \times 10^{-6} \text{ m}$$

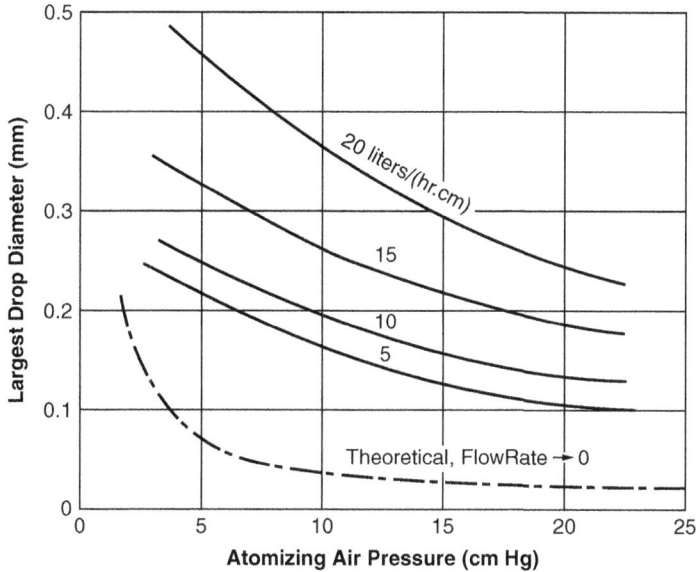

Figure 12.9 Droplet size as a function of pressure in a pneumatic atomizer. (From Marshall, W. R. Jr., 1954. Chem. Eng. Prog. Monogr. Ser. 50(20):79. AIChE, New York. Used with permission.)

Equation (12.81) in SI units:

$$t_c = \frac{4(8.09 - 2)(139.5 \times 10^{-6})^2(993)(2.399 \times 10^6)}{3(0.0314)(1 + 8.09)(175 - 43)} = 10 \, s$$

For Equation (12.84), the $\Delta \overline{T}$ is

$$\overline{\Delta T} = \frac{(175 - 43) - (104.4 - 54.4)}{\ln 132/50} = 84.5 \, K$$

$$t_f = \frac{2.399 \times 10^6(0.3 \times 1000)(193.5 \times 10^{-6})^2(2.0 - 0.087)}{3(0.0314)(84.5)} = 6.5 \, s$$

12.8 FREEZE DRYING

Dehydration carried out at low absolute pressures will allow the vaporization of water from the solid phase. Figure 12.10 shows the vapor pressure of water over ice at various temperatures below the freezing point of water. To carry out freeze drying successfully, the absolute pressure in the drying chamber must be maintained at an absolute pressure of at least 620 Pa.

Figure 12.11 is a schematic diagram of a freeze drier. The absolute pressure inside the drying chamber is determined by the temperature at which the vapor trap is maintained. This pressure corresponds to the vapor pressure over ice at the vapor trap temperature. The vacuum pump is designed primarily to exhaust the vacuum chamber at the start of the operation and to remove noncondensing gases and whatever air leaked into the system. The volume of vaporized water at the low absolute pressures in

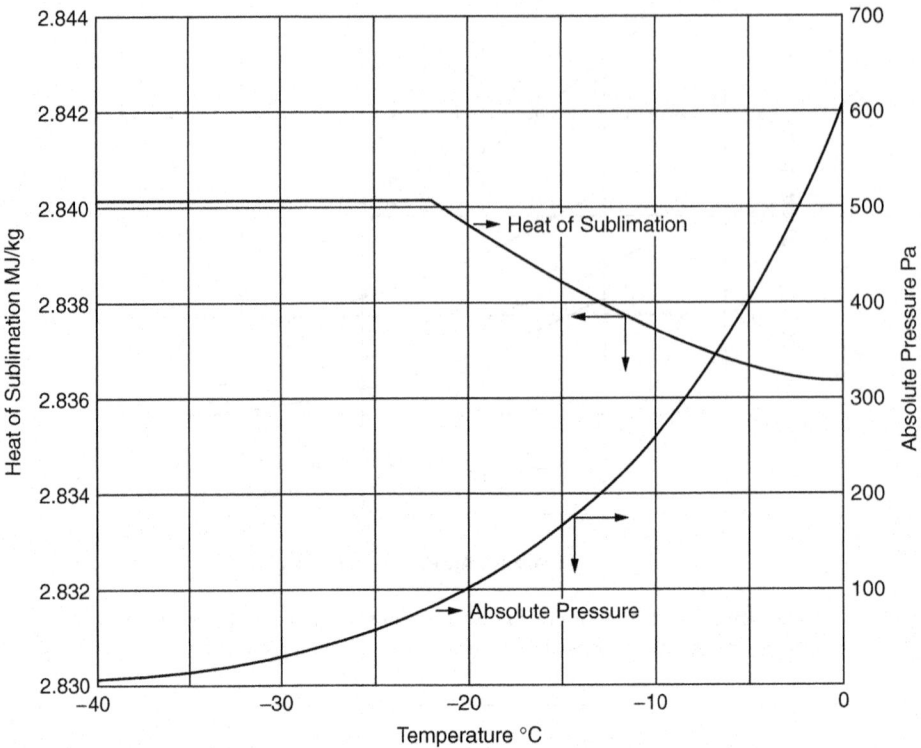

Figure 12.10 Heat of sublimation (ΔH_s) and vapor pressure of water above ice (P_i). (Based on data from Charm, S. E. 1971. Fundamentals of Food Engineering. 2nd ed. AVI Publishing Co. Westport, CT.)

freeze drying is very large; therefore, removal of the vapor by the vacuum pump alone would require a very large pump. Condensing the vaporized water in the form of ice in the vapor trap is an efficient means of reducing the volume of gases to be removed from the system by the vacuum pump.

Heat must also be supplied to the material being dried to provide the energy of vaporization. This is accomplished by the use of hollow shelves through which a heated liquid is circulated. The temperature of the shelves can be regulated by regulating either the temperature or the amount of heat transfer medium supplied to the shelves. The material to be dried rests on the top of the heated shelves. Heat transfer occurs by conduction from the heated shelves, by convection from the air inside the drying chamber to the exposed surfaces and by radiation.

12.8.1 Drying Times for Symmetrical Drying

Analysis of freeze drying is different from that in conventional drying in that drying proceeds from the exposed surfaces toward the interior. The outer layers are completely dry as the ice core recedes. Vaporization of water occurs at the surface of the ice core. Heat of sublimation is conducted to the surface of the ice core through the dried outer layer. Vaporized water diffuses through the pores of the dried outer layer before it leaves the solid and goes to the atmosphere in the drying chamber.

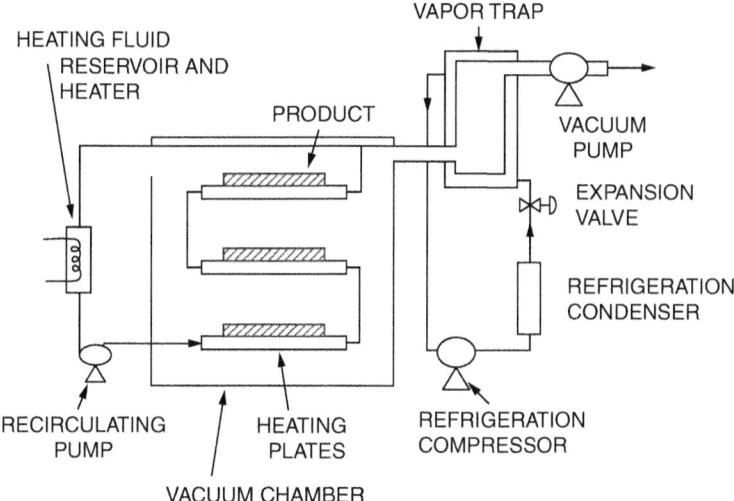

Figure 12.11   Schematic diagram of a freeze drier showing a vapor trap external to the vacuum chamber.

Symmetrical drying occurs when the rate at which the ice core recedes is equal at both top and bottom of the material. To simplify the calculations, unidirectional heat transfer is assumed.

Let W = kg water/m^3 of wet material. If ρ is the density of the wet material and X_0 is the initial moisture content on a dry basis, kg water/kg dry matter, W = $(\rho) X_0/(1 + X_0)$.

If drying is symmetrical, the mass of water evaporated, M_c, expressed in terms of a dried layer ΔL, is

$$M_c = WA\,(\Delta L)\,(2)$$

Let X' = fraction of water evaporated = M_c/total water.

$$X' = \frac{W(A)(\Delta L)(2)}{W(A)(L)} = \frac{\Delta L(2)}{L}$$

X' is also $\dfrac{X_0 - X}{X_0}$

where L = the thickness of the solid. X = moisture content, kg H$_2$O/kg dry matter.

$$\Delta L = \frac{L'X}{2}$$

The overall heat transfer coefficient calculated from the sum of the resistance to heat transfer of the dried layer and the heat transfer coefficient at the surface is

$$\frac{1}{U} = \frac{1}{h} + \frac{\Delta L}{k} = \frac{1}{h} + \frac{LX'}{2k} \tag{12.85}$$

$$U = \frac{k}{k/h + LX'/2} \tag{12.86}$$

If drying is symmetrical, heat is transferred from both sides. Let T_s = temperature of the shelf and T_f is the temperature of the frozen core surface. The heat transferred to the ice core will be

$$q = UA\Delta T = \frac{k}{k/h + LX'/2}(2A)(T_a - T_f) \tag{12.87}$$

For simultaneous heat and mass transfer, heat transferred = heat of vaporization. ΔH_s is the heat of sublimation of ice at T_f.

$$\Delta H_s(W)(A)(L)\frac{d'X}{dt} = \frac{2A(T_a - T_f)(k)}{k/h + LX'/2} \tag{12.88}$$

Simplifying and integrating Equation (12.88):

$$t = \frac{\Delta H_s(W)(L)}{2k(T_s - T_f)}\left[\frac{k'X}{h} + \frac{LX'^2}{4}\right] \tag{12.89}$$

W in Equation (12.89) can be expressed in terms of either the density of the wet material, ρ, or the density of the dried material ρ_s.

$$W = X_0\rho_s \quad \text{or} \quad W = (\rho)\frac{(X_0)}{1 + X_0}$$

$$t = \frac{\Delta H_s(X_0)(\rho_s)(L)}{2k(T_s - T_f)}\left[\frac{k'X}{h} + \frac{LX'^2}{4}\right] \tag{12.90}$$

Or:

$$t = \frac{\Delta H_s(\rho)(X_0)(L)}{2k(1 + X_0)(T_s - T_f)}\left[\frac{k'X}{h} + \frac{LX'^2}{4}\right] \tag{12.91}$$

Either Equation (12.90) or (12.91) can be used to calculate the time of drying depending on whether ρ or ρ_s is known.

Example 12.15. The density of a sample of beef is 60 lb/ft^3 (965 kg/m^3). How long will it take to dry a 1 in. (2.54 cm) thick strip of this sample from an initial moisture of 75% to a final moisture content of 4% (wb). Freeze drying is carried out at an absolute pressure of 500 microns of mercury. Assume that the shelf temperature and the air in the drying chamber are equal at 80°F (26.7°C). Assume symmetrical drying. The thermal conductivity of the dried meat is 0.0692 W/m · K. Estimate the heat transfer coefficient by assuming that a 3 mm thick of vapor (k of water = 0.0235 W/m · K) envelopes the surfaces where drying occurs and that the heat transfer coefficient is of equivalent resistance to the resistance of this vapor film.

$$h = \frac{k}{x} = \frac{0.0235}{0.003} = 7.833 \text{ W/m}^2\text{K}$$

The absolute pressure in the chamber is

$$P = 500 \times 10^{-6}\text{ m Hg}\frac{133.3 \times 10^3\text{ Pa}}{\text{m Hg}} = 66.65 \text{ Pa}$$

From Fig. 12.10, the temperature of ice in equilibrium with a pressure of 66.65 Pa is −24.5°C. The heat of sublimation H_s = 2.8403 MJ/kg. The final moisture content on a dry basis is $X = 0.04/0.96 = 0.0417$ kg water/kg dry matter. The initial moisture content on a dry basis is

$X_0 = -0.75/0.25 = 3.00$. The fraction water remaining at the completion of the drying period is $X' = X_0 - X/X_0 = 3 - 0.0417/3.0 = 0.986$.

Substituting in Equation (12.91):

$$t = \left[\frac{2.8403 \times 10^6 (965)(3.00)(0.0254)}{2(0.0692)(1+3)(26.7+24.5)} \right]$$

$$\left[\frac{0.0692(0.986)}{7.833} + \frac{0.0254(0.986)^2}{4} \right] = 109,673\,\text{s} = 30.46\,\text{h}$$

12.9 VACUUM BELT DRYER

Principle of operation: A vacuum belt dryer is used to dry heat-sensitive pasty materials and concentrated solutions. The system (Fig. 12.12) consists of a long cylinder with a continuous conveyor belt traversing the diameter the whole length of the cylinder. Product is deposited on the belt at one end of the unit and water is evaporated as the belt moves forward until the dried product falls off the belt into a receiver tank at the opposite end.

Energy transfer: Energy to vaporize water is applied to the product on the belt by conduction from the bottom and radiation from the top. A heated plate under the belt heats the product by conduction. Above the conveyor is another heated plate that transfers energy to the paste by radiation. The temperatures of the top and bottom plates are independently controlled and are set to the maximum temperature that would permit complete drying of the product when the belt makes a turn into the opposite direction. Temperatures must be optimized to avoid heat-induced degradation reactions.

Vacuum: Initial evacuation of air from the unit must be done rapidly, and a low absolute pressure must be maintained so that water vapor removal from the product is heat transfer rather than mass transfer controlled. Water vapor and noncondensible gases entering the cylinder must be evacuated using a liquid ring seal mechanical vacuum pump or a steam jet ejector. Typically, an absolute pressure of not more than 50 mm Hg is maintained inside the cylinder. Inadequate vacuum pump capacity will be manifested by condensation of water vapor on the cylinder walls. To facilitate evacuation of water vapor, dry air from the outside is continuously bled into the unit at a slow rate.

Continuous feeding of wet product: The paste to be dried is pumped into the unit and is discharged on the surface of the conveyor belt through a nozzle at the tip of a wand that slowly oscillates from one side of the conveyor belt to the other. Typically, the rate of addition of the paste and the speed of side to side movement of the wand is such that a thin uniform layer of paste is deposited on the belt. As the paste on the belt receives energy from the heated plates, vapor is generated because of the elevated temperature and low absolute pressure. Product on the belt will foam. If the foam forms a thick layer, a porous layer of dried solid forms at the top and bottom hindering heat transfer into the wet middle layer. Ideally, the thickness of the foam should be such that drying will continue throughout the whole thickness all the way to the end of the belt. Spattering of wet material on the belt must also be avoided because the spatter will be deposited on the surface of the plate above the belt and interfere with radiant heat transfer. The combination, vacuum, upper and lower heated plate temperature, and thickness of the deposited layer of wet product on the belt must be carefully regulated to dry successfully in this system.

Operating the drier: Drying time can be controlled by slowing down the speed of the belt. However, to operate at maximum drying capacity, heated plate temperatures must be elevated and too long a residence time of the wet product in the heated zone could result in scorching of the product. Conditions

a- belt; b – top heater plate; c – swivel liquid feed port; d – dried product receiver

Figure 12.12   Picture of a pilot plant size vacuum belt dryer.

must be optimized for each product dried. The paste density, foaming tendency, thermophysical properties and radiant energy absorption all play a role in the rate of dehydration.

PROBLEMS

12.1. Pork has a_w of 1.00 at a moisture content of 50% (wet basis) and higher. If pork is infused with sucrose and NaCl and dehydrated such that at the end of the dehydration process the moisture content is 60% (wet basis) and the concentration of sugar and NaCl are 10% and 3%, respectively, calculate the a_w of the cured product.

12.2. What concentration of NaCl in water would give the same water activity as a 20% solution of sucrose?

12.3. The following data were obtained on the dehydration of a food product: initial moisture content = 89.7% (wet basis).

Drying time (min)	Net weight (kg)
0	24.0
10	17.4
20	12.9
30	9.7
40	7.8
50	6.2
60	5.2
70	4.5
80	3.9
90	3.5

Draw the drying curve for this material and construct a curve for the drying rate as a function of the moisture content.

(a) What is the critical moisture content for each of the falling rate zones?

(b) What is the constant drying rate?

(c) Determine the residual moisture content for each of the falling rate stages.

(d) The dehydration was conducted at an air flow rate of 50 m/s at a dry bulb temperature of 82°C and a wet bulb temperature of 43°C. The wet material has a density of 947 kg/m³ and were dried in a layer 2.5 cm thick. If the same conditions were used but the initial moisture content was 91% (wet basis) and a thicker layer of material (3.5 cm) were used on the drying trays, how long will it take to dry this material to a final moisture content of 12% (wb)?

12.4. A continuous countercurrent drier is to be designed to dry 500 kg/h of food product from 60% (wet basis) moisture to 10% (wet basis) moisture. The equilibrium moisture content for the material is 5% (wet basis) and the critical moisture content is 30% (wet basis). The drying curve of the material in preliminary drying studies showed only one falling rate zone. Air at 66°C dry bulb and 30°C wet bulb will be used for drying. The exit air relative humidity is 40%. Assume adiabatic humidification of the air. The drying air is drawn from room temperature at 18°C and 50% RH. The wet material has a density of 920 kg/m³. The drying tunnel should use trucks that hold a stack of 14 trays, each 122 cm wide, 76 cm deep along the length of the tunnel, and 5 cm thick. The distance between trays on the stack is 10 cm. The drying tunnel has a cross-sectional area of 2.93 m². The material in the trays will be loaded at a depth of 12.7 mm. Calculate:

(a) The number of trays of product through the tunnel/h.

(b) The rate of travel by the trucks through the tunnel. Assume distance between trucks is 30 cm.

(c) The constant drying rate and the total time for drying.

(d) The length of the tunnel.

(e) If air recycling is used, the fraction of the inlet air to the drier that must come from recycled air.

(f) The capacity of the heater required for the operation with recycling.

12.5. A laboratory drier is operated with a wet bulb temperature of 115°F and a dry bulb temperature of 160°F. The air leaving the drier is at 145°F dry bulb. Assume adiabatic operation. Part of the discharge air is recycled. Ambient air at 70°F and 60% RH is heated and mixed with the recycled hot air. Calculate the proportion of fresh air and recycled hot air that must be mixed to achieve the desired inlet dry and wet bulb temperatures.

12.6. If it takes 8 hours to dry a material in a freeze drier from 80% H_2O to 10% H_2O (wet basis) at an absolute pressure of 100 Φm and a temperature of 110°F (43.3°C), how long will it take to dry this material from 80% to 40% water if the dehydration is carried out at 500 Φm and 80°F (26.7°C). The material is 25 mm thick, has a density of 950 kg/m³, and the thermal conductivity of the dried material is 0.35 W/m · K. Thermal conductivity and heat transfer coefficients are independent of plate temperature and vacuum.

12.7. Calculate the constant rate of drying in a countercurrent continuous belt dehydrator that processes 200 lb/h (90.8 kg/h) of wet material containing 80% water to 30% water. Air at 80° EF (26.7°C) and 80% RH is heated to 180°F (82.2°C) in an electric heater, enters the drier and leaves at 10% RH. The critical moisture content of the material is 28%. The drier is 4 ft (1.21 m) wide, the belt loaded to a depth of 2 in. (5.08 cm) of material, and the clearance from the top of the drier to the top of the material on the belt is 10 in. (25.4 cm). The density of the dry solids in the material is 12 lb/ft^3 (193 kg/m ^{3}).

12.8. In a spray drying experiment, a sample containing 2.15% solids and 97.8% water was fed at the rate of 6.9 lb per hour (3.126 kg/h) and this sample was dried at 392°F (200°C) inlet air temperature. The exit air temperature was 200°F (93.3°C). The dried product was 94.5% solids and the outside air was at 79°F (26.1°C) and 20% RH.

Calculate:

(a) The weight water evaporated per hour.

(b) The % RH of the exit air.

(c) The mass flow rate of air through the drier in weight dry air/h.

(d) In this same drier, if the inlet air temperature is changed to 440°F (226.7°C) and the % RH of the exit air were kept the same as in (b), weight of a sample containing 5% solids and 98% water can be dried to 2% water in 1 hour? (Air flow rate is the same as before.) What would be the exit temperature of the air from the dried under the conditions? Assume adiabatic drying.

12.9. A dehydrator when operated in the winter where the outside air was 10°F (-12.2°C) and 100% RH (H = 0.001) can dry 100 lb (45.5 kg) of fruit per hour from 90% water to 10% water. The inlet temperature of the air to the drier is 150°F (65.6°C) and leaves at 100°F (37.8°C). In the summer when the outside air is at 90°F (32.2°C) and 80% RH, determine the moisture content of the product leaving the drier if the operator maintains the same rate of 100 lb (45.4 kg) of wet fruit/hr and the exit air from the drier has the same % RH as it was in the winter.

12.10. The desorption isotherm of water in carrots at 70° EC is reported to fit Iglesias and Chirife's equation (Eq. 12.36) with the constants $B_1 = 3.2841$ and $B_2 = 1.3923$.

(a) Determine the moisture contents where a shift in drying rate may be expected in the dehydration of carrots.

(b) The following data represents the equilibrium water activity (a_w) for carrots at various moisture contents in kg water/kg dry matter (X): (a_w, X); (0.02, 0.0045), (0.04, 0.009), (0.06, 0.0125), (0.08, 0.016), (0.10, 0.019), (0.12, 0.0225), (0.14, 0.025), (0.016, 0.028), (0.18, 0.031), (0.20, 0.034). Fit this data to the BET isotherm and determine the moisture content for a unimolecular layer, X_m

(c) Fit the data to the GAB equation and determine the constants.

12.11 The diffusivity of water in scalded potatoes at 69° EC and 80% moisture (wet basis) has been determined to be 0.22×10^{-5} m^2/h. If 1 cm potato cubes are dried using air at 1.5 m/s velocity and 1% relative humidity, calculate the dry bulb temperature of the air that can be used such that the diffusion rate from the interior to the surface will be equal to the surface dehydration rate. Assume the air flows parallel to the cubes and that dehydration proceeds from all faces of each cube. The density of the potato cube is 1002 kg/m^3 at 80% moisture.

12.12 Puffing can be induced during dehydration of diced carrots if the dehydration rate at the constant rate period is of the order 1 kg water/(min kg DM). In a fluidized bed drier where the air contacts individual particles at a velocity of 12 m/s, calculate the minimum dry bulb temperature of the drying air that would induce this rate of drying at the constant rate period. Assume drying air has a humidity of 0.001 kg water/kg dry air and surface temperature under these conditions is 5° EC higher than the wet bulb temperature.

Calculate the mass transfer rate under these conditions. Is dehydration rate heat or mass transfer controlled?

SUGGESTED READING

Barbosa-Canovas, G. V. and Vega-Mercado, H. 1996. Dehydration of Foods. Chapman and Hall, New York.

Charm, S. E. 1971. Fundamentals of Food Engineering. 2nd ed. AVI Publishing Co., Westport, CT.

Foust, A. S., Wenzel, L. A., Clump, C. W., Maus, L., and Andersen, L. B. 1960. Principles of Unit Operations. John Wiley & Sons, New York.

Goldblith, S. A., Rey, L., and Rothmayr, W. W. 1975. Freeze Drying and Advanced Food Technology. Academic Press, New York.

Green, D. W. and Mahoney, J. O., Eds. 1997. Perry's Chemical Engineers Handbook. 7th ed. McGraw-Hill Book Co., New York.

Greensmith, M. 1998. Practical Dehydration. CRC Press, Boca Raton, FL.

Hartman, T. M. 1989. Waer and Food Quality. Elsevier, New York.

Heldman, D. R. 1975. Food Process Engineering. AVI Publishing Co., Westport, CT.

Hildebrand, J. and Scott, R. L. 1962. Regular Solutions. Prentice-Hall, Englewood Cliffs, NJ.

Iglesias, H. A., and Chirife, J. 1982. Handbook of Food Isotherms. Academic Press, New York.

Leniger, H. A. and Beerloo, W. A. 1975. Food Process Engineering. D. Riedel Publishing Co., Boston.

Marshall, W. R., Jr. 1954. Atomization and spray drying. Am. Inst. Chem. Eng., Prog. Mon. Ser. 50(2).

McCabe, W. L. and Smith, J. C. 1967. Unit Operations of Chemical Engineering. 2nd ed. McGraw-Hill Book Co., New York.

McCabe, W. L., Smith, J. C., and Harriott, P. 1985. Unit Operations in Chemical Engineering. 4th ed. McGraw Hill, New York.

Mujumdar, A., Ed. 2004. Dehydration of Products of Biological Origin. Science Publishers, Enfield, NH.

Norrish, R. S. 1966. An equation for the activity coefficient and equilibrium relative humidity of water in confectionery syrups. J. Food Technol. 1:25–39.

Perry, R. H., Chilton, C. H., and Kirkpatrick, S. D. 1963. Chemical Engineers Handbook. 4th ed. McGraw-Hill Book Co., New York.

Rockland, L. B. and Stewart, G. F. Eds. 1981. Water Activity: Influences on Food Quality. Academic Press, New York.

Ross, K. D. 1975. Estimation of a_w in intermediate moisture foods. Food Technol. 29(3):26–34.

Sandal, O. C., King, C. J., and Wilke, C. R. 1967. The relation-ship between transport properties and rates of freeze drying of poultry meat. AICHE J. 13(3):428–438.

Watson, E. L. and Harper, J. C. 1989. Elements of Food Engineering, 2nd ed. Van Nostrand Reinhold, New York.

CHAPTER 13

Physical Separation Processes

Food technology has evolved from the practice of preserving products in very much the same form as they occur in nature to one where desirable components are separated and converted to other forms. Separation processes have been in use in the food industry for years, but sophistication in its use is a fairly recent occurrence. Current technology makes it possible to remove haze from wine and fruit juices or nectars, separate the proteins of cheese whey into fractions having different functional properties, separate foreign matter from whole or milled grains, and concentrate fruit juices without having to employ heat. Efficient separation processes have been instrumental in making economically viable the recovery of useful components from food processing wastes.

13.1 FILTRATION

Filtration is the process of passing a fluid containing suspended particles through a porous medium. The medium traps the suspended solids producing a clarified filtrate. Filtration is employed when the valuable component of the mixture is the filtrate. Examples are clarification of fruit juices and vegetable oil. If the suspended material is the valuable component (e.g., recovery of precipitated proteins from an extracting solution), and rapid removal of the suspending liquid cannot be carried out without the addition of a filter aid, other separation techniques must be used.

Surface filtration is a process where the filtrate passes across the thickness of a porous sheet while the suspended solids are retained on the surface of the sheet. A sheet with large pores has low resistance to flow therefore filtrate flow is rapid, however, small particles may pass through resulting in a cloudy filtrate. Surface filtration allows no cake accumulation. Flow stops when solids cover the pores. If the solids do not adhere to the filter surface, the filter may be regenerated by backwashing the surface. Filtration sterilization of beer using microporous filters is a form of surface filtration.

Depth filtration is a process where the filter medium is thick, and solids penetrate the depth of the filter. Eventually, solids block the pores and stop filtrate flow, or solids may break through the filter and contaminate the filtrate. Once filtrate flow stops or slows down considerably, the filter must be replaced. In depth filtration, particle retention may occur by electrostatic attraction in addition to the sieving effect. Thus, particles smaller than the pore size may be retained. Depth filters capable of electrostatic solids retention will be ideal for rapid filtration. Cartridge, fiber, and sand filters are forms of depth filtration. Depth filtration is not very effective when suspended solids concentration is very high.

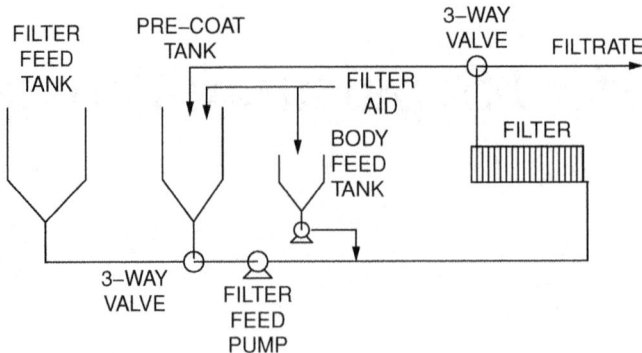

Figure 13.1   Diagram of filter system for filter aid filtration.

Filter aid filtration involves the use of agents that form a porous cake with the suspended solids. Filter aids used commercially are diatomaceous earth and perlite. Diatomaceous earth consists of skeletal remains of diatoms and is very porous. Perlite is milled and classified perlite rock, an expanded crystalline silicate. In filter aid filtration, the filter medium is a thin layer of cloth or wire screen that has little capacity for retention of the suspended solids and serves only to retain the filter aid. A layer of filter aid (0.5 to 1 kg/m^2 of filter surface) is precoated over the filter medium before the start of filtration. Filter aid, referred to as "body feed," is continuously added to the suspension during filtration. As filtration proceeds, the filter aid and suspended solids are deposited as a filter cake, which increases in thickness with increasing filtrate volume. Body feed filter aid concentration is usually in the range of one to two times the suspended solids concentration. A body feed concentration must be used that will produce a cake with adequate porosity for filtrate flow. Inadequate body feed concentration of filter aid will result in rapid decrease in filtrate flow and consequently shorten filtration cycles. Figure 13.1 is a diagram of a filtration system for filter aid filtration. The system is designed to make it easy to change from precoat operation to body feed filtration operation with the switch of a 3-way valve. Some filters such as the vertical leaf filter can be backflushed to remove the filter cake. Some, like the plate and frame filter, are designed to be easily disassembled to remove the filter cake. The body feed may be a slurry of the filter aid that is metered into the filter feed, or the dry filter aid may be added using a proportional solids feeder into the filter feed tank. In continuous vacuum filters, precoating is done only once at the start of the operation. Body feed is added directly to the vacuum filter pan. Figure 13.2 is a diagram of the most common filters used for filter aid filtration. Figure 13.2A is a vertical leaf filter, Fig. 13.2B is a plate and frame filter, and Fig. 13.2C is a continuous vacuum rotary drum filter.

13.1.1   Filtrate Flow Through Filter Cake

In filter aid filtration, filtrate flow through the pores in the cake is dependent on the pressure differential across the cake and the resistance to flow. The total resistance increases with increasing cake thickness, thus filtrate flow decreases with time of filtration. The resistance to filtrate flow across the filter cake is expressed as the specific cake resistance.

Figure 13.3 shows a section of a filter showing the filter medium, precoat, and filter aid. The total pressure drop across the filter is the sum of the pressure drop across the filter medium (the filter cloth and precoat), ΔP_m, and that across the cake, ΔP_c.

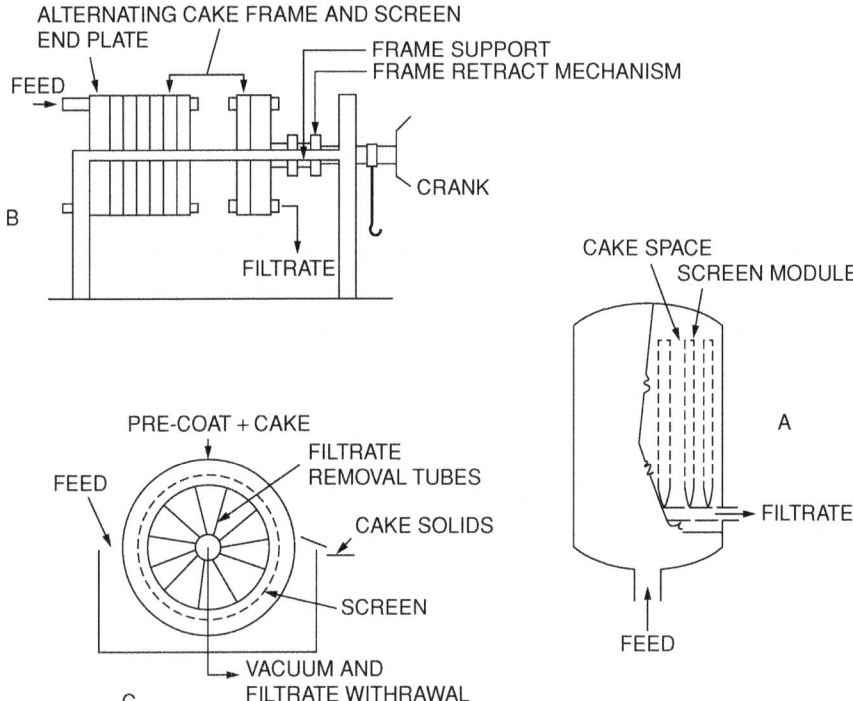

Figure 13.2 Diagram of common filters for filter aid filtration. (A) vertical leaf filter; (B) plate and frame filter; (C) rotary vacuum drum filter.

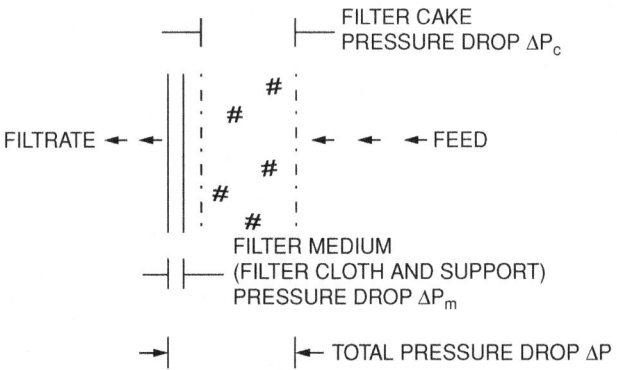

Figure 13.3 Diagram of a filter section showing filter cloth, precoat, and filter cake.

Let v = velocity of filtrate flow, μ = filtrate viscosity, A = filter area, and m = mass of filter cake. The medium resistance, R_m, is

$$R_m = \frac{\Delta P_m}{\mu v} \tag{13.1}$$

The specific cake resistance, α, is

$$\alpha = \frac{\Delta P_c}{\mu v (m/A)} \tag{13.2}$$

The units of R_m and α are m^{-1} and m/kg, respectively, in SI.

Let: ΔP = pressure differential across the filter

$$\Delta P = \Delta P_m + \Delta P_c \tag{13.3}$$

Substituting Equations (13.1) and (13.2) in Equation (13.3):

$$\Delta P = \alpha \mu v \frac{m}{A} + R_m \mu v \tag{13.4}$$

Let: V = volume of filtrate and c = concentration of cake solids in the suspension to be filtered. m = Vc. The filtrate velocity, v = (1/A) dV/dt.

Substituting for m and v in Equation (13.4):

$$\Delta P = \frac{\mu (dV/dt)}{A} \left[\frac{\alpha V c}{A} + R_m \right] \tag{13.5}$$

$$\frac{dt}{dV} = \frac{\mu}{A \cdot \Delta P} \left[\frac{\alpha V c}{A} + R_m \right] \tag{13.6}$$

Equation (13.6) is the Sperry equation, the most widely used model for filtrate flow through filter cakes. Equation (13.6) can be used to determine the specific cake resistance from filtration data. Filtration time, t, is plotted against filtrate volume, V, tangents to the curve are drawn at several values of V, and the slopes of the tangents, dt/dV, are determined. A plot of dt/dV versus V will have a slope equal to $\alpha c \Phi (A^2 \cdot \Delta P)$ and an intercept on the ordinate at V = 0 equal to $R_m \Phi (A \cdot \Delta P)$. An easier method for determining α and R_m will be shown in next section.

13.1.2   Constant Pressure Filtration

When a centrifugal pump is used as the filter feed pump, the pressure differential across the filter, ΔP, is constant, and Equation (13.6) can be integrated to give:

$$t = \frac{\mu}{\Delta P} \left[\frac{\alpha c}{2} \frac{V^2}{A^2} + R_m \frac{V}{A} \right] \tag{13.7}$$

Dividing Equation (13.7) through by V:

$$\frac{t}{V} = \frac{\mu \, \alpha c}{2 A^2 \Delta P} V + \frac{\mu R_m}{A \Delta P} \tag{13.8}$$

Equation (13.8) shows that a plot of t/V against V will be linear, and the values of α and R_m can be determined from the slope and intercept. A common problem with the use of either Equation (13.6) or Equation (13.8) is that negative values for R_m may be obtained. This may occur when R_m is much smaller than α; if finely suspended material is present that rapidly reduces medium porosity even with very small actual amount of cake solids deposited; or when α increases with time of filtration such that the least squares method of curve fitting weighs heavily the data during the later stages of filtration relative to those at the early stages. To avoid having negative values for R_m, Equation (13.7) may be used in the analysis. If filtration is carried out using only filtrate containing no suspended solids, $c = 0$ and equation 7 becomes:

$$t = \frac{\mu R_m}{\Delta PA} V \qquad (13.9)$$

Because $R_m = \Delta P_m/(\Phi \cdot v)$; and since during filtration with only the filter cloth and precoat, the pressure differential is ΔP_m, and because $A \cdot v =$ the filtrate volumetric rate of flow, q, the coefficient of V in Equation (13.9) is $1/q$. Thus, Equation (13.7) can be expressed as:

$$t = \frac{\mu \alpha c}{2 \Delta PA^2} V^2 + \left(\frac{1}{q}\right) V \qquad (13.10)$$

q is evaluated separately as the volumetric rate of filtrate flow on the precoated filter at the pressure differential used in filtration with body feed. Because ΔP across the precoated filter is primarily due to the resistance of the deposited filter aid, q is primarily a function of the type of filter aid, the pressure applied, and the thickness of the cake. For the same filter aid and filtrate, q is proportional to the thickness and the applied pressure, therefore the dependence can be quickly established. If q is known, α can be easily determined from the slope determined by a regression of the function $(t - V/q)$ against V^2 or from the intercept of a log-log plot of $(t - V/q)$ against V.

The use of filtration model equations permits determination of the filtration constants R_m and α on a small filter which can then be used to scale up to larger filtrations. An example of a laboratory filtration module is shown in Fig. 13.4. This is a batch filter with a tank volume of 7.5 L and a filter

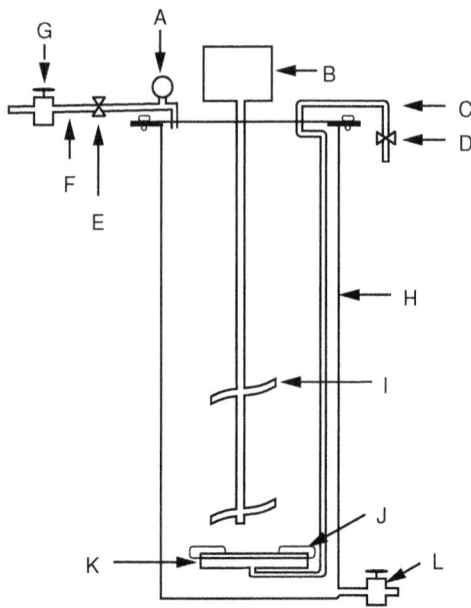

Figure 13.4 Diagram of a laboratory test filtration cell.

Table 13.1   Data for Filtration Clarification of Apple Juice.

Time t(s)	Filtrate Volume $V(m^3) \times 10^5$	$t/V \times 10^{-6}$
60	3.40	1.7647
120	5.80	2.0689
180	7.80	2.3077
240	9.60	2.5000
300	11.4	2.6316
360	13.1	2.7586
420	14.7	2.8669
480	16.1	2.9721
540	17.6	3.0769
600	19.0	3.1496
660	20.6	3.2117
720	22.0	3.2727
780	23.3	3.3476

area of 20 cm^2. An opening at the bottom allows draining of the precoat suspension at the beginning of a filtration cycle, and charging of the tank after precoating. To precoat, a suspension of filter aid is made at such a concentration that filtration of 1000 mL will give the precoating level of 1.0 kg/m^2. For example, 2.0 g dry filter aid in 1000 mL of water will be equivalent to a precoat of 1.0 kg/m^2 on a filter area of 20 cm^2 if 1000 mL of filtrate is allowed to pass through the filter from this suspension. Depending upon the fineness of the filter aid used, filter paper, or cloth may be used as the filter medium.

Example 13.1. Data in Table 13.1 were collected in filtration clarification of apple juice. The apple juice was squeezed from macerated apples, treated with pectinase, and after settling and siphoning of the clarified layer, the cloudy juice that remained in the tank was filter clarified using perlite filter aid. The cloudy juice contained 1.19 g solids/100 mL. Perlite filter aid was used which had a pure water permeability of 0.40 mL $\cdot$ s^{-1} for 1 cm^2 filter area, 1 atm pressure drop and 1 cm thick cake. Whatman no. 541 filter paper was used as the filter medium. The precoat was 0.1 g/cm^2 (1 kg/m^2). The filter shown in Fig. 13.4, which had a filtration area of 20 cm^2, was used. Flow of clear apple juice through the precoated filter was 0.6 mL/s at a pressure differential of 25 lb$_f$/in.^{2} (172.369 kPa). The filtrate had a viscosity of 1.6 cP.

Calculate α and R$_m$. Calculate the average filtration rate in m^3/(min $\cdot$ m^2 filter area) if the filtration time per cycle is 30 min and the pressure differential across the filter is 30 lb$_f$/in.^{2} (206.84 kPa). Assume the same solids and body feed concentration.

The cumulative filtrate volume with filtration time at a filter aid concentration of 2.5 g/100 mL of suspension, is shown in Table 13.1.

Solution:

Using Equation (13.8), a regression of (t/V) against V was done using the transformed data in Table 13.1.

The regression equation was:

$$(t/V) = 7.61 \times 10^9 V + 1,684,000$$

The coefficient of V is $\alpha \cdot \mu \cdot c/(2\Delta P\ A^2)$. c = concentration of filter cake solids in the feed, which is the sum of the original suspended solids, 1.19 g/100 mL or 11.9 g/L, and the body feed filter aid concentration which is 2.5 g/100 mL or 25 g/L.

$$\alpha = \frac{7.61 \times 10^9 (2)(20 \times 10^{-4})^2 (172,369)}{1.6(0.001)925 + 11.9} = 1.78 \times 10^{11} \text{m/kg}$$

The intercept is $R_m \mu/(A\Delta P)$

$$R_m = \frac{1.684 \times 10^6 (20 \times 10^{-4})(172,369)}{0.001(1.6)} = 3.63 \times 10^{11} \text{m}^{-1}$$

Similar values for α and R_m were obtained using Equation (13.10) by performing a regression on $(t - V/q)$ against V^2. q is 0.6×10^{-6} m^3/s. A slightly better fit was obtained using Equation (13.10) with $R^2 = 0.994$ compared with $R^2 = 0.964$ using Equation (13.8). The regression equation using Equation (13.10) was:

$$(t - V/q) = 7.29 \times 10^9 V^2$$

Values for R_m and α were 3.59×10^{11} and 1.7×10^{11}, respectively, in SI units. These values would be more accurate than those obtained using equation 8 because of the better R^2 value for the regression. However, the values are very close and would be considered the same for most practical purposes.

The average filtrate flow is calculated using Equation (13.7). t = 30 min = 1800 s

$$1800 = \frac{1.6(0.001)}{206,840} \left(\frac{1.78 \times 10^{11} \times (36.9)}{2} \left[\frac{V}{A} \right]^2 + 3.63 \times 10^{11} \left[\frac{V}{A} \right] \right)$$

$$32.841 \times 10^{11} (V/A)^2 + 3.63 \times 10^{11} (V/A) = 2.32698375 \times 10^{11}$$

Dividing through by 32.841×10^{11} and solving for the positive root of the quadratic equation

$$(V/A)^2 + 0.1105(V/A) - 0.070856 = 0$$

$$\frac{V}{A} = \frac{-0.1105 \pm ([0.1105]^2 + 4(1)(0.070856))^{0.5}}{2} = 0.217 \text{m}^3/\text{m}^2 \text{filter area})$$

This is the volume of filtrate which passed through a unit area of the filter for a filtration cycle of 30 min. The average rate of filtration is

$$(V/A)_{a\ vg}/t = 0.217/30 = 0.00722 \text{m}^3/(\text{min} \cdot \text{m}^2 \text{filter area})$$

The average filtration rate can be used to size a filter for a desired production rate if the same filtration cycle time is used.

Two approaches were used to evaluate R_m and α in example 13.1. In both approaches, the traditional graphical method for determining filtration rate by drawing tangents to the filtration curve was eliminated. In this example, the resistance of the filter medium was the same magnitude as the specific cake resistance, therefore, the Sperry equation worked well in determining R_m. There are instances, particularly when $R_m \ll \alpha$, when a negative intercept will be obtained using Equation (13.8). Under these conditions, R_m is best determined by filtration of solids-free filtrate to obtain q, and using

Equation (13.10) to determine α. In the above example, use of the values of α and R_m calculated using Equation (13.10), will give a filtrate volume per 30 minutes cycle of 0.221 m^3/m^2 of filter area, or an average filtration rate of 0.00737 m^3/(min $\cdot$ m^2 filter area).

13.1.3 Filtration Rate Model Equations for Prolonged Filtration When Filter Cakes Exhibit Time-Dependent Specific Resistance

Equation (13.6) and variations on it such as Equation (13.7) are used assuming that specific cake resistance is constant. These equations usually provide good fit with experimental filtration data for short filtration times as was shown in the previous example. However, when filtration time is extended, Equation (13.8) usually overestimates the filtrate flow. Some filter cakes undergo compaction or fine solids may migrate within the pores blocking flow and increase cake resistance as filtration proceeds.

13.1.4 Exponential Dependence of Rate on Filtrate Volume

The Sperry equation has been modified to account for changing resistance with increasing filtration time. de la Garza and Bouton (Am. J. Enol. Vitic. 35:189, 1984) assumed α to be constant and modified the Sperry equation such that the filtration rate is a power function of filtrate volume.

$$\frac{dt}{dV} = \frac{\mu}{\Delta P\,A}\left(\alpha c\left[\frac{V}{A}\right]^n + R_m\right) \tag{13.11}$$

Integration of Equation (13.11) gives:

$$t = \frac{\mu}{\Delta P\,A}\left(\frac{\alpha c}{n+1}\left[\frac{V}{A}\right]^{n+1} + R_m V\right) \tag{13.12}$$

Bayindirly et al. (J. Food Sci. 54:1003, 1989) tested Equation (13.12) on apple juice and found that a common n in Equation (13.11) could not be found to describe all data at different body feed concentrations. Consequently, an alternative equation was proposed which combines the specific cake resistance and solids concentration into a parameter k.

$$\frac{dt}{dV} = \frac{\mu}{A\Delta P}R_m[e]^{kV/A} \tag{13.13}$$

Integration of Equation (13.13) gives:

$$t = \frac{\mu}{k\Delta P}R_m[e]^{kV/A} \tag{13.14}$$

Taking the natural logarithm of Equation (13.14):

$$\ln(t) = \ln\left[\frac{\mu R_m}{k\Delta P}\right] + \frac{k}{A}V \tag{13.15}$$

A semi-log plot of t against (V/A) will be linear with slope k. R_m is evaluated from the intercept.

Example 13.2. Data in Table 13.2 was obtained from Bayindirly et al. on filtration of apple juice through a filter with a 30.2 cm^2 area using diatomaceous earth filter aid with an average particle size

Table 13.2 Filtration Data for Apple Juice.

Time (s)	ln(t)	Filtrate Volume $V(m^3) \times 10^5$	$t/V \times 10^{-6}$	$t - (V/q)$
84.3	4.353	9.74	8.659	14.78
112.5	4.723	12.3	9.141	24.59
140.6	4.946	14.4	9.793	38.06
196.9	5.283	17.4	11.291	72.33
281.2	5.639	20.0	14.062	138.39
421.9	6.045	22.6	18.697	260.70
759.4	6.632	27.2	27.939	565.24
1068.8	6.974	29.7	35.932	856.29
1575	7.362	32.3	48.750	1344.2
2250	7.719	24.9	64.522	2000.9
3825	8.249	39.5	96.867	3542.9

of 20 μm. The juice was said to have a viscosity just slightly greater than that of water (assume $\mu = 1.0$ cP) and suspended solids in the juice was 0.3% (3.0 kg/m^3) Precoating was 0.25 g/cm^2 (2.5 kg/m^2) and body feed was 0.005 g/mL (5.0 kg/m^3). Pressure differential was 0.65 atm. Calculate the filtration parameters, k and R_m and the average filtration rate in m^3/(min · m^2 of filter area), if a filtration cycle of 60 min is used in the filtration process.

Solution:

Table 13.2 also shows the transformed data on which a linear regression was performed. Linear regression of (t/V) against V to fit Equation (13.8) results in the following correlation equation:

$$t/V = 2.64 \times 10^{10}V - 3,107,386$$

The problem of a negative value for the term involving the medium resistance, R_m is apparent in the analysis of this data. The correlation coefficient was 0.835 which may indicate reasonable fit, however, when filtration time against filtrate volume is plotted the lack of fit is obvious. Thus, the Sperry equation could not be used on the date for this filtration.

A linear regression of ln(t) against V according to Equation (13.15) results in the following regression equation:

$$\ln(t) = 13155.92V + 3.70293$$

The correlation coefficient was 0.998 indicating very good fit. A plot of filtration time against filtrate volume also shows very good agreement between values calculated using the correlation equation and the experimental data. The exponential dependence of filtration rate with filtrate volume appropriately described the filtration data.

The value of k is determined from the coefficient of V in the regression equation.

$$k/A = 13155.92; \quad k = 13155.92(30.2 \times 10^{-4})$$
$$k = 39.73 \text{ m}^{-1}$$

The constant, 3.70293 in the regression equation is the value of $\ln(\mu R_m / k \cdot \Delta P)$

$$R_m = \frac{e^{3.70293}(39.73)(0.65)(101,300)}{0.001(1)} = 5.652 \times 10^{10} m^{-1}$$

The correlation equation will also be used to calculate the filtrate volume after a filtration time of 60 minutes.

$$V = \frac{\ln(3600) - 3.70293}{13155.92} = 0.000348 m^3$$

Average filtration rate $= (V/A)_{avg}/t$

$$\frac{(V/A)_{avg}}{t} = \frac{0.000341}{(30.2 \times 10^{-4})(60)} = 0.001882 \, m^3/(min \cdot m^2 \text{ filter area})$$

13.1.5   Model Equation Based on Time-Dependent Specific Cake Resistance

Chang and Toledo (J. Food Proc. Pres. 12:253, 1989) modified Equation (13.10), derived from the Sperry equation, by assuming a linear dependence of the specific cake resistance with time. The model fitted experimental data on body feed filtration of poultry chiller water overflow for recycling, using perlite filter aid.

$$t = (k_0 + \beta t)V^2 + (1/q)V \tag{13.16}$$

The ratio $(t - V/q)/V^2$ is $(k_0 + \beta t)$ at each filtration time t. The slope of the regression equation of $(t - V/q)/V^2$ against filtration time will have a slope of β and an intercept of k_0. The filtrate volume at time t will be the positive root of Equation (13.16).

$$V = \frac{-(1/q) + [(1/q)^2 + 4(k_0 + \beta t)t]^{0.5}}{2(k_0 + \beta t)} \tag{13.17}$$

Filtration data may be analyzed using this procedure by using the raw volume vs. time data, and the filter area is not considered until the final analysis (i.e , calculated value of filtrate volume or filtration rate is converted to a per unit area basis). Another approach is to use filtrate volume per unit area of filter in the calculations, in which case the calculated value of filtrate volume will already be on a per unit area basis. The former approach is used in the following example to avoid having to manipulate very small numbers.

The units of k_0 and β in Equation (13.16) are $s \cdot m^{-6}$ and m^{-6}, respectively, because values of V used in the analysis have not been converted to Volume/filter area.

Example 13.3. Table 13.3 shows data on filtration of poultry chiller water overflow using perlite filter aid which has a rated pure water permeability of 0.4 mL/(s $\cdot$ m^2) for a 1 cm cake and a 1 atm pressure differential. The filter has an area of 20 cm^2, water at 2°C was the filtrate. The precoat was 1.0 kg/m^2 of filter area. Whatman no. 541 filter paper was used as the filter medium. Body feed was 5 kg/m^3 and suspended solids was 5 kg/m^3. Pressure across the filter was 172 kPa. Filtrate flow across the precoated filter was 25.5 mL/s at 172 kPa pressure differential. Calculate the parameters for the time dependence of specific cake resistance and determine the fit of Equation (13.16) with experimental data. Calculate the average filtration rate if the cycle time is 20 min.

Table 13.3 Data on Filtration of Poultry Chilller Water.

Time (s)	Volume of Filtrate $V(m^3)$	$t - V/q$	$R = (t - V/q)/V^2$ $\times 10^8$
60	0.000483	41.059	1.760
180	0.000823	147.7	2.181
300	0.001031	259.6	2.442
420	0.001173	374.0	2.718
540	0.001292	489.3	2.931
660	0.001385	605.7	3.156
780	0.001466	722.5	3.362
900	0.001538	839.7	3.550
1020	0.001601	957.1	3.734
1140	0.001658	1075.0	3.910
1200	0.001660	1134.9	4.118

$q = 25.5 \times 10^{-6}\, m^3/s$

Solution:

Table 13.3 also shows the values of $(t - V/q)/V^2$ on which a regression analysis was done against time to yield the following regression equation ($R^2 = 0.9883$)

$$k_0 + \beta t = 191,347t + 1.83 \times 10^8 \qquad (13.17a)$$

Table 13.4 shows the calculation of filtrate volume against filtration time using the expression for $k_0 + \beta t$ in Equation (13.17a) and Equation (13.17). A graph of calculated filtrate volume against filtration time shows good agreement between the model and experimental value.

For a filtration time of 20 minutes, the calculated value of V in Table 13.4 is 0.001659 m^3. The average filtration rate is

$$(V/A)_{avg}/t = 0.001659/[(20)(20 \times 10^{-4})]$$
$$(V/A)_{avg}/t = 0.0415 m^3/(s \cdot m^2 \text{filter area})$$

at 172 kPa pressure differential across the filter.

13.1.6 Optimization of Filtration Cycles

Filtration cycles are generally based on obtaining the fastest average filtrate flow. As filtrate flow drops with increasing cake thickness, a very long filtration time will result in the lowered average filtration rate. However, labor involved in disassembling the filter, removal of the cake, and pre-coating will be a major expense. Because precoating is needed before actual filtration, shortened filtration cycles will necessitate increased consumption of filter aid. Thus, optimum cycles based on maximum filtrate flow, may not always be the ideal cycle time from the standpoint of economics. Filtrations with short cycle times for maximum filtrate flow per unit filter area may have to be done

Table 13.4 Calculation of Filtrate Volume from Cake Resistance Data.

Time (s)	$k_0 + \beta t \times 10^{-8}$	$1/q$	$C \times 10^{-10}$	Filtrate Volume (V) Calculated from Equation 17 (mL)
60	1.94	39215	4.82	464
180	2.17	39215	15.8	824
300	2.40	39215	29.0	1038
420	2.63	39215	44.4	1191
540	2.86	39215	62.0	1307
660	3.09	39215	81.8	1399
780	3.32	39215	104	1474
900	3.55	39215	128	1537
1020	3.78	39215	154	1591
1140	4.01	39215	183	1637
1200	4.13	39215	198	1659

$q = 2.55 \times 10^{-5}$; From equation 17: $C = (1/q)^2 + 4(k_0 + \beta t)$;
$V = [-(\text{column 2}) + (\text{column 3})^{0.5}]/[2 \text{ column 1}]$.

using a continuous rotary vacuum filter to prevent excessive filter aid use. The optimum cycle time for maximizing filtrate flow per cycle is derived as follows.

For filtrations that fit the Sperry equation (Eq. 13.7):

A cycle is the sum of filtration time, t and the time to disassemble, assemble and precoat the filter t_{DAP}.

$$\frac{V}{\text{Cycle}} = \frac{V}{t + t_{DAP}}$$

Let $k_1 = \pi \alpha c / 2\Delta P$; and $k_2 = R_m \pi /\Delta P$. Equation 7 becomes:

$$t = k_1(V/A)^2 + k_2(V/A)$$

To obtain the optimum cycle time, the filtrate volume per cycle is maximized by taking the derivative and equating to zero as follows:

$$\frac{d}{dV}\left[\frac{V}{t + t_{DAP}}\right] = \frac{d}{dV}\left[\frac{V}{k_1(V/A)^2 + k_2(V/A) + t_{DAP}}\right]$$

The derivative is equated to zero to obtain V/A for the maximum V/cycle.

$$0 = \frac{[k_1(V/A)^2 + k_2(V/A) + t_{DAP}] - V[2k_1(V/A)^2 + k_2/A]}{[k_1(V/A)^2 + k_2(V/A) + t_{DAP}]^2}$$

$$t_{DAP} - k_1(V/A)^2 = 0$$

$$\left(\frac{V}{A}\right)_{max} = \left[\frac{t_{DAP}}{k_1}\right]^{0.5}$$

The optimum cycle time, t_{opt} is

$$t_{opt} = k_1 \left[\frac{t_{DAP}}{k_1} \right] + k_2 \left[\frac{t_{DAP}}{k_1} \right]^{0.5}$$

$$t_{opt} = t_{DAP} + k_2 \left[\frac{t_{DAP}}{k_1} \right]^{0.5}$$

$$t_{opt} = t_{DAP} + R_m \left[\frac{2\mu t_{DAP}}{\alpha c \Delta P} \right]^{0.5} \tag{13.18}$$

When filtration data fits Equation (13.14), the (V/A) for maximum filtrate flow derived using the above procedure, is the root of Equation (13.19):

$$\mu R_m (e)^{k \left[\frac{V}{A} \right]} \left(\frac{k}{A^2} - \frac{V}{A} \right) - t_{DAP} = 0 \tag{13.19}$$

When filtration data fits Equation (13.16), t as a function of k_0, V, and q only, is solved as follows:

$$t = \frac{k_0 V^2 + q^{-1} V}{1 - \beta V^2}$$

The optimum V for one filtration cycle is the root of Equation (13.20).

$$V^4 (\beta k_0) + V^3 (2\beta q^{-1}) + V^2 (k_0 + 2\beta t_{DAP}) - t_{DAP} = 0 \tag{13.20}$$

Example 13.4. Calculate the optimum filtration cycle in example 13.3 for poultry chiller water overflow filtration, to maximize filtrate flow, assuming 10 min. for disassembly, assembly and pre-coating. Use the values, $k_0 = 1.83 \times 10^8$ s/m^6; $\beta = 191,347$ 1/m^6; $q = 25.5 \times 10^{-6}$ m^3/s calculated in Example 13.3.

Solution:

The values of k_0, β, q and $t_{DAP} = 600$ s are substituted in Equation (13.20), which is then solved for V. A Visual BASIC program will be used to solve for the optimum filtrate volume, V, and the filtration time. The first 5 lines in the program assigns the known values of β, k_0, q, and t_{DAP}, followed by Equation (13.20), and then the equation solves for t required for filtrate volume V to flow through the filter. Figure 13.5 shows the program and the output.

The value of F $= -1.046$ when V $= 0.001122$, and F $= 0.143$ when V $= 0.001123$. Thus, the optimum V for a cycle will be 0.001123 m^3 or 1123 mL. The optimum cycle time is 370 s.

The filtration behavior shown in this example results in very short cycle times because of very rapid loss of filter cake porosity. Thus, use of a batch filter in carrying out this filtration will result in excessive filter aid use. Use of a continuous rotary vacuum filter is indicated for this application.

13.1.7 Pressure-Driven Membrane Separation Processes

A form of filtration that employs permselective membranes as the filter medium is employed to separate solute, macromolecules, and small suspended particles in liquids. A thin membrane with small pore size, which possesses selectivity for passing solute or solvent, is used. The solvent and small molecules pass through the membrane and other solutes, macromolecules or suspended solids are

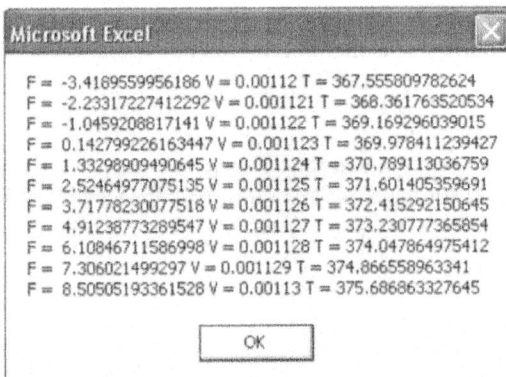

Figure 13.5 Visual BASIC program for calculating optimum filtration cycle times.

retained either by repulsive forces acting on the membrane surface or by a sieving effect. Figure 13.6 shows a typical membrane filtration system. Cross-flow filtration is the most efficient configuration, as discussed later, and retentate recycling is needed to obtain both high cross-membrane fluid velocities and the desired final solids concentration in the product. The fluid crossing the membrane is the "permeate" and the fluid retained on the feed side of the membrane is the "retentate." The fluid entering the membrane is the "feed."

Pressure driven membrane separation processes include:

Microfiltration (MF): particle size retained on the membrane is in the range of 0.02 to 10 μm. Sterilizing filtration is a MF process.

Ultrafiltration (UF): particle size retained on the membrane is in the range of 0.001 to 0.02 μm. Concentration of cheese whey and removal of lactose is a UF process.

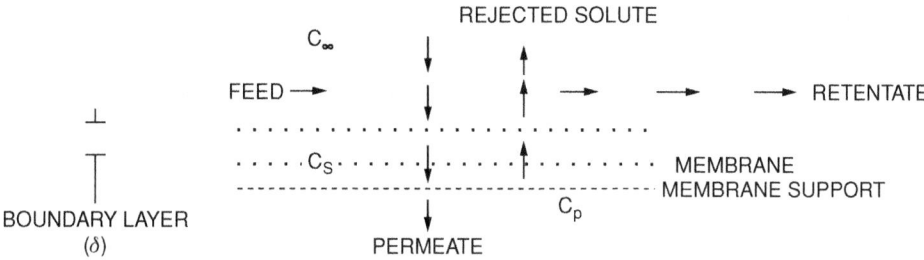

Figure 13.6 Cross-section of a membrane in ultrafiltration.

Reverse osmosis (RO): solute molecules with molecular size less than 0.001 Φm (molecular weight <1000 daltons) are retained on the membrane surface. In RO, solutes increase the osmotic pressure, therefore the transmembrane pressure is reduced by the osmotic pressure as the driving force for solvent flux across the membrane. Concentration of apple juice and desalination of brackish water are examples of RO processes.

Membranes are either isotropic or anisotropic. An isotropic membrane has uniform pores all the way across the membrane thickness, while an anisotropic membrane consists of a thin layer of permselective material on the surface and a porous backing. MF membranes may be isotropic or anisotropic, while permselective high flux membranes used in UF and RO are primarily anisotropic. Membranes are formed by casting sheets using a polymer solution in a volatile solvent followed by evaporation of the solvent. Removal of the solvent leaves a porous structure in the membrane. Larger pore size membranes are produced by subjecting a nonporous polymeric membrane to high-energy radiation. The polymer is changed at specific points of entry of the radiation into the membrane. A secondary treatment to dissolve the altered polymer produces pores within the membrane. Dynamic casting may also be employed. A porous support is coated with the membrane material by pumping a solution containing the membrane material across its surface. As the solvent filters across the porous support, the membrane material is slowly deposited. Eventually, a layer with the desired permselectivity is produced.

Two major factors are important in the design and operation of pressure driven membrane separation processes. These are transmembrane flux and solute rejection properties of the membrane. In general, transmembrane flux and rejection properties are dependent on properties of solute and suspended solids and membrane characteristics of (a) mean pore size, (b) range of pore size distribution, (c) tortuous path for fluid or particle flow across the membrane thickness or "tortuosity," (d) membrane thickness, and (e) configuration of pores. Operating conditions also affect flux and rejection of solutes, and this will be discussed in the succeeding sections.

13.1.8 Membrane System Configurations

The effectiveness of membrane systems may be measured by effectiveness in separating the component of interest at a high production rate. Regardless of whether the valuable component is in the retentate or in the permeate, transmembrane flux must be maximized. Because flux is a function of pressure and membrane area, and is constrained by fouling, the different membrane system

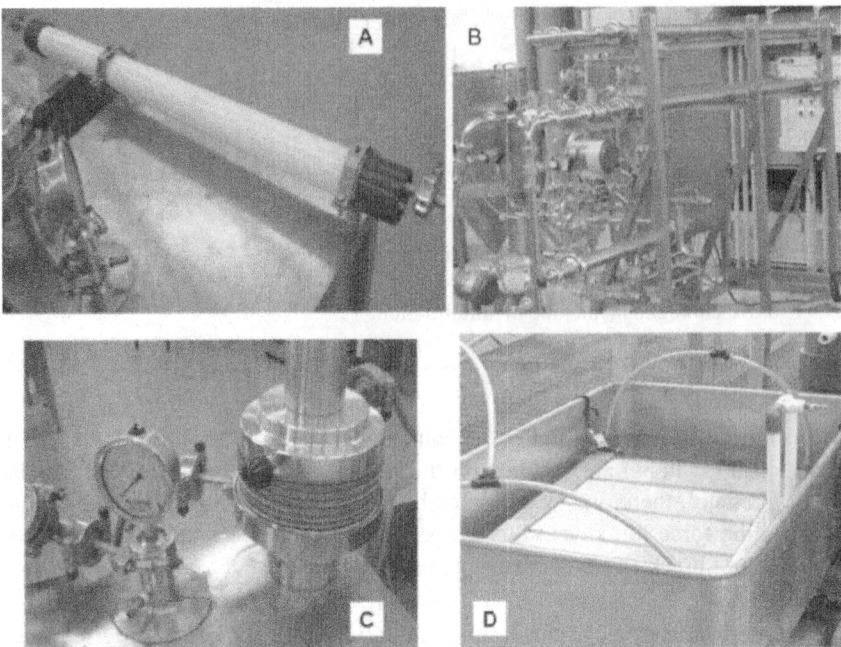

Figure 13.7   Membrane system configuration.

configurations available are designed to maximize membrane area within a small footprint, operate at high pressure, minimize fouling, and facilitate cleaning. Figure 13.7 shows four different membrane system configurations. The spiral wound membrane (Fig. 13.7A) is the most commonly used configuration. The membrane module is tubular with channels between the spiral wound sheet for feed to flow across the membrane from one end of the tube to the opposite end. A spacer sheet between the membrane layer on the permeate side permits permeate to travel around the spiral until it enters the permeate tube at the center of the spiral. The tubular spiral wound module is placed inside a tubular pressure vessel. This configuration permits the most effective packing of a large membrane area in a small space. The disadvantage is the tendency to trap solid particles in the spaces within the spiral thus making it difficult to clean. Feed to a spiral wound membrane system must be pre-filtered to remove particles that could get trapped in the module. The use of spiral wound membranes has made it possible to economically produce potable water from sea water in large desalination plants.

Tubular membranes (Fig. 13.7B) made of ceramic material or inorganic solids deposited on a porous stainless steel support require a large footprint for a given membrane surface area. They have the advantage of being able to operate at high pressure and temperature, resist cleaning compounds, and permit back-flushing to dislodge foulant that penetrated the pores. These membranes have high first cost but membrane life is considerably longer than the polymeric membranes.

Plate and frame membrane units (Fig. 13.7C) permit stacking of several membrane sheets to form a compact unit with large membrane surface area. Plates with the membrane material deposited on each surface are available and when these plates are used, the unit is easily disassembled for cleaning. The manufacturer also supplies membrane support frames which permits a user to use flat sheet membranes on the unit.

A modular submerged module membrane unit is shown in Fig. 13.7D. The module consists of stacks of multiple envelope-like membrane elements arranged in a very compact configuration. Each element consists of two sheets of membrane material with a spacer between them. The edges of the membrane sheets are welded together to prevent feed from entering the element. An opening in the center of the element permits stacking them with gaskets in-between through a permeate removal tube. Because of the large membrane surface area, these units require very low transmemrane pressure to obtain the desired flux. Air scouring of the membrane surface prevents fouling.

Other configurations such as hollow fiber bundles, vibrating membrane units, and cartridge modules are also available and are used in industry.

13.1.9   Transmembrane Flux in Pressure-Driven Membrane Separation Processes (Polarization Concentration and Fouling)

Qualitatively, similarities exists in ordinary filtration and membrane separation processes. The net transmembrane pressure, which drives solvent flow across the membrane, is the sum of the pressure drop across the filter medium, ΔP_m, and the pressure drop across the combined deposited solids on the surface and the boundary layer of concentrated suspension flowing across the membrane surface, ΔP_δ. The development of the equation for transmembrane flux will be similar to that for filtration rate in filter aid filtrations. The pure water permeability of the membrane usually given by the membrane manufacturer, is q in Equation (13.10).

Transmembrane flux when a compressible deposit is on the membrane may be expressed by modifying the Sperry equation (Eq. 13.6) by letting $\alpha = \alpha_0 \Delta P^n$ as follows:

$$\frac{dV}{dt} = \frac{Aq\Delta P}{\mu \alpha_0 \Delta P^n cq(V/A) + A\Delta P} \tag{13.21}$$

n is a compressibility factor that is 0 for a completely incompressible and 1 for a completely compressible solid deposit. q is the pure water permeability at the transmembrane pressure used in the filtration, ΔP.

If the second term in the denominator is much smaller than the first term, i.e., the resistance of the deposit and the fluid boundary layer at the membrane surface controls the flux, Equation (13.21) becomes:

$$\frac{dV}{dt} = \frac{A^2 \Delta P^{1-n}}{\mu \alpha_0 c \, V} \tag{13.22}$$

Equation (13.22) shows that if n is 1 (i.e., the solid deposit is completely compressible), the transmembrane flux is independent of the transmembrane pressure. This phenomenon has been observed in UF and RO operations. For example, in UF of cheese whey, transmembrane flux is proportional to pressure at transmembrane pressure between 30 and 70 $lb_f/in.^2$ gauge (207 to 483 kPa) and is independent of pressure at higher pressures. The phenomenon of decreasing flux rates with increased permeate throughput V, shown in Equation (13.22), has also been observed. Solids deposition on the membrane surface is referred to as "fouling."

Even in the absence of suspended solids, transmembrane flux with solutions will be different from the pure water permeability. This decrease in flux is referred to as "polarization concentration." In simplistic terms, Equation (13.22) qualitatively expresses the phenomenon of polarization concentration in membrane filtrations. As solvent permeates through the membrane, the solids concentration, c, increases in the vicinity of the membrane. The transmembrane flux is inversely proportional to the solids concentration in Equation (13.22).

The concentration on the membrane surface increases with increasing transmembrane flux, therefore, the phenomenon of flux decrease due to polarization concentration will be more apparent at high transmembrane flux. Flux decrease due to polarization concentration is minimized by increasing flow across the membrane surface by cross-flow filtration. In the absence of suspended solids which form fouling deposits, flux will be a steady state. In the presence of suspended solids which form compressible deposits, flux will continually decrease with increasing throughput of permeate, V.

Flux decrease due to polarization concentration is a function of the operating conditions, and if flow rate on the membrane surface, solids concentration in the feed, and transmembrane pressure are kept constant, flux should remain constant. Flux decline under constant operating conditions can be attributed to membrane fouling.

Porter (1979) presented theoretical equations for evaluation of polarization concentration in ultrafiltration and reverse osmosis. The theoretical basis for the equations is the mass balance for solute or solids transport occurring at the fluid boundary layer at the membrane surface.

In cross-flow filtrations, fluid flows over the membrane at a fast rate. A laminar boundary layer of thickness δ exists at the membrane surface. If C is the solids concentration in the boundary layer, and C_4 is the concentration in the liquid bulk, a mass balance of solids entering the boundary layer with the solvent and those leaving the boundary layer by diffusion will be

$$\left(\frac{\partial V}{\partial t}\right) C = -D \left(\frac{\partial C}{\partial x}\right)$$

D is the mass diffusivity of solids. Integrating with respect to x, and designating the transmembrane flux, MV/Mt = J, and using the boundary conditions: $C = C_4$ at $x = \delta$:

$$\ln \left(\frac{C}{C_\infty}\right) = \frac{J}{D}(\delta - x) \tag{13.23}$$

At the surface, $x = 0$ and $C = C_s$.

$$\ln \left(\frac{C_s}{C_\infty}\right) = \frac{J}{D}\delta \tag{13.24}$$

Equation (13.24) shows that polarization concentration expressed as the ratio of surface to bulk solids concentration increases as the transmembrane flux and thickness of boundary layer increases. Equation (13.24) demonstrates that polarization concentration can be reduced by reducing transmembrane flux and decreasing the boundary layer thickness. Since flux must be maximized in any filtration, reduced polarization concentration can be achieved at maximum flux if high fluid velocities can be maintained on the membrane surface to decrease the boundary layer thickness, δ.

Rearranging Equation (13.24):

$$J = \frac{D}{\delta} \ln \left(\frac{C_s}{C_\infty}\right)$$

The ratio D/δ may be represented by a mass transfer coefficient for the solids, k_s. Thus:

$$J = k_s[\ln (C_s) - \ln (C_4)] \tag{13.25}$$

Equation (13.25) shows that a semi-log plot of the bulk concentration C_4 against the transmembrane flux under conditions where C_s is constant, will be linear with a negative slope of $1/k_s$. Thus, flux will be decreasing with increasing bulk solids concentration and the rate of flux decrease will be inversely proportional to the mass transfer coefficient for solids transport between the surface and the fluid bulk.

Equations for estimation of mass transfer coefficients have been applied to determine the effects of cross-flow velocity on transmembrane flux, when polarization concentration alone controls flux. In Chapter 12, mass transfer coefficients were determined using the same equations as for heat transfer. The Sherwood number (Sh) corresponds to the Nusselt number; and the Schmidt number (Sc) corresponds to the Prandtl number.

The Dittus-Boelter equation can thus be used for estimation of mass transfer in turbulent flow.

$$Sh = 0.023 Re^{0.8} Sc^{0.33}$$

Because $Sh = k_s d_h / D$; $Re = d_h v \rho / \Phi$; and $Sc = \Phi / (D \cdot \rho)$:

$$k_s = 0.023 \frac{D}{d_h} \frac{d_h^{0.8} v^{0.8} \rho^{0.8}}{\mu^{0.8}} \frac{\mu^{0.33}}{D^{0.33} \rho^{0.33}} = \frac{0.023 D^{0.67} v^{0.8} \rho^{0.47}}{d_h^{0.2} \mu^{0.47}} \tag{13.26}$$

Although the Dittus-Boelter equation has been derived for tube flow, an analogy with pressure drops through non-circular conduits will reveal that equations for tube flow can be applied to noncircular conduits if the hydraulic radius is substituted for the diameter of the tube. D = mass diffusivity, m^2/s; d_h = hydraulic radius = 4 (cross-sectional area)/wetted perimeter.

If fluid flows in a thin channel between parallel plates with width W and channel depth, 2b:

$$d_h = 4(2b)(W)/(4b + 4W) = 2bW/(b + w)$$

If b << W, d_h = 2b = the channel depth.

In laminar flow, the Sieder-Tate equation for heat transfer can be used as an analog to mass transfer:

$$Sh = 1.86 [Re \cdot Pr \cdot (d/L)]^{0.33} \tag{13.27}$$

Substituting the expressions for Sh, Re, and Pr:

$$k_s = 1.86 \frac{v^{0.33} D^{0.66}}{d_h^{0.33} L^{0.33}} \tag{13.28}$$

L is the length of the channel.

Equations (13.25) and (13.28) can be used to predict transmembrane flux when polarization concentration occurs if the diffusivity D of the molecular species involved and the surface concentration are known. The form of these equations agrees with experimental data. However it is ineffective in predicting actual transmembrane flux by the inability to predict the concentration at the membrane surface, C_s. The value of these equations is in interpolating within experimentally observed values for flux, and in extrapolating within reasonable limits, fluxes when data at one set of operating conditions are known. Porter (1979) suggests using the Einstein-Stokes equation (equation. 29) to estimate D:

$$D = \frac{1.38 \times 10^{-23} T}{6 \pi \mu r} \tag{13.29}$$

where D is in m^2/s, T = K, Φ = medium viscosity in Pa · s, and r = molecular radius in meters.

Example 13.5. The molecular diameter of β-lactoglobulin having a molecular weight of 37,000 daltons is 1.2×10^{-9}m. When performing ultrafiltrations at 30°C at a membrane surface velocity of 1.25 m/s, solids concentration in the feed of 12% and transmembrane pressure of 414 kPa, the flux was 6.792 L/(m^2 · h). Calculate the concentration at the membrane surface, and the flux under the same conditions but at a higher cross-membrane velocity of 2.19 m/s. The membrane system was a thin channel with a separation of 7.6 mm. Flow path was 30 cm long. The viscosity of the solution at 30°C was 4.8 centipoise, and the density was 1002 kg/m^3.

Solution:

The diffusivity using Equation (13.29) is

$$D = \frac{1.38 \times 10^{-23}(30 + 273)}{6\pi(4.8)(0.001)(1.2 \times 10^{-9})} = 3.851 \times 10^{-11} m^2/s$$

The hydraulic radius, d_h = channel depth = 7.6×10^{-3} m. The Reynolds number is

$$Re = 7.6 \times 10^{-3}(1.25)(1002)/[(4.8)(0.001)] = 1983$$

For turbulent flow, (Eq. 13.26):

$$k_s = \frac{0.023(3.851 \times 10^{-11})^{0.66}(1.25)^{0.33}(1002)^{0.47}}{(7.6 \times 10^{-3})^{0.2}[(4.8)(0.001)]^{0.47}} = \frac{0.023(1.338 \times 10^{-7})(1.076)(25.72)}{0.376(0.081317)}$$

$$= 2.785 \times 10^{-6} m/s$$

Given: J = 6.792 L / (m^2· h) = $1.887 \times 10^{-6} m^3/m^2$· s)
Using Equation (13.25): $C_s = C_4[e]^{J/ks}$
C_4 is given as 0.12 g solids/g soln.

$$C_s = 0.12[e]^{1.88 \times 10^{-6}/2.785 \times 10^{-6}} = 0.12\,(1.969) = 0.236 \text{ g solute/g soln.}$$

For v = 2.19 m/s:

$$k_s = 2.785 \times 10^{-6} \left[\frac{2.19}{1.25}\right]^{0.8} = 4.362 \times 10^{-6} m/s$$

Using Equation (13.25):

$$J = 4.362 \times 10^{-6} \, [\ln (0.236/0.12)] = 2.95 \times 10^{-6} \text{ m}^3/(m^2 \cdot s) \text{ or } 10.62 \text{ L}/(m^2 \cdot h)$$

13.1.10   Solute Rejection

Solute rejection in membrane separations is dependent on the type of membrane and the operating conditions. A solute rejection factor, R, is used, defined as:

$$R = \frac{C_s - C_p}{C_s} \tag{13.30}$$

where C_s = concentration of solute on the membrane surface on the retentate side of the membrane, and C_p = concentration of solute in the permeate.

Equation (13.30) shows that polarization concentration and increasing solute concentration in the feed, decreases the solute rejection by membranes.

Rejection properties of membranes are specified by the manufacturer in terms of the "molcular weight cut-off," an approximate molecular size that will be retained by the membrane with a rejection factor of 0.99 in very dilute solutions. The ideal membrane with a sharp molecular weight cut-off does not exist. On a particular membrane, the curve for rejection factor against molecular weight is often sigmoidal with increasing molecular weight of solute. Low molecular weight solutes will completely pass through the membrane and R = 0. As the solute molecular weight increases, small increases in R will be observed until the molecular size reaches that which will not pass through a majority of the pores. When the molecular size exceeds the size of all the pores, the rejection factor will be 1.0. The pore size distribution therefore, determines the distribution of molecular sizes which will be retained by the membrane. Some solutes will be retained because they are repelled by the membrane at the surface. This is the case with mineral salts on cellulose acetate membrane surfaces. This property of a membrane to repel a particular solute will be beneficial under conditions where separation from the solvent is desired (e.g., concentration) because high rejection factors can be achieved with larger membrane pore size therefore allowing solute rejection at high transmembrane flux.

Interactions between solutes also affect the rejection factor. For example, rejection factor for calcium in milk or whey is higher than aqueous calcium solutions. A protein to calcium complex will bind the calcium preventing its permeation through the membrane.

13.1.11 Sterilizing Filtrations

Sterilizing filtrations employ either depth cartridge filters or microporous membrane filters in a plate and frame or cartridge configuration. These filtrations are MF processes. Microporous membrane filters have pore sizes smaller than the smallest particle to be removed and performs a sieving process. They are preferred for sterilizing filtrations on liquid because the pressure drop across the membrane is much smaller than in a depth filter and the possibility of microorganisms breaking through is minimal. Pore size of microporous membranes for sterilizing filtrations is less than 0.2 Φm.

The filtration rate in sterilizing filtrations has been found by Peleg and Brown (J. Food Sci. 41:805, 1976) to follow the following relationship:

$$\frac{dV}{dt} = k\frac{V^{-n}}{c} \tag{13.31}$$

where c is load of microorganisms and suspended material that is removed by the filter, and k and n are constants which are characteristic of the fluid being filtered, the filter medium, the suspended solids, and fluid velocity across the membrane surface. k is also dependent on the transmembrane pressure. n is greater than 1; therefore, filtration rate decreases rapidly with increase in filtrate volume. No cake accumulation occurs in this type of filtration. The time for membrane replacement and pre-sterilization must be included in the analysis of the optimum cycle time and procedures for optimization will be similar to that in the section "Optimization of Filtration Cycles."

Cycle time can be increased when fluid is in cross-flow across the membrane surface at high velocities. This configuration minimizes solids deposition and membrane fouling. Fouled membrane surfaces may be rejuvenated by occasionally interrupting filtration through reduction of transmembrane pressure while maintaining the same velocity of fluid flow across the membrane. This procedure is called "flushing," as opposed to "backwashing" where the transmembrane pressure is reversed. Some membrane configurations are suitable for backwashing, while membrane fragility may restrict removal of surface deposits by flushing in other configurations. When rejuvenation is done by flushing or backflushing, the flushing fluid must be discarded or filtered through another coarser filter to remove the solids, otherwise, mixing with new incoming feed will result in rapid loss of filtration rate.

Pre-sterilization of filter assemblies may be done using high pressure steam, or using chemical sterilants such as hydrogen peroxide, iodophores, or chlorine solutions. Care must be taken to test for filter integrity after pre-sterilization, particularly when heating of filter assemblies is used for presterilization. One method to test for filter integrity in line, is the "bubble point test," where the membrane is wetted, sterile air is introduced, and the pressure needed to dislodge the liquid from the membrane is noted. Intact membranes require specific pressures to dislodge the liquid from the surface and any reduction in this bubble pressure is an indication of a break in the membrane.

Sterilizing filtrations are successfully employed in cold pasteurization of beer and wine, in the pharmaceutical industry for sterilization of injectable solutions, and in the biotechnological industry for sterilization of fermentation media and enzyme solutions.

Example 13.6. Data on flux during membrane filtration to clarify apple juice (Mondor et al. Food Res. International 33:539–548, 2000) showed the flux to follow two different domains when plotted with time. The initial rapid flux decay was attributed to deposition of solids on the membrane pores, and the domain where flux appear to level off with time was attributed to be due to concentration polarization and formation of a gel layer by the deposited solids on the membrane surface. If J_o is the flux rate at time $= 0$, J_{41} is the flux corresponding to the point where flux levels off in the first part of the filtration curve, and J_{42} is the flux corresponding to the time when flux leveled off in the second part of the filtration curve, then the flux over the whole filtration cycle can be expressed as:

$$J = (J_o - J_{\infty 1})\exp(-\alpha t) + (J_{\infty 1} - J_{\infty 2})\exp(-\beta t) + J_{\infty 2}$$

An experiment involving cross-flow filtration of apple juice through a PVDF membrane system with a pore size of 0.2 micrometers and a total membrane surface area of $1.0\ m^2$. Cross flow velocity was 3 m/s and transmembrane pressure was 1.5 Bars. The apple juice was obtained by pressing pectinase enzyme treated ground apples with rice hulls as filtration aid through mutiple layers of cheese cloth in a bladder press. The data (time in hours, flux in liters/m^2 h) are as follows: (0.11, 32), (0.24, 28.6), (0.55, 22.7), (0.96, 18.1), (1.2, 15.1), (2.5,9.8), (5.3, 5.9), (7. 4.7), (8, 4.7), (9, 4.6).

(a) Assuming that a cleaned membrane will result in the same pattern of flux decay, how long would it take to recover 15 L of the clarified juice (permeate)?

Solution:

The first term in the flux decay equation above is good for $t = 0$ to $t = t_{s1}$ where t_{s1} corresponds to the end of the first stage of the filtration cycle. This time is determined by plotting $\ln(J - J_{42})$ against time to see where the linear portion deviates from the experimental data curve. A regression is then conducted for the range of time where the curve is linear. The second term in the above expression for J represents the flux decay from the start of the second stage in the filtration cycle. Again, the linear portion of the $\ln(J - J_{42})$ versus time curve is located and a linear regression is conducted to obtain the slope. The time variable in the second term should be the difference between the actual time and the start of the linear plot close to the actual experimental curve.

J_{42} from the above data $= 4.6\ L/m^2h$.

The transformed data is as follows:

Time (h)	J (L/m²h)	ln(J + J_{42})
0.11	32	3.31
0.24	28.6	3.18
0.55	22.7	2.89
0.96	18.1	2.60
1.2	15.1	2.35
2.5	9.8	1.65
5.3	5.9	0.26
7	4.7	−2.3
8	4.7	−2.3

The first linear segment fitted $t = 0.11$ to $t = 2.5$. This segment has a slope of -0.69 and an intercept of 3.31. The value of $J_{41} = 4.7$ L/m²h. The second line segment fitted the curve from $t = 5.3$ to 8 and has a slope of -1.0). This line segment intersects the first line segment at $t = 3.5$. Thus the equation for J is as follows: $J = (32 - 4.7) \exp(-0.69 \cdot t) + (4.7 - 4.6)\exp(-1.0(t - 3.5)) + 4.6$
$J = 27.3 \exp(-0.69 \cdot t) + 0.1 \exp(-(t - 3.5)) + 4.6$. V = volume of permeate.

$$V = \int_0^t J dt = \int_0^t \{27.3[e]^{-0.69t} + 0.1[e]^{-(t-3.5)} + 4.6$$

$$V = \frac{-27.3}{0.69}[e]^{-0.69t} - 0.1[e]^{-(t-3.5)} + 4.6t$$

This equation will be solved using Excel.

Set up volume increments in 0.1 hours. Suppose $t = 0.1$ h is in a3. The value of the integral will be: $(-27.3/0.69)* \exp(-0.79*a3) - 0.1*\exp(-(a3 - 3.5)) + 4.6*a3 + (27.3/0.69) + 0.1$. The calculated values are as follows:
$t = 0$, $V = 0$; $t = 0.1$, $V = 0.20$; $t = 0.3$, $V = 6.42$, $t = 0.4$, $V = 0.3$, $t = 0.5$, $V = 11.9$; $t = 0.6$, $V = 14.5$, $t = 0.7$, $V = 16.8$. Thus, it would take 0.62 hours of filtration to obtain 15 L of filtrate.

13.1.12 Ultrafiltration

Ultrafiltration is widely employed in the increasingly important biotechnological industry for separation of fermentation products, particularly enzymes. Its largest commercial use is in the dairy industry for recovery of proteins from cheese whey and for pre-concentration of milk for cheese making. UF systems also have potential for use as biochemical reactors, particularly in enzyme or microbial conversion processes where the reaction products have an inhibitory effect on the reaction rate.

A large potential use for UF is in the extraction of nectar from fruits which cannot be pressed for extraction of the juice. The fruit mash is treated with pectinase enzymes and the slurry is clarified by UF. UF applications in recycling of food processing waste water, and in recovery of valuable components of food processing wastes are currently in the developmental stages. More membranes for UF applications are available than for RO applications, indicating increasing commercial applications for UF in the food industry.

13.1.13   Reverse Osmosis

Reverse osmosis has potential for generating energy savings in the food processing industry, when used as an alternative to evaporation in the concentration of products containing low molecular weight solutes. The early work on RO has been devoted to desalination of brackish water, and a number of RO systems have now been developed which can purify waste water for re-use after an RO treatment. In the food industry, RO has potential for juice concentration without the need for application of heat.

In RO, the transmembrane pressure is reduced as a driving force for solvent flux by the osmotic pressure. Since solutes separated by RO have low molecular weights, the influence of concentration on osmotic pressure is significant. Equation (12.5) in Chapter 12 can be used for determining the osmotic pressure. Because the product of the activity coefficient of water and the mole fraction of water is the water activity, this equation can be written as:

$$\pi = -\frac{RT}{V} \ln (a_w) \tag{13.32}$$

where π = osmotic pressure, R = gas constant, T = absolute temperature, and V = molar volume of water. a_w = water activity.

The minimum pressure differential needed to force solvent across a membrane in RO is the osmotic pressure. Equations (12.17) and (12.21) in Chapter 12 can be used to determine a $_w$ of sugar solutions.

Example 13.7. Calculate the minimum transmembrane pressure that must be used to concentrate sucrose solution to 50% sucrose at 25°C.

Solution:

From the Example 12.1 in Chapter 12, the water activity of a 50% sucrose solution was calculated to be 0.935.

R = 8315 N m /(kgmole · K); T = 303 K, and the density of water at 25°C = 997.067 kg/m^3 from the steam tables.

$$V = \frac{18kg}{kgmole} \frac{1}{997.067kg/m^3} = 0.018053 m^3/kgmole$$

$$\pi = \frac{-[8315 Nm/(kgmole \cdot K)][303K]}{0.018053 m^3/kgmole} \ln (0.935)$$

$$= 9379 kPa$$

A minimum of 9379 kPa transmembrane pressure must be applied just to overcome the osmotic pressure of a 50% sucrose solution.

This example illustrates the difficulty in producing very high concentrations of solutes in reverse osmosis processes, and the large influence of concentration polarization in reducing transmembrane flux in RO systems.

Transmembrane flux in RO systems is proportional to the difference between the transmembrane pressure and the osmotic pressure.

$$J = \left(\frac{k_w}{x}\right) (P - \Delta\pi) \tag{13.33}$$

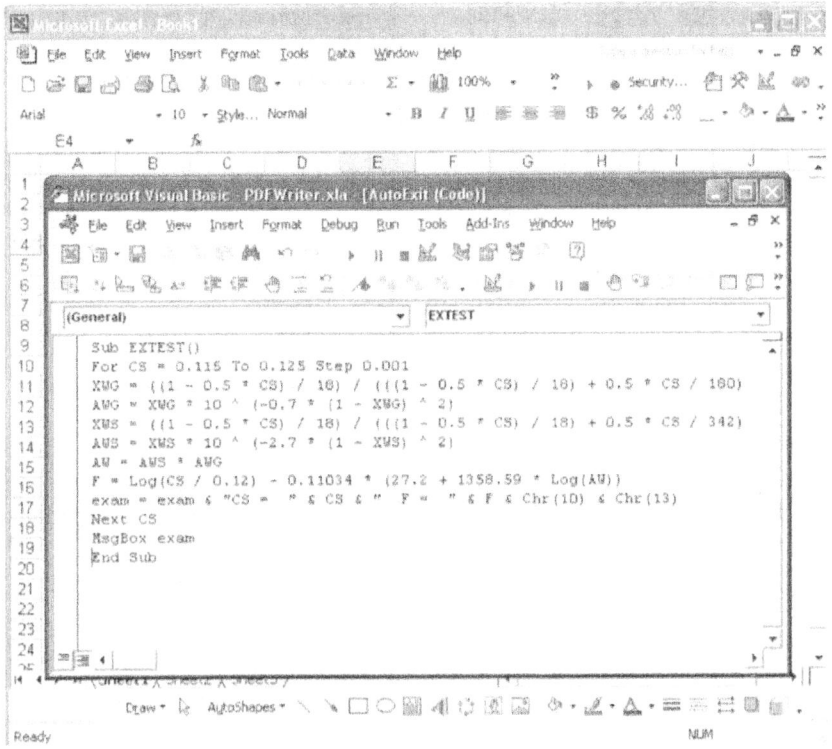

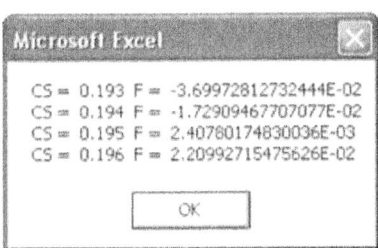

Figure 13.8 Visual BASIC program for problem involving reverse osmosis concentration of orange juice.

where J = transmembrane flux, kg water/(s · m^2); k$_w$ = mass transfer coefficient, s; x = thickness of membrane, m; P = transmembrane pressure, Pa; and π = osmotic pressure, Pa.

Example 13.8. The pure water permeability of cellulose acetate membrane having a rejection factor of 0.93 for a mixture of 50% sucrose and 50% glucose is 34.08 kg/h of water at a transmembrane pressure of 10.2 atm (1034 kPa) for a membrane area of 0.26 m^2. This membrane system had a permeation rate of 23.86 kg/h when 1% glucose solution was passed through the system at a transmembrane pressure of 10.2 atm. at 25°C.

Orange juice is to be concentrated using this membrane system. In order to prevent fouling, the orange juice was first centrifuged to remove the insoluble solids, and only the serum which contained 12% soluble solids was passed through the RO system. After centrifugation, the serum was 70% of the total mass of the juice.

Assume that the soluble solids are 50% sucrose and 50% glucose. The pump that feeds the serum through the system operates at a flow rate of 0.0612 kg/s (60 mL/s of juice having a density of 1.02 g/mL). The transmembrane pressure was 2758 kPa (27.2 atm).

Calculate the concentration of soluble solids in the retentate after one pass of the serum through the membrane system.

Solution:

This problem illustrates the effects of transmembrane pressure, polarization concentration and osmotic pressure on transmembrane flux and solute rejection in reverse osmosis systems. The governing equations are Equation (13.25) for polarization concentration and Equation (13.33) for transmembrane pressure.

The reduction in the pure water permeability when glucose solution is passed through the membrane system can be attributed to polarization concentration. Equation (13.25):
$J = k_s \ln (C_s/C_4)$. Let N_w = water permeation rate in kg/h.
$N_w = JA \, m^3/s \, (3600s/h)(1000 kg/m^3)$.

$$J = \left(\frac{N_w}{A}\right)\left(\frac{1}{3.6}\right) \times 10^{-6}$$

The osmotic pressure at the membrane surface responsible for the reduction of transmembrane flux in the 1% glucose solution is calculated using Equation (13.33).

$$N_w = \left(k_w \frac{A}{x}\right)(P - \pi)$$

The factor ($k_w A/x$) can be calculated using the pure water permeability data, where $\pi = 0$:

$$\left(k_w \frac{A}{x}\right) = \frac{N_w}{P} = \frac{34.08}{10.2} = 3.34 \, kg/(h \cdot atm)$$

For the 1% glucose solution, using Equation (13.33):

$$N_w = \left(k_w \frac{A}{x}\right)(P - \pi)$$

$$23.86 = 3.34(10.2 - \pi)$$

$$\pi = 3.05 \, atm.$$

A 1% glucose solution has a water activity of practically 1.0, therefore, the osmotic pressure must be exerted at the membrane surface. Thus, C_s can be calculated as the glucose concentration that gives an osmotic pressure of 3.05 atm. Using Equation (13.32), for $P = 3.05(101,325) = 309,041$ Pa.:

$$309,041 = -(8315)(298/0.018) \ln (a_w)$$

$$a_w = e^{-0.00225} = 0.99775$$

For very dilute solutions, $a_w = x_w$. Let C_s = solute concentration at the surface, g solute/g solution.

$$a_w = x_w = \frac{(1 - C_s)/18}{(1 - C_s)/18 - C_s/180}$$

$$a_w = \frac{(1 - C_s)(18)(180)}{18[180(1 - C_s) + 18C_s]} = 0.99775$$

$$321.271C_s + 3232.71 - 3232.71C_s = 3420 - 3420C_s$$

$$C_s = 7.29/330.561 = 0.022 \text{ g solute / g solution}$$

Solving for J from the permeation rate:

$$J = (23.86/0.26)(1/3.6) \times 10^{-6} = 2.55 \times 10^{-5} \text{m/s}$$

Using Equation (13.25):

$$k_s = 2.55 \times 10^{-5}/\ln(0.022/0.01) = 3.234 \times 10^{-5} \text{m/s}$$

For the orange juice serum: $C_4 = 0.12$ g solute/g soln. Solving for C_s using Equation (13.25):

$$J = 3.234 \times 10^{-5} \ln(C_s/0.12)$$

Solving for N_w from J:

$$N_w = J A (3.6)(10^6) = [3.234 \times 10^{-5} \ln(C_s/0.12)](0.26)(3.6 \times 10^6) = 30.27 \ln(C_s/0.12)$$

Solving for N_w from the transmembrane pressure using Equation (13.36):

$$N_w = (k_w A/x)(P - \pi) = 3.34(27.2 - \pi)$$

Equating the two equations for permeation rate:

$$30.27 \ln(C_s/0.12) = 3.34(27.2 - \pi)$$

$$\ln(C_s/0.12) = 0.1103(27.2 - \pi)$$

Solving for π in atm using Equation (13.32):

$$\pi = -(8315)(298/0.018) \ln(a_w) [1 \text{ atm}/101{,}325 \text{ Pa}] = -1358.59 \ln(a_w), \text{ atm}$$

Substituting in the equation for C_s:

$$\ln(C_s/0.12) = 0.1103[27.2 + 1358.59 \ln(a_w)]$$

a_w can be solved in terms of C_s as follows:

For sucrose: x_{ws} = mole fraction of water in the sucrose fraction of the solution; a_{ws} = water activity due to the sucrose fraction.

$$x_{ws} = \frac{(1 - 0.5C_s)/18}{(1 - 0.5C_s)/18 + 0.5C_s/342}$$

$$a_{ws} = x_{ws}[10]^{-2.7(1-x_{ws})^2}$$

For the glucose fraction of the mixture: Let x_{wg} = mole fraction of glucose; a_{wg} = water activity due to the glucose fraction.

$$x_{wg} = \frac{(1 - 0.5C_s)/18}{(1 - 0.5C_s)/18 + 0.5C_s/180}$$

$$a_{wg} = x_{wg}[10]^{-0.7(1-x_{wg})^2}$$

$$a_w = a_{ws} \cdot a_{wg}$$

C_s will be solved using a Visual BASIC program. In this program, a wide range of values of C_s from 0.1 to 0.2 was first substituted with a larger step change of 0.01. The printout was then checked for the two values of C_s between which the function F crossed zero. A narrow range of C_s between 0.115 and 0.125 was later chosen with a narrower step change in order to obtain a value of C_s to the nearest third decimal place which satisfied the value of the function F to be zero.

The value of C_s will be 0.1949. The rejection factor will be used to calculate the permeate solute concentration.

$$R = (C_s - C_p)/C_s$$
$$C_p = C_s(1 - R) = 0.1949(1 - 0.93)$$
$$C_p = 0.0136 \text{ g solute/g permeate.}$$

Total mass balance:

$$F = P + R; \quad F = 0.0612 \text{ kg/s } (3600\text{s/h}) = 220.3 \text{ kg/h}$$
$$P = N_w/(1 - C_p)$$
$$N_w = 30.27 \ln(0.1949/0.12) = 14.68 \text{ kg/h}$$
$$P = 14.68/(1 - 0.0134) = 14.87 \text{ kg/h}$$
$$R = 220.3 - 14.87 = 205.43 \text{ kg/h}$$

Solute balance: $FC_f = RC_r + PC_p$

$$0.12(220.3) = 205.43(C_r) + 14.87(0.0134)$$
$$C_r = 0.128\text{g solute/g retentate}$$

The effect of polarization concentration shown in this example problem demonstrates that increasing transmembrane pressure may not always result in a higher flux. A pressure will be reached beyond which polarization concentration will negate the effect of the applied pressure and any further increase in transmembrane pressure will result in no further increase in flux.

Fouling remains a major problem in both RO and UF applications. Optimization of operating cycles using the principles discussed in the section "Optimization of Filtration Cycles" is an aspect of the operation to which an engineer can make contributions to successful operations. Application of theories and equations on polarization concentration as discussed in the preceding sections will help in identifying optimum operating conditions, but absolute values of transmembrane flux are better obtained empirically on small systems and results used for scale-up than calculating theoretical values. Material balance calculations as discussed in Chapter 3 in the section "Multistage Processes," Example 3.21, will also help in analyzing UF systems, particularly in relation to recycling and multi-stage operations to obtain the desired composition in the retentate.

13.1.14 Temperature Dependence of Membrane Permeation Rates

Temperature affects liquid permeation rates through membranes in proportion to the change in viscosity of the fluid with temperature. If the solvent is water, and permeate flux at one temperature is given, the permeation rate at any other temperature can be calculated by:

$$J_{T1} = J_{T2} \frac{\mu_{T2}}{\mu_{T1}}$$

where J_{T1} and J_{T2} are fluxes at T1 and T2 respectively. The viscosity of water at temperature T in Celsius is

$$\mu = \frac{1795 \cdot 10^{-6}}{1 + 0.036T + 0.000185T^2}$$

where μ = viscosity in Pa s and T is temperature in °C.

13.1.15 Other Membrane Separation Processes

Membranes are also used to remove inorganic salts from protein solutions, or for recovery of valuable inorganic compounds from process wastes. Dialysis and electrodialysis are two commonly used processes.

Dialysis is a diffusion rather than pressure-driven process. A solute concentration gradient drives solute transport across the membrane. Dialysis is well-known in the medical area, for removing body wastes from the blood of persons with diseased kidneys. In the food and biochemical industries, dialysis is used extensively for removal of mineral salts from a protein solution by immersing the protein solution contained in a dialysis bag or tube in flowing water. Solute flux in dialysis is very slow.

Electrodialysis is a process where solute migration across the membrane is accelerated by the application of an electromotive potential. Anion and cation selective membranes are used. The electromotive potential forces the migration of ionic species across the membrane towards the appropriate electrode. Electro dialysis is used to demineralize whey, but food industry applications of the process are still rather limited.

13.2 SIEVING

Sieving is a mechanical size separation process. It is widely used in the food industry for separating fines from larger particles, and also for removing large solid particles from liquid streams prior to further treatment or disposal. Sieving is a gravity driven process. Usually a stack of sieves are used when fractions of various sizes are to be produced from a mixture of particle sizes.

To assist in the sifting of solids in a stack of sieves, sieve shakers are used. The shakers may be in the form of an eccentric drive that gives the screens a gyratory or oscillating motion, or it may take the form of a vibrator which gives the screens small amplitude high frequency up and down motion. When the sieves are inclined, the particles retained on a screen fall off at the lower end and are collected by a conveyor. Screening and particle size separation can thus be carried out automatically.

13.2.1   Standard Sieve Sizes

Sieves may be designated by the opening size, US-Sieve mesh or Tyler Sieve mesh. The Tyler mesh designations refer to the number of openings per inch, while the US-Sieve mesh designations is the metric equivalent. The latter has been adopted by the International Standards Organization. The two mesh designations have equivalent opening size although the sieve number designations are not exactly the same. Current sieve designations, unless specified, refer to the US-Sieve series. Size of particles are usually designated by the mesh size that retains particles that have passed through the next larger screen size. A clearer specification of particle size by mesh number would be to indicate by a plus sign before the mesh size that retains the particles and by a negative sign the mesh size that passed the particles. If a mixture of different sized particles are present, the designated particle size must be the weighted average of the particle sizes.

Table 13.5 shows US sieve mesh designations and the size of openings on the screen. A cumulative size distribution of powders can be made by sieving through a series of standard sieves and determining the mass fraction of particles retained on each screen. The particles are assumed to be spherical with a diameter equal to the mean of the sieve opening which passed the particles and that which retained the particles.

Example 13.9. Table 13.6 shows the mass fraction of a sample of milled corn retained on each of a series of sieves. Calculate a mean particle diameter which should be specified for this mixture.

Solution:

The sieve opening is obtained from Table 13.6. The mean particle size retained on each screen is the mean opening size between the screens which retained the particular fraction and the one above

Table 13.5   Standard US-Sieve Sizes.

US-Sieve Size (mesh)	Opening (mm)	US-Sieve Size (mesh)	Opening (mm)
2.5	8.00	35	0.500
3	6.73	40	0.420
3.5	5.66	45	0.354
4	4.76	50	0.297
5	4.00	60	0.250
6	3.36	70	0.210
7	2.83	80	0.177
8	2.38	100	0.149
10	2.00	120	0.125
12	1.68	140	0.105
14	1.41	170	0.088
16	1.19	200	0.074
18	1.00	230	0.063
20	0.841	270	0.053
25	0.707	325	0.044
30	0.595	400	0.037

Source: Perry, R. H., Chilton, C. H., and Kirkpatrick, S. D. 1963. *Chemical Engineers Handbook,* 4th ed. McGrawo-Hill Book Co., New York.

Table 13.6 US-Sieve Size Distribution of Power Sample and Calculation of Mean Particle Diameter.

Mass Fraction Retained	US-Sieve Screen	Sieve Opening (mm)	Particle size	Fraction × Size
0	35	0.500		
0.15	45	0.354	0.437	0.065
0.35	60	0.250	0.302	0.1057
0.45	80	0.177	0.213	0.095585
0.05	120	0.125	0.151	0.00755
				Sum = 0.2741

it. The mass fraction is multiplied by the mean particle size on each screen and the sum will be the weighted mean diameter of the particle. From Table 13.5, the mean particle diameter is 0.2741 mm.

This example clearly shows that the choice of sieves used in classifying the sample into the different size fractions, will affect the mean particle diameter calculated. The use of a series of sieves of adjacent sizes in Table 13.5 is recommended.

13.3 GRAVITY SEPARATIONS

This type of separation depends on density differences between several solids in suspension or between the suspending medium and the suspended material. Generally, the force of gravity is the only driving force for the separation. However, the same principles will apply when the driving force is increased by application of centrifugal force.

13.3.1 Force Balance on Particles Suspended in a Fluid

When a solid is in suspension, an *external force*, F_e, acts on it to move in the direction of the force. This external force may be gravitational force, in which case it will cause a particle to move downward, or it can be centrifugal force, which will cause movement toward the center of rotation. In addition, a *buoyant force*, F_b, exists that acts in a direction opposite the external force. Another force, the *drag force*, F_d, also exists, and this force is due to motion of the suspended solid and acts in the direction opposite the direction of flow of the solid. Let F_t = the net force on the particle. The component of forces in the direction of solid motion will obey the following force balance equation.

$$F_t = F_e - F_d - F_b \tag{13.34}$$

In applying Equation (13.34), the sign on F^t will determine its direction based on the designation of the direction of F_e as the positive direction.

13.3.1.1 Buoyant Force

F_b is the mass of fluid displaced by the solid multiplied by the acceleration provided by the external force. Let a_e = acceleration due to the external force, m = mass of particle, ρ_p = density of particle, ρ = density of the fluid.

506   13. PHYSICAL SEPARATION PROCESSES

The mass of displaced fluid = volume of the particle /density = m/ρ_p

$$F_b = ma_e \left(\frac{\rho}{\rho_p} \right) \tag{13.35}$$

The mass of displaced fluid = $m(\rho/\rho_p)$.

13.3.1.2 Drag Force

F_d is fluid resistance to particle movement. Dimensional analysis will be used to derive an expression for F_d.

Assume that the F_d/A_p, where A_p is the projected area perpendicular to the direction of flow; F_d is a function of the particle diameter, d; v_r is the relative velocity between fluid and particle; ρ is the fluid density; ρ nd μ is the fluid viscosity.

The general expression used in dimensional analysis is that the functionality is a constant multiplied by the product of the variables each raised to a power.

$$\frac{F_d}{A_p} = C(v_r)^a \rho^b \mu^c d^e$$

Expressing in terms of the base units, mass, M, length, L, and time, t.

$$\frac{ML}{t^2} \frac{1}{L^2} = \frac{L^a}{t^a} \frac{M^b}{L^{3b}} \frac{M^c}{L^c t^c} L^e$$

Consider the coefficients of each base unit. To be dimensionally consistent, the exponents of each unit must the opposite side of the equation.

Exponents of L:

$$-1 = a - 3b - c - e \tag{i}$$

Exponents of t:

$$2 = a + c; \quad a = 2 - c \tag{ii}$$

Exponents of M:

$$1 = b + c; \quad b = 1 - c \tag{iii}$$

Combining equations (ii) and (iii) in (i)

$$-1 = 2 - c - 3(1 - c) - c + e$$
$$e = -c$$

Thus, the expression for drag force becomes:

$$\frac{F_d}{A_p} = CV^{2-c} \rho^{1-c} \mu^c d^{-c}$$

Combining terms with the same exponent:

$$\frac{F_d}{A_p} = C \left[\frac{\mu}{dv_r \rho} \right]^n (v_r)^2 \rho \tag{13.36}$$

The first two terms on the right are replaced by a drag coefficient, C_d, defined as $C_d = C/Re^n$.

Equation (13.36) is often written in terms of C_d as follows:

$$F_d = C_d \frac{(v_r)^2 \rho}{2} \tag{13.37}$$

Consider a spherical particle settling within a suspending fluid. A sphere has a projected area of a circle; $A_p = \pi d^2/4$. If particle motion through the fluid results in laminar flow at the fluid terface with the particle, $C_d = 24/Re$. The equation for drag force becomes:

$$F_d = \left(\frac{24}{Re}\right)\left(\frac{\pi d^2}{4}\right)(v_r)^2 \rho = 3\pi\mu dv_r \tag{13.38}$$

13.3.2 Terminal Velocity

The terminal velocity of the particle is the velocity when $dv_r/dt = 0$, that is, settling will occur at a constant velocity. Substituting Equations (13.35) and (13.38) into Equation (13.34):

$$F_t = ma_e - ma_e\left(\frac{\rho}{\rho_p}\right) - 3\pi\mu dv_r$$

$F_t = ma_p = m(dv_r/dt)$. The mass of the particle is the product of the volume and the density. For a sphere: $m = \pi d^3 \rho/6$. Thus, at the terminal velocity, v_t:

$$ma_e\left(1 - \frac{\rho}{\rho_p}\right) = 3\pi\mu v_t d$$

Substituting for m:

$$v_t = \frac{a_e(\rho_p - \rho)d^2}{18\mu} \tag{13.39}$$

Equation (13.39) is *Stokes'= law* for velocity of particle settling within the suspending fluid. For gravity settling, $a_e = g$. For centrifugation, $a_e = \omega^2 r_c$, where r_c = radius of rotation and ω = the angular velocity of rotation.

13.3.3 The Drag Coefficient

C_d used in the derivation of Equation (13.39) is only good when v_r is small and laminar conditions exists at the fluid-particle interface. In general, C_d can be determined by calculating an index K as follows:

$$K = d\left[\frac{a_e\rho(\rho - \rho_p)}{\mu^2}\right]^{0.33} \tag{13.40}$$

Data by Lapple and Shepherd (Ind. Eng. Chem. 32:605, 1940) shows that the value of C_d can be determined for different values of K as follows:

If K > 0.33, or Re < 1.9; $C_d = 24/Re$
If 1.3 < K < 44 or 1.9 < Re < 500; $C_d = 18.5/Re^{0.6}$
If K > 44 or Re > 500; $C_d = 0.44$

If the Reynolds number at the particle to fluid interface exceeds 1.9, Equation (13.35) and (13.37) are combined in Equation (13.34), and under conditions when the particle is at the terminal velocity:

$$ma_e\left[1 - \left(\frac{\rho}{\rho_p}\right)\right] = C_d A_p (v_r)^2 \frac{\rho}{2}$$

Substituting $A_p = \pi d^2/4$; and $m = \pi d^3 \rho_p/6$ and solving for v_r:

$$v_r = 1.155\left[\frac{a_e d(\rho_p - \rho)}{\rho C_d}\right]^{0.5} \tag{13.41}$$

Example 13.10. In the processing of soybeans for oil extraction, hulls must be separated from the cotyledons in order that the soy meal eventually produced will have a high value as feed or as a material for further precessing for food proteins. The whole soybeans are passed between a cracking roll which splits the cotyledons and produces a mixture of cotyledons and hulls. Figure 13.9 shows a system that can be used to separate hulls from the cotyledons by air classification. The projected diameters and the densities of cotyledons and hulls are 4.76 mm and 1003.2 kg/m^3 and 6.35 and 550 kg/m^3, respectively. The process is carried out at 20°C. Calculate the terminal velocity of hulls and cotyledons in air. The appropriate velocity through the aperture will be between these two calculated terminal velocities.

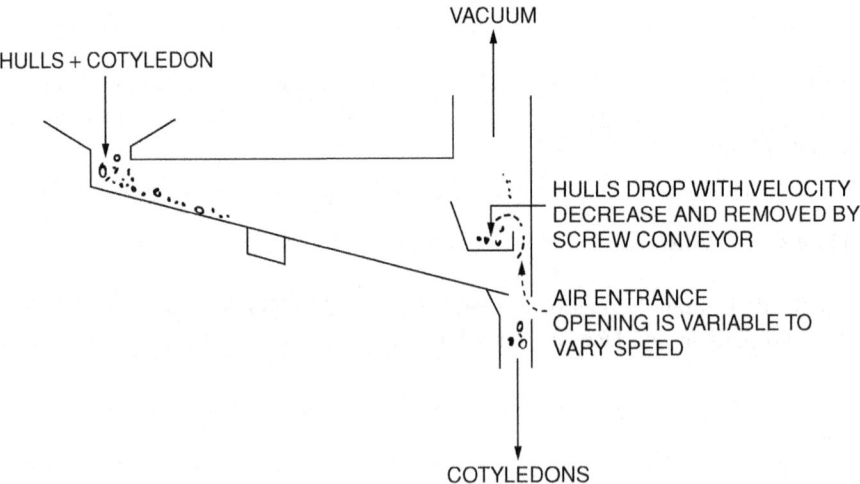

Figure 13.9 Diagram of air classification system to separate soybean hulls from the cotyledons.

Solution:

The density of air at 20°C is calculated using the ideal gas equation. The molecular weight of air is 29; atmospheric pressure = 101.325 kPa;

$$R = 8315 \ N \cdot M/(kgmole \cdot K)$$

$$\rho = \frac{p(29)}{RT} = \frac{101,325(29)}{8315(293)} = 1.206 kg/m^3$$

The viscosity of air at 20°C is 0.0175 centipoise (from *Perry's Chemical Engineers Handbook*). Using Equation (13.40):

$$a_e = \text{acceleration due to gravity} = 9.8 m/s^2.$$

For the cotyledons:

$$K = 0.00476 \left[\frac{9.8(1.206)(1003.2 - 1.206)}{[(0.0175)(0.001)]^2} \right]^{0.33} = 145$$

Because K > 44, $C_d = 0.44$.
    Using Equation (13.41): $a_e = 9.8 \ m/s^2$

$$v_r = 1.155 \left[\frac{9.8(0.00476)(1003.2 - 1.206)}{1.206(0.44)} \right]^{0.5} = 10.84 m/s$$

For the hulls:

$$K = 0.00476 \left[\frac{9.8(1.206)(550 - 1.206)}{[0.0175(0.001)]^2} \right]^{0.33} = 118.9$$

K > 444 ; therefore, $C_d = 0.44$.
    Using Equation (13.41):

$$v_r = 1.155 \left[\frac{9.8(0.00635)(550 - 1.206)}{1.206(0.44)} \right]^{0.5} = 9.266 \ m/s$$

Thus air velocity between 9.266 and 10.84 m/s must be used on the system to properly separate the hulls from the cotyledons.

Equations (13.37) to (13.40) also applies in sedimentation, flotation, centrifugation, and fluidization of a bed of solids. In flotation separation where gravity is the only external force acting on the particles, Equation (13.39) shows that solids having different densities can be separated if the flotation fluid density is between the densities of the two solids. The solid with lower density will float and that with the higher density will sink. Air classification of solids with different densities as shown in the example can be achieved by suspending the solids in air of the appropriate velocity, the denser solid will fall and the lighter solid will be carried off by the air stream. Air classification is a commonly used method for removal of lighter foreign material from grains, and in fractionation of milled wheat into flour with different protein and bran contents. Density gradient separations are employed in fractionating beans of different degrees of maturity, and crab meat from scraps after manual removal of flesh.

In fluidization, the fluid velocity must equal the terminal velocity to maintain the particles in a stationary motion in the stream of air. Fluidized bed drying and freezing are commercially practiced in the food and pharmaceutical industries.

PROBLEMS

13.1. Based on data for apple juice filtration in Table 13.2, calculate the filtration area of a rotary filter that must be used to filter 4000 L/h of juice having a suspended solids content of 50 kg/m^3. Assume k is proportional to solids content. The medium resistance increases with increasing pre-coat thickness. A 10-cm-thick precoat is usually used on rotary filters. Assume the medium resistance is proportional to the precoat thickness. The rotational speed of the filter is 2 rev/min and the diameter is 2.5 m. A filtration cycle on a rotary filter is the time of immersion of the drum in the slurry and in this particular system, 45% of the total circumference is immersed at any given time.

13.2. The following data were obtained in a laboratory filtration of apple juice using a mixture of rice hulls and perlite as the filter aid. The filtrate had a viscosity of 2.0 centipose. The suspended solids in the juice is 5 g/L, and rice hulls and perlite were added each at the same concentration as the suspended solids. The filter has an area of 34 cm^3. Calculate:
(a) The specific cake resistance and the medium resistance.
(b) The optimum filtration time per cycle and the average volume of filtrate per unit area per hour at the optimum cycle time.

Time(s)	Volume (mL)	Time (s)	Volume (mL)
100	40	400	83
200	60	500	91
300	73		

13.3. The following data were reported by Slack [Process Biochem. 17(4):7, 1982] on flux rates at different solids content during ultrafiltration of milk. Test if the flux vs. solids content follow that might be predicted by the theory on polarization concentration. What might be the reasons for the deviation? All flow rates are the same and the same membrane was used in the series of tests.

Mean solids Content (%)	Flux (L/(m$^2 \cong$h))
12.3	28.9
13.6	25.5
14.9	22.1
24.1	23.8

13.4. The following analysis has been reported for retentate and permeate in ultrafiltration of skim milk. Retentate: 0.15% fat, 16.7% protein, 4.3% lactose, 22.9% total solids. Permeate: 0% fat, 0% protein, 4.6% lactose, 5.2% total solids.
(a) Calculate the rejection factor for lactose by the membrane used in this process.
(b) If the membrane flux in this process follows the data in Problem 3, calculate the lactose content of skim milk which was subjected to a diafiltration process where the original milk containing 8.8% total solids, 4.5% lactose, 3.3% protein and 0.03% fat was concentrated to 17% total solids, diluted back to 10 % total solids and re-concentrated by UF to 20% total solids.

13.5. Calculate the average particle diameter and total particle surface area assuming spherical particles, per kg solids, for a powder which has a bulk density of 870 kg/m^3 and which has the following particle size distribution:

+14,	−10 mesh	10%
+18,	−14 mesh	14.6%
+25,	−18 mesh	22.3%
+35,	−45 mesh	30.2%
+60,	−45 mesh	22.9%

SUGGESTED READING

Cheryan, M. 1986. Ultrafiltration Handbook. Technomic Publishing Co., Lancaster, PA.

Crewspo, J. G. and Boddeker, K. W. 1994. Membrane Processes in Separation and Purification. Kluwer Academic Publishers, Boston.

Green. D. W. and Mahoney, J. O. 1997. Perry's Chemical Engineers Handbook. 7th ed. Chapter 18 and 22. McGraw Hill, New York.

Noble, R. D. and Stern, S. A. 1995. Membrane Separations Technology: Principles and Applications. Elsevier, New York.

Poole J. B. and Doyle, D. 1968. Solid-Liquid Separation. Chemical Publishing Co., New York.

Porter, M. C. 1979. Membrane Filtration. In: Handbook of Separation Techniques for Chemical Engineers. P. A. Schweitzer, Ed. McGraw-Hill, New York.

Sourirajan, S. 1970. Reverse Osmosis. Logos Press, New York.

CHAPTER 14

Extraction

One area in food processing that is receiving increasing attention is extraction. This separation process involves two phases. The solvent is the material added to form a phase different from that where the material to be separated originally was present. Separation is achieved when the compound to be separated dissolves in the solvent while the rest of the components remain where they were originally. The two phases may be solid and liquid, immiscible liquid phases, or solid and gas. Solid-liquid extraction is also called leaching. In supercritical fluid extraction, gas at supercritical conditions contacts a solid or a liquid solution containing the solute. Extraction has been practiced in the vegetable oil industry for a long time. Oil from soybean, corn, and rice bran cannot be separated by mechanical pressing, therefore, solvent extraction is used for their recovery. In the production of olive oil, the product from the first pressing operation is the extra virgin olive oil, the residue after first press may be re-pressed to obtain the virgin olive oil, and further recovery of oil from the cake is done by solvent extraction. Oil from peanuts is recovered by mechanical pressing and extraction of the pressed cake to completely remove the oil. One characteristic of solvent extracted oilseed meal is the high quality of the residual protein, suitable for further processing into food-grade powders. They may also be texturized for use as food protein extenders.

Extraction of spice oils and natural flavor extracts has also been practiced in the flavor industry. Interest in functional food additives used to fortify formulated food products has led to the development of extraction systems to separate useful ingredients from food processing waste and medicinal plants.

Extraction is also used in the beet sugar industry to separate sugar from sugar beets. Sugar from sugar cane is separated by multistage mechanical expression with water added between stages. This process may also be considered a form of extraction. Roller mills used for mechanical expression of sugar cane juice is capital intensive and when breakdowns occur, the down time is usually very lengthy. It is also an energy intensive process, therefore, modern cane sugar processing plants are installing diffusers, a water extraction process, instead of the multiple roller mills previously used.

In other areas of the food industry, water extraction is used to remove caffeine from coffee beans, and water extraction is used to prepare coffee and tea solubles for freeze or spray drying. Supercritical fluid extraction has been found to be effective for decaffeinating coffee and tea and for preparing unique flavor extracts from fruit and leaves of plants.

14.1 TYPES OF EXTRACTION PROCESSES

Extraction processes may be classified as follows.

14.1.1   Single-Stage Batch Processing

In this process, the solid is contacted with solute-free solvent until equilibrium is reached. The solvent may be pumped through the bed of solids and recirculated, or the solids may be soaked in the solvent with or without agitation. After equilibrium, the solvent phase is drained out of the solids. Examples are brewing coffee or tea, and water decaffeination of raw coffee beans.

14.1.2   Multistage Cross-Flow Extraction

In this process, the solid is contacted repeatedly, each time with solute free solvent. A good example is soxhlet extraction of fat in food analysis. This procedure requires a lot of solvent, or in the case of a soxhlet, a lot of energy is used in vaporizing and condensing the solvent for recycling, therefore, it is not used as in industrial separation process.

14.1.3   Multistage Countercurrent Extraction

This process utilizes a battery of extractors. Solute-free solvent enters the system at the opposite end from the point of entry of the unextracted solids. The solute-free solvent contacts the solids in the last extraction stage, resulting in the least concentration of solute in the solvent phase at equilibrium at this last extraction stage. Thus, the solute carried over by the solids after separation from the solvent phase at this stage is minimal. Solute-rich solvent, called the extract, emerges from the system at the first extraction stage after contacting the solids that had just entered the system. Stage to stage flow of solvent moves in a direction countercurrent to that of the solids. The same solvent is used from stage to stage, therefore solute concentration in the solvent phase increases as the solvent moves from one stage to the next, while the solute concentration in the solids decreases as the solids move in the opposite direction. A good example of a multistage countercurrent extraction process is oil extraction from soybeans using a carousel extractor. This system called the "rotocell" is now in the public domain and can be obtained from a number of foreign equipment manufacturers. A similar system produced by Extractionstechnik GmbH of Germany was described by Berk in a FAO publication. In this system (Fig. 14.1), two cylindrical tanks are positioned over each other. The top tank rotates while the lower tank is stationary. Both top and bottom tanks are separated into wedges, such that the content of each wedge are not allowed to mix. Each wedge of the top tank is fitted with a swinging false bottom to retain the solids, while a pump is installed to draw out solvent from each of the wedges except one, in the lower tank. A screw conveyor is installed in one of the wedges in the lower tank to remove the spent solids and convey them to a desolventization system. The false bottom swings out after the last extraction stage to drop the solids out of the top tank into the bottom wedge filled with the screw conveyor. The movement of the wedges on the top tank is indexed such that with each index, each wedge will be positioned directly over a corresponding wedge in the lower tank. Thus, solvent draining through the bed of solids in a wedge in the top tank will all go into one wedge in the lower tank. Solvent taken from the wedge forward of the current wedge is pumped over the bed of solids, drains through the bed, and enters the receiving tank, from which another pump transfers this solvent to the top of the bed of solids in the preceding wedge. After the last extraction stage, the swinging false bottom drops down releasing the solids, the swinging false bottom is lifted in place, and the empty wedge receives fresh solids to start the process over again. A similar system although of a different design, is employed in the beet sugar industry.

Figure 14.1   The rotating basket extractor. (Source: Berk, Z. 1992. Technology for the production of edible flours and protein products from soybeans. FAO Agricultural Services Bulletin 97. Food and Agriculture Organization of the United Nations, Rome.)

14.1.4   Continuous Countercurrent Extractors

In this system, the physical appearance of an extraction stage is not well defined. In its most simple form, an inclined screw conveyor may be pictured. The conveyor is initially filled with the solvent to the overflow level at the lower end, and solids are introduced at the lower end. The screw moves the solids upward through the solvent. Fresh solvent introduced at the highest end, will move countercurrent to the flow of solids picking up solute from the solids as the solvent moves down. Eventually, the solute-rich solvent collects at the lowermost end of the conveyor and is withdrawn through the overflow. In this type of extraction system, term "height of a transfer unit" (HTU) is used to represent the length of the conveyer where the solute transfer from the solids to the solvent is equivalent to one equilibrium stage in a multistage system.

Continuous conveyor type extractors are now commonly used in the oilseed industry. One type of extractor is a sliding cell basket extractor (Fig. 14.2A). The baskets affixed to a conveyor chain have false bottoms, which permits solvent sprayed at the top to percolate through and collect at a reservoir at the bottom of the unit. Pumps take the solvent from the reservoirs and takes them to nozzles at the top of the baskets. The discharge point of the solvent at the top of the baskets is advanced such that the solvent weak in solute is fed to the baskets forward of the baskets from where the solvent had

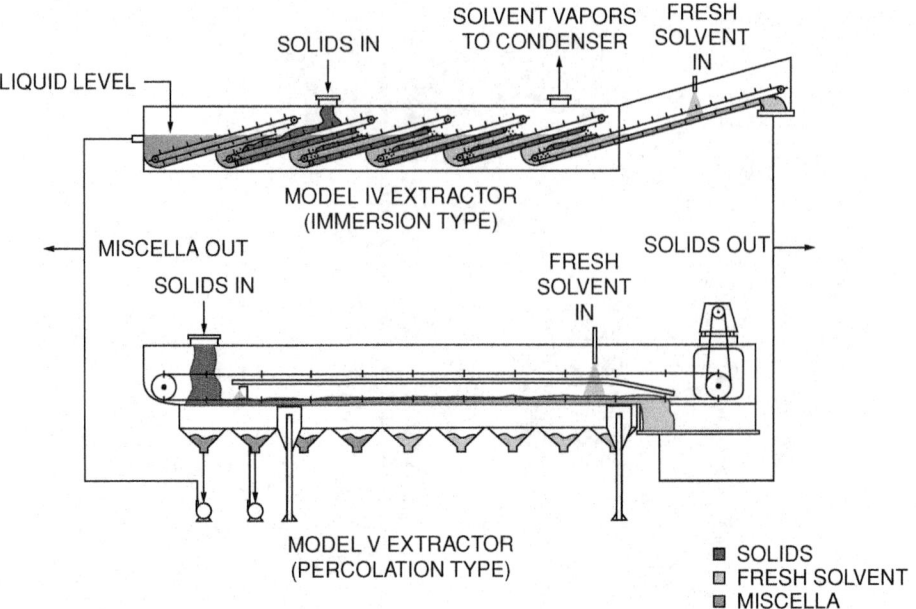

Figure 14.2   Continuous belt-type extractor. (A) An immersion-type multistage countercurrent extractor. (B) A percolation-type extractor.

previously percolated. These units have been described in Berk's article and are produced by a German firm, Lurgi, GmbH. Another extractor suitable for not only oilseed extraction but also for extraction of health-functional food ingredients from plant material, is a perforated belt extractor. Figure 14.2B shows a perforated belt extractor produced in the United States by Crown Iron Works of Minneapolis, Minnesota. This unit is made to handle as small as 5 kg of solvent/h. A single continuous belt moves the solids forward while solvent is sprayed over the solids. A series of solvent collection reservoirs underneath the conveyor evenly spaced along the length of the unit, separates the solvent forming the different extraction stages. Each collection reservoir has a pump which takes out solvent from one stage and this liquid is applied over the solids on the conveyor in such a manner that the liquid will drain through the bed of solids and collect in another collection reservoir of the preceding stage.

Most extractions in the food industry involve solid-liquid extraction, therefore, the discussion in this chapter will be limited to solid-liquid extraction.

14.2 GENERAL PRINCIPLES

The following are the physical phenomena involved in extraction processes.

14.2.1 Diffusion

Diffusion is the transport of molecules of a compound through a continuum in one phase, or through an interface between phases. In solid liquid extraction (also known as leaching), the solvent must diffuse into the solid in order for the solute to dissolve in the solvent, and the solute must diffuse out of the solvent saturated solid into the solvent phase. The rate of diffusion determines the length of time needed to achieve equilibrium between phases. In Chapter 12, section "Mass Diffusion," the solution to the differential equation for diffusion through a slab is presented in Equation (12.41) in terms of the average concentration of solute within a solid as a function of time. This equation may be used to calculate the time needed for equilibrium to be achieved when the quantity of solid is small relative to the solvent such that concentration of solute in the solvent phase remains practically constant. Qualitatively, it may be seen that the time required for diffusion to occur in order to reach equilibrium, is inversely proportional to the square of the diffusion path. Thus, in solvent extraction, the smaller the particle size, the shorter the residence time for the solids to remain within an extraction stage.

Particle size, however, must be balanced by the need for the solvent to percolate through the bed of solids. Very small particle size will result in very slow movement of the solvent through the bed of solids, and increases the probability that fines will go with the solvent phase interfering with subsequent solute and solvent recovery.

In soybean oil extraction, the soy is tempered to a certain moisture content in order that they can be passed through flaking rolls to produce thin flakes without disintegration into fine particles. The thin flakes have very short diffusion path for the oil, resulting in short equilibrium time in each extraction stage, and solvent introduced at the top of the bed of flakes percolates unhindered through the bed. The presence of small particle solids is not desirable in this system because the fine solids are not easily removed from the solvent going to the solvent/oil recovery system. The high temperature needed to drive off the solvent will result in a dark colored oil if there is a large concentration of fine particles.

In cottonseed and peanut oil extraction, a pre-press is used to remove as much oil as can be mechanically expressed. The residue comes out of the expeller as small pellets which then goes into the extractor.

Some raw materials may contain lipoxygenase, which catalyzes the oxidation of the oil. Extraction of oil from rice bran involves the use of an extruder to heat the bran prior to extraction to inactivate lipoxygenase. The extruder produces small pellets which facilitates the extraction process by minimizing the amount of fines that goes with the solvent phase.

In cane sugar diffusers, hammer mills are used to disintegrate the cane such that the thickness of each particle is not more than twice the size of the juice cells. Thus, equilibrium is almost instantaneous upon contact of the particles with water. The cane may be pre-pressed through a roller mill to crush the cane and produce very finely shredded solids for the extraction battery.

14.2.2 Solubility

The highest possible solute concentration in the final extract leaving an extraction system is the saturation concentration. Thus, solvent to solids ratio must be high enough such that, when fresh solvent contacts fresh solids, the resulting solution on equilibrium, will be below the saturation concentration of solute.

In systems where the solids are repeatedly extracted with recycled solvent (e.g., supercritical fluid extraction), a high solute solubility will reduce the number of solvent recycles needed to obtain the desired degree of solute removal.

14.2.3 Equilibrium

When the solvent to solid ratio is adequate to satisfy the solubility of the solute, equilibrium is a condition where the solute concentration in both the solid and the solvent phases are equal. Thus, the solution adhering to the solids will have the same solute concentration as the liquid or solvent phase. When the amount of solvent is inadequate to dissolve all the solute present, equilibrium is considered as a condition where no further changes in solute concentration in either phase will occur with prolonged contact time. In order for equilibrium to occur, enough contact time must be allowed for the solid and solvent phases.

The extent to which the equilibrium concentration of solute in the solvent phase is reached in an extraction stage is expressed as a stage efficiency. If equilibrium is reached in an extraction stage, the stage is 100% efficient and is designated an "ideal stage."

14.3   SOLID-LIQUID EXTRACTION: LEACHING

14.3.1   The Extraction Battery: Number of Extraction Stages

Figure 14.3 shows a schematic diagram of an extraction battery with n stages. The liquid phase is designated the overflow, the quantity of which is represented by V. The solid phase is designated the underflow, the quantity of which is represented by L. The extraction stages is numbered 1 after the mixing stage where the fresh solids first contacts the solute laden extract from the other stages, and n as the last stage where fresh solvent first entered the system and where the spent solids leave the system. The stage where extract from the first extraction stage contacts the fresh solids, is called the mixing stage. It is different from the other stages in the extraction battery because at this stage the solids have to absorb a much larger amount of solvent than in the other stages.

Each of the stages is considered an ideal stage. Solute concentration in the underflow leaving stage n, must be at the designated level considered for completeness of the extraction process. Residual solute in the solids fraction must be maintained at a low level. If it is a valuable solute, the efficiency of solute recovery is limited only by the cost of adding more extraction stages. On the other hand, if the spent solids is also valuable and the presence of solute in the spent solids affects its value, then the number of extraction stages must be adequate to reduce the solute level to a minimum desirable value. For example, in oilseed extraction, the residual oil in the meal must be very low, otherwise the meal will rapidly become rancid.

Determination of the number of extraction stages may be done using a stage by stage material balance. Because the only conditions known are those at the entrance and outlet from the

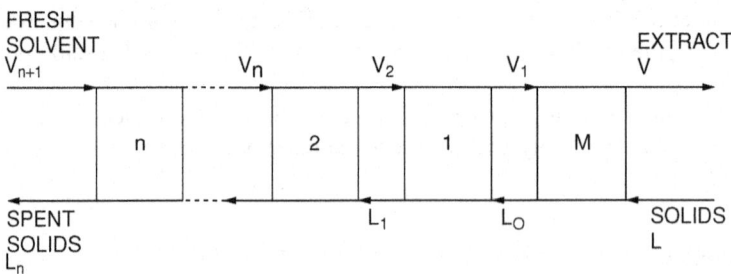

Figure 14.3   Schematic diagram of an extraction battery for multistage countercurrent extraction.

extraction system, the stage by stage material balance will involve setting up a system of equations that are solved simultaneously to determine the conditions of solute concentration in the underflow and overflow leaving each stage. The procedure will involve assumption of a number of ideal stages, solving the equations, and comparing if the calculated level of solute in the underflow from stage n, matches the specified value. The process is very tedious. A graphical method is generally used.

14.3.2 Determination of the Number of Extraction Stages Using the Ponchon-Savarit Diagram

This graphical method for determining the number of extraction stages in a multi-stage extraction process, involves the use of an X-Y diagram. The coordinates of this diagram are defined as follows:

$$Y = \frac{solid}{solute + solvent}$$

$$X = \frac{solute}{solute + solvent}$$

where solid = concentration of insoluble solids; solute = concentration of solute, and solvent = concentration of solvent. Thus, the composition of any stream entering and leaving an extraction stage can be expressed in terms of the coordinates X,Y.

Figure 14.4 shows the X-Y diagram for a solid-liquid extraction process. The overflow line represents the composition of the solvent phase leaving each stage. If no solids entrainment occurs, the overflow line should be represented by Y = 0. The underflow line is dependent on how much of the solvent phase is retained by the solids in moving form one stage to the next. The underflow line will be linear if the solvent retained is constant, and curved when the solvent retained by the solids varies with the concentration of solute. Variable solvent retention in the underflow occurs when the presence of solute increases significantly the viscosity of the solvent phase.

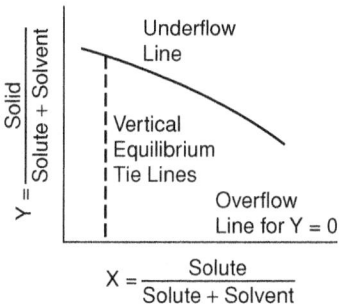

Figure 14.4 The underflow and overflow curves and equilibrium tie lines on a Ponchon-Savarit diagram.

14.3.3   The Lever Rule in Plotting Position of a Mixture of Two Streams in an X-Y Diagram

Let two streams with mass, R and S, and with coordinates X_r and Y_r, and X_s and Y_s, respectively, be mixed together to form T with coordinates X_t and Y_t. The diagram for the material balance is shown in Fig. 14.5(I). If R and S consist of only solute and solvent, solute balance results in:

$$RX_r + SX_s = (R + S)X_t \tag{14.1}$$

A solids balance gives:

$$RY_r + SY_s = (R + S)Y_t \tag{14.2}$$

Solving for R in Equation (14.1) and S in Equation (14.2) and dividing:

$$\frac{R}{S} = \frac{X_s - X_t}{X_t - X_r} = \frac{Y_s - Y_t}{Y_t - Y_r} \tag{14.3}$$

Figure 14.5(II) represents the term involving Y in Equation (14.3). The coordinate for the mixture should always be between those of its components. The ratio of the distance between the line $Y = Y_s$ and the line $Y = Y_t$, represented on the diagram by A, and the distance between the line, $Y = Y_t$ and the line $Y = Y_r$, represented on the diagram by B, will equal the ratio of the mass of R and S.

Figure 14.5(III) represents the term involving X in Equation (14.3). Again, the ratio of the masses of R and X equals the ratio of the distance C/distance D in the X-Y diagram. The composite of the material balance is drawn in Fig. 14.5(IV). The coordinates of points S and R when plotted in the X-Y diagram and joined together by a straight line, will result in the point representing T to be in the line between S and R. The ratio of the distance on the line, between S and T, represented by E, and that between R and T, represented by F is the ratio of the mass of R to S.

These principles show that a material balance can be represented in the X-Y diagram with each process stream represented as a point on the diagram and any mixture of streams can be represented

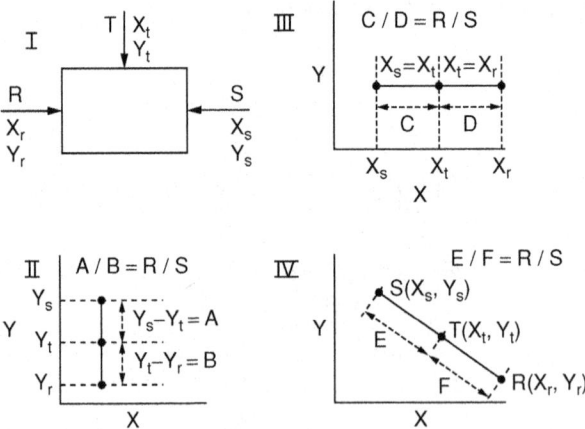

Figure 14.5   Representation of a material balance on the X-Y diagram, and the lever rule for plotting mass ratio of process streams as ratio of distances between points on the X-Y diagram.

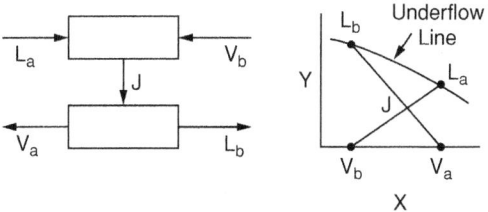

Figure 14.6 Representation of the point J as the intersection of lines connecting four process streams entering and leaving a system.

---

by a point on the line drawn between the coordinates of the components of the mixture. The exact positioning of the location of the point representing the mixture can be made using the lever rule on distances between the points as represented in Fig. 14.5(IV).

14.3.4 Mathematical and Graphical Representation of the Point J in the Ponchon-Savarit Diagram

Consider underflow, L_b mixing with overflow, V_a, and the solid and solvent phases later separated to form the overflow stream V_b and the underflow stream L_a. A total mass balance gives:

$$L_b + V_a = V_b + L_a = J \tag{14.4}$$

The mixture of L_b and V_a forms the point J, shown in Fig. 14.6. A line drawn between the coordinates of L_b and V_a and another line drawn between the point V_b and L_a will intersect at point J. The point J may be used to help in plotting points representing the incoming and exiting streams in an extraction battery from the solvent to solids ratio.

14.3.5 Mathematical and Graphical Representation of the Point P

The point P is a mixture that results when the underflow leaving a stage mixes with the overflow entering that stage. It is also the mixture of overflow leaving a stage and underflow entering that stage. The mixture representing point P is shown schematically in Fig. 14.7. A total mass balance around

---

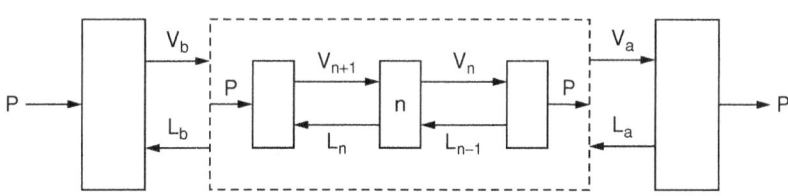

Figure 14.7 Representation of the point P for the common point through which all lines connecting the underflow leaving and the overflow entering an extraction stage must pass.

the system represented by the dotted line in Fig. 14.6 gives:

$$-V_b + L_b = -V_a + L_a = -V_{n+1} + L_n = P \tag{14.5}$$

Thus, the point P will be an extrapolation of the line that joins V_b and L_b; the line that joins V_a and L_a; and the line that joins V_{n+1} and L_n. All lines which join the underflow stream leaving a stage and the overflow stream which enters that stage, will all meet at a common point P. This is a basic principle used to draw the successive stages in a Ponchon-Savarit diagram for stage by stage analysis of an extraction process.

14.3.6   Equation of the Operating Line and Representation on the X-Y Diagram

Figure 14.8 represents an extraction battery with n cells. Each stage in the extractor may also be called an extraction cell. Cell $n + 1$ is the cell after cell n. The subscripts on the overflow stream, V, and the underflow stream, L, represent the cell from which the streams is leaving. Thus, the solvent phase entering cell n comes from cell $n + 1$ and is designated V_{n+1} and the solid phase leaving cell n is designated L_n. Let V represent the mass of solute and solvent in an overflow stream, and L represent the mass of solute and solvent in an underflow stream. A material balance around the battery of n cells is as follows:

Total mass balance:

$$V_{n+1} + L_a = L_n + V_a; \qquad V_{n+1} = L_n + V_a - L_a \tag{14.6}$$

Solute balance:

$$V_{n+1}X_{n+1} = V_aX_{Va} - L_aX_{La} + L_nX_n \tag{14.7}$$

Substituting Equation (14.6) in Equation (14.7) and solving for X_{n+1}:

$$X_{n+1} = \frac{L_n}{L_n + V_a - L_a} X_n + \frac{V_a X_{Va} + L_a X_{La}}{L_n + V_a - L_a} \tag{14.8}$$

Equation (14.8) is the equation of an operating line. It shows that the point representing stage $n + 1$, which has the coordinate X_{n+1}, is on the same line drawn through the point representing stage n, which has the coordinate X_n.

It can be shown that the coordinates of point P represented by Equation (14.5) can also satisfy Equation (14.8), thus, a line drawn from a point representing stage n to point P will allow the determination of a point representing the coordinate of stage $n + 1$.

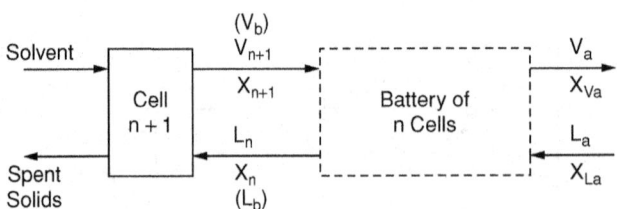

Figure 14.8   Diagram representing the system used to make a material balance for deriving the equation of the operating line in a Ponchon-Savarit diagram.

14.3.7   Construction of the Ponchon-Savarit Diagram for the Determination of the Number of Ideal Extraction Stages

Figure 14.9 shows the X-Y diagram for determination of the number of ideal stages. Known points representing the coordinates of solvent saturated solids from the mixing stage, L_a, the extract, V_a, the spent solids, L_b, and the fresh solvent entering the system, V_b, are plotted first on the X-Y diagram. A total material balance must first be made to determine the coordinates of L_b and V_a from the solvent to solids ratio.

Point P is established from the intersection of lines going through L_b and V_b and L_a and V_a.

The condition of equilibrium is represented by a vertical line. Equilibrium means that X_{Ln} and X_{Vn} are equal. The equilibrium line is also known as a "tie line" and may not be vertical if equilibrium is not achieved. However, it is easier to assume equilibrium to occur, determine the number of ideal extraction stages, and incorporate the fact that equilibrium does not occur in terms of a stage efficiency.

The succeeding stages are established by drawing a line from the point representing stage n on the underflow line to point P. The intersection of this line with the overflow curve will determine X_{n+1}.

Example 14.1. Draw a diagram for a single stage extraction process involving beef (64% water, 20% fat, 16% nonextractable solids) and isopropyl alcohol in a 1 to 5 ratio. Isopropyl alcohol and water are totally miscible, and the mixture is considered to be the total solvent. Assume all fat dissolves in this solvent. Following equilibrium, the solids fraction separated by filtration retained 10% by weight of the total solution of fat and solvent. Determine graphically the concentration of the fat in the extract and calculate the fat content in the solvent free solids.

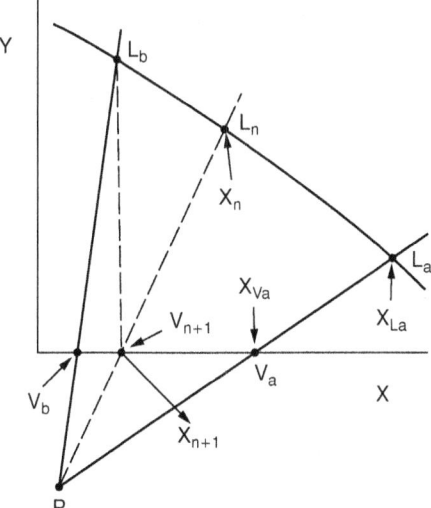

Figure 14.9 Plotting of point P, the operating line, and the equilibrium tie-line on a Ponchon-Savarit diagram.

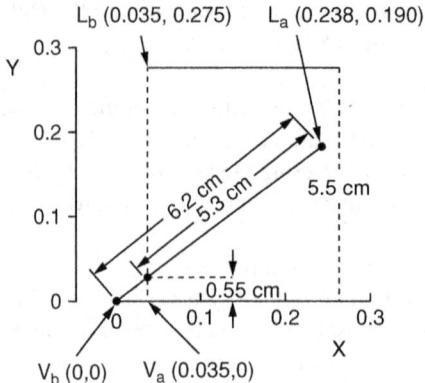

Figure 14.10   X-Y diagram for a single-stage extraction process.

Solution:

Basis: 100 kg beef. Because the X-Y diagram uses as a denominator for the X and Y coordinates only the mass of the solution phase rather than total mass, the distances on the diagram should be scaled on the basis of the mass of solute and solvent present in each stream. Let L_a represent the beef.

$X_{La} = 20/(20 + 64) = 0.238$

$Y_{La} = 16/(20 + 64) = 0.190$

The solvent, V_b, will have the coordinates (0,0) because it contains no solute nor solids. L_a and V_b are plotted in Fig. 14.10.

Point J, the point representing the mixture before filtration is determined using the lever rule as follows: The ratio of the length of line segment $L_a J$/length of line segment $V_b L_a$ equals the ratio of the mass of solvent and solute in V_b to the total mass of solvent and solute in V_b and L_a.

Let: $V_b L_a = 100$ units

The ratio $V_b/(L_a + V_b) = 500/(20 + 64 + 500) = 0.856$. Therefore, the distance $V_b J$ should be 85.6 units. For example, when plotting on a scale of 1 cm = 0.5 units for X and Y, length of $V_b L_a = 6.2$ cm. Thus, $L_a J$ should be 6.2(0.856) = 5.31 cm long. These distances, when plotted, establish point J, which has coordinates (0.035, 1.1275). The vertical line X = 0.035 is the equilibrium line. Assuming no insoluble solids in the extract, V_a has coordinates (0.035, 0). Point L_b is determined by plotting the distance $V_a L_b$ (5.5 cm), which is 10 times $V_a J$ (0.55 cm) because 10% of the solution is retained in the underflow. The ratio X_{Lb}/Y_{Lb} = solute/solids in L_b = mass fraction solute in the solvent free solids = 0.127.

Example 14.2. Table 14.1 represents the amount of solution retained in soybean meal as a function of the oil concentration.

(a) Draw the underflow curve.
(b) If a solvent to soy ratio of 0.5 to 1 is used for extraction, and the original seed contains 18% oil, determine the number of extraction stages needed such that the meal after final desolventization will have no more than 0.01 kg oil/kg oil free meal.

Table 14.1 Data for Overflow Composition During Extraction of Oil from Soybeans.

$X = \dfrac{kg\ oil}{kg\ soln.}$	Soln. retained kg/kg solids	$Y = \dfrac{Solids}{kg\ soln.}$
0	0.5	2.0
0.1	0.505	1.980
0.2	0.515	1.942
0.3	0.530	1.887
0.4	0.550	1.818
0.5	0.571	1.751
0.6	0.595	1.680
0.7	0.620	1.613

Solution represents the solvent phase, mass of solvent plus mass of dissolved solute.

Solution:

(a) The coordinates of points representing the underflow curve are given in Table 14.1 The parameter Y is the reciprocal of the solution retained/kg solids. The underflow curve is linear except for the last three points at high oil concentrations where there was a slight deviation from linearity. A regression analysis results in the equation of the best fit line as:

$$Y = -0.5775\,X + 2.036$$

Figure 14.11 shows the Ponchon-Savarit diagram for this problem. The underflow curve is plotted in Fig. 14.12.

(b) The composition of the final extract and spent solids stream is calculated by performing a material balance around the whole system. Basis: 1 kg soybean; solvent = 0.5 kg From the specified level of oil in the extracted meal:

$$\text{kg oil in spent solids} = \frac{0.01\ \text{kg oil}}{\text{kg solids}} 0.82\ \text{kg solids} = 0.0082$$

From the data in Table 14.1, at very low oil contents, the amount of solution retained by the solids is 0.5 kg/kg solids.

$$\text{kg soln. in spent solids} = \frac{0.5\ \text{kg soln.}}{\text{kg solids}} 0.82\ \text{kg solids} = 0.41$$

$$\text{kg solvent} = \frac{1\ \text{kg solvent}}{1.0082\ \text{kg soln.}} 0.41\ \text{kg soln.} = 0.4018\ \text{kg solvent in spent solids}$$

Mass of L_n, the spent solids = 0.82 + 0.0082 + 0.4018 = 1.23 kg

Coordinates of spent solids, L_n:

$X = 0.0082/0.41 = 0.02$ kg solute/soln.

$Y = 1\,\text{kg solids}/0.5\,\text{kg soln.} = 2.0\,\text{kg solids/kg soln.}$

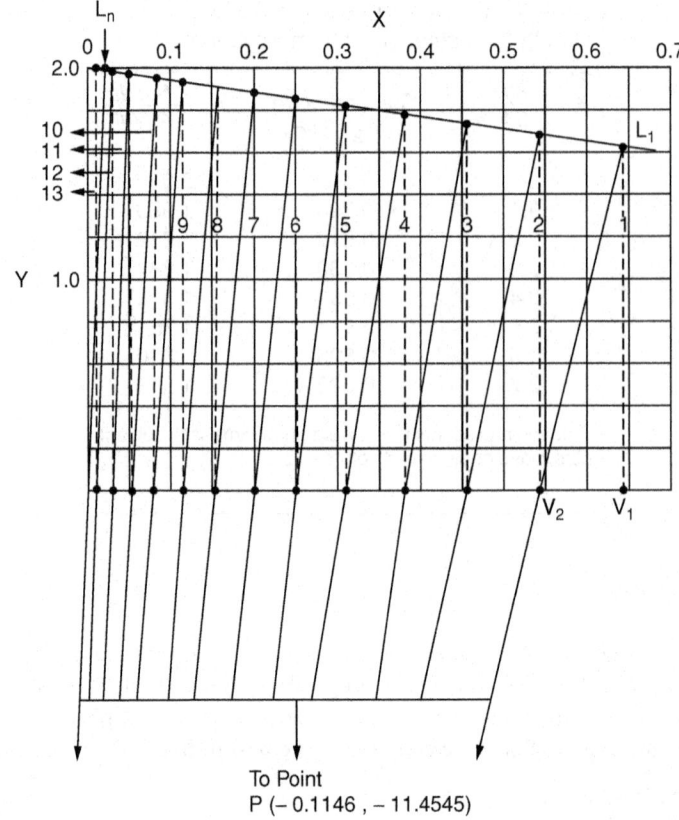

Figure 14.11   The underflow curve and operating lines establishing the number of equilibrium stages needed to solve the example problem on continuous countercurrent oil extraction from soybeans.

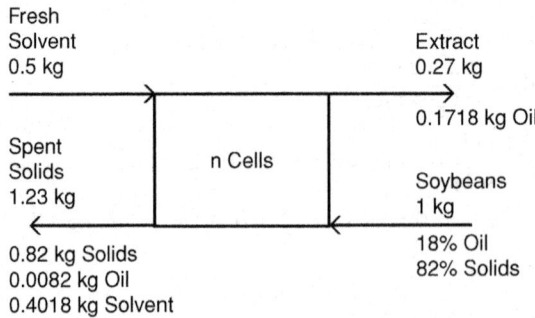

Figure 14.12   Overall material balance to determine the extract mass and composition and spent solids mass and composition for the example problem on soybean extraction.

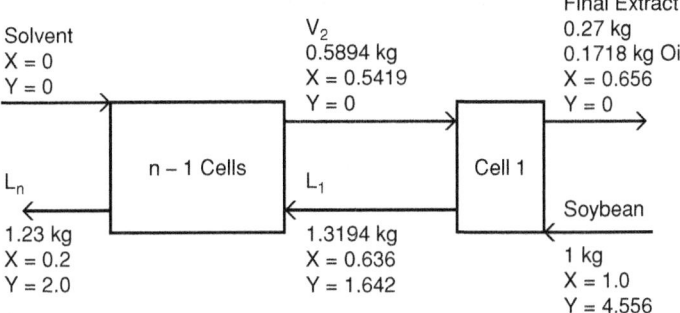

Figure 14.13 Diagram showing material balance around the mixing stage, Cell 1, to establish the composition of overflow stream leaving and underflow stream entering the extraction battery after the mixing stage.

Figure 14.13 shows the overall material balance with the mass and components of each stream entering and leaving the extraction battery.

Mass of extract $V_a = 1.5 - 1.23 = 0.27\,\text{kg}$

Mass of oil $V_a = 0.82 - 0.0082 = 0.1718\,\text{kg}$

The Ponchon-Savarit diagram can be constructed only for the cells following the mixing stage. Figure 14.14 shows the material balance for the mixing stage which is considered as cell 1 in this case,

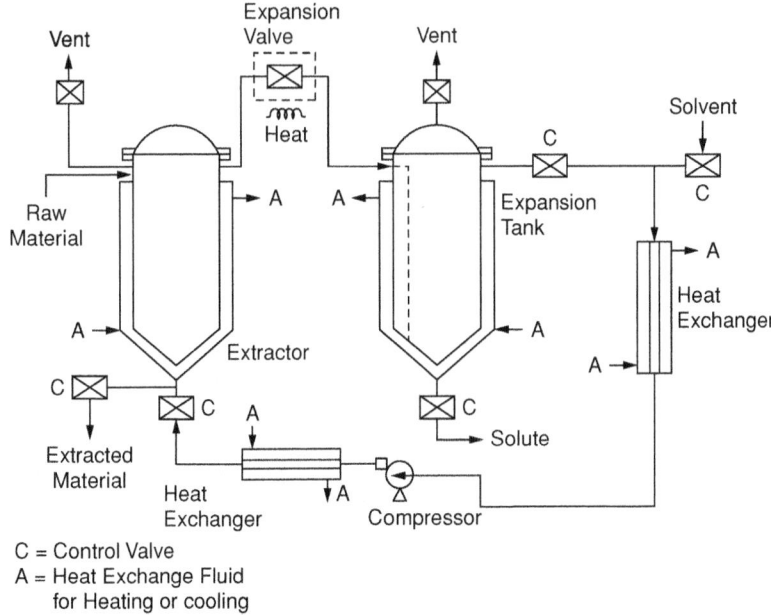

C = Control Valve
A = Heat Exchange Fluid
for Heating or cooling

Figure 14.14 Schematic diagram of a supercritical fluid extraction system.

and the stream entering and leaving stage 2. Material balance around cell 1:

Final extract, V_a: X = 0.1718/0.27 = 0.636 kg oil/kg soln.

Because the last three data points deviated slightly from linearity, the amount of solution retained by the solids in stream L_1 will be calculated by interpolation from the tabulated data instead of the regression equation. Because final extract V_a is in equilibrium with the solids in stream L_1, the X coordinate of stream L_1 must be the same as for the final extract. Thus, X for stream L_1 = 0.636 kg oil/kg soln. From Table 14.1 by interpolation, the solution retained by the solids if the solute concentration is 0.636 kg oil/kg soln. is

$$= 0.595 + (0.620 - 0.595)(0.656 - 0.0.6)/(0.7 - 0.6) = 0.609 \text{ kg soln/kg solids.}$$

Total mass of L_1 = 0.82 kg solids + 0.609(0.82) = 1.3194 kg

Y for stream L_1 = 1/0.609 = 1.642 kg solids/kg soln. The mass and composition of stream V_2 can now be calculated.

Mass of $V_2 = L_1 + V_a - 1 = 1.3194 + 0.27 - 1 = 0.5894$ kg

$$\text{Mass oil in } L_1 = \left[\frac{0.609 \text{ kg soln.}}{\text{kg solids}} (0.82 \text{ kg solids}) \right] \left(\frac{0.656 \text{ kg oil}}{\text{kg soln.}} \right) = 0.3276 \text{ kg}$$

Oil balance around cell 1:

Oil in V_2 = 0.1718 + 0.3276 − 0.18 = 0.3194 kg

X coordinate for V_2:

X = 0.3194/0.5894 = 0.5419

The streams to be considered to start plotting of the Ponchon-Savarit diagram are streams V_2 and L_1, and the solute-free solvent stream and the spent solids stream, L_n.

V_2 and L_1 are first plotted in Fig. 14.11. A line is drawn through the two points and extended beyond the graph. The equation of the line that connects the two points may also be determined. Points representing the solute-free solvent and the spent solids stream, L_n, are plotted next. The solute free solvent is represented by the origin. A line is drawn through the two points and extended until it intersects the previous line drawn between L_1 and V_2. The equation for the line may also be determined and the point of intersection, representing the point, P, can be calculated by solving the two equations simultaneously. The coordinate of point P is −0.1146, −11.4545.

An equilibrium line drawn from V_2 to the underflow curve will locate a point that represents L_3. Drawing a line from this point to connect with point P will locate an intersection on the overflow curve, for the point representing V_3. The process is repeated, until the last line drawn from the underflow curve to point P will intersect the overflow curve at a point that is equal to or less than the specified X for the spent solids, L_n. The number of extraction stages is the number of equilibrium tie-lines drawn in the diagram in Fig. 14.11. In this problem, 13 equilibrium stages are required.

14.4 SUPERCRITICAL FLUID EXTRACTION

Supercritical fluid extraction may be done on solids or liquids. The solvent is a dense gas at conditions of temperature and pressure where further increase in pressure or a reduction in temperature will not result in a phase change from gas to liquid. The density of a supercritical fluid, however, is almost that

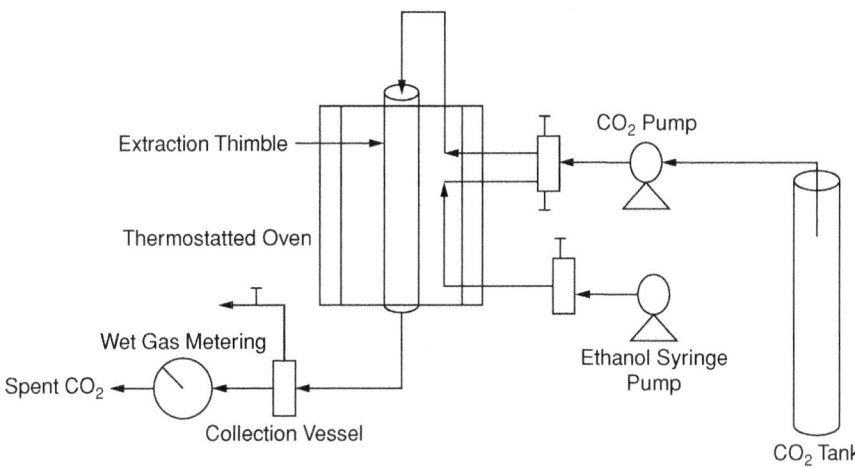

Figure 14.15 Schematic diagram of a supercritical fluid extraction system using entrained ethanol in supercritical carbon dioxide.

of a liquid, but it is not a liquid. In addition, the solubility of solutes in a supercritical fluid approaches the solubility in a liquid. Thus, the principle of solute extraction from solids using a supercritical fluid is very similar to that for solid-liquid extractions.

14.4.1 Extraction Principles

Supercritical fluid extraction is done in a single-stage contractor with or without recycling of the solvent. When recycling is used, the process involves a reduction of pressure to allow the supercritical fluid to lose its ability to dissolve the solute, after which the solid is allowed to separate by gravity, and the gas at low pressure is compressed back to the supercritical conditions and recycled. Temperature reduction may also be used to drop the solute and the solvent is reheated for recycling without the need for recompression.

Figure 14.14 shows a schematic diagram of a supercritical fluid extraction system. The basic components are an extractor tank and an expansion tank. Supercritical fluid conditions are maintained in the extractor. Temperature is usually maintained under controlled conditions in both tanks. Charging and emptying the extractor is a batch operation. The pressure of the supercritical fluid is reduced by throttling through a needle valve or orifice after which it enters an expansion tank where the supercritical fluid becomes a gas. Because solute solubility in the gas is much less than in the supercritical fluid, solute separates from the gas in the expansion tank. The spent gas is then recompressed and recycled. Heat exchanges are needed to maintain temperatures and prevent excessive cooling at the throttling valve due to the Joule-Kelvin effect.

Two of the major problems of supercritical fluid extraction are channeling of solvent flow through the bed of solids, and entrainment of the nonextractable component by the solvent. Time of solid-solvent contact is the quotient of extraction vessel volume divided by the solvent volumetric flow rate. The volume is calculated at the temperature and pressure inside the extraction vessel. Normally, volume of the solvent is measured at atmospheric pressure after the gas exited the expansion tank. From this measured volume, the number of moles of gas is calculated and the volume of the supercritical fluid in the extraction vessel is then calculated using the equations of state for gases. The contact time should be adequate to permit solvent to penetrate solid particles and permit diffusion of solute from inside the solid particles into the solvent phase. To achieve equilibrium between the solution inside solid particles and the solvent phase, solvent flow must be adjusted to achieve the necessary contact time and to provide enough solvent such that concentration of dissolved solutes in the solvent phase will be below the solubility of solute in the solvent. A large quantity of solute to be extracted would require a larger rate of solvent flow to permit thorough solute extraction within a reasonable length of time. Supercritical fluid penetration into the interior of a solid is rapid, but solute diffusion from the solid into the supercritical fluid may be slow thus requiring prolonged contact time in the extraction vessel. Solvent flow rate, pressure, and temperature in the extraction vessel are the major supercritical fluid extraction process parameters.

14.4.2   Critical Points of Supercritical Fluids Used in Foods

Carbon dioxide is the most widely used supercritical fluid. The compound is nontoxic, nonflamable, and is inexpensive and readily available. The critical point of carbon dioxide is 31.1°C and 7.39 MPa (74 Bars). At temperature and pressure above the critical point, the fluid is supercritical. Ethanol may also be added in small amounts to supercritical carbon dioxide to change its polarity in some extrations. The critical point of ethanol is 243°C and 6.38 MPa (64 Bars). Water may also be transferred from the solid to the solvent phase therefore the supercritical fluid in the extraction vessel may contain water. The critical point of water is 364°C and 22.1 MPa (221 Bars).

14.4.3   Critical Point of Mixtures

The critical point of a mixture of compounds may be calculated as a mole fraction weighted average of the individual components.

$$T_c = 3(x_1 T_{c1} + x_2 T_{c2} \ldots + x_n T_{cn}$$
$$P_c = 3(x_1 P_{c1} + x_2 P_{c2} \ldots + x_n P_{cn}$$

14.4.4   Properties of Supercritical Fluids Relative to Gases

The relative density of a supercritical fluid is in the range 0.1 to 1 s compared with a density of 1 for liquids and 0.001 for gases. The relative viscosity is 0.1 to 1 compared with 1 for liquids and 0.01 for gases. The relative diffusivity is 10 to 100 compared with 1 for liquids and 10^4 for gases.

14.4.5   Supercritical Fluid Extraction Parameters

Separation of multiple solutes from a solid may be the objective of a supercritical fluid extraction process. For example, components of a fat or oil high in polyunsaturated fatty acids may be removed from the mixture leaving triglycerides with saturated fatty acids with higher melting points and higher resistance to oxidation. Another example is the separation of phospholipids from the triglycerides in soybean oil to prevent gumming of the oil. Solubility of solutes in the supercritical solvent may be a function of pressure and/or temperature. Table 14.2 shows the solubility of some compounds in supercritical carbon dioxide. These data show that the right conditions may be selected to maximize

Table 14.2 Solubility of various compounds in supercritical CO_2 (In mg/g).

Pressure(Bars)/ Temperature °C	80/40	90/40	100/40	200/40	80/50	90/50	100/50	200/50	162/45	146/45
Limonine	5.8	CM			17.1	CM				
Lauric acid			6.8	11.6						
Palmitic acid			0.8	4.4						
Trilaurin			0.1	26.8						
Tripalmitin				0.3						
Trimyristin			0.5	5.6						
β-carotene				0.0021				0.0028		
Stearic acid[1]									23.8	23.8
Water					1.6		1.9	2.8		

[1] Solvent is CO_2 with acetic acid at 0.03 mole fraction as entrainer.
CM = completely miscible

solute solubility. Another method for changing solubility is the use of entrainers. For example, in Table 14.2, tristearin solubility in carbon dioxide with an acetic acid entrainer is very high at relatively low pressure. Data in Table 14.2 also shows that a pressure of 200 Bars and 40°C should be good for extracting neutral lipids. Another example is the enhancement of the solubility of lecithin when ethyl alcohol is an entrainer in supercritical carbon dioxide. Other compounds with improved solubility in supercritical carbon dioxide-ethanol include limonoids from citrus seeds and pospholipids such as phosphatidyl ethanolamine and phosphatidyl choline from dried egg yolk.

Another technique used to separate solutes is by using a two-stage process. For example, one group of compounds may easily dissolve in the solvent under a given set of conditions while leaving the less soluble ones in the solid. A second extraction under conditions that favor the dissolution of the remaining solute in the solid in the solvent will result in the isolation of these group of compounds from the other group removed in the first extraction. The following examples are optimized extraction conditions reported in the literature:

1. Preparation of ginger flavor extract from dried ginger. CO_2 at 15 g/min. First-stage extraction at 79 Bars and 30°C; second stage at 246 Bars and 40°C. (Yonei, Y, et al J. Sup. Fl. 8:156–161, 1995).
2. Flavor extracts from coriander seed: 250 Bars and 40°C. Celery powder and sage: 70 Bars and 20°C (Catchpoole, O.J. et al. J. Sup. Fl. 9:273–279, 1996).
3. Egg phospholipids from dried egg yolk. First stage: CO_2 at 414 Bars and 45°C. Second stage: $CO_2 - 5\%$ ethanol at 414 Bars and 45°C (Shah A. et al. J. Sup. Fl. 30:303–313, 2004).

PROBLEMS

14.1. Ground roasted coffee contains 5% soluble solids, 3% water, and 92% inert insoluble solids. In order to obtain an extract with high soluble solids content without having to concentrate it for spray drying, a counter-current extraction process is to be used to prepare this extract.

It is desired that the final extract contain 0.1 kg solubles/kg water and that the spent coffee grounds should have solubles not to exceed 0.005 kg/kg dry inert solids.

(a) Determine the water to coffee ratio to be used in the extraction.

(b) The coffee grounds carry 1 kg water/kg of solubles-free inert solids, and this quantity is constant with solute concentration in the extract. Calculate the number of extraction stages needed for this process.

14.2. A process for extracting sugar from sweet sorghum involves pressing the cane through a three-roll mill followed by shredding the fibrous residue (bagasse) and extracting the sugar out with water. The sorghum originally contained 20% fiber, 16% sugar, and 64% water. After milling, the moisture content of the bagasse is 55%. Because the fiber is used for fuel, after the extraction battery, the solids is squeezed to remove the absorbed solution and the squeezed solution is added to the last stage of the extractor. The following are the constraints: The sugar recovery must be a minimum of 99%, and the concentration of sugar in the final extract must be 10%. The bagasse carries a constant amount of solution, 1.22 kg solution/ kg fiber. Calculate:

(a) The water to solids ratio needed.

(b) The number of ideal extraction stages.

(c) The final sugar content if the extract is mixed with the juice first pressed out of the cane.

SUGGESTED READING

Berk, Z. 1992. Technology for the production of edible flours and protein products from soybeans. FAO Agricultural Services Bulletin 97. Food and Agriculture Organization of the United Nations, Rome.

Green, D. W. and Mahoney, J. O. 1997. Perry's Chemical Engineers Handbook. 7th ed. McGraw-Hill, New York.

Johnston, K. P. and Penninger, J. M. L., Eds. 1989. Supercritical Fluid Science and Technology. ACS Symposium Series 406. American Chemical Society, Washington, DC.

McCabe, W. L., Smith, J. C., and Harriott, P. 1985. Unit Operations in Chemical Engineering. 4th ed. McGraw-Hill, New York.

Perry, R. H., and Chilton, C. H., Eds. 1973. The Chemical Engineers Handbook. 5th ed. McGraw-Hill, New York.

Prabhudesai, R. K. 1979. Leaching. In: Handbook of Separation Techniques for Chemical Engineers. Schweitzer, P. A., Ed. McGraw-Hill, New York.

Conversion Factors Expressed as a Ratio

Denominator	Numerators		Denominator	Numerators	
I	II	III	I	II	III
Acre	0.4046873	Hectare	cPoise	0.001	Pa · s
Acre	43560.0	$(ft)^2$	$(cm)^3$	3.531* E − 5	$(ft)^3$
Atm (std)	101.325.0	Pascal	$(cm)^3$	0.061023	$(in)^3$
Atm (std)	14.696	lb_f/in^2	$(cm)^3$	2.642*E − 4	Gal
Atm (std)	29.921	in Hg	$(ft)^3$	7.48052	Gal
Atm (std)	76.0	cm Hg @ 0°C	$(ft)^3$	28.316	Liter
			Cup	2.36588* E − 4	$(m)^3$
Atm (std)	33.899	ft H_2O @ 32.2°F	Dyne	1.000* E − 5	Newton
			Dyne·cm	1.000* E − 7	Newton·m
BTU	1.0550* E + 10	Erg	Dyne	2.248* E − 6	lg_f
BTU	778.3	ft·lb_f			
BTU	3.9292* E − 4	Hp·h	Erg	9.486* E − 11	BTU
BTU	1054.8	Joule	Erg	2.389* E − 8	Cal
BTU	252	Cal	Erg	1.000* E − 7	Joule
BTU	2.928* E − 4	kW·h			
BTU/min	0.023575	Hp	Foot	0.3048	Meter
BTU/min	0.01757	kW	$(ft)^3$	2.83168* E − 2	$(m)^3$
BTU/min	17.5725	Watt	$(ft)^2$	9.29030* E − 2	$(m)^2$
BTU/s	1054.35	Watt	ft·lb_f	1.355818	Joule
BTU/h	0.29299	Watt			
$\dfrac{BTU}{h(ft^2)(°F)}$	5.678263	W/m²· K	Gal (U.S.)	0.13368	$(ft)^3$
			Gal (U.S.)	3.78541	Liter
			Gal (U.S.)	3.78541*E−3	$(m)^3$
$\dfrac{BTU}{h(ft)(°F)}$	1.730735	W/m· K	Gram	2.2046* E − 3	Pound

(Continued)

533

Denominator I	Numerators II	Numerators III
BTU/lb	2326.0	J/kg
$\dfrac{BTU}{lb(°F)}$	4186.8	J/kg· K
Bushel	1.2445	ft^3
Bushel	0.035239	m^3
Cal	4.1868	Joule
Cal	3.9684* E − 3	BTU
Cal	4.1868* E + 7	Erg
centimeter	0.3937	Inch
cm Hg@ 0°C	1333.33	Pascal
cm H$_2$O@ 4°C	98.0638	Pascal
cPoise	0.01	g/cm·s
cPoise	3.60	kg/m·h
cPoise	6.72* E − 4	lb/ft·s
Joule	10.000* E + 7	Erg
Joule	0.73756	ft·lb$_f$
Joule	2.77* E − 4	W·h
kg	2.2046	Pound
km	3281	Foot
km	0.6214	Mile
kW	3413	BTU/h
kW·h	3.6* E + 6	Joule
Liter	0.03532	(ft)^3
Liter	0.2642	Gal (U.S.)
Liter	2.113	Pint
Meter	3.281	Foot
Meter	39.37	Inch
Newton	1.000* E+5	Dyne
Oz (liq)	29.57373	(cm)^3
Oz (liq)	1.803	(in)^3
Oz (av)	28.3495	Gram
Oz (av)	0.0625	Pound

Denominator I	Numerators II	Numerators III
Hectare	2.471	Acre
Hp	42.44	BTU/min
Hp	33,000	ft · lb$_f$/min
Hp	0.7457	kW
Hp(boiler)	33,480	BTU/h
Inch	2.5400* E − 2	Meter
in Hg @ 0°C	3.38638* E + 3	Pascal
in Hg@ 0°C	0.4912	lb$_f$/in^2
Joule	9.48* E − 4	BTU
Joule	0.23889	Cal
Pound	453.5924	Gram
Pound	0.45359	kg
lb$_f$	4.44823	Newton
lb$_f$/in.^{2}	0.068046	Atm (std)
lb$_f$/in.^{2}	68947	dynes/cm^2
lb$_f$/in.^{2}	2.3066	ft H$_2$O@39.2°F
lb$_f$/in.^{2}	2.035	in Hg@0°C
lb$_f$/ft^2	47.88026	Pascal
lb$_f$/in.^{2}	6894.757	Pascal
Qt (U.S.)	9.4635* E + 4	(m)^3
Qt (U.S.)	946.358	(cm)^3
Qt (U.S.)	57.75	(in.)^3
Qt (U.S.)	0.9463	liter
Qt (U.S.)	0.25	Gallon
Ton (metric)	1000	kg
Ton (metric)	2204.6	Pound
Ton (short)	2000	Pound
Ton (refri)	12,000	BTU/h
Torr (mmHg @0°C)	133.322	Pascal
Watt	3.413	BTU/h

(Continued)

Denominator	Numerators		Denominator	Numerators	
I	II	III	I	II	III
Pascal	1.4504* E − 4	$lb_f/in.^2$	Watt	44.27	$ft \cdot lb_f/min$
Pascal	1.0197*E − 5	kg_f/cm^2	Watt	1.341* E − 3	Hp
Pint	28.87	$(in.)^3$	Watt·h	3.413	BTU
Poise	0.1	Pa·s	Watt·h	860.01	Cal
lb_f	444823	Dyne	Watt·h	3600	Joule

To use: multiply quantities having units under Column 1 with the factors under Column II to obtain quantities having the units under Column III. Also use as a ratio in a dimensional equation Example: 10 acres = 10 × 0.4.046.873 hectares. The dimensional ratio is (0.4046873) hectare/acre. The symbol *E represents exponents of 10.9.486* $E − 11 = 9.486 \times 10^{-11}$.

Properties of Superheated Steam

	Absolute Pressure lb_f/in^2 (psi)					
	1 psi $T_s = 101.74°F$		5 psi $T_s = 162.24°F$		10 psi $T_s = 193.21°F$	
Temp. °F	v	h	v	h	v	h
200	392.5	1150.2	78.14	1148.6	38.84	1146.6
250	422.4	1172.9	84.21	1171.7	41.93	1170.2
300	452.3	1195.7	90.24	1194.8	44.98	1193.7
350	482.1	1218.7	96.25	1218.0	48.02	1217.1
400	511.9	1241.8	102.24	1241.3	51.03	1240.6
450	541.7	1265.1	108.23	1264.7	54.04	1264.1
500	571.5	1288.6	114.21	1288.2	57.04	1287.8
600	631.1	1336.1	126.15	1335.9	63.03	1335.5

	Absolute Pressure lb_f/in^2 (psi)					
	14.696 psi $T_s = 212.00°F$		15 psi $T_s = 213.03°F$		20 psi $T_s = 227.96°F$	
Temp. °F	v	h	v	h	v	h
250	28.42	1168.8	27.837	1168.7	20.788	1167.1
300	30.52	1192.6	29.889	1192.5	22.356	1191.4
350	32.60	1216.3	31.939	1216.2	23.900	1215.4
400	34.67	1239.9	33.963	1239.9	25.428	1239.2
450	36.72	1263.6	35.977	1263.6	26.946	1263.0
500	38.77	1287.4	37.985	1287.3	28.457	1286.9
600	42.86	1335.2	41.986	1335.2	31.466	1334.9

(Continued)

Temp. °F	Absolute Pressure lb$_f$/in^2 (psi)					
	25 psi $T_s = 240.07°F$		30 psi $T_s = 250.34°F$		35 psi $T_s = 259.29°F$	
	v	h	v	h	v	h
250	16.558	1165.6				
300	17.829	1190.2	14.810	1189.0	12.654	1187.8
350	19.076	1214.5	15.589	1213.6	12.562	1212.7
400	20.307	1238.5	16.892	1237.8	14.453	1237.1
450	21.527	1262.5	17.914	1261.9	15.334	1261.3
500	22.740	1286.4	18.929	1286.0	16.207	1285.5
600	25.153	1334.6	20.945	1334.2	17.939	1333.9

v = specific volume in ft^3/lb; h = enthalpy in BTU/lb.
T_s = saturation temperature at the designated pressure.

Source: Abridged from ASME. 1967. *Steam Tables. Properties of Saturated and Superheated Steam*—from 0.08865 to 15,500 lb per sq in. absolute pressure. American Society of Mechanical Engineers. NY. Used with permission.

Saturated Steam Tables: English Units

Temp. °F	Abs. pressure lb/in²	Specific Volume (ft³/lb)			Enthalpy (BTU/lb)		
		Sat. liquid v_f	Evap v_{fx}	Sat. vapor v_x	Sat. liquid h_f	Evap. h_{fg}	Sat. vapor h_g
32	0.08859	0.016022	3304.7	3304.7	−0.0179	1075.5	1075.5
35	0.09998	0.016020	2950.5	2950.5	3.002	1073.8	1076.8
40	0.12163	0.016019	2445.8	2445.8	8.027	1071.0	1079.0
45	0.14753	0.016020	2039.3	2039.3	13.044	1068.2	1081.2
50	0.17796	0.016023	1704.8	1704.8	18.054	1065.3	1083.4
55	0.21404	0.016027	1384.2	1384.2	23.059	1062.5	1085.6
60	0.25611	0.016033	1207.6	1207.6	28.060	1059.7	1087.7
65	0.30562	0.016041	1022.8	1022.8	33.057	1056.9	1089.9
70	0.36292	0.016050	868.3	868.4	38.052	1054.0	1092.1
75	0.42985	0.016061	740.8	740.8	43.045	1051.3	1094.3
80	0.50683	0.016072	633.3	633.3	48.037	1048.4	1096.4
85	0.59610	0.016085	543.9	543.9	53.028	1045.6	1098.6
90	0.69813	0.016099	468.1	468.1	58.018	1042.7	1100.0
95	0.81567	0.016114	404.6	404.6	63.008	1039.9	1102.9
100	0.94924	0.016130	350.4	350.4	67.999	1037.1	1105.1
105	1.10218	0.016148	304.6	304.6	72.991	1034.3	1107.2
110	1.2750	0.016165	265.4	265.4	77.98	1031.4	1109.3
115	1.4716	0.016184	232.03	232.0	82.97	1028.5	1111.5
120	1.6927	0.016204	203.25	203.26	87.97	1025.6	1113.6
125	1.9435	0.016225	178.66	178.67	92.96	1022.8	1115.7
130	2.2230	0.016247	157.32	157.33	97.96	1019.8	1117.8
135	2.5382	0.016270	138.98	138.99	102.95	1016.9	1119.9
140	2.8892	0.016293	122.98	123.00	107.95	1014.0	1122.0
145	3.2825	0.016317	109.16	109.18	112.95	1011.1	1124.1
150	3.7184	0.016343	97.05	97.07	117.95	1008.2	1126.1

(Continued)

Temp. °F	Abs. pressure lb/in²	Specific Volume (ft³/lb)			Enthalpy (BTU/lb)		
		Sat. liquid v_f	Evap v_{fx}	Sat. vapor v_x	Sat. liquid h_f	Evap. h_{fg}	Sat. vapor h_g
155	4.2047	0.016369	86.53	86.55	122.95	1005.2	1128.2
160	4.7414	0.016395	77.27	77.29	127.96	1002.2	1130.2
165	5.3374	0.016423	69.19	69.20	132.97	999.2	1132.2
170	5.9926	0.016451	62.04	62.06	137.97	996.2	1134.2
175	6.7173	0.016480	55.77	55.79	142.99	993.2	1136.2
180	7.5110	0.016510	50.21	50.22	148.00	990.2	1138.2
185	8.3855	0.016543	45.31	45.33	153.02	987.2	1140.2
190	9.340	0.016572	40.941	40.957	158.04	984.1	1142.1
195	10.386	0.016605	37.078	37.094	163.06	981.0	1144.1
200	11.526	0.016637	33.622	33.639	168.09	977.9	1146.0
205	12.776	0.016707	30.567	30.583	173.12	974.8	1147.8
210	14.132	0.016705	27.822	27.839	178.16	971.6	1149.8
212	14.696	0.016719	26.782	26.799	180.17	970.3	1150.5
220	17.186	0.016775	23.131	23.148	188.23	965.2	1153.4
225	18.921	0.016812	21.161	21.177	193.28	961.9	1155.2
230	20.791	0.016849	19.379	19.396	198.33	958.7	1157.1
235	22.804	0.016887	17.766	17.783	203.39	956.5	1158.9
240	24.968	0.016926	16.304	16.321	208.45	952.1	1160.6
245	27.319	0.016966	14.998	15.015	213.52	948.8	1162.4
250	29.840	0.017006	13.811	13.828	218.59	945.5	1164.1
255	32.539	0.017047	12.729	12.747	223.67	942.1	1165.8
260	35.427	0.017089	11.745	11.762	228.76	938.6	1167.4
265	38.546	0.017132	10.858	10.875	233.85	935.2	1169.0
270	41.875	0.017175	10.048	10.065	238.95	931.7	1170.7
275	45.423	0.017219	9.306	9.324	244.06	928.2	1172.2
280	49.200	0.017264	8.627	8.644	249.17	924.6	1173.8
285	53.259	0.017310	8.0118	8.0291	254.32	920.9	1175.3
290	57.752	0.017360	7.4468	7.4641	259.45	917.3	1176.8

Source: Abridged from: ASME 1967. *Steam Tables. Properties of Saturated and Superheated Steam.* American Society of Mechanical Engineers, NY. Used with permission.

APPENDIX A.4

Saturated Steam Tables: Metric Units

Temperature °C	Absolute pressure kPa	Saturated liquid h_f	Enthalpy (MJ/kg) (MJ/kg) Evaporation h_{fg}	Saturated vapor h_g
0	0.6108	-0.00004	2.5016	2.5016
2.5	0.7314	0.01049	2.4956	2.5061
5	0.8724	0.02100	2.4897	2.5108
7.5	1.0365	0.03151	2.4839	2.5153
10	1.2270	0.04204	2.4779	2.5200
12.5	1.4489	0.05253	2.4720	2.5245
15	1.7049	0.06292	2.4661	2.5291
17.5	2.0326	0.07453	2.4595	2.5342
20	2.3366	0.08386	2.4544	2.5381
22.5	2.7248	0.09780	2.4484	2.5428
25	3.1599	0.10477	2.4425	2.5473
27.5	3.6708	0.11522	2.4367	2.5518
30	4.2415	0.12566	2.4307	2.5563
32.5	4.8913	0.13611	2.4246	2.5609
35	5.6238	0.14656	2.4188	2.5653
37.5	6.4488	0.15701	2.4129	2.5699
40	7.3749	0.16745	2.4069	2.5744
42.5	8.4185	0.17789	2.4009	2.5788
45	9.5851	0.18834	2.3949	2.5832
47.5	10.8868	0.19880	2.3889	2.5877
50	12.3354	0.20925	2.3829	2.5921
52.5	13.9524	0.21971	2.3769	2.5966
55	15.7459	0.23017	2.3705	2.6000
57.5	17.7295	0.24062	2.3648	2.6054
60	19.9203	0.25109	2.3586	2.6098
62.5	22.3466	0.26155	2.3525	2.6140

(Continued)

			Enthalpy (MJ/kg)	
Temperature °C	Absolute pressure kPa	Saturated liquid h_f	(MJ/kg) Evaporation h_{fg}	Saturated vapor h_g
65	25.0159	0.27202	2.3464	2.6184
67.5	27.9479	0.28249	2.3402	2.6226
70	31.1622	0.29298	2.3339	2.6270
72.5	34.6961	0.30345	2.3276	2.6312
75	38.5575	0.31394	2.3214	2.6354
77.5	42.7706	0.32442	2.3151	2.6395
80	47.3601	0.33492	2.30879	2.64373
82.5	52.5777	0.34542	2.30251	2.64792
85	57.8159	0.34659	2.29611	2.65199
87.5	63.7196	0.36643	2.28971	2.65606
90	70.1059	0.37693	2.28320	2.66025
92.5	77.0489	0.38747	2.27669	2.66420
95	84.5676	0.39799	2.27023	2.66821
97.5	92.6379	0.40853	2.26349	2.67214
100	101.3250	0.41908	2.25692	2.67606
102.5	110.7410	0.42962	2.25035	2.67996
105	120.8548	0.44017	2.24354	2.68368
107.5	131.7114	0.45074	2.23674	2.68752
110	143.3489	0.46132	2.22994	2.69129
112.5	155.8051	0.47190	2.22313	2.69508
115	169.1284	0.48249	2.21615	2.69874
117.5	183.3574	0.49309	2.20929	2.70241
120	198.5414	0.50372	2.20225	2.70607
122.5	214.8337	0.51434	2.19519	2.70949
125	232.1809	0.52499	2.18807	2.71311
127.5	250.6391	0.53565	2.18083	2.71651
130	270.2538	0.54631	2.17365	2.71991
132.5	291.0837	0.55698	2.11632	2.72331
135	313.1771	0.56768	2.15899	2.72654

Source: Calculated from ASME 1967. *Steam Tables. Properties of Saturated and Superheated Steam*—from 0.08865 to 15.500 lb per sq in. absolute pressure. American Society of Mechanical Engineers, NY. Used with permission.

Flow Properties of Food Fluids

Product	% Solids	Temperature °C	Flow constants	
			n (dimensonless)	k dyne - dyne · s^n/cm^2
Applesauce	11	30	0.34	116
		82	0.34	90
Apple juice	50–65.5 Brix	30	0.65	–
	10.5–40 Brix	30	1.0	–
Apricot puree	16	30	0.30	68
		82	0.27	56
Apricot concentrate	26	4.5	0.26	860
		25	0.30	670
		60	0.32	400
Banana puree	—	24	0.458	65
Grape juice	64 Brix	30	0.9	—
	15–50 Brix	30	1.0	—
Orange juice concentrate	—–	15	0.584	11.9
		0	0.542	18.0
Orange juice concentrate	30 Brix	30	0.85	—
	60 Brix	30	0.55	15.5
	65 Brix	30	0.91	2.6
Pear puree	18.3	32	0.486	22.5
		82	0.484	14.5
	26	32	0.450	62
		82	0.455	36
	31	32	0.450	109
		82	0.459	56
	37	32	0.479	355
		82	0.481	160

(Continued)

Product	% Solids	Temperature °C	Flow constants	
			n (dimensonless)	*k dyne - dyne · s^n/cm^2*
Peach puree	12	30	0.28	72
		82	0.27	58
Plum juice	14	30	0.34	22
		82	0.34	20
Tomato juice	12.8	32	0.43	20
		82	0.345	31.2
	25	32	0.405	129
		82	0.43	61
	30	32	0.40	187
		82	0.445	79
Tomato catsup (0.15 g tomato solids/g catsup)	36	30	0.441	81

Source: Holdsworth, S. D. Applicability of theological models to the interpretation of flow and processing behavior of fluid food products. J. Tex. Studies 2(4): 393–418, 1971.

Psychrometric Chart: English Units

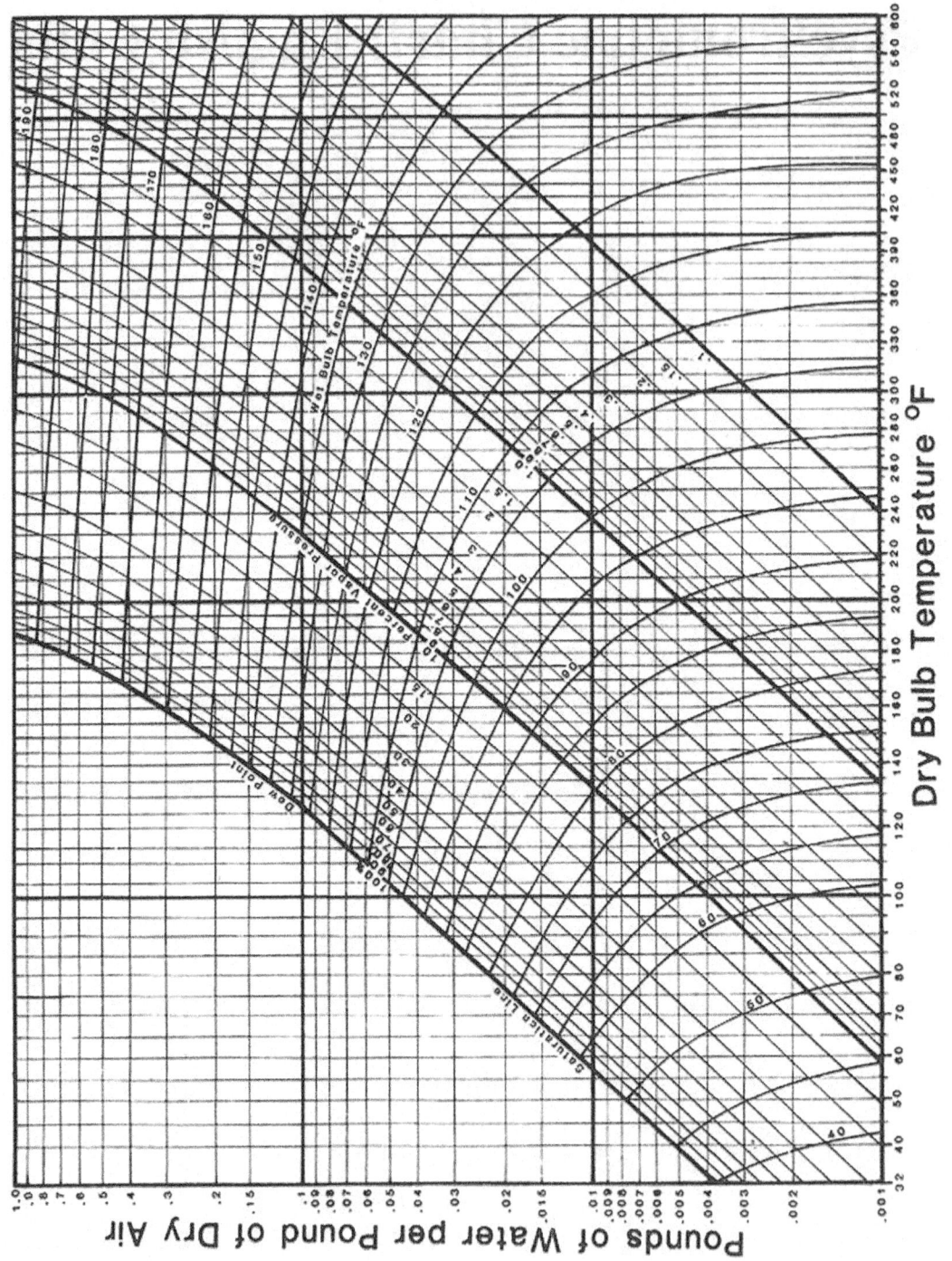

Psychrometric Chart: Metric Units

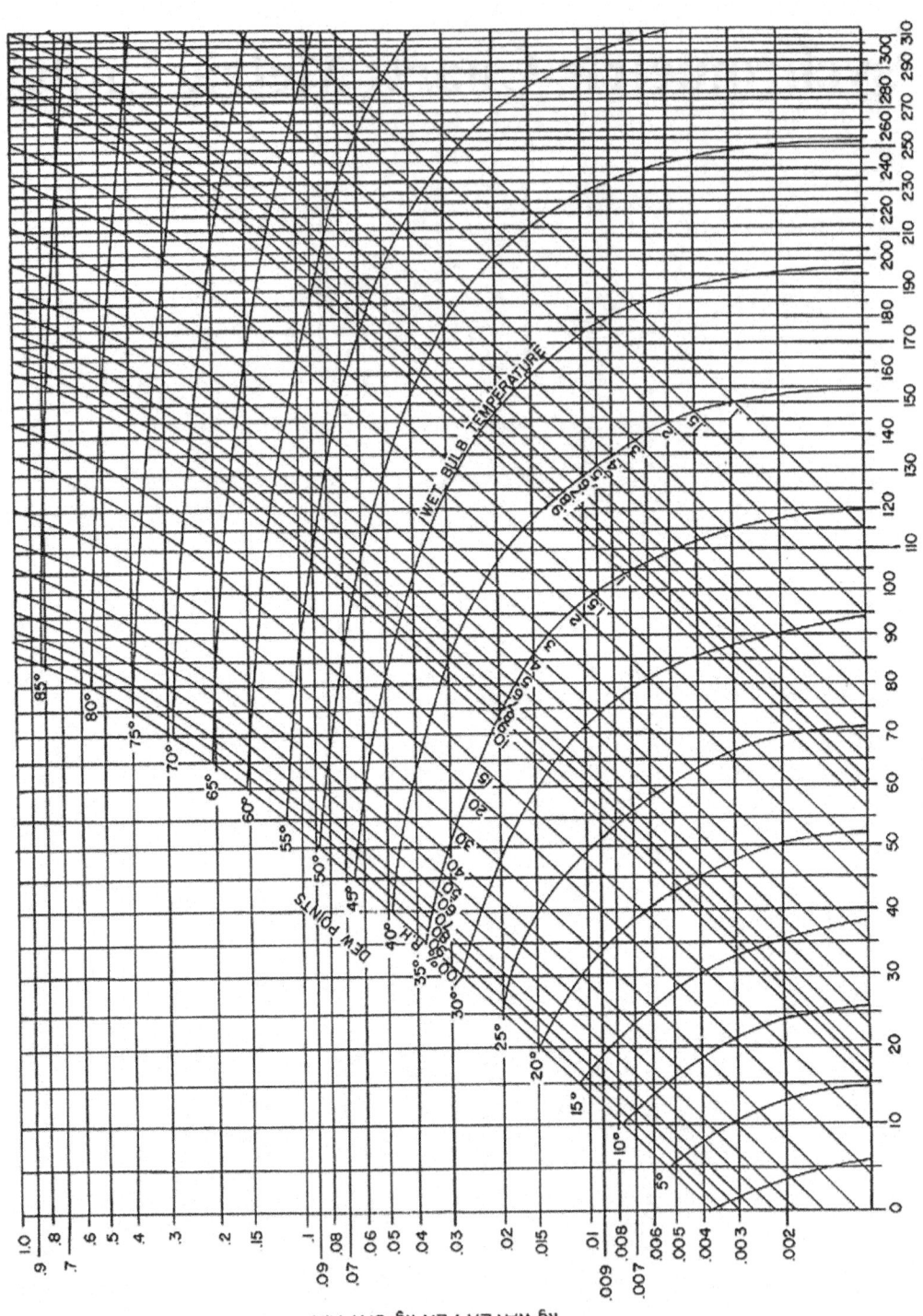

Average Composition of Foods, In Per Cent (From USDA Handbook 8)

Product	Water	Protein	Fat	Carbohydrate	Fiber	Ash
Sausage						
Beef Bologna	55.3	12.2	28.5	0.8	0	3.2
B/P Bologna	54.3	11.7	28.3	2.8	0	3
Pork Bologna	60.6	15.3	19.9	0.73	0	3.5
Beef Franks	54.7	12	28.5	1.8	0	2.9
B/P Franks	53.9	11.3	29.2	2.6	0	3.2
Chick. Franks	57.5	12.9	19.5	6.8	0	3.3
Turkey Franks	63	14.3	17.7	1.5	0	3.5
Canned Chopped Ham,	60.8	16.1	18.8	0.3	0	4.1
Sliced Ham (Ex. Lean)	70.5	19.4	5	1	0	4.2
Sliced Ham (Reg.)	64.6	17.6	10.6	3.1	0	4.1
Salami, Beef (Cooked)	59.3	14.7	20.1	2.5	0	4.4
Salami, Pork, Dry	36.2	22.6	33.7	1.6	0	5.9
Salami, B/F, Dry	34.7	22.9	34.4	2.6	0	5.5
Smoked Link Sausage	52.2	13.4	30.3	1.4	0	2.7
Turkey Roll (light meat)	71.6	18.7	7.2	0.5	0	2
Beef						
Brisket (cooked)	44.8	23.5	31.6	0	0	0.85
Brisket (raw)	55.1	16.9	26.5	0	0	0.8
Lean, Prime, Raw	68.6	21.2	9.7	0	0	1
Lean, Prime, Cooked	58.1	29	13	0	0	1.2
Chuck, Raw	61.8	18.5	18.4	0	0	0.9
Chuck, Braised	52.5	29.7	17	0	0	1
Round , Raw	65.9	20.4	11.6	0	0	1
Round , Cooked	58.1	27.4	11.7	0	0	1.3
Top Sirloin, Ch. Raw	62.7	19	16.2	0	0	0.9
Top Sirloin, Ch. Cooked	55.6	27.6	16.7	0	0	1.2
Top Sirloin, Sel. Raw	64.2	19.3	13.8	0	0	1
Top Sirloin, Sel. Cooked	57.6	28	13.9	0	0	1.3

Product	Water	Protein	Fat	Carbohydrate	Fiber	Ash
Gound, Ex-lean, Raw	63.2	18.7	17.1	0	0	0.9
Ground, Ex-lean, Cooked	58.6	24.5	16.1	0	0	0.8
Ground, Reg. Raw	48.9	28.8	21.5	0	0	1.2
Ground, Reg. Cooked	54.2	24.1	20.7	0	0	1
Pork						
Fresh Ham, Raw	64.7	18.7	15.7	0	0	0.9
Fresh Ham, Cooked	56.8	28.9	14.3	0	0	1.1
Loin, Raw	72.2	21.4	5.7	0	0	1.1
Loin, Cooked	61.4	28.6	9.1	0	0	1.4
Poultry						
Chicken Dark, Raw, no Skin	76	20.1	4.3	0	0	0.9
Chicken Dark, Cooked, no Skin	63.1	27.4	9.7	0	0	1
Chicken LT, Raw, no Skin	74.9	23.2	1.7	0	0	1
Chicken, LT, Cooked, no Skin	64.8	30.9	4.5	0	0	1
Turkey, Dark, Raw, no Skin	74.5	20.1	4.4	0	0	0
Turkey, Dark, Cooked, no Skin	63.1	28.6	7.2	0	0	1
Turkey, LT, Raw, no Skin	73.8	23.6	1.6	0	0	1
Turkey, LT, Cooked, no Skin	66.3	29.9	3.2	0	0	1.1
Dairy/Eggs						
Butter	15.9	0.9	81.1	0.1	0	2.1
Cheese, Cheddar	36.8	24.9	33.1	1.3	0	3.9
Cream, Half&Half	80.6	3	11.5	4.3	0	0.7
Cream, Whipping, Lt	63.5	2.2	30.9	3	0	0.5
Cream Whipping, Heavy	57.7	2.1	37	2.8	0	0.5
Egg, Whole	75.3	12.5	10	1.2	0	0.9
Egg, White	87.8	10.5	0	1	0	0.6
Egg, Yolk	48.8	16.8	30.9	1.8	0	1.8
Milk, Whole	88	3.3	3.4	4.7	0	0.7
Milk, Skim	90.8	3.4	0.2	4.9	0	0.8
Whey, acid	93.4	0.8	0.1	5.1	0	0.6
Whey, sweet	93.1	0.9	0.4	5.1	0	0.5
Fish/Shellfish						
Catfish Raw	75.4	15.6	7.6	0	0	1
Catfish Cooked	71.6	18.7	8	0	0	1.2
Cod, Raw	81.2	17.8	0.7	0	0	1.2
Cod, Cooked	75.9	22.8	0.9	0	0	1.5
Halibut, Raw	77.9	20.8	2.3	0	0	1.4
Halibut, Cooked	71.7	26.7	2.9	0	0	1.7
Mackerel, Raw	63.6	18.6	13.9	0	0	1.4
Mackerel, Cooked	53.3	23.9	17.8	0	0	1.5
Salmon, Farmed, Raw	68.9	19.9	10.9	0	0	1.1
Salmon, Farmed, Cooked	64.8	22.1	12.4	0	0	1.2
Shrimp, Raw	75.9	20.3	1.7	0.9	0	1.2
Shrimp, Steamed	77.3	20.9	1.1	0	0	1.6
Oyster, Raw	85.2	7.1	2.5	3.9	0	1.4

Product	Water	Protein	Fat	Carbohydrate	Fiber	Ash
Oyster, Steamed	70.3	14.1	4.9	7.8	0	2.8
Vegetables/Fruits						
Beans, Lima, Raw	70.2	6.8	0.9	20.2	1.9	1.9
Beans, Lima, Boiled	67.2	6.8	0.3	23.6	2.1	2.1
Beans, Snap, Raw	90.3	1.8	0.1	7.1	1.1	0.7
Beans, Snap, Boiled	89.2	1.9	0.3	7.9	1.4	0.7
Beets, Raw	87.6	1.6	0.2	0.6	0.8	1.1
Beets, Boiled	87.1	1.7	0.2	10	0.8	1.1
Carrots, Raw	87.8	1	0.2	10.1	1	0.9
Carrots, Boiled	87.4	1.1	0.2	10.5	1.5	0.9
Potatoes, Raw (Flesh)	79	2.1	0.1	18	0.4	0.9
Potatoes, Baked (Flesh)	75.4	2	0.1	21.6	0.4	1
Fruits/Juices						
Apples	83.9	0.2	0.4	15.3	0.8	0.3
Apple Juice, Bottled	87.9	0.1	0.1	11.7	0.2	0.2
Apricots	86.4	1.4	0.4	11.1	0.6	0.8
Avocados	72.6	2.1	17.3	6.9	2.1	1.1
Bananas	74.3	1	0.5	23.4	0.5	0.8
Cherries, Sour	86.1	1	0.3	12.2	0.2	0.4
Cherries, Sweet	80.8	1.2	1	16.6	0.4	0.5
Grapefruit, white	90.5	0.7	0.1	8.4	0.2	0.3
Grapefruit juice	90	0.5	0.1	9.2	0	0.2
Grape	81.3	0.6	0.4	17.2	0.8	0.6
Grape Juice	84.1	0.6	0.1	15	0	0.3
Peach	87.7	0.7	0.1	11.1	0.6	0.5
Pears	83.8	0.4	0.4	15.1	1.4	0.3
Pineapple	86.5	0.4	0.4	12.4	0.5	0.3
Strawberries	91.6	0.6	0.4	7	0.5	0.4

Thermal Conductivity of Construction and Insulating Materials

	$\dfrac{BTU}{h(ft)(^\circ F)}$	$\dfrac{W}{m(K)}$
Building materials		
Asbestos cement boards	0.43	0.74
Building brick	0.40	0.69
Building plaster	0.25	0.43
Concrete	0.54	0.93
Concrete blocks		
Two oval core, 8 in. thick	0.60	1.04
Two rectangular core, 8 in. thick	0.64	1.11
Corkboard	0.025	0.043
Felt (wool)	0.03	0.052
Glass	0.3–0.61	0.52–1.06
Gypsum or plasterboard	0.33	0.57
Wood (laminated board)	0.045	0.078
Wood (across grain, dry)		
Maple	0.11	0.19
Oak	0.12	0.21
Pine	0.087	0.15
Wood (plywood)	0.067	0.12
Rubber (hard)	0.087	0.15
Insulating materials		
Air		
32°F (0°C)	0.014	0.024
212°F (100°C)	0.0183	0.032
392°F (200°C)	0.0226	0.039

(Continued)

	$\dfrac{BTU}{h(ft)(°F)}$	$\dfrac{W}{m(K)}$
Fiberglass (9 lb/ft density)	0.02	0.035
Polystyrene		
2.4 lb/ft density	0.019	0.032
2.9 lb/ft density	0.015	0.026
1.6 lb/ft density	0.023	0.040
Polyurethane (5–8.5 lb/ft density)	0.019	0.033
Hog hair with asphalt binder		
(8.5 lb/ft density)	0.028	0.048
Mineral wool with binder	0.025	0.043
Metals		
Aluminum		
32°F (0°C)	117	202
212°F (100°C)	119	205
572°F (300°C)	133	230
Cast iron		
32°F (0°C)	32	55
212°F (100°C)	30	52
572°F (300°C)	26	45
Copper		
32°F (0°C)	294	509
212°F (100°C)	218	377
572°F(300°C)	212	367
Steel (carbon)		
212°F (100°C)	26	45
572°F (300°C)	25	43
Steel, stainless type 304 or 302	10	17
Steel, stainless type 316	9	15

Thermal Conductivity of Foods

Food	Temp. °C	Thermal conductivity W/(m · K)	Food	Temp. °C	Thermal conductivity W/(m · K)
Apple juice	80	0.6317	Lemon	—	1.817
Applesauce	29	0.5846	Limes		
Avocado	—	0.4292	Peeled	—	0.4900
Banana	—	0.4811	Margarine	—	0.2340
Beef	5	0.5106	Milk		
	10	0.5227	3% fat	—	0.5296
Beets	28	0.6006	2.5% fat	20	0.05054
Broccoli	−6.6	0.3808	Oatmeal, dry	—	0.6404
Butter	—	0.1972	Olive oil	5.6	0.1887
Butterfat	−10.6			100	0.1627
	to 10	0.1679	Onions	8.6	0.5746
Cantaloupe	—	0.5711	Oranges		
Carrots			peeled	28	0.5800
Fresh	—	0.6058	Orange juice	−18	2.3880
Puree	—	1.263	Peaches	28	0.5815
Corn			Peanut oil	3.9	0.1679
Yellow	—	0.1405	Pear	8.7	0.5954
Dent	—	0.577	Pear juice	20	0.4760
Egg white	—	0.338		80	0.5365
Egg yolk	2.8	0.5435	Peas	2.8−	
Fish	−10	1.497	Blackeye	16.7	0.3115
Cod	3.9	0.5019	Pineaple	—	0.5486
Salmon	−2.5	1.2980	Plums	—	0.5504
			Pork	6	0.4881
Gooseberries	—	0.2769		59.3	0.5400
Dry	—	0.3288	Potato, raw	—	0.554
Wet	—	0.0277	Poultry,	—	0.4119
Frozen			broiler		

(Continued)

555

Food	Temp. °C	Thermal conductivity W/(m · K)	Food	Temp. °C	Thermal conductivity W/(m · K)
Grapefruit	—	1.3500	Sesame oil	—	0.1755
Mashed			Strawberries	13.3	0.6750
Honey		0.5019		−12.2	1.0970
80% water	2	0.5019	Tomato	—	0.5279
80% water	69	0.6230	Turkey	2.8	0.5019
14.8% water	69	2.4230		−10	1.461
Ice	−25	0.4500	Turnips	—	0.5625
Lamb	5.5	0.4777			
	61.1				

Source: Excerpted from *Food Technol.* 34(11): 76–94, 1980.

APPENDIX A.11

Spreadsheet Program for Calculating Thermophysical Properties of Foods from Their Composition

Microsoft Excel - app11

File   Edit   View   Insert   Format   Tools   Data   Window   Help

Arial   10   Style... Normal   B I U

C18   fx   =1.9842+(1.4733*10^-3*B$1)-(4.8008*10^-6*(B$1^2))

	A	B	C	D	E
1	TEMPERATURE (°C)	19			
2	Thermal	Major	Group models		
3	property	component	temperature function		
4	k	water	=5.7109*10^-1+(1.7625*10^-3*B$1)-(6.7036*10^-6*(B$1^2))		
5	W/m° C	protein	=1.7881*10^-1+(1.1958*10^-3*B$1)-(2.7178*10^-6*(B$1^2))		
6		fat	=1.8071*10^-1-(2.7604*10^-3*B$1)-1.7749*10^-7*(B$1^2)		
7		carbohydrate	=2.014*10^-1+(1.3874*10^-3*B$1)-(4.3312*10^-6*(B$1^2))		
8		fiber	=1.8331*10^-1+(1.2497*10^-3*B$1)-(3.1683*10^-6*(B$1^2))		
9		ash	=3.2962*10^-1+(1.4011*10^-3*B$1)-(2.9069*10^-6*(B$1^2))		
10	ρ	water	=9.9718*10^2+(3.1439*10^-3*B$1)-(3.7574*10^-3*(B$1^2))		
11	kg/m³	protein	=1.3299*10^3-5.184*10^-1*B$1		
12		fat	=9.2559*10^2-4.1757*10^-1*B$1		
13		carbohydrate	=1.5991*10^3-3.1046*10^-1*B$1		
14		fiber	=1.3115*10^3-3.6589*10^-1*B$1		
15		ash	=2.4238*10^3-2.8063*10^-1*B$1		
16	Cp	water	=4.1762-(9.0864*10^-5*B$1)+(5.4731*10^-6*(B$1^2))		
17	kJ/kg°C	protein	=2.0082+(1.2089*10^-3*B$1)-(1.3129*10^-6*(B$1^2))		
18		fat	=1.9842+(1.4733*10^-3*B$1)-(4.8008*10^-6*(B$1^2))		
19		carbohydrate	=1.5488+(1.9625*10^-3*B$1)-(5.9399*10^-6*(B$1^2))		
20		fiber	=1.8459+(1.8306*10^-3*B$1)-(4.6509*10^-6*(B$1^2))		
21		ash	=1.0926+(1.8896*10^-3*B$1)-(3.6817*10^-6*(B$1^2))		
22	x mass fraction	water	0.71		
23		protein	0.19		
24		fat	0.078		
25		carbohydrate	0		
26		fiber	0		
27		ash	0.015		
28		xi/ρi	cp	xvi	k = Σ[(ki)(xvi)]
29		=C22/C10	=C16*C22	=C22*B$36/C10	=C4*D29
30		=C23/C11	=C17*C23	=C23*B$36/C11	=C5*D30
31		=C24/C12	=C18*C24	=C24*B$36/C12	=C6*D31
32		=C25/C13	=C19*C25	=C25*B$36/C13	=C7*D32
33		=C26/C14	=C20*C26	=C26*B$36/C14	=C8*D33
34		=C27/C15	=C21*C27	=C27*B$36/C15	=C9*D34
35		=SUM(B29:B34)	=SUM(C29:C34)		=SUM(E29:E34)
36	ρ	=1/B35			
37	α = k/(ρCp)	=E35/(B36*C35*10^3)			
38					

Sheet1 / Sheet2 / Sheet3 /

Ready

start   Microsoft Excel - app11

Correlation Equations for Heat Transfer Coefficients

Correlation equations for heat transfer coefficients:

Very viscous fluids flowing inside horizontal tubes

$$Nu = 1.62 \left[prRe\frac{d}{L} \right]^{0.33} [1 + 0.015(Gr)^{0.33}] \left[\frac{\mu_f}{\mu_n} \right]^{0.33}$$

Very viscous fluids flowing inside vertical tubes

$$Nu = 0.255 Gr^{0.25} Re^{0.07} Pr^{0.37}$$

Fluids in laminar flow inside bent tubes

$$h = h_r \left[\frac{1 + 21}{Re^{0.14}} \right] \left[\frac{d}{D} \right]$$

$h_r = h$ in straight tube, $d =$ tube diameter, $D =$ diameter of curvature of bend.

Evaporation from heat exchange surfaces

$$\frac{q}{A} = 15.6 P^{1.156}(T_w - T_s)^{2.30/p^{0.24}}$$

$q/A =$ heat flux, $T_w =$ wall temperature, $T_s =$ saturated temperature of vapor at pressure P.

Condensing vapors outside vertical tubes

$$Nu_L = 0.925 \left[\frac{L^3 \rho^2 g \lambda}{\mu k \Delta T} \right]^{0.25}$$

Condensation outside horizontal tubes

$$Nu_{D0} = 0.73 \left[\frac{Do^3 \rho^2 g \lambda}{k \mu \Delta T} \right]^{0.25}$$

Condensation inside horizontal tubes

$$Nu_{Di} = 0.612 \left[\frac{Di^3 \rho_1 (\rho_1 - \rho_v) g}{k \mu \Delta T} \right]^{0.25}$$

Condensation inside horizontal tubes

$$Nu_{Di} = 0.024(Re)^{0.8}(Pr)^{0.43}\rho_{corr}$$

Re is based on the total mass of steam entering the pipe. $\rho_{corr} = [0.5/\rho_i](\rho_1 - \rho_v)(x_t - x_0).x = $ steam quality, subscripts *i* and *o* refer to inlet and exit, and 1 and *v* refer to condensate and vapor.

Fluids in cross-flow to a bank of tubes

Re is based on fluid velocity at the entrance to the tube bank.

Tubes in line: $a = $ diameter to diameter distance between tubes.

$$Nu = [1.517 + 205Re^{0.38}]^2 \left[\frac{4a}{(4a - \pi)} \right]$$

Tubes staggered, hexagonal centers; $a = $ distance between tube rows.

$$Nu = [1.878 + 0.256Re^{0.36}] \left[\frac{4a}{(4a - \pi)} \right]$$

Laminar flow in annuli

$$Nu = 1.02Re^{0.45} Pr^{0.5} \left(\frac{De}{L} \right)^{0.4} \left(\frac{D_2}{D_1} \right)^{0.8} Gr^{0.05} \left(\frac{\mu}{\mu_1}^{0.14} \right)$$

$De = $ hydraulic diameter; subscripts 1 and 2 refer to outside diameter of inner cylinder and inside diameter of outer cylinder, respectively.

Turbulent flow in annuli

$$Nu = 0.02Re^{0.8} Pr^{0.33} \left(\frac{D_2}{D_1} \right)^{0.53}$$

Finned tubes

Nu_d, Re_d, and A_0 use the outside diameter of the bare tube. $A = $ total area of tube wall and fin.

Tubes in line: $$Nu_d = 0.3 Re_d^{0.625} \left(\frac{A_0}{A} \right)^{0.375} Pr^{0.33}$$

Staggered tubes: $$Hu_d = 0.45 Re_d^{0.625} \left(\frac{A_0}{A} \right)^{0.375} Pr^{0.33}$$

Swept surface heat exchangers

$N = $ rotational speed of blades, $D = $ inside diameter of heat exchanger, $V = $ average fluid velocity, $L = $ swept surface length.

$$Nu = 4.9Re^{9.57} Pr^{0.47} \left(D\frac{N}{V} \right)^{0.17} \left(\frac{D}{L} \right)^{0.37}$$

Individual particles

The Nusselt number and the Reynolds number are based on the particle characteristic diameter and the fluid velocity over the particle.

Particles in a packed bed: $Nu = 0.015 Re^{1.6} Pr^{0.67}$

Particles in a gas stream: $Nu = 2 + 0.6 Re^{0.5} Pr^{0.33}$

Sources: Perry and Chilton, *Chemical Engineers Handbook*, 5th ed., McGraw-Hill Book Co., New York; Rohsenow and Hartnett, *Handbook of Heat Transfer*, McGraw-Hill Book Co., New York; Hausen, *Heat Transfer in Counterflow, Parallel Flow and Cross Flow*, McGraw-Hill Book Co., New York; Schmidt, *Kaltechn.* 15:98, 1963 and 15:370, 1963; Ranz and Marshall, *Chem. Eng. Prog.* 48(3):141 1952.

Visual BASIC Program for Evaluating Temperature Response of a Brick-Shaped Solid

```
Option Explicit
Dim X
Dim i As Integer
Dim BI(3) As Single
Dim Delta1(6) As Single
Dim Delta2(6) As Single
Dim Delta3(6) As Single
Dim YS, YC, TS, TC

Const TM = 177
Const T0 = 4
Const L1 = 0.0245
Const L2 = 0.0256
Const L3 = 0.0254
Const H1 = 6.5
Const H2 = 6.5
Const H3 = 6.5
Const K = 0.455
Const Rho = 1085
Const Cp = 4100

Dim TIMX
Dim d1, d2, d3
Dim YS1, YC1, YS2, YC2, YS3, YC3
Dim YXn, YCN, YCXn

Dim test
Dim Numx(6), DeNumx(6)
Dim YX(6), YCX(6)
```

(Continued)

```
Dim j As Integer
Dim fh
Dim LPHA As Variant

Sub TEMPSLAB()

LPHA = K / Rho / Cp
Debug.Print "alpha=" & LPHA

Worksheets("sheet1").Cells.Clear
BI(1) = H1 * L1 / K
Debug.Print "BI(1)" & BI(1)

BI(2) = H2 * L2 / K
Debug.Print "bi = " & BI(2)
BI(3) = H3 * L3 / K
Debug.Print "bi =" & BI(3)
Call toexc
j = 1

For TIMX = 0 To 4800 Step 600
Call DELN(BI(1), L1)
YS1 = YXn
YC1 = YCXn
d1 = Delta1(1)

Call DELN(BI(2), L2)
YS2 = YXn
YC2 = YCXn
d2 = Delta1(1)

Call DELN(BI(3), L3)
YS3 = YXn
YC3 = YCXn
d3 = Delta1(1)

YS = YS1 * YS2 * YS3
YC = YC1 * YC2 * YC3

TS = TM - YS * (TM - T0)
TC = TM - YC * (TM - T0)
If TIMX <> 0 Then Debug.Print "TIME =" & TIMX, "TS =" & TS, "TC =" & TC

Call toexcel(j, TIMX / 60, Format(TS, "0.000"), Format(TC, "0.000"))
j = j + 1

Next TIMX

fh = 1 / (LPHA * 0.4343 * ((d1 ^ 2 / L1 ^ 2) + (d2 ^ 2 / L2 ^ 2) + (d3 ^ 2 / L3 ^ 2)))
Debug.Print "FH=", fh
Call toexc1(fh, TM, T0, L1, L2, L3)
```

```
End Sub
Sub DELN(BI, L) 'DETERMINE ROOTS OF DELTA1
Delta1(1) = DntanDn(0, 1.57, BI)
Delta1(2) = DntanDn(3.14, 4.71, BI)
Delta1(3) = DntanDn(6.28, 7.85, BI)
Delta1(4) = DntanDn(9.42, 11, BI)
Delta1(5) = DntanDn(12.5, 14.1, BI)
Delta1(6) = DntanDn(15.7, 17.3, BI)

For i = 1 To 6
Numx(i) = 2 * Sin(Delta1(i)) * Cos(Delta1(i))
DeNumx(i) = Delta1(i) + Sin(Delta1(i)) * Cos(Delta1(i))
YX(i) = (Numx(i) / DeNumx(i)) * Exp(-LPHA * TIMX * Delta1(i) ^ 2 / L ^ 2)
YCX(i) = YX(i) / Cos(Delta1(i))

Next i

YXn = YX(1) + YX(2) + YX(3) + YX(4) + YX(5) + YX(6)
YCXn = YCX(1) + YCX(2) + YCX(3) + YCX(4) + YCX(5) + YCX(6)

If YXn > 1 Then YXn = 1
If YCXn > 1 Then YCXn = 1

End Sub
Function DntanDn(Lo, Hi, BI)
Do
X = 0.5 * (Lo + Hi)
test = X * Tan(X) - BI
If test > 0 Then Hi = X Else Lo = X
If Abs(Hi - Lo) < 0.00001 Then GoTo LABEL
Loop While Abs(test) > 0.001
LABEL:
DntanDn = X
End Function

Sub toexc()
With Sheets("sheet1")
.Cells(1, 1).Value = "alpha= "
.Cells(1, 2).Value = LPHA
.Cells(2, 1).Value = "bi(1)"
.Cells(2, 2).Value = BI(1)
.Cells(3, 1).Value = "bi(2)"
.Cells(3, 2).Value = BI(2)
.Cells(4, 1).Value = "(bi(3)"
.Cells(4, 2).Value = BI(3)
.Cells(5, 1).Value = "time": .Cells(5, 2).Value = "tsurf": .Cells(5, 3).Value = "tcent"
End With
End Sub
```

(Continued)

```
Sub toexcel(j, tx, surf, cent)
With Sheets("sheet1")
.Cells(j + 5, 1).Value = tx
.Cells(j + 5, 2).Value = surf
.Cells(j + 5, 3).Value = cent
End With
End Sub
Sub toexc1(fh, TM, T0, L1, L2, L3)
With Sheets("sheet1")
.Cells(j + 5, 1).Value = "fh = ": .Cells(j + 5, 2).Value = fh
.Cells(j + 6, 1).Value = "Tm = ": .Cells(j + 6, 2).Value = TM
.Cells(j + 7, 1).Value = "T0 = ": .Cells(j + 7, 2).Value = T0
.Cells(j + 8, 1).Value = "L1 =": .Cells(j + 8, 2) = L1
.Cells(j + 9, 1).Value = "L2 = ": .Cells(j + 9, 2).Value = L2
.Cells(j + 10, 1).Value = "L3 = ": .Cells(j + 10, 2).Value = L3
End With
End Sub
```

Visual BASIC Program for Evaluating Local Heat Transfer Coefficient from Temperature Response of a Brick-Shaped Solid

```
Sub localh()
'calculate h from Fh
Fh = 33596
L1 = 0.0254
L2 = 0.0508
L3 = 0.1016
k = 0.455
RHO = 10855
Cp = 4100
ALPHA = k / (RHO * Cp)
h = 375
Do
Bi1 = h * L1 / k
Bi2 = h * L2 / k
Bi3 = h * L3 / k
Delta1 = getDelta(Bi1)
Delta2 = getDelta(Bi2)
Delta3 = getDelta(Bi3)
testfh2 = testfh
testfh = 1 − (Fh * ALPHA * 0.4343 * ((Delta1 ^ 2 / L1 ^ 2) + (Delta2 ^ 2 /
L2 ^ 2) + (Delta3 2 ^ L3 ^ 2)))
If testfh > 0 Then dh = 0.001 Else dh = −0.001
h = h + dh
Debug.Print h, testfh
If testfh2 <> Empty Then If testfh / testfh2 < 0 Then Exit Do
```

```
Loop While Abs(testfh) > 0.01
MsgBox "Heat transfer coef, H Test" & Chr(13) & h − dh & " " &
testfh2 & Chr(13) & h & " " & testfh

End Sub
Function getDelta(Bi)
Lo = 0
Hi = 2
Do
x = 0.5 * (Lo + Hi)
Test = x * Tan(x) − Bi
If Test > 0 Then Hi = x Else Lo = x
If Abs(Hi - Lo) < 0.00001 Then Exit Do
Loop
getDelta = x
End Function
```

Thermal Conductivity of Water as a Function of Temperature

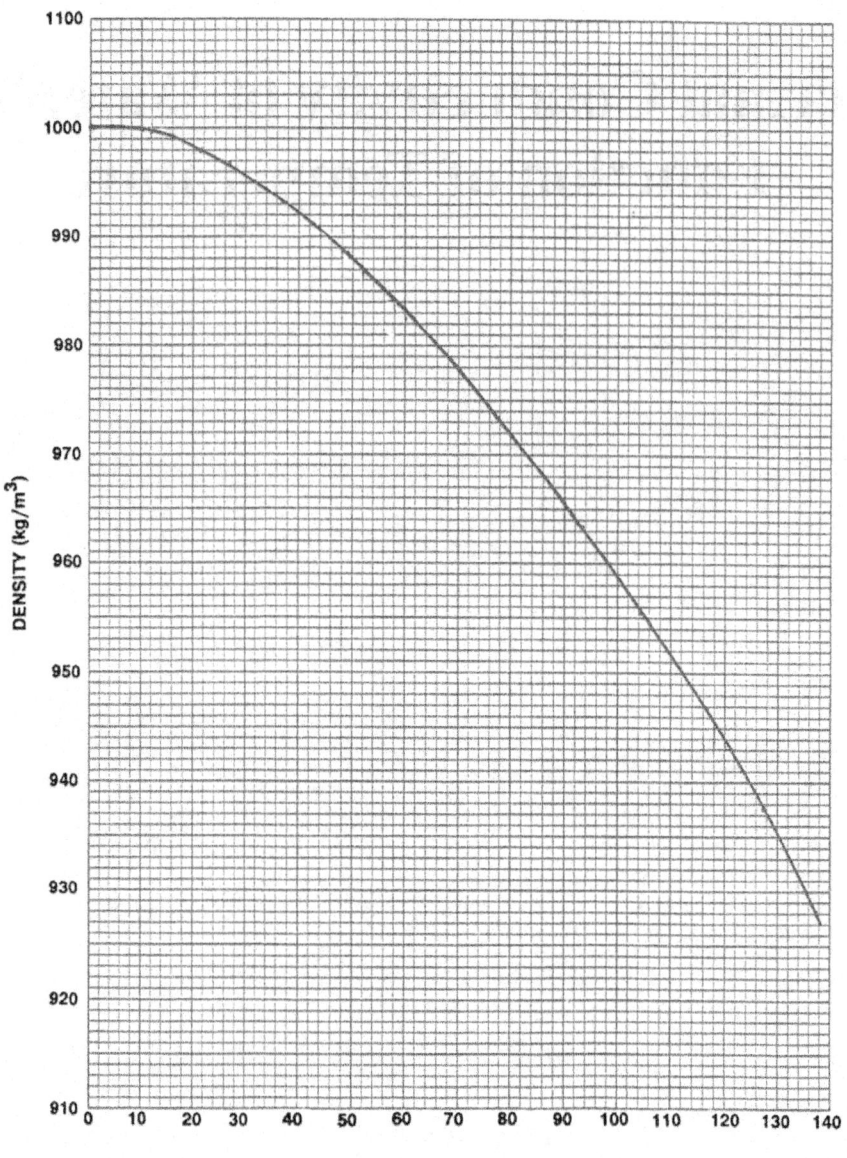

TEMPERATURE(°C)

Density of Water as a Function of Temperature

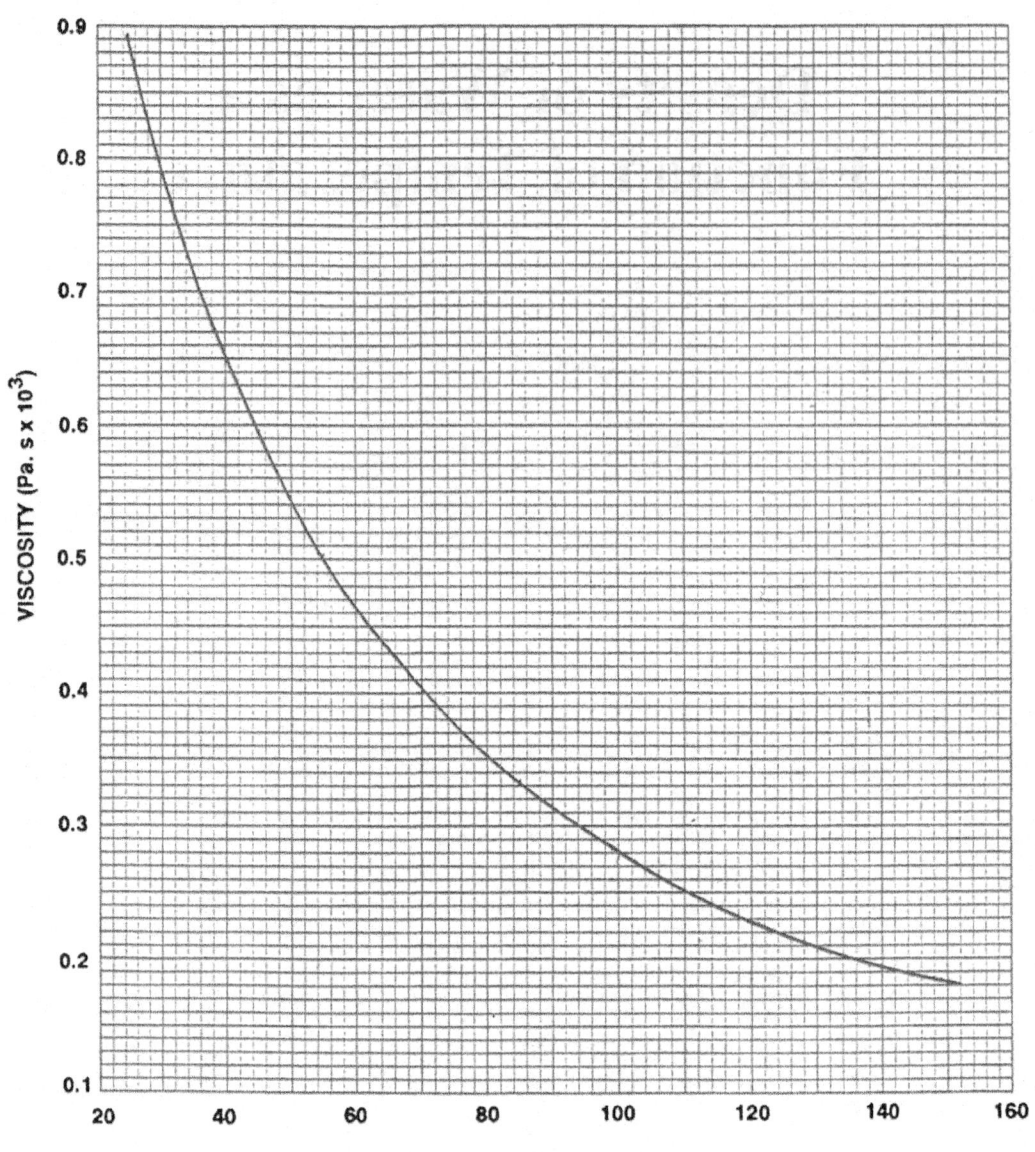

Viscosity of Water as a Function of Temperature

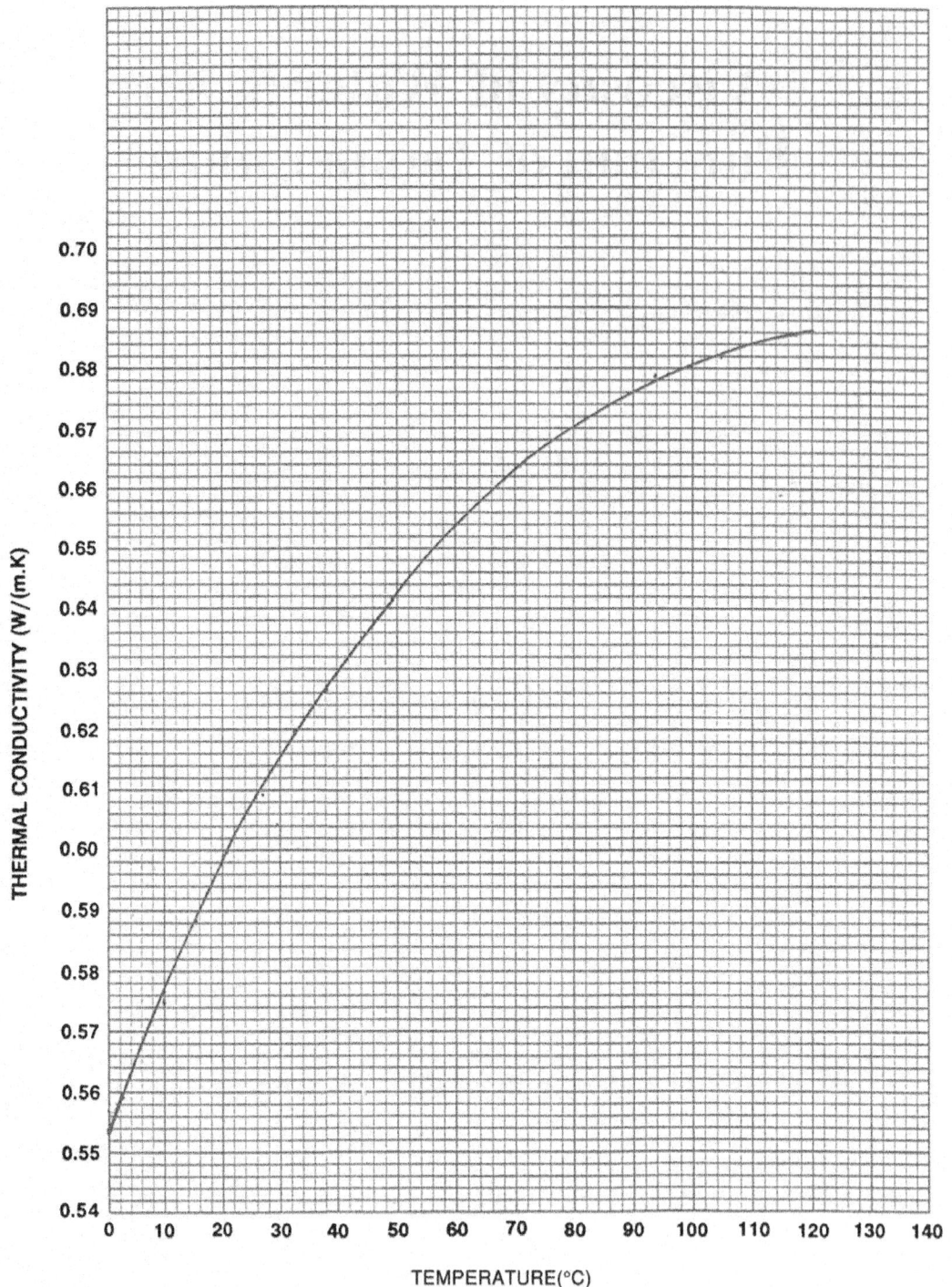

THERMAL CONDUCTIVITY (W/(m.K))

TEMPERATURE(°C)

Index

CPSIA information can be obtained
at www.ICGtesting.com
Printed in the USA
LVOW04*0921181117
556811LV00013B/304/P

9 780387 290195